Active tectonics and seismic hazards of Puerto Rico, the Virgin Islands, and offshore areas

Edited by

Paul Mann
Institute for Geophysics
John A. and Katherine G. Jackson School of Geosciences
University of Texas at Austin
4412 Spicewood Springs Road, Building 600
Austin, Texas 78759
USA

THE
GEOLOGICAL
SOCIETY
OF AMERICA

Special Paper 385

3300 Penrose Place, P.O. Box 9140 ▪ Boulder, Colorado 80301-9140 USA

2005

Published by The Geological Society of America, Inc.
3300 Penrose Place, P.O. Box 9140, Boulder, Colorado 80301-9140, USA
www.geosociety.org

Printed in U.S.A.

GSA Books Science Editor: Abhijit Basu

Library of Congress Cataloging-in-Publication Data

Active tectonics and seismic hazards of Puerto Rico, the Virgin Islands, and offshore areas / edited by Paul Mann.
p. cm. — (Special paper ; 385)
Includes bibliographic references and index.
ISBN 0-8137-2385-X (pbk.)
1. Earthquake zones--Puerto Rico. 2. Earthquake zones--Virgin Islands. 3. Fault zones--Puerto Rico. 4. Fault zones--Virgin Islands. I. Mann, Paul, 1956- II. Special papers (Geological Society of America) ; 385.

QE535.2.P9A28 2005
551.8'72'097295--dc22

2004063581

Cover: Shaded relief of the northern margin of Puerto Rico and the Virgin Islands. View is looking from north to south, illumination is from the northeast. Note two large (up to 55 km wide) amphitheater-shaped scarps less than 60 km north of the island of Puerto Rico that are believed to be sites of past submarine landslides. See Grindlay et al. (this volume, Chapter 2) for discussion of the neotectonic setting of the northern Puerto Rico–Virgin Islands margin. Digital elevation model generated from cruise EW96-05 Hydrosweep bathymetry, NOAA hydrographic data and U.S. Geological Survey topography. Grid interval = 6 arc sec.

10 9 8 7 6 5 4 3 2 1

Contents

Dedication

I dedicate this volume on the geology and tectonics of the Puerto Rico and Virgin Islands to Betty Bunce and Carl Bowin, two long-term researchers on the tectonics and marine geophysics of this area. Carl Bowin is presently an emeritus researcher at the Woods Hole Oceanographic Institution. Betty Bunce was his long-time colleague at that same institute, but unfortunately passed away in 2003 at the age of 88. Both came to the study of Caribbean geology through interesting career paths and contributed in unique and important ways to the understanding of the area described in this volume.

Paul Mann, editor

Elizabeth "Betty" Bunce. Photograph courtesy of the Woods Hole Oceanographic Institution.

BETTY BUNCE

Elizabeth "Betty" Bunce was born in 1915 in Mineola, New York, where she attended elementary and high school. She attended Smith College, where she graduated with a bachelor's degree in physical education in 1937 and a certificate in physical education in 1941. Her love of sports, particularly field hockey and lacrosse, led her to graduate work in physical education and a four-year position as a physical education teacher at the Kent Place School in New Jersey. She was a member of the All-American Reserve team and a national honorary judge and umpire in both sports.

Betty's long affiliation with Woods Hole Oceanographic Institution (WHOI) began with a chance summer vacation in 1944 to visit a friend working at the institution. That visit led to a job in the underwater explosives research group, known as Navy 7, headed by J. Brackett Hersey. She worked off and on during the summers while teaching and then attending graduate school.

Interested in the theory behind the research she was a part of at WHOI, Betty returned to Smith College and received a master's degree in physics in 1949, spending summers at WHOI as a technician. She worked as an instructor in physics at Smith from 1949 to 1951, leaving in July 1952 to join the WHOI staff full time as a research associate, a position she thought would last no more than two years. She was appointed an associate scientist in 1964, a senior scientist in 1975, and a scientist emeritus in 1980. Betty was the first woman to serve as a department chair at WHOI, serving as acting chair of the Geology and Geophysics Department a number of times during the 1970s and 1980s.

Betty sailed aboard many WHOI ships, including *Bear* and *Asterias*, but her cruises aboard *Chain* and *Atlantis II* were perhaps the most noteworthy. Betty's colleagues note that she was a jack-of-all-trades, a trail blazer and straight talker, and a very loyal friend. She appeared on the popular television program "To Tell the Truth" in the early 1960s and fooled three of the four panelists, who did not select her as an oceanographer. Famous for working out with a punching bag aboard ship, Betty maintained her love of sports throughout her life. She played squash and was often seen rowing in the waters around Woods Hole. She also enjoyed bird watching. She lived in a house of her own design overlooking Buzzard's Bay.

Her research interests were crustal structure, marine seismology, reflection and refraction, and underwater acoustics associated with seafloor studies. She was the author or co-author of more than 25 papers in marine geophysics, including several pioneer works on the marine geology and geophysics of the Puerto Rico trench and Lesser Antilles island arc. In 1971, she was awarded an honorary doctor of science degree from Smith College.

The first American woman to serve as chief scientist on a major oceanographic expedition, Betty opened the door for other woman desiring to enter oceanography. She subsequently served as chief scientist on many cruises, including a major portion of the Indian Ocean expedition in 1964, and in 1971 she headed a team on the *Chain* conducting site surveys in the Indian Ocean for future scientific drilling. She was a member of several scientific societies and served on many scientific committees.

She was a fellow of the Geological Society of America, a member of the American Geophysical Union, and she served on JOIDES site survey panels. In March 1995, she was honored at the Institution's second "Woman Pioneers in Oceanography" seminar, during which her friends and colleagues recounted some of her many contributions to oceanography. She was a woman of many firsts: the first woman chosen as a chief scientist on a deep-sea drilling cruise aboard *Glomar Challenger*, the first woman scientist to go to sea routinely, and the first woman to dive in *Alvin* (1965). She also laid claim as the first western woman to present a paper in marine geophysics at the International Union of Geology and Geophysics in Finland in 1960.

ACKNOWLEDGMENTS

Much of the information presented here was derived from an obituary written by the Media Relations Office of the Woods Hole Oceanographic Institution.

Selected Publications of Betty Bunce

Bunce, E.T., and Fahlquist, D.A., 1962. Geophysical Investigation of the Puerto Rico Trench and Outer Ridge: Journal of Geophysical Research, v. 67, p. 3955–3972.

Heezen, B.C., Bunce, E.T., Hersey, J.B., and Tharp, M., 1964, Chain and Romanche Fracture Zones: Deep-Sea Research, v. 11, p. 11–33.

Bunce, E.T., Crampin, S., Hersey, J.B., and Hill, M.N., 1964, Seismic Refraction Observations on the Continental Boundary West of Britain: Journal of Geophysical Research, v. 69, p. 3853–3863.

Bunce, E.T., Emery, K.O., Gerard, R.D., Knott, S.T., Lidz, L., Saito, T., and Schlee, J., 1965, Ocean Drilling on the Continental Margin: Science, v. 150, p. 709–716.

Bunce, E.T., 1966, The Puerto Rico Trench, *in* Continental Margins and Island Arcs: Geological Survey of Canada, Paper 66-15, p. 165–175.

Knott, S.T., Bunce, E. T., and Chase, R.L., 1966, Red Sea Seismic Reflection Studies, *in* World Rift System: Geological Survey of Canada, Paper 66-14, p. 33–61.

Bunce, E.T. and Hersey, J.B., 1966, Continuous Seismic Profiles of the Outer Ridge and Nares Basin North of Puerto Rico: Geological Society of America Bulletin, v. 77, p. 803–811.
Bunce, E.T., Bowin, C.O., and Chase, R.L., 1966, Preliminary Results of the 1964 Cruise of R/V *Chain* to the Indian Ocean: Philosophical Transactions of the Royal Society, A, v. 259, p. 218–226.
Bunce, E.T., Langseth, M.G., Chase, R.L., and Ewing, M., 1967, Structure of the Western Somali Basin: Journal of Geophysical Research, v. 72, p. 2547–2555.
Knott, S.T., and Bunce, E.T., 1968, Recent Improvement in Technique of Continuous Seismic Profiling: Deep-Sea Research, v. 15, p. 633–636.
Chase, R.L., and Bunce, E.T., 1969, Underthrusting of the Eastern Margin of the Antilles by the Floor of the Western North Atlantic Ocean, and Origin of the Barbados Ridge: Journal of Geophysical Research, v. 74, p. 1413–1420.
Bunce, E.T., Fahlquist, D.A., and Clough, J.W., 1969, Seismic Refraction and Reflection Measurements—Puerto Rico Outer Ridge: Journal of Geophysical Research, v. 74, p. 3082–3094.
Bunce, E.T., 1969, Seismic Refraction Measurements in the Baltic Sea: Geophys. Prospecting, v. 17, p. 28–35.
Emery, K.O., Uchupi, E., Phillips, J.D., Bowin, C.O., Bunce E.T., and Knott, S.T., 1970, The Continental Rise off Eastern North America: American Association of Petroleum Geologists Bulletin, v. 54, p. 44–108.
Bunce, E.T., Phillips, J.D., Chase R.L., and Bowin, C.O., 1970, The Lesser Antilles Arc and the Eastern Margin of the Caribbean Sea, *in* Maxwell, A.E., ed., The Sea: New York, Wiley Interscience, v. 4, p. 359–385.
Pimm, A.C., Burroughs, R.H., and Bunce, E.T., 1972, Oligocene Sediments near Chain Ridge, Northwest Indian Ocean Structural Implications: Marine Geology, v. 13, p. M14–M18.
Fisher, R.L., Bunce, E.T., Cernock, P.J., Cronan, D.S., Damiani, V.C., Dmitriev, L., Kinsman, D.J.J., Roth, P.H., Thiede, J., and Vincent, E.S., 1972, Deep Sea Drilling Project in Dodoland, Leg 24. Geotimes, v. 17, no. 19, p. 17–21.
Luyendyk, B.P., Bunce, E.T., 1973, Geophysical Study of the Northwest African Margin off Morocco: Deep-Sea Research, v. 20, p. 537–549.
Bunce, E.T., Phillips, J.D., and Chase, R.L., 1974, Geophysical Study of Antilles Outer Ridge, Puerto Rico Trench, and Northeast Margin of Caribbean Sea: American Association of Petroleum Geologists Bulletin, v. 58, p. 106–123.
Fisher, R.L., Bunce, E.T., et al., 1974, Initial Reports of the Deep Sea Drilling Project: Washington D.C., U.S. Government Printing Office, v. 24, 1183 p.
Hobart, M.A., Bunce E.T., and Sclater, J.G., 1975, Bottom water Flow through the Kane Gap, Sierra Leone Rise, Atlantic Ocean: Journal of Geophysical Research, v. 80, p. 5083–5088.
Chase, R.L., Bunce, E.T., and Phillips, J.D., 1975, Geophysical Study of Antilles Outer Ridge, Puerto Rico Trench, and Northeast Margin of Caribbean Sea: Reply: American Association of Petroleum Geologists Bulletin, v. 59, p. 2323–2324.
Ben-Avraham, Z., and Bunce, E.T., 1977, Geophysical Study of the Chagos-Laccadive Ridge, Indian Ocean: Journal of Geophysical Research, v. 82, p. 1295–1305.
Bunce, E.T., and Molnar, P., 1977, Seismic Reflection Profiling and Basement Topography in the Somali Basin: Possible Fracture Zones between Madagascar and Africa: Journal of Geophysical Research, v. 82, p. 5305–5311.

Carl Bowin. Photograph courtesy of the Woods Hole Oceanographic Institution.

CARL BOWIN

Carl Bowin was born in 1934 in Los Angeles, California, to working-class parents at the height of the Great Depression. The family lived in a tenement in central Los Angeles while Carl's father supported the family on a salary of $55 a month from the WPA (Works Progress Administration) and collected scrap metal for additional income. In 1938, the family was able to move to a better neighborhood near the downtown civic center. When in junior high, Carl became interested in insect collecting, and his parents encouraged him to contact the entomologist, Dr. Pearce, at the Los Angeles County Museum. On Saturdays, his parents would drive him to the museum where he would help Dr. Pearce mount insect specimens while they waited outside. Insect collecting eventually led to fossil collecting, which led to Saturday afternoons helping the museum paleontologist, Dr. Kanakoff, sort fossil samples. This led Carl to meetings of a group of high school students from all over Los Angeles who met at the museum to pursue their scientific interests.

In 1948, Carl applied to and was accepted at the California Institute of Technology. By living at home and commuting in the family car, his parents were able to afford his tuition of about $200 per month. He became a geology major and spent one summer at a field camp mapping deformed rocks in Wyoming. At a school dance organized for the all-male student body at that time, Carl met his future wife, Jean, a student from nearby Occidental College. Following his B.S. degree in geology from Caltech, Carl attended Northwestern University, where he obtained his M.S. degree mapping metamorphic and igneous rocks.

Following graduation from Northwestern, Carl continued his migration eastward and to the Ph.D. program in geology at Princeton University, working under the renowned tectonicist H.H. Hess. Carl's project was a study of serpentinized peridotite in the Dominican Republic, which formed part of a much larger study Hess was conducting on Caribbean peridotites at that time. Part of the study involved an early gravity survey of the country as a way to "look inside" the tall mountains he had studied from outcrops. Completion of the geologic mapping-gravity study led to his graduation in 1960 and an offer by Hess to continue on at Princeton as an instructor in geology. He developed new interests during this time, including the use of a prototype IBM computer at Princeton to study heat flow and crustal evolution. A visit between J. Brackett Hersey of Woods Hole Oceanographic Institution and Hess revealed that Hersey was interested in starting a marine gravity program using U.S. research vessels. Along with geoscience research, the motivation for this program was linked to the interest of the U.S. military in improving intercontinental missile guidance systems. Carl quickly realized the utility of computers for processing gravity data while at sea and helped design the first seagoing computer system. The marine gravity data program prospered and led to many interesting surveys, including that of the extreme free-air anomalies in the Puerto Rico trench.

In 1981, Carl recognized the potential of the fledgling Global Positioning System for measurements of plate motions. A proposal to NASA to measure Caribbean plate motions laid the framework for measurements now routinely made on the Caribbean islands and larger landmasses. Carl continues to be active in his retirement producing video documentaries about science for the local Public Access Television studio near his home in Falmouth, Massachusetts. These programs are available on the Web at www.whoi.edu/scc.

In commemoration of the many contributions of Betty and Carl, we have named two submarine faults for them: the Bunce fault in the deep part of the Puerto Rico trench (ten Brink and Lian, 2004) and the subparallel Bowin fault at shallower depths in the trench (Grindlay et al., this volume, Chapter 2).

REFERENCE CITED

ten Brink, U., and Lin, J., 2004, Stress interaction between subduction earthquakes and forensic strike-slip faults: Modeling and application to the northern Caribbean plate boundary: Journal of Geophysical Research, v. 109, B12310, doi: 10.1029/2004JB003031.

Selected Publications of Carl Bowin

Bowin, C., 1960, Geology of Central Dominican Republic: Geological Society of America Bulletin, v. 71, p. 1831 (Abstract, Ph.D. dissertation, Princeton University).

Bowin, C., 1962, A New System for the Automatic Measurement of Gravity at Sea: Oceanus, v. 4, no. 1, p. 22–23.

Bowin, C., 1963, Gravity Anomaly Maps of the Puerto Rico Trench Region from New Continuous Measurements (Abstract). Vol. III, International Assoc. of Seismology and Physics of the Earth's Interior, XII General Assembly, International Union of Geodesy and Geophysics, Berkeley, California.

Bowin, C., and Bernstein, R., 1963, A New Automatic Sea Gravity System [abs.]: Transactions, American Geophysical Union, v. 44, no. 1, p. 30–31.

Bowin, C., and Nalwalk, A., 1963, Serpentinized Peridotite Dredged from the North Wall, Puerto Rico Trench [abs.]: Transactions, American Geophysical Union, v. 44, no. 1, p. 120.

Bernstein, R., and Bowin, C.O., 1963, Real-Time Digital Computer Acquisition and Computation of Gravity Data at Sea: IEEE Transactions on Geoscience Electronics, v. GE-1, no. 1, p. 2–10.

Bowin, C.O., 1965, A Shipboard Navigational and Geophysical Processing System: Proceedings of the Second U.S. Navy Symposium on Military Oceanography, 5–7 May 1965, v. 1, p. 369–374.

Bowin, C.O., 1966, Geology of Central Dominican Republic (A Case History of Part of an Island arc), *in* Hess, H.H., ed., Caribbean Geological Investigations: Geological Society of America Memoir 98, p. 11–84.

Bowin, C., 1966, Gravity over Trenches and Rifts, *in* Poole, W.H., ed., Geological Survey Paper 66-15, p. 430–439.

Bowin, C.O., and Vogt, P.R., 1966, Magnetic Lineation between Carlsberg Ridge and Seychelles Bank, Indian Ocean: Journal of Geophysical Research, v. 71, p. 2625–2630.

Bowin, C.O., Nalwalk, A.J., and Hersey, J.B., 1966, Serpentinized Peridotite from the North Wall of the Puerto Rico Trench: Geological Society of America Bulletin, v. 77, p. 257–270.

Bowin, C.O., 1967, Five Years Experience with a Shipboard Oceanographic Data Processing System: Hindsight and Foresight: Fourth U.S.N. Symposium on Military Oceanography, The Proceedings of the Symposium, v. I, NRL, p. 253–264.

Bowin, C.O., Bernstein, R., Ungar, E., and Madigan, J.R., 1967, A Shipboard Oceanographic Data Processing and Control System: IEEE Transactions on Geoscience Electronics, v. GE-5, no. 2, p. 41–50.

Bowin, C.O., Chase, R.L., and Hersey, J.B., 1967, Geological Application of Sea-Floor Photography. *in* Hersey, J.B., ed.,

Deep-Sea Photography: Johns Hopkins University Press, p. 117–140.
Bowin, C., 1967, Geology of the Caribbean, *in* Runcorn, K., ed., International Dictionary of Geophysics: Pergamon Press, v. 1, p. 605–609.
Bowin, C.O., 1968, Geophysical Study of the Cayman Trough: Journal of Geophysical Research, v. 73, no. 16, p. 5159–5173.
Bunce, E.T., Bowin, C.O., and Chase, R.L., 1968, Structure of the Lesser Antillean Arc [abs.]: Transactions, American Geophysical Union, v. 49, no. 1, p. 197.
Bowin, C.O., 1968, Shipboard Computer Data Processing; Some Results at the Woods Hole Oceanographic Institution: Second FAO Tech. Conference on Fishery Research Craft, Seattle, Washington, 18–24 May 1968: Rome, Italy, Food and Agriculture Organization of the United Nations, v. II, 4 p.
Bowin, C.O., 1969, Experience with a Sea-Going Computer System: Lessons, Recommendations, and Predictions: Transactions, Marine Tech. Society Symposium on Application of Sea-Going Computers, p. 141–157.
Ruppert, G.N., Bernstein, R., and Bowin, C.O., 1969, Precise Positioning of a Ship at Sea Utilizing VLF Transmissions—Methods and Results, Navigation: Journal Inst. Nav., v. 16, no. 2, p. 11–128.
Bowin, C., Wing, C.G., and Aldrich, T.C., 1969, Test of the MIT Vibrating String Gravimeter, 1967: Journal of Geophysical Research, v. 74, p. 3278–3280.
Bowin, C., 1971, Some Aspects of the Gravity Field and Tectonics of the Northern Caribbean Region: Transactions, Fifth Caribbean Geological Conference, Geological Bulletin, no. 5, Queens College Press, May 1971, p. 1–6.
Bunce, E.T., Phillips, J.D., Chase, R.L., and Bowin, C., 1971, The Lesser Antilles Arc and the Eastern Margin of the Caribbean Sea, *in* Maxwell, A.E., ed., The Sea: Wiley Interscience, vol. 4, part II, p. 359–385.
Uchupi, E., Milliman, J.D., Luyendyk, B.P., Bowin, C.O., and Emery, K.O., 1971, Structure and Origin of Southeastern Bahamas: American Association of Petroleum Geologists Bulletin, v. 55, no. 5, (May 1971), p. 687–704.
Bowin, C., Aldrich, T.C., and Folinsbee, R.A., 1972, VSA Gravity Meter System: Tests and Recent Developments: Journal of Geophysical Research, v. 77, p. 2018–2033.
Bowin, C., 1972, The Puerto Rico Trench Negative Gravity Anomaly Belt: Geological Society of America Memoir 132, p. 339–350.
Ben-Avraham, Z., Bowin, C., and Segawa, J., 1972, An Extinct Spreading Centre in the Philippine Sea: Nature, v. 240, p. 453–455.
Bowin, C., 1974, Migration of a Pattern of Plate Motion: Earth and Planetary Science Letters, no. 21, p. 400–404.
Grow, J.A., and Bowin, C.O., 1975, Evidence for High-Density Crust and Mantle Beneath the Chile Trench due to the Descending Lithosphere: Journal of Geophysical Research, v. 80, p. 1449–1458.
Bowin, C., 1975, The Geology of Hispaniola, *in* Nairn, A.E.M., and Stehli, F.G., eds., Ocean Basins and Margins: New York, Plenum Press, v. 3, p. 501–552.
McKenzie, D., and Bowin, C., 1976, The relationship between bathymetry and gravity in the Atlantic Ocean: Journal of Geophysical Research, v. 81, no. 11, p. 1903–1915.
Sclater, J.G., Bowin, C., Hey, R., Hoskins, H., Pierce, J., Phillips, J. and Tapscott, C., 1976, The Bouvet Triple Junction: Journal of Geophysical Research, v. 81, p. 1857–1869.
Bowin, C., 1976, Caribbean Gravity Field and Plate Tectonics: Geological Society of America Special Paper 169, 79 p.
Bowin, C., Lu, R.S., Chao-Shing Lee, and Schouten, H., 1978, Plate Convergence and Accretion in the Taiwan-Luzon Region: American Association of Petroleum Geologists Bulletin, v. 62, p. 1645–1672.
Bowin, C., Purdy, G.M., Johnston, C., Shore, G., Lawver, L., Hartono, H.M.S., and Jezek, P., 1980, Arc-Continent Collision in the Banda Sea Region: American Association of Petroleum Geologists Bulletin, v. 64, p. 868–915.
Johnson, C., and Bowin, C., 1981, Crustal Reactions Resulting from the Mid-Pliocene to Recent Continent Island Arc Collision in the Timor Region: Australian Bureau of Mineral Resources Journal, v. 6, p. 223–243.
Bowin, C., 1982, Gravity and Geoid Anomalies of the Caribbean: Transactions of the Ninth Caribbean Geological Conference, Santo Domingo, Dominican Republic, v. 12, p. 527–538.
Bowin, C., and Nagle, F., 1982, Igneous and Metamorphic Rocks of Northern Dominican Republic: Uplifted Subduction Zone: Transactions of the Ninth Caribbean Geological Conference, Santo Domingo, Dominican Republic, v. 1, p. 39–45.
Bowin, C., Warsi, W., and Milligan, J., 1982, Free-Air Gravity Anomaly Map and Atlas of the World: Geological Society of America Map and Chart Series, MC-45 and MC-46.
Bowin, C., Scheer, E., and Smith, W., 1986, Depth estimates from ratios of gravity, geoid, and gravity gradient anomalies: Geophysics, v. 51, no. 1, p. 123–136.
Bowin, C., 1987, Historical Note on "Evolution Ocean Basins" preprint by H.H. Hess (1906–1969): Geology, v. 15, p. 475–476.
Lewis, J.B., Draper, G., Bowin, C., Bourdon, L., Maurrasse, F. and Nagle, F., 1990, Hispaniola section, *in* Lewis, J.F., and Draper, G., eds., The Caribbean region: Geological Society of America, Geology of North America, v. H., p. 97–112 (Chapter 4).
Bowin, C., Earth's Gravity Field and Plate Tectonics, 1991, Tectonophysics issue with contributions to Texas A&M Geodynamics Silver Anniversary Symposium, v. 187, p. 69–89.
Bowin, C., 1997, Mass Anomalies of Earth, Venus and Mars: Initial Estimates, *in* Segawa, et al., eds., Gravity, Geoid and Marine Geodesy: International Symposium Tokyo, Springer-Verlag, v. 117, p. 257–264.
Bowin, C., 2000, Mass anomalies and the structure of the Earth, *in* Physics and Chemistry of the Earth (A): v. 25, no. 4, p. 343–353.
Bowin, C., 2000, Mass anomaly structure of the Earth: Reviews of Geophysics, v. 38, no. 3, p. 355–387.
Bowin, C., 2004, Why Plates Move [abs.]: 2004 Joint assembly program and abstracts CGU, AGU, SEG, and EUG, 17–24 May 2004, Montreal, Canada, Session T33A-05.

Geological Society of America
Special Paper 385
2005

Introduction

Paul Mann*
Institute for Geophysics, Jackson School of Geosciences, University of Texas at Austin, 4412 Spicewood Springs Road, Building 600, Austin, Texas 78759, USA

PURPOSE AND SIGNIFICANCE OF VOLUME

The purpose of this volume is to present 14 well-integrated chapters on the complex active tectonic setting, seismicity, and paleoseismicity of the Puerto Rico and Virgin Islands area astride the northeastern segment of the 3200-km-long, active North America–Caribbean plate boundary zone (Fig. 1). The faults producing seismic hazards in Puerto Rico and the Virgin Islands are formed in this plate boundary setting and extend across the northern Caribbean to Central America. The zone of interplate faulting as indicated by earthquake activity ranges from 100 to 250 km in width (Fig. 1). This dominantly left-lateral strike-slip plate boundary ranks with the great seismogenic, strike-slip fault boundaries of the world, including the San Andreas fault of California (1500 km in length), the Alpine fault of New Zealand (600 km in length), the North Anatolian fault of Turkey (1000 km in length), and the Dead Sea fault of the Middle East (900 km in length) (Yeats et al., 1997). One obvious difference with these other better-studied subaerial strike-slip faults is that ~75% of the North America–Caribbean plate boundary fault zone lies offshore and therefore cannot be studied using direct observations and paleoseismic methods based on detailed logging of trenched exposures and radiometric dating of the trench stratigraphy (Sieh, 1981). Nonetheless, the fault system is a major source of destructive earthquakes for 54 million people living in 11 countries along its 3200 km length (Fig. 1).

The two most extensive landfalls of the plate boundary fault shown in Figure 1 include the 500-km-long Motagua and Polochic faults of northern Central America (Schwartz et al., 1979) and the 320-km-long Septentrional fault zone of Hispaniola, the island that includes the countries of Dominican Republic and Haiti (Mann et al., 1998; Prentice et al., 2003). Offshore segments of the active plate boundary faults have been mapped using marine geophysical methods in the Cayman trough (Rosencrantz and Mann, 1991; Dillon et al., 1996; Leroy et al., 1996); southern Cuba (Calais et al., 1998), northern Hispaniola (Dillon et al., 1992; Dolan et al., 1998), and the Puerto Rico trench (Masson and Scanlon, 1991; Grindlay et al., this volume, Chapter 2). A summary of names of various segments of the North America–Caribbean plate boundary fault zone is given in Figure 1C. Although the fault zones have different names, the maps in Figure 1 show that the plate boundary faults are remarkably linear and continuous for hundreds of kilometers.

TECTONIC SETTING OF PUERTO RICO AND THE VIRGIN ISLANDS

More than a decade of Global Positioning System (GPS)–based geodetic results from the plate boundary zone has clarified the rates and direction of interplate motion of the Caribbean plate that is bounded mainly by subduction or strike-slip plate boundaries where plate rates and directions are difficult to precisely estimate from geologic structures and earthquakes alone (Fig. 1). GPS-determined motion of sites within the stable, nondeforming interior of the Caribbean plate relative to the North America plate by DeMets et al. (2000) shows that the Caribbean plate moves to the east-northeast (070°) at a rate of 18–20 ± 3 mm/yr. Puerto Rico and the Virgin Islands move with the larger Caribbean plate while the area of Hispaniola to the west moves independently as a detached part of the Caribbean plate (Jansma et al., 2000; Mann et al., 2002; Jansma and Mattioli, this volume, Chapter 1) (Fig. 1).

Types of faulting and structural styles shown from west to east on Figure 1B are related to the angle between the direction of plate motion and the plate boundary fault in that area and include zones of (1) transtension in Central America; (2) "pure" strike-slip in the western and central Cayman trough; (3) transpression in the eastern Cayman trough and southern Cuba; (4) oblique collision between Hispaniola and the Bahama carbonate platform; (5) oblique oceanic subduction of Atlantic oceanic crust beneath Puerto Rico, and (6) orthogonal subduction at the Lesser Antilles island arc (DeMets et al., 2000; Mann et al., 2002; Calais et al., 2002, Grindlay et al., this volume, Chapter 2). GPS studies in Puerto Rico and the Virgin Islands by Jansma et al. (2000)

*paulm@ig.utexas.edu

Mann, P., 2005, Introduction, *in* Mann, P., ed., Active tectonics and seismic hazards of Puerto Rico, the Virgin Islands, and offshore areas: Geological Society of America Special Paper 385, p. 1–12. For permission to copy, contact editing@geosociety.org.

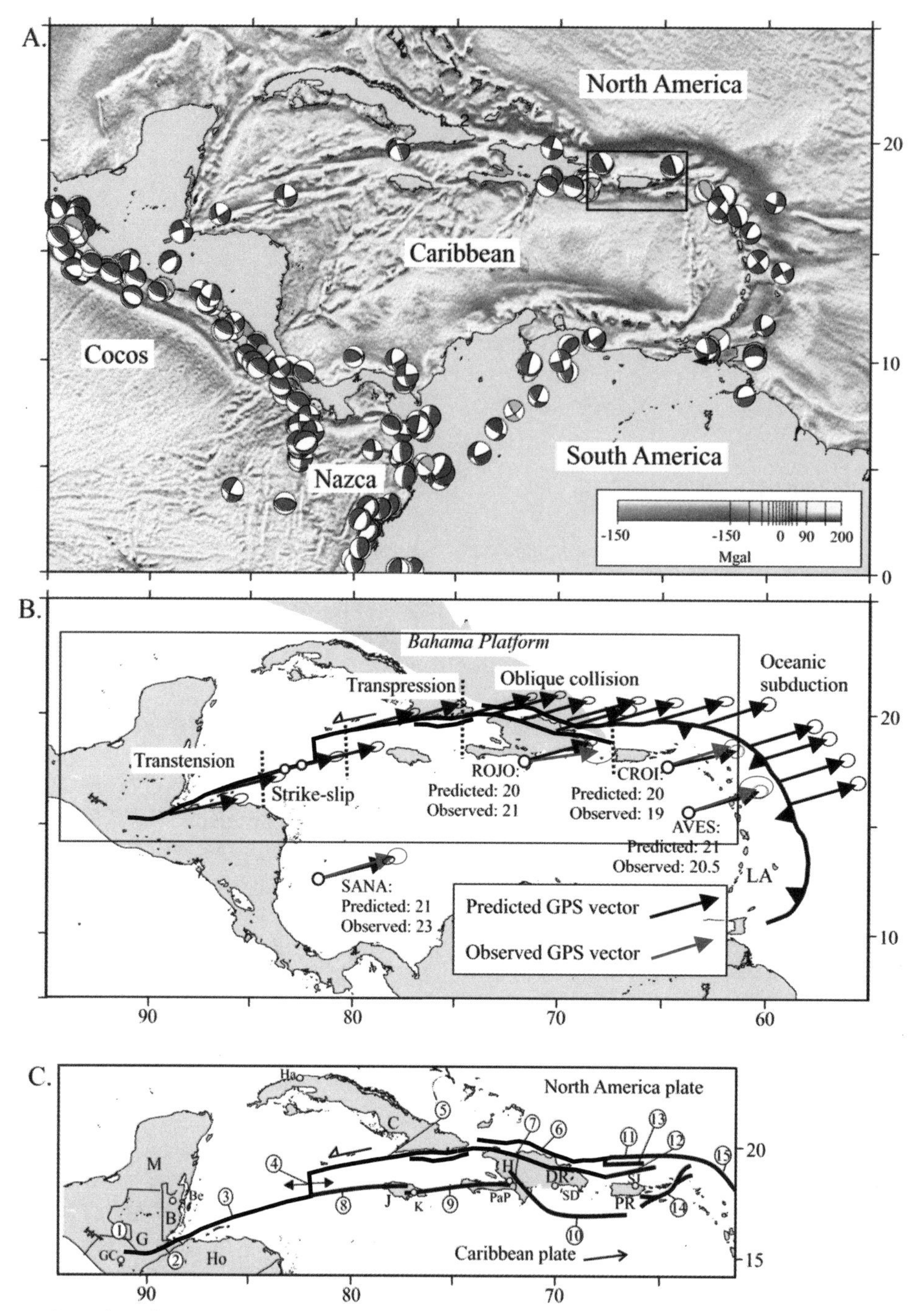

Figure 1. (A) Plate tectonic setting of Puerto Rico and the Virgin Islands in the northeastern Caribbean (boxed area) (modified from Mann et al., 2002). Base map is a satellite-derived gravity map compiled by Smith and Sandwell (1997). Earthquake focal mechanisms in red are 0–75 km in depth; blue mechanisms are 75–150 km in depth; and green mechanisms are >150 km in depth. Earthquakes indicate the Caribbean plate moves eastward relative to North and South America. (B) Caribbean–North America velocity predictions of DeMets et al. (2000) (black arrows) based on GPS velocities at four localities in the stable interior of the plate (red vectors) and two fault strike measurements in the strike-slip segment of the North America–Caribbean strike-slip boundary (open circles). The predicted velocities are consistent with the along-strike transition in structural styles from transtension in the northwestern corner of the plate in Central America to oblique collision and oceanic subduction in the area of the Puerto Rico and the Virgin Islands in the northeastern corner of the plate. (C) Map showing political divisions of the North America–Caribbean plate boundary zone with largest cities labeled. Most large cities have been damaged several times by earthquakes over the past four or five centuries. Key to country and city abbreviations: M—Mexico; B—Belize; Be—Belmopan, Belize; G—Guatemala; GC—Guatemala City; Ho—Honduras; C—Cuba; Ha—Havana; J—Jamaica; K—Kingston; H—Haiti; PaP—Port au Prince; DR—Dominican Republic; SD—Santo Domingo; PR—Puerto Rico; SJ—San Juan. Dominican Republic and Haiti together occupy the island of Hispaniola. Key to major faults of the North America-Caribbean plate boundary zone: 1—Polochic fault zone, Guatemala; 2—Motagua fault zone, Guatemala; 3—Swan Islands fault zone; 4—Mid-Cayman spreading center; 5—Oriente fault zone; 6—North Hispaniola deformed belt; 7—Septentrional fault zone; 8—Walton fault zone; 9—Enriquillo-Plaintain Garden fault zone; 10—Muertos trench; 11—Puerto Rico trench; 12—Bowin fault zone; 13—Bunce fault zone; 14—Anegada Passage fault zone; 15—Lesser Antilles frontal thrust.

and Jansma and Mattioli (this volume, Chapter 1) show that this region moves to the east-northeast at the same rate as the larger Caribbean plate and that differential motion with the adjacent island of Hispaniola is accommodated by slow (~5 mm/yr) rifting in the Mona Passage separating western Puerto Rico and eastern Hispaniola.

NEOGENE TECTONIC EVOLUTION OF ACTIVE FAULT SYSTEMS WITHIN THE PUERTO RICO–VIRGIN ISLANDS AREA

The Neogene tectonic evolution of the North America–Caribbean plate boundary in the region of Puerto Rico and the Virgin Islands appears to have been closely controlled by oblique west-southwestward subduction of the Bahama carbonate platform beneath the Virgin Islands, Puerto Rico, and Hispaniola (Dolan et al., 1998; Mann et al., 2002; Grindlay et al., this volume, Chapter 2). Reconstructions shown in Figure 2 and discussed in detail by Grindlay et al. (this volume, Chapter 2) show that the locus of oblique collision moved from west to east over the past 10 m.y. Detachment of the Hispaniola area is probably related to its oblique collision with the Bahama carbonate platform as first noted by Vogt et al. (1976). The reconstructions shown in Figure 2A–D illustrate how the oblique collision of the Bahama platform forms a series of three microplates on the northeastern edge of the Caribbean plate. Each of these microplates (color coded in Fig. 2D) are bounded by complex zones of late Quaternary faults with potential to generate earthquakes. Because of oblique plate convergence, these microplates are pushed or "backthrust" to the southwest as a result of the inability of the trench to consume the thicker-than-oceanic crust of the southeastern Bahama Platform (Mann et al., 2002) (Fig. 2D).

EARTHQUAKES AND POPULATION GROWTH IN THE NORTHERN CARIBBEAN

Mapping and detailed studies of the plate boundary faults using geologic methods in onshore areas and geophysical methods in offshore areas is a high priority for research because of the extreme population density in the northern Caribbean region, (Fig. 1; Table 1). Because of the relatively small land areas on islands, population densities in Puerto Rico (433 persons/km^2) and the U.S. Virgin Islands (348 persons/km^2) are similar to better-known, densely populated areas such as Japan (338 persons/km^2) and Taiwan (627 persons/km^2).

Earthquake risk is especially high in developing countries like Haiti, the Dominican Republic, Guatemala, and Honduras for several reasons: (1) the population is becoming increasingly urbanized; (2) many people live in poorly constructed housing (especially concrete structures lacking proper reinforcement); and (3) low-income, makeshift housing is commonly found on steep slopes that are susceptible to collapse or sliding during major earthquakes. Moreover, populations on islands and in countries like Belize, Puerto Rico, and the Virgin Islands are also largely concentrated in low-lying coastal areas and confined harbors, making them vulnerable to tsunamis related to motions on offshore faults and earthquake-triggered submarine slumps.

The historic record of earthquakes and tsunamis in the northeastern Caribbean is the longest in the western hemisphere and dates back to the arrival of Europeans in the New World in 1492. The compilation of historic earthquake information in Table 1 reveals that rapidly growing cities such as Kingston, Jamaica (present population 922,900), Port-au-Prince, Haiti (population 3,066,700) and Santo Domingo, Dominican Republic (population 1,891,700) have experienced repeated damaging earthquakes in past centuries, some of which were devastating (e.g., Kingston, Jamaica, in 1692 and 1907). All of these large cities are likely to experience future shocks under much more crowded conditions in the 21st century. Of particular concern for earthquake safety is the proliferation of high-rise apartment and office buildings, one- and two-story buildings constructed with unreinforced concrete blocks, and elevated highways necessitated by the rapid increase and densification of urban populations.

A growth rate of 1.5% averaged for all countries in the North America–Caribbean plate boundary zone indicates that the present population of 50 million persons will double to 100 million persons by the year 2050. At its most basic, the earthquake cycle on active faults of the North America–Caribbean plate boundary is an ongoing process of strain buildup related to plate motion followed by a sudden and catastrophic release. GPS data has better quantified the rates at which strain is accumulating (Calais et al., 2002). Earthquake cycles, or the time between major fault ruptures, are measured in hundreds or sometimes thousands of years, as shown by fault trenching or "paleoseismology" studies in the northern Caribbean and other areas (Prentice et al., 1993, 2003; Prentice and Mann, this volume, Chapter 10).

Because these earthquakes cannot be prevented and major population centers near earthquake faults cannot be relocated, earthquakes will strike increasingly larger populations again and again into the future. The best protection against these future events is identifying the locations of active faults, conducting paleoseismic studies to analyze their past amounts of slip (and therefore range of earthquake size), and to appropriately design buildings that can withstand the predicted size range of these recurring events.

Earthquakes will be most damaging when their stored energy is released near cities, even when those cities are of no great size (Bilham, 1988). A recent example of the deadly effects of a nearby earthquake on a relatively small city was the 2003, M6.5 shallow earthquake near Bam, Iran, where 26,000 inhabitants of the city were killed, 9,000 were injured, and twice as many were left homeless in a city with a population of 100,000. The proximity of the fault rupture to the city and collapse of unreinforced mud brick dwellings was the main reason the death toll was so high. The most recent devastating earthquake of the North America–Caribbean

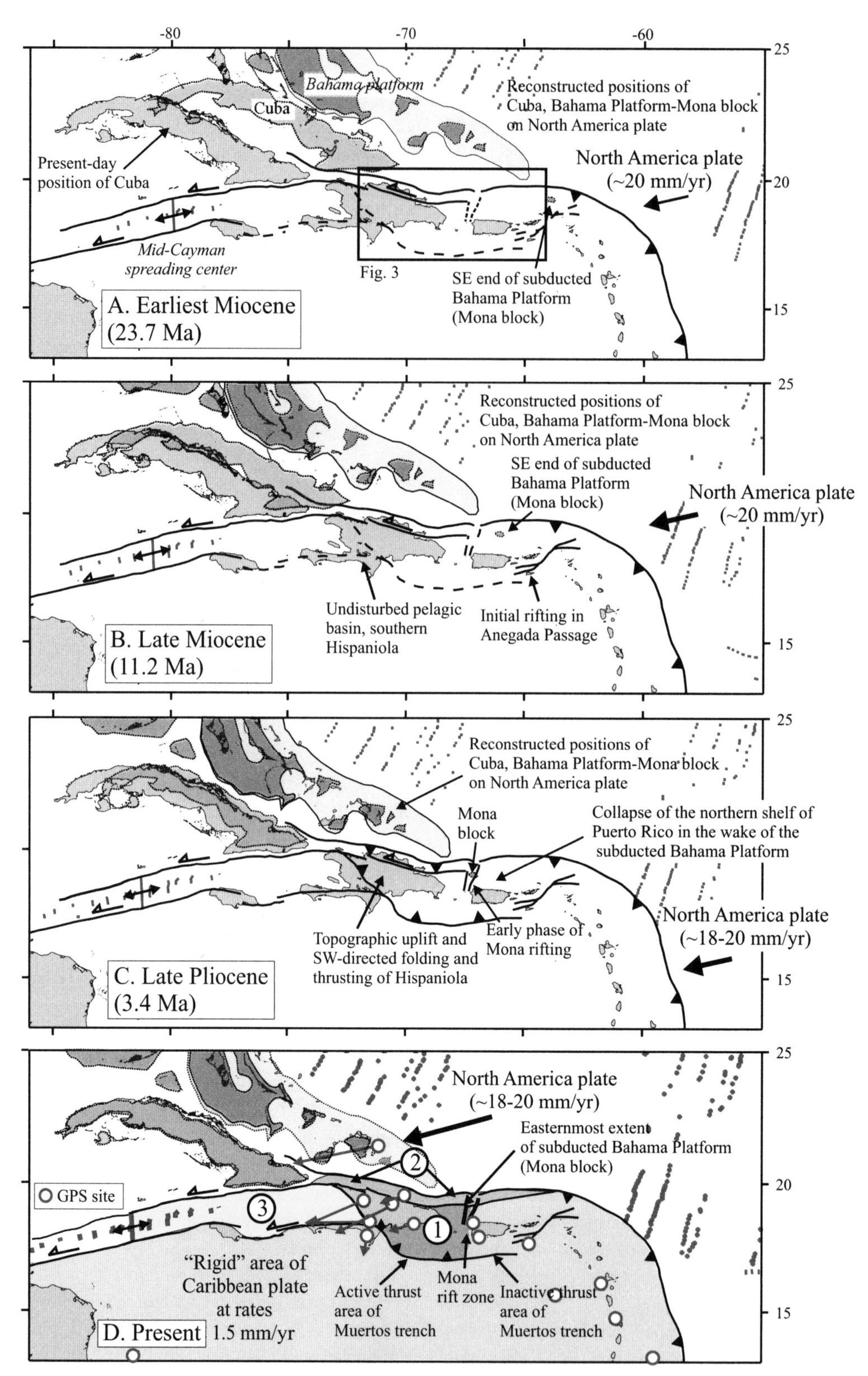

Figure 2. Plate tectonic reconstructions of the Hispaniola-Bahamas collision zone for four intervals (modified from Mann et al., 2002). The Caribbean plate is held fixed and land areas of Cuba (pink) and the Bahama carbonate platform (high-standing banks are dark blue, deeper bank areas are light blue) are moved with North America. An uncolored outline of the present-day land area of Cuba is shown on the North America plate for reference. Red dots are magnetic anomaly picks in oceanic crust of the Atlantic Ocean and Cayman trough. Oblique collision between the Bahama Platform and the northeastern corner of the plate is proposed to be the process for forming the microplates shown on Figure 2D. (A) Earliest Miocene (23.7 Ma) reconstruction. (B) Late Miocene (11.2 Ma). (C) Late Pliocene (3.4 Ma). (D) Present. Key to numbered microplates in D: 1—Hispaniola microplate; 2—Septentrional microplate; 3—Gonave microplate. Red dots and arrows in D show GPS site velocities relative to a fixed Caribbean plate. See text for discussion.

TABLE 1: COMPARISON OF POPULATIONS AND HISTORICAL SEISMICITY OF THE NORTH AMERICA–CARIBBEAN PLATE BOUNDARY ZONE

Country	Total population in 2004	Number of cities with populations in 2004 over 1,000,000 inhabitants	Population (per km^2)	Growth Rate (est. 2004)	Largest historical earthquakes and estimated loss of life	References
Belize	282,500	0	12	2.39%	–	–
British Virgin Islands	22,700	0	150	2.06%	1867	McCann (1985)
Cayman Islands	48,800	0	188	2.71%	–	–
Cuba	11,265,100	1	98	0.34%	1578, 1675, 1677, 1678, 1755, 1766, 1852, 1887	Calais et al. (1998)
Dominican Republic	8,545,300	1	175	1.33%	1564, 1783, 1842, 1887, 1899	Scherer (1912); Kelleher et al. (1973)
Guatemala	11,917,800	0	109	2.61%	1785, 1816, 1976	Plafker (1976); White and Harlow (1993)
Haiti	8,666,200	1	312	1.71%	1751, 1770, 1842, 1860, 1887	Scherer (1912); Kelleher et al. (1973)
Honduras	6,530,300	0	58	2.24%	1849, 1853, 1854, 1856, 1859, 1862, 1873	Osieki (1981)
Jamaica	2,658,600	0	242	0.66%	1692, 1712, 1722, 1744, 1812, 1907, 1957	Shepherd and Aspinall (1980); Van Dusen and Doser (2000)
Puerto Rico	3,942,100	0	433	0.49%	1787, 1867, 1918	McCann (1985); Doser et al. (this volume)
U.S. Virgin Islands	122,600	0	348	–.05%	1867	McCann (1985); Doser et al. (this volume)
TOTAL	54,002,000	3	118 (avg.)	–1.5% (avg.)	40 events	

plate boundary occurred in 1976 and affected mainly rural areas of eastern and central Guatemala adjacent to the plate boundary fault in that region (Motagua fault zone). The sudden, one meter, left-lateral slip of a 230-km-long segment of the plate boundary fault produced a catastrophic M7.2 earthquake. The resulting shock and subsequent strong aftershocks killed ~26,000 Guatemalans and left more than a million homeless in a country with a total population of 5.5 million (Plafker, 1976). Most of these victims were living in small towns or in rural settings. A similar scenario of a deadly event affecting similar, non-urban settings is possible in the dense populations living along the 3200-km-long plate boundary shown on Figure 1.

Presently, there are only three cities with populations greater than one million inhabitants in the North America–Caribbean plate boundary zone (Table 1). Two of these cities, Port-au-Prince, Haiti, and Santo Domingo, Dominican Republic, are less than 100 km from major plate boundary fault zones (Fig. 1). These cities represent only a small fraction of the 54 million persons living near the plate boundary zone. Although both cities are rapidly growing, most inhabitants of the plate boundary zone are living in cities in the size range of 100,000 persons or fewer. Some of these cities, including Santiago in the Dominican Republic (population 753,000), are within 10 km of a major plate boundary fault and are therefore subject to more potential damage due to their proximity to seismogenic plate boundary faults (Figs. 3 and 4).

ORGANIZATION AND HISTORY OF THIS VOLUME

This volume deals with seismic hazards of the relatively small region of Puerto Rico and the Virgin Islands (Fig. 1) that encompasses a relatively small subset (~4 million persons) of the total population of the North America–Caribbean plate boundary zone (Table 1). Puerto Rico and the Virgin Islands have long been studied for their pre-Quaternary bedrock geology (cf. volume by Lidiak and Larue, 1998), but studies of Quaternary geology and active tectonics have been relatively rare (Table 2). This volume attempts to fill that large gap in the knowledge base of Quaternary geology and its record of recent tectonic activity and to improve our overall understanding of the seismic hazard in this region.

The 14 papers in this volume that follow this introduction are divided into four sections: (1) two papers of regional scope that better constrain the active tectonic setting of Puerto Rico and the Virgin Islands; (2) three studies of instrumental and historical seismicity of Puerto Rico; (3) five papers on the identification of late Quaternary faults in Puerto Rico and shallow coastal areas; and (4) four papers on seismic source, ground amplification, and paleoliquefaction.

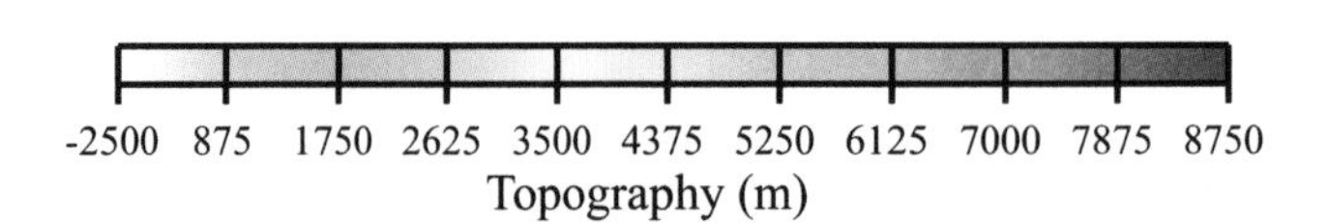

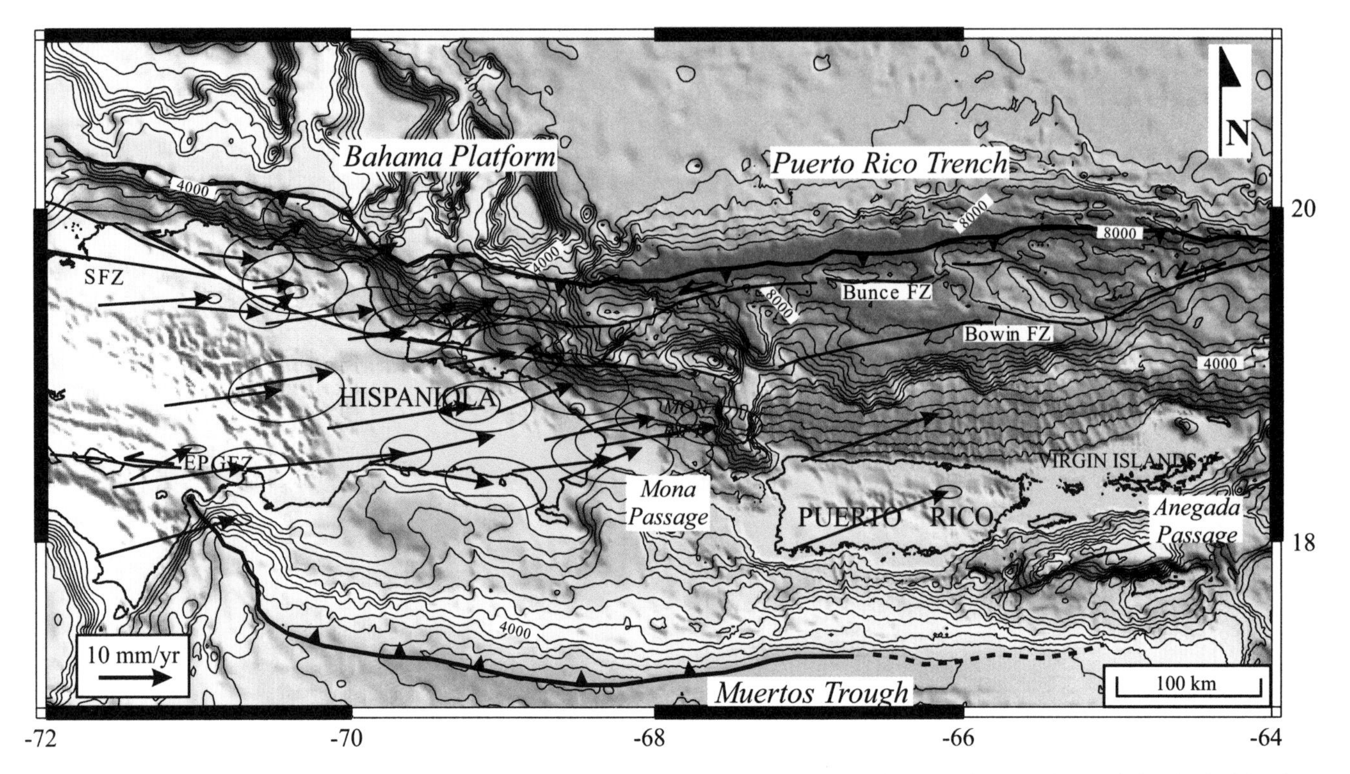

Figure 3. Topographic map of the Hispaniola, Puerto Rico, and Virgin Islands area showing main active faults and GPS-derived velocities from Calais et al. (2002) and Mann et al. (2002) with respect to the North America plate. The arrow length is proportional to the displacement rate and ellipses. Error ellipses for each vector represent two-dimensional, 95% confidence limit.

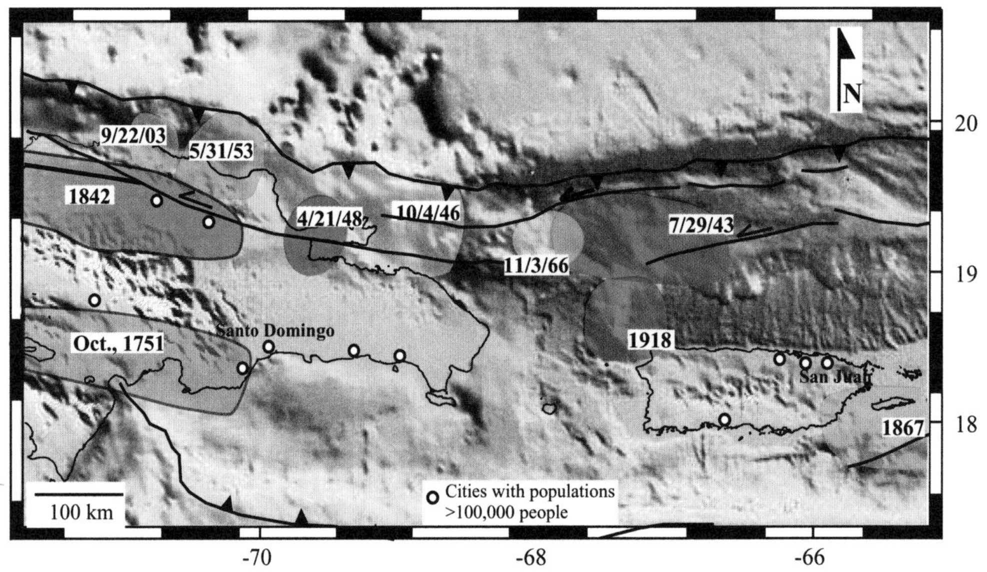

Figure 4. Map of northeastern Caribbean showing rupture zones of major earthquakes compiled from Kelleher et al. (1973), Dolan et al. (1998), and Dolan and Bowman (2004). Major cities are indicated; open circles denote cities with populations over 100,000. The area shown that includes Haiti, Dominican Republic, Puerto Rico, and the Virgin Islands has one of the densest and most rapidly growing populations in the world (Table 1). Note east to west progression of historical earthquakes across the northern margin of Hispaniola beginning in 1943 and culminating in the September 22, 2003, event in the northern Dominican Republic (M6.3).

TABLE 2. LIST OF UNPUBLISHED THESES CONCERNING THE ACTIVE TECTONICS OF THE NORTHERN CARIBBEAN PLATE BOUNDARY ZONE

1. Almy, C.C., Jr., 1965, Parguera Limestone, Upper Cretaceous Mayaguez Group, southwest Puerto Rico [Ph.D. thesis]: Houston, Texas, Rice University 203 p. (mapping of Neogene faults).
2. Cooey, J.C., 1978, Structure and stratigraphy of the offshore margin of the southern Dominican Republic [M.S. thesis]: Hattiesburg, Mississippi, University of Southern Mississippi, 53 p. (mapping study of active features associated with the Muertos trench and Beata Ridge).
3. Curet, A.F., 1976, Geology of the Cretaceous-Tertiary rocks of the southwest quarter of the Monte Guillarte quadrangle, west central Puerto Rico [M.Sc. thesis]: University of Minnesota, Duluth, 121 p.
4. **DelGreco, L.,** Identification of Holocene(?) faulting on the shallow insular shelf of Puerto Rico: Implications for seismic hazards [M.S. thesis]: University of Carolina Wilmington, 79 p. (mapping of coastal faults).
5. Detrich, C., 1995, Characterization of faults off the western coast of Puerto Rico using seismic reflection profiles [M.S. thesis]: Lehigh University, 122 p. (mapping of coastal faults).
6. Erikson, J.P., 1993, Neogene sedimentation and tectonics in the Cibao basin and northern Hispaniola [Ph.D. thesis]: Hanover, New Hampshire, Dartmouth College (study included mapping of young faults in Puerto Rico and gravity modeling of the Septentrional fault in northern Hispaniola).
7. Feliciano, J., 1985, Stratigraphic and structural analysis of the Middle to Late Tertiary sequence on the southwest coast of Puerto Rico [senior thesis]: Mayaguez, Puerto Rico, University of Puerto Rico, 12 p. plus appendix (mapping of late Neogene faults in the youngest exposed section in southern Puerto Rico).
8. Feliciano, J., 1993, Paleomagnetic study of mid-Tertiary rocks in the northeastern Caribbean: Investigations of suspected tectonic block rotations. [M.S. thesis]: Syracuse, New York, Syracuse University, 61 p. (paleomagnetic study of Neogene rocks on Antigua showing 25° of counterclockwise rotation since Oligocene time).
9. Fernandez, S., 1988, Recent faulting on southeastern Puerto Rico [senior thesis]: Mayaguez, Puerto Rico, University of Puerto Rico–Mayaguez, 20 p. (mapping of young faults in bedrock and alluvium in southwestern Puerto Rico).
10. Feuillet, N., 2000, Sismotectonique des Petites Antilles: Liaison entre activite sismique et volcanique [Ph.D. dissertation]: University of Paris, VII, 234 p. (fault striation studies in Lesser Antilles).
11. Fischer, K.M., 1984, Velocity modeling and earthquake relocation in the northeast Caribbean [Ph.D. thesis]: New York, Columbia University (study of the seismotectonics of the northeastern Caribbean region).
12. Frankel, A., 1982, Earthquake source parameters and seismic attenuation in the northeast Caribbean [Ph.D. thesis]: New York, Columbia University, 159 p. (seismological study of source parameters, scaling and attenuation relationships, precursors, and a composite focal mechanism of small earthquakes in the Virgin Islands area).
13. Gill, I.P., 1989, The evolution of Tertiary St. Croix [Ph.D. thesis]: Baton Rouge, Louisiana, Louisiana State University, 320 p. (biostratigraphic study of Neogene rocks on St. Croix documenting fault movements along Anegada Passage fault zone and transtensional separation of St. Croix and Puerto Rico).
14. Hearne, M., 2004, Three-dimensional investigation of slope failure along the northern margin of Puerto Rico [MS Thesis]: University of Carolina at Wilmington, 87 p. (study of tsunamigenic slumping along the north coast of Puerto Rico).
15. Houlgatte, E., 1983, Etude d'une Partie de la Frontière Nord-est de la Plaque Caraïbe [M.A. thesis]: Brest, Université de Bretagne Occidentale: 69 p. (marine geophysical study of active faulting in the Anegada Passage using seismic reflection profiles).
16. **Huerfano, V.,** 1996, Modelo de Corteza y Mecanica del Bloque de Puerto Rico (Crustal model and Stress Regime of the Puerto Rico block) [M.S. thesis]: Mayaguez, Puerto Rico, University of Puerto Rico–Mayaguez (Physics Dept.), 90 p. (mechanical modeling of the Puerto Rico microplate).
17. Jany, I., 1989, Neotectonique au sud des Grandes Antilles, collision (ride de Beata, Presqu'ile de Bahoruco); subduction (fosse de Muertos); transtension (passage d'Anegada) [Ph.D. thesis]: Paris, Université Pierre-et-Marie-Curie (Paris-VI), 306 p. (marine geophysical mapping of active faulting at the northern end of the Beata Ridge, the Muertos Trench, and the Anegada Passage).
18. Joyce, J.J., 1985, High pressure–low temperature metamorphism and tectonic evolution of the Samana Peninsula, Dominican Republic (Greater Antilles) [Ph.D. thesis]: Evanston, Illinois, Northwestern University, 246 p. (mapping study included the eastern part of the Septentrional fault zone, Dominican Republic).
19. Kafka, A.L., 1980, Caribbean tectonic processes, seismic surface wave source and path property analysis [Ph.D. thesis]: Stonybrook, New York, State University of New York at Stonybrook, 121 p. (early study of earthquake focal mechanisms in northern Caribbean region; includes compilation of focal mechanisms in the Hispaniola, Puerto Rico, and central Caribbean areas).
20. Kazcor, L, 1987, Petrology of the Mid-Cretaceous Aguas Buenas and Rio Maton, limestones in southeast-Central Puerto Rico [M.Sc. thesis]: Chapel Hill, University of North Carolina, 121 p.
21. Ladd, J., 1974, South Atlantic seafloor spreading and Caribbean tectonics [Ph.D. thesis]: New York, Columbia University, 251 p. (includes marine geophysical data that showed active accretionary prism development in the Los Muertos Deformed Belt south of Hispaniola and Puerto Rico, as well as other major tectonic features along the southern and northern margins of the Caribbean basin).
22. **Laó Dávila,** D., 2002, Tertiary tectonics of western Puerto Rico: paleostress and paleomagnetic study [M.S. thesis]: Florida International University, Miami, Florida, 189 p. (fault striation and paleomagnetic study).
23. Lee, V., 1974, Petrography, metamorphism, and geochemistry of the Bermeja Complex and related rocks in southwestern Puerto Rico and their significance in the evolution of the eastern Greater Antillian Island Arc [M.S. thesis]: State University of New York at Albany, 120 p.
24. Leroy, S., 1995, Structure et origine de la plaque Caraibe: Implications geodynamiques [Ph.D. thesis (Doctoral d'Universite)]: Paris, Université Pierre-et-Marie-Curie, 250 p. (marine geophysical mapping of active, intraplate faults along the Beata Ridge, Cayman Trough, and adjacent areas in the Colombian and Venezuelan basins).
25. Longshore. J.D., 1965. Chemical and mineralogical variations in the Virgin Island Batholith and its associated wall rocks [Ph.D. thesis]: Houston, Texas, Rice University. 94 p.
26. Lopez, A., 2000, Microplate behavior in the northeastern Caribbean as constrained by Global Positioning System (GPS) geodesy [M.S. thesis]: University of Puerto Rico–Mayaguez, 67 p. (GPS study of fault motions).

(*continued*)

TABLE 2. LIST OF UNPUBLISHED THESES CONCERNING THE ACTIVE TECTONICS OF THE NORTHERN CARIBBEAN PLATE BOUNDARY ZONE (*continued*)

27. Matthews, J.E., 1970, Geology of the northeastern Caribbean [M.A. thesis]: Salt Lake City, Utah, University of Utah (pioneer study of the marine geology of the Beata Ridge and Los Muertos trench).
28. **McCann, W.**, 1980, Large and moderate-size earthquakes: Their relationships to the tectonics of subduction [Ph.D. thesis]: New York, Columbia University, 189 p. (seismological and marine geophysical study that included analysis of earthquakes and active faults of the northern and eastern margins of the Caribbean plate).
29. **Moya, J.**, 1998, The neotectonics of western Puerto Rico [Ph.D. thesis]: University of Colorado at Boulder, 132 p. (wide-ranging study of neotectonic features).
30. Muszala, S., 1999, Aeromagnetics of the North Slope of Alaska and magnetics of the Puerto Rico trench [M.S. thesis]: University of Texas at Austin, 101 p. (used aeromagnetic data to map faults).
31. Otalora, G., 1961, Geology of the Barranquitas quadrangle. Central Puerto Rico [Ph.D. thesis]: Princeton, New Jersey, Princeton University, 152 p.
32. Reid, J.A., 1992, Paleomagnetism and rock magnetics of the middle Tertiary carbonate sequence. North Coast, Puerto Rico, with tectonic implications [M.S. thesis]: Syracuse. New York. Syracuse University, 117 p., with five appendices (paleomagnetic study of Neogene limestones that revealed 24.5° of counterclockwise rotation of Puerto Rico during late Miocene–early Pliocene time).
33. Schellekens, J.H., 1993, Geochemical evolution of volcanic rocks in Puerto Rico [Ph.D. thesis]: Syracuse, New York, University of Syracuse, 289 p.
34. Slodowski, T.R., 1956, Geology of the Yauco area, Puerto Rico [Ph.D. thesis]: Princeton, New Jersey, Princeton University, 130 p.
35. Taggart, B.E., 1991, Radiometric dating of marine terraces and coral reefs on northwestern Puerto Rico [Ph.D. thesis]: Mayaguez, Puerto Rico, University of Puerto Rico–Mayaguez, 252 p. (mapping and radiometric dating study of young coral reef terraces in Puerto Rico).
36. Treadgold, G.E., 1975, Modeling and interpretation of the oceanic crustal structure north of the Puerto Rico trench [M.A. thesis]: Austin, Texas, The University of Texas at Austin, 190 p. (seismic reflection and refraction study of normal faults along the outer bulge of the subducting North America plate adjacent to the Puerto Rico trench).
37. Tzeng, S.-Y., 1976, Low-grade metamorphism in east-central Puerto Rico [Ph.D. thesis]: Pittsburgh, Pennsylvania, University of Pittsburgh, 159 p.
38. **van Gestel, J-P.**, 2000,1, Tectonics of the Puerto Rico trench and Puerto Rico–Virgin Islands Platform as Determined from Studies of Seismic Reflection Data; II, Migration of Ground Penetrating Radar Data [Ph.D. dissertation]: University of Texas at Austin, 213 p. (used seismic data to map faults).
39. Velez, R., 1985, A survey of seismic reflection data for Puerto Rico and its environs [senior thesis): Mayaguez, Puerto Rico, University of Puerto Rico–Mayaguez, 25 p. (compilation of petroleum industry seismic reflection profiles and maps of late Neogene structures around Puerto Rico).
40. Weiland, T.J., 1988, Petrology of volcanic rocks in lower Cretaceous formations of northeastern Puerto Rico [Ph.D. thesis]: Chapel Hill, North Carolina, University of North Carolina. 205 p.
41. While, P.M., 1993, Structural evolution of the San Pedro basin: Implications for the paleogeographic position of the Dominican Republic and Puerto Rico in a plate-tectonic framework, University of Texas at Arlington [M.S. thesis]: 46 p. (used industry seismic data for basin analysis).

*Note: Authors contributing to this volume are in bold type.

The multidisciplinary approach of this volume grew out of a workshop held March 23–24, 1999, in San Juan, Puerto Rico, sponsored by the U.S. Geological Survey (USGS). (Workshop convenors were U. ten Brink, W. Dillon, A. Frankel, C. Mueller, and R. Rodriguez.) The host institution for the workshop was Colegio de Ingenieros y Agrimensores of San Juan, Puerto Rico. The complete workshop report was published as USGS Open-file Report 99-353 (available online at: http://pubs.usgs.gov/of/of99-353/). The workshop was historic in that it brought together for the first time all groups concerned with seismic hazard in Puerto Rico and the Virgin Islands, including scientists, structural engineers, and public and private officials charged with the establishment of building codes, probabilistic assessment of seismic hazard, and educating the public on seismic hazard.

Almost 100 workshop participants were divided into six working groups that prepared specific recommendations for promoting an increased understanding of the seismic hazards of the region. These groups included: (1) marine geology and geophysics; (2) paleoseismology and active faults; (3) earthquake seismology; (4) earthquake engineering; (5) tsunami hazards; and (6) societal concerns. All information was to be incorporated into a new USGS seismic hazard map of the island that would in turn help influence the establishment of revised building codes.

The contributors to this volume followed up on many of the recommendations put forward by the 1999 USGS workshop (cf. open-file report for a complete listing of all of the USGS workshops recommendations):

Marine geology and geophysics: conduct high-resolution sonar and seismic reflection studies around possible active faults in the nearshore region (Grindlay et al., Chapter 7).

Paleoseismology and active faults: (1) conduct field reconnaissance aimed at identifying Quaternary faults and determining their paleoseismic chronology and slip rates, as well as identifying and dating paleoliquefaction features from large earthquakes; (2) Quaternary mapping of marine terraces, fluvial terraces and basins, beach ridges, etc., to establish a framework for understanding neotectonic deformation of the island; and (3) interpretation of aerial photography to identify possible Quaternary faults (Hippolyte et al., Chapter 8; Mann et al., Chapter 6; Mann et al., Chapter 9; Prentice and Mann, Chapter 10; Tuttle et al., Chapter 13).

Earthquake seismology: (1) determine an empirical seismic attenuation function using observations from local seismic networks; (2) evaluate existing earthquake catalogs from local networks and regional stations and complete the catalogs; and (3) use GPS measurements to constrain deformation rates used in seismic-hazard maps (Motazedian and Atkinson, Chapter 3; Huerfano et al., Chapter 4; Doser et al., Chapter 5).

Earthquake engineering: (1) prepare liquefaction susceptibility maps for urban areas; (2) update and improve databases for types of site conditions; (3) collect site effect observations and near-surface geophysical measurements for future urban area hazard maps (LaForge and McCann, Chapter 11; Hengesh and Bachhuber, Chapter 12; Tuttle et al., Chapter 13; Macari and Hoyos, Chapter 14).The group reporting in this volume did not directly address the final two themes of the USGS workshop: tsunami hazard and societal concerns.

Several of the papers of the volume (Mann et al., Chapter 6; Grindlay et al., Chapter 7; Hippolyte et al., Chapter 8; Mann et al., Chapter 9; Prentice and Mann, Chapter 10; Hengesh and Bachhuber, Chapter 12; Tuttle et al., Chapter 13) grew out of several proposals funded in 2000 and 2001 by the U.S. National Earthquake Hazards Reduction Program of the USGS. The data reported in this volume will be integrated by the USGS in its revised earthquake hazards map for Puerto Rico and the U.S. Virgin Islands.

RESOURCES FOR NEOTECTONIC RESEARCH IN THE PUERTO RICO–VIRGIN ISLANDS AREA

There are increasing resources for neotectonic research in this region as the awareness becomes greater for the high seismic risk for the region. Basic resources for researchers include almost complete topographic and geologic map coverage of the islands compiled by the USGS (Figs. 5 and 6). These maps can be used in conjunction with aerial photographs at various scales available at cost from the Puerto Rico highway department in San Juan. In our work, we have found the 1:18,500 scale aerial photographs taken in 1936 in Puerto Rico to be particularly useful for studies

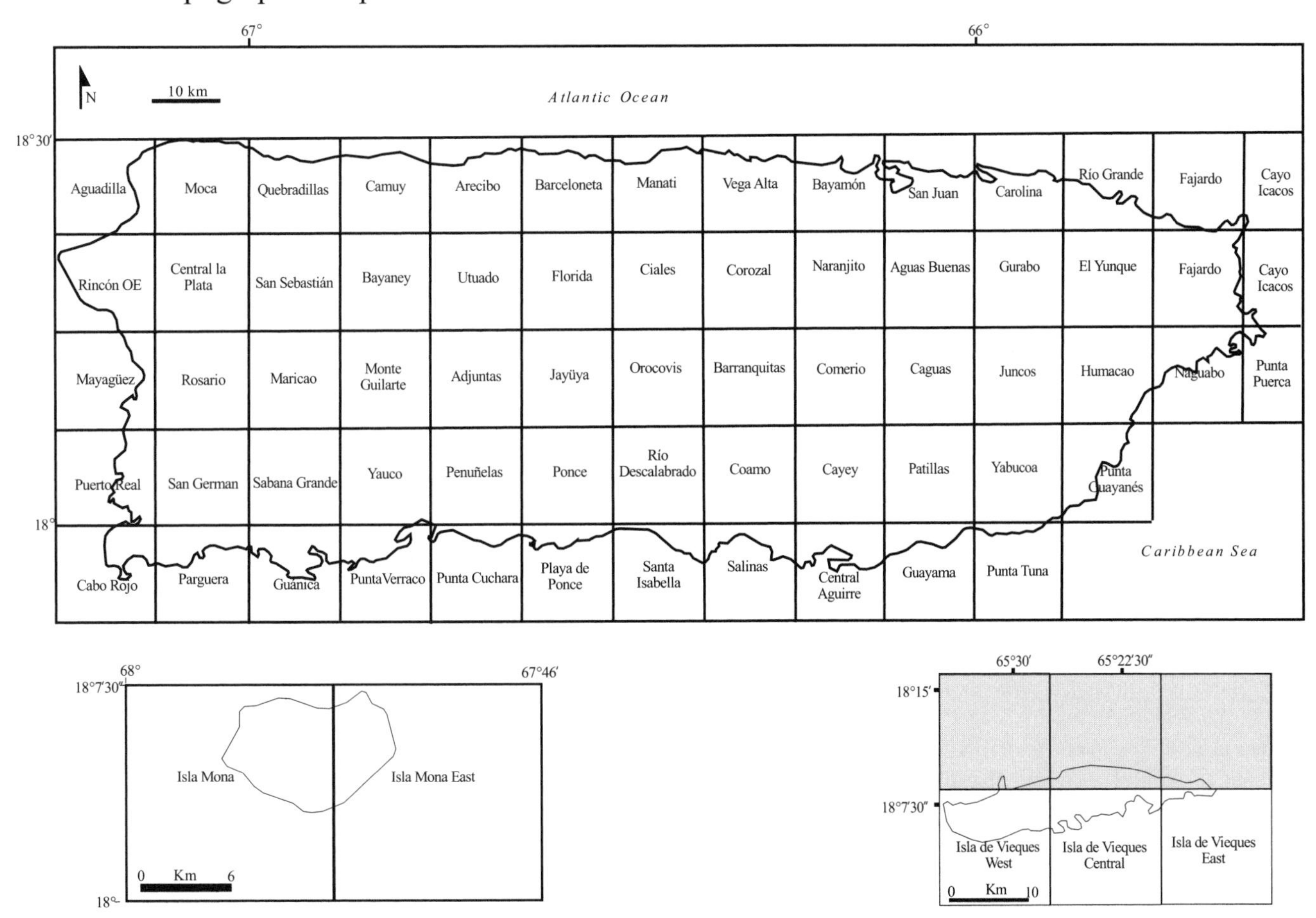

Figure 5. Index map showing 1:25,000-scale U.S. Geological Survey topographic maps used for field studies in Puerto Rico, Mona Island, Vieques Island, and St. Croix (U.S. Virgin Islands). No topographic maps are available for gray boxes.

USGS geologic quadrangle maps of Puerto Rico

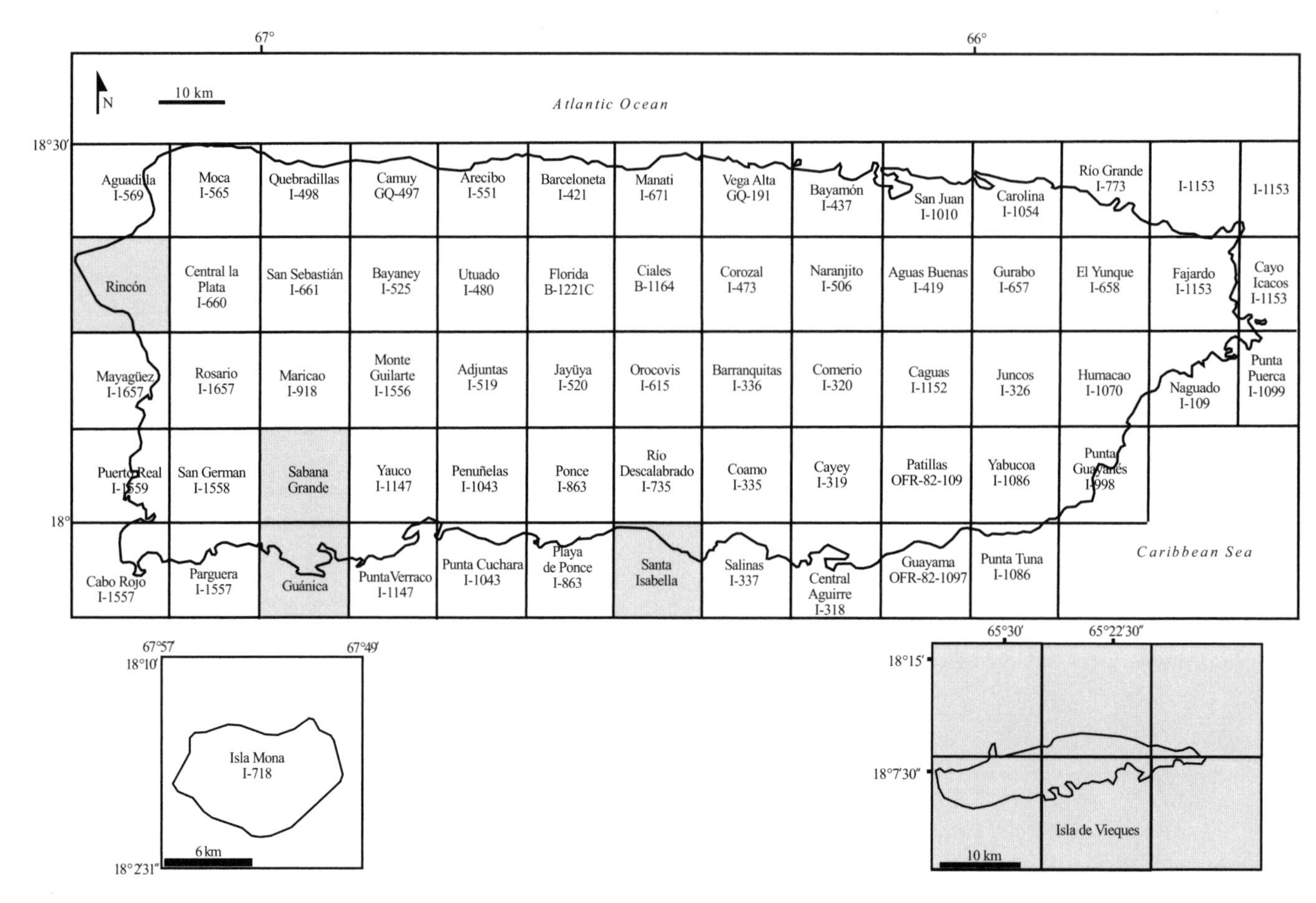

Figure 6. Index map showing 1:25,000-scale U.S. Geological Survey geologic maps used for field studies in Puerto Rico, Mona Island, Vieques Island, and St. Croix (U.S. Virgin Islands). Grey quadrangles have not been mapped.

of fault scarps (cf. Mann et al., Chapter 6; Mann et al., Chapter 9; Prentice and Mann, Chapter 10).

On Figure 7 and in Table 2, we have compiled unpublished thesis studies at the undergraduate and graduate level in the region which commonly provide useful sources of field observations and map information. Many of these studies have been carried out by scientists associated with the Department of Geology of the University of Puerto Rico in Mayaguez, which offers both B.S. and M.S. degrees in geology and B.S., M.S., and Ph.D. degrees in marine sciences. Most U.S.-based thesis studies can be accessed through University Microfilms International (ProQuest) whereas foreign-based theses need to be obtained through interlibrary loan.

Finally, digital data products are becoming more widespread and useful for neotectonic studies. Local earthquake data recorded by the Puerto Rico Seismic Network based at the Department of Geology in Mayaguez are available digitally (cf. Huerfano et al., Chapter 4). Topographic digital elevation models and compilations of offshore bathymetric data are also available through the University of Puerto Rico or the U.S. Geological Survey (cf. Grindlay et al., Chapter 7). Gravity data as shown in Figure 1A in this paper is also available over the Internet (Smith and Sandwell, 1997).

CLOSING RECOMMENDATION

We are entering a productive period for neotectonic studies in Puerto Rico and the Virgin Islands as demonstrated by the multidisciplinary studies in this volume as well as the many other previous works cited in these papers. However, for effective seismic hazard mitigation in this vulnerable region, earthquake research must be accompanied by progress in earthquake engineering and in public outreach and education. The USGS workshop in 1999 and this volume is a solid start, but much more work lies ahead in integrating earthquake research, earthquake engineering, and public outreach and earthquake education.

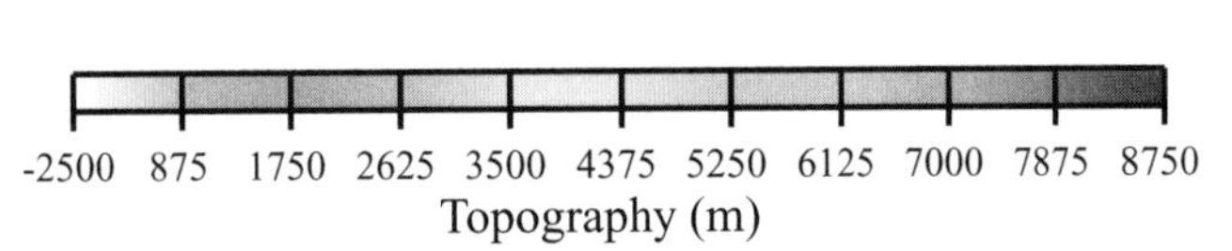

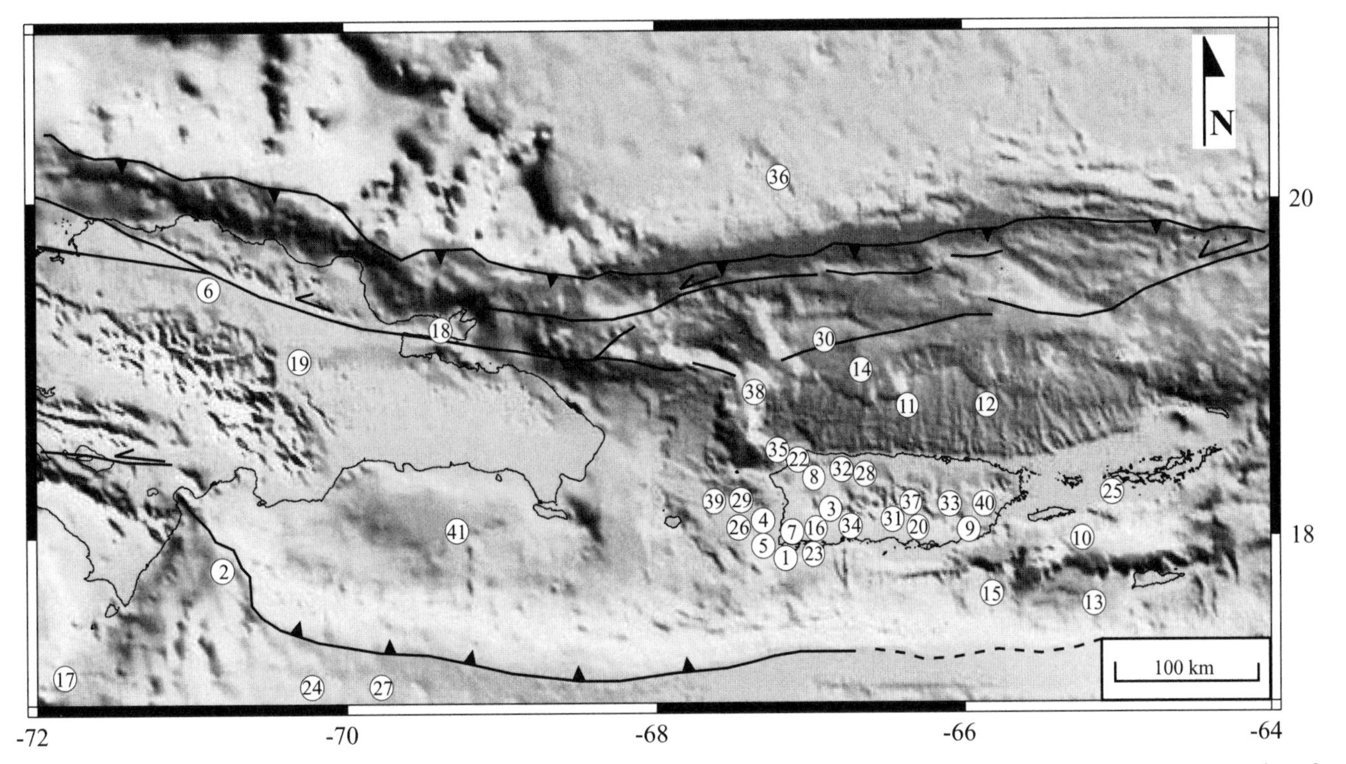

Figure 7. Map showing approximate area of thesis studies compiled on Table 2 (numbers are keyed to numbers on table). Note concentration of theses in the area of western Puerto Rico, partly because this area is the location of the Department of Geology of the University of Puerto Rico at Mayaguez.

ACKNOWLEDGMENTS

Field work by Mann in Puerto Rico was supported by the USGS National Earthquake Hazards Reduction Program (grants 1434-00-G-2341 and 1434-00-G-2342). I thank M. Hearne and N. Grindlay for kindly providing the topographic base used in Figures 3–5. Key long-term collaborators in Puerto Rico include C. von Hildebrandt and V. Huérfano of the Puerto Rico Seismic Network of the University of Puerto Rico at Mayaguez and J. Joyce, and H. Schellekens of the Department of Geology of the University of Puerto Rico at Mayaguez. I would like to thank the following colleagues for volunteering their time to review the papers presented in this volume: E. Asencio, G. Atkinson, E. Calais, C. DeMets, J.W. Dewey, J. Dolan, G. Draper, C. Fenton, A. Fernandez, A. Frankel, M. Gordon, N. Grindlay, T. Holcombe, K. Knudsen, R. LaForge, W. McCann, P. McCrory, C. Mueller, O. Perez, T. Pratt, C. Prentice, E. Schweig, D. Scholl, F. Taylor, U. ten Brink, J. Tinsley, T. Tuttle, and A. Villesenor. Special thanks to A. Basu for his patience and editorial efforts and to C. Watson for GIS, computer illustration, and editing work. University of Texas at Austin Institute for Geophysics (UTIG) contribution no. 1742. The authors acknowledge the financial support for this publication provided by The University of Texas at Austin's Geology Foundation and Jackson School of Geosciences.

REFERENCES CITED

Bilham, R., 1988, Earthquakes and urban growth: Nature, v. 336, p. 625–626, doi: 10.1038/336625A0.

Calais, E., Perrot, J., and Mercier de Lepinay, B., 1998, Strike-slip tectonics and seismicity along the northern Caribbean plate boundary from Cuba to Hispaniola, *in* Dolan, J., and Mann, P., eds., Active Strike-slip and Collisional Tectonics of the Northern Caribbean Plate Boundary Zone: Geological Society of America Special Paper 326, p. 125–141.

Calais, E., Mazabraud, Y., Mercier de Lepinay, B., Mann, P., Mattioli, G., and Jansma, P., 2002, Strain partitioning and fault slip rates in the northeastern Caribbean from GPS measurements: Geophysical Research Letters, v. 29, 1856, doi:10.1029/2002GL015397.

DeMets, C., Jansma, P., Mattioli, G., Dixon, T., Farina, F., Bilham, R., Calais, E., and Mann, P., 2000, GPS geodetic constraints on Caribbean-North America plate motion: Geophysical Research Letters, v. 27, p. 437–440, doi: 10.1029/1999GL005436.

Dillon, W., Austin, J., Jr, Scanlon, K., Edgar, N., and Parson, L., 1992, Accretionary margin of northwestern Hispaniola: Morphology, structure, and development of part of the northern Caribbean plate boundary: Marine and Petroleum Geology, v. 9, p. 70–88, doi: 10.1016/0264-8172(92)90005-Y.

Dillon, W., Edgar, N., Scanlon, K., and Coleman, D., 1996, A review of the tectonic problems of the strike-slip northern boundary of the Caribbean plate and examination by GLORIA, *in* Gardner, J., Field, M., and Twichell, D., eds., Geology of the United States' Seafloor: The View from GLORIA: Cambridge, U.K., Cambridge University Press, p. 135–164.

Dolan, J., and Bowman, D., 2004, Tectonic and seismologic setting of the 22 September, 2003, Puerto Plata, Dominican Republic earthquake: Implications for earthquake hazard in northern Hispaniola: Seismological Research Letters, v. 75, p. 587–597.

Dolan, J., Mullins, H., and Wald, D., 1998, Active tectonics in the north-central Caribbean: Oblique collision, strain partitioning, and opposing subduction slabs, *in* Dolan, J., and Mann, P., eds., Active Strike-slip and Collisional Tectonics of the Northern Caribbean Plate Boundary Zone: Geological Society of America Special Paper 326, p. 143–169.

Jansma, P., Mattioli, G., Lopez, A., DeMets, C., Dixon, T., Mann, P., and Calais, E., 2000, Neotectonics of Puerto Rico and the Virgin Islands, northeastern Caribbean from GPS geodesy: Tectonics, v. 19, p. 1021–1037.

Kelleher, J., Sykes, L., and Oliver, J., 1973, Possible criteria for predicting earthquakes and their application to major plate boundaries of the Pacific and the Caribbean: Journal of Geophysical Research, v. 78, p. 2547–2585.

Leroy, S., Mercier de Lepinay, B., Mauffret, A., and Pubellier, M., 1996, Structural and tectonic evolution of the eastern Cayman trough (Caribbean Sea) from seismic reflection data: AAPG Bulletin, v. 80, p. 222–247.

Lidiak, E., and Larue, D., editors, 1998, Tectonics and Geochemistry of the Northeastern Caribbean: Geological Society of America Special Paper 322, 194 p.

Mann, P., Prentice, C., Burr, G., Pena, L., and Taylor, F., 1998, Tectonic geomorphology and paleoseismology of the Septentrional fault system, Dominican Republic, *in* Dolan, J., and Mann, P., eds., Active Strike-slip and Collisional Tectonics of the Northern Caribbean Plate Boundary Zone: Geological Society of America Special Paper 326, p. 63–123.

Mann, P., Calais, E., Ruegg, J.-C., DeMets, C., Jansma, P., and Mattioli, G., 2002, Oblique collision in the northeastern Caribbean from GPS measurements and geological observations: Tectonics, v. 21, doi: 10.1029?2001TC001304, 2002.

Masson, D., and Scanlon, K., 1991, The neotectonic setting of Puerto Rico: Geological Society of America Bulletin, v. 103, p. 144–154, doi: 10.1130/0016-7606(1991)103<0144:TNSOPR>2.3.CO;2.

McCann, W., 1985, On the earthquake hazards of Puerto Rico and the Virgin Islands: Bulletin of the Seismological Society of America, v. 75, p. 251–262.

Osieki, P., 1981, Estimated intensities and probably tectonic sources of historic (pre-1898) Honduran earthquakes: Bulletin of the Seismological Society of America, v. 71, p. 865–881.

Plafker, G., 1976, Tectonic aspects of the Guatemalan earthquake of 4 February, 1976: Science, v. 193, p. 1201–1208.

Prentice, C., Mann, P., Taylor, F., Burr, G., and Valastro, S., Jr., 1993, Paleoseismicity of the North American-Caribbean plate boundary (Septentrional fault), Dominican Republic: Geology, v. 21, p. 49–52.

Prentice, C., Mann, P., Pena, L., and Burr, G., 2003, Slip rate and earthquake recurrence along the central Septentrional fault, North American–Caribbean plate boundary, Dominican Republic: Journal of Geophysical Research, v. 108, 2149, doi: 10.1029/2001JB000442.

Rosencrantz, E., and Mann, P., 1991, SeaMARC II mapping of transform faults in the Cayman trough: Geology, v. 19, p. 690–693, doi: 10.1130/0091-7613(1991)0192.3.CO;2.

Scherer, J., 1912, Great earthquakes in the island of Haiti: Bulletin of the Seismological Society of America, v. 2, p. 161–180.

Schwartz, D., Cluff, L., and Donnelly, T., 1979, Quaternary faulting along the Caribbean–North American plate boundary in Central America: Tectonophysics, v. 52, p. 431–445, doi: 10.1016/0040-1951(79)90258-0.

Shepherd, J., and Aspinall, W., 1980, Seismicity and seismic intensities in Jamaica, West Indies: A problem in risk assessment: Earthquake Engineering and Structural Dynamics, v. 8, p. 315–335.

Sieh, K., 1981, A review of geological evidence for recurrence times of large earthquakes, *in* Simpson, D., and Richards, P., eds., Earthquake Prediction—An International Review, Maurice Ewing Series: American Geophysical Union, Washington, D.C., v. 4, p. 181–207.

Smith, W., and Sandwell, D., 1997, Global sea floor topography from satellite altimetry and ship depth soundings: Science, v. 277, p. 1956–1962, doi: 10.1126/SCIENCE.277.5334.1956.

Vogt, P., Lowrie, A., Bracey, D., and Hey, R., 1976, Subduction of aseismic oceanic ridges: Effects on shape, seismicity, and other characteristics of consuming plate boundaries: Geological Society of America Special Paper 172, 59 p.

Van Dusen, S., and Doser, D., 2000, Faulting processes of historic (1917–1962) $M \geq 6.0$ earthquakes along the north-central Caribbean margin: Pure and Applied Geophysics, v. 157, p. 719–736.

White, R., and Harlow, D., 1993, Destructive upper-crustal earthquakes of Central America since 1900: Bulletin of the Seismological Society of America, v. 83, p. 1115–1142.

Yeats, R., Sieh, K., and Allen, C., 1997, The Geology of Earthquakes: Oxford University Press, Oxford and New York, 568 p.

Manuscript Accepted by the Society August 18, 2004

Printed in the USA

Geological Society of America
Special Paper 385
2005

GPS results from Puerto Rico and the Virgin Islands: Constraints on tectonic setting and rates of active faulting

Pamela E. Jansma
Glen S. Mattioli
Department of Geosciences, University of Arkansas, Fayetteville, Arkansas 72701, USA

ABSTRACT

Puerto Rico and the northern Virgin Islands define the eastern terminus of the Greater Antilles, which extend eastward from offshore eastern Central America to the Lesser Antilles volcanic arc and mark the boundary between the Caribbean and North America plates. In Hispaniola, Puerto Rico, and the northern Virgin Islands, the Puerto Rico trench and the Muertos trough define the northern and southern limits of the plate boundary zone, respectively. Three microplates lie within the boundary zone: (1) the Gonave in the west; (2) the Hispaniola in the center; and (3) the Puerto Rico–northern Virgin Islands in the east. Results from Global Positioning System (GPS) geodesy conducted in the region since 1994 confirm the presence of an independently translating Puerto Rico–northern Virgin Islands microplate whose motion is 2.6 ± 2.0 mm/yr toward N82.5°W ± 34° (95%) with respect to the Caribbean. Geodetic data are consistent with east-west extension of several mm/yr from eastern Hispaniola to the eastern Virgin Islands. Extension increases westward with the most, 5 ± 3 mm/yr, accommodated in the Mona rift, confirming earlier GPS geodetic results. East-west extension of 3 ± 2 mm/yr also is observed across the island of Puerto Rico, consistent with composite focal mechanisms and regional epicentral distributions. Although the loci of extension are not known, similarity of GPS-derived velocities among sites in eastern Puerto Rico suggests the active structures lie west of the San Juan metropolitan area. Reactivation of the Great Northern and Southern Puerto Rico fault zones as oblique normal faults with right-lateral slip is a possibility. East-west extension of 2 ± 1 mm/yr also must exist between eastern Puerto Rico and Virgin Gorda, which likely is attached to the Caribbean plate. These extensional belts allow eastward transfer of slip between North America and the Caribbean from the southern part of the plate boundary zone in the west to the northern segment in the east. Motions along or across any of the individual subaerial structures of Puerto Rico are ≤2 mm/yr. The Lajas Valley in the southwest, where microseismicity is greatest, is the locus of highest permissible on-land deformation. Northwest-southeast to east-west extension of 2 ± 1 mm/yr is also observed across the Anegada Passage.

Keywords: microplate tectonics, Caribbean, GPS geodesy, extension.

Jansma, P.E., and G.S. Mattioli, G.S., 2005, GPS results from Puerto Rico and the Virgin Islands: Constraints on tectonic setting and rates of active faulting, *in* Mann, P., ed., Active tectonics and seismic hazards of Puerto Rico, the Virgin Islands, and offshore areas: Geological Society of America Special Paper 385, p. 13–30. For permission to copy, contact editing@geosociety.org.

INTRODUCTION

One of the fundamental questions concerning lithospheric behavior in zones of active tectonics is how relative motion between blocks or plates is accommodated. Is motion taken up along a few major faults or narrow belts of deformation separating discrete crustal blocks, or is strain broadly distributed throughout the deforming region as similar displacements occur on many closely spaced faults of comparable size? The two end members have different implications for seismic hazard. Larger earthquakes within narrow zones are likely to occur in the former, whereas smaller events over a broad region are more probable in the latter. Which end member more closely characterizes Puerto Rico and the northern Virgin Islands has long been controversial, i.e., do Puerto Rico and the northern Virgin Islands define a discrete, rigid block within the North America–Caribbean plate boundary within which little motion occurs, or do several faults capable of accommodating significant displacement cross the region? Recent results from Global Positioning System (GPS) geodesy provide upper bounds on the potential intrablock displacement.

Puerto Rico and the northern Virgin Islands define the eastern terminus of the Greater Antilles, which extend eastward from offshore eastern Central America to the Lesser Antilles volcanic arc and mark the boundary between the Caribbean and North America plates. Tectonic models for the northern Caribbean (e.g., Byrne et al., 1985; Mann et al., 1995) propose active microplates within the boundary zone on the basis of geologic and earthquake evidence. Seismicity along the EW-trending boundary between the North America and Caribbean plates is consistent with evolution of the boundary zone from a relatively simple set of transform faults in the west to a more complex deformation zone ~250 km wide in the east (Fig. 1A). Motion along the predominantly east-west striking major structures of the northern Caribbean is primarily left-lateral. In the west, the Swan and Oriente transform faults define the EW-trending Cayman trough and bound the short (~100 km), NS-trending Mid-Cayman spreading center (Fig. 1B). In Hispaniola, Puerto Rico, and the Virgin Islands, the Puerto Rico trench and the Muertos trough define the northern and southern limits of the plate boundary zone, respectively. Three microplates lie within this diffuse boundary zone (Fig. 1B): (1) the Gonave in the west (Mann et al., 1995); (2) the Hispaniola in the center (Byrne et al., 1985); and (3) the Puerto Rico–northern Virgin Islands in the east (Masson and Scanlon, 1991). Such a microplate model assumes that nearly all of the deformation associated with North America–Caribbean motion is concentrated along the faults that bound the three rigid blocks: the Oriente, Septentrional, Enriquillo-Plantain Garden, and Anegada faults; the Muertos trough and North Hispaniola deformed belt; and the Mona rift faults northwest of Puerto Rico (Fig. 1).

Recent results from GPS geodesy support the presence of an independently translating Puerto Rico–northern Virgin Islands microplate within the northeastern Caribbean (Jansma et al., 2000) with ~85% of the relative motion occurring between the Puerto Rico–northern Virgin Islands and North America and ~15% between the Puerto Rico–northern Virgin Islands and the Caribbean. (This distribution of interplate slip is our current best estimate, given that rigorous quantitative modeling is not yet completed, and potential elastic strain accumulation along the Puerto Rico trench has not been considered). The geodetic data are not consistent with models that advocate either counterclockwise rotation of the Puerto Rico–northern Virgin Islands about a nearby vertical axis (Masson and Scanlon, 1991; Reid et al., 1991) or eastward tectonic escape of the block within the Caribbean–North America plate boundary zone (Jany et al., 1987). Because velocities of the Puerto Rico–northern Virgin Islands microplate with respect to the Caribbean plate are small and the errors were significant, both as a consequence of the paucity of data for a Caribbean reference frame and the geographically restricted GPS network, Jansma et al. (2000) were limited in their interpretation of interplate deformation along the Puerto Rico–northern Virgin Islands–Caribbean boundary and intrablock deformation of the Puerto Rico–northern Virgin Islands. In this paper, we update the analysis of Jansma et al. (2000) with the inclusion of additional geodetic data from both existing and new continuous and campaign GPS sites in the Puerto Rico–northern Virgin Islands block. We also use ITRF00 (International Terrestrial Frame 2000) and an improved Caribbean reference frame (DeMets et al., 2000, and 2002, personal commun.). Our objectives are to refine the velocity of the Puerto Rico–northern Virgin Islands block with respect to the Caribbean plate, to constrain maximum permissible displacements along potentially active faults on the island of Puerto Rico and immediately offshore, and to examine potential diffuse extension between the Virgin Islands and eastern Puerto Rico.

TECTONIC SETTING OF PUERTO RICO AND THE VIRGIN ISLANDS

Hundreds of earthquakes per year occur within and around Puerto Rico and the Virgin Islands (Fig. 1). The majority of events are small and located offshore. Several large events have occurred during historic time, including the 1916, 1918, and 1943 Mona Passage earthquakes (Ms = 7.2, 7.3, and 7.5, respectively), the 1867 Anegada earthquake (Ms = 7.3), the 1787 Puerto Rico trench earthquake (M = 7.5?) and the 1670 San German earthquake (M = 6.5?) (Pacheco and Sykes, 1992). Noting the concentration of both current seismicity and historic events offshore (Fig. 1C), several workers proposed a rigid Puerto Rico–northern Virgin Islands block in the northeastern corner of the Caribbean (Byrne et al., 1985; Masson and Scanlon, 1991).

The EW-striking Puerto Rico trench and Muertos trough define the northern and southern limits of the Puerto Rico–northern Virgin Islands, respectively (Fig. 1). Both are characterized by diffuse zones of earthquakes that dip below the island of Puerto Rico (McCann and Pennington, 1990; McCann, 2002), consistent with subduction of Atlantic lithosphere below northern Puerto Rico and Caribbean lithosphere below the southwestern coast of the island. Interaction of the two slabs at depth has

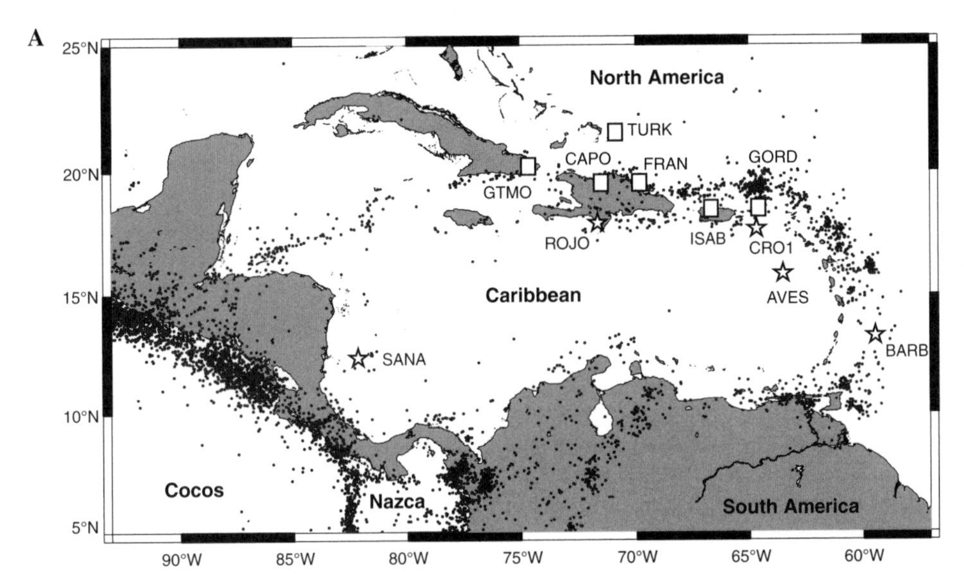

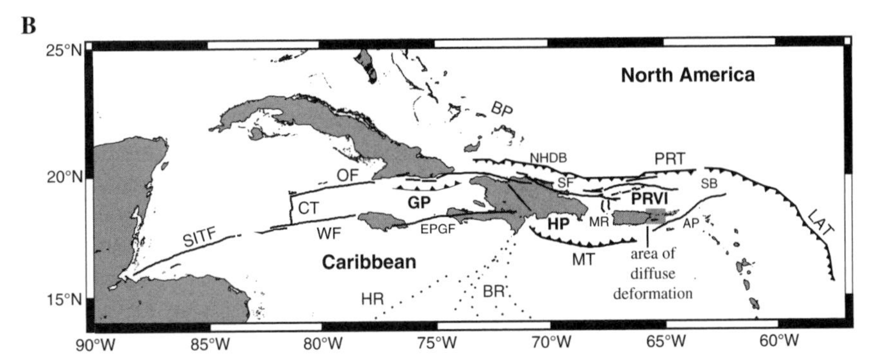

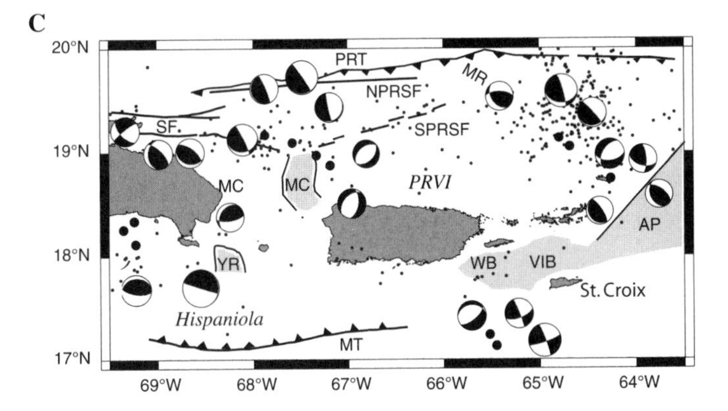

Figure 1. (A) Map of Caribbean plate and regional seismicity. Epicenters are for earthquakes above depths of 60 km with magnitudes >3.5 from 1 January 1967 until 28 April 1999 (U.S. Geological Survey). The five GPS sites used to constrain Caribbean reference frame are plotted as stars: SANA—San Andres Island; ROJO—Cabo Rojo, Dominican Republic; CRO1—St. Croix, U.S. Virgin Islands; AVES—Aves Island; and BARB—Barbados. Original CANAPE Global Positioning System sites are ROJO and CRO1 plus the squares: CAPO—Capotillo, Dominican Republic; FRAN—Cabo Frances Viejo, Dominican Republic; GORD—Virgin Gorda; GTMO—Guantanamo Bay, Cuba; ISAB—Isabela, Puerto Rico; and TURK—Grand Turk, Turks and Caicos. (B) Map of northern Caribbean plate boundary showing microplates and structures. AP—Anegada Passage; BP—Bahamas Platform; BR—Beata Ridge; CT—Cayman trough spreading center; EPGF—Enriquillo-Plantain Garden fault; GP—Gonave platelet; HP—Hispaniola platelet; HR—Hess Rise; LAT—Lesser Antilles Trench; MR—Mona Rift; MT—Muertos trough; NHDB—North Hispaniola deformed belt; OF—Oriente fault; PRT—Puerto Rico Trench; PRVI—Puerto Rico–Virgin Islands block; SB—Sombrero Basin; SITF—Swan Islands transform fault; SF—Septentrional fault; WF—Walton fault. (C) Focal mechanisms for depth <35 km for eastern Hispaniola, Puerto Rico, and Virgin Islands. Sources are the Harvard CMT (Centroid-Moment Tensor) catalogue, the Puerto Rico Seismic Network, Deng and Sykes (1995), and Molnar and Sykes (1969). Dots are USGS epicenters for earthquakes above depths of 60 km with magnitudes >3.5 from 1 January 1967 until 28 April 1999 (USGS). Abbreviations as in Figure 1B. MC—Mona Canyon; NPRSF—North Puerto Rico Slope fault; SPRSF—South Puerto Rico Slope fault; VIB—Virgin Islands Basin; WB—Whiting Basin; YR—Yuma Rift.

been postulated (Dolan and Wald, 1998; Dolan et al., 1998) with resultant arching of the island of Puerto Rico about an EW-oriented axis above (van Gestel et al., 1998). The Puerto Rico trench lies ~100 km offshore and reaches a water depth >8 km. Water depth in the Muertos trough is ~5 km. GPS-derived velocities of western Puerto Rico with respect to North America are consistent with oblique convergence across the western Puerto Rico trench (Jansma et al., 2000). Geologic evidence to support convergence includes earthquake slip vectors and low-angle thrusts imaged in seismic profiles (Sykes et al., 1982; Deng and Sykes, 1995; Larue and Ryan, 1998). The Puerto Rico trench and additional mapped offshore faults between the north coast of Puerto Rico and the Puerto Rico trench, such as the South Puerto Rico Slope fault (Grindlay et al., 1997) accommodate ~85% of the current highly oblique North America–Caribbean relative plate motion. The remaining 15% must be taken up along onland faults in Puerto Rico or along the Muertos trough. These results are preliminary and await confirmation by quantitative models of slip distribution. GPS-derived velocities of Puerto Rico with respect to the Caribbean are small, making interpretations from 1999 and earlier geodetic data alone tenuous. Evidence for ongoing deformation along the western Muertos trough includes seismicity and the existence of an accretionary prism along the lower slope south of southeastern Hispaniola and southwestern Puerto Rico along which 40 km of underthrusting is thought to have occurred (Ladd et al., 1977; Ladd and Watkins, 1978; Byrne et al., 1985; Larue and Ryan, 1990; McCann and Pennington, 1990; Masson and Scanlon, 1991; Deng and Sykes, 1995). The accretionary prism narrows eastward along the southern boundary of the Puerto Rico–northern Virgin Islands microplate and disappears near 65°W, southwest of St. Croix (Mauffret and Jany, 1990; Masson and Scanlon, 1991), implying an eastward decrease in normal convergence across the Muertos trough.

The Mona passage, which bounds the Puerto Rico–northern Virgin Islands microplate to the west, contains the NS-trending Mona rift in the north and the similarly oriented Yuma rift in the south (Figs. 1 and 2). The Mona rift formed as a result of east-west extension between Puerto Rico and Hispaniola during the last few million years (Larue and Ryan, 1990, 1998; van Gestel et al., 1998). The 1918 earthquake (Ms = 7.3) in the Mona rift likely occurred along a NS-trending normal fault (Mercado and McCann, 1998) and indicates extension in the Mona rift is ongoing. Focal mechanisms of other historic and recent events offshore northwestern Puerto Rico are consistent with normal faulting along NNE-striking planes (Fig. 1C) (Doser et al., this volume). The kinematics of the Yuma rift are not well constrained, and thus the nature of the boundary between the Puerto Rico–northern Virgin Islands and Hispaniola microplates south of the Mona rift is unclear.

The eastern boundary of the Puerto Rico–northern Virgin Islands block is marked by the ENE-trending Anegada passage, which connects the Neogene Virgin Island and Whiting basins in the southwest with the Sombrero basin in the northeast (Fig. 1). Transtension is likely across the Anegada passage, which is dominated by northeast-southwest–striking and east-west–striking faults with a significant extensional component (Jany et al., 1987; Holcombe et al., 1989; Masson and Scanlon, 1991). Normal displacements along the major faults decrease southwestward to the Muertos trough from more than 4 km to zero over a distance of 50 km, implying rapidly decreasing extension westward (Masson and Scanlon, 1991). North of the Virgin Islands, where most large historic earthquakes have occurred (Murphy and McCann, 1979; Frankel et al., 1980; McCann, 1985; Doser et al., this volume), focal mechanisms record a combination of reverse and sinistral motion along roughly east-west–striking faults, reflecting ENE–directed relative motion of the Caribbean with respect to North America.

ON-LAND FAULTING ON PUERTO RICO

Most studies of on-land faults in Puerto Rico have focused primarily on structures that cut Tertiary and older rocks (Glover and Mattson, 1960; Glover, 1971; Erikson et al., 1990, 1991). Prior to the work of Prentice and Mann (this volume) and Mann et al. (this volume) documentation of features that offset Quaternary units has been limited (e.g., McCann, 1985; Meltzer et al., 1995), leaving most workers to model Puerto Rico as a rigid block (e.g., Masson and Scanlon, 1991). Shallow microseismic-

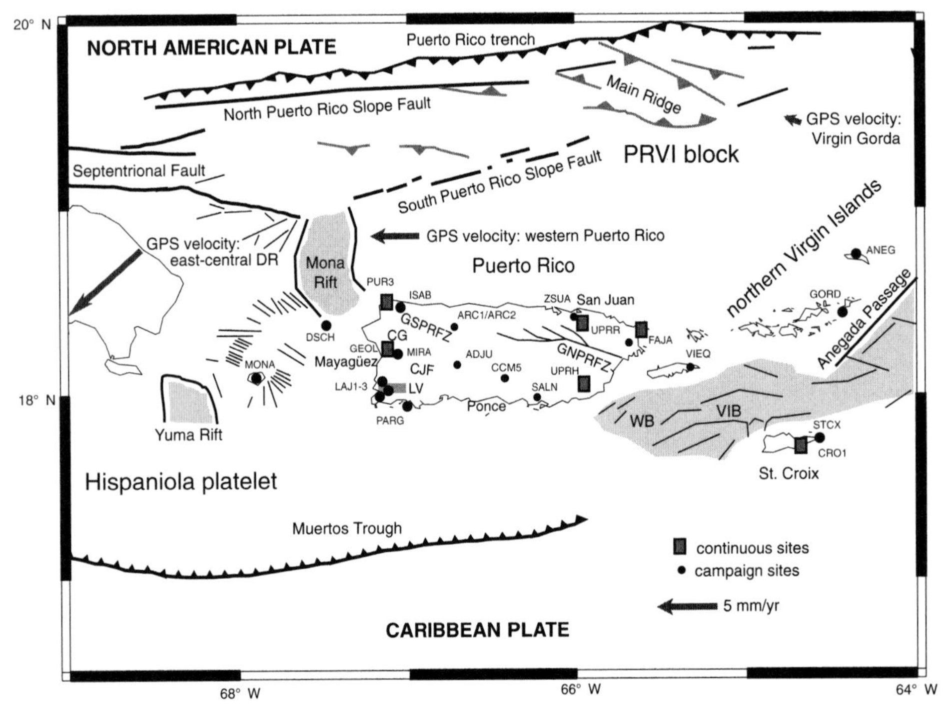

Figure 2. Current mixed-mode Global Positioning System (GPS) geodetic network in the northeastern Caribbean. GNPRFZ—Great Northern Puerto Rico fault zone; GSPRFZ—Great Southern Puerto Rico fault zone; CG—Cerro Goden fault; CJF—postulated Cordillera-Joyuda faults; LV—Lajas Valley (medium gray shaded rectangle in southwestern Puerto Rico). Major offshore structures also are shown. Light gray shaded regions are zones of inferred extension. WB—Whiting Basin; VIB—Virgin Islands Basin. Arrow in Dominican Republic is GPS-derived velocity relative to the fixed Caribbean for central Hispaniola, south of Septentrional fault. Arrow north of the island of Puerto Rico is average GPS-derived velocity relative to the Caribbean for sites in western Puerto Rico. Arrow north of the northern Virgin Islands is GPS-derived velocity relative to the Caribbean for site in Virgin Gorda. Length of arrow in lower left corresponds to 5 mm/yr for scale. Error ellipses are not shown for clarity.

ity does occur onshore, but the historic record is consistent with major events limited to the offshore region (McCann, 2002). The recognition of Quaternary displacements when coupled with observations of shallow seismicity in western Puerto Rico, however, appears to argue against truly rigid-block behavior. The important questions to be answered are: what are the permissible displacement rates along the mapped faults given the GPS geodetic results, do these rates agree with those derived from the geologic data, and are these rates consistent with significant and therefore possibly destructive earthquakes occurring along subaerial faults within the Puerto Rico–northern Virgin Islands microplate?

The highest levels of onshore seismicity are in the Lajas Valley (Asencio, 1980), an EW-trending feature in southwestern Puerto Rico, which continues offshore to the west and passes south of the southern termination of the Mona Canyon, where offshore faults are mapped (Fig. 2). Quaternary offsets along the southern edge of the Lajas Valley were interpreted from seismic reflection profiles (Meltzer et al., 1995; Meltzer, 1997), and active faulting was inferred from the "basin-and-range" style topography of the region (Joyce et al., 1987). Displacements along the EW-striking faults are inferred to be normal with components of strike-slip (Meltzer, 1997; Almy et al., 2000; Prentice et al., 2000). In addition, the island of Puerto Rico is traversed by two northwest-southeast striking fault zones: (1) the Great Northern Puerto Rico fault zone, and (2) the Great Southern Puerto Rico fault zone (Fig. 2). The fault zones were active during the Eocene and record predominantly thrust and left-lateral displacement in response to amalgamation of discrete island arc terranes at the leading edge of the Caribbean plate into the Puerto Rico–northern Virgin Islands microplate (Glover and Mattson, 1960; Glover, 1971; Erikson et al., 1990, 1991). Both the Great Northern Puerto Rico fault zone and the southern end of the Great Southern Puerto Rico fault zone are covered by little-deformed Neogene strata. The two fault zones, however, represent large areas of weakness within the Puerto Rico–northern Virgin Islands along which active displacements may be localized. Indeed, the southern end of the Great Southern Puerto Rico fault zone immediately offshore may cut and disturb Recent shelf sediments (McCann, 1985; J. Joyce, 2000, personal commun.). The projection of the northern end of the Great Southern Puerto Rico fault zone, which continues offshore into Mona Canyon, is subparallel to faults of similar orientation (NW-SE), which are seismically active (McCann, 1985; Joyce et al., 1987). In addition, an EW-striking splay of the Great Southern Puerto Rico fault zone, the Cerro Goden fault, cuts across to the west coast of Puerto Rico ~10 km north of the city of Mayagüez (Fig. 2). Whether Quaternary motion occurred along the Cerro Goden fault is unknown, although the offshore projection of the fault merges with other mapped structures that presumably are Quaternary in age. Lao-Davila et al. (2000) infer Recent displacement with components of normal motion and left-lateral strike-slip on the basis of offset stream drainages and terraces. Some workers have identified the surficial expressions of the Cordillera and Joyuda faults, which they argue may correspond to a WNW-ESE trend across southwestern Puerto Rico that is defined by a series of epicenters of small earthquakes that were recorded by the Puerto Rico Seismic Network in 1995 (J. Moya, 1999, personal commun.). Geological estimates for displacements along the fault are unconstrained.

PRIOR GPS RESULTS FOR PUERTO RICO AND THE NORTHERN VIRGIN ISLANDS

Jansma et al. (2000) provided a detailed discussion and interpretation of GPS geodetic data collected in the Puerto Rico–northern Virgin Islands microplate from 1994 until 1999. With the exception of two sites (ZSUA in San Juan and GORD in Virgin Gorda), the network consisted of stations in western Puerto Rico. Their results, therefore, emphasized the western half of the island and are summarized below.

To assess if Puerto Rico was attached to the Caribbean plate, Jansma et al. (2000) compared the predicted velocity of the Caribbean with respect to the North America plate at the longitude of western Puerto Rico from the model of DeMets et al. (2000) against the velocity of western Puerto Rico relative to North America derived from GPS geodetic data. The predicted and GPS-derived velocities of the Caribbean with respect to North America for western Puerto Rico are 19.4 ± 1.2 mm/yr toward N79°E ± 3° (1σ) and 16.9 ± 1.1 mm/yr with an azimuth of N68°E ± 3 (1σ), respectively, with the latter slightly slower and more northerly than that for Caribbean–North America plate motion as a whole. Thus nearly 85% of the relative motion between the Caribbean and North America plates is accommodated offshore northern Puerto Rico. Relative motion of western Puerto Rico with respect to the Caribbean as constrained by GPS geodesy through 1999 is 2.4 ± 1.4 mm/yr to the west (Jansma et al., 2000). The slow velocity of Puerto Rico with respect to the Caribbean has led some authors to assume that Puerto Rico is part of the Caribbean plate and to include sites in Puerto Rico in the formulation of the Caribbean reference frame (Weber et al., 2001). Although this interpretation was permissible given the errors associated with the geodetic data included in the inversion of the Euler pole, we believe that Puerto Rico is a discrete microplate. The current geodetic data set allows us to infer this unequivocally (see below). Additional compelling evidence is seismicity associated with the Muertos trough, which defines a pattern consistent with overriding of Caribbean lithosphere by southwestern Puerto Rico (Byrne et al., 1985; Dolan and Wald, 1998; McCann et al., this volume).

To assess whether the Puerto Rico–northern Virgin Islands block is rigid, Jansma et al. (2000) used the methodology of Ward (1990) to examine the dispersion of geodetic velocities about predictions of an angular velocity that best-fits those velocities. The results were only applicable to western Puerto Rico where the majority of sites were located. The average rate misfit to the 14 horizontal velocity components of the seven sites that were included (ARC1, GEOL, ISAB, MIRA, PARG, PUR3 AND ZSUA) was 1.2 mm/yr, with only two velocity components misfit by more than 2 mm/yr and one misfit at a level exceeding one standard deviation. The data were fit within their estimated uncertainties of 1–3 mm/yr. The approximate upper bound on the level of internal deformation of western Puerto Rico thus is 1–3 mm/yr. Velocity uncertainties were on the order of several mm/yr for GPS sites in eastern Puerto Rico.

In contrast to western Puerto Rico where GPS velocities of the Puerto Rico–northern Virgin Islands microplate with respect to the Caribbean are significant above error, the GPS-derived velocity at Virgin Gorda (GORD) at the eastern end of a presumed Puerto Rico–northern Virgin Islands block (Fig. 2) is zero within error to that predicted for the Caribbean plate at that location (DeMets et al., 2000; Jansma et al., 2000). The eastern Virgin Islands, therefore, may be attached to the Caribbean plate, implying no displacement along or across the eastern Anegada passage and a zone of EW extension of a few millimeters per year between Virgin Gorda and western Puerto Rico (Fig. 2). The limited time series and geographic distribution of the network made these interpretations only tentative in Jansma et al. (2000). We reexamine these conclusions here with the additional geodetic data from new campaign and continuous sites, additional occupations for preexisting campaign sites and the accumulation of two more years of data for continuous GPS (CGPS) sites.

The GPS-derived velocities relative to the Caribbean plate were consistent with W to SW motion of Hispaniola within the plate boundary zone at a rate faster than that of Puerto Rico, yielding EW extension of 5 ± 3 mm/yr across the NS-trending Mona passage (Fig. 2). The large velocity of Hispaniola with respect to the Caribbean (and, therefore, lower relative motion with respect to North America) within the plate boundary zone likely reflects collision with the Bahama Bank (Mann et al., 1995), a carbonate platform that extends northwest from the northern Dominican Republic to offshore Florida (Fig. 1B). In contrast, Puerto Rico has bypassed the collision and translates eastward with respect to North America at nearly the Caribbean–North America plate rate.

THE CARIBBEAN REFERENCE FRAME

One of the critical issues for GPS geodesy in the northeastern Caribbean is the development and potential limitations of a Caribbean reference frame. The published version is that of DeMets et al. (2000), which uses GPS-derived velocities from Aves Island (AVES) in the east, San Andres Island (SANA) in the west, St. Croix (CRO1) in the U.S. Virgin Islands, and Cabo Rojo (ROJO) in the southern Dominican Republic (Fig. 1). To supplement the limited geodetic data, azimuths of the eastern Swan Islands transform fault also are included to constrain Caribbean motion. CRO1 is the only continuous site of the original four and has been recording data since late 1995. Time series for the other sites range from 4 to 8 years. An updated version of the reference frame incorporates the CGPS site in Barbados (BARB, Fig. 1) and is the one we use in this paper (DeMets et al., 2000, and 2002, personal commun.). The best-fit angular velocity for the five Caribbean GPS sites yields an average residual velocity of 1.5 mm/yr. Another realization of the Caribbean reference frame by Weber et al. (2001) uses Barbados (BARB, Fig. 1) and also includes Puerto Rico (ISAB, Fig. 1). As we will show below, the latter likely sits on a separate tectonic block within the plate boundary zone and sites there will introduce increased uncertainties.

NEW GPS DATA ACQUISITION

GPS measurements were first collected in the northeastern Caribbean in 1986 at six locations (Fig. 1) (Dixon et al., 1991), which were reoccupied subsequently as part of CANAPE (Caribbean–North American Plate Experiment) in 1994. The original sites were: Grand Turk (TURK), Turks and Caicos; Guantanamo (GTMO), Cuba; Cabo Rojo (ROJO), Capotillo (CAPO), and Cabo Frances Viejo (FRAN) in the Dominican Republic, St. Croix (STCX), U.S. Virgin Islands, and Isabela (ISAB), Puerto Rico (Dixon et al., 1991). The network was densified during CANAPE and each subsequent year (for details, see Dixon et al., 1998; Jansma et al., 2000; Calais et al., 2002; Mann et al., 2002). Since 1994, measurements have been made on subsets of the entire network each year. A permanent IGS (International GPS Service) station was established in St. Croix in 1995 (CRO1) and a vector tie to the original 1986 site, STCX, was established (Dixon et al., 1998), which extended the time series by nearly a decade and therefore improved the CRO1 velocity estimate.

The GPS network in Puerto Rico and the Virgin Islands (Fig. 2) consists of the original 1994 CANAPE locations (ISAB, PARG, and GORD) plus campaign sites MIRA (Miradero-Mayagüez), ZSUA (San Juan), MONA (Mona island), DSCH (Desecheo island), ADJU (Adjuntas), ARC2 (Arecibo), CCM5 (Ponce), FAJA (Fajardo), LAJ1, LAJ2, and LAJ3 (Lajas Valley), SALN (Salinas), VIEQ (Vieques), and ANEG (Anegada, British Virgin Islands) and continuous sites GEOL in Mayagüez, FAJA in Fajardo, UPRR in Rio Piedras, and UPRH in Humacao operated by the Department of Geosciences, University of Arkansas, and PUR3 in Aguadilla maintained by the U.S. Coast Guard.

Details of the campaign observations from 1994 to 1999 can be found in Jansma et al. (2000) and will not be repeated here. All campaign observations after 1999 were obtained with the following hardware: Trimble 4000 SSi 12-channel, dual-frequency code-phase receivers and Dorn-Margolin type choke ring antennae. Data were collected using either 0.5 m spike mounts or standard tripod and/or rotating optical plummet setups at a 30 s observation epoch and 10° elevation mask. An individual campaign occupation usually obtained between 10–24 h of continuous data for 2–3 consecutive observation days. The descriptions of the continuous sites remain unchanged from that reported in Jansma et al. (2000).

Previously, we chose to report and analyze CGPS data from PUR3, GEOL, and CRO1. Here we include data from FAJA, a CGPS site in northeastern Puerto Rico for which over two years of observations now exist. Two additional new CGPS stations, UPRH (University of Puerto Rico, Humacao) and UPRR (University of Puerto Rico, Rio Piedras) had been operating for approximately one year at the time of our analysis for this report.

Accordingly, these sites have not yet accumulated enough data to merit serious interpretation.

DATA ANALYSIS

Data from continuous sites PUR3, GEOL, FAJA, and CRO1 (Fig. 3) and campaign sites ISAB, PARG, ADJN, ARC2, ZSUA, SALN, and GORD (Fig. 4) are considered in this paper. Although several additional campaign sites have been installed on the Puerto Rico–northern Virgin Islands block since 2000, most have only had initial occupations completed to date. Others that have had a second occupation have only 1.5–1.8 yr between observations, making estimates of their velocities too uncertain at this time. Data from a total of 11 sites are reported here.

Data analysis procedures have changed slightly from those reported in Jansma et al. (2000) and only a brief description and the significant departures will be reviewed here. All data were processed as free-network point positions using GIPSY-OASISII (version 2.5.8a) (Lichten, 1990; Zumberge et al., 1997). Free-network solutions were transformed, scaled, and rotated into

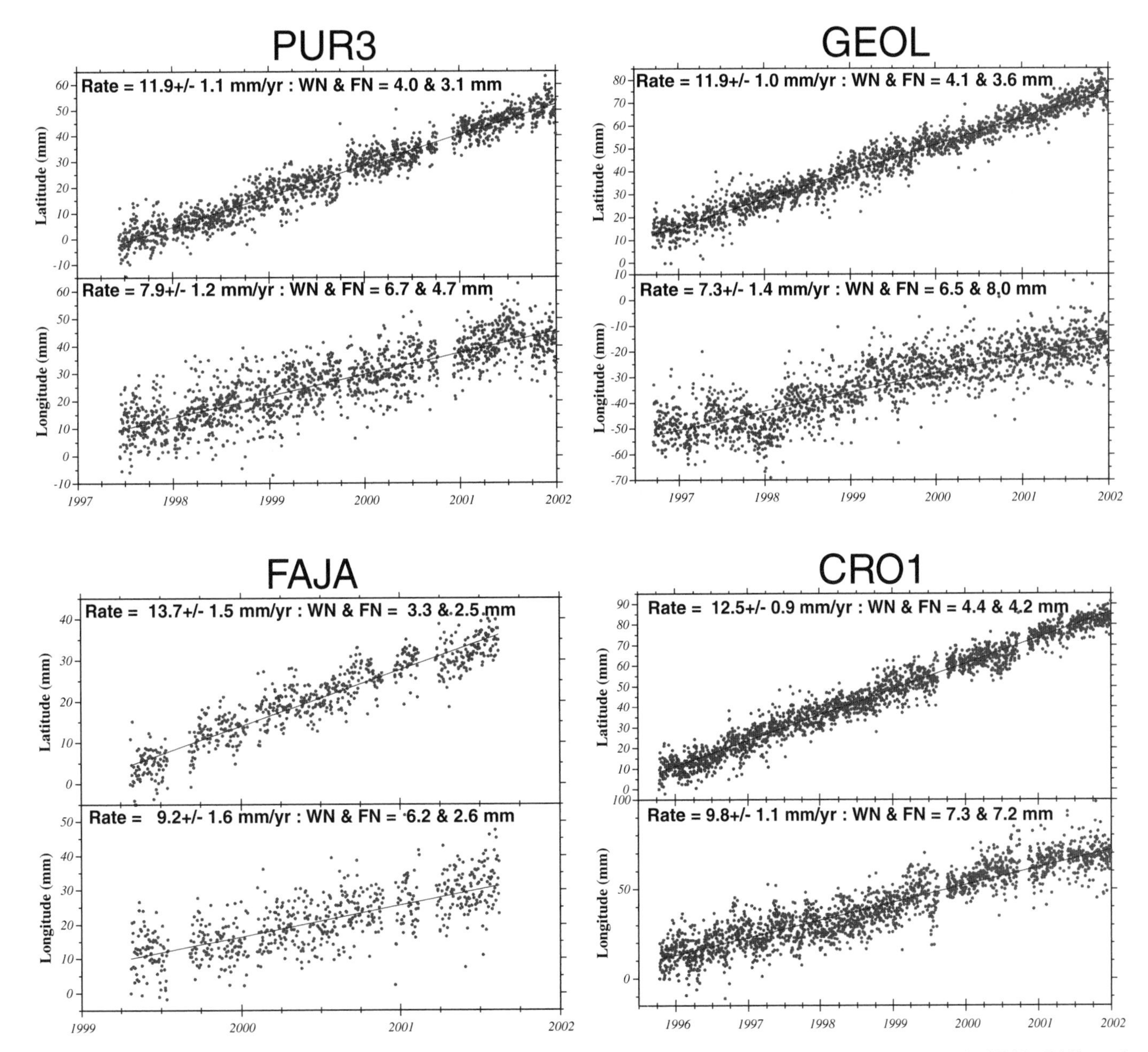

Figure 3. Global Positioning System station latitude and longitude coordinate time series for continuous stations PUR3, GEOL, FAJA, and CRO1. Daily point positions are in ITRF00 (International terrestrial reference frame, 2000). Formal solution errors are not shown for clarity. Note correlated noise among the three sites. WN—white noise; FN—flicker noise.

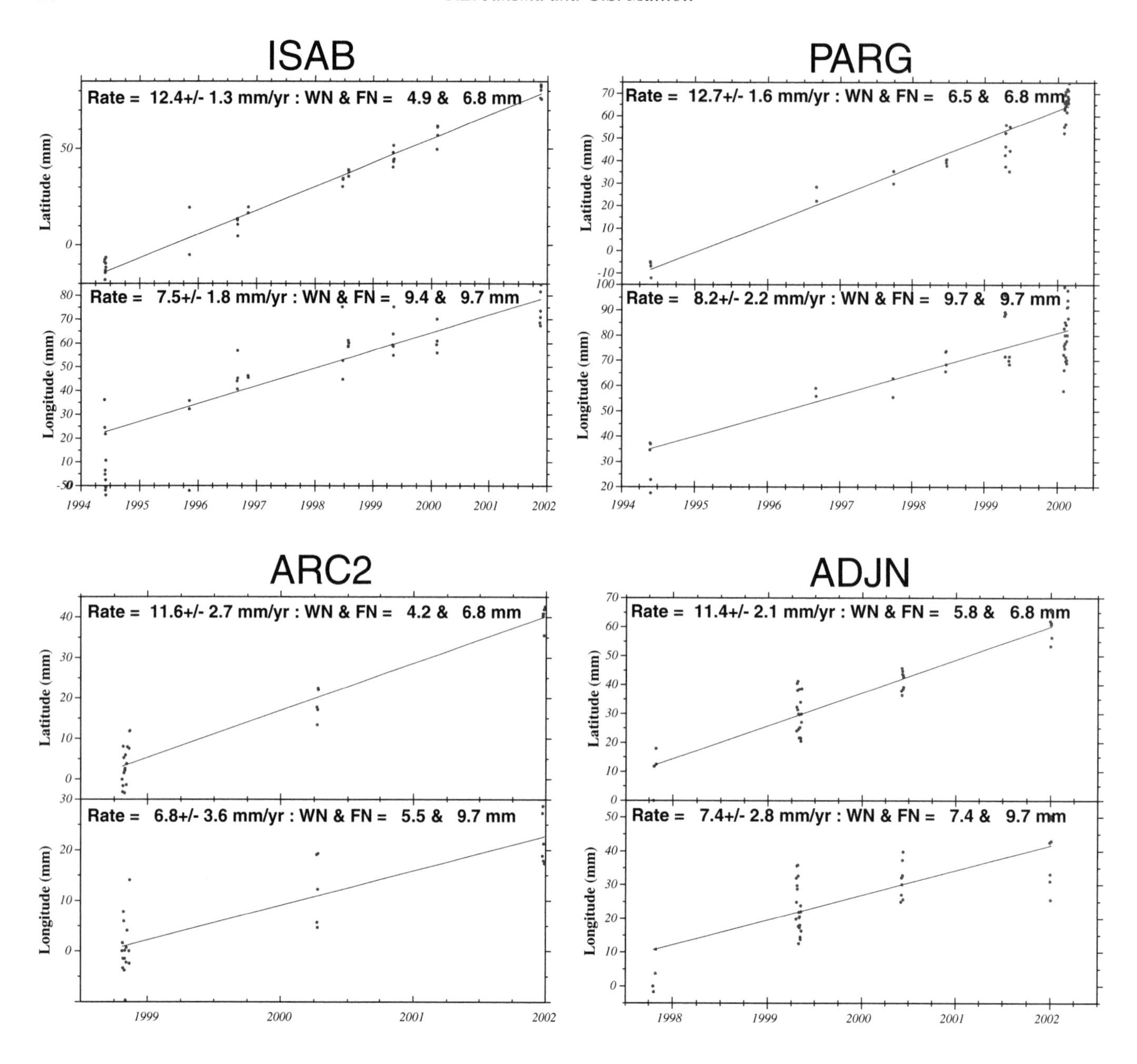

ITRF00 (the current realization of the International Terrestrial Reference Frame, epoch 2000) using x-files from the Jet Propulsion Laboratory (JPL; Blewitt et al., 1992; Heflin et al. 1992). All processing used final, precise non-fiducial orbit, earth-orientation, and GPS clock files from JPL (300 s epoch). Time series velocity errors were calculated using the formulation of Mao et al. (1999), which includes both colored (time-correlated) and white noise contributions and an assumed estimate of $\sqrt{2}$ mm/yr of random monument noise at each site. Scaled formal error estimates for individual site positions are not shown on the time series plots as their utility is recognized as limited. Component site velocities were calculated in ITRF00 using post-processing software modules, which allow estimation of component offsets along with estimates of velocities in a rapid and internally consistent way that includes the full-covariance of each daily site position in the analysis.

After obtaining an ITRF00 velocity for the latitude, longitude, and radial component and their associated errors, final velocities and errors in the Caribbean reference frame were calculated using the current best-fit model for Caribbean motion with respect to the ITRF00. Caribbean fixed velocities include the full covariance of the individual sites velocities and the predicted motion of the Caribbean plate at that location (DeMets et al., 2000, and 2002, personal commun.). Velocities and their

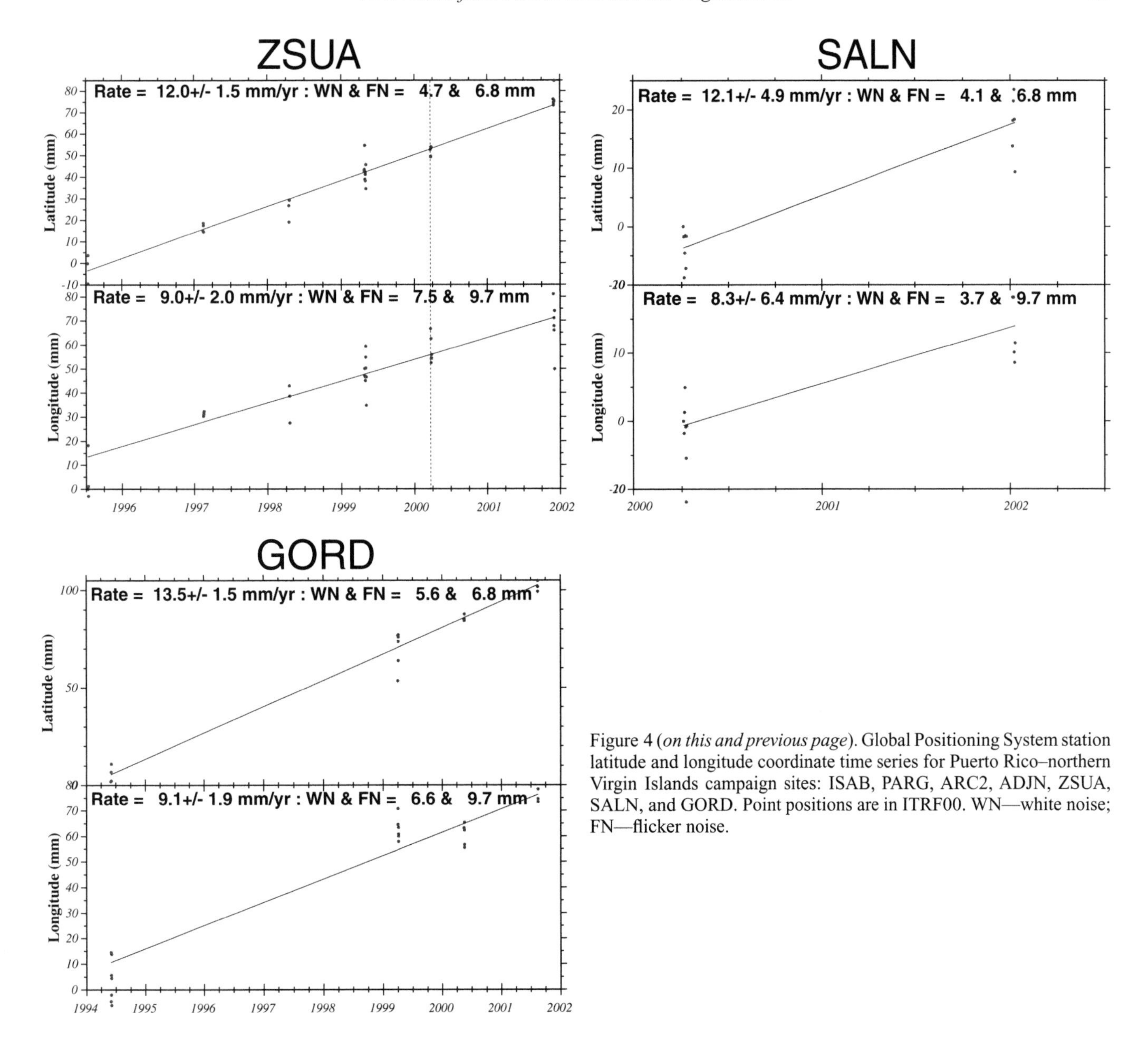

Figure 4 (*on this and previous page*). Global Positioning System station latitude and longitude coordinate time series for Puerto Rico–northern Virgin Islands campaign sites: ISAB, PARG, ARC2, ADJN, ZSUA, SALN, and GORD. Point positions are in ITRF00. WN—white noise; FN—flicker noise.

errors are shown with respect to ITRF00 and the Caribbean reference frame in Table 1. Error ellipses in Figure 5A are 1σ and include both the error of the site time series and the Caribbean reference frame formulation.

In addition to the point positions and velocities derived from them, baseline lengths and components (north, east, and up) were determined to evaluate the internal deformation of the Puerto Rico–northern Virgin Islands block. The derived baseline data have been calculated from the ITRF00 site positions by geometrical differencing using standard GIPSY-OASISII routines to obtain relative changes in station position of the second station with respect to fixed position of the first station. Phase ambiguities were not resolved and therefore the baseline lengths and component velocities are simply a recasting of the absolute point position results described above into a fixed-site reference frame. Such estimates, although useful, can potentially yield spurious results because of spatially correlated noise and unmodeled monument motion (Langbein and Johnson, 1997). No attempt was made during most campaign occupations to assure that several other sites were operating simultaneously, which limits the available data for this type of analysis primarily to the four CGPS sites (PUR3, GEOL, FAJA, and CRO1), and the PARG and ISAB campaign sites, which fortuitously have several synchronous observations over a six year period. The Caribbean-fixed and Puerto Rico–northern Virgin Islands site-fixed velocities and their errors are discussed in detail below.

TABLE 1. VELOCITIES OF GPS SITES IN PUERTO RICO AND THE VIRGIN ISLANDS IN ITRF2000 AND WITH RESPECT TO THE STABLE CARIBBEAN

	Latitude north	Longitude east	Time span (years)	Velocity north (ITRF2000) (mm/yr)	Velocity east (ITRF2000) (mm/yr)	Velocity north (Caribbean) (mm/yr)	Velocity east (Caribbean) (mm/yr)
ADJN	18.17	292.20	4.211	11.4 ± 2.1	7.4 ± 2.8	–0.6 ± 2.2	–3.2 ± 2.9
ARC2	18.34	293.25	3.181	11.6 ± 2.7	6.8 ± 3.6	–0.4 ± 2.7	–3.8 ± 3.7
CRO1	17.76	295.42	6.216	12.5 ± 0.9	9.8 ± 1.1	–0.3 ± 1.5	–1.2 ± 1.4
FAJA	18.38	294.38	2.301	13.7 ± 1.5	9.2 ± 1.6	1.4 ± 1.7	–1.4 ± 1.8
GEOL	18.21	292.86	5.290	11.9 ± 1.0	7.3 ± 1.4	0.1 ± 1.2	–3.3 ± 1.6
GORD	18.43	295.56	7.214	13.5 ± 1.5	9.1 ± 1.9	0.6 ± 1.6	–1.6 ± 2.1
ISAB	18.47	292.95	7.488	12.4 ± 1.3	7.5 ± 1.8	0.5 ± 1.4	–3.0 ± 2.0
PARG	17.97	292.96	5.750	12.7 ± 1.6	8.2 ± 2.2	0.8 ± 1.7	–2.5 ± 1.7
PUR3	18.46	292.93	4.556	11.9 ± 1.1	7.9 ± 1.2	0.1 ± 1.2	–2.6 ± 1.5
SALN	18.03	293.77	1.768	12.1 ± 4.9	8.3 ± 6.4	–0.1 ± 5.0	–2.4 ± 6.5
ZSUA	18.43	294.01	6.406	12.0 ± 1.5	9.0 ± 2.0	–0.3 ± 1.6	–1.6 ± 2.2

Note: For definition of stable Caribbean, see text. Uncertainties are 1σ and include white noise, flicker noise, and monument error. ITRF2000—International terrestrial reference frame—2000.

GPS-DERIVED VELOCITIES AND BASELINES FOR PUERTO RICO AND THE VIRGIN ISLANDS

In our discussion of GPS-derived velocities and baselines, we separate the island of Puerto Rico from the neighboring Virgin Islands to the east. Our reasoning is that the potential attachment of Virgin Gorda (GORD) to the Caribbean plate requires small relative motion between the eastern Virgin Islands and eastern Puerto Rico. To assess whether the island of Puerto Rico and the Virgin Islands as a whole are each rigid, we examine GPS-derived velocities of sites in Puerto Rico and the Virgin Islands relative to the Caribbean. To constrain the permissible upper limits for displacement rates along active faults in Puerto Rico, we include in our analysis calculated component velocities relative to continuous GPS sites that cross major structures: PUR3-GEOL; PUR3-PARG; PUR3-FAJA; GEOL-FAJA; FAJA-CRO1; and PUR3-CRO1. PARG, a campaign site, also is used to create baselines that cross the seismically active Lajas Valley (PARG-PUR3; PARG-GEOL).

Velocities of Puerto Rico and the Virgin Islands Relative to the Caribbean Plate: Implications for a Separate Puerto Rico–Northern Virgin Islands Microplate

GPS-derived velocities relative to the Caribbean for the Puerto Rico–northern Virgin Islands block appear consistent with a linear increase in velocity from near zero (1.7 ± 2.1 mm/yr) at Virgin Gorda (GORD) in the east to ~3 mm/yr (3.0 ± 0.3 mm/yr mean and standard deviation of the average of five sites) toward the west along the west coast of Puerto Rico (Fig. 5A, Table 1). To assess whether motion of the Puerto Rico–northern Virgin Islands microplate with respect to the Caribbean was significant, we tested for the existence of a separate microplate using the F-ratio test of Stein and Gordon (1984). This method compares the weighted least-squares fits of models that use two angular velocities (six parameters) and one angular velocity (three parameters), respectively, to fit a set of kinematic observations, in our case, five Caribbean and 15 Puerto Rico–northern Virgin Islands sites. The F-test is insensitive to systematic overestimates or underestimates of velocity uncertainties, an important consideration with GPS velocities. We undertook a similar analysis in Jansma et al. (2000) using the geodetic data collected between 1994 and 1998. Our previous attempt resulted in significance of a separate Puerto Rico–northern Virgin Islands block only at the 80% confidence level. We estimated that an additional two years of data would be required before the GPS-derived velocities would be sufficiently precise to detect motion of the Puerto Rico–northern Virgin Islands relative to the Caribbean of a few mm/yr at the 95% confidence level. The F-test is defined here as:

$$F = \frac{\left[\chi^2_{(2\text{ plates})} - \chi^2_{(1\text{ plate})}\right]/3}{\chi^2_{(2\text{ plates})}/\nu}$$

where χ^2 is defined as the weighted root mean square misfit (chi-squared) of the kinematic model to the GPS site velocities and ν represents the number of degrees of freedom ($N - p$), which in this case is 40 – 6 or 34.

Fitting the Puerto Rico–northern Virgin Islands and Caribbean velocities with a single angular velocity gives a summed weighted least-squares misfit of 20.78. Fitting the two sets of velocities with separate angular velocities gives a summed misfit of 13.16. The improvement in fit from the 2-plate model, which stems from using three additional adjustable parameters, gives F = 6.56, which is significant at the 99.9% confidence level for 3 versus 34 degrees of freedom. The inclusion of the additional data thus constrains the unequivocal existence of a Puerto Rico–northern Virgin Islands microplate within the northeastern Caribbean. This conclusion is robust despite the still significant error in the individual time series of Puerto Rico–northern Virgin Islands

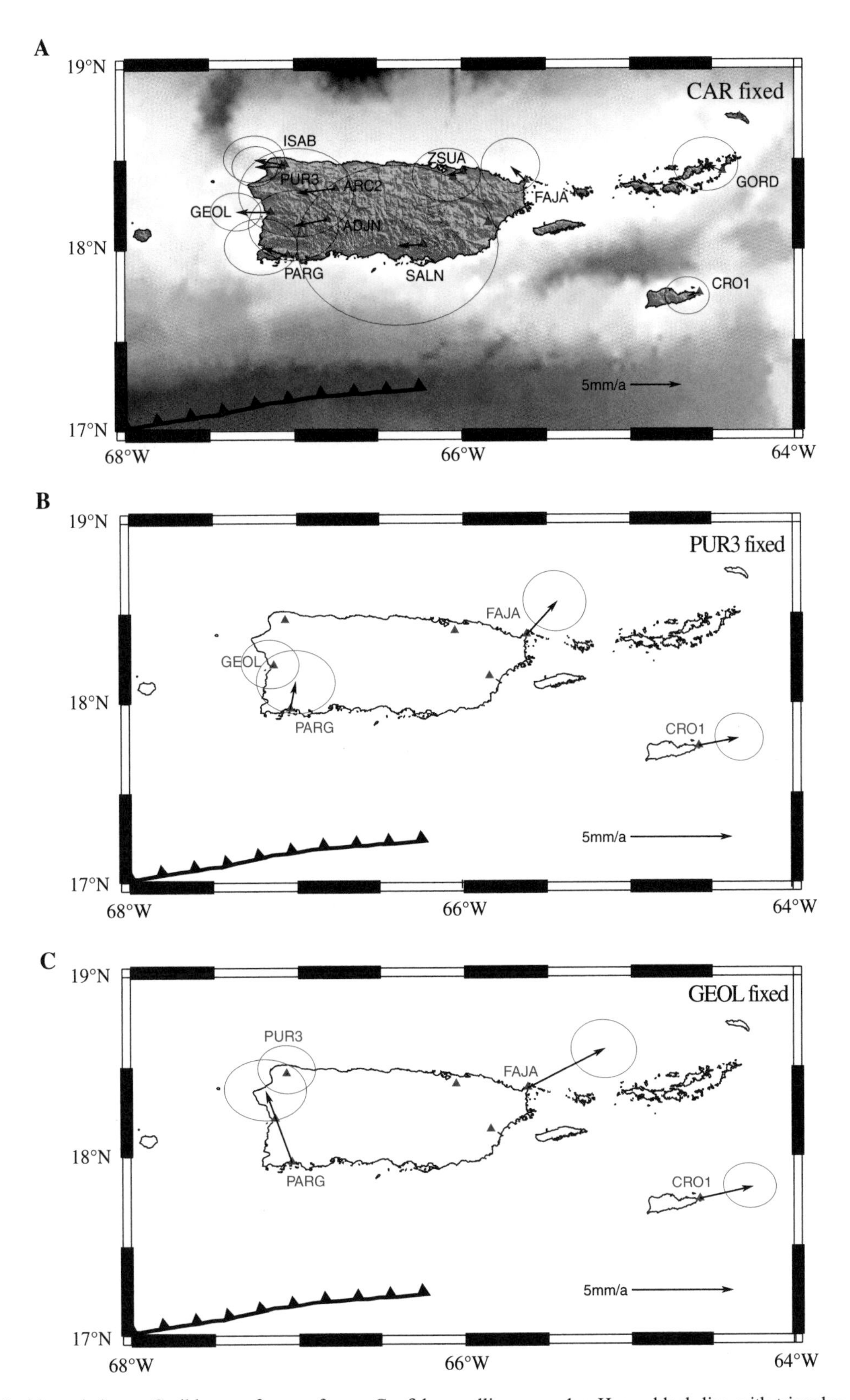

Figure 5. (A) Velocities relative to Caribbean reference frame. Confidence ellipses are 1σ. Heavy black line with triangles offshore south of Puerto Rico represents the Muertos trough. (B) Velocities relative to fixed GEOL. For magnitude of changes along baselines, see Table 2. (C) Velocities relative to fixed PUR3. Table 2 lists velocities and 1σ for sites. Note that calculated 1σ errors were determined by propagating errors from the ITRF2000 time series component errors and likely are overestimates; therefore, error ellipses shown have been scaled to 0.5 1σ.

TABLE 2. BASELINE VELOCITY COMPONENTS FOR SELECTED PAIRS OF GPS SITES IN PUERTO RICO AND THE VIRGIN ISLANDS.

Baseline	Fixed site	Velocity north (mm/yr)	Velocity east (mm/yr)	Correlation coefficient
GEOL-PUR3	GEOL	0.02 ± 1.49	0.10 ± 1.84	–0.0328
GEOL-PARG	GEOL	3.34 ± 1.89	–1.26 ± 2.61	–0.0123
GEOL-FAJA	GEOL	1.85 ± 1.80	3.74 ± 2.10	–0.0368
GEOL-CRO1	GEOL	0.56 ± 1.38	2.66 ± 1.75	–0.0485
PUR3-PARG	PUR3	1.31 ± 1.94	–0.26 ± 2.51	–0.0274
PUR3-FAJA	PUR3	1.58 ± 1.86	1.49 ± 2.00	–0.0426
PUR3-CRO1	PUR3	0.36 ± 1.45	1.99 ± 1.60	–0.0538
FAJA-CR01	FAJA	–2.13 ± 1.77	0.96 ± 1.92	–0.0508
ISAB-PARG	ISAB	0.47 ± 2.06	–1.07 ± 2.84	–0.0027

Note: Changes are expressed in terms of velocity north and velocity east in mm/yr. Stated error is 1σ propagated from International terrestrial reference frame—2000 time series component errors without any provision for common mode errors. It is likely that these stated uncertainties substantially overestimate the real uncertainties.

sites. We note that site velocities for all 15 Puerto Rico–northern Virgin Islands sites are included in our analysis, despite the fact that four of these sites (ARC1, MIRA, UPRR, and ZSUB) are not reported in Table 1 because their time series component errors remain too large for meaningful geologic interpretation. The total contribution of these site velocities to the final best fit (2-plate) kinematic model is 0.36 out of 6, or only 6%, however, and thus they have little impact on the final model. Their low contribution arises because these site velocities are more imprecise than other sites located nearby. Additional kinematic modeling and further details will be reported elsewhere.

Velocities of Puerto Rico and the Virgin Islands Relative to the Caribbean Plate: Implications for Internal Puerto Rico–Northern Virgin Islands Microplate Deformation

Although all Puerto Rico–northern Virgin Islands velocities are similar within error, raising the possibility that no deformation occurs across the microplate, we believe that the systematic pattern of the velocities across the Puerto Rico–northern Virgin Islands block suggests that the small differences are likely real, although they are close to the limits of detection in the current geodetic data. The observed residual velocities are not likely a result of bias in the location of the Caribbean Euler pole because the location of the pole is much farther away than the distance between the sites in the Puerto Rico–northern Virgin Islands microplate. Changes in the location or rotation rate for the Caribbean would cause a systematic shift in the observed Puerto Rico–northern Virgin Islands residual velocities, but would not eliminate the observed differences among the sites within the Puerto Rico–northern Virgin Islands block. The new observations support earlier interpretations that the Puerto Rico–northern Virgin Islands block may be attached to the Caribbean plate at GORD on its eastern end, requiring east-west extension across Puerto Rico and the western Virgin Islands (Jansma et al., 2000). The preliminary results were based on only two epochs (1994 and 1999) of data at GORD. The new data set includes additional epochs at GORD in 2000 and 2001, such that the time series reported here spans seven years (Fig. 4 and Table 1). The inclusion of data from FAJA in the northeastern corner of Puerto Rico and from SALN in south-central Puerto Rico also strengthens this conclusion: the GPS-derived velocities of FAJA and SALN (average of 2.2 ± 0.3 mm/yr) relative to the Caribbean are faster than that of GORD (1.72 ± 2.1 mm/yr), but slower than that of GEOL, PUR3, PARG, ADJN, and ARC2 (average of 3.0 ± 0.3 mm/yr). The magnitude of the velocity of FAJA also is similar to the magnitude of that at ZSUA, ~35 km to the west. The velocity azimuths, however, vary by ~45°.

Comparison of the GPS-derived velocities relative to the Caribbean at GORD, SALN, and FAJA provides important constraints for the neotectonics of the Anegada passage. At the longitude of GORD, where the velocity is zero within error, little if any motion occurs across the Anegada passage. The possibility does exist that CRO1 and GORD velocities are affected by strain accumulation along a locked, fast-slipping, eastern Anegada passage fault. We believe this is unlikely, however, because of the similarity of CRO1's velocity with the other rigid Caribbean sites (AVES, SANA, ROJO, BARB) (DeMets et al., 2000). We note that CRO1 is closer to the Anegada passage faults than GORD and therefore should be more affected by any possible strain accumulation along these structures. Low levels of microseismicity south of the eastern Virgin Islands (Frankel et al., 1980) also support little motion between GORD and CRO1. To the west, however, where microseismicity is greater and motions of FAJA and SALN with respect to the Caribbean are toward the northwest and west, respectively (Fig. 5A), displacement across the northeast-trending structures of the Whiting and Virgin Island basins must have an extensional component, a conclusion supported by evidence of normal faulting documented in seismic profiles acquired in the Anegada passage (Jany et al., 1987; Holcombe et

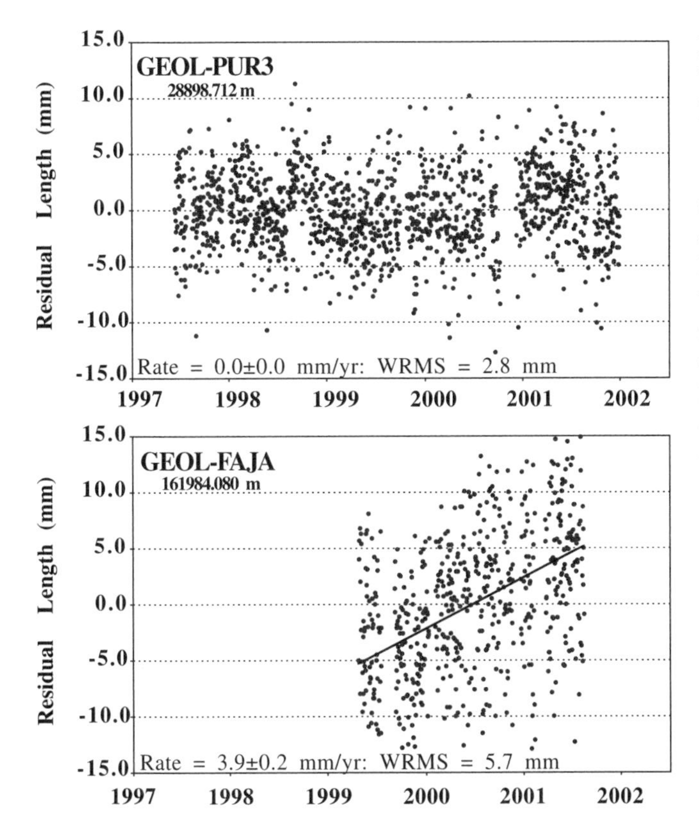

Figure 6. Evolution of residual baseline length between continuous sites PUR3 and GEOL (upper) and GEOL and FAJA (lower). Rate of change of baseline length between PUR3 and GEOL is 0.0 ± 0.0 mm/yr. Motion transverse to the baseline of –0.1 ± 0.1 also occurs. WRMS (weighted root mean square) is 2.8 mm. Median baseline length is 28898.712 m. Rate of change of baseline length between GEOL and FAJA is 3.9 ± 0.2 mm/yr. Motion transverse to the baseline of 1.4 ± 0.1 also occurs. WRMS is 5.7 mm. Median baseline length is 161984.080 m. Stated error is 1σ formal error with white noise only and no provision for monument noise.

al., 1989; Masson and Scanlon, 1991). The change from near zero motion at GORD to 2.0 ± 1.8 mm/yr relative to the Caribbean at FAJA (Fig. 5A, Table 1) requires a small component of northwest-southeast to east-west extension between the eastern Virgin Islands and eastern Puerto Rico. The northwest-oriented velocity of FAJA relative to the Caribbean yields northwest-southeast extension across the Anegada passage at the longitude of 65°W, whereas the west-directed velocity relative to the Caribbean of SALN produces near east-west extension across the Whiting and Virgin Islands basins south of Vieques. The westward motion of western Puerto Rico relative to the Caribbean of 3.0 ± 0.3 mm/yr (average of five sites) implies a component of left-lateral strike-slip motion along either the east-west–trending Muertos trough or offshore faults south of Puerto Rico at the longitude of western and central Puerto Rico. Up to 2 mm/yr of convergence also is permitted along the Muertos trough within error (Fig. 5A).

Puerto Rico–Northern Virgin Islands–Fixed Component Velocities and Baseline Evolution for Selected Site Pairs in Puerto Rico

Although the GPS-derived velocities of sites in Puerto Rico and the Virgin Islands relative to the Caribbean are essential to constrain the kinematics of the Puerto Rico–northern Virgin Islands microplate, the 1σ uncertainties associated with the individual site velocities frequently exceed geologically based estimates of slip rates along those faults. To assess whether the small differences among the GPS-derived velocities for sites in Puerto Rico were geologically significant, we examined the evolution of component velocities and baseline lengths between several pairs of CGPS stations in the northeastern Caribbean. The campaign sites, PARG and ISAB, also were included in our analysis. The CGPS sites were selected because they have the longest time series, which generally should minimize the error associated with the calculations, although as we will discuss below, CGPS data also may be biased. PARG and ISAB were included to provide baseline coverage of structures in southwest Puerto Rico. Relative velocity components (north, east, and up) (Table 2) and length were calculated among the CGPS sites and the ISAB and PARG campaign sites holding either GEOL, PUR3, or ISAB fixed in the ISAB-PARG baseline. Vectors shown in Figures 5B and 5C were determined by holding the ITRF00 position of one site fixed as a reference, GEOL in the case of the upper diagram, and determining the relative change for the north and east components for the other sites. Derived time series of the baseline components were then used to calculate component velocities. The full covariance for each component velocity in ITRF00 was included in the calculation of the standard deviation in fixed site component velocities reported in Table 2. The estimated component velocity errors already included our estimate of white time-correlated and monument noise. They do not include any provision, however, for error reduction as a result of common-mode noise in the time series and therefore are likely to be larger than the real errors for these site pairs.

The derived residual baseline length change for GEOL-FAJA is 3.9 ± 0.2 mm/yr, for example, appears to be inconsistent with velocities reported in Table 1 in either the ITRF00 or Caribbean reference frame. Using the ITRF00 component velocities from Table 1, the square root of the sum of the squared differences for GEOL-FAJA is 2.4 ± 3.2 mm/yr rather than the 3.9 ± 0.2 mm/yr determined by the baseline evolution shown in Figure 6. In principle, the coordinate transformation whereby GEOL was held fixed should not have biased the calculated baseline length change, if the linear model and associated errors for the component velocities accurately represents the temporal evolution of the GEOL and FAJA point positions. This apparently was not the case here. The FAJA time series is much shorter and started much later than the GEOL time series. To test whether this affected our results, we have recalculated the GEOL component velocities using only position estimates overlapping in time with FAJA. The result is V_{north} = 11.6 ± 1.6 mm/yr and V_{east} = 5.2 ± 1.6 mm/

yr (2.31 yr from April 24, 1999, to August 15, 2001, relative to ITRF00). This is essentially the same in the north component, but substantially different in the east component (7.3 ± 1.4 and 5.2 ± 1.6 mm/yr for the full versus the abbreviated time series, respectively). The implication of this is that between April 1999 and August 2001, the GEOL site motion did not closely approximate its site velocity for the much longer 5.3 yr period. Using these rates for GEOL and recalculating the expected baseline rate of change as well as propagating the errors in the point position time series yields 4.2 ± 3.4 mm/yr for the motion of FAJA relative to fixed GEOL. The motion of FAJA relative to fixed GEOL can now be recast into baseline and transverse components changes for comparison with the baseline evolution diagram shown in Figure 6. The resulting V_{length} = 4.0 ± 3.1 mm/yr is now in excellent agreement with the 3.9 ± 0.2 calculated by solely fitting the baseline length residuals. Taking a very conservative approach, we can safely estimate that the baseline length change between FAJA and GEOL is between the 2.3 ± 3.0 and 4.0 ± 3.1 mm/yr estimates. In contrast, the PUR3-GEOL baseline length change of 0.0 ± 0.0 mm/yr shown in Figure 6 closely approximates the calculated difference based on the velocity time series of 0.1 ± 1.5 mm/yr. In this case, either long period noise does not affect this northeast-oriented baseline significantly or the 4.6 yr period over which we have calculated the velocities and residual baseline changes is sufficient to effectively average them out. Because of these considerations, real errors remain relatively large for the baseline estimates and care must be exercised in their interpretation. Nevertheless, we believe that an internally consistent tectonic model can be developed based upon both the point position estimates in the Caribbean fixed frame and the calculated baseline changes. This is presented below.

GEOLOGICAL AND TECTONIC IMPLICATIONS

From Northwestern to Southwestern Puerto Rico: Displacement across Subaerial Faults and Zone of High Microseismicity

To constrain the upper bounds of displacements along potentially active faults in western Puerto Rico, we examined two baselines: PUR3-GEOL, which extends generally north-south from the northwestern corner of the island to Mayagüez in the central western coast, and PUR3-PARG, which joins the northwestern and southwestern corners of Puerto Rico and crosses the seismically active Lajas Valley. PUR3-GEOL crosses the Great Southern Puerto Rico fault zone and the Cerro Goden fault (Fig. 2). The length of the baseline between PUR3 and GEOL has remained constant within error during the 4.5 yr interval for which we have data (Fig. 6) (Table 2). The total integrated displacement across the Great Southern Puerto Rico fault zone and the Cerro Goden fault, therefore, is less than the 0.0–1.6 mm/yr error estimated along the baseline.

In contrast, the length of the baseline between PUR3 and PARG has decreased 1.6 ± 0.3 mm/yr during the same time interval. (The baselines ISAB-PARG and GEOL-PARG yielded similar results.) The velocity of PARG relative to a fixed PUR3 is oriented NNE (Figs. 5B and 5C), a direction perpendicular to the major structures of the plate boundary and consistent with a component of convergence across the Caribbean-Puerto Rico boundary zone. Because PARG is a campaign site with only six epochs of observations, the errors associated with this baseline are larger than those between two continuous sites. Nevertheless, the implication is that deformation ≤2 mm/yr is likely across southwestern Puerto Rico, a scenario that is not unreasonable given the high microseismicity in the Lajas Valley (Asencio, 1980) (Fig. 2). The decrease in line length implies shortening between PARG and GEOL. East-west–trending faults that bound the Lajas Valley, therefore, likely have oblique-reverse motion. Because the degree of obliquity is small, the sense of the component of strike-slip displacement predicted along the edges of the Lajas Valley and the postulated Cordillera and Joyuda faults is highly sensitive to slight changes in the azimuth of the displacement between PUR3 and PARG and these remain uncertain. The sense implied by the baseline change, however, is right-lateral. Trenching across one of the bounding faults of the Lajas Valley revealed components of normal (valley-side down) and strike-slip displacement (Prentice and Mann, this volume). Because a three-dimensional excavation of channels and other features was not attempted, no quantitative estimate of the sense of fault offset could be determined. Almy et al. (2000) argued for components of normal motion and left-lateral strike-slip. Other evidence for normal faulting in southwestern Puerto Rico along east-west to northwest-southeast trends is cited from seismic reflection profiling of the Lajas Valley (Meltzer et al., 1995), fault striation analysis of Miocene rocks (Hippolyte et al., this volume; Mann et al., this volume), and marine geophysical surveying of faults offshore western Puerto Rico (Grindlay et al., this volume; Mann et al., this volume). The opposite sense of dip-slip observed in the trench relative to the decrease in the baseline length between PUR3 and PARG may arise from unmodeled signals in the relative velocities, such as seasonal effects, ocean loading and monument noise that yield potentially spurious results or may reflect changes in the local kinematics between the time of faulting and the present. We think it unlikely given the 6 yr time span and the consistency of other baseline measurements across Puerto Rico, however, that these results are unrelated to tectonic processes. For example, it has been demonstrated that velocity transients in GPS time series may be related to the earthquake cycle (Dragert et al., 2001). Although verification of such transients would require mechanical modeling to demonstrate their cause unequivocally, the possibility exists that the geodetically measured velocity field is recording strain accumulation related to the regional geodynamics of the plate boundary. Variations in the local strain field and the existence of additional unmapped structures between PARG and the Lajas Valley are other possible alternatives. Resolution of this discrepancy will require additional GPS geodetic data and a further analysis to remove regionally correlated noise. The campaign sites that span the Lajas Valley (LAJ1, LAJ2, and

LAJ3) (Fig. 2), which were first measured in early 2000, will provide critical constraints in the near future.

From Western to Eastern Puerto Rico: Distributed Deformation across the Island

The relative velocities and baseline evolution for GEOL-FAJA and PUR3-FAJA provide constraints on the overall deformation within and across Puerto Rico. GEOL-FAJA crosses both the Great Southern and Great Northern Puerto Rico fault zones, whereas PUR3-FAJA traverses only the latter (Fig. 2). The time interval considered is April 1999 until December 2001, reflecting when the site at Fajardo was installed. During this period, the baseline length between GEOL-FAJA increased by 3.9 ± 0.2 mm/yr (Fig. 6) (Table 2). The length change for PUR3-FAJA is smaller, but still positive at 1.5 ± 0.2 mm/yr, providing additional support for little displacement between GEOL and PUR3. While caution must be exercised when interpreting these data (see discussion above), the increasing baseline length and relative velocity (Fig. 5B) between fixed GEOL or fixed PUR3 and FAJA imply a component of approximately ENE-WSW–oriented extension of between 1.5 and 3.9 mm/yr across the island of Puerto Rico. Whether this extension is accommodated along a few major structures (e.g., the Great Northern Puerto Rico fault zone where the extension would generate dextral transtension) or distributed across several smaller features is not clear. Higher-than-average levels of seismicity do occur along the central segment of the Great Northern Puerto Rico fault zone (Asencio, 1980). The GPS-derived velocity relative to the Caribbean at ZSUA, a site between PUR3 and FAJA, is identical within error to that at FAJA, tentatively suggesting that the structures that take up displacement may be west of the San Juan metropolitan area. The errors associated with the GPS-derived velocities are still too large to conclude this definitively. We do note that there are several north-south–trending valleys that channel the major rivers along the north coast of Puerto Rico; these may be manifestations of east-west–oriented extension. One focal mechanism for the northwest corner of Puerto Rico is consistent with slip along north-south–striking normal faults (Fig. 1C). Hippolyte et al. (this volume) infer east-west extension across north-south–striking faults from fault striation studies in the Miocene carbonates of the northwestern corner of the island. The observation of likely ENE-WSW extension across Puerto Rico, therefore, is not surprising. Results from previous GPS studies and marine geophysical surveys document east-west extension between western Puerto Rico and eastern Hispaniola (van Gestel et al., 1998; Jansma et al., 2000; Calais et al., 2002). In addition, potential attachment of Virgin Gorda to the Caribbean at the eastern end of the Puerto Rico–northern Virgin Islands microplate requires near east-west extension between the eastern Virgin Islands and eastern Puerto Rico. In summary, the GPS geodetic data suggest that east-west extension is not limited to the offshore regions, but also likely affects the island of Puerto Rico at the level of 2–3 mm/yr. Orientations of T-axes from composite focal mechanisms also indicate east-west extension across eastern Puerto Rico (McCann, 2000).

From Eastern Puerto Rico and the Northern Virgin Islands to St. Croix and the Northern Lesser Antilles: Deformation across the Anegada Passage

The GPS-derived velocity relative to the Caribbean for FAJA in the northeastern corner of Puerto Rico is 2.0 ± 1.8 mm/yr (Fig. 5A and Table 1), which is very nearly no motion within error. We believe that no motion is unlikely and that after several more years of data accumulation the error estimate will likely decrease, yielding a small but statistically significant residual velocity. Evidence to support active slip in the region includes high seismicity (Frankel et al., 1980) and the occurrence in 1867 of a large ($7 < M < 7.75$) tsunamogenic earthquake along the north wall of the Virgin Islands basin that caused extensive damage in St. Croix and St. Thomas (Reid and Taber, 1920). To investigate potential motion across and along the Anegada passage in more detail, we examined the change in baseline length between FAJA in the northeastern corner of Puerto Rico and CRO1 on the island of St. Croix within the stable interior of the Caribbean plate. This has the advantage of eliminating the errors introduced by uncertainties in the Caribbean reference frame, related to variations in velocities among the five stable Caribbean sites. Baseline length increased at 1.9 ± 0.2 mm/yr (white noise; fit residuals) to 2.0 ± 2.3 mm/yr (relative velocities with errors propagated) (Table 2) in a direction perpendicular to the northeast-trending structures of the Whiting and Virgin Island basins, implying active extension across the Anegada passage, which is consistent with mapped structures in the offshore (Jany et al., 1987; Holcombe et al., 1989; Masson and Scanlon, 1991). A component of left-lateral strike-slip motion (1.2 ± 1.3 mm/yr) is also permissible within error from the baseline component and fixed site velocity analysis. This result is consistent given the westward-directed velocities of other sites in Puerto Rico (SALN, ADJN, PARG) relative to the Caribbean. A pair of focal mechanisms immediately south of the Anegada passage is consistent with left-lateral strike-slip faulting along east-west–oriented vertical planes (Fig. 1C).

From Western Puerto Rico to the Northeastern Dominican Republic: Opening of the Mona Rift

The new GPS observations from western Puerto Rico that incorporate 2000 and 2001 data are consistent with earlier results: velocities of sites in western Puerto Rico relative to the Caribbean remain significantly slower and more westerly trending (Fig. 5A) than velocities for sites in eastern Hispaniola (Fig. 2), implying ENE-WSW extension between the two islands. Comparing the total slip along major faults north and west of the Mona rift estimated from two-dimensional elastic models with the velocity of northwestern Puerto Rico, Jansma et al. (2000) inferred 5 ± 4 mm/yr extension across the Mona rift. New GPS geodetic data from the Dominican Republic and results from three-dimensional elastic modeling of strain accumulation across Hispaniola and the immediate offshore region (Calais et al., 2002) lend additional support to near east-west extension of a few millimeters per year across the Mona rift.

Two major EW- to WNW-trending fault zones cut the island of Hispaniola: the Septentrional fault zone in the north and the Enriquillo fault zone in the south (Fig. 1B). The offshore continuation of the Septentrional fault projects to the South Puerto Rico Slope fault, which lies north of the Mona rift (Fig. 2) (Grindlay et al., 1997). Deformation also occurs immediately offshore Hispaniola to the north along the North Hispaniola deformed belt (Fig. 1B), a submarine zone of folds and thrusts, which was the locus of a series of large thrust earthquakes in 1946 and 1948 (Russo and Villaseñor, 1995; Dolan and Wald, 1998) and to the south along the Muertos trough. The Enriquillo fault disappears in central Hispaniola and slip in eastern Hispaniola may be transferred to the Muertos trough (Mann et al., 1999; Mauffret and Leroy, 1999). From their three-dimensional elastic model, Calais et al. (2002) estimate slip rates of 12.9 mm/yr, 9.0 mm/yr, 10.0 mm/yr, and 7.7 mm/yr along the North Hispaniola deformed belt, the Septentrional fault, the Enriquillo fault, and the Muertos trough, respectively. The 9 mm/yr rate for the Septentrional fault was constrained by Holocene rates derived from paleoseismological studies (Prentice et al., 2003). Their model is consistent with slip transfer from the Enriquillo fault to the Muertos trough south of eastern Hispaniola. If this is correct, GPS sites in central eastern Hispaniola are little affected by elastic strain accumulation, lying at substantially greater distance than the 15 km locking depth from the Septentrional fault and more than 100 km from the Muertos trough. GPS-derived velocities relative to the Caribbean of the central eastern Hispaniola sites are 3–5 mm/yr faster toward the WSW than the velocity at PARG (Calais et al., 2002), consistent with earlier estimates for opening across the Mona rift (Jansma et al., 2000).

CONCLUSIONS

Using GPS geodetic data collected throughout the northeastern Caribbean from 1994 until 2001, we are able to refine the velocities of Puerto Rico and the Virgin Islands with respect to the Caribbean plate, to constrain maximum permissible displacement rates along potentially active faults on the island of Puerto Rico and immediately offshore, and to examine potential diffuse extension between the Virgin Islands and eastern Puerto Rico. The existence of an independent Puerto Rico–northern Virgin Islands microplate was demonstrated at the 99.9% confidence level. At a location near the center of Puerto Rico (18.25°N, 66.5°W), motion of the Puerto Rico–northern Virgin Islands microplate with respect to the Caribbean is 2.6 ± 2.0 mm/yr (95%) directed toward N82.5°W ± 34° (95%). The geodetic data from the Puerto Rico–northern Virgin Islands block are consistent with east-west extension of several mm/yr in the boundary zone between the North America and Caribbean plates from eastern Hispaniola to the eastern Virgin Islands. The amount of extension increases westward with the most accommodated in the Mona rift between the Dominican Republic and western Puerto Rico, confirming earlier results from GPS geodesy. East-west extension of 1.5–3.9 mm/yr per year also potentially acts across the island of Puerto Rico. Whether this extension is localized along a few structures or is broadly distributed is unknown. We note that reactivation of the Great Northern and Great Southern Puerto Rico fault zones would yield oblique-normal slip with right-lateral sense along these structures. A zone of east-west extension of 1–2 mm/yr also must exist between eastern Puerto Rico and Virgin Gorda, which likely is attached to the Caribbean plate. These extensional belts allow eastward transfer of slip between North America and the Caribbean from the southern part of the plate boundary zone (Enriquillo fault in the southern Dominican Republic) to the northern (Puerto Rico trench offshore northern Virgin Islands). Increased seismicity in the Sombrero trend north of the northern and easternmost Virgin Islands is consistent with this interpretation.

The GPS-derived velocities of sites in Puerto Rico relative to the Caribbean, coupled with baseline changes and calculated relative velocities between selected stations, constrain maximum permissible displacement rates along active faults on the island and immediately offshore. Motions along or across any of the subaerial structures of Puerto Rico are ≤2 mm/yr. The Lajas Valley in the southwest, where microseismicity is greatest, is the locus of highest permissible on-land deformation. Northwest-southeast to east-west extension of 1.9 ± 0.2–2.0 ± 2.3 mm/yr across the Anegada Passage is also inferred from our GPS-derived velocities.

ACKNOWLEDGMENTS

This project could not have been completed without the support of numerous people: A. Eaby, A. Lopez, D. Martinez, H. Rodriguez, and S. Matson for assistance in data collection and E. Calais, C. DeMets, T. Dixon, and P. Mann for extensive discussions throughout the tenure of CANAPE. Thoughtful and comprehensive reviews by E. Calais and C. DeMets improved the final version of the manuscript. We also thank the Department of Marine Sciences, University of Puerto Rico, Mayagüez; the Agricultural Experiment Station in Isabela, Puerto Rico; the University of Puerto Rico, Humacao; B. Wiener; the Federal Aviation Administration Headquarters in San Juan, Puerto Rico; and the Guavaberry for continued access to their facilities. Work was supported by National Science Foundation grants EAR-9316215, EAR-9628553, EAR-9807289, EAR9806456, and HRD-9353549, National Aeronautics and Space Administration grant NCCW-0088, U.S. Geological Survey grant 01HQGR0041, and Puerto Rico Sea Grant award 535868.

REFERENCES CITED

Almy, C., Meltzer, A., and Dietrich, C., 2000, Faulting in the Lajas Valley and on the adjacent shelf, southwestern Puerto Rico: Eos (Transactions, American Geophysical Union), F1181.

Asencio, E., 1980, Western Puerto Rico seismicity: U.S. Geological Society Open-File Report, 80-192, p. 135.
Blewitt, G., Heflin, M.B., Webb, F.H., Lindqwister, U.J., and Malla, R.P., 1992, Global coordinates with centimeter accuracy in the International Terrestrial Reference Frame using GPS: Geophysical Research Letters, v. 19, p. 853–856.
Byrne, D.B., Suarez, G., and McCann, W.R., 1985, Muertos Trough subduction—Microplate tectonics in the northern Caribbean?: Nature, v. 317, p. 420–421.
Calais, E., Mazabraud, Y., Mercier de Lepinay, B., Mann, P., Mattioli, G., and Jansma, P., 2002, Strain partitioning and fault slip rates in the Caribbean from GPS measurements: Geophysical Research Letters, v. 29, no. 18, 1856, doi: 10.1029/2002G1015397.
DeMets, C., Jansma, P., Mattioli, G., Dixon, T., Farina, F., Bilham, R., Calais, E., and Mann, P., 2000, GPS geodetic constraints on Caribbean–North America plate motion: Implications for plate rigidity and oblique plate boundary convergence: Geophysical Research Letters, v. 27, p. 437–440, doi: 10.1029/1999GL005436.
Deng, J., and Sykes, L.R., 1995, Determination of Euler pole for contemporary relative motion of Caribbean and North American plates using slip vectors of interplate earthquakes: Tectonics, v. 14, p. 39–53, doi: 10.1029/94TC02547.
Dixon, T., G. Gonzales, E. Katsigris, and S. Lichten, 1991, First epoch geodetic measurements with the Global Positioning System across the northern Caribbean plate boundary zone: Journal of Geophysical Research, v. 96, p. 2,397–2,415.
Dixon, T.H., Farina, F., DeMets, C., Jansma, P., Mann, P., and Calais, E., 1998, Caribbean–North American plate relative motion and strain partitioning across the northern Caribbean plate boundary zone from a decade of GPS observations: Journal of Geophysical Research, v. 103, p. 15,157–15,182, doi: 10.1029/97JB03575.
Dolan, J.F., and Wald, D.J., 1998, The 1943–1953 north-central Caribbean earthquakes: active tectonic setting, seismic hazards and implications for Caribbean–North American plate motions, *in* Dolan, J., and P. Mann, eds., Active strike-slip and collisional tectonics of the northern Caribbean Plate boundary zone: Geological Society of America Special Paper 326, p. 143–161.
Dolan, J., Mullins, H., and Wald, D., 1998, Active tectonics of the north-central Caribbean: oblique collision, strain partitioning, and opposing subducted slabs, *in* Dolan, J., and Mann, P., eds., Active strike-slip and collisional tectonics of the northern Caribbean Plate boundary zone: Geological Society of America Special Paper 326, p. 1–61.
Dragert, H., Wang, K., and James, T., 2001, A silent slip event on the deeper Cascadia subduction interface: Science, v. 292, p. 1,525–1,528.
Erikson, J., Pindell, J., and Larue, D., 1990, Tectonic evolution of the south-central Puerto Rico region: Evidence for transpressional tectonism: Journal of Geology, v. 98, p. 365–368.
Erikson, J., Pindell, J., and Larue, D., 1991, Fault zone deformational constraints on Paleogene tectonic evolution in southern Puerto Rico: Geophysical Research Letters, v. 18, p. 569–572.
Frankel, A., McCann, W.R., and Murphy, A.J., 1980, Observations from a seismic network in the Virgin Islands region: Tectonic structures and earthquake swarms: Journal of Geophysical Research, v. 85, p. 2,669–2,678.
Glover, L., III, 1971, Geology of the Coama area, Puerto Rico and its relation to the volcanic arc-trench association: U.S. Geological Survey Professional Paper 636, p. 102.
Glover, L., III, and Mattson, P., 1960, Successive thrust and transcurrent faulting during the early Tertiary in south-central Puerto Rico: U.S. Geological Survey Professional Paper 400-B, 363–365.
Grindlay, N.R., Mann, P., and Dolan, J.F., 1997, Researchers investigate submarine faults north of Puerto Rico: Eos (Transactions, American Geophysical Union), v. 78, p. 404.
Heflin, M., Bertiger, W., Blewitt, G., Freedman, A., Hurst, K., Lichten, S., Lindqwister, U., Vigue, Y., Webb, F., Yunck, T., and Zumberge, J., 1992, Global geodesy using GPS without fiducial sites: Geophysical Research Letters, v. 19, p. 131–134.
Holcombe, T.L., Fisher, C.G., and Bowles, F.A., 1989, Gravity-flow deposits from the St. Croix Ridge; depositional history: Geo-Marine Letters, v. 9, p. 11–18.
Jansma, P., Mattioli, G., Lopez, A., DeMets, C., Dixon, T., Mann, P., and Calais, E., 2000, Neotectonics of Puerto Rico and the Virgin Islands, northeastern Caribbean from GPS geodesy: Tectonics, v. 19, p. 1021–1037.
Jany, I., Mauffret, A., Bouysse, P., Mascle, A., Mercier de Lépinay, B., Renard, V., and Stephan, J.F., 1987, Relevé bathymétrique Sea beam et tectonique en décrochement au sud des Iles Vierges (Nord-Est Caraibes): Comptes Rendus de l'Académie des Sciences, Serie II, Mecanique, Physique, Chimie, Sciences de l'Univers: Sciences de la Terre, v. 304, p. 527–532.
Joyce, J., McCann, W., and Lithgow, C., 1987, Onland active faulting in the Puerto Rico platelet: Eos (Transactions, American Geophysical Union), v. 68, p. 1,483.
Ladd, J.W., Worzel, J.L., and Watkins, J.S., 1977, Multifold seismic reflection records from the northern Venezuela Basin and the north slope of Muertos Trench, *in* Talwani, M., and Pitman, W.C., eds., Island arcs, deep-sea trenches and back-arc basins: Washington, D.C., American Geophysical Union, p. 41–56.
Ladd, J.W., and Watkins, J.S., 1978, Active margin structures within the north slope of the Muertos Trough: Geologie en Mijnbouw, v. 57, p. 225–260.
Langbein, J., and Johnson, H., 1997, Correlated errors in geodetic time series: implications for time-dependent deformation: Journal of Geophysical Research, v. 102, p. 591–603, doi: 10.1029/96JB02945.
Lao-Davila, D., Mann, P., Prentice, C., and Draper, G., 2000, Late Quaternary activity of the Cerro Goden fault zone, transpressional uplift of the La Cadena Range, and their possible relation to the opening of the Mona rift, western Puerto Rico: Eos (Transactions, American Geophysical Union), v. 81, p. F1181.
Larue, D.K., and Ryan, H.F., 1990, Extensional tectonism in the Mona Passage, Puerto Rico and Hispaniola: a preliminary study: Transactions of the Caribbean Geological Conference, v. 12, p. 301–313.
Larue, D.K., and Ryan, H.F., 1998, Seismic reflection profiles of the Puerto Rico Trench; shortening between the North American and Caribbean plates, *in* Lidiak, E.G., and Larue, D.K., eds., Tectonics and geochemistry of the northeastern Caribbean: Geological Society of America Special Paper 322, p. 193–210.
Lichten, S.M., 1990, Estimation and filtering for high precision GPS applications: Manual Geodesy, v. 15, p. 159–176.
Mann, P., Taylor, F., Edwards, L., and Ku, T., 1995, Actively evolving microplate formation by oblique collision and sideways motion along strike-slip faults: an example from the northeastern Caribbean plate margin: Tectonophysics, v. 246, p. 1–69, doi: 10.1016/0040-1951(94)00268-E.
Mann, P., McLaughlin, P., Jr., van den Bold, W., Lawrence, S., and Lamar, M., 1999, Tectonic and eustatic controls on Neogene evaporitic and siliciclastic deposition in the Enriquillo basin, Dominican Republic, *in* Mann, P., ed., Caribbean Basins, Sedimentary Basins of the World, 4: Elsevier: Amsterdam, p. 287–342.
Mann, P., Calais, E., Ruegg, J., DeMets, C., Jansma, P., and Mattioli, G., 2002, Oblique collision in the northeastern Caribbean from GPS measurements and geological observations, Tectonics, v. 21, no. 6, 1058, doi: 10.1029/2001TC001363.
Mao, A., Harrison, C.G.A., and Dixon, T.H., 1999, Noise in GPS coordinate time series: Journal of Geophysical Research, v. 104, p. 2,797–2,816.
Masson, D.G., and Scanlon, K.M., 1991, The neotectonic setting of Puerto Rico: Geological Society of America Bulletin, v. 103, p. 144–154, doi: 10.1130/0016-7606(1991)1032.3.CO;2.
Mauffret, A., and Jany, I., 1990, Collision et tectonique d'expulsion le long de la frontiere Nord-Caraibe: Oceanologica Acta, v. 10, p. 97–116.
Mauffret, A., and Leroy, S., 1999, Neogene intraplate deformation of the Caribbean plate at the Beata Ridge, *in* Mann, P., ed., Caribbean Basins, Sedimentary Basins of the World, 4: Elsevier: Amsterdam, p. 627–669.
McCann, W.R., 1985, On the earthquake hazards of Puerto Rico and the Virgin Islands: Bulletin of the Seismological Society of America, v. 75, p. 251–262.
McCann, W.R., 2002, Microearthquake data elucidate details of Caribbean subduction zone: Seismological Research Letters, v. 73, p. 25–32.
McCann, W.R., and Pennington, W.D., 1990, Seismicity, large earthquakes, and the margin of the Caribbean plate, *in* Dengo, G., and Case, J., eds., The Geology of North America, The Caribbean Region: Geological Society of America: Boulder, Colorado, Geology of North America, v. H, p. 291–306.
McCann, W. R., 2000, Characterization of active submarine faults near U.S. Caribbean territories: U.S. Geological Survey Final report for 99HQGR0067, 8 p.
Meltzer, A.S., Schoemann, M.L., Dietrich, C., Almy, C., and Schellekens, H., 1995, Characterization of faulting: southwest Puerto Rico: Geological Society of America Abstracts with Programs, v. 27, no. 6, p. 227.

Meltzer, A., 1997, Fault structure and earthquake potential of the Lajas Valley, SW Puerto Rico: U.S. Geological Survey Technical Abstract.

Mercado, A., and McCann, W., 1998, Numerical simulation of the 1918 Puerto Rico tsunami: Journal of Natural Hazards, v. 18, p. 57–76, doi: 10.1023/A:1008091910209.

Molnar, P., and Sykes, L.R., 1969, Tectonics of the Caribbean and Middle America regions from focal mechanisms and seismicity: Geological Society of America Bulletin, v. 80, p. 1639–1684.

Murphy, A.J., and McCann, W.R., 1979, Preliminary results from a new seismic network in the northeastern Caribbean: Bulletin of the Seismological Society of America, v. 69, p. 1497–1513.

Pacheco, J.F., and Sykes, L.R., 1992, Seismic moment catalog of large shallow earthquakes, 1900 to 1989: Bulletin of the Seismological Society of America, v. 82, p. 1306–1349.

Prentice, C., Mann, P., Peña, L., and Burr, G., 2003, Slip rate and earthquake recurrence along the central Septentrional fault, North American–Caribbean plate boundary, Dominican Republic: Journal of Geophysical Research, v. 108(B3), 2149, doi: 10.1029/2001JB00042.

Prentice, C., Mann, P., and Burr, G., 2000, Prehistoric earthquakes associated with a Late Quaternary fault in the Lajas Valley, southwestern Puerto Rico: Eos (Transactions, American Geophysical Union), v. 81, p. F1182.

Reid, H., and Taber, S., 1920, The Virgin Islands earthquakes of 1867–1868: Bulletin of the Seismological Society of America, v. 10, p. 9–30.

Reid, J., Plumley, P., and Schellekens, J., 1991, Paleomagnetic evidence for late Miocene counterclockwise rotation of the North Coast carbonate sequence, Puerto Rico: Geophysical Research Letters, v. 18, p. 565–568.

Russo, R.M., and Villaseñor, A., 1995, The 1946 Hispaniola earthquake and the tectonics of the North America–Caribbean plate boundary zone, northeastern Hispaniola: Journal of Geophysical Research, v. 100, p. 6,265–6,280.

Stein, S., and Gordon, R.G., 1984, Statistical tests of additional plate boundaries from plate motion: Earth and Planetary Science Letters, v. 69, p. 401–412, doi: 10.1016/0012-821X(84)90198-5.

Sykes, L.R., McCann, W.R., and Kafka, A.L., 1982, Motion of Caribbean plate during last 7 million years and implications for earlier Cenozoic movements: Journal of Geophysical Research, v. 87, p. 10,656–10,676.

van Gestel, J., Mann, P., Dolan, J., and Grindlay, N., 1998, Structure and tectonics of the upper Cenozoic Puerto Rico–Virgin Islands carbonate platform as determined from seismic reflection studies: Journal of Geophysical Research, v. 103, p. 30,505, doi: 10.1029/98JB02341.

Ward, S.N., 1990, Pacific–North America plate motions: new results from very long baseline interferometry: Journal of Geophysical Research, v. 95, p. 21,965–21,981.

Weber, J.C., Dixon, T., DeMets, C., Ambeh, W.B., Jansma, P., Mattioli, G., Saleh, J., Sella, G., Bilham, R., and Pérez, O., 2001, A GPS estimate of relative motion between the Caribbean and South American plates and geologic implications for Trinidad and Venezuela: Geology, v. 29, p. 75–78, doi: 10.1130/0091-7613(2001)0292.0.CO;2.

Zumberge, J.F., Heflin, M., Jefferson, D., Watkins, M., and Webb, F., 1997, Precise point positioning for efficient and robust analysis of GPS data from large networks, Journal of Geophysical Research, v. 102, p. 5,005–5,017.

Manuscript Accepted by the Society 18 August 2004

Geological Society of America
Special Paper 385
2005

Neotectonics and subsidence of the northern Puerto Rico–Virgin Islands margin in response to the oblique subduction of high-standing ridges

Nancy R. Grindlay*
Center for Marine Science and Department of Earth Sciences, University of North Carolina at Wilmington, Wilmington, North Carolina, 28409, USA

Paul Mann
Institute for Geophysics, Jackson School of Geosciences, University of Texas at Austin, Austin, Texas 78759-8500, USA

James F. Dolan
Department of Earth Sciences, University of Southern California, Los Angeles, California 90089-0740, USA

Jean-Paul van Gestel*
Department of Geological Sciences and Institute for Geophysics, University of Texas at Austin, Austin, Texas, 78759-8500, USA

ABSTRACT

High-resolution single-channel seismic reflection profiles, bathymetry and sidescan sonar imagery from the Puerto Rico trench document the present-day and post-collisional effects of the obliquely subducting southeastern extension of the Bahama Province and the Main Ridge fracture zone on the northern Puerto Rico–Virgin Islands margin. In contrast to an orthogonal system, where it is unlikely that two high-standing ridges will impact the same section of margin, along the Puerto Rico trench convergence is highly oblique and the deformational effects of the two ridges are superimposed and often difficult to isolate. A middle–upper Miocene margin-wide unconformity in the Oligocene–lower Pliocene shallow-water carbonate platform of the Virgin Islands, Puerto Rico, and eastern Hispaniola provides an excellent horizontal reference frame for the timing and impact of the high-standing ridges on the margin. During the past 10 m.y., trenchward tilting (4–6°) of the carbonate platform including the margin-wide erosional surface to >5000 m water depths provides evidence that the margin has experienced significant subsidence. In this paper we present a model of accelerated subduction erosion and diachronous margin subsidence triggered by the ridges sweeping from east to west beneath the Puerto Rico forearc. Evidence for ongoing and past subduction erosion include zones of enhanced seismicity, oversteepening and mass wasting of the forearc slope, and landward migration of the inner trench-slope break. We document over 3 km of Neogene subsidence presumed to be the result of rapid crustal thinning associated with the tunneling of the

*E-mail, Grindlay: grindlayn@uncw.edu. Present address, van Gestel: BP/Amoco, Houston, Texas, USA.

Grindlay, N.R., Mann, P., Dolan, J.F., and van Gestel, J-P, 2005, Neotectonics and subsidence of the northern Puerto Rico–Virgin Islands margin in response to the oblique subduction of high-standing ridges, *in* Mann, P., ed., Active tectonics and seismic hazards of Puerto Rico, the Virgin Islands, and offshore areas: Geological Society of America Special Paper 385, p. 31–60. For permission to copy, contact editing@geosociety.org.

buoyant, thick (~20-km-thick crustal/sediment section) southeastern Bahama Province beneath the forearc. During the past 3.5–5 m.y., the volume of material eroded from the overriding plate is estimated to be ~210 km^3/km of margin, equivalent to an erosion rate of 42–60 km^3/m.y./km of margin. A significant reduction in the rate of margin subsidence in the eastern part of the survey area suggests that the Main Ridge fracture zone has had a relatively small erosional impact on the margin, and that under normal conditions, this highly oblique convergent boundary is characterized by relatively slow rates of subduction erosion. Three strike-slip fault zones are imaged in the forearc: the East Septentrional fault zone, the Bunce fault zone, and the Bowin fault zone. The Bunce and Bowin fault zones trend N80°–90°E, subparallel to the predicted North America–Caribbean relative plate motion vector (N70°E) determined by global positioning system (GPS) studies. Seismic reflection profiles across the Puerto Rico trench in the western section of the survey area, where the Bunce fault zone is <10 km from the deformation front, reveal thick (0.75–1.5 km) accumulations of relatively undeformed trench sediment fill. This lack of shortening deformation suggests that convergence approaches pure strike-slip and that strain within the forearc is largely accommodated by the strike-slip fault zones. Strike-slip faulting appears to be progressively partitioned to the east where deformation in the trench is more clearly contractional. Although not a typical plow mark, the Bowin fault zone appears to represent the long-term (~10 m.y.) track of the underthrusting and colliding southeastern Bahama Province across the forearc area that is consistent with the convergence direction exhibited by GPS results.

Keywords: subduction erosion, Puerto Rico trench, oblique convergence, Bahama Province, North America–Caribbean plate boundary.

INTRODUCTION

This paper explores the morphotectonic features of the Puerto Rico trench and northern Puerto Rico and Virgin Islands insular margins, from the Bahamas oblique collision zone at the northeastern coast of Hispaniola eastward to the Virgin Islands (McCann and Sykes, 1984; Dolan et al., 1998) (Fig. 1). A high-resolution sidescan sonar, bathymetric and single-channel seismic data set is used to unravel the complex Neogene deformation of the plate boundary zone within this region. Previous tectonic models have proposed that rapid late Neogene margin subsidence occurred in response to "deep plate collision" or mantle interaction between a south-dipping North America slab depressed by a north-dipping Caribbean slab as shown by serial profiles of seismicity through the Mona Passage and Hispaniola (Dillon et al., 1996: van Gestel et al., 1999). In these models, subsidence occurs synchronously and is ongoing with active plate convergence. This paper presents evidence for accelerated subduction erosion and subsidence of the forearc region of the Puerto Rico trench associated with the oblique subduction of high-standing ridges on the down-going North America plate. The high-standing ridges include the southeastern extension of the Bahama Province, a carbonate platform above thinned continental or thickened oceanic crust, and the Main Ridge fracture zone, a fracture zone formed in oceanic crust of early Cretaceous age.

Studies of obliquely subducting ridges and seamounts along the Tonga Trench (Ballance et al., 1989), and orthogonally subducted seamounts and ridges at the Japan and Peru convergent margins (von Huene and Lallemand, 1990) show accelerated subduction erosion of the forearc due to the fracturing and shearing of arc substrate rocks as they are lifted up by the colliding bathymetric high, and then left to collapse as the base of the forearc is removed and the ridge is subducted. This wave of vertical tectonism travels along the landward trench slope lifting it and then letting it subside again leaving an extended or collapsed terrain in its "wake." Von Huene et al. (1995) and Ranero and von Huene (2000) have shown spectacular examples using multibeam bathymetry of furrows up to 55 km long in the Costa Rica landward trench slope that mark the collapsing paths of subducting seamounts.

As is the case along other margins bordered by poorly sedimented trenches, (e.g., the Guatemala to Costa Rica sector of the Middle American margin and northern and central Chile margin), along the northeastern North America–Caribbean plate boundary, subducted bathymetric highs are tracked by subtle and difficult-to-image uplift and deformational response of crystalline and sedimentary rock in the overriding plate. The Mona Block and the Main Ridge areas of the Puerto Rico trench were selected as the focus of this study because these areas appear to represent the zones of active collision with the southeastern Bahama Province and the Main Ridge fracture zone, respectively, along this segment of the plate boundary (McCann and Sykes, 1984; Dolan et al., 1998) (Fig. 1). Documenting the nature of these collision zones is critical for understanding patterns of ancient

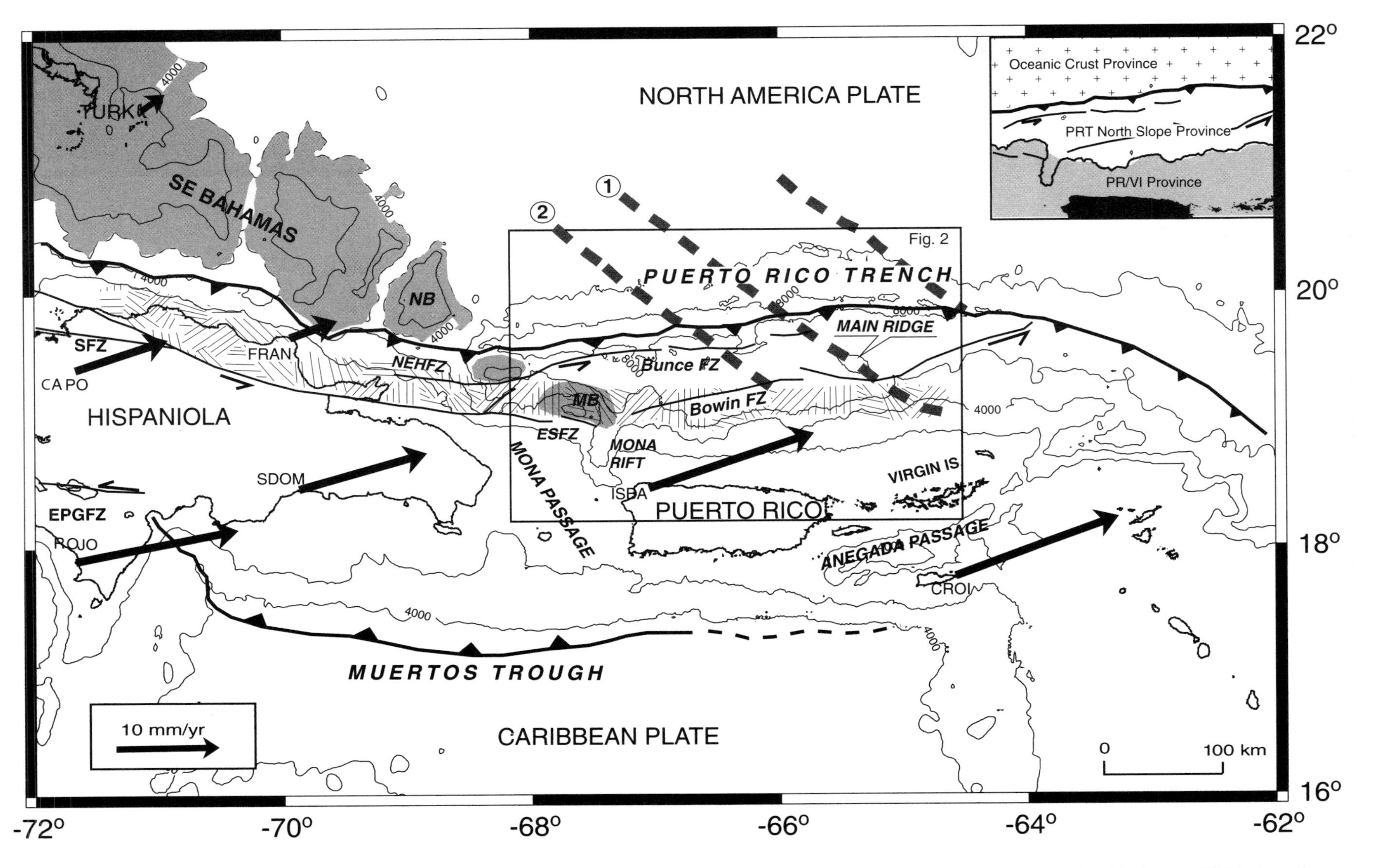

Figure 1. Regional tectonic setting of study area. Bathymetry contour interval is 2000m. MB—Mona Block; NB—Navidad Bank; SFZ—Septentrional fault zone; ESFZ—East Septentrional fault zone; EPGFZ—Enriquillo–Plantain Garden fault zone; NEHFZ—Northeast Hispaniola fault zone. The area covered by striped pattern corresponds to a metamorphic blue schist belt that has been identified in outcrop (Joyce, 1991) seafloor dredges (Perfit et al., 1980), and as a zone of low magnetization (Muszala et al., 1999). Thick dashed dark gray lines indicate the linear magnetic highs identified by Muszala et al., (1999) that are believed to be associated with fracture zones in the downgoing North America plate. 1—Main Ridge fracture zone; 2—Dos Ninos fracture zone. Also shown are the GPS plate motion vectors relative to a fixed North America plate (Mann et al., 2002). CAPO—Capotillo, Dominican Republic; ROJO—Cabo Rojo, Dominican Republic; TURK—Turk island, Bahamas; FRAN—Cabo Frances Viejo, Dominican Republic; SDOM—Santo Domingo, Dominican Republic; ISBA—La Isabela, Puerto Rico; CROI—St. Croix, U.S. Virgin Islands; PR-VI—Puerto Rico–Virgin Islands.

deformation in a post-collisional zone that extends from the southeastern Bahamas Province to the Lesser Antilles islands, as well as understanding the present-day zones of deformation and related seismicity in the Puerto Rico trench area.

REGIONAL TECTONIC AND GEOLOGIC SETTING OF THE PUERTO RICO TRENCH

Vening Meinesz et al. (1934) first recognized the Puerto Rico trench as a unique oceanic feature. With depths reaching 8450 m, the trench is the deepest point in the Atlantic Ocean and is associated with the largest (–350 mgal) free-air gravity minimum in the world (Bowin, 1972; 1976). The trench forms an ~150 km wide, and ~500 km long submarine valley that is located ~70 km north of the island of Puerto Rico and delineates the leading edge of the Caribbean plate (Figs. 1 and 2). The trench is continuous with the North Hispaniola deformed belt, a zone of active southward-dipping thrust faults at the base of the Hispaniola island slope (Austin, 1983; Dillon et al., 1992; Mullins et al., 1992; Dolan et al., 1998). Both the Puerto Rico trench and the North Hispaniola deformed belt lie within the part of the plate boundary that appears to be transitional between orthogonal oceanic subduction at the Lesser Antilles, oblique collision in the Hispaniola area, and pure strike-slip at the Cayman trough (Calais et al., 1992; DeMets et al., 2000). North of the Puerto Rico trench, ~100-m.y.-old Atlantic oceanic crust is moving westward with the North America plate predominantly parallel to the orientation of the trench. The Puerto Rico trench east of the northern Virgin Islands curves southward into the north-south Lesser Antilles subduction zone and disappears as a bathymetric feature as it becomes filled with southward-increasing amounts of terrigenous sediment derived from South America.

Over the past three decades, numerous kinematic models have been put forth that predict significantly different rates and directions of motion between the North America and Caribbean plates, from 1 to 4 cm/yr of EW strike-slip motion (e.g., Jordan, 1975) to ENE convergence (Deng and Sykes, 1995). The most recent GPS-geodetic measurements in the region have led to a robust model for the entire Caribbean plate that is supportive of ENE convergence (070°) in the NE Caribbean at rates of 19–20 mm/yr (DeMets et al., 2000; Jansma et al., 2000; Mann et al., 2002). This direction of convergence is subparallel to the overall trend (080°) of the Puerto Rico trench, thus relative motion along this sector of the North America–Caribbean boundary is largely strike-slip with a small component of convergence. Subducted slabs beneath eastern Hispaniola indicate at least 150–200 km of Neogene plate convergence (Dillon et al., 1996; Dolan et al., 1998).

The Puerto Rico trench and its southern slope can be divided into three distinct geologic provinces (Fig. 1 inset). On the basis of dredge hauls and submersible observations (Fox and Heezen, 1975; Weaver et al., 1975; Perfit et al., 1980), the northernmost province is associated with Lower Cretaceous ocean crust with sedimentary cover. It is also characterized by northwest to southeast trending magnetic highs (Muszala et al., 1999). This province extends to the surface trace of the Puerto Rico trench thrust fault and is subducted along this fault beneath the northern slope of Puerto Rico. The North Slope province extends south of the trench axis to the base of the northern Puerto Rico insular slope. Dredge hauls reveal that this central terrane corresponds to a region of mostly indurated rocks, including Cretaceous though Miocence volcanic and carbonate rocks and a blueschist-grade metamorphic belt of Cretaceous age (Fox and Heezen, 1975; Perfit et al., 1980). The blueschist belt is characterized by an E-W–trending zone of low-amplitude magnetic anomalies extends from north of the Dominican Republic across the southern slope of the Puerto Rico trench (striped area in Fig. 1) (Muszala et al., 1999). West of the Puerto Rico trench, the blueschist belts outcrop in northeastern Hispaniola (Joyce, 1991). The southernmost geologic province, the Puerto Rico–Virgin Islands province, is characterized by high-amplitude magnetic anomalies and is associated with Oligocene-Cretaceous arc-related rocks overlain by a cap of shallow-water limestone. This carbonate platform is comprised of a lower Oligocene–middle Miocene carbonate sequence that was deposited horizontally in a shallow-water environment (Monroe, 1980). By middle–late Miocene a rapid decline in sea level had exposed the shallow water limestone producing a regional subaerial unconformity (Monroe, 1980). The unconformity is clearly imaged by single-channel (SCS) and multichannel seismic (MCS) profiles and is a superb horizontal reference frame from which the timing and amount of margin subsidence can be estimated. Northward tilting (4–6°) of the carbonate platform including the margin-wide erosional surface to >5000 m water depths provides evidence that the margin has experienced significant subsidence during the past 10 m.y.

The Quebradillas formation (upper Miocene–lower Pliocene) was deposited during a rapid marine transgression/regression sequence (Moussa et al., 1987). The formation was not deposited uniformly on the platform. In the eastern area of the platform, it was deposited after the initial tilting and is mainly confined to nearshore areas within ~5–10 km of the coast, <1000–1500 m water depths (Larue et al., 1998)(Fig. 2A). The Quebradillas deposition is more extensive in the western part of the platform and includes onland areas of northwest Puerto Rico (Moussa et al., 1987) (Fig. 2A). Pleistocene corals and shell hash have been recovered on the eastern flanks of the Mona rift and in the Mona Passage (Fox and Heezen, 1975; Weaver et al., 1975; Perfit et al., 1980; Heezen et al., 1985).

Prior to 1996, the morphology and structure of the seafloor near the Puerto Rico trench was poorly constrained owing to limited coverage and insufficient resolution of bathymetric features by previous geophysical studies. GLORIA sidescan sonar data collected in 1985 by the U.S. Geological Survey in the Exclusive Economic Zone (EEZ) around Puerto Rico, in combination with widely spaced SCS and MCS reflection profiles, reveal pronounced and continuous seafloor lineations in the Puerto Rico trench that have been interpreted by previous workers as active, left-lateral strike-slip faults accommodating North America–Caribbean relative plate motion (e.g., EEZ SCAN 85 Scientific

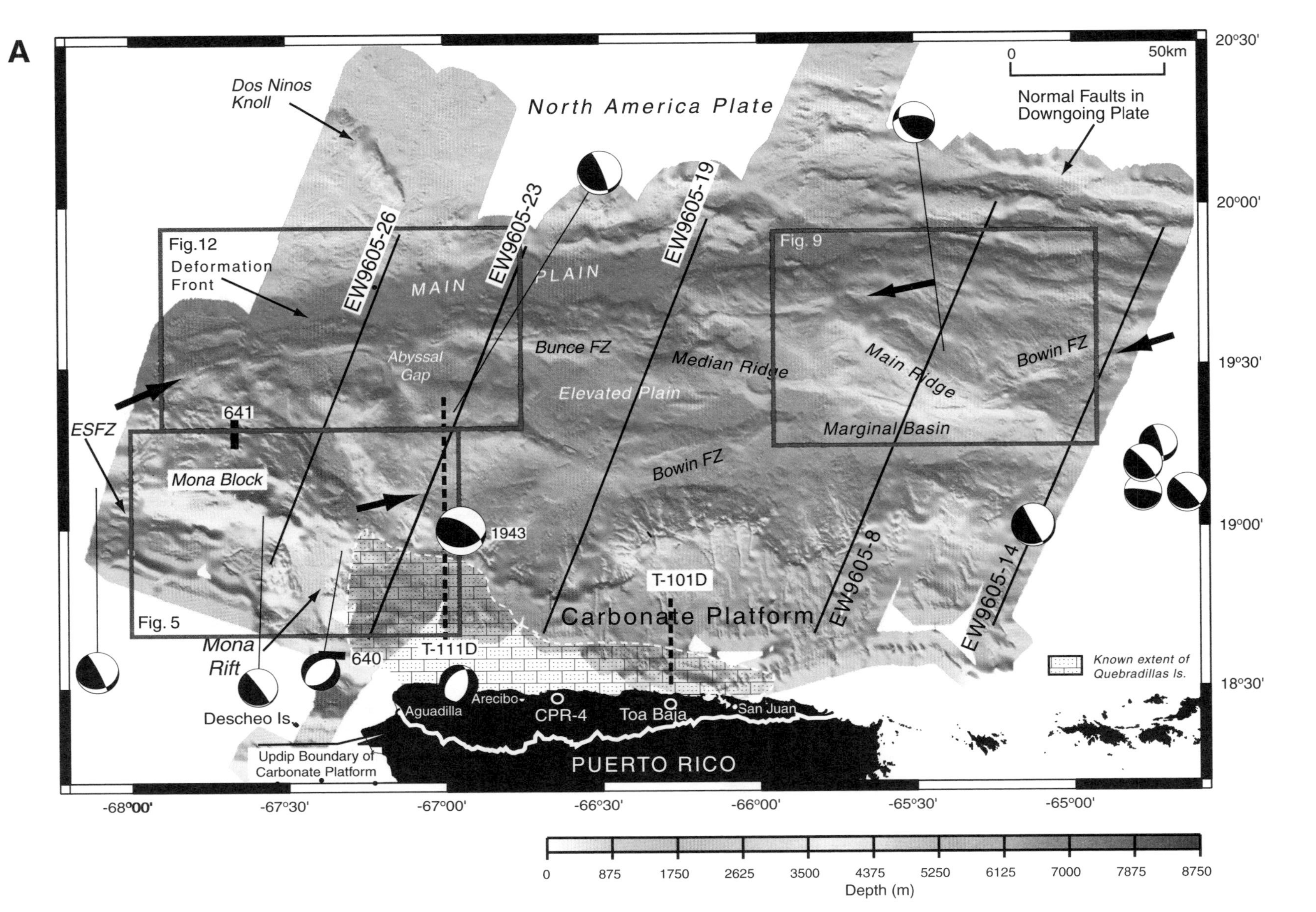

Figure 2 (*continued on following page*). (A) Colored shaded relief of EW9605 Hydrosweep bathymetry showing detailed structural features. Illumination is from the southwest. Long thick black lines indicate the locations of EW96–05 single-channel seismic reflection profiles discussed in the paper. Long thick dashed black lines show location of multichannel seismic profiles collected by Western Geophysical and used to constrain the known extent of Quebradillas deposition (Moussa et al., 1987; Larue et al., 1998). Location of ALVIN dives 640 and 641 (Heezen et al., 1985) are shown by short thick black lines. The updip limit of the north coast carbonate platform on land is shown by a white line on the island of Puerto Rico. Boxes show the location of the sidescan sonar images and contoured bathymetry shown in Figures 5, 9, and 12. Shallow (<50 km) focal mechanisms from Harvard CMT catalog, 1976–2003. Focal mechanism of the 1943 Ms 7.8 earthquake from Dolan and Wald (1998). ESFZ—East Septentrional fault zone.

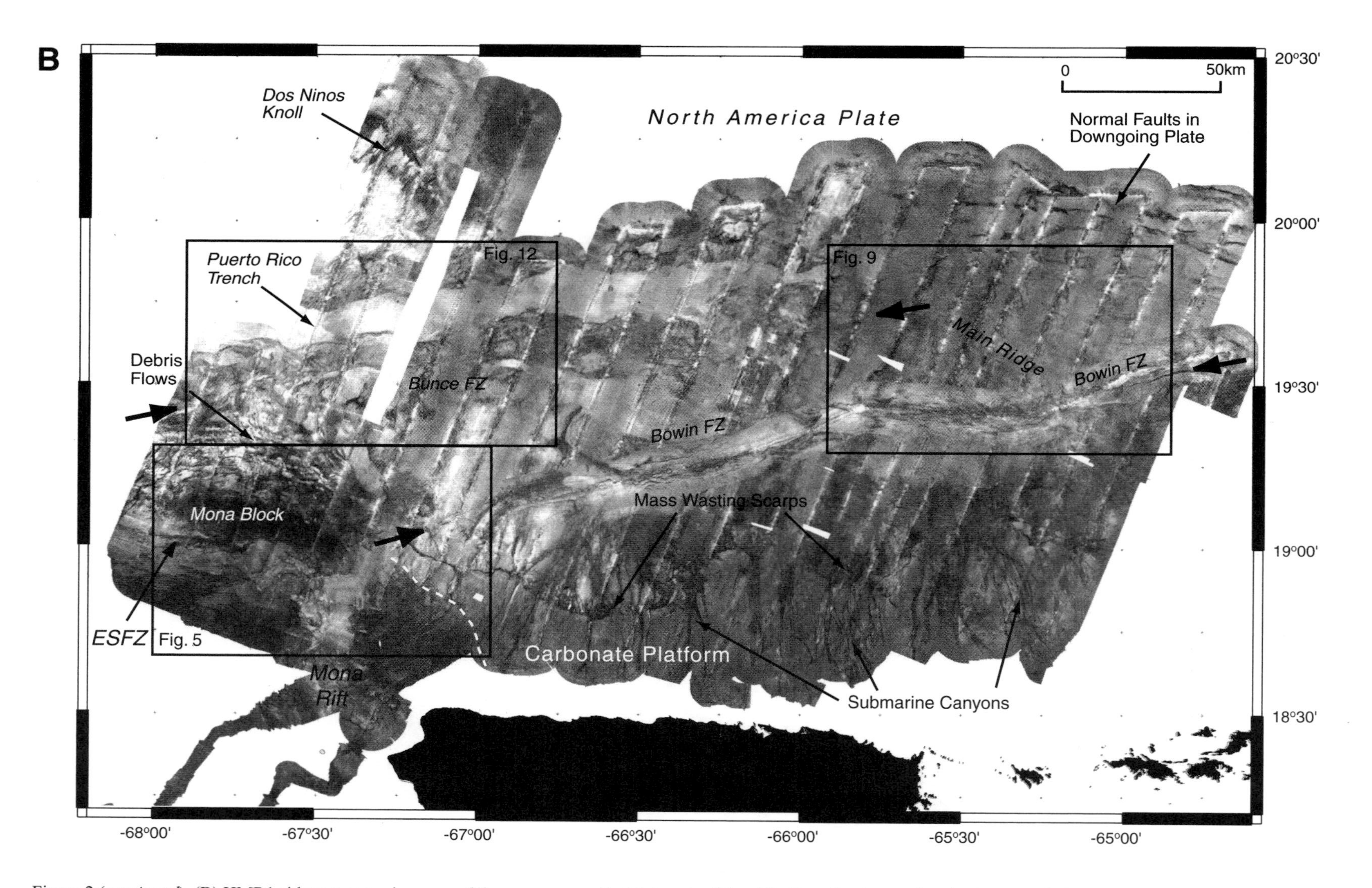

Figure 2 (*continued*). (B) HMR1 sidescan sonar imagery of the survey area. Grid interval = 17 m. Highly reflective seafloor is shown as dark returns, seafloor with low reflectivity as light returns. Boxes show the location of the sidescan sonar images and contoured bathymetry in Figures 5, 9. and 12.

Staff, 1987; Masson and Scanlon, 1991; Larue et al., 1989; Larue and Ryan, 1998). Low-angle overthrusting of trench turbidite beds by slope rocks and areas of possible accreted sediment imaged in MCS lines indicate a small convergent component of plate motion in addition to the strike-slip component suggested by these seafloor faults (Larue and Ryan, 1998). Seismic reflection profiles and GLORIA sidescan imagery also reveal horst-and-graben structures in North America oceanic lithosphere that are attributed to bending of a subducting slab (Masson and Scanlon, 1991, Dillon et al., 1998). In addition, several prominent bathymetric highs have been imaged with SeaMARC II sidescan sonar on the North America plate: the southeasternmost, unsubducted extension of the southeastern Bahama Province, including, Navidad Bank (Mullins et al., 1992; Dolan et al., 1998). On the northern slope of Puerto Rico, anomalously shallow bathymetric features include the Mona Block, which is collinear across the Puerto Rico trench with the Navidad Bank, and the Main Ridge (Figs. 1 and 2A; Ewing et al., 1968; McCann and Sykes, 1984; Masson and Scanlon, 1991; Dolan et al., 1998).

NEW DATA: SINGLE-CHANNEL REFLECTION SEISMIC PROFILES, SIDESCAN SONAR, AND BATHYMETRY

In June and July of 1996, an extensive marine geophysical survey (EW96-05) of the Puerto Rico trench, Mona Passage, and northern insular margins of Puerto Rico and the Virgin Islands was conducted from the R/V *Maurice Ewing*. Over a 21-day period, Hawaii-MR1 (HMR1) sidescan sonar imagery, Hydrosweep multibeam bathymetry, gravity, magnetics, and SCS data were systematically collected along 5600 km of closely spaced (12–14 km) tracklines.

An area of ~50,000 km^2 was imaged by the HMR1 bathymetry and sidescan sonar system (Figs. 2A and 2B). With the exception of the shallow areas in the southern section of the survey area, sidescan sonar coverage of the survey is almost continuous. An electrical problem in the HMR1 tow fish resulted in noise contamination of the swath bathymetry and sidescan sonar data during the majority of the cruise. The noise problem rendered the HMR1 bathymetry data unusable. In the sidescan sonar data, the noise appears in the starboard half of the swath as a track parallel band of muted data. The data degradation was particularly apparent in swaths covering the deeper (>4500 m) areas of the survey. Despite these shortcomings, these sidescan sonar data are far superior to previous efforts including the GLORIA survey (EEZ SCAN 85 Scientific Staff, 1987) and hull-mounted surveys.

Additional bathymetry were collected along all survey tracks using the hull-mounted multibeam Atlas Hydrowseep echosounder. Hydrosweep swath coverage of 90° is typically twice water depths (vertical resolution of 10–15 m) and, thus, provided nearly continuous coverage in areas >6000 m. In shallower areas, the Hydrosweep bathymetry have been interpolated between swaths to produce a continuous contoured map. The Hydrosweep bathymetry have been processed to produce two- and three-dimensional shaded relief images to enhance visualization of seafloor features (Figs. 2A, 3A, and 3B).

An array of six air guns with a1385 ci capacity was towed simultaneously with the HMR1 system along most survey tracks. The seismic reflection data were recorded by a four-channel streamer with an active section length of 137.5 m. The SCS data presented in this paper have been processed through migration (processing details in van Gestel et al.; 1998; 1999). Interpreted lines drawings of representative seismic profiles acquired in the survey area are show in (Fig. 4).

DATA DESCRIPTION AND INTERPRETATION

The sidescan sonar, bathymetry, and SCS data collected during cruise EW96-05 allow a much more detailed view of the morphology and structure of the seafloor within and near the Puerto Rico trench. Below, we discuss the major features of the northern Puerto Rico slope and flanking trench incorporating existing and newly acquired data.

Mona Block

Mona Block is located in the southwest portion of our survey area ~40 km offshore the northwest tip of the island of Puerto Rico (Figs. 2A, 3A, and 3B). The entire Mona Block region (Mona Block, adjacent carbonate platform, and Mona rift) is characterized by frequent and intense seismic activity (Molnar and Sykes, 1969; McCann and Sykes, 1984; Dolan et al., 1998; Dolan and Wald, 1998) (Fig. 2A). Hydrosweep bathymetry indicate Mona Block is an anomalous high in the forearc region whose crest lies at ~1000 m depth and strikes N75°W parallel to the overall trend of the southeastern Bahama Province (Fig. 5B). Mona Block is ~30 km in the east-west direction and 20 km in the north-south direction with a very steep slope on its southern side and a gentler northward-dipping convex slope on its northern side. The sidescan sonar imagery shows a highly reflective NW-trending elongate zone associated with the shallowest portions of the block (Fig. 5A). We interpret this to be a cap of platform carbonate because it has a similar reflectivity and seismic layering to the carbonate platform in the adjacent Mona Passage (Figs. 5A and 6).

EW96-05 SCS profile 26 (Figs. 4 and 6) and sidescan sonar imagery (Fig. 5A) also show evidence of large blocks of carbonate material and slumps of debris being shed down the northern slope of Mona Block toward the trench. During DSRV *Alvin* dive 641 Heezen et al. (1985) observed and recovered in situ Pleistocene-aged shallow-water corals on this slope at depths of 3600 m. Heezen et al. (1985) suggested that presence of young corals at these depths indicated extremely recent subsidence of the forearc in this region. On the same dive at depths ranging from 3345 to 3000 m, Cretaceous blueschist metamorphic rocks were recovered from the northern slope of Mona Block. In the sidescan sonar imagery, an east-west–trending, discontinuous fabric of alternating light and dark bands appear to encircle the north slope of the Mona Block (Fig. 5A). We interpret these features to be a

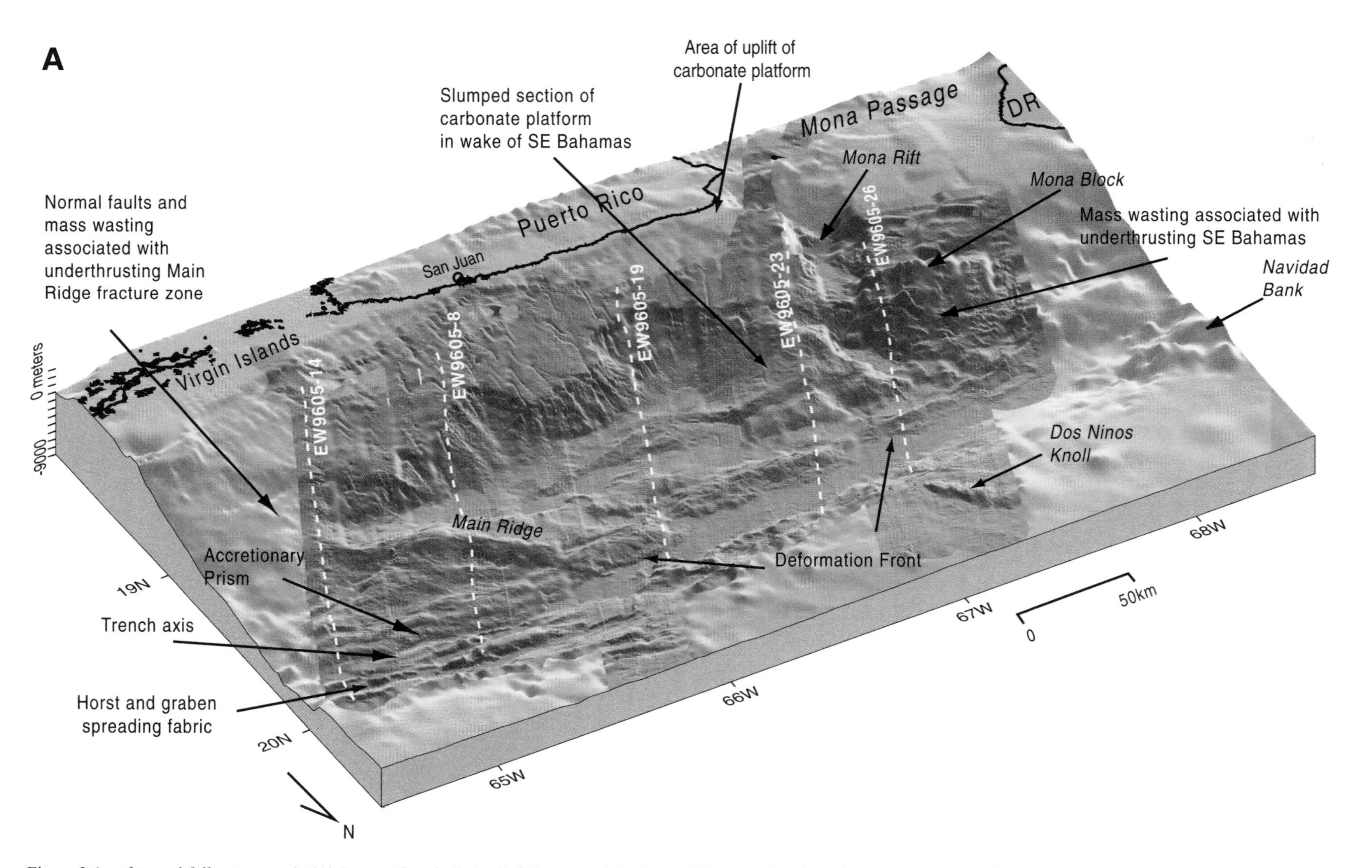

Figure 3 (*on this and following page*). (A) Perspective shaded-relief diagram of the Puerto Rico trench and northern Puerto margin from east to west, identifying major structural features. The EW9605 high-resolution Hydrosweep bathymetry is draped over a coarser grid of predicted bathymetry (Smith and Sandwell, 1997), National Ocean Survey bathymetry and USGS digital elevation models of Puerto Rico, the Virgin Islands, and Hispaniola. Location of EW9605 single-channel seismic reflection profiles discussed in the paper are shown as dashed white lines.

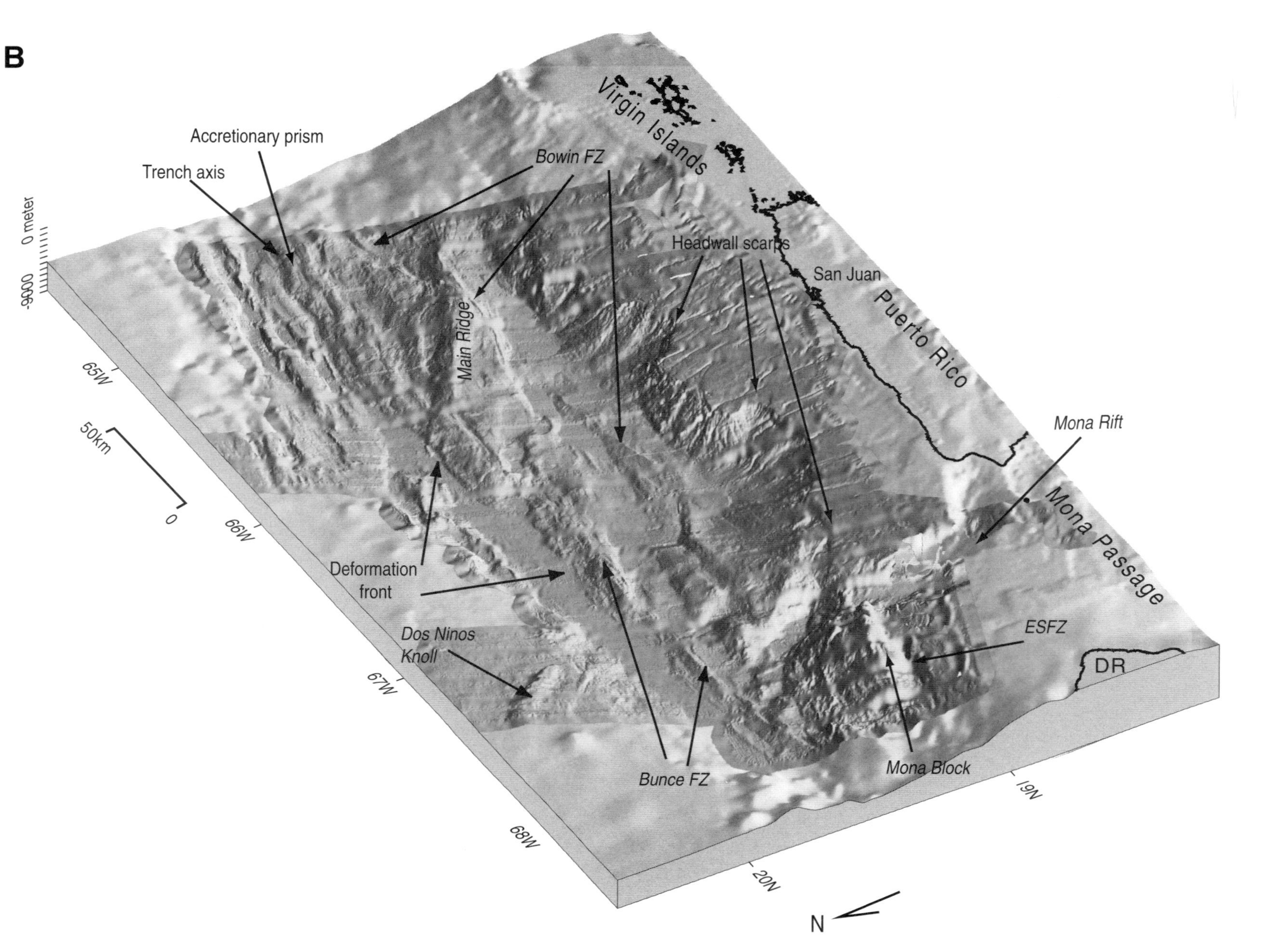

Figure 3 (*continued*). (B) Perspective shaded-relief diagram of the Puerto Rico trench and northern margin looking from west to east, identifying major structural features. ESFZ—East Septentrional fault zone.

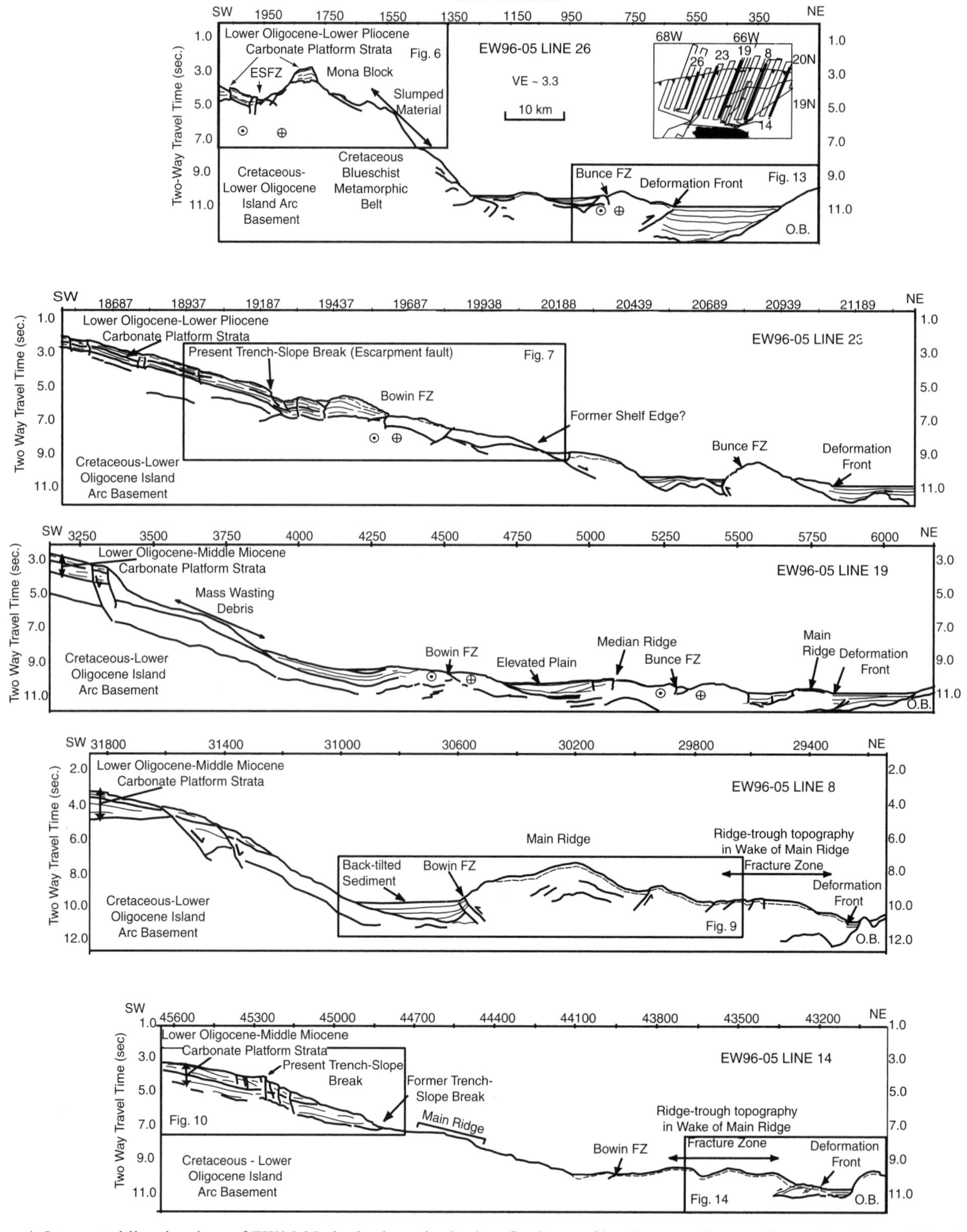

Figure 4. Interpreted line drawings of EW96-05 single-channel seismic reflection profiles that extend across the northern Puerto Rico–Virgin Islands margin and the Puerto Rico trench. Inset and Figures 2A and 3A show the location of profiles. Vertical exaggeration is 1:3.3. Boxes indicate location of seismic data panels shown in Figures 6, 7, 9, 13 and 14. Location of the Puerto Rico trench deformation front is marked on all profiles. ESFZ—East Septentrional fault zone. These profiles show that Puerto Rico forearc is characterized by a wide, low-relief lower slope and a steep upper slope. The trench-slope break is marked by large-throw, trenchward-dipping normal faults that cut the carbonate platform and underlying Cretaceous-Eocene basement.

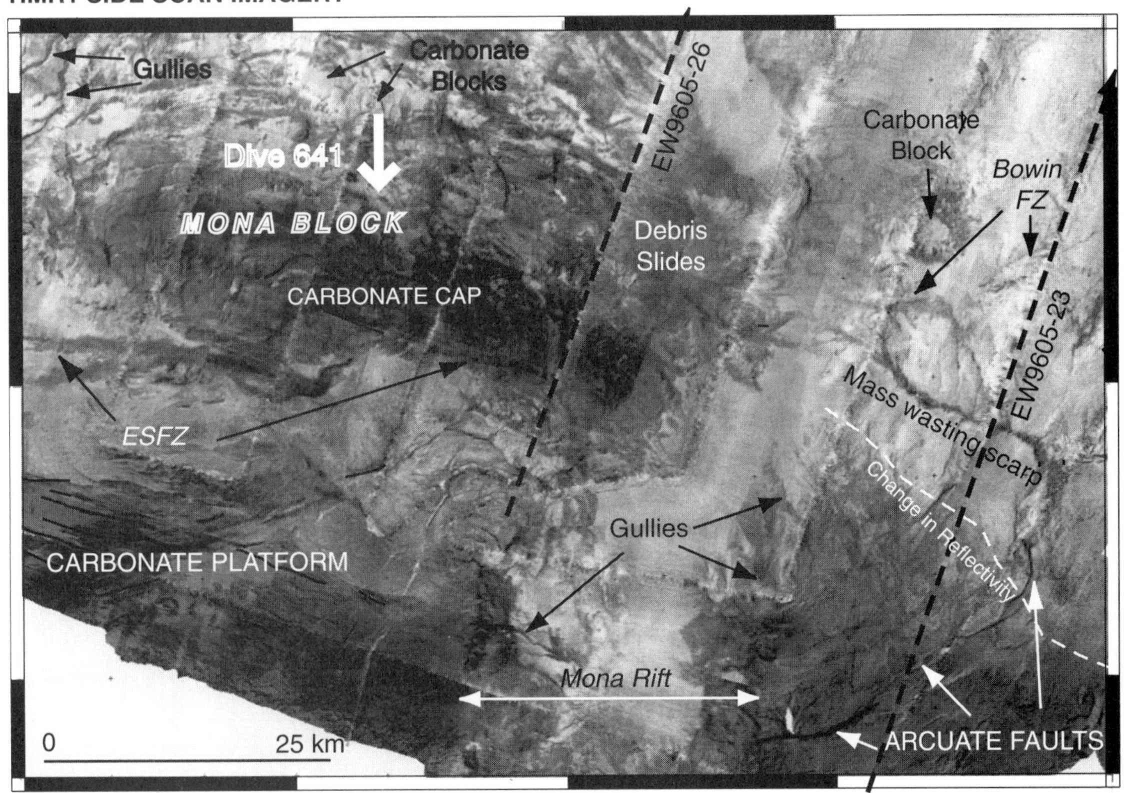

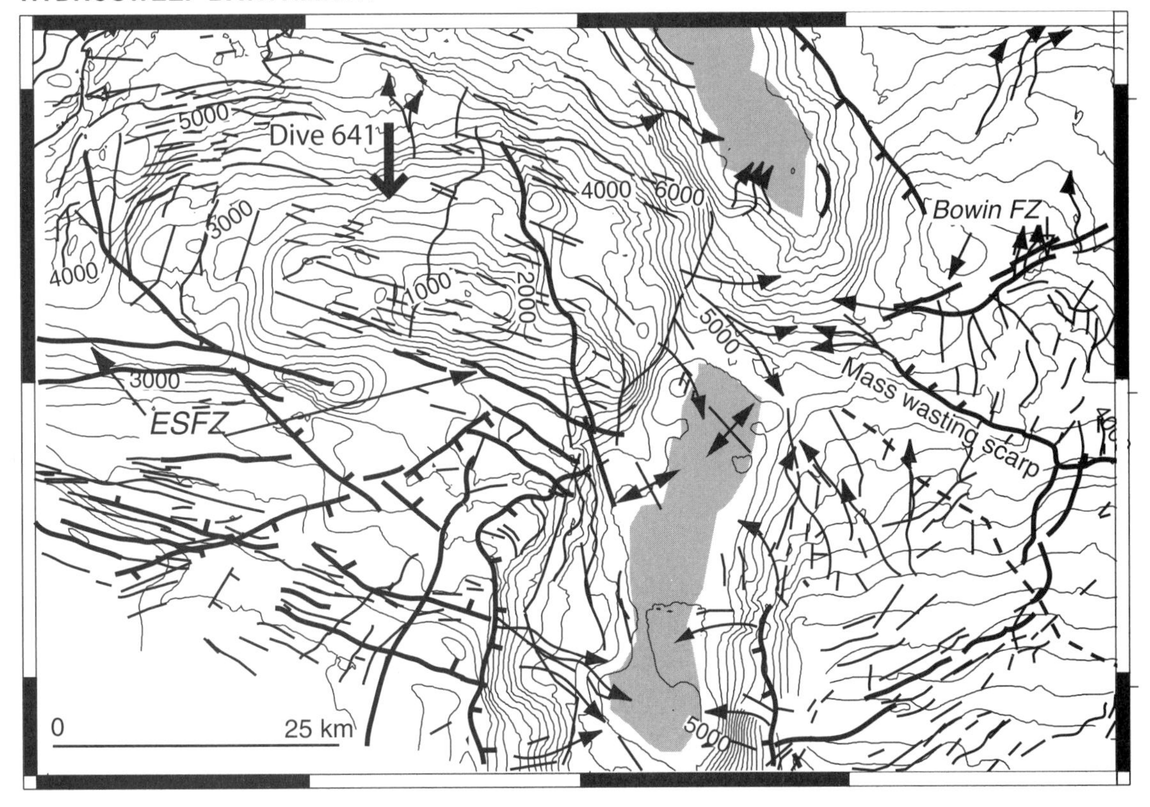

Figure 5. (A) HMR1 sidescan imagery of southeastern Bahamas collision point with carbonate platform in the vicinity of the Mona rift. Dashed white line indicates location of change in platform reflectivity. Dark areas characterized by high reflectivity are interpreted to be covered by Pleistocene-Recent corals. Location of EW96-05 SCS profiles shown by black dashed lines. ESFZ—East Septentrional fault zone. (B) Bathymetry of same region at 250 m contour interval with structures interpreted from sidescan sonar imagery. Arcuate faults and enhanced seismicity in the carbonate platform on the eastside of Mona rift suggest that the southeastern tip of the underthrust Bahamas extends southeast of the Mona Block.

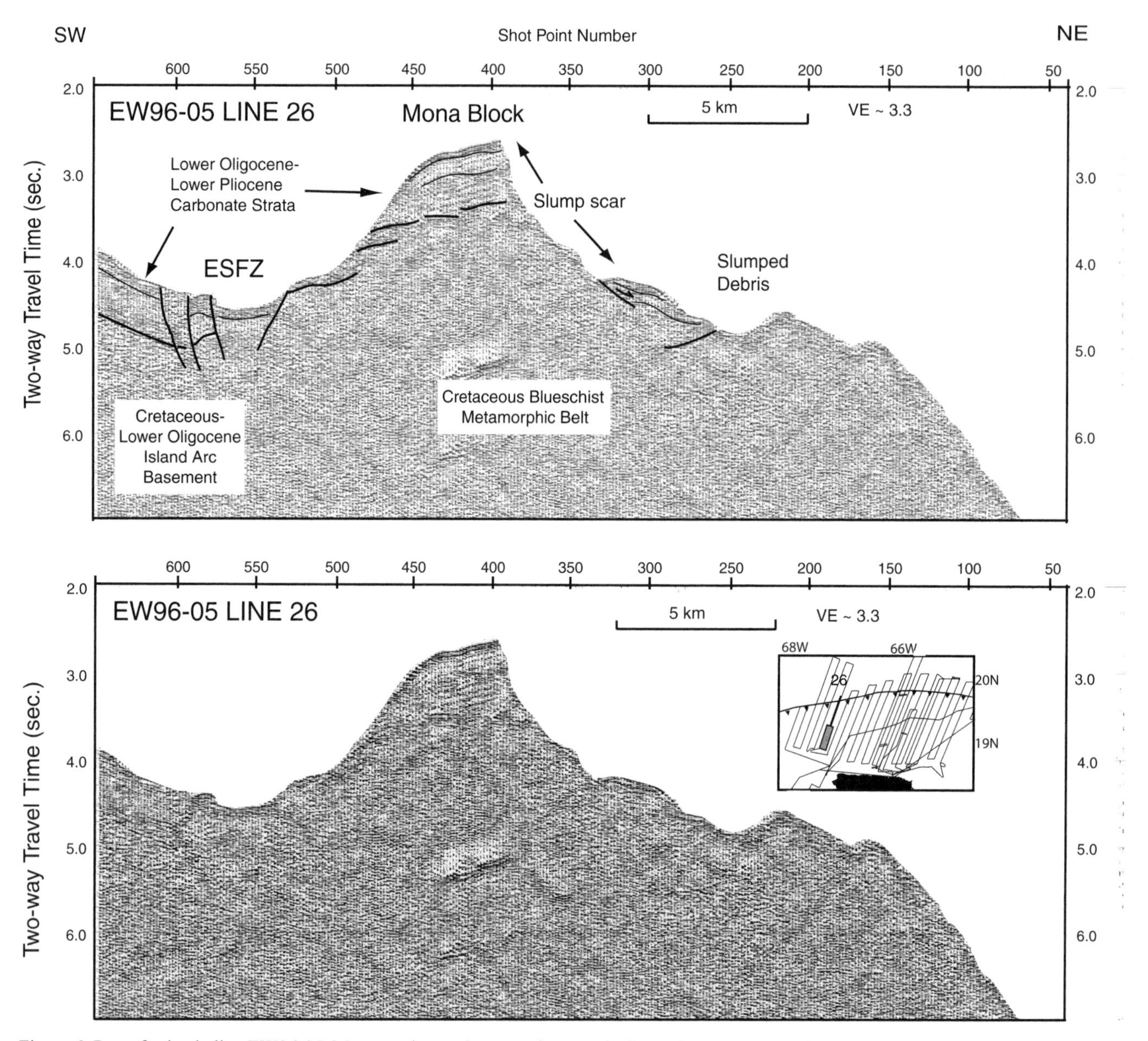

Figure 6. Part of seismic line EW96-05-26 across the northwest carbonate platform, the East Septentrional fault zone (ESFZ) and Mona Block. Note the parallel reflector in the upper section of Mona Block interpreted to be a cap of carbonate platform strata uplifted by the underthrusting SE Bahama Province. Location of the line is shown in the inset.

series of ledges formed by normal faulting (down to the north) of the metamorphic unit observed by Heezen et al. (1985) during dive 641. Although not imaged by the EW96-05 SCS profiles, it likely that the décollement surface has been deflected up over the top of the underthrusting southeastern Bahamas and outcrops on the north flank of Mona Block (cf. Lallemand et al., 1994; Dominguez et al., 1998a). The debris flows, however, obscure any surface expression of this feature. North-south–trending sinuous lineaments are also observed to extend downslope to the north (Figs. 3A, 3B, and 5A) and are interpreted to be gullies and headless channels, similar in size and shape to those observed in the accretionary complex offshore northern Hispaniola and believed to be associated with submarine seeps (Mullins et al., 1992; Orange et al., 1994; Dolan et al., 1998). The steep scarp that defines the southern edge of Mona Block is interpreted to be the eastward extension of the East Septentrional strike-slip fault zone. Directly south of Mona Block the sidescan imagery shows a series of crossing lineaments forming a broad "X" in the carbonate platform (Fig. 5A). EW96-05 SCS profiles 26 (Fig. 6), 29 and 30 (van Gestel et al., 1998) show that many of these lineaments correspond to normal faults or, in several cases where no seafloor offset is apparent, potential strike-slip faults.

Southern slope of the Puerto Rico trench

East of the Mona Block

To the northeast of the Mona Block, the southern slope of the Puerto Rico trench deepens sharply by ~6.5 km into a flat-bottom deep, the lower Mona rift (Figs. 2A, 5A, and 5B). The curved shape and orientation of the deep is similar to tectonically controlled canyons observed in areas of highly oblique subduction along the Aleutian Island forearc (Geist et al., 1988; Dobson et al., 1996). These canyons represent the trailing edge of a rotating block of forearc material and act as conduits for sediment transport from the upper and lower forearc to the trench.

Alternatively, this curved deep could be a reentrant related to the subduction of the southeastern Bahamas. The reentrant curves around the northeast base of the Mona Block and intersects forearc material (Zone 2 of Muszala et al., 1999). Its shape is similar to the margin reentrant and "shadow zone" identified by Dominguez et al. (1998b) along the Ryukyu margin where the basal décollement is deflected upward over the subducting Gagua Ridge to the upper part of the frontal margin. As a result, frontal accretion is inhibited in the wake of the subducting ridge.

Immediately to the east of the Mona Block, the slope deepens by 3.5 km into Mona rift (5000 m) a north-south–trending half-graben (Figs. 2A, 5A, and 5B). EW96-05 SCS profiles show that the bounding walls of the Mona rift cut through the carbonate platform and underlying arc material (van Gestel et al., 1998, 1999; Mann et al., 2002). Submersible work (dive 640) by Heezen et al. (1985) identified Cretaceous greenschist-grade metavolcanic rocks and metaplutonic rocks overlain by shallow water limestone on the eastern wall of the rift. The low reflectively of the floors of these deeps seen in the sidescan sonar imagery indicates the presence of fine-grained materials such as fine sand and mud (Fig. 5A). The Mona rift has been linked to ongoing extension between Hispaniola and Puerto Rico in the Mona Passage (van Gestel et al. 1998, 1999; Jansma et al., 2000; Mann et al., 2002).

The carbonate platform on the eastern flank of the Mona rift shows that distinctive, arcuate lineaments in sidescan sonar imagery (Fig. 5A). EW96-05 SCS line 23 (Figs. 4 and 7) crosses this area and indicates that these are normal faults that cut through the platform into the underlying Eocene volcaniclastic rock (also see van Gestel et al., 1998, their Fig. 7). Large, isolated blocks of carbonate material positioned at the top of a narrow ridge that extends from the carbonate platform across the forearc region are visible both in the bathymetry and sidescan sonar imagery (Figs. 3A, 3B, and 5A) and on EW96-05 SCS profile 23 (Fig. 7). EW96-05 SCS profile 23 reveals the continuation of the Oligocene-Pliocene strata, including the middle–late Miocene unconformity across the mid-slope to water depths of ~5200 m indicating massive subsidence of an eroded paleoseafloor (Figs. 4 and 7).

The sidescan imagery shows patches of higher reflectivity on the carbonate platform on the eastern flanks of Mona rift and in the Mona Passage. The change in reflectively across the platform was also observed in the SeaMARC II imagery by Dolan et al. (1998). In addition to middle-upper Miocene aged limestone, dredge hauls and *Alvin* sampling recovered Pleistocene corals (Fox and Heezen, 1975; Weaver et al., 1975; Perfit et al., 1980; Heezen et al., 1985) in this area of the platform. Van Gestel et al. (1998) noted that these areas of the platform are associated with the thickest sequences of Miocene-Pliocene carbonate beds.

The SCS data from EW96-05 have increased coverage of the carbonate platform on the northern margin of Puerto Rico over an along-strike distance of 200 km. To the eastern edge of our survey area at ~65°W, the platform has along-strike uniform northward dip of 4°–6° (van Gestel et al., 1998, 1999). Van Gestel et al. (1998) discuss in detail the evolution of the sedimentary units that form the carbonate platform. Carved out of the northern edge of the platform are two huge (up to 55 km across) amphitheater-shaped scarps (Figs. 2A and 2B). The headwall scarps occur at a water depth of ~3000–3500m. Schwab et al., (1991) first recognized them in the GLORIA survey and estimated that 1500 km^3 of material slumped down the south slope of the Puerto Rico trench. EW96-05 SCS profile 19 (Fig. 4) shows that a section of the platform has slumped several thousand meters to the base of the slope.

Above the headwall scarp, the sidescan sonar imagery shows evidence of submarine channels recorded as high-reflectivity (dark) north-south–trending lineaments (Fig. 2B). Other than these channels and uniform dip, the platform appears to be relatively undeformed. Some of the channels that extend to dark streaks in the head-wall scarp have presumably been cut into the slope and carry debris from the upper slope platform. Many of the dark (high reflectivity) streaks, however, are not associated with submarine channels. These we interpret to be headless canyons or gullies similar to those observed along other convergent margins, e.g., Oregon (Kulm and Suess, 1990) and Hispaniola (Mullins et al., 1992; Orange et. al; 1994; Dolan et al., 1998).

The Main Ridge and Vicinity

The Main Ridge

The Main Ridge was first identified by Ewing et al. (1968) in seismic profiler records as a bathymetric high on the southern wall of the Puerto Rico trench. McCann and Sykes (1984) noted that the strike of this feature is about N50W, nearly parallel to the inferred strike of fracture zone orientations on the down-going North America plate. Mainly on the basis of teleseismic data and widely spaced SCS lines, McCann and Sykes (1984) proposed that the Main Ridge represents the zone of deformation on the overriding Caribbean plate produced by the subduction of the northwestern extension of the aseismic Barracuda fracture zone transverse ridge of the Lesser Antilles where it is one of several west-northwest–trending aseismic ridges that extend beneath the Barbados accretionary prism. In support of the McCann and Sykes hypothesis, Muszala et al., (1999) found that the Main Ridge lines up with a northwest-trending linear magnetic high that extends beneath the trench slope. This linear magnetic high is interpreted to be a fracture zone (Main Ridge

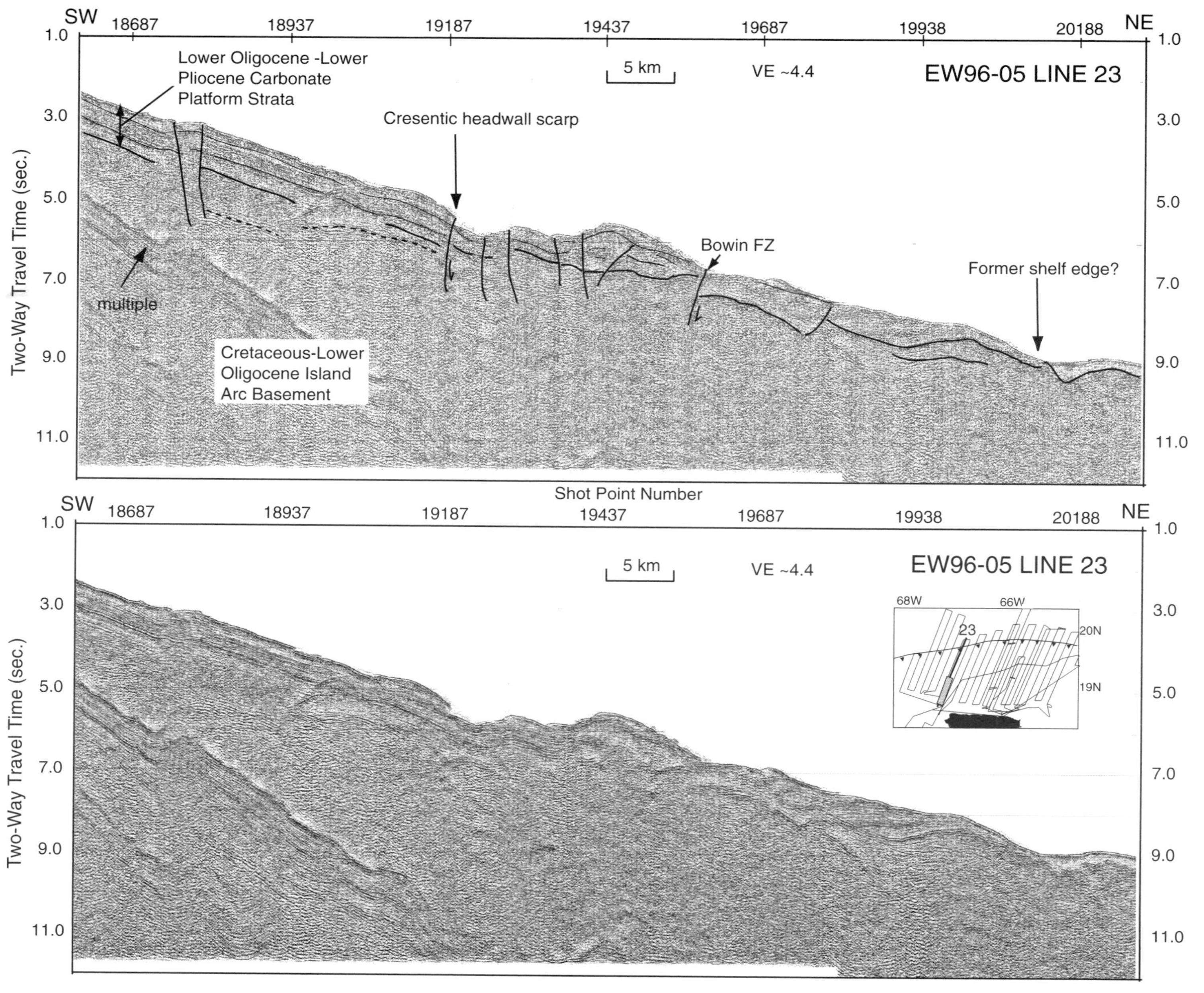

Figure 7. Part of seismic line EW96-05-23 showing the subsided Lower Oligocene-Lower Pliocene carbonate platform strata to water depths of ~5200 m. The position of the Bowin fault zone on the mid-slope is also shown. Dashed lines mark location of a mid-Cretaceous unconformity in the island arc basement. Location of the line is shown in the inset.

fracture zone) on the down-going North America plate north of Puerto Rico (Fig. 1).

The Main Ridge is ~15 km wide and rises ~2000–1500 m above the surrounding seafloor at its highest point (5450 m) (Figs. 2A, 8A, 8b, and 9). EW96-05 SCS profiles 14, 8, and 19 (Fig. 4) image cross sections of the ridge from its easternmost expression on the mid-slope of the Puerto Rico trench at ~65°W to its westernmost expression at ~66°20′W, where it intersects the floor of the trench. The orientation of the crest of the Main Ridge is N60°W, ~10° different than reported by McCann and Sykes (1984). The easternmost expression of the Main Ridge is the rise on the slope of the Virgin Islands platform seen on EW96-05 SCS profile 14 (Fig. 4) at the eastern limit of the study area. Immediately above the mid-slope rise the carbonate platform shows extensional normal faulting extending ~20 km northeast from the trench-slope break (Fig. 10).

The EW96-05 cruise data indicate that the ridge is defined to the southwest and northeast by faults (Figs. 3A, 3B, 8A, 8B, and 9). Beneath the ridge, EW96-05 SCS profile 8 images the top of a seamount/transverse ridge, as indicted by rounded or domed reflectors at ~9 stwt. (Fig. 9). The unit above the subducted ridge has poorly defined chaotic internal reflectors, suggesting it is uplifted forearc debris. This is evidence supporting McCann and Sykes' (1984) hypothesis that the Main Ridge represents the zone of deformation on the overriding Caribbean plate produced by the subduction of a bathymetric high on the underthrusting North America plate. On EW96-05 SCS profile 19 (Fig. 4), the Main Ridge is imaged as a discrete topographic high protruding from sediment within the Puerto Rico trench floor.

East and West of the Main Ridge

As noted by McCann and Sykes (1984) using SCS and 3.5 kHz lines, we see a dramatic difference in the shaded relief images of the texture or roughness of the forearc region east and west of the Main Ridge (Figs. 3A and 3B). West of the Main Ridge lies relatively smooth topography that has been draped by pelagic, hemipelagic sediment and turbidite deposits shed from the island (Fig. 3B) (Connolly and Ewing, 1967; Pilkey and Cleary, 1986). EW96-05 SCS profiles 19 and 8 (Figs. 4 and 9) show ponded turbidite beds with accumulations of up to 0.5 stwt in the Marginal Basin and Elevated Plain areas that have been tilted downward to the southwest away from the Median and Main ridges. The Marginal Basin acts as a channel for turbidity currents derived from the Puerto Rican shelf (Pilkey and Cleary, 1986). Turbidity currents flow from the Elevated Plain into the Main Plain of the Puerto Rico trench through an "Abyssal Gap" identified by Ewing et al. (1968) at the western end of the Median Ridge (Fig. 2A).

East of the Main Ridge, the forearc topography is characterized by closely spaced ridges and troughs. On EW96-05 SCS profile 8 (Fig. 9) directly northeast of the Main Ridge and on EW96-05 SCS profile 14 (Fig. 10) northeast of where the Main Ridge fracture zone is underthrusting the mid-slope area, we identify folds and closely spaced thrusts with little or no ponded sediment in low-lying areas.

Strike-Slip Faults in the Forearc Region

The Bowin Fault Zone

East of Hispaniola, the Septentrional strike-slip fault extends offshore and has been mapped with SeaMARC II from Samana Bay to ~15 km west of the Mona rift (Dolan et al., 1998). The offshore extension of the Septentrional fault zone has been previously referred to as the 19° Latitude fault by Larue et al. (1989) and Larue and Ryan (1989).

Larue et al. (1989) tentatively traced the East Septentrional (19° Latitude) fault zone farther east of the Mona rift with widely spaced SCS and MCS lines along the northern coast of Puerto Rico where the East Septentrional (19° fault zone) as south-dipping listric fault. In addition, Larue et al. (1989) identified a large-throw, northward-dipping normal fault, the Escarpment fault zone, which is parallel to and thought to be separated from the East Septentrional (19° Latitude) fault zone by a narrow horst.

The EW96-05 cruise provided complete coverage of the fault zone from the eastern tip of the Samana Peninsula in Hispaniola, through the Mona Passage, north of Puerto Rico and the Virgin Islands, to close to its intersection with the Puerto Rico trench around 64°W (Fig. 1). The final trackline of the cruise followed the eastward continuation of the fault zone within the Marginal Basin along the base of the northern Puerto Rico slope. Because of the distinct break and abrupt change in trend of the East Septentrional fault zone across the Mona rift, we have opted to name this strike-slip fault zone separately as the Bowin fault zone (previously identified to as the South Puerto Rico Slope fault zone in Grindlay et al., 1997). The fault zone is characterized by linear scarps that range from 5 to 15 km in length and extend from the Mona Block to the eastern edge of the survey. The EW96-05 survey data reveal that the Escarpment fault zone is an E-W series of overlapping, crescentic normal fault lines that define the northern edge of the carbonate platform, whereas the Bowin fault zone is an ENE-trending, linear feature roughly 5 km in width. Only a small part (~10 km) of this lineament was recognized on the GLORIA mosaic of Masson and Scanlon (1991).

The Bowin fault zone trends N75°E to N80°E east and west of the Main Ridge, subparallel to the N70°E relative plate motion vectors determined by geodesy studies for North America–Caribbean (Jansma et al., 2000; Mann et al., 2002) (Fig. 2A). What we interpret as a pull-apart basin at 65°50′W (Figs. 8A and 8B), suggests a transtensional component. This configuration is what would be expected along a left-lateral shear, given the slightly oblique fault orientation relative to the GPS plate motion vector. With the exception of the area near the Main Ridge, the Bowin fault zone appears to be steeply dipping with small amounts of oblique dip-slip motion (Figs. 4, 7, and 10). Between 65°45′W and 65°15W, where the fault lies directly south of the Main Ridge, the trend changes from N90°E to N100°E, and the fault displays a component of reverse dip or a transpressional character (Fig. 9). Thus, some amount of the relief associated with the Main Ridge may be attributed to uplift in a restraining bend of left-lateral strike-slip faulting. A slight (<5 km) left lateral offset

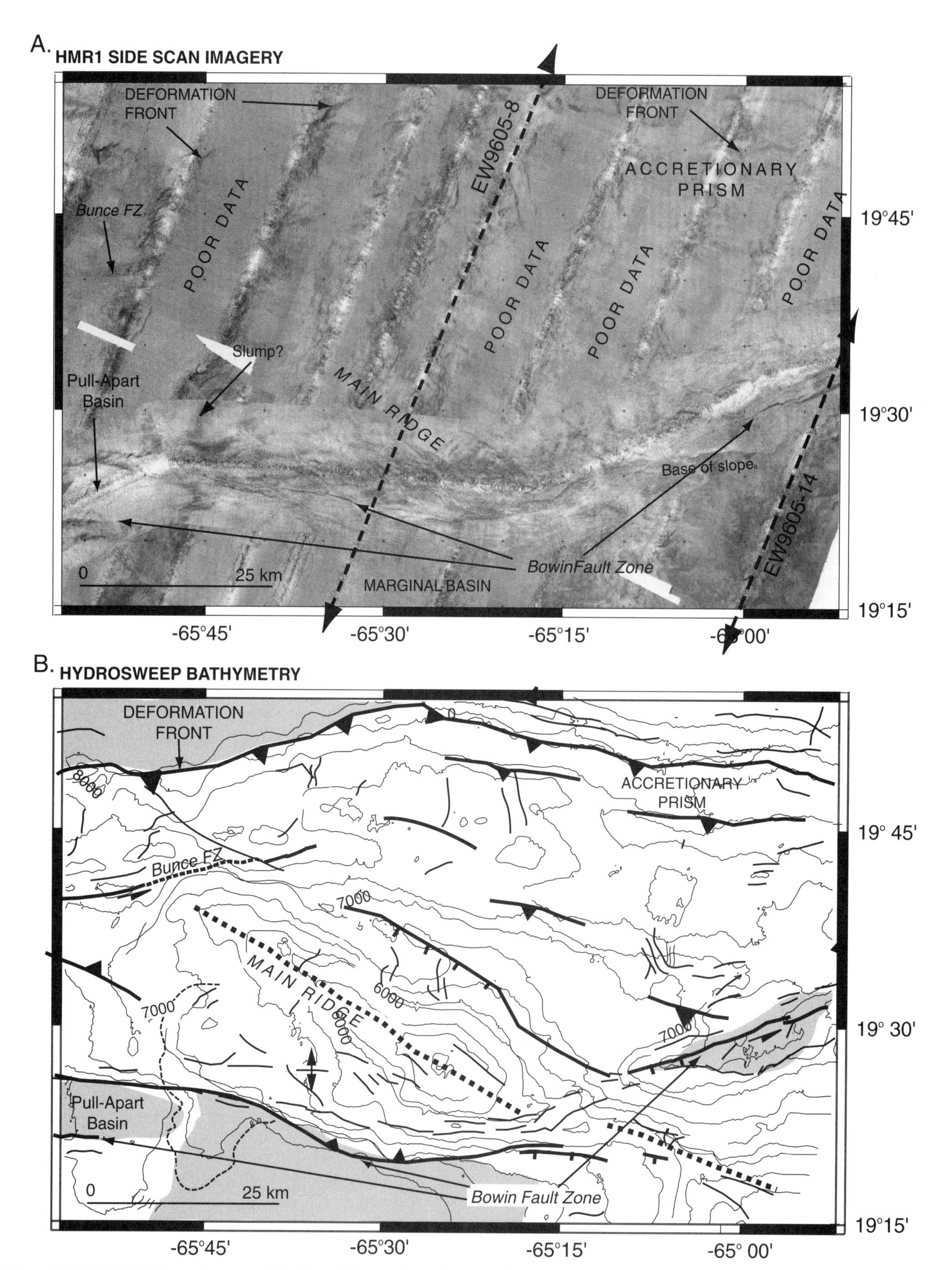

Figure 8. (A) HMR1 sidescan imagery of the Main Ridge and accretionary prism of the Puerto Rico trench. Location of single-channel seismic profiles shown as black dashed lines. White areas are data gaps. (B) Bathymetry of same section at a 250 m contour interval with structures interpreted from sidescan imagery. Dashed black line marks the apex of the Main Ridge. Note the ~5 km left-lateral offset of the Bowin fault zone across the Main Ridge.

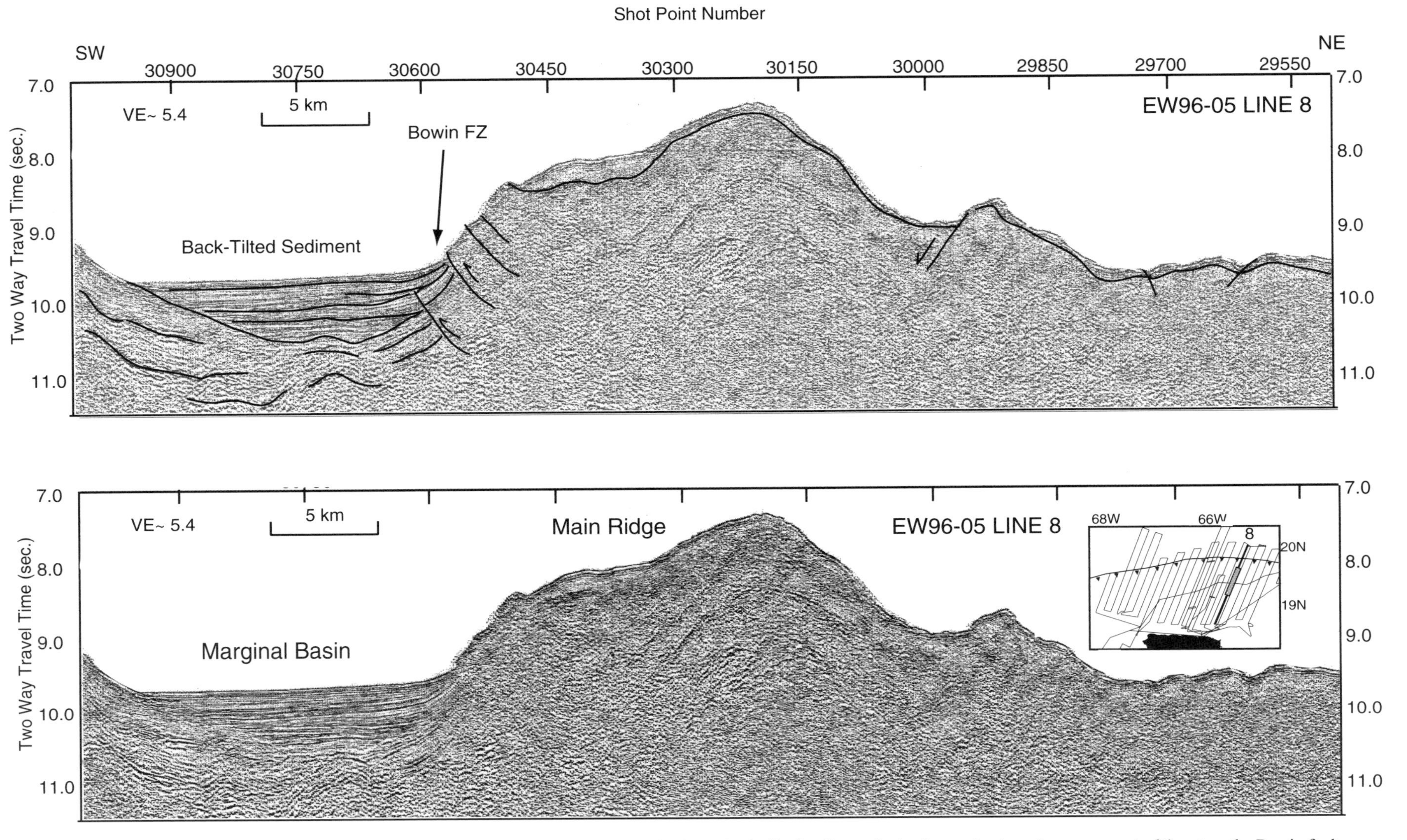

Figure 9. Part of seismic line EW9505-8 across the apex of the Main Ridge. Profile 8 clearly shows back-tilted sediment in the forearc basin and a component of thrust on the Bowin fault zone. Note the domed reflectors beneath the crest of the Main Ridge believed to be the top of the underthrust Main Ridge fracture zone.

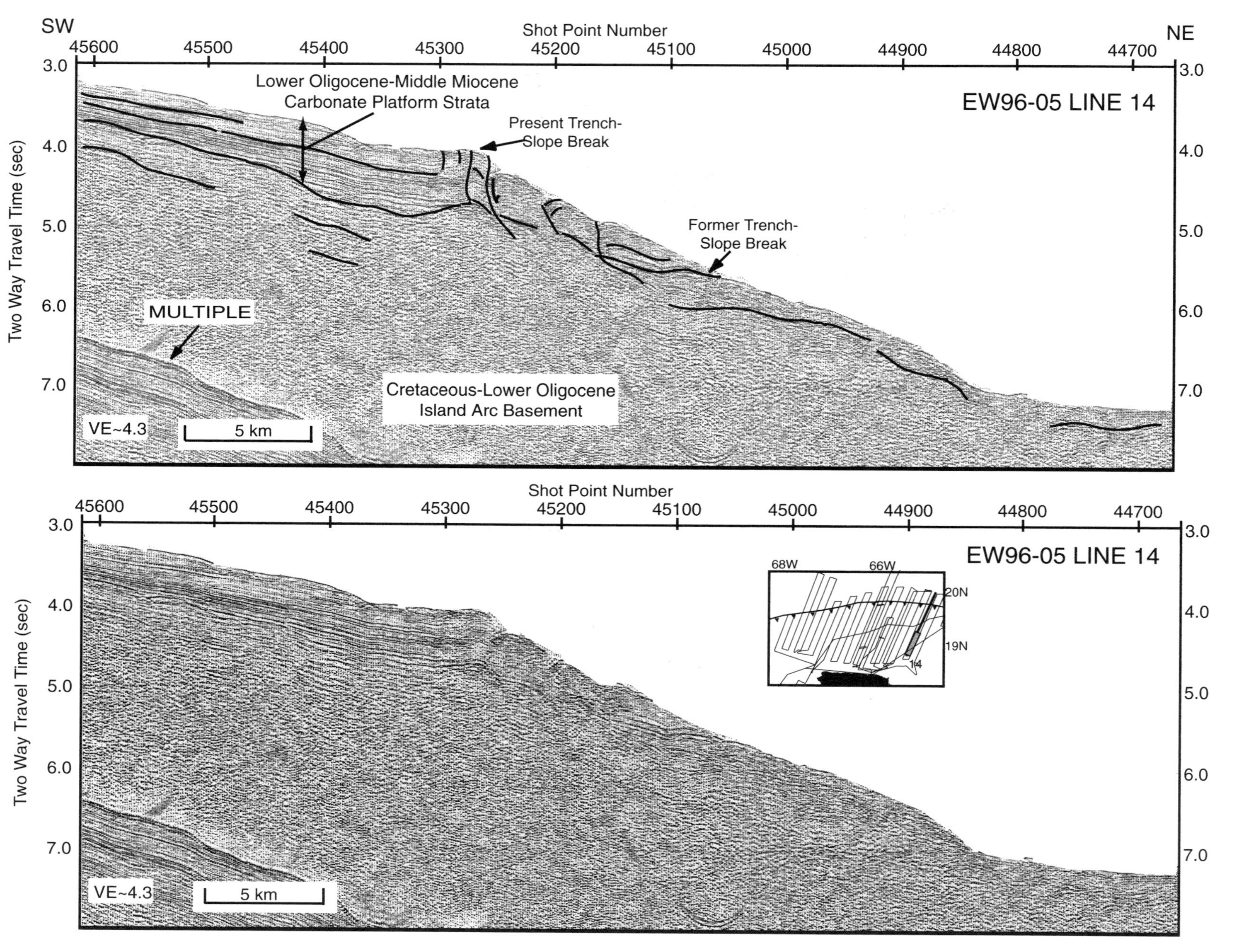

Figure 10. Part of seismic line EW9605–14 showing the fractured nature of the carbonate platform, and the positions of the former and present trench-slope break. Location of the line is shown in the inset.

of the bathymetric trend of the Main Ridge and the uplifted section of the mid-slope is observed (Fig. 8) and suggests that (1) either the fault is young, and offset has not been significant, or (2) that the Main Ridge is a topographic reflection of a high standing feature (fracture zone transverse ridge?) on the underthrust plate, a part of which has been accreted or underplated to the island arc and a part of which continues to move westward on the obliquely subducting North America plate.

The Bunce Fault Zone

In the area off the northeast coast of Hispaniola, Dolan et al., (1998) identified the northeastern strand of the Septentrional fault zone as a wide zone of highly reflective, parallel lineaments on SeaMARCII sidescan sonar imagery. The Northeast Septentrional fault zone trends N50°E and extends northward from the main Septentrional Fault to at least ~19°11′N where it forms the southeastern edge of a deep (~7 km) rhomb-shaped basin that lies between two bathymetric highs, the Navidad bank and Mona Block (Fig. 11). The fault can be traced northeastward from the basin as the Bunce fault zone (previously identified to as the North Puerto Rico Slope fault zone in Grindlay et al., 1997). North of the Mona Block the trend of the Bunce fault zone becomes subparallel to the trend (N80°–90°E) of the trench and approaches to within 5–10 km of the Puerto Rico trench deformation front (Figs. 2A, 3B, and 12). Thus, the Bunce fault zone isolates a remarkably narrow (5–10 km) and thin (4–5 km) forearc sliver underlain by the low-angle subduction zone thrust imaged on MCS lines presented by Larue and Ryan (1998). The fault zone, also identified in GLORIA imagery by Masson and Scanlon (1991), can be traced in the bathymetry and sidescan imagery more than 40 km along a narrow ridge eastward to where it intersects the Main Ridge (Figs. 2A and 3B). It is unclear if the fault continues eastward to merge with the fold-thrust belt west of the Main Ridge. Because of its extreme along-strike continuity, we interpret the Bunce fault zone as a left-lateral strike slip fault. Little seismicity is associated with this shear, probably because of its limited depth penetration into sedimentary rocks on the lower forearc slope.

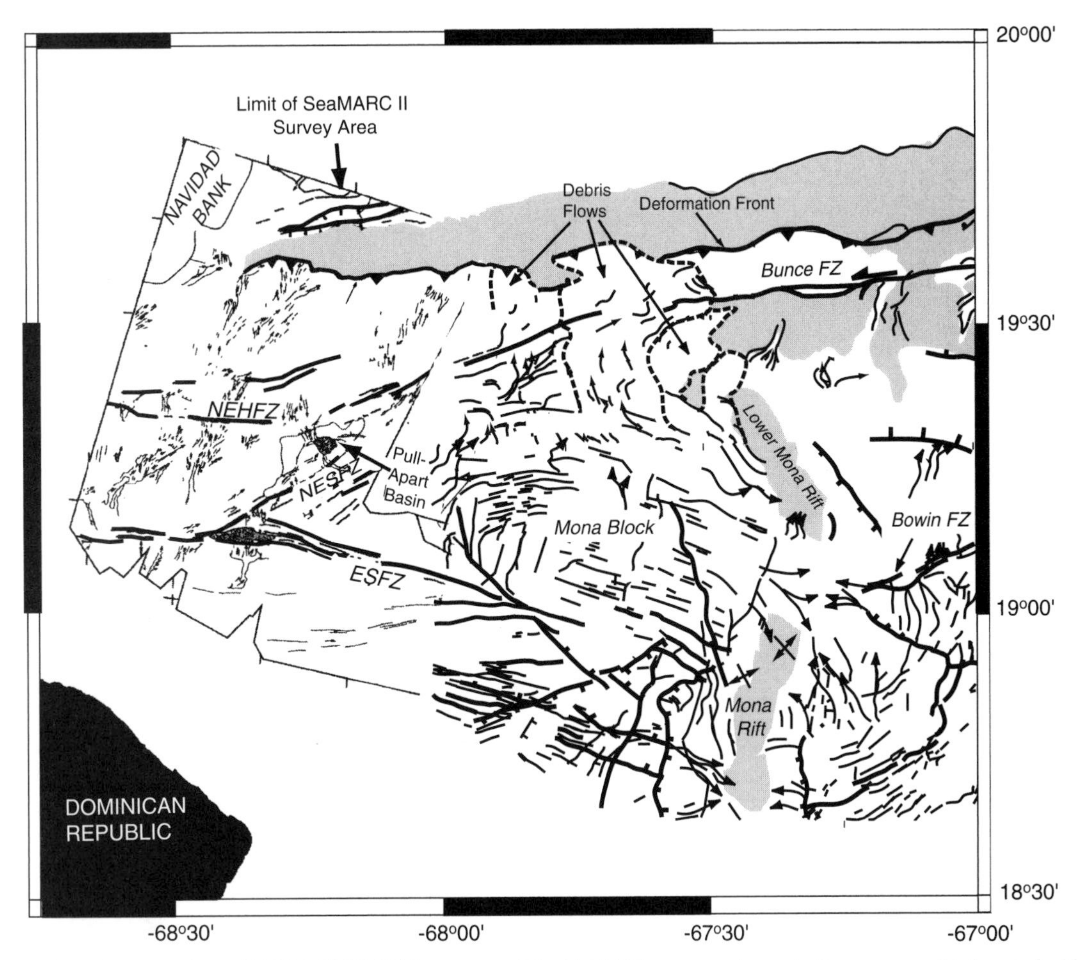

Figure 11. Combined interpretation of EW96-05 HMR1 imagery and SeaMARCII coverage to west of survey area (Dolan et al., 1998). This figure illustrates the continuation of the northeast Septentrional fault zone (NESFZ) with the Bunce fault zone. Note that the fault zones lie between two bathymetry highs, Navidad Bank and Mona Block. NEHFZ—Northeast Hispaniola fault zone; ESFZ—East Septentrional fault zone.

A. **HMR1 SIDESCAN IMAGERY**

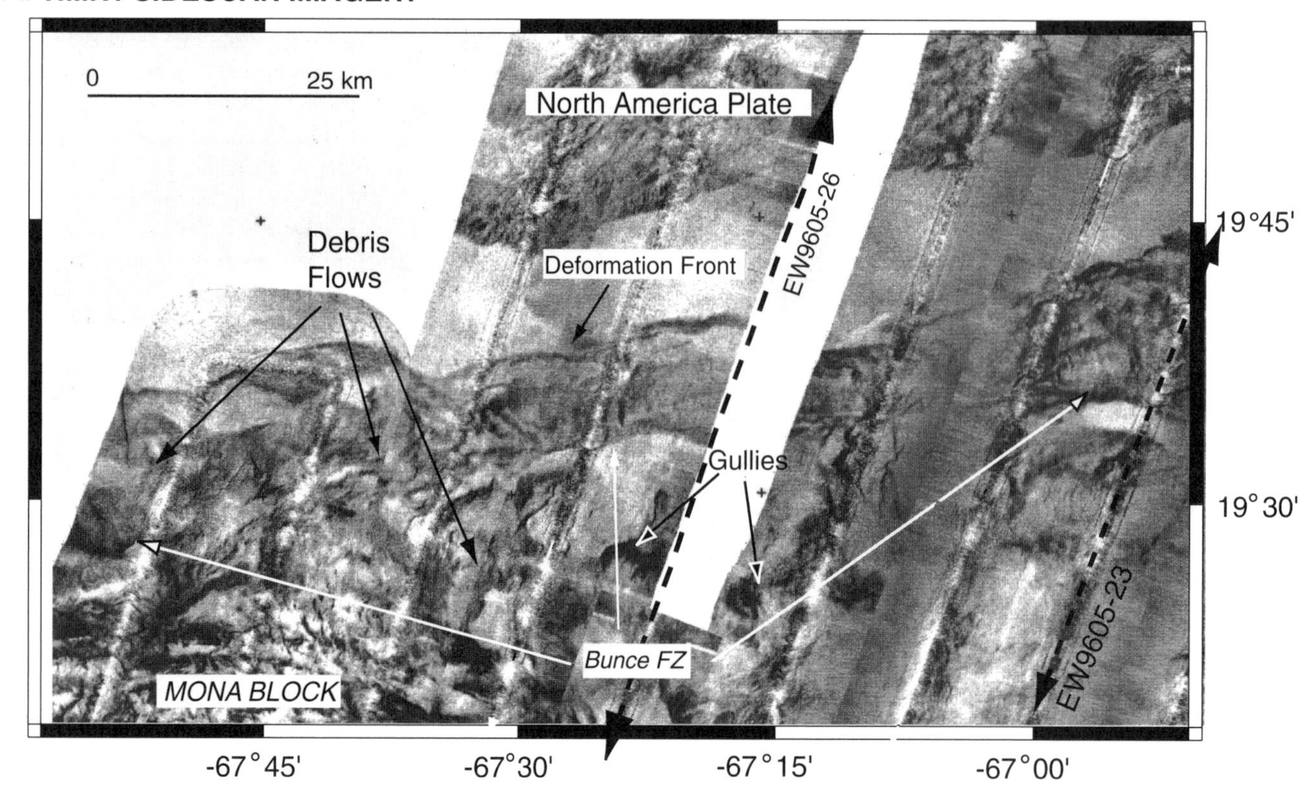

B. **HYDROSWEEP BATHYMETRY**

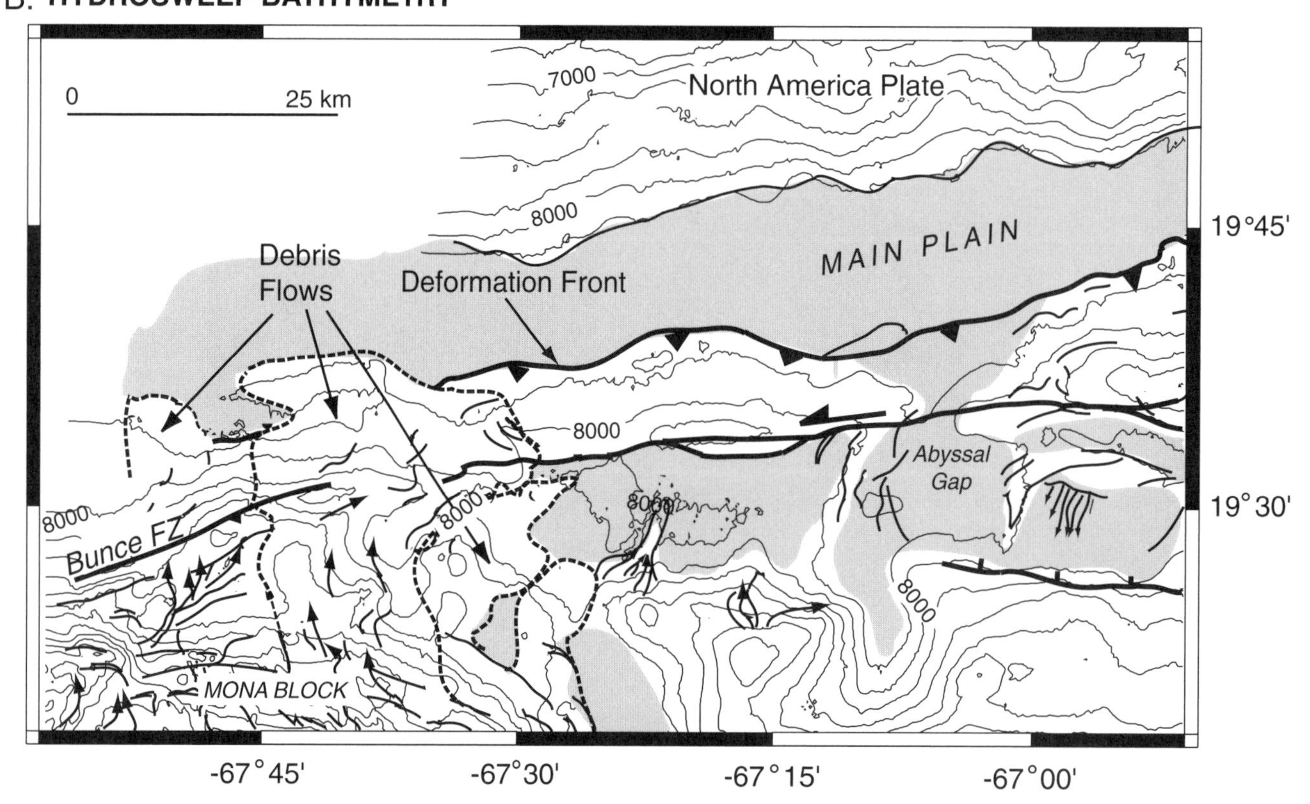

Figure 12. (A) HMR1 sidescan imagery of the Bunce fault zone. EW96-05 SCS profiles shown by dashed lines. White areas within the mosaic are data gaps. (B) Bathymetry of same region at a 250 m contour interval with structures interpreted from sidescan imagery.

Puerto Rico Trench and Adjacent North America Plate Structures

The flat, sediment-covered floor of the Puerto Rico trench is easily recognized as a band of low reflectivity trending ~N85°E and crossing the surveyed area at 19°45′N. The trench floor reaches a maximum depth of 8450 m at ~67°W, northeast of the reentrant that is adjacent to the Mona Block. Between 68°W and 66°15W the trench floor is broad (15–20 km) and then narrows, shallows, and changes trend eastward to N90°E at 65°40W. EW96-05 SCS profiles 26, 23, and 19 show thick (1–2 stwt, 0.75–1.5 km) accumulations of sediment in the western section of the survey area (Figs. 4 and 13). Sediment thickness decrease abruptly to less than 0.5 stwt. north and northeast of the Main Ridge (SCS profiles 8 and 14; Figures 4 and 14). The western EW96-05 SCS profiles (26, 23, and 19) show a sharp deformation front (Figs. 3B and 4). Almost no deformation of the trench sedimentary sequence is observed seaward of the deformation front, nor is earthquake activity associated with this boundary (Figs. 2A and 13). This lack of deformation in trench sediment could reflect several factors: (1) relatively young fill; (2) the low rate of convergence across the deformation front; (3) the accommodation of the majority of plate motion by strike-slip fault(s) within the forearc; and/or (4) recent roll back of the slab resulting in limited coupling with the overriding plate. Evidence for contractional deformation is not entirely lacking as Larue and Ryan (1998) note the presence of what appears to be an underriding slab and thrust faulting in several MCS profiles that cross the trench. These features are also observed in the SCS profiles that were collected during cruise EW96-05 (Fig. 4). A small, narrow (~5 km) wedge of accreted sediment is observed on the easternmost SCS line EW96-05-14 (Fig. 14)

The subducting North America plate is offset by normal faults interpreted to be the result of the bending of the plate as it enters the subduction zone (Figs. 3A and 3B). The throw of these normal faults increases toward the trench and along the trench to the east. The general trend of the faults west of 66°00W is N85°E, east of 66°W the trend changes to 090°–100°E and roughly corresponds to the location where the trench floor begins to narrow and change its trend. Dillon et al. (1996, 1998) have interpreted this to reflect breakup of the downgoing slab into segments having differing dips.

DISCUSSION

Present-Day Collisional Impacts of the Southeastern Bahama Province

Numerous refraction and reflection seismics and gravity studies have helped to constrain the structure of the southeastern Bahamas as 100-km-thick lithosphere that includes a 10-km-thick limestone and sediment section capping 10 km of transitional rift crust. (Uchupi et al., 1971; Sheridan et al., 1981; Ball et al., 1985; Sheridan et al., 1988; Masaferro and Eberli, 1999). The response of the northern Puerto Rico margin to the underthrusting of the southeastern Bahama Province has been uplift, oversteepening, mega-slumps, and rapid subsidence. Areas of the forearc that have experienced recent uplift followed by rapid subsidence due to subduction erosion include the Mona Block, the Mona Passage directly south of the Mona Block and the carbonate platform off the northwest tip of Puerto Rico. These areas are characterized by highly reflective seafloor (Figs. 2B and 5A) relative to the carbonate platform to the east and based on dredge hauls and sampling from submersible (Fox and Heezen, 1975; Perfit et al., 1980; Heezen et al., 1985) are interpreted to be Pleistocene corals that grew during a recent pulse of uplift. Bathymetric profiles oriented perpendicular to the trench show zones of uplift on the forearc slope associated with presumed position of the underthrust southeastern Bahamas beneath the forearc (Fig. 15A). This area of the forearc is characterized by frequent seismic events that continue along strike of the southeastern extension of the Bahama Province beneath the northern Puerto Rico margin (McCann, 2002). Focal mechanisms in this region show mainly thrust motions along shallowly south- and southwest-dipping planes as well as some component of oblique, left-lateral motion (Fig. 2A). The shallow focal depths (15–23 km) of these events are consistent with the earthquakes having occurred along the top of the underthrust southeastern Bahama Province on the North America slab.

We interpret the arcuate faults on the east flank of the Mona rift to be faults in the upper plate that have been produced by crustal warping above a subcircular underthrust high (cf. Dominguez et al., 1998b). This upwarping is related to the subducted southeast extension of the Bahama Province. Dominguez et al., (1998a; 1998b) show in experimental modeling that when a seamount underthrusts the cohesive part of the margin, subsidence initiates above the seamount trailing flank. Former back thrusts, in this case the arcuate faults, are then reactivated as steep normal faults that control the margin collapse in the wake of the seamount (Figs. 2A, 3A, and 7). The arcuate faults have similar spatial dimensions to the headwall scarps to the east and potentially represent the location of an incipient landslide (Figs. 2A, and 2B).

In addition to the localized forearc impacts discussed above, the collision zone has a regional signature. On the basis of EW96-05 SCS profiles, bathymetry, and topography, van Gestel et al. (1998) have documented that a prominent feature of the Puerto Rico–Virgin Islands platform is an arch that extends from the eastern Hispaniola to the western central part of Puerto Rico. The arch changes trend from dominantly northeast in Hispaniola to EW in the area of the Mona Passage directly south of the collision area of the southeastern Bahamas (Mann et al., 2002). On the basis of regional geologic data and GPS results, Mann et al. (2002) have suggested that this region represents the indentation of a buttress of ancient arc rocks by the southeastern Bahama Province. Mann et al. (2002) also suggest that the collision resulted in back-thrusting (initiation of subduction?) in the Muertos trough.

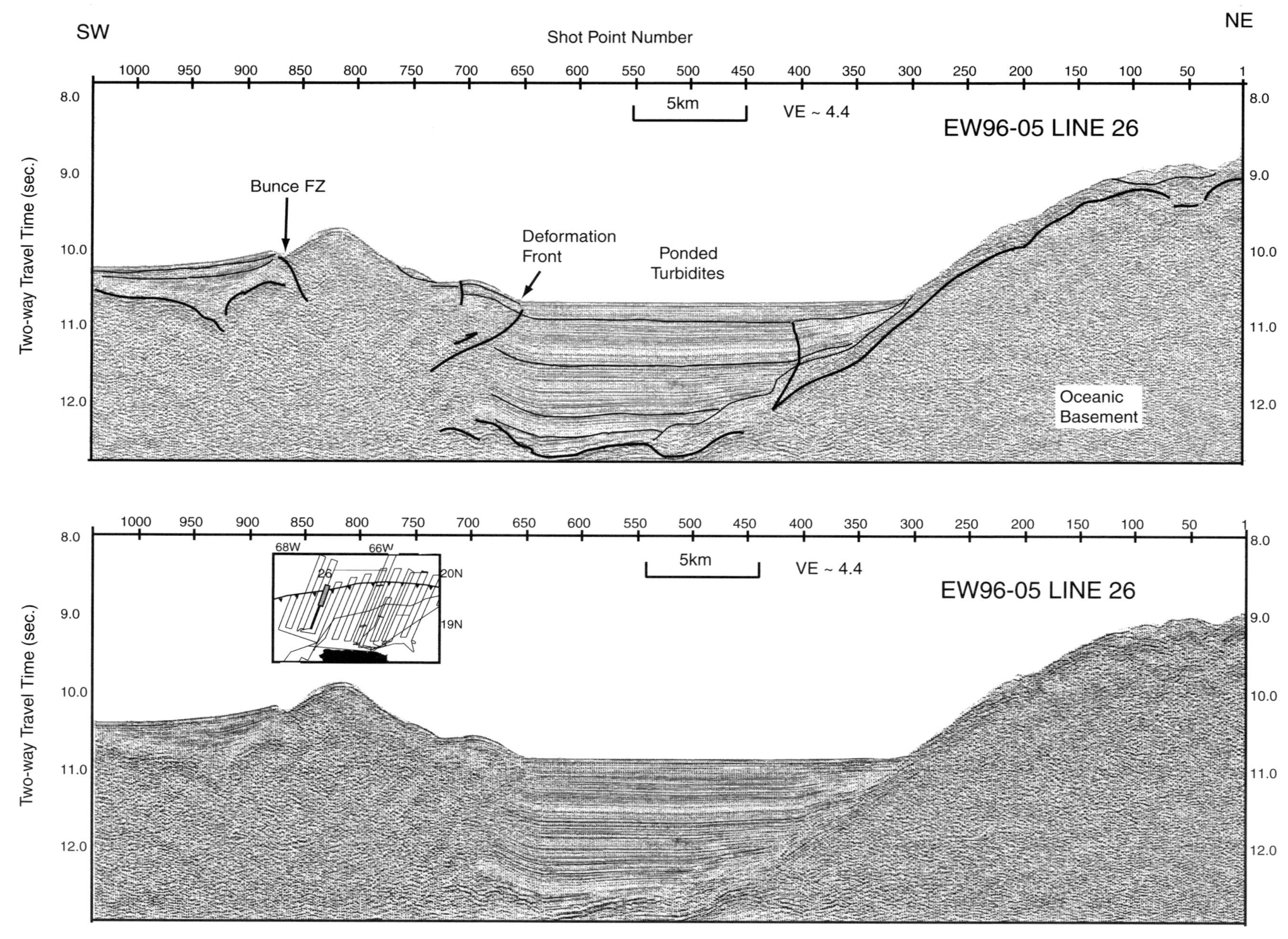

Figure 13. Part of seismic line EW9605–26 showing thick accumulation of relatively undeformed trench fill and the close proximity of the strike-slip Bunce fault zone to the deformation front in the western section of the survey area. Location of line is shown in the inset.

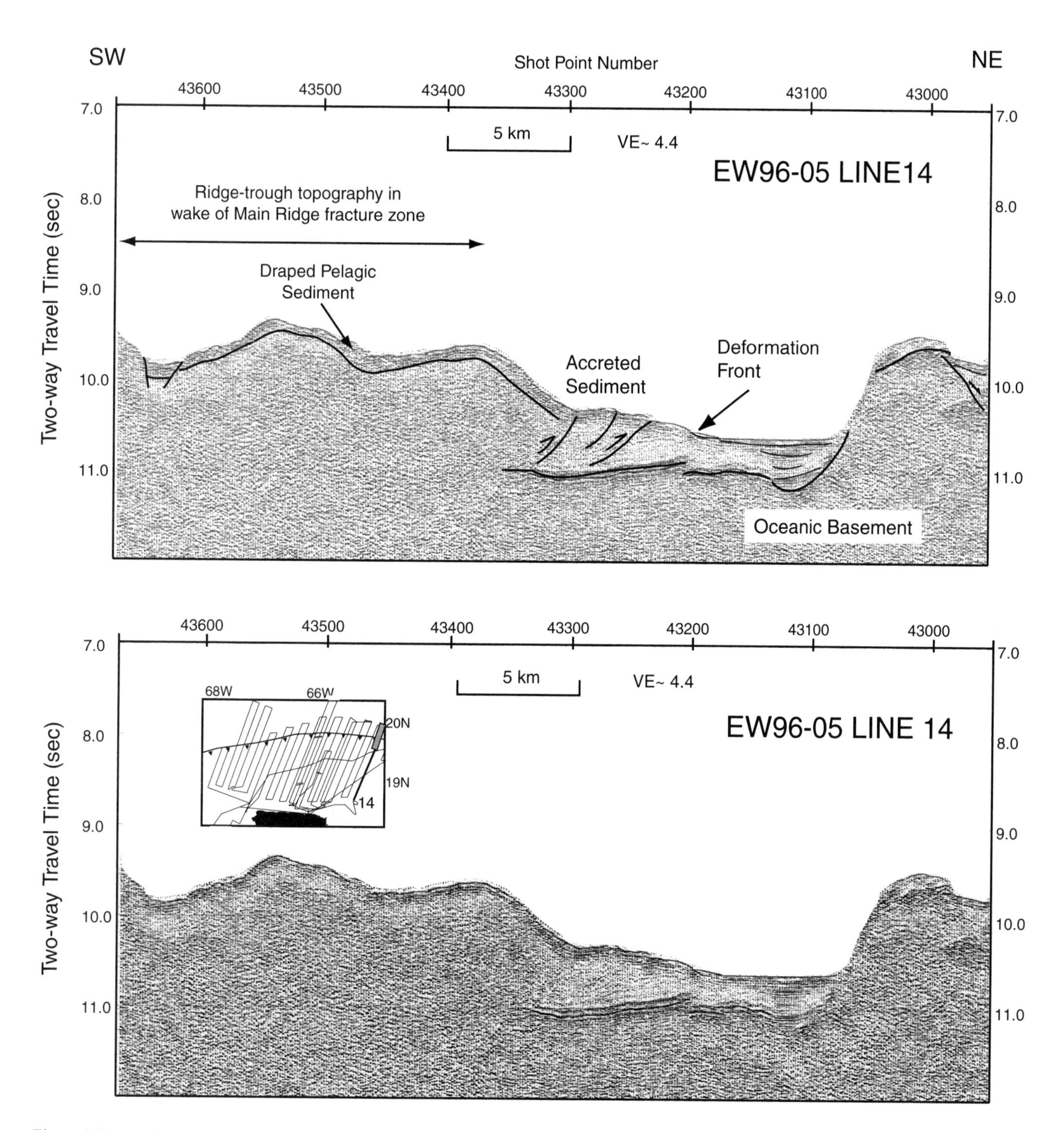

Figure 14. Part of seismic line EW9605–14 showing the narrow (~5 km) zone of accreted sediment and relatively thin accumulation of trench fill in the eastern section of the survey area. Location of the line is shown in the inset.

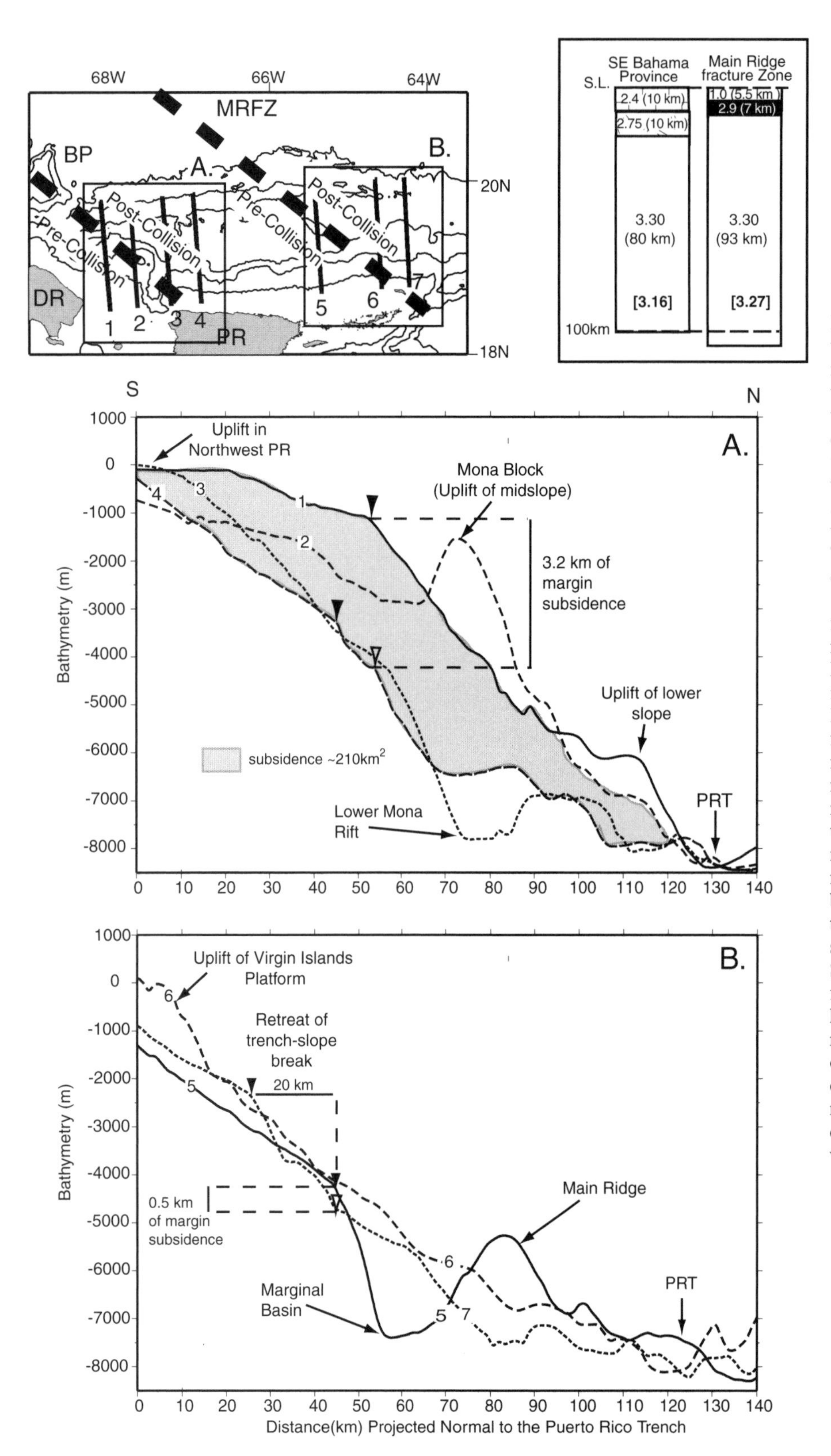

Figure 15. A series of bathymetric profiles taken perpendicular to the trench and margin. Using these profiles, we attempt to quantify the amount of subsidence that the northern margin of Puerto Rico has experienced over the past 10 m.y. Top left inset shows location of profiles in relation to the pre- and post-collision areas of the southeastern Bahamas Province (BP) and the Main Ridge fracture zone (MRFZ). Top right inset shows isostatic columns of southeastern Bahamas lithosphere and 100 Ma oceanic lithosphere (Main Ridge fracture zone). Density values for units making up columns are given in gm/cm^3. Lithosphere bulk density for each column is given in brackets. Open triangles mark location of former inner-trench slope break, filled triangles mark location of present inner-trench slope break. (A) We document a minimum of 3.2 km of subsidence of the margin during the past 3.5–5 m.y. due to the passage of the SE Bahamas. The shaded region (difference between profiles 1 and 4) represents the volume of material removed by subduction erosion during the same time interval, equivalent to a rate of erosion of 46–67 km^3/m.y./km of margin. (B) Subsidence associated with the Main Ridge fracture zone is estimated to be ~0.5 km, however, the inner-trench slope break has moved landward ~20 km due to the passage of the fracture zone. Note that the gradient of the mid-slope is reduced post collision, mostly likely the result of the transport of material from the upper-slope by mass wasting. PRT—Puerto Rico trench.

Present-Day Collisional Impacts of the Main Ridge Fracture Zone

The collisional impact of the Main Ridge fracture zone with the Puerto Rico trench slope provides a clear picture of "before," "during," and "after" tectonism of the margin. Most obvious is the change in the profile and structuring of the forearc from smooth, ponded sediment masses that are gently tilted to the southwest in the "before" snapshot (i.e., west of the Main Ridge) compared to a terrain characterized by closely spaced folds and thrust faults to the east in the aftermath area of the collision (Figs. 3B and 9). McCann and Sykes (1984) emphasized a "ridge-trough topography" between the Main Ridge and the trench whose troughs are inferred to be young in age because they do not contain young turbidite bodies. This area is inferred to be the chaotic area formed in the wake of the obliquely subducting Main Ridge fracture zone. One should note, however, that the turbidite fill of the Marginal Basin and Elevated Plain is dependent on the location of this area downslope from the Puerto Rico shelf (Ewing et al., 1966; Connolly and Ewing, 1967). The Median and Main ridges shield the area along their northeastern flanks from turbidite sedimentation. Although it is possible that this northeastern flank lacks sediment because of its post-collisional setting, it is also possible that these differences are related only to the blockage of turbidite sedimentation by the Median Ridge itself.

This area of the forearc is similar in appearance to that of the northern Chile margin, where von Huene and Ranero (2003) attribute the "riffled" nature of the forearc to the draping of slope debris over underthrusting seafloor relief. Given the significant horst and graben structures imaged in the down-going North America plate, a similar explanation might apply to the ridge and trough topography of the Puerto Rico trench forearc. However, since neither the EW96-05 SCS profiles nor previous MCS lines have imaged the top of the underthrusting plate any distance beneath the forearc, a direct link between underthrusting seafloor relief and forearc topography cannot yet be made.

The "during" picture shows oversteepening and slope gravity failure on the margin in the impact zone. Bathymetric profiles oriented perpendicular to the trench show evidence of uplift on the mid-slope of the forearc and Virgin Islands platform associated with presumed position of the underthrust Main Ridge fracture zone beneath the forearc (Fig. 15B). Slumping, mass wasting of the forearc slope, and faulting and thinning of the carbonate platform and migration of the trench-slope break landward by ~20 km characterize the region where the Main Ridge fracture zone is underthrusting the margin north of the Virgin Islands (EW96-05 SCS profile 14, Fig. 10). This area of the forearc is characterized by frequent seismic events that continue along strike of the southeastern extension of the Main Ridge fracture zone for more than 50 km beneath the margin (McCann, 2002). Focal mechanisms indicate that many are shallow (~15 km) south-dipping thrust events (Fig. 2A). In addition, an earthquake swarm was recorded by the Atlantic Sound Surveillance System (SOSUS) network in this area (Bryan and Nishimura, 1995).

Neogene Subsidence of the Margin: The Result of Accelerated Subduction Erosion

Birch (1986) was the first to propose tectonic erosion of the island arc and removal of topographic support for the northern Puerto Rico margin to explain the rapid late Neogene subsidence. Elaborating on Birch's hypothesis, we propose that much of the subsidence occurred in response to removal of upper-plate material by the underthrusting southeastern Bahama Province and overlying carbonate banks as the colliding ridges swept westward through time. This implies a time-transgressive deformation beginning in the east and moving to the west along the northern Puerto Rico margin.

Reconstruction of the past position of the southeastern extension of the Bahama Province, assuming constant velocities and directions similar to the present, place the carbonate province offshore the Virgin Islands and colliding obliquely with the margin beginning ca. 10 Ma (Mann et al., 2002) (Fig. 16). In this reconstruction, the Bowin fault zone appears to represent the approximate track of the southern edge of the southeastern Bahama Province along the forearc, with thinner disrupted forearc material to the north and thicker, older and more rigid metamorphic and volcanic arc material south of the Bahama Province. Divergence in the tracks of the Bowin fault zone and southeastern Bahama Province is most likely due to a recent change in trend of the Bowin fault zone caused by the underthrusting Main Ridge fracture zone. The East Septentrional fault shows a similar but more exaggerated deflection southward in the vicinity of the southeastern Bahama Province collision zone. Although not a typical plow mark, in this interpretation, the Bowin fault zone represents the long-term (10 m.y.) track of the southeastern Bahamas across the forearc area that is consistent with the convergence direction taken from GPS results. This model predicts time-transgressive subduction erosion and subsidence of the margin tied to the passage of specific fracture zones and the southeastern extension of the Bahama Province.

Subduction erosion is commonly thought to occur by a combination of frontal erosion and basal erosion. Frontal erosion refers to material eroded from the lower forearc slope, deposited in the trench, and then subducted, whereas basal erosion refers to transfer of material from the upper plate to the lower plate at the base of the forearc. Given the thick accumulation of turbidite fill in the trench adjacent to and in the wake of the southeastern Bahamas collision zone, material has, and most likely currently is, being transferred from the mid- and lower-slope into the trench. However, considering the lack of contractional deformation of the trench fill and the extremely low convergence rates, we postulate that little of material being eroded from the lower forearc slope is actually being subducted. Therefore basal erosion must significantly exceed frontal erosion in this area.

Figure 15 shows seven bathymetric profiles that cross the northern Puerto Rico margin and trench. The profiles are oriented perpendicular to the present-day direction of relative motion. The westernmost bathymetric profile 1 represents that

A.

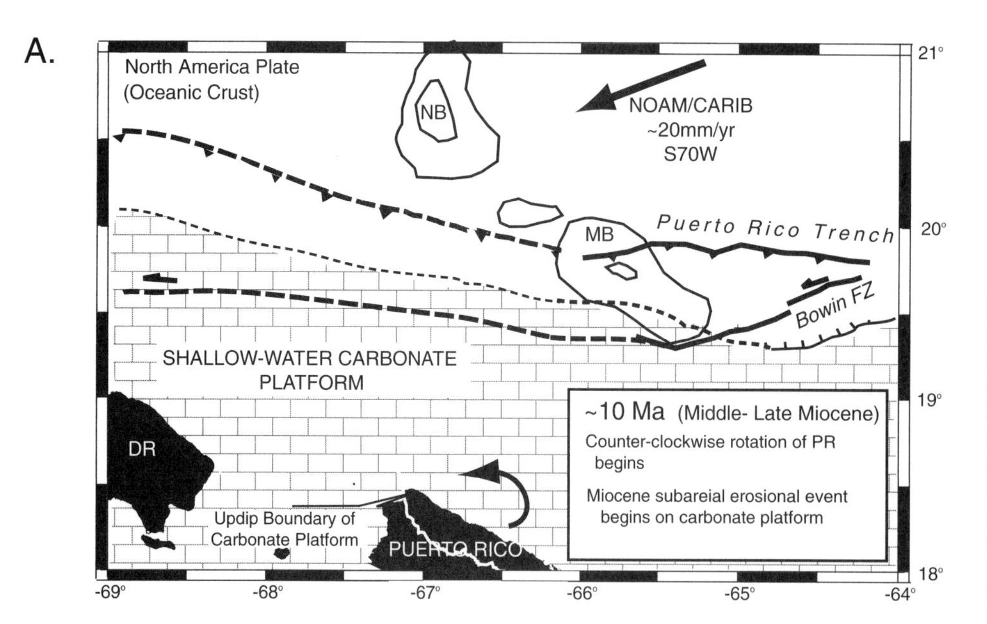

B.

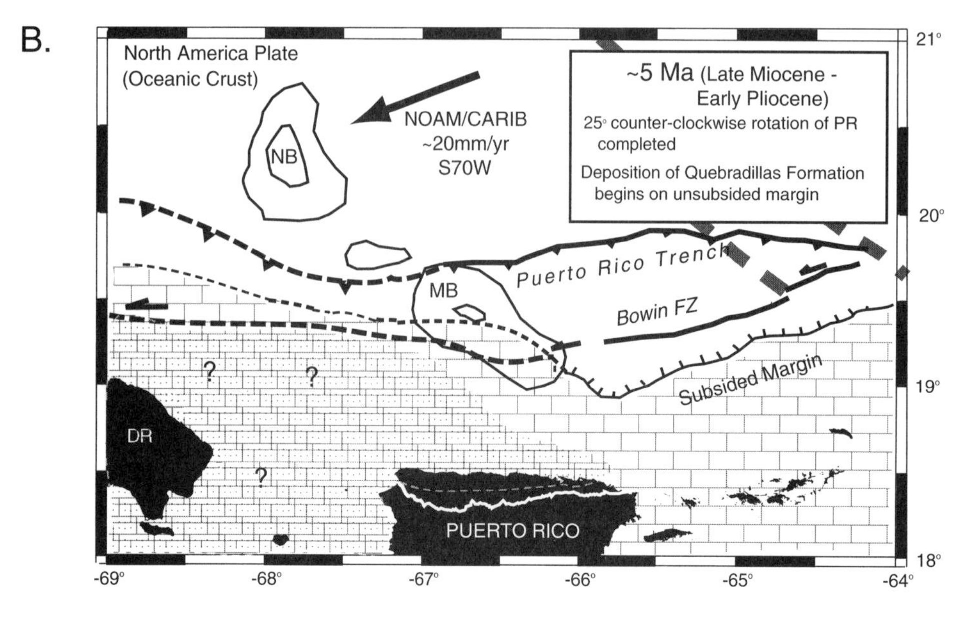

C.

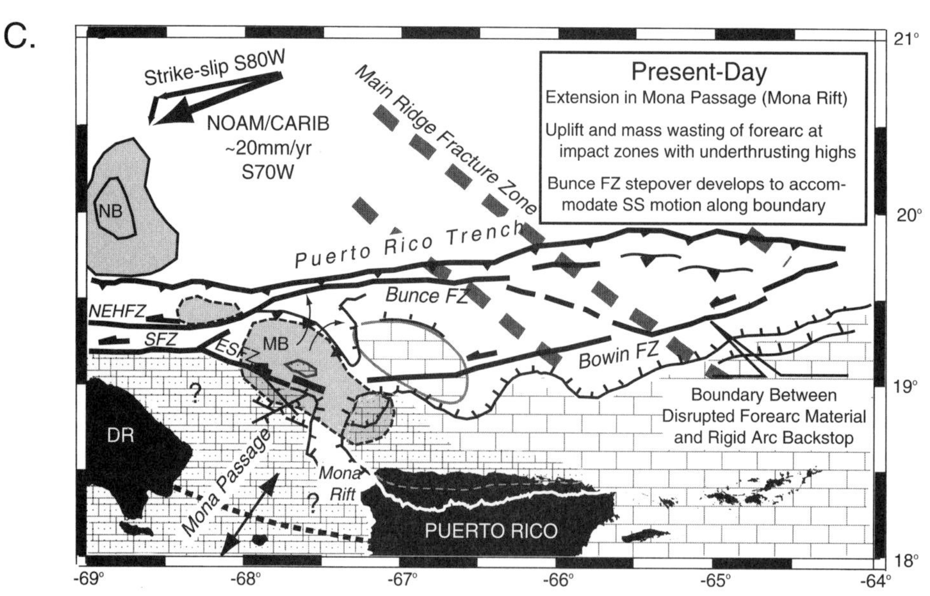

Figure 16. Time-transgressive schematic showing relative location of southeastern Bahama Province, indicated by NB (Navidad Bank) and MB (Mona Block) to the northern margin of Puerto Rico and the Virgin Islands during the past 10 m.y. GPS-derived vector for North America–Caribbean relative motion shown as large black arrow (Mann et al., 2002). Dashed boundaries indicate large uncertainties in exact location. (A) Amount and timing of counter-clockwise rotation of Puerto Rico from Reid et al. (1991). (B) The Quebradillas formation is indicated by hachured pattern on top of the existing carbonate pattern. The Quebradillas extends on land in northern Puerto Rico as indicated by the thin dashed white line. The extent of the formation on the northern Puerto Rico margin from Moussa et al., 1987. (C) Vectors representing strain partitioning within the forearc shown as small black arrows. Additional evidence of the time-transgressive nature of the tectonic erosion comes from the positions and locations of the headwall scarps. A line drawn through the apex of the two 40–55-km-wide, 2-km-high amphitheater-shaped mass wasting scarps at the trench slope break and the arcuate faults in the carbonate platform adjacent to the eastern flank of the Mona rift parallels the projected track of the southeastern Bahama Province along the margin. Regional arching of eastern Dominican Republic (DR) and Puerto Rico shown as dashed black line through Mona Passage (van Gestel et al., 1998; Mann et al., 2002). Note the change in orientation of the arch from NW-SE to EW south of the SE Bahamas collision zone. ESFZ—East Septentrional fault zone; NEHFZ—Northeast Hispaniola fault zone; SFZ—Septentrional fault zone.

of a slightly tilted precollision platform, profiles 2 and 3 show the uplifted collision zone on the forearc and northwest Puerto Rico, respectively, and profile 4 shows the post-collisional and subsided platform of late Miocene–early Pliocene age (5–3.5 Ma Quebradillas formation used as a time marker) (Fig. 15A). By comparing the depth to the inner trench slope break on profiles 1 and 4, we calculate that a minimum of 3.2 km of subsidence has occurred as a result of the passage of the high-standing bank of the southeastern Bahama Province and the short-term rate of subsidence to be 0.64–0.91 km/m.y. The short-term rates of the post-collision subsidence along the northern Puerto Rico margin are comparable to rates observed at other convergent plate boundaries where accelerated subduction erosion is thought to be the cause of subsidence. For example, during the middle to late Eocene, the Tonga forearc subsided by more than 2 km over a time span of only 3 m.y. (Clift and MacLeod, 1999). An ~210 km^2 change in margin cross section is calculated between profiles 1 and 4 and assumed to represent material loss as the result of subduction erosion, or equivalent to a rate of erosion of 42–60 km^3/m.y./km of margin.

In contrast to the relatively buoyant ~20-km-thick southeastern Bahamas, transverse ridges that are often associated with fracture zones do not represent over-thickened crust; they lack a crustal root and are uncompensated at depth (Abrams. et al., 1988; Pockalny et al., 1996). Transverse ridges are relatively small features (20–20 km wide, 1–2 km high) and are typically discontinuous along-strike. Because the Main Ridge fracture zone transverse ridge is not formed by over-thickened crust, it is readily subducted and causes relatively minimal subduction erosion. In Figure 15B, we calculate the amount of margin subsidence related to the passage of the Main Ridge fracture zone transverse ridge to be ~0.5 km. The most dramatic difference observed between profiles 4 and 6 is the location of the inner-trench slope break. The trench-slope break has moved ~20 km landward caused by the passage of the Main Ridge fracture zone. Mass wasting of the upper slope has transferred a significant amount of material down slope thereby reducing the gradient of the mid-slope in the post-collision area compared to areas east of the collision zone. Subsidence of the margin has slowed significantly or stopped to the west of the Main Ridge fracture collision point as shown by the Pliocene-Recent deposition of flat-lying carbonates in the Virgin Island Platform (van Gestel et al., 1998). This suggests that basal erosion caused by the underthrusting Main Ridge fracture zone is minimal. Moreover, the normal subduction erosion rates along this plate boundary must also be relatively low. This is not unexpected given the very slow and highly oblique convergence that characterizes this plate boundary (Lallemand et al., 1994).

Subduction zones are dynamic settings. We therefore cannot rule out that a non-isostatic component contributes or has contributed to the overall margin subsidence. These forces may be responsible for the small amount platform tilting observed in the northwestern Mona Passage, an area yet to be underthrust by the southeast Bahamas. Dillon et al. (1996) and van Gestel et al. (1999) propose that the margin subsidence occurred in response to "deep plate collision" or mantle interaction between a south-dipping North America slab depressed by a north-dipping Caribbean slab as shown by serial profiles of seismicity through the Mona Passage and Hispaniola. Calais and Lepinay (1991) propose that an eastward lateral migration of slab detachment within the subducting North America plate reaches its easternmost extent beneath the Mona Passage. At this position, excess pull exerted by the detached lithosphere could potentially cause forearc subsidence (van der Meulen et al., 1998) and rollback. Both of these mechanisms, however, predict synchronous and ongoing subsidence of the entire margin, which the majority of our data do not support.

Strain Regime within the Forearc Region

The unfolded nature of the units that make up the northern Puerto Rico margin is not consistent with a model of fold-thrust deformation of the Puerto Rico arc during its Neogene collision with the southeastern Bahama Province. We suggest that owing to the highly oblique nature of the collision and accelerated basal subduction erosion, little in the way of compressional deformation would be evident. Instead, we would anticipate that the regional strain regime would have a significant extensional component. Normal faulting along the trench-slope break revealed by the sidescan sonar imagery and seismic profiles suggest that extension is occurring at and trenchward of the trench-slope break. In addition, fault analyses on land in northern Puerto Rico (Hippolyte et al., this volume) document only extensional faulting, mainly trending N-S, from late Oligocene to recent times.

In addition to gravity induced extensional faulting in the Puerto Rico trench forearc, strain associated with highly oblique convergence appears to be largely accommodated along a series of strike-slip faults that trend sub-parallel to the N70°E GPS-derived direction of relative motion for the North America–Caribbean plates. This is consistent with recent GPS studies that find that a minimum of 80% of the North America–Caribbean plate motion occurs north of the island of Puerto Rico (Jansma et al., 2000). At least three strike slip faults offset the forearc in the western section of the survey area, the East Septentrional fault zone, the Bunce fault zone, and the Bowin fault zone (Fig. 4A). The Bunce fault zone is in close proximity (<10 km) to the deformation front of the trench. Little convergent deformation is observed in the trench sediment seaward of the front, suggesting limited coupling between the subducting plate and overriding plate. We speculate that the Bunce fault zone is a relatively recent feature that represents a "step-over" from the East Septentrional fault zone and potentially the westernmost section of the Bowin fault zone. The majority of forearc strain had been accommodated along these strike-slip fault zones; however, motion along these fault zones has recently become restricted and/or locked due to the collision of southeastern Bahama Province. The step-over point of the Bunce fault zone lies in a trough between the unsubducted Navidad bank and the subducted section of southeastern Bahama Province, presumably a relatively thin area of the forearc that is more easily faulted than the adjacent area being uplifted by relatively thick carbonate bank (Fig. 11).

In the eastern extent of the Puerto Rico trench, forearc strain is largely accommodated along the Bowin fault zone, as the Bunce fault zone appears to terminate at the Main Ridge around 65°35W. Here the floor of the Puerto Rico trench narrows eastward and changes trend. Contractional deformation of the trench sediment and the development of a narrow (<5 km) accretionary prism indicates a small component of strain partitioning east of the Main Ridge. The Main Ridge is where the Bowin fault zone displays a transpressional character and changes trend from N80°E to N90–100°E.

SUMMARY AND CONCLUSIONS

The systematic marine geophysical cruise EW96-05, conducted in 1996 along the northeastern North America–Caribbean plate boundary, provided the SCS reflection, bathymetry, and sidescan imagery that enables us to document the ongoing and post-collisional effects of obliquely subducting high-standing ridges underthrusting the northern Puerto Rico–Virgin Islands margin. In many aspects, the study was reconnaissance in nature and thus, several unsolved questions remain that require MCS reflection data, refraction studies, and ultimately deep ocean drilling. Nevertheless, we are able to describe the main structural elements of the margin and put forward a reasonable interpretation on the basis of existing, albeit incomplete data. On the basis of these observations and related interpretations, we conclude:

1. The oblique underthrusting of the southeastern extension of the Bahama Provinces has had a significant and diachronic impact on the tectonic evolution of northern Puerto Rico–Virgin Islands margin during the Neogene. Evidence of tectonic erosion in the wake of the obliquely subducting high-standing ridge includes increased seismicity, oversteepening, mass wasting, landward migration of the trench-slope break, and rapid subsidence of the margin. We document over 3.2 km of margin subsidence associated with the east-west passage of the southeastern Bahamas province beneath the margin. Calculated short-term (3.5–5 m.y.) subsidence rates are 0.64–0.91 km/m.y.

2. Accelerated basal subduction erosion of the overriding plate, although difficult to document directly, undoubtedly plays a significant role as the buoyant 20-km-thick southeastern Bahama Province tunnels beneath the margin. During the past 3.5–5 m.y., we estimate that the volume loss attributable to accelerated subduction erosion is 210 km^3/km of margin, or equivalent to a rate of 42–60 km^3/m.y./km of margin of material that has been eroded significantly thinning the overriding plate.

3. Subsidence of the margin associated with the passage of the Main Ridge fracture zone is relatively minimal (~0.5 km), nevertheless, the trench-slope break has retreated ~20 km landward. The near cessation of margin subsidence in the eastern section of the survey area and the rapid development of a small accretionary prism in the wake of the underthrusting Main Ridge fracture zone suggests that, under normal conditions, subduction erosion rates are minimal. This is not unexpected given the slow and highly oblique convergence that characterizes this plate boundary.

4. In the western section of the survey area, convergence approaches pure strike-slip. Here thick (0.75–1.5 km) accumulations of trench fill sediment are relatively undeformed. This lack of shortening deformation indicates that strain within the forearc is mainly accommodated along at least three strike-slip fault zones; the East Septentrional fault zone, the Bunce fault zone and the Bowin fault zone. Some strain partitioning, however, occurs in the eastern section of the survey area where trench sediment show evidence of folding and a narrow (<5 km wide) accretionary prism is apparent.

5. The Bowin fault zone appears to represent the track of the southern edge of the underthrusting and colliding southeastern Bahama Province along the forearc with thinner forearc debris to the north and thicker, more rigid volcanic arc material to the south. Although not a typical plow mark, in this interpretation the Bowin fault zone represents the long-term track (~10 m.y.) of the southeastern Bahamas across the forearc area that is consistent with the convergence direction taken from GPS results.

6. The underthrust high-standing ridges appear to alter the trends and geometries of the strike-slip faults within the forearc. We speculate that the Bunce fault zone is a relatively recent feature that developed when strike-slip motion became restricted along the East Septentrional and Bowin fault zones due to the collision of the southeastern Bahama Province. The Main Ridge is where the Bowin fault zone displays a transpressional character and changes trend from N80°E to N90–100°E.

NOTE IN PRESS

Our interpretation of the location and geometries of strike-slip faults within the Puerto Rico Trench differs from that of ten Brink et al. (2004). On the basis of multibeam bathymetry alone, ten Brink et al. link a possible splay of the Bunce fault at 66°45′W with a fault south of the Main Ridge, east of 65°30′W. Our interpretation is that the Bunce fault continues north of the Main Ridge trending approximately due east from 67°W. A separate strike-slip fault, the Bowin fault (referred to as the South Puerto Rico Slope Fault in Grindlay et al., 1997), defines the southern boundary of the Main Ridge. We mapped the Bowin fault eastward from the Main Ridge to the edge of our survey area at 64°45′W and westward across the southern trench slope to the Mona rift at 67°15′W.

ACKNOWLEDGMENTS

The dedication and hard work of Captain, crew, science officer and technicians aboard the R/V *Ewing* during cruise EW96-05 are gratefully acknowledged. Reviews by David Scholl and Troy Holcombe significantly improved the manuscript. We would also like to thank the HMR1 group, especially Bruce

Applegate, and Wilfredo Rodriguez, Frances Delano, Araceli Munoz, John Charles, and Steve Muszala, who participated as watchstanders during the cruise. Amy Miller helped to post-process the bathymetry data, and Neva Godwin and Luke Del Greco helped with figure making. GMT software (Wessel and Smith, 1995) was used to generate many of the figures in this manuscript. This research is supported by collaborative grants from the National Science Foundation Grant to our respective institutions: no. OCE-9796189 (NRG), no. OCE-9504118 (PM). University of Texas at Austin Institute for Geophysics (UTIG) contribution no. 1744.

REFERENCES CITED

Abrams, L.J., Detrick, R.S., and Fox, P.J., 1988, Morphology and crustal structure of the Kane fracture zone transverse ridge: Journal of Geophysical Research, v. 93, p. 3195–3210.

Austin, J.A., Overthrusting in a deep-water carbonate terrane, 1983, *in* Bally, A.W., ed., Seismic Expression of Structural Styles, American Association of Petroleum Geologists Studies in Geology. v. 15, p. 3.4.2-167–3.4.2-172.

Ball, M.M., Martin, R.G., Bock, W.D., Sylvester, R.E., Bowles, R.M., Taylor, D., Coward, E.L., Dodd, J.E., and Gilbert, L., 1985, Seismic structure and stratigraphy of northern edge of Bahaman-Cuban collision zone: American Association of Petroleum Geologists Bulletin, v. 69, p. 1275–1294.

Ballance, P., Scholl, D., Vallier, T., Stevenson, A., Ryan, H., and Herzer, R., 1989, Subduction of a late Cretaceous seamount of the Louisville Ridge at the Tonga Trench: A model of normal and accelerated tectonic erosion: Tectonics, v. 8, p. 953–962.

Birch, F.S., 1986, Isostatic, thermal and flexural models of the subsidence of the north coast of Puerto Rico: Geology, v. 14, p. 427–429.

Bowin, C., 1972, Puerto Rico trench negative anomaly belt: Geological Society of America Memoir 132, p. 339–350.

Bowin, C., 1976, Caribbean gravity field and plate tectonics: Geological Society of America Special Paper 169, p. 79.

Bryan, C.J., and Nishimura, C.E., 1995, Monitoring oceanic earthquakes with SOSUS: an example from the Caribbean: Oceanography, v. 8, p. 4–10.

Calais, E., and Lepinay, B. Mercier de, 1991, From transpression to transtension along the northern Caribbean plate boundary: Implications for the recent motion of the Caribbean plate: Tectonophysics, v. 186, p. 329–350, doi: 10.1016/0040-1951(91)90367-2.

Calais, E., Béthoux, N., and Lépinay, B. Mercier de, 1992, From transcurrent faulting to frontal subduction: A seismotectonic study of the northern Caribbean plate boundary from Cuba to Puerto Rico: Tectonics, v. 11, p. 114–123.

Clift, P., and MacLeod, C., 1999, Slow rates of subduction erosion estimated from subsidence and tilting of the Tonga forearc: Geology, v. 27, p. 411–414, doi: 10.1130/0091-7613(1999)0272.3.CO;2.

Connolly, F., and Ewing, M., 1967, Sedimentation in the Puerto Rico Trench: Journal of Sedimentary Petrology, v. 37, p. 44–59.

DeMets, C., Jansma, P., Mattioli, G., Dixon, T., Farina, F., Bilham, R., Calais, E., and Mann, P., 2000, GPS geodetic constraints on Caribbean–North America plate motion: Geophysical Research Letters, v. 27, p. 437–440, doi: 10.1029/1999GL005436.

Deng, J., and Sykes, L.R., 1995, Determination of Euler pole for contemporary relative motion of Caribbean and North America plates using slip vectors of interplate earthquakes: Tectonics, v. 14, p. 39–53, doi: 10.1029/94TC02547.

Dillon, W., Acosta, J., Uchupi, E., and ten Brink, U., 1998, Joint Spanish-American research uncovers fracture pattern in northeastern Caribbean: Eos (Transactions, American Geophysical Union), v. 79, p. 336.

Dillon, W.P., Austin, J., Scanlon, K., Edgar, N., and Parson, L., 1992, Accretionary margin of northwestern Hispaniola: Morphology structure and development as part of the northern Caribbean plate boundary: Marine and Petroleum Geology, v. 9, p. 70–88, doi: 10.1016/0264-8172(92)90005-Y.

Dillon, W.P., Edgar, N.T., Scanlon, K.M., and Coleman, D.F., 1996, A review of the tectonic problems of the strike-slip northern boundary of the Caribbean plate and examination by GLORIA, *in* Gardner, J.V., Field, M.E., and Twichell, D.C., eds., Geology of the United States Seafloor: The View from GLORIA, p. 135–164.

Dobson, M.R., Karl, H.A., and Vallier, T.L., 1996, Sedimentation along the fore-arc region of the Aleutian Island Arc, Alaska, *in* Gardner, J.V., Field, M.E., and Twichell, D.C., eds., Geology of the United States Seafloor: The View from *Gloria*, p. 279–304.

Dolan, J., Mullins, H.T., and Wald, D.J., 1998, Active tectonics of the north-central Caribbean: oblique collision, strain partitioning, and opposing subducted slabs, *in* Dolan, J.F., and Mann, P., eds., Active Strike-slip and Collisional Tectonics of the Northern Caribbean Plate boundary Zone, eds.: Geological Society of America Special Paper 326, p. 1–61.

Dolan, J.F., and Wald, J.D., 1998, The 1943–1953 north central Caribbean earthquakes: Active tectonic setting, seismic hazards and implications for Caribbean–North America plate motions, *in* Dolan, J.F., and Mann, P., eds., Active Strike-slip and Collisional Tectonics of the Northern Caribbean Plate Boundary Zone: Geological Society of America Special Paper 326, p. 143–169.

Dominguez, S., Lallemand, S., Malavieille, J., and von Huene, R., 1998a, Upper plate deformation associated with seamount subduction: Tectonophysics, v. 293, p. 207–224, doi: 10.1016/S0040-1951(98)00086-9.

Dominguez, S., Lallemand, S., Malavieille, J., and Schnurle, P., 1998b, Oblique subduction of the Gagua Ridge beneath the Ryukyu accretionary wedge system: Insights from marine observations and sandbox experiments: Marine Geophysical Researches, v. 20, p. 383–402, doi: 10.1023/A:1004614506345.

EEZ SCAN 85 Scientific Staff, 1987, Atlas of the US Exclusive Economic Zone, eastern Caribbean: U.S. Geologic Survey Miscellaneous Investigator Series, I-1864-B, p. 58.

Ewing, M., Leonardi, A., and Ewing, J., 1968, The sediments and topography of the Puerto Rico trench and outer ridge: Transactions, Fourth Caribbean Geological Conference, 28 March–12 April 1965, Port-of-Spain, Trinidad and Tobago, Caribbean Printers, Arima, Trinidad and Tobago, p. 325–334.

Fox, J., and Heezen, B., 1975, Geology of the Caribbean crust, *in* Nairn, A., and Stehli, F., eds., The Ocean Basins and Margins, vol. 3: New York, Plenum Press, p. 421–466

Geist, E., Childs, J. and Scholl, D., 1988, The origin of summit basins of the Aleutian Ridge: Implications for block rotation of an arc massif: Tectonics, v. 7, p. 327-341.

Grindlay, N.R., Mann, P., and Dolan, J.F., 1997, Researchers investigate submarine faults north of Puerto Rico: Eos (Transactions, American Geophysical Union), v. 78, p. 404.

Heezen, B.C., Nesteroff, V., Rawson, M., and Freeman-Lynde, R.P., 1985, Visual evidence for subduction in the western Puerto Rico trench, Geodynamiques des Caribes, Symposium Paris, 5–8 Fevrier: Editions Technip, 27, Rue GINOUX, 75015, Paris, France.

Jansma, P., Lopez, A., Mattioli, A., DeMets, C., Dixon, T., Mann, P., and Calais, E., 2000, Neotectonics of Puerto Rico and the Virgin Islands, northeastern Caribbean, from GPS geodesy: Tectonics, v. 19, p. 1021–1037, doi: 10.1029/1999TC001170.

Jordan, T., 1975, The present-day motions of the Caribbean plate: Journal of Geophysical Research, v. 80, p. 4433–4439.

Joyce, J., 1991, Blueschist metamorphism and deformation on the Samana Peninsula; a record of subduction and collision on the Greater Antilles, *in* Mann, P., Draper, G., and Lewis, J.F. eds., Geologic and Tectonic Development of the North America–Caribbean Plate Boundary in Hispaniola: Geological Society of America Special Paper 262, p. 47–76.

Kulm, L.V., and Suess, E., 1990, Relationship of carbonate deposits and fluid venting: Oregon accretionary prism: Journal of Geophysical Research, v. 95, p. 8899–8916.

Lallemand, S.E., Schnurle, P., and Malavieille, J., 1994, Coulomb theory applied to accretionary and nonaccretionary wedges: Possible causes for tectonic erosion and/or frontal accretion: Journal of Geophysical Research, v. 99, p. 12,033–12,055, doi: 10.1029/94JB00124.

Larue, D.K., and Ryan, H.F., 1989, Extensional tectonism in the Mona Passage, Puerto Rico and Hispaniola: A preliminary study: Twelfth Caribbean Geological Conference Transactions, p. 223–230.

Larue, D.K., and Ryan, H.F., 1998, Seismic reflection profiles of the Puerto Rico Trench: Shortening between the North America and Caribbean Plates, *in* Lidiak, E., and Larue, D., eds., Tectonics and Geochemistry of the Northeastern Caribbean: Geological Society of America Special Paper 322, p. 193–210.

Larue, D.K., Joyce, J., and Ryan, H., 1989, Neotectonics of the Puerto Rico Trench: Extensional tectonism and forearc subsidence: Twelfth Caribbean Geological Conference Transactions p. 231–247.
Larue, D., Torrini, R., Smith, A., and Joyce, J., 1998, North Coast Tertiary basin of Puerto Rico: From arc basin to carbonate platform to arc-massif slope, *in* Lidiak, E., and Larue, D., eds., Tectonics and Geochemistry of the Northeastern Caribbean: Geological Society of America Special Paper 322, p. 155–176.
Mann, P., Calais, E., Ruegg, J.-C., DeMets, C., Dixon, T., Jansma, P., and Mattioli, G., 2002, Oblique collision in the northeastern Caribbean from GPS measurements and geological observations: Tectonics, v. 21, no. 6, p. 1057, doi: 10.1029/2002TC001304.
Masaferro, J.L., and Eberli, G.P., 1999, Jurassic-Cenozoic structural evolution of the southern Great Bahama Bank, *in* Mann, P., ed., Sedimentary Basins of the World: Caribbean Basins, v. 4, p. 167–196.
Masson, D.G., and Scanlon, K.M., 1991, The neotectonic setting of Puerto Rico: Geological Society of America Bulletin, v. 103, p. 144–154, doi: 10.1130/0016-7606(1991)1032.3.CO;2.
McCann, W., 2002, Microearthquake data elucidate details of Caribbean subduction zone: Seismological Research Letters, v. 73, p. 25–32.
McCann, W., and Sykes, L., 1984, Subduction of aseismic ridges beneath the Caribbean plate: Implications for the tectonics and seismic potential of the Northeastern Caribbean: Journal of Geophysical Research, v. 89, p. 4493–4519.
Molnar, P., and Sykes, L.R., 1969, Tectonics of the Caribbean and Middle America regions from focal mechanisms and seismicity: Geological Society of America Bulletin, v. 80, p. 1639–1684.
Monroe, W.H., 1980, Geology of the middle Tertiary formations of Puerto Rico: U.S. Geological Survey Professional Paper 953, 93 p.
Moussa, M.T., Seiglie, G.A., Meyerhoff, A.A., and Taner, I., 1987, The Quebradillas Limestone (Miocene-Pliocene), northern Puerto Rico, and tectonics of the northeastern Caribbean margin: Geological Society of America Bulletin, v. 99, p. 427–439.
Mullins, H.T., Breen, N., Dolan, J., Wellner, R., Petruccione, J., Gaylord, M., Andersen, B., Mellillo, A., Jurgens, A., and Orange, D., 1992, Carbonate platforms along the Southeast Bahamas-Hispaniola collision zone: Marine Geology, v. 105, p. 169–209, doi: 10.1016/0025-3227(92)90188-N.
Muszala, S., Grindlay, N.R., and Bird, R., 1999, Three–dimensional Euler deconvolution and tectonic interpretation of marine magnetic anomaly data in the Puerto Rico trench: Journal of Geophysical Research, v. 104, p. 29,175–29,187, doi: 10.1029/1999JB900233.
Orange, D., Anderson, R., and Breen, N., 1994, Regular canyon spacing in the submarine environment: The link between hydrology and geomorphology: GSA Today, v. 4, p. 30–39.
Perfit, M.R., Heezen, B.C., Rawson, M., and Donnelly, T., 1980, Chemistry, origin and tectonic significance of metamorphic rocks from the Puerto Rico Trench: Marine Geology, v. 34, p. 125–156, doi: 10.1016/0025-3227(80)90069-9.
Pilkey, O., and Cleary, W., 1986, Turbidite sedimentation in the northwestern Atlantic Ocean basin, *in* Vogt, P., and Tucholke, B., eds., The Western North Atlantic Region: Geological Society of America, Geology of North America, v. M, p. 437–450.
Pockalny, R., Gente, P., and Buck, R., 1996, Oceanic transverse ridges: A flexural response to fracture-zone-normal extension: Geology, v. 24, p. 71–74, doi: 10.1130/0091-7613(1996)0242.3.CO;2.
Ranero, C.R., and von Huene, R., 2000, Subduction erosion along the Middle America convergent margin: Nature, v. 404, p. 748–752, doi: 10.1038/35008046.
Reid, J., Plumley, P.W., and Schellekens, J.H., 1991, Paleomagnetic evidence for Late Miocene counterclockwise rotation of north coast carbonate sequence, Puerto Rico: Geophysical Research Letters, v. 18, p. 565–568.
Schwab, W.C., Danforth, W.W., Scanlon, K.M., and Masson, D.G., 1991, A giant submarine slope failure on the northern insular slope of Puerto Rico: Marine Geology, v. 96, p. 237–246, doi: 10.1016/0025-3227(91)90149-X.
Sheridan, R.E., Crosby, J.T., Bryan, G., and Stoffa, P.L., 1981, Stratigraphy and structure of southern Blake Plateau, northern Florida Straits, and northern Bahama Platform from multichannel seismic reflection data: American Association of Petroleum Geologists Bulletin, v. 65, p. 2571–2593.
Sheridan, R.E., Mullins, H.T., Austin, J.A., Ball, M.M., and Ladd, J.W., 1988, Geology and geophysics of the Bahamas, *in* Sheridan, R.E., and Grow, J.A., eds., The Atlantic Continental Margin, U.S.: Geological Society of America, Geology of North America, v. I-2, p. 329–364.
Smith, W.H., and Sandwell, D.T., 1997, Global seafloor topography from satellite altimetry and ship depth soundings: Science, v. 277, p. 1956–1962, doi: 10.1126/SCIENCE.277.5334.1956.
ten Brink, U., Danforth, W., Polloni, C., Andrews, B., Lianes, P., Smith, S., Parker, E., and Uozumi, T., 2004, New seafloor map of the Puerto Rico Trench helps assess earthquake and tsunami hazards: Eos (Transactions, American Geophysical Union), v. 85, p. 349.
Uchupi, E., Milliman, J.D., Luyendyk, B.P., Bowin, C.O., and Emery, K.O., 1971, Structure and Origin of southeastern Bahamas: American Association of Petroleum Geologists Bulletin, v. 55, p. 687–704.
van der Meulen, M.J., Meulenkamp, J.E., and Wortel, M.J.R., 1998, Lateral shifts of Apenninic foredeep depocenteres reflecting detachment of subducted lithosphere: Earth and Planetary Science Letters, v. 154, p. 203–219, doi: 10.1016/S0012-821X(97)00166-0.
van Gestel, J.-P., Mann, P., Dolan, J.F., and Grindlay, N.R., 1998, Structure and tectonics of the upper Cenozoic Puerto Rico–Virgin Islands carbonate platform as determined from seismic reflection studies: Journal of Geophysical Research, v. 103, p. 30,505–30,530, doi: 10.1029/98JB02341.
van Gestel, J.-P., Mann, P., Grindlay, N.R., and Dolan, J.F., 1999, Three-phase tectonic evolution of the northern margin of Puerto Rico as inferred from an integration of seismic reflection, well and outcrop data: Marine Geology, v. 161, p. 257–286, doi: 10.1016/S0025-3227(99)00035-3.
Vening Meinesz, F.A., Umbgrove, J.H.F., and Kuenen, P.H., 1934, Gravity expeditions at sea 1923–1932, Delft, 2d Netherlands Geodetic Commission, 139 p.
von Huene, R., and Lallemand, S., 1990, Tectonic erosion along the Japan and Peru convergent margins: Geological Society of America Bulletin, v. 102, p. 704–720, doi: 10.1130/0016-7606(1990)1022.3.CO;2.
von Huene, R., Bialas, J., Flueh, E., Cropp, B., Csernok, T., Fabel, E., Hoffmann, J., Emeis, K., Holler, P., Jeschke, G., Leandro, C., Peréz Fernandéz, I., Chavarria, J., Florez, A., Escobedo, D., León, R., and Barrios, O., 1995, Morphotectonics of the convergent margin of Costa Rica, *in* Mann, P., ed., Geologic and Tectonic Development of the Caribbean Plate boundary in Southern Central America, Geological Society of America Special Paper 295, p. 291–308.
von Huene, R. and Ranero, C., 2003, Subduction erosion and basal friction along the sediment-starved convergent margin off Antofagasta, Chile: Journal of Geophysical Research, v. 108, no. 2079, doi:10.1029/2001JB001569.
Weaver, J.D., Smith, A.L., and Sieglie, G., 1975, Geology and tectonics of the Mona Passage: Eos (Transactions, American Geophysical Union), v. 56, p. 45.
Wessel, P., and Smith, W.H.F., 1995, New version of the Generic Mapping Tools released: EOS, Transactions, American Geophysical Union, v. 76, p. 329.

Manuscript Accepted by the Society 18 August 2004

Printed in the USA

Geological Society of America
Special Paper 385
2005

Ground-motion relations for Puerto Rico

Dariush Motazedian
Gail Atkinson
Dept. of Earth Sciences, Carleton University, 1125 Colonel By Drive, Ottawa, Ontario K1S 5B6, Canada

ABSTRACT

Fourier amplitudes and response spectra of more than 3000 waveforms from ~300 Puerto Rico earthquakes of magnitude 3–5.5 have been analyzed. Due to a paucity of data at small distances and large magnitudes, the ground-motion data cannot be used directly to obtain ground-motion relations for magnitudes and distances of most engineering interest. Instead, the data were used to determine key attenuation parameters, such as regional anelastic attenuation, duration behavior and generic site amplification that are input to seismological ground-motion models. The data were also used to validate ground-motion model predictions.

To overcome the incompleteness of the data set, we applied the stochastic method to simulate waveforms for different magnitudes and distances. The stochastic method has been applied for other regions such as California, Cascadia, and eastern North America, and on average reproduces empirical attenuation relationships that can be obtained by direct use of enough data. The input parameters for the simulations are based on the attenuation parameters obtained from the recorded waveforms.

We simulated 1950 acceleration time series for magnitudes from M3.0 to M8.0 and distances from 2 km to 500 km. Simulation was performed in magnitude steps of 0.2 units. In order to provide a good database, we simulated data for both backward and forward directivity cases, as well as data for azimuths with minimal directivity effects. The response spectra of the simulated time series have been calculated. The maximum likelihood method has been applied in order to derive ground-motion relations for a generic soft rock site condition for frequencies from 0.1 to 20Hz. The stochastic-model ground-motion relations for Puerto Rico are validated using available seismographic data, and compared to ground-motion relations for other regions. These are the first region-specific ground-motion relations developed for seismic hazard analysis of Puerto Rico.

Keywords: Puerto Rico, seismicity, ground-motion relations, stochastic finite fault modeling, response spectra.

INTRODUCTION

Puerto Rico has a high level of seismic activity due to its location on the boundary between the North America and the Caribbean plates. At least four destructive earthquakes are documented in the historical records before 1700. There was a possible great earthquake in 1787 (**M**8–8.2)[1] and a major earthquake in 1918 (**M**7.3) (McCann, 1985, 2002). About 9000 earthquakes of M > 3 have been recorded since the inception of the Puerto Rico Seismic Network in 1974 (McCann, 2002). A felt seismic event occurs about once per month. Thus, Puerto

[1]In this paper, boldface **M** is moment magnitude and regular M is catalogue magnitude.

Motazedian, D., and Atkinson, G., 2005, Ground-motion relations for Puerto Rico, *in* Mann, P., ed., Active tectonics and seismic hazards of Puerto Rico, the Virgin Islands, and offshore areas: Geological Society of America Special Paper 385, p. 61–80. For permission to copy, contact editing@geosociety.org.

Rico's 3.8 million inhabitants are exposed to a significant earthquake hazard.

Despite the hazard, there are to date no region-specific ground-motion relations for Puerto Rico. Ground-motion relations, describing the expected amplitudes of ground motions as functions of magnitude and distance, are a key component of seismic hazard analyses. In order to provide accurate seismic hazard assessments for Puerto Rico—a prerequisite for making informed seismic design decisions—it is important to develop such ground-motion relations. In this paper, we develop ground-motion relations for Puerto Rico using a combination of seismological and empirical modeling. We use data recorded on regional broadband and local seismic networks to determine the underlying attenuation parameters and validate the predictions of the ground-motion relations. First, we describe the overall method that is employed to define ground-motion relations. Then we describe the Puerto Rico ground-motion data and how they are used to obtain underlying attenuation parameters for the model. We then present the simulated ground-motion database and use it to derive ground-motion relations. Finally, we compare the derived relations to recorded data and to ground-motion relations for other regions.

METHOD

The stochastic model is a widely used tool to simulate acceleration time series and develop ground-motion relations (Hanks and McGuire, 1981; Boore, 1983; Atkinson and Boore, 1995, 1997a; Toro et al., 1997; Atkinson and Silva, 2000). The stochastic method begins with the specification of the Fourier spectrum of ground motion as a function of magnitude and distance. Often the acceleration spectrum is modeled by a spectrum with an ω^2 shape, where ω = angular frequency (Aki, 1967; Brune, 1970, 1971; Boore 1983). The "Brune model" spectrum is derived for an instantaneous shear dislocation at a point. The acceleration spectrum of the shear wave, $A(f)$, at hypocentral distance R from an earthquake is given by

$$A(f) = CM_0\,(2\pi f)^2/[1+f/f_0)^2]\ \exp(-\pi f\kappa)\ \exp(-\pi f R/Q\beta)/R, \quad (1)$$

where M_0 is seismic moment and f_0 is corner frequency, which is given by $f_0 = 4.9*10^6\beta\,(\Delta\sigma/M_0)^{1/3}$, where $\Delta\sigma$ is stress drop in bars; M_0 is in dyne-cm; and β is shear wave velocity in km/s (Boore, 1983). The constant $C = \Re_{\theta\phi}\ F\ V/(4\pi\rho\beta^3)$, where $\Re_{\theta\phi}$ = radiation pattern (average value of 0.55 for shear waves); F = free surface amplification (2.0); V = partition onto two horizontal components (0.71); ρ = density; and R = hypocentral distance (Boore, 1983). The term $\exp(-\pi f\kappa)$ is a high-cut filter to account for near-surface "kappa" effects, which describe the commonly observed rapid spectral decay at high frequencies (Anderson and Hough, 1984). In the above equation, the power of R in the denominator of the attenuation term, $\exp(-\pi f R/Q\beta)/R$, is considered equal to 1, which is appropriate for body-wave spreading in a whole space. This value can be changed as needed in order to account for the presence of postcritical reflections from the Moho discontinuity or multiple reflected waves traveling in the crustal waveguide. The quality factor, $Q(f)$, is an inverse measure of anelastic attenuation. Through this equation, the spectrum is diminished with distance to account for empirically defined attenuation behavior.

Finite fault modeling has been an important tool for the prediction of ground motion near the epicenters of large earthquakes (Hartzell, 1978; Irikura, 1983; Joyner and Boore, 1986; Heaton and Hartzell, 1986; Somerville et al., 1991; Tumarkin and Archuleta, 1994; Zeng et al., 1994; Beresnev and Atkinson, 1998a). One of the most useful methods to simulate ground motion for a large earthquake is based on the simulation of a number of small earthquakes as subfaults that comprise a big fault. A large fault is divided into N number of subfaults, and each subfault is considered to be a small point source (introduced by Hartzell, 1978). Ground motions of subfaults, each of which may be calculated by the stochastic point-source method as described above, are summed with a proper delay time in the time domain to obtain the ground motion from the entire fault, $a(t)$:

$$a(t) = \sum_{i=1}^{nl}\sum_{j=1}^{nw} a_{ij}\left(t + \Delta t_{ij}\right), \quad (2)$$

where nl and nw are the number of subfaults along the length and width of the main fault, respectively ($nl*nw = N$); Δt_{ij} is the relative delay time for the radiated wave from the ijth subfault to reach the observation point. The $a_{ij}(t)$ are each calculated by the stochastic point source method.

The stochastic method has also been used to derive ground-motion relations for many different regions. Atkinson and Boore (1995) derived ground-motion relations for eastern North America, using a stochastic point-source model with an empirical two-corner source model. Toro et al. (1997) developed similar relations for eastern North America using a Brune point-source model. Atkinson and Silva (2000) developed ground-motion relations for California using a stochastic method that exploits the equivalence between the finite fault model and a two-corner point-source model of the earthquake spectrum. In each of these cases, region-specific input parameters derived from seismograms were used to specify the model parameters that drive the ground-motion relations for that region. For California, where there is a good empirical strong-motion database, it was shown that the stochastic relations agree well with empirical regression equations (e.g., Abrahamson and Silva, 1997; Boore et al., 1997; Sadigh et al., 1997; Atkinson and Silva, 2000). The stochastic ground-motion relations provide a sound basis for estimating peak ground motions and response spectra for earthquakes of magnitude 4 through 8, at distances from 1 to 200 km over the frequency range 0.2–20 Hz.

In this study, we use a stochastic finite-fault approach. A modified version of the computer program FINSIM (Beresnev and Atkinson, 1998b) was used for the simulations. The modifications to FINSIM introduce the new concept of a "dynamic corner frequency" to more closely model real finite-fault dynamic behavior (see Appendix). The modified program has been renamed EXSIM (extended earthquake fault simulation program). EXSIM model parameters that represent the earthquake source processes have been calibrated for general applications, using data from 27 moderate to large, well-recorded earthquakes in California. For use in Puerto Rico, the model requires region-specific attenuation and generic site parameters, which are derived from recordings of small to moderate earthquakes.

EXSIM is used to simulate a ground-motion database from which to develop ground-motion equations. We take this approach because there are not enough real data in the magnitude-distance ranges of engineering interest. We can use the empirical data to establish the underlying parameters and validate the model predictions. The region-specific parameters needed for simulations are the following:

1. Attenuation of Fourier amplitudes with distance (geometric spreading and anelastic attenuation);
2. Duration of ground motion as a function of magnitude and distance; and
3. Regional generic crust/site amplification.

With these parameters established, we can use the calibrated EXSIM model to extend our prediction to the magnitude-distance range of interest. We then compare predictions with Puerto Rico data and with ground-motion relations for other regions.

GROUND-MOTION DATA FOR PUERTO RICO

The Puerto Rico Seismic Network (PRSN), consisting of 13 vertical component short-period 1 Hz natural frequency seismometers, was installed in 1974. Digital time series have been recorded by the PRSN since 1991 in International Association of Seismology and Physics of the Earth's Interior (IASPEI) format (16-bit analogue to digital converter, 100 samples/s). More than 2000 time series recorded by the PRSN from more than 300 events with M ≥ 3.0 have been compiled; however, many records were clipped and could not be used. (Note: M refers to catalog magnitude, to be discussed in a later section.) Time series recorded before 1991 cannot be used due to lack of information about the instrument responses. Nine new three-component broadband stations were installed as of the year 2000 and will soon provide additional data.

The Puerto Rico Strong Motion Network (PRSMN) has been installed gradually since 1994 and now comprises 32 strong motion stations. The number of recorded acceleration time series is limited, since the earthquake ground motion must be strong enough to trigger these accelerographs. About 30 acceleration time series from 4 events are available from this network.

There has been a single Incorporated Research Institutions for Seismology (IRIS)/U.S. Geological Survey station (SJG) operating in Puerto Rico since 1993; it includes five broadband seismometers (three-component), and one short period seismometer (three-component). The sampling rate of the short period three-component seismometers is 80 samples/s. The highest sampling rate for the broadband seismometer data is 20 samples/s. From the SJG station, 1289 time series from more than 300 events with M ≥ 3.0 have been compiled. These are the data with highest reliability and well-known instrument response.

In order to study ground-motion processes, we need to remove all instrument response effects from the records so that we can obtain the actual motion of the ground. Instrument-corrected data allow compilation of a regional ground-motion database in a uniform format for ground-motion studies. The instrument response of all stations was obtained from a detailed calibration program that included in situ field calibrations of instruments using a portable seismograph with well-known response and by calibration against the well-known response of the SJG station. The stability over time of the obtained instrument responses of the PRSN stations was also examined by plotting the residuals (e.g., deviations from calculated model values) of the Fourier spectra versus time for all of the recorded data. Data for which the instrument response could not be reliably determined were deleted from the database.

Fourier amplitudes and response spectra of more than 3000 time series (of known and/or calibrated instrument response) from more than 300 earthquakes in the Puerto Rico region were analyzed. Figure 1 shows the distribution of the ground-motion database by magnitude and distance. The databases can be obtained by sending a request to the first author (dariush@ccs.carleton.ca). Table 1 lists the moment magnitude for each event. The determination of these magnitudes is discussed later.

The shear-wave portion of signal and pre-event noise was windowed for all of the time series. We compiled data for which

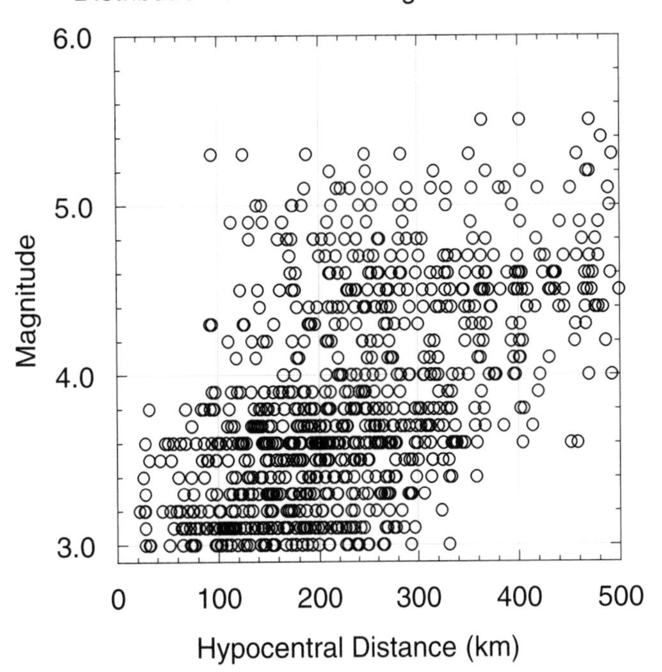

Figure 1. Distribution of study events in magnitude and distance.

TABLE 1. EARTHQUAKES USED IN THIS STUDY

	Date Y	M	D	Time h	Lat.	Long.	Depth (km)	Catalogue magnitude		Moment magnitude
1	1993	7	22	20	18.58	–69.00	109	4.8	mb	4.5
2	1993	8	1	19	17.40	–65.72	25	5.3	Mw	5.3
3	1993	8	10	7	19.35	–64.89	36	4.5	mb	4.3
4	1993	9	30	0	18.75	–62.80	33	4.3	mb	3.5
5	1993	10	3	13	17.80	–62.75	10	4.2	Md	3.8
6	1993	10	18	9	18.66	–64.45	30	4.3	mb	4.0
7	1993	10	24	17	19.68	–70.38	33	4.0	mb	4.1
8	1993	11	5	2	19.02	–66.01	48	4.9	mb	4.4
9	1993	11	8	3	19.20	–68.08	10	4.6	mb	4.1
10	1994	1	6	17	18.05	–68.37	87	4.0	mb	4.2
11	1994	1	8	22	18.22	–64.33	103	4.8	mb	4.6
12	1994	1	13	20	18.84	–66.18	47	4.0	mb	3.7
13	1994	1	18	12	18.58	–68.82	163	4.6	mb	4.3
14	1994	1	21	18	19.67	–64.43	10	4.0	mb	3.8
15	1994	2	25	16	19.25	–64.23	33	4.3	mb	4.2
16	1994	2	25	17	19.25	–64.33	32	4.9	mb	4.7
17	1994	3	1	18	19.39	–65.16	10	4.3	mb	4.1
18	1994	3	10	3	17.81	–65.35	10	4.1	mb	3.8
19	1994	4	21	17	18.00	–62.88	75	5.0	mb	4.8
20	1994	5	1	4	17.93	–64.70	162	4.3	mb	3.9
21	1994	6	25	6	19.01	–66.83	46	4.7	mb	4.3
22	1994	6	25	8	18.94	–66.68	33	4.0	Md	3.4
23	1994	7	11	3	19.23	–66.76	10	4.4	mb	4.1
24	1994	7	17	23	17.63	–62.94	131	4.2	mb	4.0
25	1994	9	23	23	18.43	–61.53	31	4.4	Md	4.3
26	1994	10	12	3	18.24	–68.37	107	4.4	mb	4.3
27	1994	11	17	3	17.92	–68.69	33	4.5	mb	4.3
28	1994	11	17	14	18.62	–68.34	81	4.7	mb	4.7
29	1994	11	30	2	19.53	–64.60	19	4.5	mb	4.5
30	1995	1	1	9	19.22	–69.42	42	4.8	mb	4.9
31	1995	1	18	20	18.89	–70.33	91	5.0	mb	5.1
32	1995	2	1	10	18.19	–68.36	179	4.6	mb	4.7
33	1995	2	16	11	18.88	–64.18	33	4.1	mb	4.3
34	1995	2	16	14	19.48	–65.79	33	4.0	mb	4.3
35	1995	5	4	20	18.89	–64.29	47	4.5	Md	4.3
36	1995	6	20	13	17.88	–62.85	100	4.2	mb	4.3
37	1995	7	9	18	19.62	–67.14	33	4.6	mb	4.1
38	1995	7	26	9	19.19	–64.67	52	4.6	mb	4.5
39	1995	7	28	0	19.57	–69.62	33	4.2	mb	4.3
40	1995	8	13	9	19.41	–69.36	10	4.4	mb	4.3
41	1995	9	19	3	18.79	–62.53	10	4.7	mb	4.1
42	1995	10	8	4	19.01	–66.96	46	4.8	Md	4.4
43	1995	10	31	1	19.70	–69.75	62	4.8	mb	4.2
44	1995	12	9	5	18.89	–65.71	50	4.2	mb	3.8
45	1995	12	31	0	18.44	–64.60	33	4.7	Ml	4.2
46	1996	1	2	18	18.79	–62.70	33	4.5	Ml	3.9
47	1996	1	24	1	18.19	–69.99	50	4.4	mb	4.1
48	1996	3	20	12	19.22	–66.75	33	4.1	mb	3.9
49	1996	4	9	19	18.88	–69.72	88	4.1	mb	4.2
50	1996	5	11	2	19.31	–64.96	35	5.1	Mw	5.1
51	1996	5	12	4	18.48	–63.84	33	4.7	mb	4.4
52	1996	5	14	11	18.94	–65.08	33	4.2	mb	3.9
53	1996	5	29	11	18.06	–69.64	52	4.0	mb	4.1
54	1996	6	11	16	17.25	–68.28	33	5.5	mb	4.7
55	1996	6	12	3	20.03	–70.19	33	4.5	mb	4.3
56	1996	7	21	9	18.30	–62.41	60	4.5	mb	4.2
57	1996	10	17	17	19.02	–69.11	33	4.4	mb	4.3
58	1996	11	1	1	18.60	–64.28	33	4.4	mb	3.8
59	1996	11	6	2	18.85	–64.32	21	5.1	mb	4.7
60	1996	11	8	7	18.04	–68.53	73	4.8	mb	4.5

continued

TABLE 1. EARTHQUAKES USED IN THIS STUDY (*continued*)

	Date			Time	Lat.	Long.	Depth	Catalogue		Moment
	Y	M	D	h			(km)	magnitude		magnitude
61	1996	12	4	19	19.04	–69.26	108	4.2	mb	4.3
62	1997	2	24	4	19.32	–69.23	90	4.5	mb	4.1
63	1997	3	17	5	19.01	–62.84	33	5.5	Mw	5.5
64	1997	4	5	21	19.08	–63.12	33	4.3	mb	4.0
65	1997	5	14	2	19.63	–70.29	54	4.7	mb	4.7
66	1997	7	30	16	18.00	–70.32	10	4.9	mb	4.1
67	1997	10	12	15	18.58	–66.22	100	3.7	Md	3.3
68	1997	10	15	19	18.67	–67.44	10	3.3	Md	3.5
69	1997	10	24	0	17.99	–65.31	5	3.2	Md	3.1
70	1997	11	2	11	19.24	–66.34	33	4.3	mb	4.1
71	1997	11	21	7	18.56	–67.01	100	3.2	Md	3.2
72	1997	12	6	17	18.75	–67.34	80	3.5	Md	3.2
73	1997	12	21	2	18.58	–66.54	80	3.1	Md	3.0
74	1997	12	26	0	18.18	–68.44	94	4.7	mb	4.3
75	1997	12	27	13	17.82	–66.10	5	3.0	Md	2.8
76	1997	12	31	3	17.84	–66.10	10	3.0	Md	2.8
77	1997	12	31	21	18.50	–66.13	100	3.1	Md	3.2
78	1998	1	16	12	18.55	–66.12	106	3.1	Md	3.0
79	1998	1	17	23	18.96	–64.61	72	3.5	Md	3.3
80	1998	1	18	14	19.17	–64.66	25	4.4	mb	4.0
81	1998	1	18	13	18.98	–64.64	52	3.6	Md	3.3
82	1998	1	19	22	18.20	–67.17	13	3.7	Md	3.3
83	1998	1	20	12	19.03	–65.09	97	3.5	Md	3.3
84	1998	2	13	3	19.09	–66.09	27	3.0	Md	3.1
85	1998	2	13	18	18.05	–65.46	14	3.7	Md	3.3
86	1998	2	18	16	18.37	–64.97	24	3.3	Md	3.1
87	1998	2	19	5	19.15	–66.43	37	3.0	Md	3.1
88	1998	3	5	7	20.01	–63.15	33	5.1	Mw	5.1
89	1998	3	10	23	18.04	–65.60	1	3.2	Md	3.0
90	1998	3	12	6	18.14	–66.70	60	3.5	Md	3.3
91	1998	3	19	1	18.86	–66.02	56	3.1	Md	3.1
92	1998	3	25	8	19.38	–67.09	25	4.9	mb	4.4
93	1998	3	27	3	18.66	–64.25	50	3.4	Md	3.4
94	1998	4	1	6	18.15	–67.34	0	3.1	Md	3.1
95	1998	4	10	12	18.15	–66.85	16	3.0	Md	2.9
96	1998	4	15	6	18.18	–64.20	25	3.5	Md	3.3
97	1998	4	15	17	17.92	–65.54	2	3.8	Md	3.5
98	1998	4	16	18	17.78	–65.61	2	3.1	Md	3.1
99	1998	4	16	20	18.02	–65.63	0	3.6	Md	3.0
100	1998	4	16	22	18.07	–65.54	0	3.0	Md	2.9
101	1998	4	17	6	19.08	–67.36	22	3.1	Md	3.2
102	1998	4	18	9	18.65	–67.49	1	3.6	Md	3.3
103	1998	4	20	15	18.67	–66.72	79	3.4	Md	3.1
104	1998	4	25	5	19.00	–67.49	25	3.3	Md	3.2
105	1998	4	26	10	18.22	–67.09	20	3.6	Md	3.1
106	1998	4	26	14	19.07	–66.41	30	3.3	Md	3.0
107	1998	4	29	15	18.16	–65.87	87	3.8	Md	3.5
108	1998	4	30	0	18.71	–65.04	71	3.2	Md	3.2
109	1998	5	12	4	18.04	–65.55	0	3.1	Md	2.8
110	1998	5	14	6	17.95	–64.67	25	3.6	Md	3.2
111	1998	5	15	21	19.14	–66.49	49	3.5	Md	3.3
112	1998	5	22	4	19.22	–66.70	36	3.8	Md	3.4
113	1998	5	24	5	19.11	–67.17	25	3.6	Md	3.2
114	1998	5	27	19	19.23	–66.65	24	3.7	Md	3.4
115	1998	5	29	15	19.66	–66.95	10	3.6	Md	3.5
116	1998	6	8	11	18.19	–66.54	66	3.6	Md	3.2
117	1998	6	13	4	17.75	–64.14	30	3.6	Md	3.3
118	1998	6	13	9	19.48	–66.36	48	3.9	Md	3.4
119	1998	6	14	13	18.62	–65.27	48	3.9	Md	3.5
120	1998	6	18	22	18.08	–65.53	0	3.3	Md	2.9

continued

TABLE 1. EARTHQUAKES USED IN THIS STUDY (*continued*)

	Date Y	M	D	Time h	Lat.	Long.	Depth (km)	Catalogue magnitude		Moment magnitude
121	1998	6	21	2	18.97	–64.30	58	3.6	Md	3.4
122	1998	6	23	19	19.42	–65.28	49	3.8	Md	3.5
123	1998	6	24	22	17.74	–66.33	17	3.6	Md	3.2
124	1998	6	25	4	18.68	–65.99	68	3.1	Md	3.1
125	1998	7	4	23	18.32	–65.95	135	3.7	Md	3.1
126	1998	7	5	7	18.85	–67.23	18	3.0	Md	3.0
127	1998	7	19	9	18.30	–65.10	131	4.6	mb	4.2
128	1998	7	25	22	19.12	–66.11	15	3.1	Md	3.2
129	1998	7	27	1	18.71	–66.51	25	3.1	Md	3.1
130	1998	7	28	8	18.69	–65.00	25	3.3	Md	3.2
131	1998	8	2	2	18.84	–67.27	13	3.1	Md	3.1
132	1998	8	3	21	18.61	–67.44	12	3.3	Md	3.2
133	1998	8	4	15	19.23	–64.66	60	3.8	Md	3.8
134	1998	8	4	21	19.13	–64.51	56	3.6	Md	3.3
135	1998	8	5	23	19.23	–64.66	60	3.7	Md	3.7
136	1998	8	8	7	18.01	–66.61	13	3.0	Md	2.8
137	1998	8	9	11	19.74	–70.00	33	4.5	mb	4.0
138	1998	8	9	21	18.97	–64.93	22	3.5	Md	3.2
139	1998	8	9	23	18.85	–64.55	95	3.6	Md	3.4
140	1998	8	10	21	18.65	–70.54	58	5.2	Mw	5.2
141	1998	8	10	0	19.16	–64.77	72	3.7	Md	3.5
142	1998	8	10	8	19.30	–64.74	25	3.6	Md	3.3
143	1998	8	11	6	19.21	–66.14	25	3.0	Md	3.0
144	1998	8	26	10	18.75	–65.99	69	3.6	Md	3.3
145	1998	8	30	20	18.69	–70.27	33	4.2	mb	4.0
146	1998	9	2	8	17.96	–66.34	9	3.3	Md	2.8
147	1998	9	3	15	17.94	–66.33	8	3.2	Md	2.7
148	1998	9	12	4	17.94	–66.32	6	3.2	Md	2.7
149	1998	10	10	2	18.25	–66.29	6	3.2	Md	2.6
150	1998	10	15	21	18.49	–70.47	68	4.4	mb	4.4
151	1998	10	15	4	18.86	–65.16	53	3.6	Md	3.2
152	1998	10	18	0	17.96	–65.68	5	3.2	Md	2.9
153	1998	10	22	7	18.92	–65.14	59	3.6	Md	3.4
154	1998	10	23	18	18.87	–64.38	33	4.0	Md	3.7
155	1998	10	23	19	19.50	–64.55	65	3.8	Md	3.9
156	1998	10	24	1	18.86	–64.32	56	4.6	mb	4.2
157	1998	10	24	14	18.62	–66.73	19	3.4	Md	3.0
158	1998	10	29	4	18.20	–65.94	14	3.0	Md	2.7
159	1998	11	1	21	19.04	–65.23	45	3.6	Md	3.3
160	1998	11	1	22	19.14	–65.13	67	3.6	Md	3.4
161	1998	11	3	19	18.84	–65.07	25	3.4	Md	3.1
162	1998	11	4	2	18.75	–65.93	73	3.1	Md	3.0
163	1998	11	9	5	19.30	–65.25	75	3.7	Md	3.4
164	1998	11	11	8	18.24	–67.04	17	3.9	Md	3.4
165	1998	11	15	0	18.82	–66.25	35	3.1	Md	3.0
166	1998	11	16	12	18.90	–67.38	58	3.1	Md	3.1
167	1998	11	21	1	18.84	–65.23	72	3.1	Md	3.2
168	1998	11	23	7	18.72	–66.12	25	3.1	Md	2.9
169	1998	11	24	6	19.14	–67.78	33	4.4	mb	4.1
170	1999	1	18	19	18.86	–67.22	33	5.0	Mw	5.0
171	1999	1	25	10	16.89	–62.50	140.3	4.6	mb	4.0
172	1999	1	25	17	19.50	–66.69	50.1	4.6	mb	4.0
173	1999	1	27	12	18.94	–63.26	33	4.1	mb	4.2
174	1999	4	20	5	18.58	–65.37	91.8	4.2	mb	3.6
175	1999	8	5	15	18.88	–67.18	71.6	4.3	mb	4.0
176	1999	8	7	12	18.76	–66.86	63.4	4.5	mb	4.1
177	1999	10	28	12	18.72	–67.25	33	4.1	mb	3.8
178	1999	12	20	10	17.31	–61.71	58.8	5.4	mb	5.0
179	2000	1	6	11	18.26	–68.32	178	4.0	Md	3.8
180	2000	1	9	2	17.94	–67.00	28	3.1	Md	3.4

continued

TABLE 1. EARTHQUAKES USED IN THIS STUDY (*continued*)

	Date			Time	Lat.	Long.	Depth	Catalogue		Moment
	Y	M	D	h			(km)	magnitude		magnitude
181	2000	1	9	5	18.77	–67.25	68	3.2	Md	3.2
182	2000	1	13	18	18.97	–66.58	45	3.1	Md	3.0
183	2000	1	14	6	18.14	–67.91	131	3.4	Md	3.4
184	2000	1	18	11	17.93	–65.66	4	3.1	Md	2.9
185	2000	1	18	23	18.93	–68.10	10	3.8	Md	3.4
186	2000	1	22	12	18.63	–66.88	12	3.0	Md	2.8
187	2000	1	25	14	19.61	–68.06	69	4.1	Md	3.7
188	2000	1	26	15	19.00	–67.76	34	3.0	Md	3.2
189	2000	1	29	3	18.89	–64.30	50	4.4	mb	4.1
190	2000	2	2	9	18.28	–66.23	74	3.0	Md	2.9
191	2000	2	3	11	19.17	–66.52	9	3.6	Md	3.2
192	2000	2	4	0	19.48	–68.05	15	3.7	Md	3.4
193	2000	2	4	23	17.86	–67.04	28	3.2	Md	3.1
194	2000	2	6	21	17.99	–65.55	4	3.1	Md	2.9
195	2000	2	9	4	18.69	–67.52	2	3.5	Md	3.3
196	2000	2	10	4	18.07	–65.86	7	3.8	Md	3.3
197	2000	2	21	16	18.32	–67.88	108	4.4	mb	4.0
198	2000	2	26	12	18.93	–65.90	76	3.9	Md	3.3
199	2000	2	27	23	17.83	–60.99	16	5.0	Md	4.6
200	2000	3	2	14	18.89	–65.18	42	3.7	Md	3.3
201	2000	3	5	6	18.91	–66.85	10	3.2	Md	2.9
202	2000	3	5	16	19.04	–66.86	40	3.3	Md	2.9
203	2000	3	5	20	18.95	–66.91	46	3.0	Md	3.0
204	2000	3	6	1	18.89	–66.40	46	3.0	Md	2.9
205	2000	3	6	2	18.13	–66.92	13	3.2	Md	3.1
206	2000	3	8	2	18.08	–67.11	18	3.0	Md	2.9
207	2000	3	8	9	19.16	–66.48	41	3.6	Md	3.1
208	2000	3	9	0	17.83	–65.66	8	3.4	Md	2.9
209	2000	3	14	18	18.01	–67.07	5	3.5	Md	3.1
210	2000	3	15	9	19.72	–66.04	72	3.9	Md	3.5
211	2000	3	20	15	18.41	–66.62	119	3.9	Md	3.7
212	2000	3	22	2	18.52	–67.56	9	3.0	Md	3.2
213	2000	3	25	7	19.58	–68.15	25	3.8	Md	3.4
214	2000	3	25	13	19.26	–67.55	14	3.4	Md	3.3
215	2000	3	26	3	18.61	–67.70	33	4.5	Md	4.0
216	2000	3	30	9	18.82	–68.01	46	4.3	Md	3.8
217	2000	4	4	23	18.97	–67.31	4	3.3	Md	3.1
218	2000	4	7	2	18.27	–67.58	6	3.3	Md	3.0
219	2000	4	8	7	18.16	–67.36	18	3.4	Md	3.2
220	2000	4	10	11	18.95	–64.23	84	4.0	Md	3.6
221	2000	4	10	22	18.67	–66.80	25	3.8	Md	3.0
222	2000	4	13	18	19.64	–63.29	33	4.2	mb	3.9
223	2000	4	18	2	19.08	–69.52	89	4.5	mb	3.9
224	2000	4	30	3	18.46	–67.97	22	3.6	Md	3.4
225	2000	5	3	1	17.67	–62.86	200	3.6	Md	3.6
226	2000	5	4	2	17.80	–65.62	3	3.3	Md	3.1
227	2000	5	8	5	19.45	–68.50	33	4.5	mb	3.9
228	2000	5	15	5	18.24	–66.35	35	3.5	Md	2.9
229	2000	5	20	22	19.38	–67.26	56	3.8	Md	3.5
230	2000	5	26	9	19.04	–63.88	10	3.8	Md	3.6
231	2000	6	1	21	19.42	–65.42	49	4.6	Md	4.3
232	2000	6	11	20	19.91	–70.85	33	4.4	Md	4.3
233	2000	6	18	10	17.93	–66.84	16	3.1	Md	3.1
234	2000	6	23	8	19.41	–65.19	76	3.6	Md	3.6
235	2000	6	24	2	18.08	–65.73	3	3.2	Md	3.0
236	2000	6	25	10	19.39	–65.11	66	3.8	Md	3.3
237	2000	7	5	12	18.40	–65.89	34	3.5	Md	3.0
238	2000	7	7	14	18.93	–66.76	53	3.3	Md	3.0
239	2000	7	24	15	18.15	–67.01	21	3.3	Md	3.0
240	2000	7	26	2	17.85	–61.51	33	4.0	mb	3.9

continued

TABLE 1. EARTHQUAKES USED IN THIS STUDY (*continued*)

	Date			Time	Lat.	Long.	Depth	Catalogue		Moment
	Y	M	D	h			(km)	magnitude		magnitude
241	2000	7	28	9	18.70	–64.29	71	4.1	Md	3.6
242	2000	7	28	15	18.93	–64.68	62	4.4	mb	4.2
243	2000	7	29	7	17.84	–68.70	33	4.4	mb	4.3
244	2000	7	30	18	18.79	–69.42	87	3.9	Md	3.7
245	2000	7	31	11	18.77	–64.65	25	4.0	Md	3.3
246	2000	8	1	23	17.46	–68.24	117	4.2	Md	3.7
247	2000	8	2	1	18.39	–66.65	13	3.6	Md	2.9
248	2000	8	12	9	18.03	–66.60	15	3.6	Md	2.9
249	2000	8	15	9	19.16	–66.84	57	3.3	Md	3.2
250	2000	8	16	8	18.40	–65.91	105	3.2	Md	3.1
251	2000	8	17	6	18.11	–64.74	60	3.3	Md	3.2
252	2000	8	19	0	18.06	–65.51	0	3.6	Md	3.0
253	2000	8	19	8	17.92	–66.94	3	3.8	Md	3.3
254	2000	8	31	1	18.75	–64.00	74	4.1	Md	3.5
255	2000	9	3	12	18.70	–66.73	12	3.5	Md	2.7
256	2000	9	4	6	19.27	–65.99	107	3.5	Md	3.2
257	2000	9	4	15	19.05	–66.32	43	3.6	Md	3.4
258	2000	9	5	20	19.00	–68.09	113	4.1	Md	3.4
259	2000	9	7	0	18.87	–65.26	35	3.4	Md	3.1
260	2000	9	12	16	17.97	–65.93	5	3.6	Md	2.8
261	2000	9	15	6	18.87	–67.47	32	3.4	Md	3.3
262	2000	9	18	10	20.06	–70.07	33	4.4	mb	4.3
263	2000	9	22	23	19.02	–64.93	25	4.0	Ml	3.4
264	2000	9	23	22	19.26	–64.15	21	4.0	Md	3.3
265	2000	9	24	7	19.47	–66.07	64	3.7	Md	3.2
266	2000	9	25	5	19.23	–62.56	33	4.7	Md	4.2
267	2000	9	25	6	19.19	–62.52	33	4.9	mb	4.0
268	2000	9	25	23	18.12	–67.29	18	3.3	Md	3.0
269	2000	9	27	7	19.11	–64.79	50	3.7	Md	3.2
270	2000	9	28	6	19.26	–66.68	48	3.8	Md	3.2
271	2000	9	29	19	19.36	–66.27	27	3.7	Md	3.1
272	2000	9	30	2	19.42	–65.12	28	3.9	Md	3.5
273	2000	10	1	19	18.88	–64.66	25	3.5	Md	3.2
274	2000	10	2	15	18.90	–68.31	41	4.0	Md	3.7
275	2000	10	2	17	18.17	–67.41	7	3.7	Md	3.4
276	2000	10	8	11	18.78	–64.26	37	3.7	Md	3.2
277	2000	10	9	5	19.19	–65.40	90	3.9	Md	3.5
278	2000	10	11	1	18.93	–64.41	25	3.7	Md	3.2
279	2000	10	11	4	18.98	–66.60	68	3.1	Md	3.0
280	2000	10	13	0	18.64	–66.47	71	3.3	Md	3.1
281	2000	10	14	2	19.04	–67.42	30	3.2	Md	3.1
282	2000	10	15	0	18.55	–65.11	20	3.6	Md	3.0
283	2000	10	16	12	19.21	–64.48	45	4.0	Md	3.5
284	2000	10	16	13	19.25	–65.85	33	3.7	Md	3.0
285	2000	10	17	3	18.81	–67.34	6	3.5	Md	3.1
286	2000	10	17	8	19.65	–68.68	13	4.0	Md	3.6
287	2000	10	22	3	19.03	–64.96	68	3.8	Md	3.2
288	2000	10	23	5	18.79	–65.03	53	3.8	Md	3.2
289	2000	10	25	16	17.85	–66.91	41	3.8	Md	3.1
290	2000	10	25	21	17.86	–66.91	39	3.9	Md	3.4
291	2000	10	27	18	17.65	–61.17	42	4.7	Md	4.5
292	2000	10	27	19	17.60	–61.19	37	5.6	Md	5.2
293	2000	10	29	3	17.98	–66.94	16	3.6	Md	3.2
294	2000	10	30	14	18.71	–67.34	21	3.7	Md	3.4
295	2000	11	1	10	18.44	–67.05	102	3.4	Md	3.2
296	2000	11	6	3	18.00	–68.81	94	3.6	Md	3.5
297	2000	11	16	19	19.84	–65.13	24	4.0	Md	3.9
298	2000	11	25	11	17.78	–65.97	46	3.2	Md	2.8
299	2000	11	30	16	18.63	–66.71	85	4.1	Md	3.8
300	2000	12	1	11	18.56	–66.68	86	3.1	Md	3.0
301	2000	12	1	15	19.30	–67.95	55	4.3	Md	4.1
302	2000	12	8	8	17.86	–68.55	35	3.7	Md	3.5

Note: mb—body wave magnitude; Md—duration magnitude; Ml—local magnitude; Mw—moment magnitude.

the signal to noise ratio was at least a factor of two. For each record, the time series were processed as follows:

- Define shear wave window (S-window) and pre-event noise (a minimum 10 s noise window was used for calculation of signal to noise ratio);
- Remove any glitches;
- Taper the windowed time series using a 5% taper on each end of the signal;
- Zero-Pad the time series to the next greatest power of 2;
- Transfer to frequency domain by Fast Fourier Transform;
- Calculate instrument response based on the poles, zeros, and constant for that specified component of the seismographic station;
- Remove the instrument response in the frequency domain by dividing the recorded spectrum by the complex instrument transfer function;
- Convert Fourier spectra of velocity to Fourier spectra of acceleration (done in the frequency domain in the same step as removal of instrument response);
- Transfer to time domain by applying the inverse Fourier Transform;
- Calculate response spectra for 5% damping from corrected acceleration time series;
- Discard the frequencies with a signal to noise ratio <2 (Note: the noise window is normalized to the same length as the signal window to check the signal/noise ratio);
- Smooth the Fourier spectra by using a weighted 9-point smoothing algorithm.

DETERMINATION OF REGIONAL ATTENUATION PARAMETERS

The key to the stochastic simulation model is the specification of the Fourier acceleration spectrum as a function of magnitude and distance. In order to specify the Fourier spectrum, we need the parameters that prescribe the earthquake source level, attenuation of ground motion, and site response. Previous studies have determined that the overall source parameters for stochastic finite fault modeling are generic and do not vary significantly by region (Atkinson and Boore, 1998; Beresnev and Atkinson, 2002). Thus, the basic model is already well established by the calibration of EXSIM to the California strong motion database. The necessary regional attenuation parameters are the geometric spreading and the coefficient of anelastic attenuation. Other region-specific parameters describe amplification through the crustal velocity profile, including any regional generic soil responses. We also need to know the duration of shaking.

In this section, we determine the duration of seismic waves and the geometric and anelastic attenuation parameters, all based on the recorded ground-motion data in Puerto Rico. This sets the stage for stochastic simulations to provide regional ground-motion relations, satisfying the immediate and most significant needs of the engineering community for seismic design applications.

GROUND-MOTION ATTENUATION MODEL

The frequency-dependent attenuation of Fourier amplitudes can be determined from ground-motion data by regression analysis, as described by Atkinson and Mereu (1992). These regressions may also be used to determine source amplitudes for specific events, and relative site terms for all stations.

Fourier amplitudes can be fit to an equation of the form:

$$\log A_{ij}(f, R) = \log A_{i0}(f) - b \log R_{ij} - c(f) R_{ij} + \log S_j(f), \qquad (3)$$

where $A_{ij}(f)$ is the observed spectral amplitude of earthquake i at station j, for frequency f; $A_{i0}(f)$ is the source amplitude of earthquake i; R is hypocentral distance; b is the geometric spreading coefficient; $c(f)$ is the coefficient of anelastic attenuation; and $S_j(f)$ is the site response term for station j (Shin and Herrmann, 1987; Atkinson and Mereu, 1992). The anelastic coefficient, $c(f)$, is inversely related to the quality factor, Q:

$$Q = [\log(e)\, \pi f]/(c\, \beta), \qquad (4)$$

where β is the shear-wave velocity and $e = 2.718$. Equation (3) may be applied to any ground-motion phase; the shear-wave phases are of most engineering interest, since they have the largest amplitudes (typically about five times larger than the P-wave amplitudes).

In developing the attenuation model, we must determine whether the ground-motion database can be adequately fit with a single geometric spreading rate b, or whether this coefficient will take on different values depending on the epicentral distance range. For example, in eastern North America the attenuation is described by a "hinged-trilinear" form (Atkinson and Boore, 1995) in which geometric attenuation goes as R^{-1} to 70 km, then as R^{0} for 70–130 km, then $R^{-0.5}$ beyond 130 km. This behavior is explained by the presence of strong post-critical reflections from the Moho discontinuity that cause "flattening" of the attenuation curve, leading to almost no apparent geometric spreading between ~70 and 130 km. It follows that the details of the empirical attenuation provide valuable clues as to the regional crustal structure and the presence of major velocity discontinuities that affect ground-motion amplitudes.

In order to find the hinge points of the attenuation curve, representing the transition from direct S-wave to post-critical Moho reflection, we applied a trial profile of different hinge points to Equation (3) using the ground-motion database to determine the best-fit model. The first hinge point is incremented from 40 to 105 km by 5 km steps. The second hinge, representing the transition to surface-wave spreading, is incremented from 40 to 150 km with the same step distance. For each pair of hinge points, the maximum likelihood method (Joyner and Boore, 1993) has been applied to Equation (3) to determine the value of the coefficients. Next, the residuals for all records, for frequencies from 0.1 to 8 Hz, have been calculated. [Note: residual = (log of observed A_{ij}) − (log of predicted A_{ij}) by Equation (3)]. Hinge points at 75 and

100 km produce the lowest averaged residuals. As Figure 2 shows, with hinge points at 75 and 100 km, the regression residuals have a good distribution versus distance, with no trend. Thus, we adopt a geometric spreading model of R^{-1}, R^{0}, $R^{-0.5}$ for the simulations, with hinge points at 75 and 100 km.

Q-VALUE

Equation (3) has been solved by the maximum likelihood method (Joyner and Boore, 1993) to obtain the geometric spreading coefficient b and the frequency-dependent attenuation c, providing a description of regional path effects. In this section, we are principally interested in the frequency-dependent attenuation coefficient c, which is related to the regional Q-value through Equation (4). Figure 3 shows the determined Q-values for shallow (h ≤ 30 km) and deep (h > 30 km) earthquakes in Puerto Rico. The standard error of the mean Q-values spans a wider range than the difference between Q-values for the deep and shallow earthquake subsets. Thus, we can fit both deep and shallow earthquakes using a single Q model: when the data are combined, the Q-values are described by the equation

$$Q = 359 f^{0.59}, \tag{5}$$

as shown in Figure 3. Figure 3 also compares the Q-values obtained for Puerto Rico to Q-values for eastern North America (Atkinson and Boore, 1995) and California (Raoof et al., 1999). The Puerto Rico Q-values are intermediate to those for eastern North America and California.

DURATION OF GROUND MOTION

The duration (T) of an earthquake signal at hypocentral distance R can generally be represented as (Atkinson and Boore, 1995):

$$T(R) = T_0 + dR,$$

where T_0 is the source duration and d is the coefficient controlling the increase of duration with distance; d is derived empirically. The value d can be a single coefficient describing all distances of interest (e.g., Atkinson, 1995), or it can take different values depending on the distance range (e.g., Atkinson and Boore, 1995).

Using the Puerto Rico ground-motion data, we developed an empirical model of the distance-dependence of duration. For each record, the measured duration, including the strong part of the shaking, was assumed to be represented by $T(R) = T_0 + dR$. The source duration (T_0) is short (<1 s) for these records due to their small magnitudes and can thus be neglected. Figure 4 compares the duration behavior with distance for Puerto Rico with that of eastern North America (using the eastern North America data of Atkinson and Boore 1995, which are defined using a root mean square duration definition consistent with the stochastic model). In Figure 4, the Y-axis is $T(R) - T_0$, where the source duration is estimated as $T_0 = 1/2 f_A$, with $\log(f_A) = 2.41 - 0.533$ **M** (Atkinson and Boore, 1995). The change of slope at ~75 and ~100 km again suggests a trilinear attenuation curve for

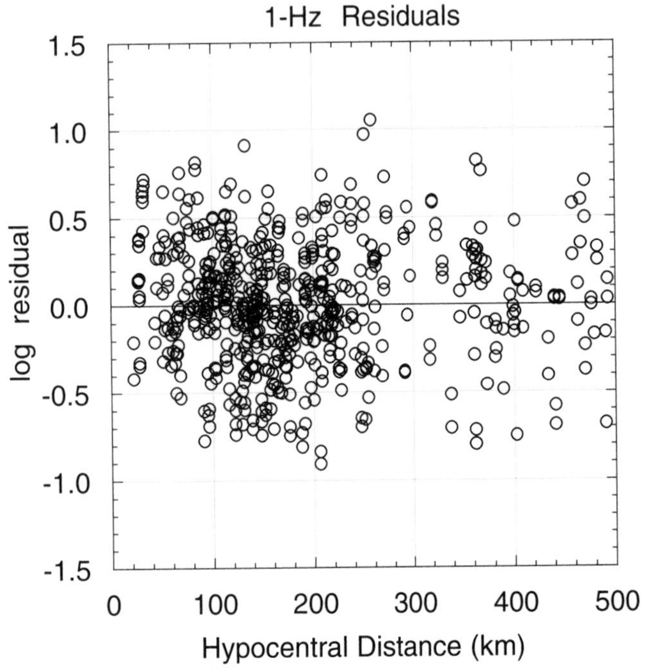

Figure 2. Distribution of residuals versus distance for hinge points 75 and 100 km and f (frequency) = 1.0 Hz.

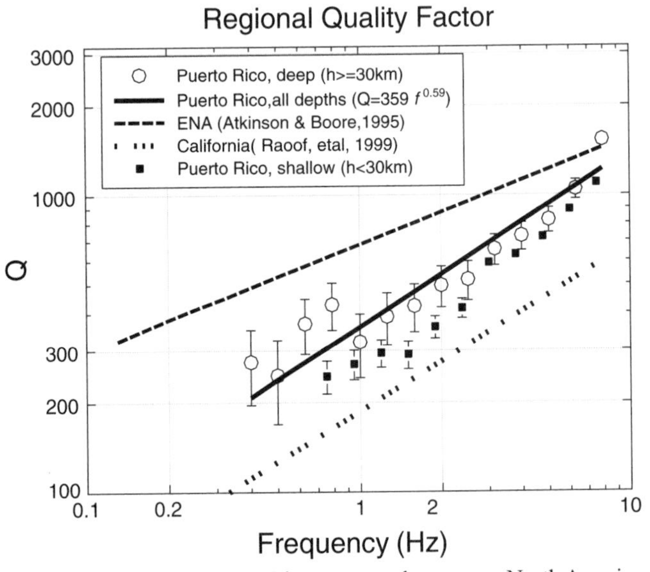

Figure 3. Q-value for Puerto Rico compared to eastern North America (Atkinson and Boore, 1995) and California (Raoof et al., 1999). Error bars show standard error of Q. The differences between Q for deep and shallow events in Puerto Rico are not significant. ENA—eastern North America.

the Puerto Rico region, similar to observations for eastern North America (Atkinson and Boore, 1995). The larger amount of scatter in the Puerto Rico data in Figure 4, compared to eastern North America data, is due to several factors. The first factor could be the greater variability of site conditions in Puerto Rico. Soil sites cause more scattered duration data than rock sites. Differences in the wave propagation behavior in the complex tectonic regime of Puerto Rico, especially for deeper events, could be another reason for the large amount of scatter. In addition in Puerto Rico there is a mixture of different types of seismic sources, including in-slab, interface, and crustal earthquakes. Finally, the durations for Puerto Rico were measured subjectively from the records, while those for eastern North America were calculated from an analytical duration definition. Based on Figure 4, we adopt the eastern North America duration model of Atkinson and Boore (1995) with the obtained hinge points for Puerto Rico (75 and 100 km) for the simulations.

H/V RATIO (GENERIC REGIONAL AMPLIFICATION EFFECTS)

The ratio of the horizontal to vertical component of ground motion (H/V) is an important parameter since most of the seismographic stations in Puerto Rico record only the vertical component, whereas the horizontal component is of primary engineering interest. The H/V ratio is believed to reflect amplification effects, both from the crustal velocity gradient and from near-surface soils (Nakamura, 1989; Atkinson and Cassidy, 2000; Lermo and Chavez-Garcia, 1993). Earthquake waves propagate from the source region, where the shear wave velocity is typically ~3.6 km/s, toward the surface, where the average shear wave velocity may be around 620 m/s for generic soft rock sites (Boore and Joyner, 1997). The spectrum will be amplified through this velocity gradient by seismic impedance effects. Amplification effects are observed strongly on the horizontal component but only weakly on the vertical component (due to the turning of rays toward the vertical by refraction). Hence, the H/V ratio is a rough measure of site amplification.

Most of the PRSN data collected to date are vertical-component recordings. However, there are also a limited number of three-component data available. We used the three-component data to establish an empirical relationship between horizontal and vertical component amplitudes. This makes it possible to use the entire PRSN database in the development of regional ground-motion relations.

Figure 5 compares H/V ratios for the IRIS 3-component broadband station in Puerto Rico (SJG) with the observed H/V ratios for various U.S. National Earthquake Hazards Reduction Program (NEHRP) site classes in other regions. The NEHRP site classes describe the soil conditions by the average shear-wave velocity (v_s) over the top 30 m. NEHRP A = v_s > 1500 m/s; NEHRP B = 760–1500 m/s; NEHRP C = 360–760 m/s; NEHRP D = 180–360 m/s; and NEHRP E = <180 m/s (Borcherdt, 1994). Information on H/V ratios for these other regions is given in Chen (2000) and Chen and Atkinson (2002). We make the assumption

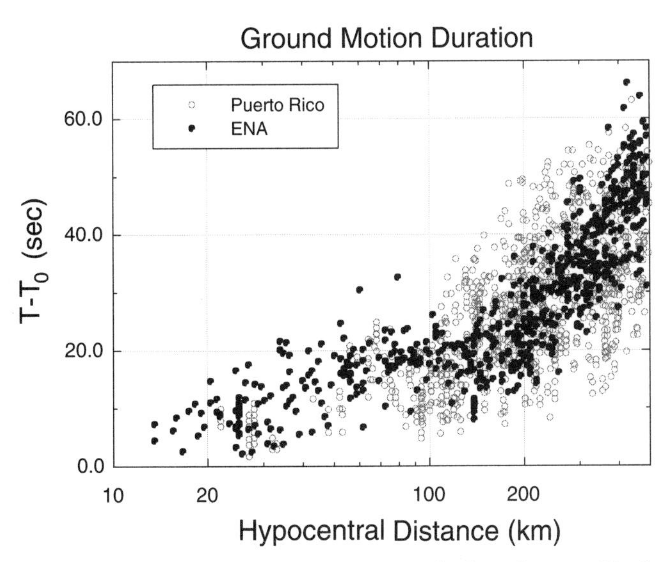

Figure 4. Comparison between duration distribution of eastern North America (ENA; Atkinson and Boore 1995) and Puerto Rico. Duration measure is total duration minus source duration, where source duration is small (<1 s) for these events. The change of slope at around 75 km and 100 km suggests a trilinear attenuation curve for the Puerto Rico region, similar to observations for eastern North America.

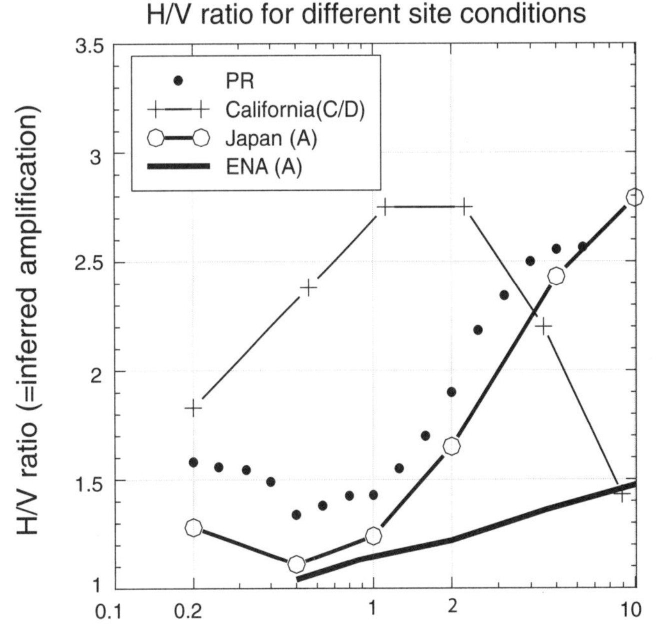

Figure 5. Horizontal/vertical (H/V) ratio for Puerto Rico (site SJG) in comparison to that for other regions. Letters (A, B, C, and D) give U.S. National Earthquake Hazards Reduction Program site classes.

that the observed H/V ratio is an estimate of the amplification of the horizontal component due to regional site effects. The observed H/V at SJG suggests an amplification at high frequencies, which probably means that there is a thin soft layer under the site (e.g., the frequency of amplification is inversely proportional to the thickness of the amplifying layer, while the degree of amplification depends on the seismic impedance contrast). The trend of H/V at the SJG station in Puerto Rico looks very much like the trend of the H/V ratios for shallow soil sites in Japan that may be classified as NEHRP B or C. For the 45 stations in the Japan data set that are so classified (by Chen, 2000), the average shear wave velocity over the top 30 m is ~738 m/s (NEHRP C), and the average layer thickness (over rock) is ~11 m (Chen and Atkinson, 2002).

For simulation of ground motions in Puerto Rico, the obtained H/V at SJG will be applied as a generic site amplification for all sites. Based on the observations made at sites, all appear to be similar in terms of site conditions. They can be characterized as soft rock sites and likely have a weathered layer that overlies harder rock. They are probably NEHRP B or NEHRP C in terms of shear-wave velocity. (Note: an improvement on this assumption will be possible in future applications, as studies being conducted by J. Martinzez-Cruzado at the University of Puerto Rico, Mayagüez, are establishing the shear-wave velocity profiles of sites across Puerto Rico.)

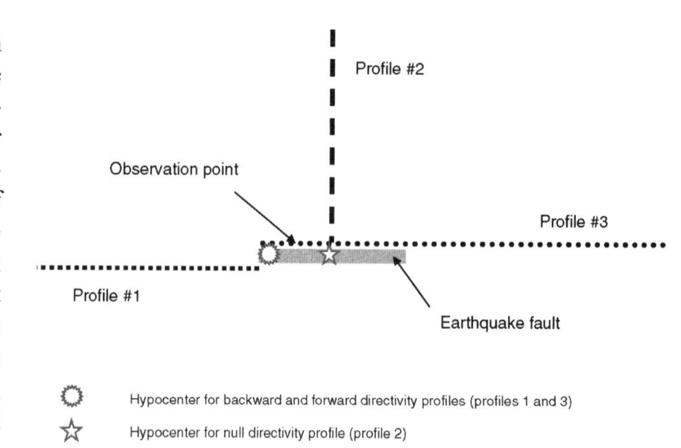

Figure 6. Profile of observation points around the earthquake fault used in simulations. See text for further explanation.

SIMULATION OF ACCELERATION TIME SERIES

Acceleration time series for earthquakes with magnitudes from **M**3.0 to **M**8.0 (where **M** is moment magnitude) and fault distances from 2 km to 500 km are simulated. Simulation is performed in magnitude steps of 0.2 units. For each simulated earthquake fault plane, we perform simulations for 25 different locations of the observation point. In order to have a uniform distribution of distances in log units, the distances are 2.2, 3.1, 3.8, 4.4, 5.4, 7, 9, 11, 16, 21, 27, 33, 42, 52, 66, 82, 103, 129, 162, 203, 254, 318, 399, and 500 km, where distance is defined as the shortest distance from the fault to the observation point. For small earthquakes, the fault distance is equivalent to the hypocentral distance.

Directivity effects, which depend on the azimuth from the fault rupture direction to the observation point, are important for larger earthquakes (Somerville et al., 1997). A complete database should provide data for both backward and forward directivity cases as well as data for azimuths with minimal directivity effects. In order to provide a good simulated database, three profiles for the above-mentioned distances have been considered and are shown in Figure 6. For the backward directivity profile (#1), the hypocenter is located at the nearest subfault. For the null-directivity profile (#2), the hypocenter is located at the middle of the fault plane. For the forward-directivity profile (#3), the hypocenter is located at the farthest subfault to maximize directivity.

There is little information about the specific fault geometry to be modeled, as Puerto Rico is affected not only by the subduction zone (~200 km away), but also by many other local faults of unknown location and orientation. Therefore, the dip has been considered to be 90°. The length and width of faults are calculated from the target moment magnitude based on the Wells and Coppersmith (1994) empirical formulas between magnitude and fault size. The sampling rate, near-source shear-wave velocity, and density have been considered as 100 samples/s, 3.6 km/s, and 2.8 g/cm^3, respectively. The attenuation, regional crustal and/or site amplification, and duration models are as given in the previous section (e.g., determined from the empirical data). Based on the calibrated generic source parameters for EXSIM, the source model corresponds to an input stress parameter given by $\Delta\sigma$ = 130 bars, with a maximum of 50% of the fault actively pulsing at any moment in time. This is the EXSIM source model that reproduces the California strong-motion database used by Atkinson and Silva (2000) in their stochastic ground-motion model development. A random slip distribution is assumed. High frequency amplitudes are reduced through the kappa operator, which models near surface attenuation, by applying the factor $\exp(-\pi f \kappa)$ (Anderson and Hough, 1984). Kappa generally varies between 0.02 and 0.04 for soft rock sites (Anderson and Hough, 1984; Boore et al., 1992; Atkinson and Silva, 1997; Boore and Joyner, 1997); κ = 0.03 is adopted for Puerto Rico as a compromise between regional estimates.

We simulated 1950 acceleration time series for magnitudes from **M**3.0 to **M**8.0 and distances from 2 km to 500 km. The response spectra (pseudoacceleration [PSA], 5% damped) of the simulated time series have been calculated for frequencies from 0.1 Hz to 20 Hz. Response spectra show the response of a simple oscillator to the time series and are used in many engineering applications. Figures 7 and 8 show the PSA of the simulated acceleration time series at 0.5 Hz and 5 Hz versus distance for different magnitudes. Each of the three directivity cases are plotted, which results in the observed scatter for each magnitude.

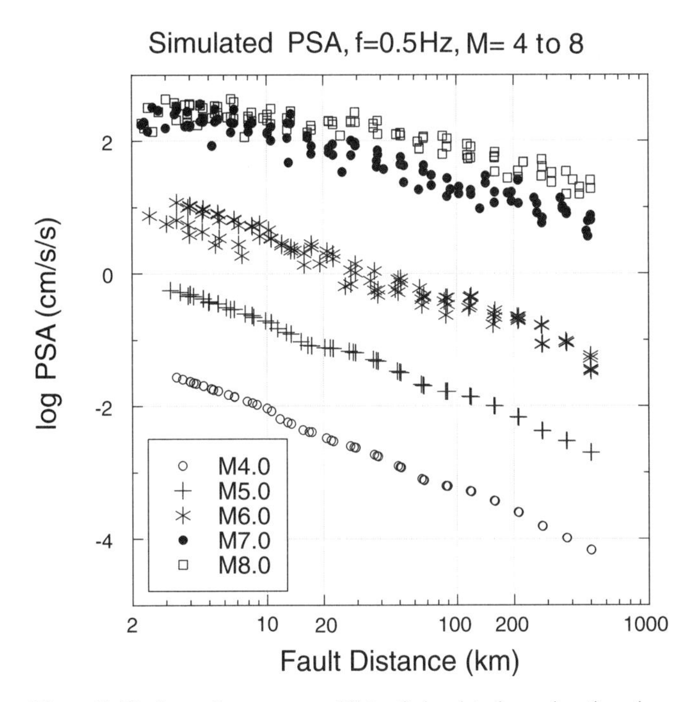

Figure 7. Horizontal-component PSA of simulated acceleration time series at 0.5 Hz versus distance for different magnitudes. Each of three directivity cases is plotted.

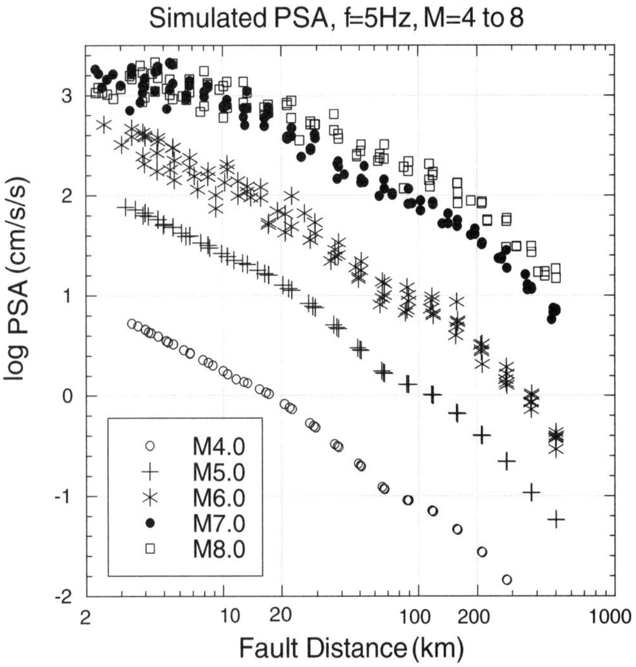

Figure 8. Horizontal-component PSA of simulated acceleration time series at 5 Hz versus distance for different magnitudes. Each of the three directivity cases is plotted.

We note that this scatter does not represent all contributions to random variability in ground-motion amplitudes, as we have not modeled variability in source parameters or attenuation; the scatter shows directivity effects only. The obtained PSA values show a well-behaved trend versus distance.

COMPARISON OF SIMULATED ACCELERATION TIME SERIES WITH DATA

In this section, we compare response spectra calculated from simulated records for Puerto Rico with those calculated from actual seismographic records. For a proper comparison, we need to have the same type of magnitude for both the simulated time series and the recorded data. The simulated time series are based on moment magnitude, **M**. The reported magnitudes for the real records include **M**, mb, and local magnitudes based on duration. Thus, we first need to obtain the moment magnitudes for the study earthquakes in Puerto Rico.

Based on the Brune point source model, which is appropriate for the small earthquakes of our empirical database, the spectra of the recorded acceleration time series can be modeled using Equation (1). The spectra of displacement at low frequencies ($f \ll f_o$) at a distance of R = 10 km would be $D(f) = C\mathbf{M}_0$ [using R = 10 km in the constant C; see Equation (1)]. If we play back the attenuation effects (including anelastic and trilinear hinged geometric behaviors) to R = 10 km, then $\mathbf{M}_0$ and hence **M** can be calculated. These calculated moment magnitudes are subjected to significant uncertainty, as they are mostly single-station estimates, based on the records at SJG.

Figures 9–12 show some examples of the distribution of simulated PSA in comparison to PSA from the recorded data. The good agreement between the simulations and records at small to moderate magnitudes, at both low and high frequencies, gives us confidence in our simulation model parameters for Puerto Rico.

GROUND-MOTION RELATIONS

Regression analysis is used to obtain ground-motion relations that describe the simulated response spectra as simple functions of magnitude and distance. The maximum likelihood method (Joyner and Boore, 1993) has been applied for frequencies from 0.1 Hz to 20 Hz. The ground-motion relations are for a generic soft rock site condition (assumed NEHRP B or C). The response spectra are fit to:

$$\log PSA\,(f, R) = c_1 + c_2(\mathbf{M} - b) + c_3\,(\mathbf{M} - b)^n + hingeFunction + c_4(f)R, \qquad (6)$$

where PSA is the random horizontal-component acceleration response spectra with 5% damping; **M** = moment magnitude; $R = (D^2 + \Delta^2)^{0.5}$; D = closest distance to fault surface, in km; $\Delta = c_5 + c_6\,\mathbf{M}$; $c_4(f)$ = the coefficient of anelastic attenuation; $hingeFunction = (c_7 + c_8\,\mathbf{M})\,\log(R)$ for $R \le 75$ km; $hingeFunction = (c_7 + c_8$

M = 3 to 4, f = 0.5 Hz
Simulated (M=3, 3.5, 4)
Recorded M3-4
log PSA (cm/s/s)
Fault Distance (km)
0
-1
-2
-3
-4
2
10
20
100
200
1000

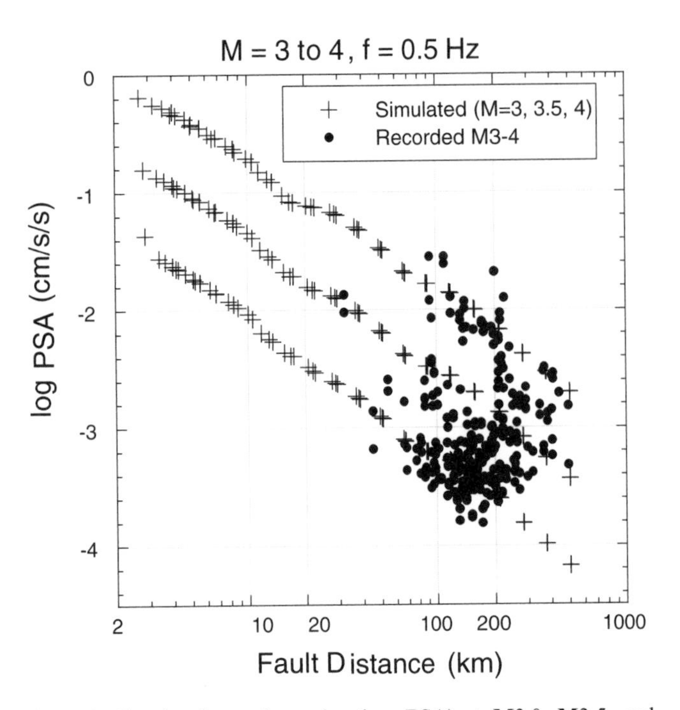

Figure 9. Simulated pseudoacceleration (PSA) at **M**3.0, **M**3.5, and **M**4.0, in comparison to PSA from the recorded data, M3.0–M4.0; *f* (frequency) = 0.5 Hz.

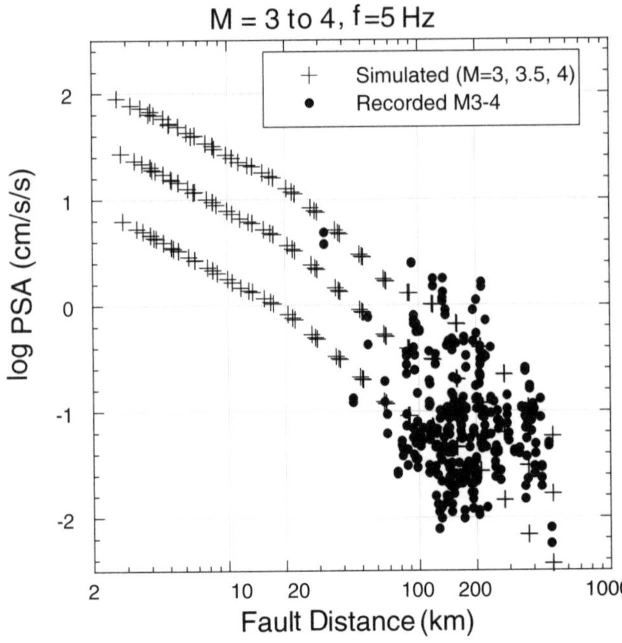

M = 3 to 4, f=5 Hz
Simulated (M=3, 3.5, 4)
Recorded M3-4
log PSA (cm/s/s)
Fault Distance (km)
2
1
0
-1
-2
2
10
20
100
200
1000

Figure 10. Simulated pseudoacceleration (PSA) at **M**3.0, **M**3.5, and **M**4.0, in comparison to PSA from the recorded data, M3.0–M4.0; *f* (frequency) = 5 Hz.

M = 4 to 5, f = 0.5 Hz
Simulated (M=4, 4.5, 5)
Recorded M4-5
log PSA (cm/s/s)
Fault Distance (km)
1
0
-1
-2
-3
2
10
20
100
200
1000

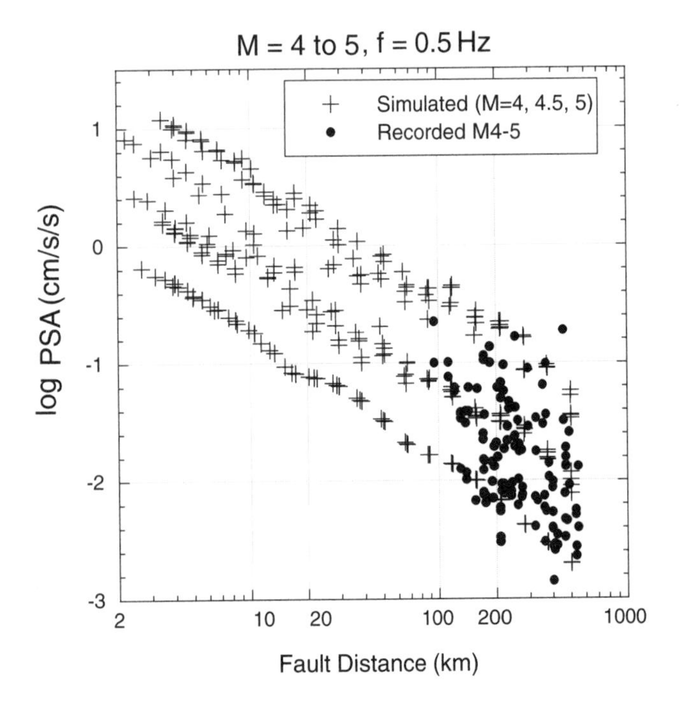

Figure 11. Simulated pseudoacceleration (PSA) at **M**4.0, **M**4.5, and **M**5.0, in comparison to PSA from the recorded data, M4.0–M5.0; *f* (frequency) = 0.5 Hz.

M = 4 to 5, f = 5 Hz
Simulated (M=4, 4.5, 5)
Recorded M4-5
log PSA (cm/s/s)
Fault Distance (km)
3
2
1
0
-1
2
10
20
100
200
1000

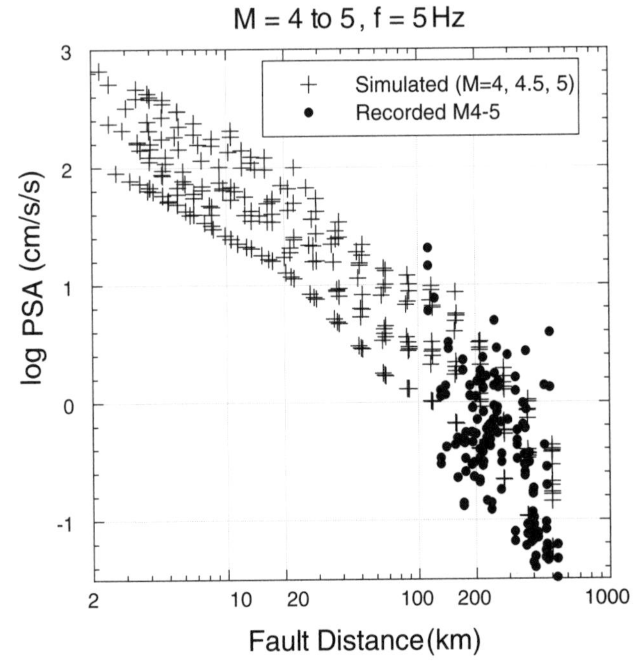

Figure 12. Simulated pseudoacceleration (PSA) at **M**4.0, **M**4.5, and **M**5.0, in comparison to PSA from the recorded data, M4.0–M5.0; *f* = 5.0 Hz.

M) $log(75)$ for 75 km $< R \le$ 100 km; and $hingeFunction = (c_7 + c_8 \mathbf{M}) \log(75) - 0.5 \log(R/100)$ for $R > 100$ km.

Some of the main seismic sources are located more than 200 km to the north of Puerto Rico. Therefore, extending the attenuation formula to large distances is important. The *hingeFunction* defines the hinged behavior of ground-motion curves for different distances.

The maximum likelihood regression algorithm of Joyner and Boore (1993) requires a linear equation for regression. Thus, some of the constants must be preset to linearize the final regression. Specifically, b, n, c_5, c_6, c_7, and c_8 are constants that are preset based on initial analysis to determine the best fit and distribution of residuals. Atkinson and Boore (1995, 1997b) applied $b = 6$ and $n = 2$, while Abrahamson and Silva (1997) applied $b = 8.5$ and $n =$ 2 or 3. In order to find the best b and n values to set in the source terms, we applied a set of different trial values for both constants to Equation (6). Values for b from 1 to 10 were considered, while n was varied from 1 to 4. For each pair of b and n, the maximum likelihood method was applied to Equation (6), and the average absolute value of residuals for each magnitude for frequencies from 0.1 to 20 Hz was calculated. This analysis showed that the best combination (lowest residuals) is $b = 6$ and $n = 2$.

It was determined that a linear relation between **M** and Δ provides a good distribution of residuals. Therefore, the parameter space was sampled to find the best linear relation (coefficients c_5 and c_6 that provide the lowest absolute value of residuals). These coefficients control the flat portion of the attenuation relations in the near-source region. The lowest residuals are obtained for the function $\Delta = -7.333 + 2.333\,\mathbf{M}$. Thus, we fix $c_5 = -7.333$ and $c_6 = 2.333$ **M.**

The slope of the attenuation curve as given in *hingeFunction* is magnitude-dependent, with the lowest residuals being obtained for $c_7 = -1.8$ and $c_8 = 0.1$. The above-mentioned procedures to determine b, n, c_5, c_6, c_7, and c_8 have been iterated many times to find the final values of the parameters that produce the lowest residuals. Therefore, Equation (6) becomes as follows:

$$\log PSA\,(f, R) = c_1 + c_2(\mathbf{M} - 6) + c_3(\mathbf{M} - 6)^2 + hingeFunction + c_4(f)R, \quad (7)$$

where PSA is the observed acceleration response spectra with 5% damping; f = frequency in hertz; **M** = moment magnitude; $R = (D^2 + \Delta^2)^{0.5}$; D = closest distance to fault surface, in km; $\Delta = -7.333 + 2.333\mathbf{M}$; $hingeFunction = (-1.8 + 0.1\mathbf{M}) \log(R)$ for $R \le 75$ km; $hingeFunction = (-1.8 + 0.1\mathbf{M}) \log(75)$ for 75 km $\le R \le$ 100 km; $hingeFunction = (-1.8 + 0.1\mathbf{M}) \log(75) - 0.5 \log(R/100)$ for $R \ge$ 100 km; and $c_4(f)$ is the coefficient of anelastic attenuation.

Table 2 shows the final results of maximum likelihood regression analysis of the simulated data to Equation (7). Figures 13–15 show the predicted PSA values by this ground-motion relation [Equation (7)] in comparison to the simulated PSA values input to the regression. The equations provide a good prediction of the simulated PSA values over all magnitudes and distances, although there is a slight tendency to overestimate near-fault PSA for earthquakes of **M**7 (the curves tend to track the "forward-directivity" or higher of the plotted values for **M**7).

For hazard calculations, we are interested in both the median ground-motion predictions, given by Equation (7) and their standard deviation. We calculated the residuals of the recorded Puerto Rico data versus the Puerto Rico attenuation relations [e.g., residual = (log of observed A_{ij}) for real recorded Puerto Rico data minus (log of predicted A_{ij}) by Equation (7)] for our study events. The study events have smaller magnitudes than the range of interest for seismic hazard calculations, but nevertheless

TABLE 2. GROUND MOTION RELATIONS FOR PUERTO RICO (HORIZONTAL COMPONENT, GENERIC SOFT ROCK SITE, NEHRP C)

f (Hz)	c_1	c_2	c_3	c_4
0.10	1.62	0.91212	−0.10486	−0.00092
0.13	1.80	0.90635	−0.11886	−0.00081
0.16	1.98	0.89009	−0.13157	−0.00064
0.20	2.16	0.87177	−0.14444	−0.00052
0.25	2.36	0.84583	−0.15306	−0.00048
0.32	2.55	0.81112	−0.16625	−0.00044
0.40	2.74	0.78035	−0.17792	−0.0005
0.50	2.89	0.73416	−0.1706	−0.00056
0.63	3.04	0.67664	−0.15973	−0.00061
0.79	3.20	0.63441	−0.15706	−0.0008
1.00	3.35	0.56986	−0.14377	−0.00086
1.26	3.47	0.497	−0.11945	−0.00105
1.59	3.58	0.47303	−0.11486	−0.00118
2.00	3.68	0.44246	−0.10831	−0.00126
2.51	3.74	0.40472	−0.08864	−0.00139
3.16	3.83	0.38087	−0.09045	−0.00159
3.98	3.88	0.35932	−0.07932	−0.00185
5.01	3.94	0.33077	−0.06816	−0.00204
6.31	3.97	0.33046	−0.07344	−0.00219
7.94	3.98	0.32515	−0.07216	−0.00234
10.00	3.96	0.32088	−0.06542	−0.00244
12.59	3.94	0.32165	−0.06523	−0.00253
15.85	3.88	0.33249	−0.06818	−0.00251
PGA	3.60	0.35181	−0.06926	−0.00201
PGV	2.35	0.54828	−0.06350	−0.00107

Note: Pseudoaccelaration (PSA) is 5% damped horizontal component pseudoacceleration in cm/s²; f in hertz; **M**—Moment magnitude; D—closest distance to fault surface in km. All logs are in Base 10. Standard deviation 0.28 in log10 units is suggested for all frequencies.

$\log PSA\,(f, R) = c_1 + c_2(\mathbf{M} - 6) + c_3(\mathbf{M} - 6)^2 + hingeFunction + c_4R$

$R = (D^2 + \Delta^2)^{0.5}$

$\Delta = -7.333 + 2.333\,\mathbf{M}$

$hingeFunction = (-1.8 + 0.1\,\mathbf{M}) \log(R)$ for $R \le 75$ km

$hingeFunction = (-1.8 + 0.1\,\mathbf{M}) \log(75)$ for 75 km $\le R \le$ 100 km

$hingeFunction = (-1.8 + 0.1\,\mathbf{M}) \log(75) - 0.5 \log(R/100)$ for $R \ge 100$ km

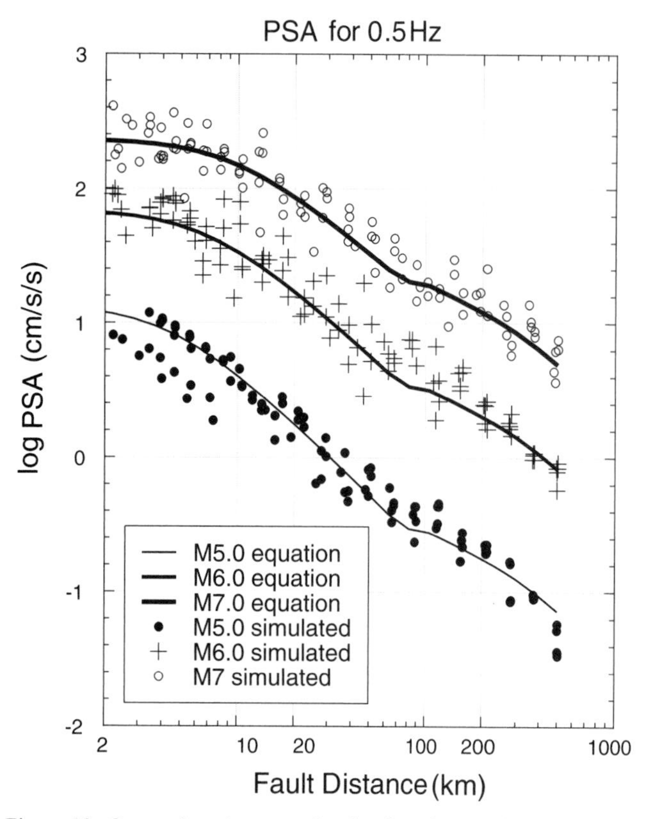

Figure 13. Comparison between the developed ground-motion equations and the simulated data; $f = 0.5$ Hz.

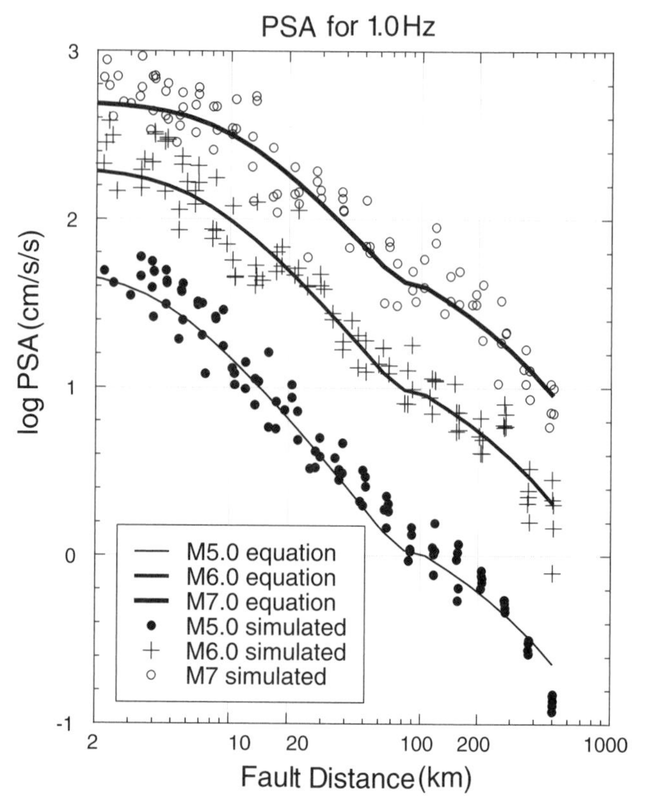

Figure 14. Comparison between the developed ground-motion equations and the simulated data; $f = 1.0$ Hz.

serve as a guide to actual variability in the recorded data about the prediction equations. The standard deviation of the distribution of the residuals for intermediate frequencies on average is ~0.28 in log10 units. We suggest a standard deviation of the relations of 0.28 log(10) units for all frequencies. This standard deviation is typical for ground-motion relations in many regions.

COMPARISON OF PUERTO RICO GROUND-MOTION RELATIONS WITH RELATIONS FOR OTHER REGIONS

In Figures 16–19, we compare our ground-motion relations for Puerto Rico with ground-motion relations for other regions. The other regions are eastern North America (Atkinson and Boore, 1995), California (Atkinson and Silva, 2000), and the global database for subduction zones (Atkinson and Boore, 2003). The eastern North America relations of Atkinson and Boore (1995) are based on a stochastic point source model with a two-corner source spectrum to mimic finite fault effects. The California relations are based on a two-corner stochastic source model calibrated against the California strong motion database. The global subduction relations of Atkinson and Boore (2003) are strictly empirical, based on regression analysis of thousands of records; separate regressions were performed for in-slab and interface events. The Atkinson and Boore subduction relations are plotted for **M**7.0 only.

The ground-motion relations for bedrock for eastern North America have been multiplied by the generic factors of Adams et al. (1999) to convert them from hard rock to a "firm ground" (NEHRP C) site condition. The generic soft rock site condition (NEHRP C) applies for the California ground-motion relations ("rock" relations of Atkinson and Silva, 2000) and is provided for the global subduction zones (Atkinson and Boore, 2003). Recall that the Puerto Rico curves are assumed to represent soft rock (NEHRP C). The plotted curves are thus comparable to the ground-motion relations of the other regions in terms of site conditions.

The Puerto Rico ground-motion relations are similar to stochastic ground-motion relations obtained for other parts of North America. Differences are attributable to regional variations in ground-motion parameters. There are three main parameters that control the behavior of PSA, which may vary regionally. The first one is the crustal and near-surface amplification. For Puerto Rico, this is modeled by the H/V ratio. The other ground-motion relations each represent slightly different overall site effects. Another parameter that controls the behavior of PSA, especially at large distances, is the regional Q-value. At large distances, Puerto Rico ground-motion curves lie between those of eastern North America and California, since the Puerto Rico Q-value is

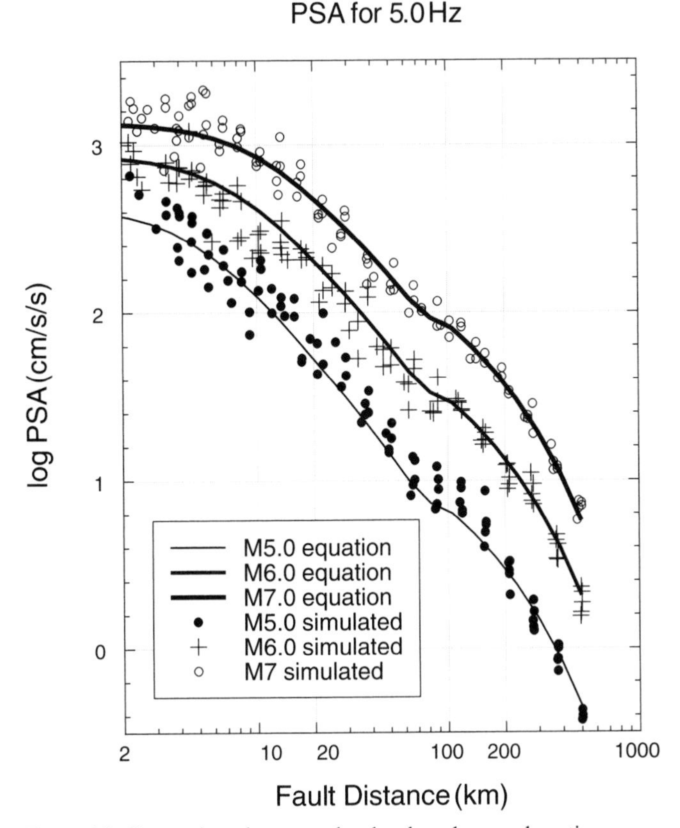

Figure 15. Comparison between the developed ground-motion equations and the simulated data; $f = 5.0$ Hz.

intermediate to those for eastern North America and California. The third parameter that makes the PSA in Puerto Rico relations deviate from the other relations for North America is differences in the hinge points of the attenuation curve, reflecting regional differences in crustal structure. Overall, the Puerto Rico relations agree well with stochastic relations for California and eastern North America.

The motions calculated by the global subduction relations of Atkinson and Boore (2003) are quite different from those suggested by this study. The global subduction relations are mostly applicable for **M** > 6.0 (interface and in-slab events) and R > 40 km (for in-slab events); thus, the Puerto Rico PSA should not be compared with the global subduction relations outside these limits. The global subduction relations are entirely empirical. The database of interface events was separated from the database of in-slab events in deriving the empirical regressions. By contrast, the Puerto Rico relations are based on the combination of events of all types. Crustal events have contributions to the Puerto Rico relations as well as interface and in-slab events. Due to the limited data precision for Puerto Rico events, it is not feasible to distinguish between event types in most cases. Thus, the Puerto Rico ground motions are region-specific, but the attenuation is averaged over all types of events. By contrast, the global subduction relations separate out event types, but combine data globally over several subduction zones. These factors contribute to the observed differences between global subduction ground-motion relations and Puerto Rico ground-motion relations. For seismic hazard analysis, differences in results obtained using global empirical relations, in comparison to those obtained using the region-specific relations presented here, are measures of our current level of uncertainty.

CONCLUSION

We derived the first region-specific ground-motion relations for Puerto Rico, based on evaluation of regional seismographic data from small to moderate earthquakes, coupled with application of a stochastic simulation model to extend the data to larger magnitudes. The relations are in good agreement with recorded ground-motion data in Puerto Rico, and are intermediate to relations for California and eastern North America. Future study is required to address such factors as the effect of event type (crustal, interface or in-slab) on ground motions and to improve the knowledge of regional site effects.

APPENDIX: EXSIM, A STOCHASTIC FINITE FAULT MODELING PROGRAM INTRODUCING THE CONCEPT OF DYNAMIC CORNER FREQUENCY

The stochastic model is a commonly used tool for ground-motion simulation. The method models ground motion as band limited Gaussian noise whose amplitude spectrum is given by a seismological model (Boore, 1983). The most commonly used seismological model for stochastic simulations has been the Brune (1970, 1971) point source model (e.g., Toro et al., 1997); however, point source models are inappropriate for large earthquakes. The effects of a large finite source, including fault geometry, heterogeneity of slip on the fault plane, and directivity, can profoundly influence the amplitudes, frequency content, and duration of ground motion. Finite-fault effects in ground motions become important for earthquakes with magnitudes exceeding ~6.0.

Finite fault modeling has been an important tool for the prediction of ground motion near the epicenters of large earthquakes (Hartzell, 1978; Irikura, 1983; Joyner and Boore, 1986; Heaton and Hartzell, 1986; Somerville et al., 1991; Tumarkin and Archuleta, 1994; Zeng et al., 1994; Beresnev and Atkinson, 1998a). One of the most useful methods to simulate ground motion for a large earthquake is based on the simulation of a number of small earthquakes as subfaults that comprise a larger fault. A large fault is divided into N number of subfaults and each subfault is considered as a small point source (introduced by Hartzell, 1978). The rupture spreads radially from the hypocenter. In our implementation, the ground motions of subfaults, each of which are calculated by the stochastic point-source method, are summed with a proper delay time in the time domain to obtain the ground motion from the entire fault, $a(t)$:

$$a(t) = \sum_{i=1}^{nl} \sum_{j=1}^{nw} a_{ij}\left(t + \Delta t_{ij}\right), \tag{1}$$

where nl and nw are the number of subfaults along the length and width of main fault, respectively ($nl*nw = N$), Δt_{ij} is the relative delay time for the radiated wave from the ijth subfault to reach the observation point. The $a_{ij}(t)$ are each calculated by the stochastic point source method. The acceleration spectrum for a subfault at a distance R_{ij} is modeled as a point source with an ω^2 shape (Aki, 1967; Brune, 1970;

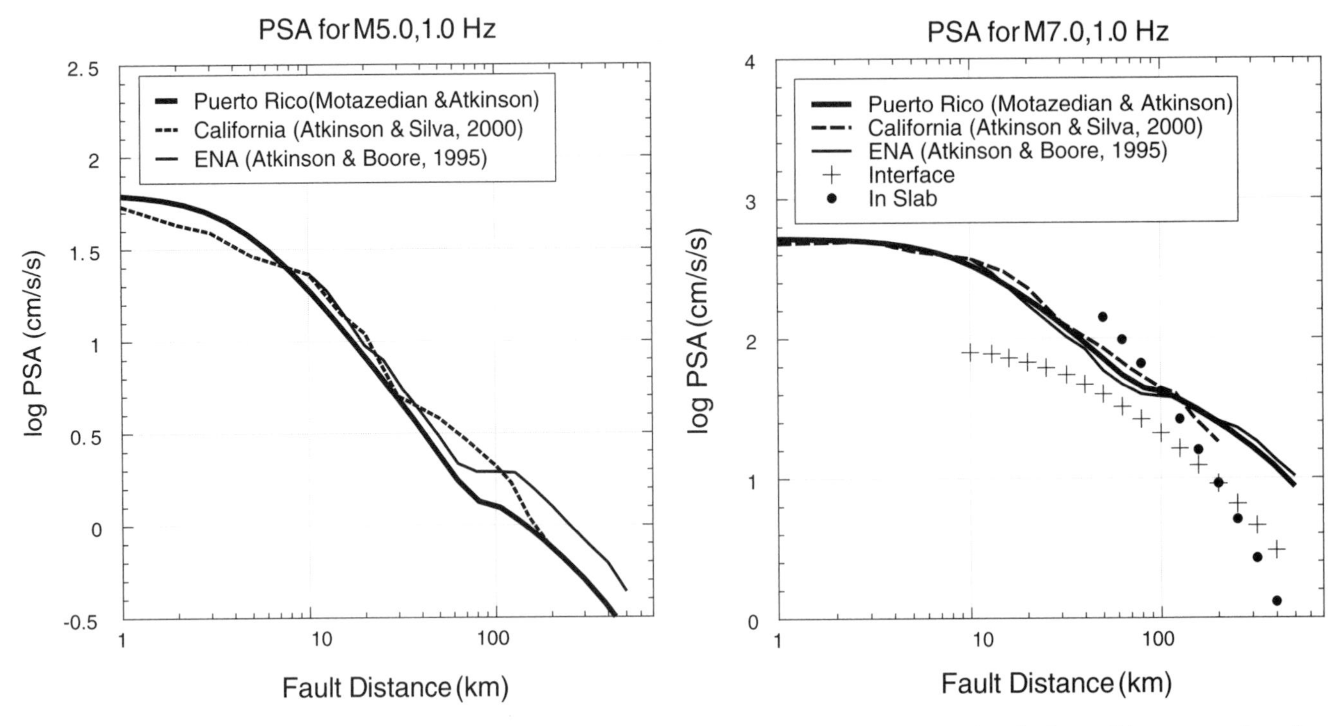

Figure 16. Ground motion relations for Puerto Rico compared to other regions. Pseudoacceleration (PSA) for **M**5.0, *f* = 1.0 Hz. All relations plotted for U.S. National Earthquake Hazards Reduction Program (NEHRP) C site conditions. ENA—eastern North America.

Figure 17. Ground motion relations for Puerto Rico compared to other regions. Pseudoacceleration (PSA) for **M**7.0, *f* = 1.0 Hz. All relations plotted for U.S. National Earthquake Hazards Reduction Program (NEHRP) C site conditions. ENA—eastern North America.

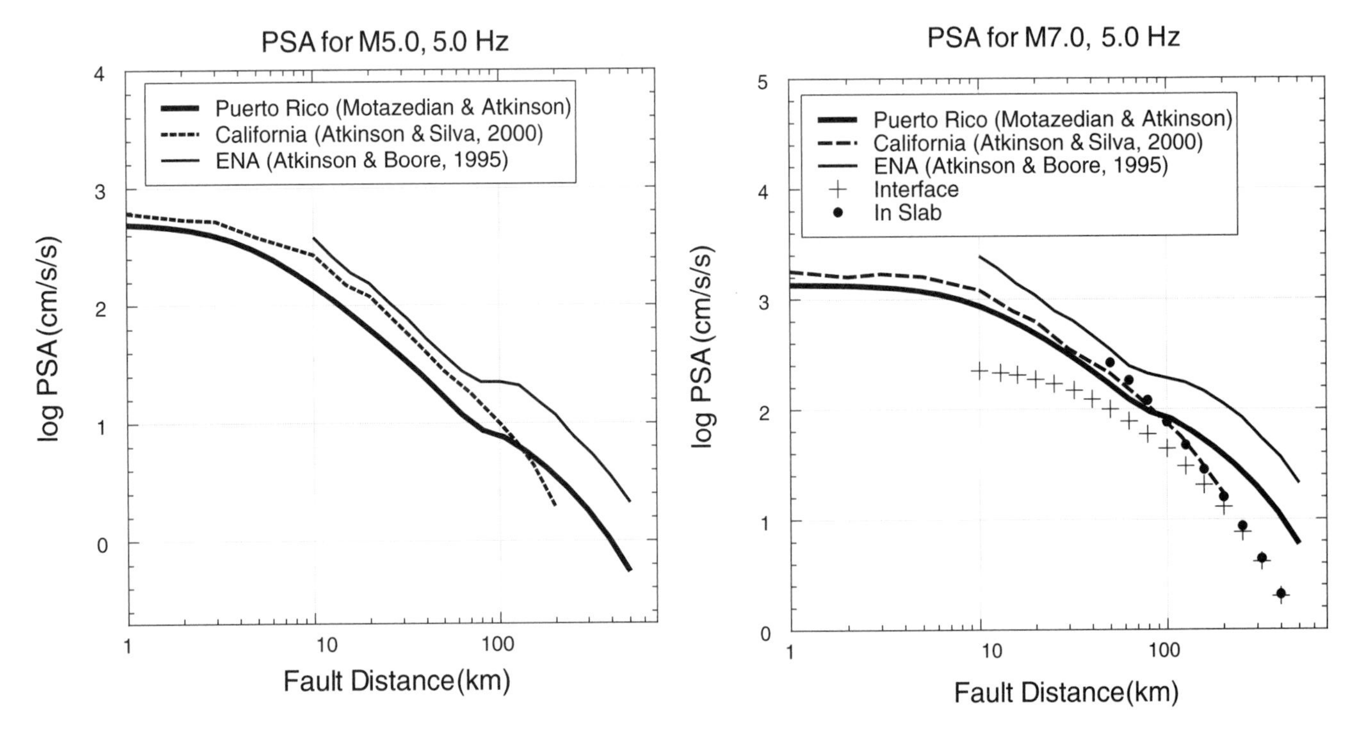

Figure 18. Ground motion relations for Puerto Rico compared to other regions. Pseudoacceleration (PSA) for **M**5.0, *f* = 5.0 Hz. All relations plotted for U.S. National Earthquake Hazards Reduction Program (NEHRP) C site conditions. ENA—eastern North America.

Figure 19. Ground motion relations for Puerto Rico compared to other regions. Pseudoacceleration (PSA) for **M**7.0, *f* = 5.0 Hz. All relations plotted for U.S. National Earthquake Hazards Reduction Program (NEHRP) C site conditions. ENA—eastern North America.

Boore, 1983). The acceleration spectrum of shear wave of the *ij*th subfault, $A_{ij}(f)$, is

$$A_{ij}(f) = CM_{0ij}(2\pi f)^2/[1 + (f/f_{0ij})^2]\ exp\ (-\pi f\kappa)\ exp\ (-\pi f R/Q\beta)/R_{ij}, \quad (2)$$

where M_{0ij}, f_{0ij}, and R_{ij} are the *ij*th subfault seismic moment, corner frequency, and distance from the observation point, respectively. Corner frequency, f_{0ij}, is given by $f_{0ij} = 4.9e + 6\beta(\Delta\sigma/M_{0ij})^{1/3}$, where $\Delta\sigma$ is stress drop in bars; M_{0ij} is in dyne-cm; and β is shear wave velocity in km/s. The constant $C = 4\pi^2\Re_{\theta\phi}/4\pi\rho\beta^3R_{ij}$, where ρ = density and $\Re_{\theta\phi}$ = radiation pattern (Boore, 1983). The $exp\ (-\pi f\kappa)$ is a high cut filter to model near-surface "kappa" effects; this is the commonly-observed rapid spectral decay at high frequencies. The quality factor, Q, is inversely related to anelastic attenuation.

The moment of each subfault is controlled by the ratio of its area to the area of the main fault ($M_{0ij} = M_0/N$ where M_0 is the seismic moment of the entire fault). If the subfaults are not identical, we can express the seismic moment of each subfault as

$$M_{0ij} = (M_0 S_{ij}) / (\sum_{l=1}^{nl} \sum_{k=1}^{nw} S_{kl}) \quad (3)$$

where S_{ij} is the relative slip weight of the *ij*th subfault. Earthquake time history simulation deals with the time from the beginning of rupture to the time the rupture stops. In our finite fault model, we deal with a ruptured area that is time dependent; it is initially zero and is finally equal to the entire fault area. Corner frequency is inversely proportional to the ruptured area. Therefore, in time history simulation, the corner frequency may be considered to be a function of time. The rupture begins with high corner frequencies and progresses to lower corner frequencies. We suppose that, during an earthquake, at each moment of time, the corner frequency is dependent on the cumulative ruptured area. The rupture history controls the frequency content of the simulated time series.

In our dynamic approach, the corner frequency of the first subfault (near the beginning of rupture) is $f_{011} = S\ 4.9e + 6\beta\ (\Delta\sigma/M_{011})^{1/3}$, where M_{011} is the seismic moment of the first subfault. The dynamic corner frequency of the *ij*th subfault, $f_{0ij}(t)$, can be defined as a function of $N_R(t)$, the cumulative number of ruptured subfaults at time t.

$$f_{0ij}(t) = N_R(t)^{-1/3}\ S\ 4.9e + 6\ \beta\ (\Delta\sigma/M_{0ave})^{1/3}, \quad (4)$$

where $M_{0ave} = M_0/N$ is the average seismic moment of subfaults and S is a constant. As the rupture proceeds toward the end of the fault, the number of ruptured subfaults increases and, based on Equation (4), the corner frequency of the subfaults decreases. The dynamic corner frequency concept will tend to decrease the high-frequency level of the spectrum of subfaults as the rupture progresses ($A_{ij}(f)_{f>>fijo} \propto f_{0ij}^{\ 2}$). We therefore introduce a scaling factor to conserve the high-frequency spectral level of the radiation produced by each subfault. The acceleration spectrum of the *ij*th subfault, $A_{ij}(f)$, is thus

$$A_{ij}(f) = CM_{0ij}\ H f^2/[1 + (f/f_{0ij})^2] \quad (5)$$

$$H = (N\int \{f^2/[1 + (f/f_0)^2]\}^2\ df / \int \{f^2/[1 + (f/f_{0ij})^2]\}^2\ df)^{1/2}, \quad (6)$$

where H is the scaling factor. Thus, the diminishing effect of f_{0ij} on the high-frequency spectral level from the subfault is compensated for by the scaling factor. The high-frequency spectral contribution from the *ij*th subfault is equal to that from the first subfault, but the calculation of corner frequency, which controls the shape of the spectrum, comes from the total ruptured area. Consequently, as the rupture progresses, there is more low-frequency energy produced, and thus the distribution of energy tends to shift toward lower frequencies.

The new simulation approach is implemented in a modified version of the computer program FINSIM (Beresnev and Atkinson, 1998b). FINSIM is a well-known stochastic finite-fault simulation program that has been validated using data from many earthquakes, including Michoacan, Mexico (1985); Lama Prieta (1989); Northridge (1994); Valparaiso, Chile (1985); and Saguenay, Quebec (1988) (Beresnev and Atkinson 1997, 1998a, 1998b, 1999, 2001, 2002). The modified program has been renamed EXSIM (extended earthquake fault simulation program). The modifications include the following features:

1. Inclusion of the new concept of "dynamic corner frequency."
2. Elimination of multiple triggering of subfaults.
3. Variability of pulsing area, or the number of subfaults that are considered as active subfaults in the calculation of dynamic corner frequency. This allows for self-healing slip models.
4. Calculation of subfault seismic moment based on dividing the total seismic moment of the earthquake fault by the number of subfaults.

In EXSIM, the percentage of pulsing area on the fault will affect the relative amplitudes of low-frequency motion in finite fault modeling. Variation of the stress drop parameter can be used to adjust the relative amplitudes of high-frequency motion. By increasing the stress parameter, the amplitude of high frequencies increases.

A generic EXSIM model was derived by finding the model parameters that best reproduce, on average, the California strong-motion database (as used by Atkinson and Silva, 2000). For this exercise, the regional attenuation parameters used in previous finite-fault simulations for California were adopted (Beresnev and Atkinson, 2002), and the calibration exercise determined the best-fit stress drop and pulsing percentage. These parameters were found to be independent of magnitude, with average values of $\Delta\sigma$ = 130 bars and a pulsing percentage of 50%. EXSIM produces similar results to FINSIM over the magnitude range from 5.5 to 7.5, but unlike FINSIM, it can be easily extended to smaller magnitudes. The advantage of EXSIM is that it introduces conceptual improvements, such as independence of results from subfault size and conservation of radiated energy, and allows simulation of self-healing slip pulses.

ACKNOWLEDGMENTS

This study was supported by the U.S. National Earthquake Hazards Reduction Program. We thank Christa von Hillebrandt, Martinez Cruzado, and the PRSN staff for their effort in providing seismic data and their help in calibrating the PRSN seismic network. We thank Igor Beresnev for stimulating discussion and comments on finite fault modeling. The comments of Bill McCann, Roland Laforge, Art Frankel, and Chuck Mueller are also gratefully acknowledged.

REFERENCES CITED

Abrahamson, N., and Silva, W., 1997, Empirical response spectral attenuation relations for shallow crustal earthquakes: Seismological Research Letters, v. 68, p. 94–127.

Adams, J., Weichert, D.H., and Halchuk, S., 1999, Trial Seismic Hazard Maps of Canada—1999: 2% / 50 year values for selected Canadian cities: Geological Survey of Canada Open File Report 3724.

Aki, K., 1967, Scaling law of seismic spectrum: Journal of Geophysical Research, v. 72, p. 1217–1231.

Anderson, J., and Hough, S., 1984, A model for the shape of the Fourier amplitude spectrum of acceleration at high frequencies: Bulletin of the Seismological Society of America, v. 74, p. 1969–1993.

Atkinson, G., 1995, Ground motion relations in eastern North America: Ottawa, Ontario, Proceedings of the Atomic Energy Control Board Workshop on Seismic Hazards, June 21–23.

Atkinson, G., and Boore, D., 1995, New ground motion relations for eastern North America: Bulletin of the Seismological Society of America, v. 85, p. 17–30.

Atkinson, G., and Boore, D., 1997a, Stochastic point-source modeling of ground motions in the Cascadia region: Seismological Research Letters, v. 68, p. 74–85.

Atkinson, G., and Boore, D., 1997b, Some comparisons of recent ground motion relations: Seismological Research Letters, v. 68, p. 24–40.

Atkinson, G., and Boore, D., 1998, Evaluation of models for earthquake source spectra in eastern North America: Bulletin of the Seismological Society of America, v. 88, p. 917–934.

Atkinson, G., and Boore, D., 2003, Empirical ground motion relation for subduction zone earthquakes and their application to Cascadia and other regions. Bulletin of the Seismological Society of America, v. 93, p. 1703–1729.

Atkinson, G., and Cassidy, J., 2000, Integrated use of seismograph and strong motion data to determine soil amplification in the Fraser Delta: results from the Duvall and Georgia Strait earthquakes: Bulletin of the Seismological Society of America, v. 90, p. 1028–1040.

Atkinson, G., and Mereu, R., 1992, The shape of ground motion attenuation curves in southeastern Canada: Bulletin of the Seismological Society of America, v. 82, p. 2014–2031.

Atkinson, G., and Silva, W., 1997, Empirical source spectra for California earthquakes: Bulletin of the Seismological Society of America, v. 87, p. 97–113.

Atkinson, G., and Silva, W., 2000, Stochastic modeling of California ground motions: Bulletin of the Seismological Society of America, v. 90, p. 255–274.

Beresnev, I., and Atkinson, G., 1997, Modeling finite fault radiation from the w^n spectrum: Bulletin of the Seismological Society of America, v. 87, p. 67–84.

Beresnev, I., and Atkinson, G., 1998a, Stochastic finite-fault modeling of ground motions from the 1994 Northridge, California earthquake. I. Validation on rock sites: Bulletin of the Seismological Society of America, v. 88, p. 1392–1401.

Beresnev, I., and Atkinson, G., 1998b, FINSIM—a FORTRAN program for simulating stochastic acceleration time histories from finite faults: Seismological Research Letters, v. 69, p. 27–32.

Beresnev, I., and Atkinson, G., 1999, Generic finite-fault model for ground-motion prediction in eastern North America: Bulletin of the Seismological Society of America, v. 89, p. 608–625.

Beresnev, I., and Atkinson, G., 2001, Subevent structure of large earthquakes—A ground motion perspective: Geophysical Research Letters v. 28, p. 53–56, doi: 10.1029/2000GL012157.

Beresnev, I., and Atkinson, G., 2002, Source parameters of earthquakes in eastern and western North America based on finite-fault modeling: Bulletin of the Seismological Society of America, v. 92, p. 696–710.

Boore, D., 1983, Stochastic simulation of high-frequency ground motions based on seismological models of the radiated spectra: Bulletin of the Seismological Society of America, v. 73, p. 1865–1894.

Boore, D., and Joyner, W., 1997, Site amplifications for generic rock sites: Bulletin of the Seismological Society of America, v. 87, p. 327–341.

Boore, D., Joyner, W., and Wennerberg, L., 1992, Fitting the stochastic omega-squared source model to observed response spectra in western North America: Trade-offs between stress drop and kappa: Bulletin of the Seismological Society of America, v. 82, p. 1956–1963.

Boore, D., Joyner, W., and Fumal, T., 1997, Equations for estimating horizontal response spectra and peak acceleration from western North American earthquakes: A summary of recent work: Seismological Research Letters, v. 68, p. 128–153.

Borcherdt, R., 1994, Estimates of site-dependent response spectra for design: Earthquake Spectra, v. 10, p. 617–653, doi: 10.1193/1.1585791.

Brune, J., 1970, Tectonic stress and the spectra of seismic shear waves from earthquakes: Journal of Geophysical Research, v. 75, p. 4997–5009.

Brune, J., 1971, Correction: Journal of Geophysical Research, v. 76, p. 5002.

Chen, S., 2000, Global comparisons of earthquake source spectra [Ph.D. thesis]: Ottawa, Ontario, Carleton University.

Chen, S., and Atkinson, G., 2002, Global comparisons of earthquake source spectra: Bulletin of the Seismological Society of America, v. 92, p. 885–895.

Hanks, T., and McGuire, R., 1981, The character of high-frequency strong ground motion: Bulletin of the Seismological Society of America, v. 71, p. 2071–2095.

Hartzell, S., 1978, Earthquake aftershocks as Green's functions: Geophysical Research Letters, v. 5, p. 1–14.

Heaton, T., and Hartzell, S., 1986, Source characteristics of hypothetical subduction earthquakes in the northwestern United States: Bulletin of the Seismological Society of America, v. 76, p. 675–708.

Irikura, K., 1983, Semi-empirical estimation of strong ground motions during large earthquakes: Bulletin of the Disaster Prevention Research Institute, Kyoto University, v. 33, p. 63–104.

Joyner, W., and Boore, D., 1986, On simulating large earthquakes by Green's function addition of smaller earthquakes, *in* Das, S., Boatwright, J., and Scholz, C.H., eds., Earthquake Source Mechanics: American Geophysical Union Geophysical Monograph 37, p. 269–274.

Joyner, W., and Boore, D., 1993, Methods for regression analysis of strong motion data: Bulletin of the Seismological Society of America, v. 83, p. 469–487.

Lermo, J., and Chavez-Garcia, F., 1993, Site effect evaluation using spectral ratios with only one station: Bulletin of the Seismological Society of America, v. 83, p. 1574–1594.

McCann, W., 1985, On the earthquake hazard of Puerto Rico and the Virgin Islands: Bulletin of the Seismological Society of America, v. 75, p. 251–262.

McCann, W.R., 2002, Catalog of felt earthquakes for Puerto Rico and neighboring islands 1492–1899 with additional information for some 20th-century earthquakes [abs.]: Seismological Research Letters, v. 74, p. 251.

Nakamura, Y., 1989, A method for dynamic characteristics estimation of subsurface using microtremor on the ground surface: Quarterly Report of Railway Technical Research Institute, v. 30, no. 1, p. 25–33.

Raoof, M., Herrmann, R., and Malagnini, L., 1999, Attenuation and excitation of three-component ground motion in southern California: Bulletin of the Seismological Society of America, v. 89, p. 888–902.

Sadigh, K., Chang, C., Egan, J., Makdisi, F., and Youngs, R., 1997, Attenuation relationships for shallow crustal earthquakes based on California strong motion data: Seismological Research Letters, v. 68, p. 180–189.

Shin, T., and Herrmann, R., 1987, Lg attenuation and source studies using 1982 Miramichi data: Bulletin of the Seismological Society of America, v. 77, p. 384–397.

Somerville, P., Sen, M., and Cohee, B., 1991, Simulations of strong ground motions recorded during the 1985 Michoacan, Mexico, and Valparaiso, Chile, earthquakes: Bulletin of the Seismological Society of America, v. 81, p. 1–27.

Somerville, P., Smith, N.C., and Graves, R.W., 1997, Modification of empirical strong ground motion attenuation relations to include the amplitude and duration effects of rupture directivity: Seismological Research Letters, v. 68, p. 199–221.

Toro, G., Abrahamson, N., and Schneider, J., 1997, Model of strong ground motion in eastern and central North America: Best estimates and uncertainties: Seismological Research Letters, v. 68, p. 41–57.

Tumarkin, A., and Archuleta, R., 1994, Empirical ground motion prediction: Annali Di Geofisica, v. 37, p. 1691–1720.

Wells, D., and Coppersmith, K., 1994, New empirical relationships among magnitude, rupture length, rupture width, rupture area, and surface displacement: Bulletin of the Seismological Society of America, v. 84, p. 974–1002.

Zeng, Y., Anderson, J., and Yu, G., 1994, A composite source model for computing realistic synthetic strong ground motions: Geophysical Research Letters, v. 21, p. 725–728.

Manuscript Accepted by the Society 18 August 2004

Geological Society of America
Special Paper 385
2005

Microseismic activity reveals two stress regimes in southwestern Puerto Rico

Victor Huérfano*
Christa von Hillebrandt-Andrade
Gisela Báez-Sanchez
Puerto Rico Seismic Network, Geology Department, University of Puerto Rico at Mayagüez, P.O. Box 9017, Mayagüez, Puerto Rico 00681

ABSTRACT

The seismic activity in the local region of Puerto Rico, lat 17°N–20°N, long 63.5°W–69°W, is the result of the interaction of the North America and Caribbean plates and the response of the Puerto Rico–Virgin Islands microplate. This seismic activity is distributed over a broad region with slightly higher activity concentrated around known fracture zones and prominent topographic features. Within the Island of Puerto Rico, the highest level of onshore shallow seismicity (≤25 km depth) is concentrated to the south of the Great Southern Puerto Rico Fault zone in the southwestern part of the island. We conducted a detailed analysis of hypocenters of 828 shallow quakes (depth ≤50 km) provided by the Puerto Rico Seismic Network (PRSN) and located within the southwestern Puerto Rico seismic zone to determine the stress patterns of this complex area and their relationship with tectonism and local geology.

A suite of first motion composite focal mechanisms was computed based on the distribution of seismicity, geology, and topographic features. On average, our solutions resemble the expected stress patterns obtained from regional plate motion and/or geophysical studies that indicate NW-SE stress fields associated with the regime of extension in the Mona Passage and the oblique convergence of the major plates. However, there are local topographic and geologic features like Cordillera Central, Monte Grande, and Sierra Bermeja that are best explained in terms of internal deformations of the southwestern Puerto Rico seismic zone.

Keywords: Puerto Rico, southwestern Puerto Rico, microseismicity, stress, focal mechanism.

INTRODUCTION

The Puerto Rico–Virgin Islands microplate (Asencio, 1980; Huérfano, 1995) is located in the northeastern corner of the Caribbean in the zone of convergence between the North America and Caribbean plates (Fig. 1). This seismically active area is characterized by complex fault systems, known topographic features and unresolved stress patterns. The Puerto Rico–Virgin Islands region has a history of several large damaging earthquakes and tsunamis, including the Virgin Islands earthquake of 1867 and the Mona Canyon earthquake of 1918. Both of these are considered

*victor@rmsismo.uprm.edu

Huérfano, V., von Hillebrandt-Andrade, C., and Báez-Sanchez, G., 2005, Microseismic activity reveals two stress regimes in southwestern Puerto Rico, *in* Mann, P., ed., Active tectonics and seismic hazards of Puerto Rico, the Virgin Islands, and offshore areas: Geological Society of America Special Paper 385, p. 81–101. For permission to copy, contact editing@geosociety.org.

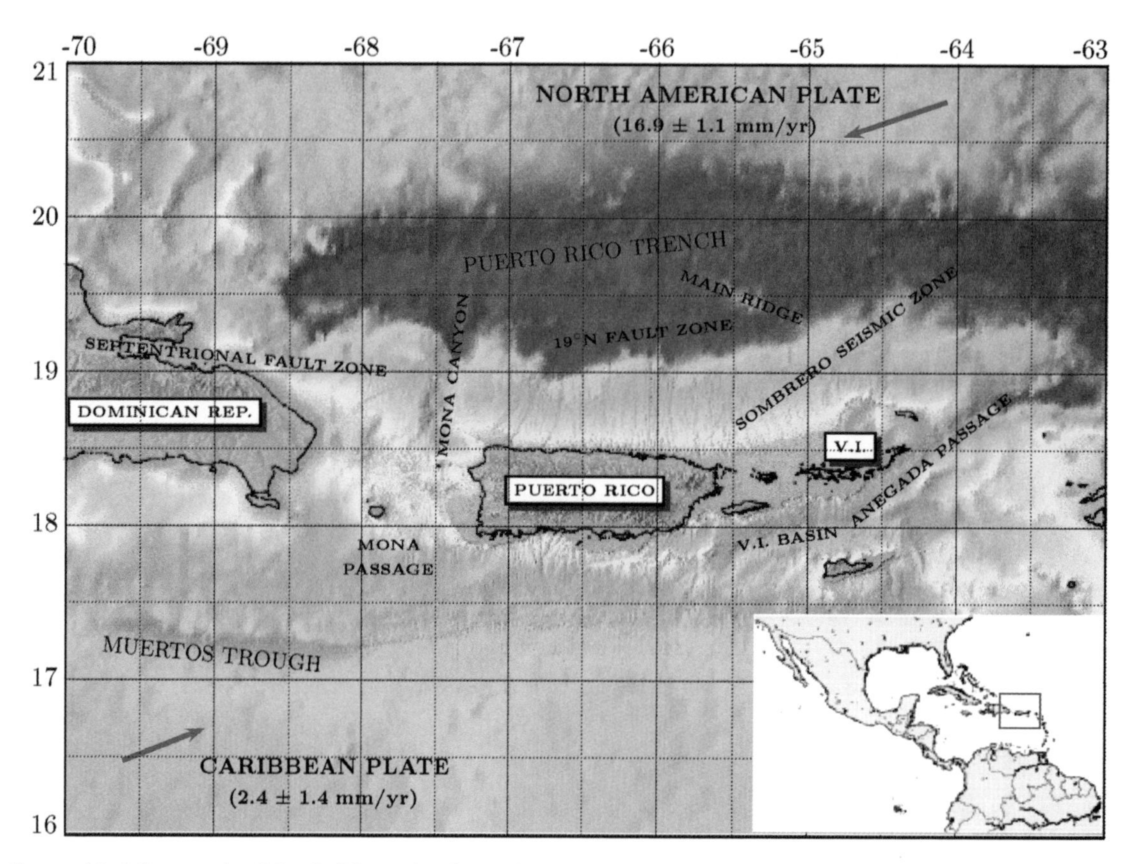

Figure 1. Geographical framework of the Caribbean showing major regional tectonics features in the local region of Puerto Rico and Virgin Islands (V.I. on figure). Vectors represent the relative plate motion and the speeds were taken from Jansma et al., (2000).

to have had the same magnitude, which was calculated to be 7.3 (Ms) for the 1918 event (Pacheco and Sykes, 1992).

The northern boundary between the Caribbean and the North America plates is marked by the Puerto Rico Trench, which is the deepest trench in the Atlantic and produces the world's largest negative free-air gravity anomaly (Dillon et al., 1999). The Puerto Rico–Virgin Islands region is bounded to the north by the 19°N fault zone, which dips south and is apparently characterized by normal motions according to Larue et al. (1990). The extensional Virgin Island Basin and the Anegada Passage follow a southwest-northeast orientation and mark the eastern boundary of the Puerto Rico–Virgin Islands. The western boundary of the Puerto Rico–Virgin Islands is the Mona Canyon, which is also under extensional dynamics (Hippolyte et al., this volume). The southern boundary of the Puerto Rico–Virgin Islands is unclear although can be related with the seismically active southwestern area of Puerto Rico. The Muertos trough, which is located to the southwest of the Puerto Rico–Virgin Islands region, is a convergence zone characterized by low rate of seismic activity. Recent global positioning system (GPS) studies (Jansma et al., 2000) indicate that the Puerto Rico–Virgin Islands region is behaving as part of the Caribbean plate and is moving ENE at a rate of 19–20 mm/yr. López et al. (1999), Jansma et al. (2000), and Mann et al. (2002) suggest from GPS measurements that the motion between Hispaniola and the Puerto Rico–Virgin Islands is ~5 mm/yr.

Figure 2A shows the local seismicity ($M_{PR} \geq 3.0$)[1] in the Puerto Rico–Virgin Islands as recorded by the Puerto Rico Seismic Network (PRSN, 2003b). This epicentral map reflects the concentration of quakes to the west in the Mona Passage and Mona Canyon, to the north and northeast of the island in the convergence zone of the Caribbean and North America plates. Inside the island, the seismicity is highly concentrated in the southwestern Puerto Rico seismic zone. The A–A′ vertical cross projection (Fig. 2B) highlights the subducted North America slab and the shallow (depth ≤25 km) seismicity in the southwestern Puerto Rico seismic zone (limited by a S-N transition zone from lat 17.75°N trending to Puerto Rico). Historically, the southwestern Puerto Rico seismic zone has been the source of at least 25 felt earthquakes in the last century (PRSN, 2003a) with maximum intensities of VI (modified Mercalli scale). The

[1] M_{PR} is the Puerto Rico Network magnitude scale; equation from von Hillebrandt and Bataille (1 995).

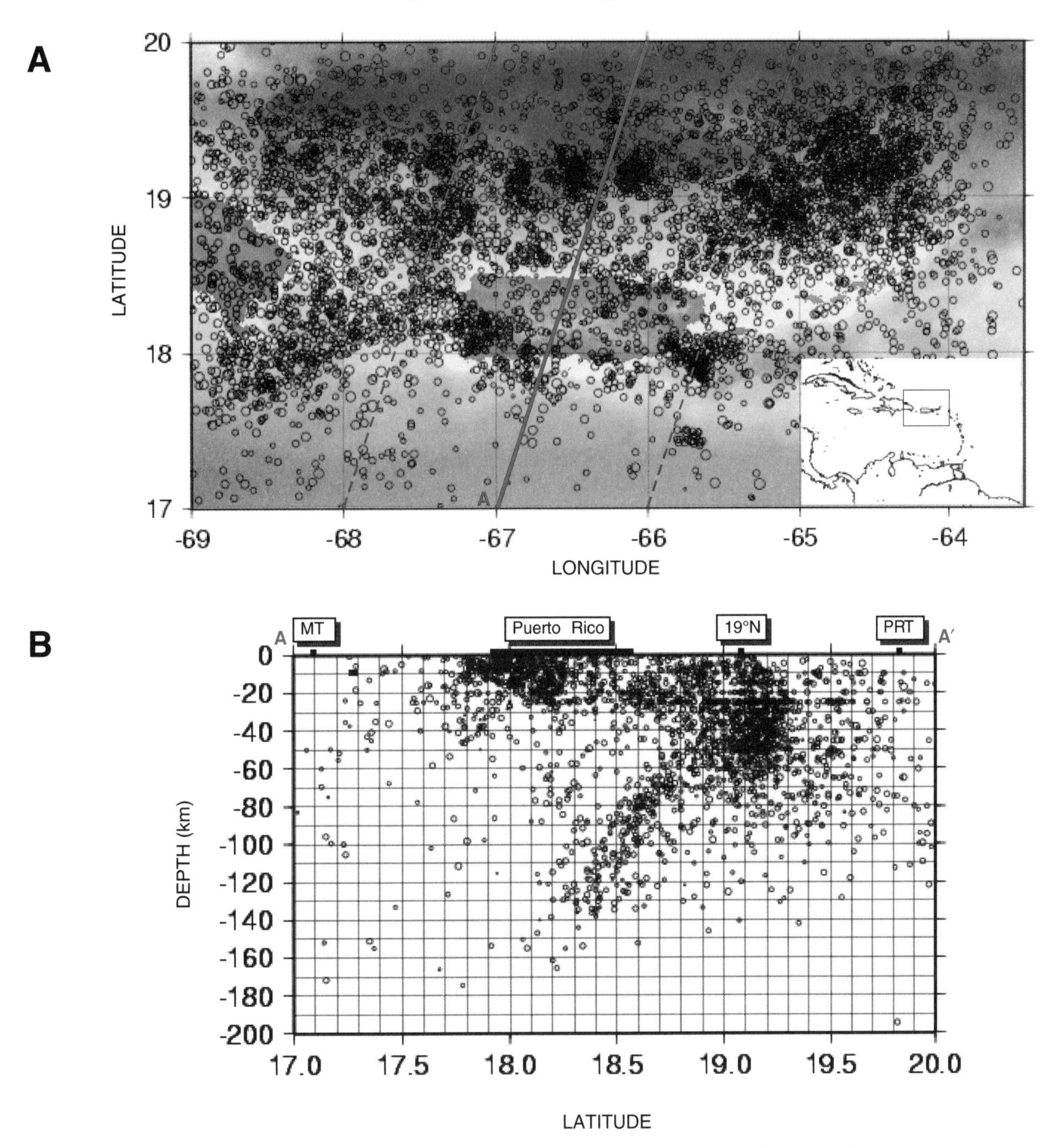

Figure 2. (A) Epicentral map of earthquakes recorded by the Puerto Rico Seismic Network with magnitudes $M_{PR} \geq 3.0$. Magenta dashed lines indicate the region used to plot the AA′ vertical cross projection (B).

events of May 1987 and March 1988, located in southwestern Puerto Rico, caused minor damage in this region.

The objective of this study is to determine the stress patterns in the southwestern Puerto Rico active seismic zone and relate it to the local geology and topographic features where clusters of seismic activity have been identified. These stress fields could help in the understanding of the kinematics and can provide a greater insight on the active faults in the area.

GEOLOGICAL SETTING AND SEISMICITY

The geology of Puerto Rico (Fig. 3) is characterized by Late Jurassic and early Tertiary metamorphic, volcanic, volcaniclastic, and sedimentary rocks. These were intruded by felsic plutonic rocks during the Late Cretaceous and early Tertiary. Slightly tilted Oligocene and younger sedimentary rocks overlay the older formations (Briggs and Akers, 1965). The island is divided

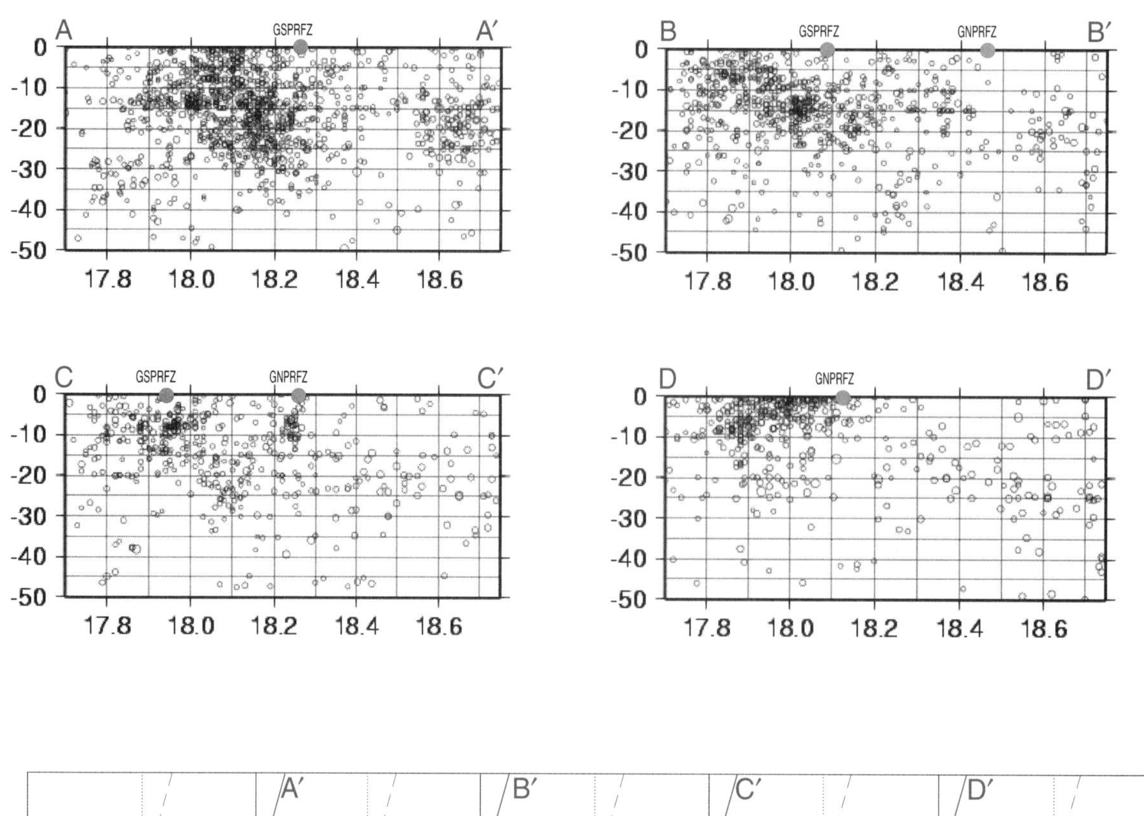

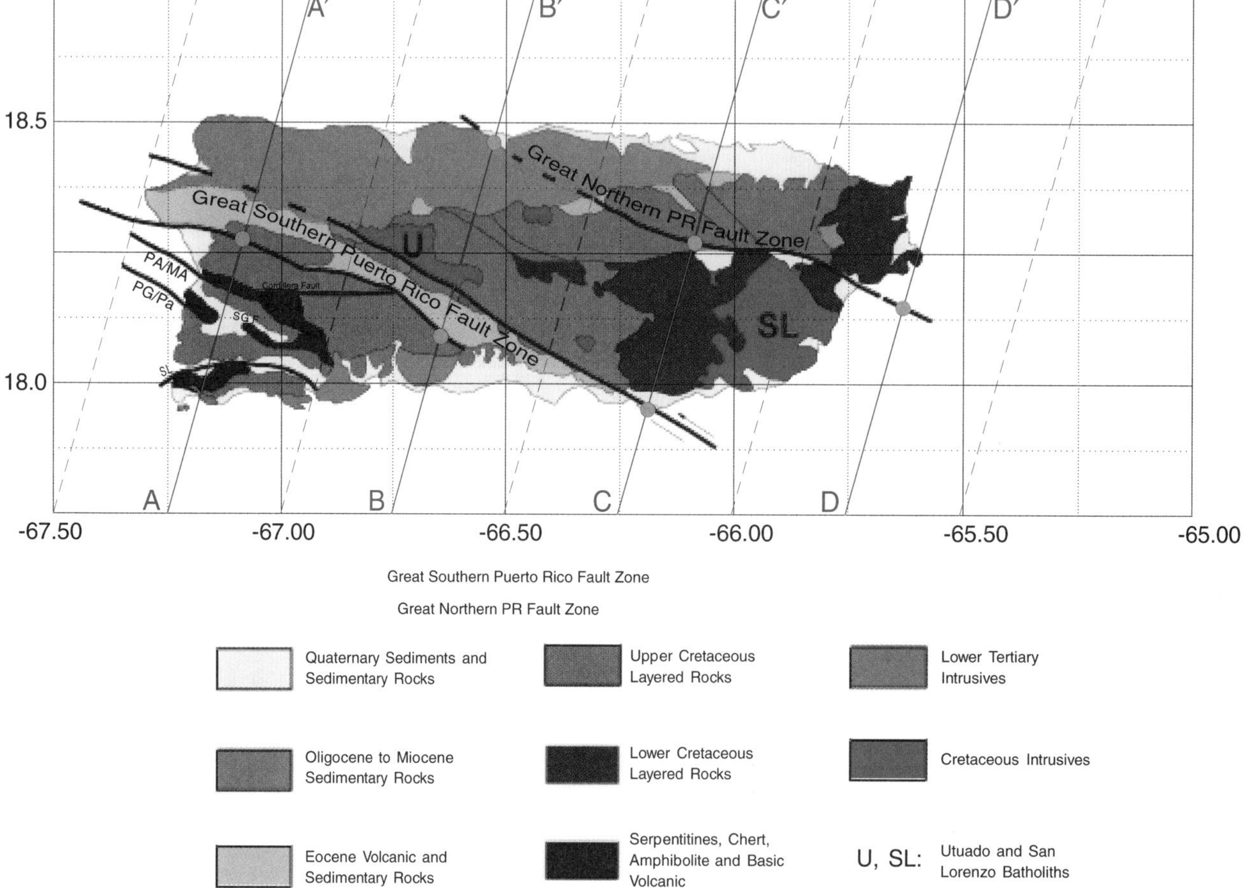

Figure 3. Geological sketch map of Puerto Rico (based on Schellekens, 1998a). Top vertical cross sections are normal to the great southern Puerto Rico fault zone, cyan points indicate the intersection between the projection line and the great southern Puerto Rico fault zone (GSPRFZ) or great northern Puerto Rico fault zone (GNPRFZ).

into three igneous provinces separated by the Great Northern and Southern Puerto Rico Fault zones.

The southwestern igneous province (southwestern Puerto Rico) is distinguished by the presence of the Bermeja Complex, which includes the oldest rocks in Puerto Rico and is located to the south of the great southern Puerto Rico fault zone. It occurs chiefly at the center of most anticlines (Mattson, 1960). Schellekens (1998a, 1998b) and Jolly et al. (1998) identified three major serpentinite belts within the Bermeja Complex: (1) the northernmost belt, known as the Monte del Estado Peridotite along the Cordillera Fault, (2) the Río Guanajibo Serpentinite Belt, and (3) the Sierra Bermeja, the southernmost group of serpentinites, which are exposed in the southwestern corner of the island.

The major tectonic feature of the southwestern Puerto Rico seismic zone is the great southern Puerto Rico fault zone, which separates an area of greater and shallower seismicity to the south as compared to the north. It extends SE-NW inland from Salinas to Rincón and trends offshore toward Isla Desecheo. The great southern Puerto Rico fault zone was active during the Eocene and recorded predominantly thrust and left-lateral strike slip displacement (Glover and Mattson, 1960). Mattson (1960) mapped and described two major faults in the southwestern Puerto Rico, Cordillera, and San Germán fault zones. He suggested a left lateral displacement along these faults. Recently, Grindlay et al. (this volume) used high-resolution seismic and sidescan sonar surveys on the insular shelf of western and southern Puerto Rico to identify three zones of active deformation: (1) the Cerro Goden, (2) the Punta Algarrobo and Mayagüez, and (3) the Punta Guanajibo and Punta Arenas. Prentice et al. (2000) proposed that the linear nature and internal drainage in the Lajas Valley is controlled by active faulting. She trenched across a linear escarpment on the south side of the Lajas Valley and found Holocene displacement along the South Lajas fault.

A compilation of 828 well constrained earthquakes located by the PRSN between 1987 and 2002 indicates that the seismicity in the southwestern Puerto Rico seismic zone is characterized by two sets of events (Fig. 3, vertical cross sections). The intermediate seismicity (depth ≥25 km) occurs along the north-dipping slab in the Muertos trough and therefore can be related to Caribbean plate motion. The shallow seismicity is related to internal deformation within the southwestern Puerto Rico seismic zone. The vertical cross sections also show that the earthquakes in Puerto Rico can be associated with the two major fault systems that transverse the island, the Great Southern Puerto Rico Fault zone and Great Northern Puerto Rico Fault zone. In the southwestern Puerto Rico seismic zone, the seismicity is mainly concentrated south of the great southern Puerto Rico fault zone and at depths <30 km. Moving to the east, the vertical cross projections (normal to the great southern Puerto Rico fault zone) show that the seismicity migrates to the south, from lat 18.1°N to 17.8°N, following a parallel surface trend of the great southern Puerto Rico fault zone.

The most striking observation is the distribution of shallow epicenters around the prominent topographic features and the linear belts of serpentinites in the southwestern Puerto Rico seismic zone (Fig. 4A). In the Sierra Bermeja, hypocenters at depths of 5–15 km are apparently concentrated in an EW plane with an inclination of ≈45°. Intermediate earthquakes (depths from 25 to 50 km) are located offshore, south of the Lajas Valley and southwest of Mayagüez (Fig. 4B). The seismic events located southwest of Mayagüez show a striking NE-SW vertical distribution. Inland intermediate seismic events occur in the area surrounding Mayagüez-Maricao.

METHOD AND RESULTS

The earthquakes used in this study were recorded by the PRSN between January 1987 and December of 2002, and were located using the typical Geiger's method as programmed in the PR-HYPO computer program. This system uses the flat and homogeneous five layer crustal model of Huérfano and Bataille (1995). Single hypocenter selection was based on the root mean square (<0.3 s), location errors (horizontal and vertical <5 km), number of readings (minimum four P-wave and two S-wave arrival times), and a magnitude >2.0 M_{PR} (von Hillebrandt and Bataille, 1995). The detection capability of the PRSN is relatively uniform for this time period, so our results avoid network-induced bias.

The data were divided into nine groups (Figs. 5A and 6A). The first five groups focused on those hypocenters located along the topographic features: (1) Group one was constrained to those earthquakes located around the Sierra Bermeja; (2) earthquakes of group two were located in the vicinity of Monte Grande; (3) group three includes events recorded along western Cordillera Central, including the Montañas de Uroyán; (4) group four includes earthquakes located at the southeastern end of the Lajas Valley, Guánica Hills and near shore; and (5) earthquakes of group five were located at the head of the Guayanilla Canyon and the hills to the north. The second set of intermediate hypocenters was selected based on their depth distribution, between 25 km and 50 km. Three intermediate subsets were identified: (1) Group six includes those events located offshore of Mayagüez–Cabo Rojo, lat 18.1°N to lat 18.2°N; (2) group seven includes earthquakes detected inland around Mayagüez; and (3) group eight was defined for a set of hypocenters located off the insular shelf south of Lajas Valley between latitudes 17.18°N and 17.89°N. Finally, group nine includes those shallow earthquakes located between latitudes 17.75°N and 17.89°N that were not included in previous groups.

For each subset of hypocenters (Table 1), the double couple fault plane solution that best fit the first motions polarities was computed. The method was based on a two stage grid search procedure (FPFIT) as explained in Reasenberg and Oppenheimer (1985). As an additional constraint, the method developed by Huérfano and Bataille (1995) was applied to select the best solution when FPFIT reported multiple solutions. Only P-wave (first arrival) readings for which the polarity was clearly Up(up)/Down(down) and weighted as 1 were used. The summary of the inversions is presented in Table 2.

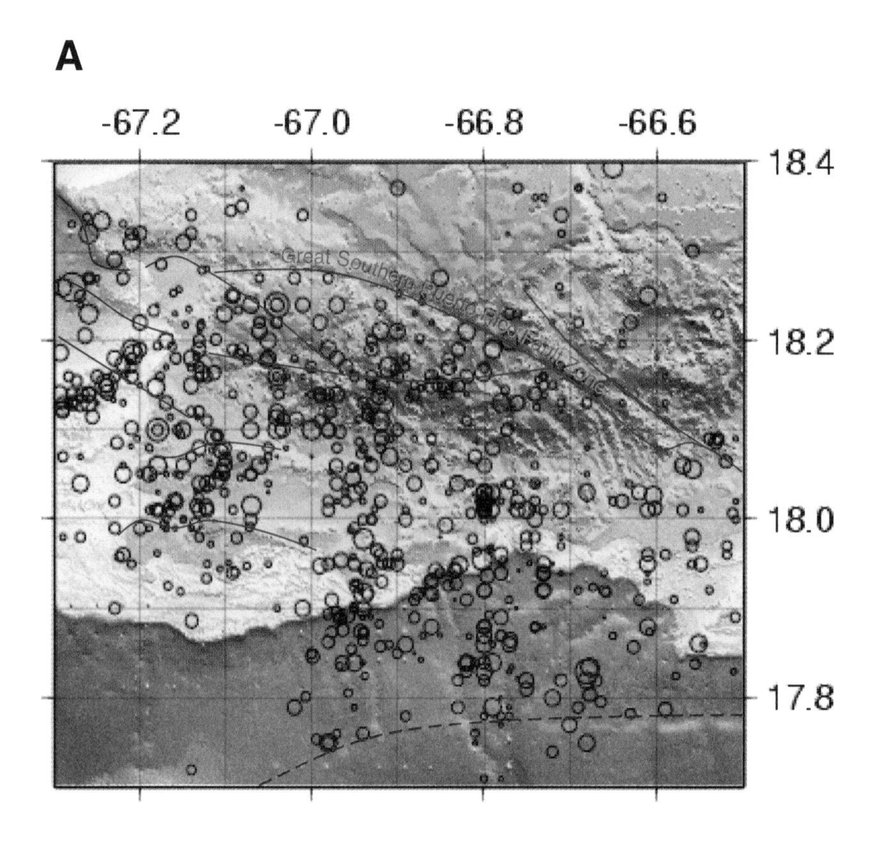

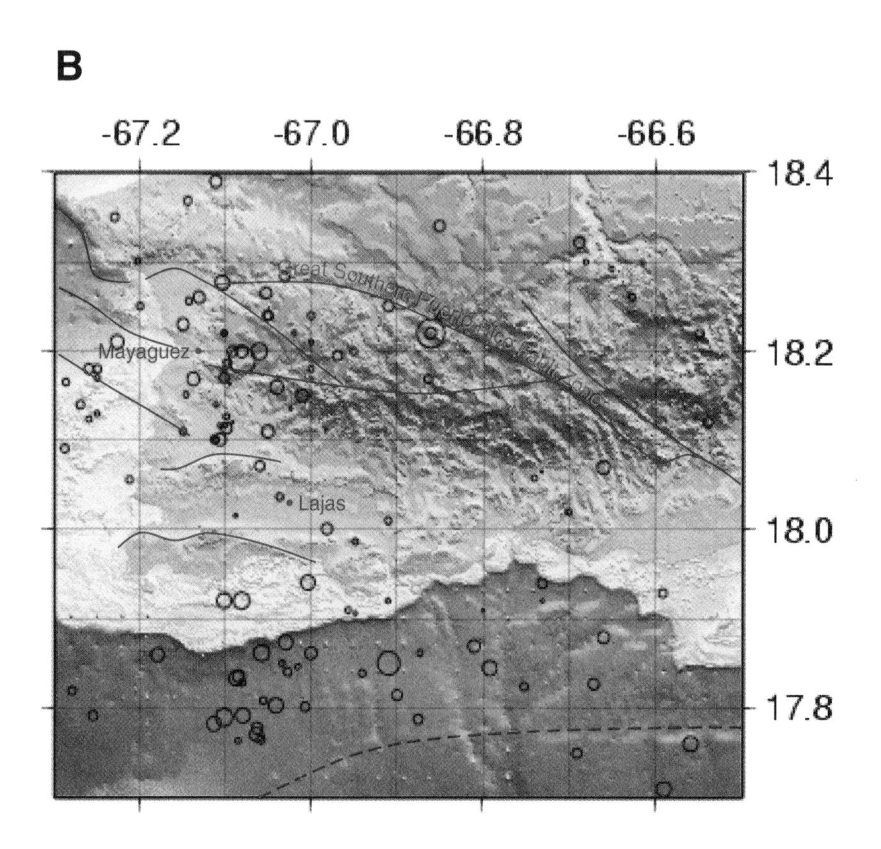

Figure 4. Well-constrained epicentral maps. (A) Shallow seismicity in the southwestern Puerto Rico seismic zone (depth ≤25 km). (B) Southwestern Puerto Rico seismicity with intermediate depths between 25 km and 50 km.

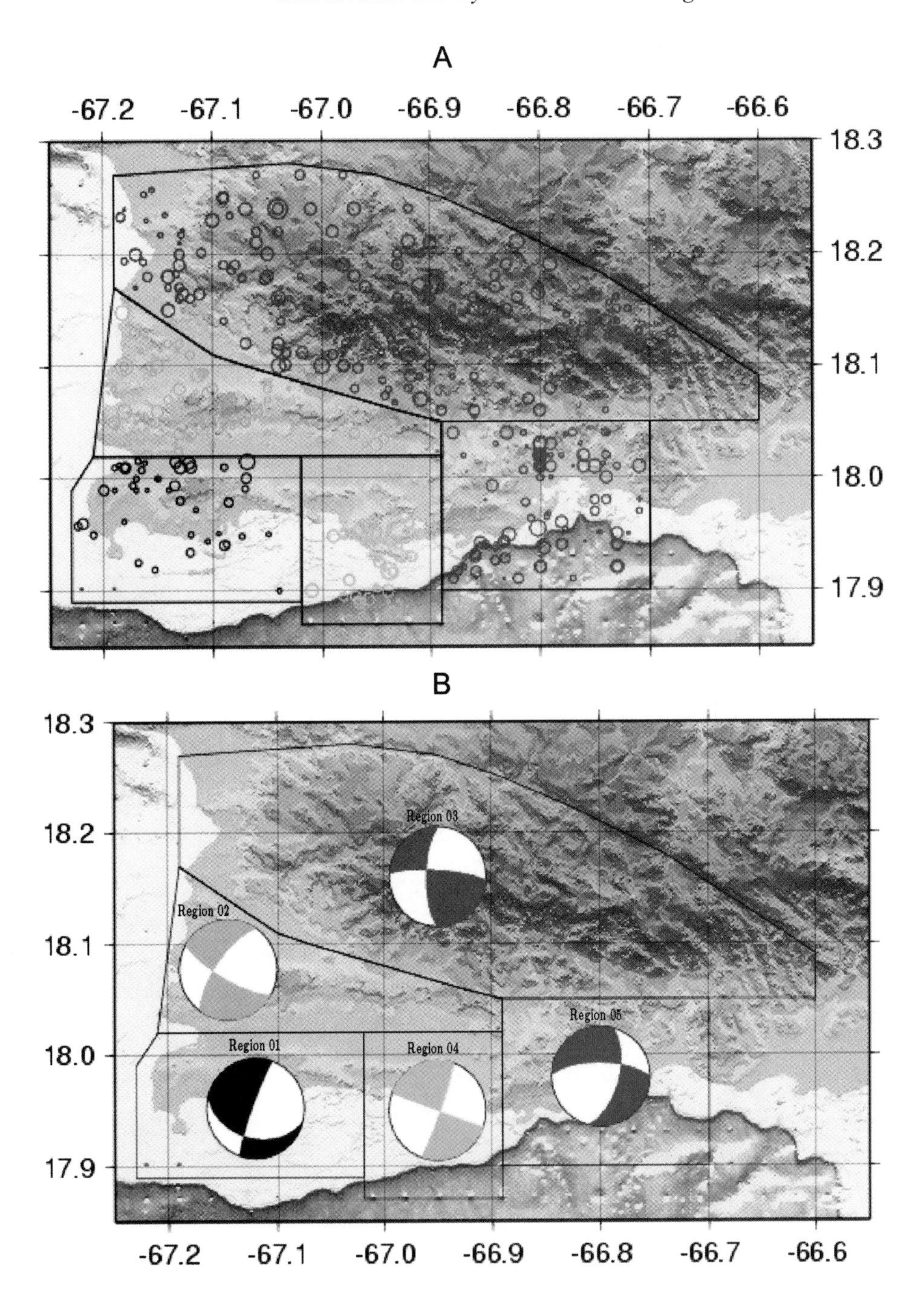

Figure 5. (A) Epicentral map showing the data used in the first five regions. (B) Composite focal mechanism computed for each region in (A). Colors associate focal spheres with earthquakes in their particular region.

THE MODEL

Because composite focal mechanisms are representative of the average stress in the source region, the slip vector as well as the P and T (compression and tension) eigenvectors indicate the pattern of deformation that generates the earthquakes. We found two major stress fields governing the southwestern Puerto Rico seismic zone (Figs. 5B and 6B). Shallow groups are consistent with the proposed extensional regime in the Mona Rift (Mann et al., 2002; Grindlay et al., this volume). We suggest that the deformation is absorbed by the great southern Puerto Rico fault zone, allowing the southwestern Puerto Rico seismic zone to extend in a NW-SE direction under lateral stresses (T axes) and a small component of uplift (P axes). An important note is that one nodal line is nearly parallel to the direction of the elongated exposure of serpentinite in the southwestern Puerto Rico seismic zone, suggesting a relationship between serpentinite remobilization and tectonic activity. In this case, a small

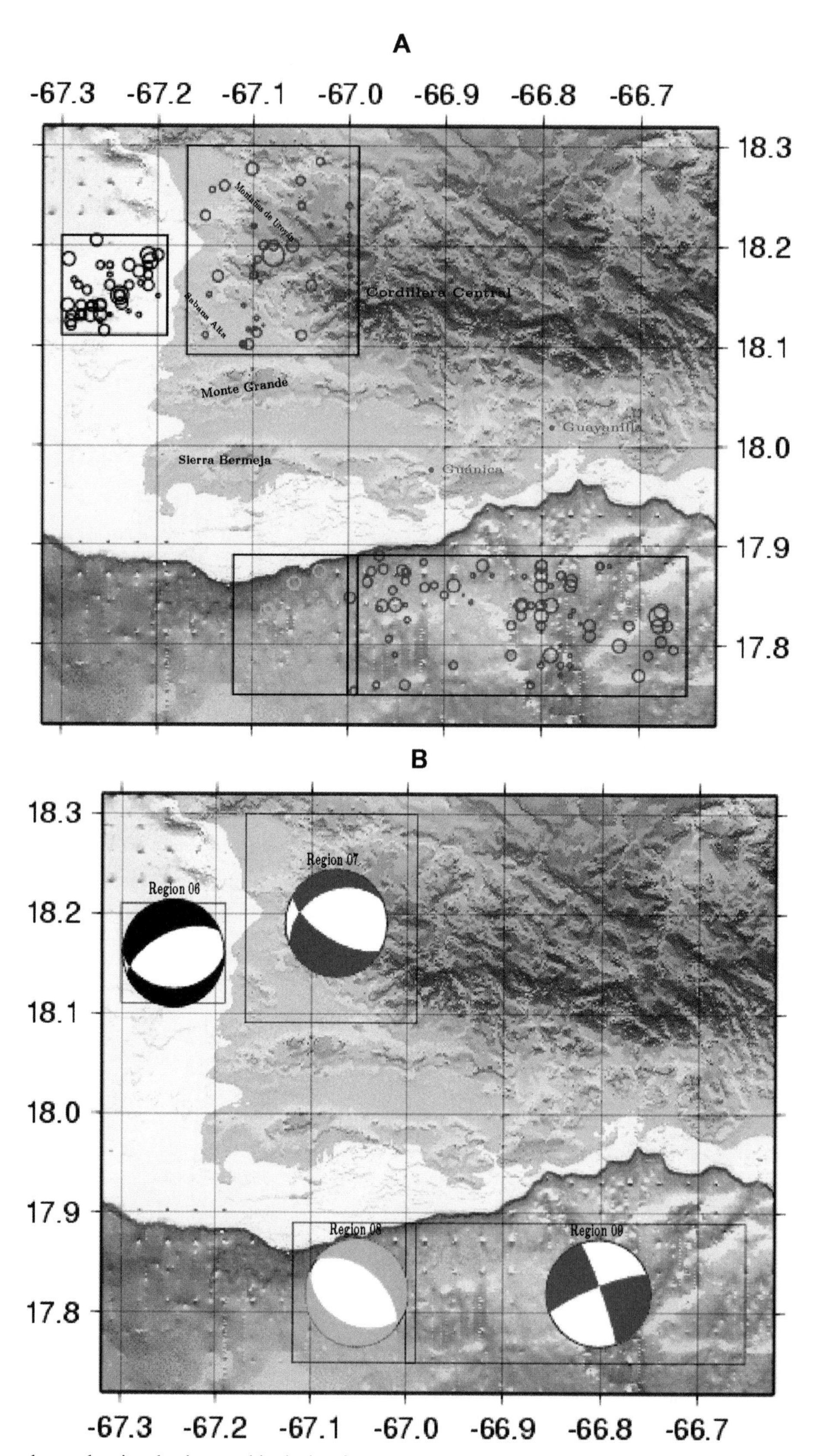

Figure 6. (A) Epicentral map showing the data used in the last four regions. (B) Composite focal mechanism computed for each region in (A). Colors associate focal spheres with earthquakes in their particular region.

TABLE 1. HYPOCENTERS USED TO COMPUTE COMPOSITE FOCAL MECHANISMS

Date	Time (GMT)	Lat	Long	Depth (km)	M_{PR}	rms	erh	erz	Gap	Q	Ps	Ss
(A) Region 01												
2/27/1987	24:37.5	18	–67.07	14.5	2.8	0.3	3.13	3.19	212	C	6	4
3/5/1988	16:21.5	17.96	–67.22	9.3	2.9	0.11	1.31	1.32	291	B	7	4
3/26/1988	51:20.4	18.01	–67.18	0.6	2.9	0.08	1.04	1.03	287	B	4	4
5/23/1993	07:31.7	17.95	–67.05	16.4	2.3	0.09	0.62	0.59	257	B	5	6
9/13/1994	04:57.6	17.99	–67.19	12.3	2.3	0.13	1.85	0.75	305	B	5	3
9/13/1994	37:49.6	17.99	–67.14	11.4	2.3	0.08	1.88	0.79	293	B	4	2
9/13/1994	01:03.0	18	–67.17	12.8	2.4	0.07	1.45	0.77	300	B	5	3
9/13/1994	23:27.9	17.99	–67.2	15.1	2.8	0.11	1.89	0.74	306	B	4	2
9/13/1994	57:06.3	17.99	–67.17	12.5	2.3	0.09	1.38	0.68	299	B	5	3
9/15/1994	03:32.4	17.95	–67.21	13.7	2.5	0.05	1.3	0.64	309	B	4	2
10/21/1995	40:26.9	17.95	–67.12	7.1	2.4	0.29	3.81	0.95	291	C	4	2
12/30/1995	56:59.0	17.94	–67.09	7.5	2.8	0.09	4.66	1.29	350	B	4	4
1/29/1996	42:37.3	18.01	–67.12	14.1	2.9	0.16	0.97	0.84	218	C	7	5
4/6/1996	58:51.3	18.01	–67.13	13.9	3	0.25	1.04	0.44	223	C	10	9
4/6/1996	27:15.0	17.99	–67.16	14.5	2.2	0.1	1.14	0.65	297	B	4	4
4/6/1996	37:13.7	18.01	–67.09	12.4	2.5	0.15	1.26	0.91	279	B	5	5
5/16/1996	34:29.1	18.01	–67.18	12.8	2.6	0.15	1.3	0.55	312	B	5	4
6/27/1996	35:23.4	17.98	–67.13	11.8	2.6	0.1	0.66	0.31	233	B	7	7
6/9/1997	49:48.0	18	–67.15	13.8	2.4	0.13	1.62	0.76	236	B	4	3
10/3/1997	52:29.8	18.01	–67.19	15.8	2.4	0.2	1.53	0.8	301	C	5	5
4/2/1998	37:46.2	17.9	–67.04	12.9	2.3	0.22	1.58	0.94	269	C	7	6
4/22/1998	24:46.5	18	–67.15	14.2	2.2	0.12	1.02	0.73	293	B	5	5
9/1/1999	13:47.2	18.014	–67.162	12.3	2.3	0.04	0.93	0.39	277	B	4	4
9/2/1999	38:46.4	17.958	–67.224	1	2.6	0.19	1.42	2.43	296	C	5	3
9/29/1999	09:18.5	17.925	–67.168	1.1	2.5	0.01	1.1	0.47	302	B	4	2
10/18/1999	07:59.9	17.948	–67.074	9.7	2.4	0.17	1.62	1.25	272	C	4	3
3/14/2000	59:45.2	18.015	–67.069	5.4	3.5	0.19	1.49	1.25	162	C	9	5
4/5/2000	29:30.3	17.962	–67.181	13.8	2.3	0.16	1.3	0.42	285	C	6	5
4/23/2000	58:19.6	18.016	–67.168	12.4	2.4	0.06	0.73	0.28	267	B	5	5
5/15/2000	48:46.7	17.994	–67.135	13.7	2.8	0.28	1.95	1.53	292	C	6	6
5/23/2000	03:08.6	17.994	–67.173	12.8	2.5	0.07	0.98	0.29	301	B	4	4
6/8/2000	26:31.6	18.014	–67.122	14.8	3	0.06	0.79	0.66	270	B	5	4
6/12/2000	20:33.3	17.934	–67.121	20.3	2.6	0.2	1.47	3.19	292	C	5	4
3/9/2001	18:30.4	18.012	–67.186	15.4	2.3	0.13	1.12	0.8	297	B	4	4
3/22/2001	06:19.8	17.991	–67.071	14.2	2.5	0.08	0.84	0.91	233	B	5	5
5/18/2001	29:08.4	18.015	–67.134	7.5	2.9	0.1	3.2	4.75	290	B	5	5
9/7/2001	15:19.8	17.979	–67.086	8.5	2.7	0.07	2.81	1.54	275	B	6	3
10/7/2001	56:01.8	17.919	–67.153	11.4	2.4	0.07	1.36	0.58	299	B	4	4
5/4/2002	00:55.0	17.972	–67.116	19.8	2.3	0.12	3.88	1.67	288	B	5	3
7/7/2002	14:25.5	18.008	–67.165	13.9	2.5	0.14	0.96	0.59	232	B	5	4
11/15/2002	43:04.9	17.951	–67.095	12.8	2.2	0.09	1.88	0.51	328	B	4	4
11/15/2002	11:41.7	17.941	–67.088	12.2	2.5	0.05	2.19	0.59	326	B	4	3
11/15/2002	15:54.5	17.944	–67.105	13.2	2.3	0.03	1.8	0.33	329	B	4	4
(B) Region 02												
4/2/1988	52:41.1	18.05	–67.19	9.8	3.1	0.09	1.64	0.77	284	B	5	3
4/2/1988	53:41.2	18.06	–67.18	11.1	3.4	0.11	1.37	0.49	306	B	5	4
4/2/1988	27:46.0	18.06	–67.1	6	3.1	0.23	3.91	2.3	205	C	7	4
4/3/1988	16:28.2	18.06	–67.15	20.6	3	0.14	1.17	1.27	296	B	6	4
4/29/1988	49:43.3	18.12	–67.14	22.1	2.6	0.2	1.14	1.25	247	C	6	6
9/19/1988	54:18.4	18.07	–67.08	8.2	2.5	0.25	3.42	2.56	211	C	4	2
9/30/1990	42:14.7	18.1	–67.15	3	3.2	0.09	1.28	0.44	266	B	6	3
10/10/1990	12:07.6	18.1	–67.18	1.7	3.7	0.11	4.73	2.71	344	B	5	3
1/10/1992	49:11.8	18.08	–67.1	0.5	2.9	0.18	1.33	0.51	234	C	6	5
1/21/1992	35:13.2	18.08	–67.13	4	3	0.21	1.24	0.82	251	C	8	5
8/5/1992	46:36.2	18.08	–67.05	17.7	2.6	0.08	0.41	0.46	142	B	9	7
8/25/1992	43:58.7	18.04	–66.97	7.9	2.9	0.18	1.26	1.71	164	C	9	6
2/11/1993	00:17.6	18.07	–67.14	21.7	2.7	0.1	0.89	1.15	267	B	5	5
7/5/1993	12:40.6	18.12	–67.13	14.7	2.7	0.16	1.01	0.83	235	C	7	6
11/15/1993	34:32.2	18.03	–66.95	7.8	3.2	0.15	0.7	1.19	229	B	9	8
7/19/1994	32:01.1	18.05	–67.07	1.2	2.8	0.09	2.17	0.76	265	B	5	3
9/13/1994	54:38.6	18.04	–67.12	7.3	3.2	0.22	1.6	0.84	214	C	10	6
9/18/1994	47:17.4	18.1	–67.18	20.1	2.7	0.12	1.07	1.27	271	B	4	5
10/23/1994	30:07.2	18.03	–67.18	12.3	2.1	0.08	1.55	0.82	285	B	4	3

continued

TABLE 1. HYPOCENTERS USED TO COMPUTE COMPOSITE FOCAL MECHANISMS (*continued*)

Date	Time (GMT)	Lat	Long	Depth (km)	M_{PR}	rms	erh	erz	Gap	Q	Ps	Ss
(B) Region 02												
12/4/1994	47:27.6	18.03	–67.05	22.4	2	0.15	1.14	1.52	157	B	5	5
9/26/1995	20:40.6	18.06	–67.06	0.8	3.2	0.23	1.15	1.57	122	C	10	6
11/8/1995	24:42.4	18.07	–67.04	17.7	2	0.08	0.7	1.36	124	B	4	4
12/8/1995	39:05.8	18.13	–67.17	8.7	2.9	0.09	0.87	0.86	215	B	4	4
2/9/1996	31:30.3	18.11	–67.12	3.8	2.7	0.05	1.65	2.38	148	B	4	3
4/2/1996	02:50.9	18.05	–67.17	6.5	2.2	0.14	1.19	0.38	270	B	5	5
4/13/1996	02:54.3	18.03	–66.96	14	2.4	0.16	1.14	0.93	176	C	6	4
9/29/1997	55:39.8	18.05	–67.07	8.1	2.7	0.29	0.85	1.27	145	C	9	8
10/19/1997	04:36.8	18.07	–67.05	12.6	2.4	0.21	1.42	1.7	127	C	6	4
10/19/1997	01:31.4	18.08	–67.03	8	2.2	0.16	1.74	2.14	127	C	5	5
2/26/1998	22:41.9	18.03	–67.1	13.8	2.5	0.15	1.24	0.69	216	B	6	5
3/10/1998	32:19.3	18.08	–67.19	9.3	2.4	0.14	1.12	1.35	280	B	4	4
7/28/1998	09:52.8	18.07	–67	19.4	2	0.08	0.83	0.58	142	B	5	4
7/29/1998	38:12.1	18.05	–67.16	16.6	2.2	0.06	0.76	0.43	248	B	6	5
8/5/1998	00:01.0	18.04	–67.05	13.3	2.3	0.12	1.52	0.97	190	B	4	3
12/19/1998	35:39.7	18.04	–67.13	7.7	3	0.21	1.43	0.56	233	C	8	6
1/9/1999	59:48.2	18.05	–67.11	24.9	2.4	0.1	2.18	1.41	280	B	4	2
3/9/1999	38:02.5	18.04	–67.12	2.3	2.2	0.1	1.73	2.19	285	B	4	3
6/20/1999	32:48.4	18.044	–67.109	7.5	3.2	0.19	1.44	0.77	212	C	8	6
7/11/1999	32:41.1	18.059	–66.971	7.3	2.9	0.11	0.65	0.97	155	B	8	7
10/2/1999	53:21.2	18.1	–67.157	7.8	2.6	0.07	1.16	0.63	256	B	5	3
12/20/1999	08:35.3	18.16	–67.177	17.2	2.5	0.27	2.47	0.96	276	C	5	4
3/8/2000	23:03.6	18.077	–67.115	18.8	2.3	0.1	1.17	0.7	215	B	5	5
3/24/2000	11:55.3	18.114	–67.132	17.8	2.2	0.11	1.71	0.62	258	B	4	3
9/7/2000	21:24.3	18.035	–66.943	5.1	2.8	0.22	1.3	2.65	170	C	5	5
9/23/2000	33:15.9	18.148	–67.182	24.2	3.4	0.25	1.56	2.13	267	C	8	7
9/25/2000	22:25.6	18.035	–67.121	16.4	2.4	0.04	0.71	1.24	242	B	4	3
12/4/2000	16:30.9	18.046	–67.188	20.6	2.2	0.08	1.26	2.42	307	B	4	3
12/4/2000	46:54.6	18.064	–67.057	24	2.2	0.09	1.05	1.5	138	B	4	4
2/9/2001	00:32.8	18.094	–67.109	24.7	2.7	0.07	1.11	2.01	279	B	4	4
2/11/2001	52:28.8	18.076	–67.104	22.6	3.1	0.13	1.29	2.05	277	B	6	4
2/15/2001	09:07.7	18.092	–67.113	23.7	3.2	0.05	0.94	1.11	282	B	4	4
3/3/2001	18:08.9	18.063	–67.117	22.3	2.2	0.13	1.36	1.68	220	B	4	4
3/7/2001	19:48.4	18.114	–67.165	13.8	2.3	0.25	2.22	0.68	264	C	6	6
3/29/2001	37:43.6	18.094	–67.107	24.5	2.6	0.07	0.98	1.18	278	B	4	4
4/24/2001	17:13.4	18.051	–67.2	22.8	2.4	0.05	0.99	1.38	311	B	5	4
5/15/2001	37:55.6	18.021	–67.159	12.9	2.7	0.09	1.27	0.51	283	B	6	5
5/16/2001	42:47.7	18.024	–67.131	10.7	2.6	0.09	0.85	0.52	263	B	6	5
6/18/2001	27:54.2	18.058	–66.947	17.5	2.2	0.07	0.51	2.54	160	B	5	5
6/18/2001	53:34.7	18.058	–66.952	20.4	2.5	0.1	1.13	3.77	159	B	5	5
6/30/2001	07:18.1	18.095	–67.153	11.7	2.5	0.08	1.17	0.41	252	B	6	5
9/7/2001	07:03.6	18.052	–66.977	11.5	2.2	0.04	0.37	1.82	159	B	5	5
1/13/2002	30:52.3	18.031	–67.134	23.3	2.2	0.12	1.46	1.74	259	B	5	5
1/26/2002	06:35.4	18.049	–67.154	10.9	2.4	0.03	1.01	0.49	287	B	4	4
3/13/2002	22:53.5	18.046	–66.952	3.9	3.2	0.18	0.71	0.61	227	C	9	7
3/23/2002	57:39.1	18.067	–67.099	15	2.8	0.08	0.75	0.92	196	B	6	6
4/3/2002	16:24.9	18.046	–67.082	22.3	2.3	0.11	1.07	1.59	168	B	4	6
4/22/2002	12:17.0	18.052	–67.103	5.4	3	0.26	2.3	1.58	204	C	9	6
5/7/2002	33:34.8	18.048	–67.099	5.6	2.5	0.2	2.8	1.85	199	C	6	3
(C) Region 03												
4/13/1987	24:48.0	18.08	–66.79	4.3	3	0.19	1	1.88	164	C	6	5
8/7/1987	37:47.9	18.24	–66.97	18.2	3.2	0.15	0.95	1.58	82	B	7	2
9/24/1987	20:09.4	18.19	–66.79	11.9	3.4	0.1	0.58	0.99	103	B	5	2
11/2/1987	45:23.0	18.19	–66.93	19.4	2.9	0.2	3.91	4.39	131	C	6	4
5/10/1988	17:04.6	18.19	–66.93	14.4	2.1	0.05	0.7	1.28	177	B	4	2
5/10/1988	40:01.2	18.15	–66.92	23.5	3.3	0.08	0.68	1.76	161	B	6	3
5/15/1988	52:42.3	18.16	–66.9	17.2	2.8	0.13	1.04	0.98	182	B	4	3
5/16/1988	07:06.8	18.21	–66.9	19.9	3.2	0.14	0.61	0.79	171	B	5	4
7/8/1988	31:16.1	18.11	–66.82	22.7	2.6	0.11	1.46	1.02	244	B	4	4
9/5/1988	11:27.4	18.27	–67.02	18.8	2.8	0.16	0.95	0.8	221	C	5	5
3/5/1989	47:31.4	18.21	–66.82	19.5	3.4	0.14	0.55	0.74	135	B	6	5
3/5/1989	01:02.1	18.17	–66.8	8.4	2.6	0.11	1.09	2.42	145	B	5	4
3/24/1989	11:53.5	18.16	–67.04	10.2	2.5	0.1	0.7	1.13	148	B	5	4

continued

TABLE 1. HYPOCENTERS USED TO COMPUTE COMPOSITE FOCAL MECHANISMS (*continued*)

Date	Time (GMT)	Lat	Long	Depth (km)	M_{PR}	rms	erh	erz	Gap	Q	Ps	Ss
(C) Region 03												
5/22/1989	04:28.3	18.14	–66.74	19.5	2.9	0.1	0.53	0.44	112	B	5	5
11/20/1989	20:08.5	18.1	–67.04	0.7	3.4	0.18	2.49	1	307	C	5	3
11/20/1989	42:31.5	18.11	–66.99	0.2	2.8	0.24	2.71	1.05	297	C	4	4
12/31/1989	32:07.3	18.16	–67.12	8.4	2.7	0.25	1.96	1.53	321	C	4	4
3/18/1990	27:34.9	18.11	–66.92	4.9	3.5	0.27	2.59	0.86	278	C	5	3
8/6/1990	48:54.9	18.22	–67.06	14.6	2.7	0.07	0.57	0.99	161	B	4	3
5/2/1991	03:10.3	18.06	–66.8	12	3	0.16	0.68	1.21	168	C	6	3
5/13/1991	43:28.4	18.16	–66.93	5.5	2.8	0.09	4.66	3.45	165	B	4	4
9/21/1991	16:38.3	18.15	–67.14	5.1	3.2	0.14	3.19	2.49	204	B	6	2
12/16/1991	29:33.5	18.18	–67.02	18.7	2.4	0.14	0.63	0.85	203	B	7	4
12/20/1991	52:16.8	18.23	–66.88	13.9	2.2	0.11	1.07	0.9	166	B	5	5
5/12/1992	13:43.9	18.21	–67.06	5.1	3.1	0.22	2.86	2.73	222	C	6	4
7/18/1992	12:59.0	18.15	–66.93	1.7	2.5	0.13	1.79	0.96	227	B	4	4
8/7/1992	10:49.6	18.24	–67.07	16.4	3.2	0.12	0.89	1.37	231	B	5	2
9/5/1992	51:33.7	18.07	–66.81	14.4	2.9	0.15	0.64	0.85	164	B	9	7
9/16/1992	25:04.6	18.09	–66.88	10.6	2.3	0.19	0.54	0.87	151	C	7	7
9/19/1992	40:08.4	18.12	–67.07	4.8	2.9	0.14	0.88	0.63	215	B	8	5
11/9/1992	44:42.7	18.2	–66.93	0	2.7	0.26	0.8	1.85	137	C	5	4
12/23/1992	24:41.4	18.14	–66.99	10.5	2.9	0.09	0.89	0.56	294	B	7	6
1/21/1993	48:57.2	18.25	–67.09	12.6	3	0.1	0.85	0.48	240	B	6	6
1/22/1993	48:57.2	18.25	–67.09	11.9	2.7	0.18	1.56	0.77	237	C	5	5
2/12/1993	30:45.9	18.19	–67.08	14.5	2.8	0.05	0.51	0.66	228	B	7	6
4/24/1993	16:50.0	18.27	–66.98	23.4	2.6	0.13	0.82	1.25	184	B	5	4
5/30/1993	53:35.2	18.2	–67.17	0.9	3.1	0.24	1.34	1.21	218	C	10	7
5/31/1993	56:19.6	18.14	–66.97	6.8	3	0.24	1.07	1	173	C	9	8
5/31/1993	54:47.5	18.14	–66.98	11.3	2.8	0.23	0.66	0.8	114	C	9	9
6/26/1993	34:55.1	18.14	–67.09	18	2.5	0.05	0.8	0.51	239	B	4	4
6/26/1993	13:48.6	18.19	–67.13	11.4	3	0.23	0.96	0.6	198	C	9	8
7/21/1993	51:01.8	18.13	–66.94	11.8	2.7	0.18	0.62	0.64	124	C	9	7
10/6/1993	18:30.9	18.18	–67.05	22.2	3	0.14	0.43	0.96	87	B	9	10
12/24/1993	14:08.5	18.24	–67.04	16.9	2.9	0.24	1.31	0.7	197	C	9	8
1/11/1994	45:55.4	18.08	–66.92	7.5	2.6	0.16	0.4	0.51	148	C	8	8
2/5/1994	16:59.2	18.14	–67	11.8	2.5	0.03	0.99	2.43	162	B	4	2
3/9/1994	12:11.5	18.14	–66.82	24.6	2.2	0.28	2.36	2.15	187	C	5	5
3/19/1994	46:41.8	18.2	–67.13	24.2	2.8	0.2	0.67	0.87	131	C	11	10
4/5/1994	16:27.6	18.16	–66.86	17.9	2.9	0.13	0.45	0.54	97	B	8	10
5/23/1994	19:49.8	18.17	–67	0	2.7	0.1	0.47	1.49	130	B	4	2
10/8/1994	55:23.6	18.2	–67.05	11.5	3.2	0.11	0.57	0.87	155	B	8	3
11/7/1994	51:51.4	18.24	–67.01	9.1	3.1	0.15	0.4	0.86	104	B	11	6
11/29/1994	05:33.5	18.16	–67.04	16.8	3.4	0.22	0.81	1.27	94	C	10	5
12/9/1994	46:10.8	18.15	–66.72	17.8	2.5	0.07	0.65	1.39	159	B	4	4
1/30/1995	39:05.6	18.24	–66.92	15.2	2.7	0.15	0.57	0.58	109	B	10	8
2/20/1995	21:16.4	18.12	–67.04	3.3	3.1	0.1	1.27	0.69	298	B	7	4
3/14/1995	45:24.2	18.22	–66.99	18.2	3.1	0.27	1.77	1.45	161	C	8	5
4/27/1995	28:03.1	18.18	–67.14	8	3.2	0.08	0.43	0.54	201	B	7	3
5/2/1995	35:26.0	18.1	–66.98	8.3	3.1	0.07	0.45	0.81	132	B	8	3
5/19/1995	19:19.2	18.18	–66.97	15.7	3.1	0.26	0.74	1.22	91	C	12	7
5/20/1995	39:09.7	18.09	–66.91	6	2.5	0.19	0.91	1.79	151	C	6	2
5/21/1995	15:18.3	18.17	–66.91	11.5	3.6	0.07	0.39	0.6	98	B	11	2
5/27/1995	12:10.9	18.06	–66.89	13.3	2.8	0.1	0.49	0.33	165	B	10	6
7/31/1995	05:00.0	18.16	–66.82	11.8	2.9	0.11	0.47	0.5	152	B	9	6
8/8/1995	28:11.9	18.17	–67.13	8.7	2.7	0.1	0.72	1.24	259	B	5	4
10/22/1995	58:49.5	18.1	–67	8.6	3.5	0.19	0.67	1.09	122	C	10	5
10/23/1995	18:20.0	18.1	–66.98	5.6	2.9	0.05	0.4	0.55	134	B	8	6
10/28/1995	33:26.2	18.19	–66.98	11.8	3.3	0.2	0.54	0.73	135	C	9	7
11/23/1995	17:22.0	18.16	–67.01	13.5	2.2	0.06	1.34	1.16	204	B	4	4
1/27/1996	57:44.9	18.17	–66.96	8	3	0.13	0.51	0.89	180	B	7	6
2/2/1996	21:01.2	18.1	–66.9	17.4	2.8	0.06	0.34	0.55	134	B	9	6
3/8/1996	03:39.9	18.14	–66.84	15.7	2.6	0.13	0.47	0.73	127	B	8	7
3/14/1996	33:10.7	18.12	–66.71	13.9	2	0.09	1.1	1.53	145	B	4	3
3/31/1996	02:19.4	18.15	–66.83	15.5	2.9	0.23	0.59	0.88	94	C	11	8
4/3/1996	22:50.6	18.21	–67.13	19.4	2.1	0.27	2.58	1.08	295	C	5	4
5/6/1996	13:53.6	18.11	–66.98	23.1	2.1	0.14	1.24	2.32	150	B	4	4

continued

TABLE 1. HYPOCENTERS USED TO COMPUTE COMPOSITE FOCAL MECHANISMS (*continued*)

Date	Time (GMT)	Lat	Long	Depth (km)	M_{PR}	rms	erh	erz	Gap	Q	Ps	Ss
(C) Region 03												
7/14/1996	40:14.9	18.19	–66.83	22.9	2.9	0.18	0.58	1.51	80	C	11	9
7/30/1996	32:59.1	18.22	–67.04	20.8	2.7	0.25	0.92	2.2	154	C	7	7
8/11/1996	48:09.9	18.09	–66.86	17.9	2.6	0.26	1.01	1.34	140	C	7	5
8/30/1996	43:05.1	18.15	–66.73	20.9	2.6	0.12	1.93	1.8	160	B	6	5
9/27/1996	42:52.0	18.16	–66.73	20.4	2.5	0.11	1.11	1.63	158	B	5	4
11/5/1996	36:31.6	18.08	–66.97	12.5	2.6	0.16	0.56	0.78	147	C	7	7
11/12/1996	19:27.7	18.13	–66.78	7.8	3.4	0.15	0.65	2.57	155	B	6	3
11/14/1996	01:28.7	18.16	–67.13	14.9	2.6	0.3	2.24	1.09	258	C	5	4
1/3/1997	16:32.8	18.17	–66.89	17.7	2.5	0.12	0.52	0.6	106	B	6	4
2/9/1997	38:48.7	18.24	–67.18	19.2	2.1	0.26	2.34	0.66	263	C	5	5
4/15/1997	32:31.8	18.13	–66.76	24.1	3.1	0.09	0.91	1.26	146	B	5	5
5/25/1997	55:01.7	18.27	–67.06	23.3	2.5	0.05	0.7	1.51	174	B	4	2
7/2/1997	50:49.6	18.16	–66.95	22.9	2.6	0.12	0.58	1	119	B	7	5
8/22/1997	16:56.9	18.23	–67.16	22.5	2.2	0.16	0.79	1.9	142	C	5	4
9/6/1997	26:44.9	18.17	–66.84	19.6	3	0.19	0.6	0.75	115	C	9	8
9/22/1997	43:43.6	18.16	–66.73	14	2.3	0.13	0.48	0.86	115	B	7	6
10/1/1997	40:59.1	18.17	–66.9	19.6	2.3	0.1	0.45	0.71	113	B	8	5
12/15/1997	40:01.9	18.17	–67.14	21.4	2.5	0.2	1.7	1.41	278	C	5	4
1/19/1998	12:10.0	18.13	–66.68	18.7	2.6	0.2	0.63	0.37	80	C	9	10
3/22/1998	09:59.0	18.18	–66.89	18	2.5	0.26	0.57	0.91	95	C	9	9
4/10/1998	31:38.4	18.15	–66.85	16.1	3	0.3	0.76	1.13	99	C	12	11
4/10/1998	33:36.1	18.15	–66.85	21.9	2.3	0.17	0.58	1.58	98	C	8	6
5/9/1998	17:31.9	18.22	–66.86	16	2.2	0.3	1.32	2.68	130	C	5	4
5/22/1998	09:36.0	18.18	–66.84	19.7	2.8	0.25	0.64	0.85	84	C	12	10
6/18/1998	42:23.4	18.13	–66.71	19.3	2.5	0.16	0.68	0.36	111	C	10	9
6/30/1998	14:39.7	18.08	–66.68	2.5	2.2	0.14	1.39	1.46	142	B	4	3
10/16/1998	19:46.5	18.14	–66.88	18.9	2.8	0.11	0.44	1.1	134	B	5	6
11/11/1998	29:20.8	18.24	–67.04	16.7	3.9	0.06	0.87	1	131	B	6	2
1/13/1999	14:54.7	18.15	–66.88	13.2	2.5	0.11	0.45	0.62	123	B	7	5
1/21/1999	38:55.0	18.15	–66.86	18.6	3	0.13	0.57	0.79	126	B	8	5
1/21/1999	18:11.1	18.15	–66.85	19.6	2.2	0.11	1.3	1.2	144	B	5	3
4/1/1999	18:05.4	18.16	–67.03	9.3	2.4	0.12	0.8	1.92	212	B	5	4
4/10/1999	20:42.5	18.06	–66.86	19	2.9	0.11	0.49	0.38	164	B	9	7
4/30/1999	22:24.5	18.07	–66.86	2.3	2.1	0.22	0.79	2.2	163	C	5	3
5/1/1999	44:37.1	18.06	–66.74	7	2.4	0.07	0.8	0.48	169	B	5	2
5/8/1999	17:36.9	18.23	–67.1	13.3	3.3	0.24	1.18	0.94	191	C	9	7
5/20/1999	18:07.2	18.19	–67.09	22.9	2.7	0.16	1.02	2.05	181	C	7	5
5/26/1999	33:37.1	18.17	–67.17	12.5	2.2	0.13	0.82	0.91	225	B	5	5
6/17/1999	35:10.9	18.07	–66.909	10.8	3.4	0.21	1.03	1.26	212	C	8	5
6/17/1999	53:14.6	18.1	–66.773	12.3	2.8	0.12	0.65	0.71	156	B	9	4
7/23/1999	44:41.0	18.145	–66.84	16.6	2.4	0.12	0.41	0.78	135	B	8	6
7/26/1999	08:47.3	18.14	–67.038	19.9	2.6	0.19	1.03	0.97	109	C	8	5
7/26/1999	09:40.0	18.147	–67.047	21.3	2	0.09	0.59	1.61	113	B	6	3
10/3/1999	15:08.2	18.14	–67.006	20.6	2	0.09	1.04	2.05	108	B	4	3
12/3/1999	25:55.9	18.091	–66.918	13.2	3	0.12	0.86	0.62	202	B	6	7
12/4/1999	45:01.3	18.112	–66.945	9.7	2.4	0.08	0.59	0.88	195	B	4	4
2/9/2000	53:23.3	18.087	–66.944	23.4	2.5	0.09	0.52	1.48	145	B	6	7
3/6/2000	58:37.7	18.129	–66.914	13.5	3.2	0.23	1.04	0.74	182	C	8	7
4/3/2000	08:53.7	18.18	–67.159	9.4	2.7	0.1	0.44	0.9	206	B	6	5
4/20/2000	12:11.0	18.125	–66.738	23.2	2.4	0.17	0.51	1.58	125	C	6	7
4/28/2000	54:33.4	18.054	–66.86	14.8	2.2	0.16	0.77	1.13	212	C	5	5
5/23/2000	32:46.6	18.145	–66.775	16.8	2.5	0.1	0.45	0.53	122	B	6	5
7/2/2000	12:28.0	18.139	–66.729	21.1	2.3	0.06	0.79	2.06	143	B	4	4
7/5/2000	32:34.4	18.155	–66.833	20	2.3	0.08	0.85	1.62	158	B	4	4
7/8/2000	15:42.7	18.119	–66.847	23.6	2	0.07	0.58	1.35	177	B	4	4
8/11/2000	56:58.5	18.234	–67.084	22.2	2.5	0.17	0.66	1.17	124	C	4	7
9/17/2000	52:07.5	18.139	–66.848	5.9	2.3	0.12	0.71	1.08	129	B	4	3
9/19/2000	08:05.0	18.168	–66.903	4.8	2.6	0.18	0.86	1.38	115	C	5	4
10/20/2000	31:56.6	18.164	–66.801	16	3.3	0.08	2.06	1.46	165	B	5	6
11/13/2000	29:52.6	18.177	–67.052	13	2.9	0.14	1.72	0.99	215	B	4	5
12/22/2000	47:43.1	18.135	–66.886	15.6	2.5	0.17	0.59	1.18	129	C	7	7
12/31/2000	43:20.4	18.185	–67.083	18.6	2.6	0.05	1.19	0.75	223	B	4	4
1/1/2001	15:10.6	18.217	–67.128	23.4	2.5	0.14	1.49	0.74	264	B	4	5

continued

TABLE 1. HYPOCENTERS USED TO COMPUTE COMPOSITE FOCAL MECHANISMS (*continued*)

Date	Time (GMT)	Lat	Long	Depth (km)	M_{PR}	rms	erh	erz	Gap	Q	Ps	Ss
(C) Region 03												
1/1/2001	15:10.7	18.182	–67.133	23.2	2.5	0.12	1.12	1.02	289	B	4	5
1/1/2001	51:05.6	18.221	–67.127	23.3	2.2	0.09	1.05	0.83	264	B	4	4
1/18/2001	20:33.7	18.086	–66.808	22	2	0.08	0.86	2.12	189	B	4	4
1/28/2001	30:18.7	18.154	–66.864	16.4	2.1	0.13	0.65	0.88	163	B	4	6
2/9/2001	19:07.8	18.101	–67.034	7.9	3.1	0.3	2.68	2.36	231	C	8	8
2/9/2001	53:25.5	18.12	–66.977	1.2	2.8	0.3	0.98	1.9	199	C	7	5
3/6/2001	46:01.1	18.193	–67.163	21.6	2.5	0.08	0.66	1.19	258	B	5	5
3/8/2001	29:03.4	18.117	–66.934	13.4	2.8	0.08	0.37	0.64	132	B	6	7
3/8/2001	15:53.6	18.121	–66.927	11.9	2.3	0.21	0.65	4.46	131	C	4	6
3/10/2001	38:02.8	18.079	–66.939	14.9	2.5	0.06	0.62	3.01	154	B	4	5
4/15/2001	35:02.9	18.253	–67.162	23	2.4	0.15	1.61	1.02	306	B	4	4
5/11/2001	23:14.7	18.156	–66.736	19.5	2.4	0.11	0.77	0.51	113	B	7	5
6/18/2001	00:21.0	18.067	–66.933	10.5	2.3	0.1	0.62	0.86	157	B	6	4
7/8/2001	55:46.5	18.233	–67.184	18.7	2.7	0.16	1.37	0.63	264	C	6	6
8/31/2001	19:44.4	18.098	–66.967	5.6	2.8	0.08	0.46	1	136	B	6	6
9/2/2001	03:34.8	18.116	–66.933	14.7	2.1	0.13	0.85	4.26	132	B	5	5
10/1/2001	37:13.9	18.112	–67.018	8.9	3	0.19	0.62	1.19	114	C	9	7
10/1/2001	05:26.8	18.112	–67.033	13	2.8	0.2	1.19	1.21	115	C	8	8
10/9/2001	51:53.4	18.073	–66.942	18.6	2.6	0.02	0.45	1.49	152	B	5	5
11/2/2001	58:40.9	18.114	–67.041	13.8	3.1	0.19	2.18	2.32	123	C	8	6
12/3/2001	06:57.6	18.124	–66.936	9.4	2.6	0.3	0.83	2.04	129	C	9	5
12/4/2001	15:48.8	18.111	–66.931	7.7	2.4	0.1	0.46	1.51	135	B	6	6
12/6/2001	19:08.1	18.211	–66.92	13.8	3.3	0.22	1.16	1.07	137	C	8	6
1/2/2002	47:46.6	18.147	–66.872	11.5	2.3	0.12	0.97	4.83	124	B	5	4
1/31/2002	53:41.8	18.217	–67.147	21.8	2.3	0.3	1.99	3.63	270	C	4	4
2/28/2002	39:13.2	18.257	–67.155	15.2	2.3	0.12	1.3	1.14	278	B	5	4
3/7/2002	13:27.1	18.136	–66.99	10.4	2.6	0.05	0.72	2.38	114	B	5	5
4/6/2002	06:00.2	18.201	–67.11	23.1	2.7	0.15	0.95	1.28	240	B	7	7
5/25/2002	41:56.8	18.213	–66.898	16.9	2.1	0.09	0.44	0.55	105	B	5	8
5/30/2002	50:55.8	18.164	–67.112	24.7	2.9	0.25	1.6	1.25	130	C	5	6
6/7/2002	10:24.9	18.137	–66.922	16.7	3	0.12	0.41	0.77	125	B	9	8
6/12/2002	40:42.6	18.168	–66.834	23.8	2.1	0.19	1.04	1.65	152	C	5	5
6/17/2002	15:49.1	18.235	–67.143	17	2.3	0.15	1.35	1.08	159	B	6	2
7/28/2002	40:02.2	18.178	–67.073	21.5	2.6	0.09	0.48	0.66	111	B	7	7
9/21/2002	26:02.6	18.164	–67.127	22.1	3.1	0.15	0.53	1.08	136	B	13	10
9/28/2002	05:10.8	18.124	–66.765	14.6	2.3	0.08	0.87	0.57	131	B	5	4
10/9/2002	56:54.5	18.202	–66.873	23.7	2.5	0.07	0.45	1.03	104	B	6	6
11/18/2002	31:07.3	18.194	–67.18	22.5	2.5	0.18	0.91	1.94	148	C	8	5
(D) Region 04												
4/15/1987	00:13.8	17.96	–66.92	3.2	2.8	0.18	2.15	2.47	203	C	4	6
1/16/1993	38:14.6	17.89	–66.94	15.2	2.2	0.13	1.19	3.76	248	B	4	4
4/20/1993	10:28.3	17.95	–66.95	18.3	2.8	0.1	0.67	0.66	218	B	6	5
1/16/1994	04:03.8	17.9	–66.94	7.2	3.1	0.25	1.15	1.33	227	C	11	9
1/20/1994	57:21.4	17.95	–66.98	12.7	2.8	0.2	1.2	0.76	251	C	7	7
10/4/1995	36:16.3	17.9	–67.01	7.5	3.2	0.12	1.22	0.74	251	B	6	4
6/16/1996	07:02.9	17.95	–66.9	8.3	2.9	0.25	0.95	1.2	204	C	10	9
8/16/1996	15:20.3	17.94	–66.93	16.5	2.6	0.26	1.72	1.22	210	C	6	6
10/24/1996	23:01.8	18	–66.94	2.6	2.6	0.22	0.78	1.18	193	C	6	5
10/3/1997	28:01.4	18	–66.93	9.6	2.6	0.19	0.71	0.98	185	C	9	7
10/3/1997	23:34.4	18	–66.93	7.4	2.6	0.24	0.95	0.8	187	C	10	7
10/13/1997	43:45.8	17.99	–66.95	6	2.7	0.28	0.78	1.17	193	C	10	8
12/8/1997	47:24.2	17.96	–66.9	7.2	2.7	0.2	0.66	1.41	199	C	9	7
11/15/1998	05:42.8	18.01	–66.98	3.1	2.8	0.04	0.39	0.57	186	B	7	3
11/24/1998	23:45.1	17.95	–66.91	4.9	2.3	0.13	1.7	1.35	222	B	4	4
4/28/1999	51:53.8	17.95	–66.91	1.6	2.8	0.11	0.62	0.94	206	B	7	3
6/11/1999	14:11.1	17.988	–66.954	5.5	2.3	0.09	2.05	2.89	202	B	4	5
11/13/1999	50:31.9	17.876	–66.963	12.2	2.8	0.05	0.68	0.51	257	B	5	6
11/13/1999	51:17.5	17.901	–66.95	6.2	2.5	0.24	2.34	1.93	246	C	5	5
11/13/1999	56:58.3	17.895	–66.97	14.6	2.8	0.08	0.99	2.2	252	B	5	5
11/14/1999	20:36.3	17.944	–66.936	0.2	3	0.1	0.46	0.71	210	B	7	6
11/14/1999	54:29.1	17.891	–66.968	15	2.6	0.09	0.81	2.18	253	B	5	4
11/14/1999	49:56.7	17.894	–66.981	17.7	2.6	0.1	0.92	2.11	255	B	5	4
11/24/1999	10:03.7	17.889	–66.967	18.2	2.8	0.09	1.01	1.73	253	B	5	4

continued

TABLE 1. HYPOCENTERS USED TO COMPUTE COMPOSITE FOCAL MECHANISMS (*continued*)

Date	Time (GMT)	Lat	Long	Depth (km)	M_{PR}	rms	erh	erz	Gap	Q	Ps	Ss
(D) Region 04												
11/29/1999	21:55.2	17.925	–66.948	5	2.9	0.14	1.11	1.18	219	B	7	6
12/14/1999	51:09.2	17.954	–66.965	14.2	3	0.12	0.94	2.37	223	B	5	4
12/19/1999	20:01.6	18.011	–66.927	6	2.3	0.07	0.88	1.67	188	B	5	5
6/13/2000	45:21.1	17.958	–66.951	15.8	2.3	0.07	0.83	1.69	219	B	4	4
8/19/2000	15:51.1	17.917	–66.939	3.9	3.6	0.08	0.83	0.58	228	B	8	3
8/19/2000	50:56.4	17.892	–66.958	8.8	3.1	0.11	0.73	1	239	B	7	7
9/1/2000	24:14.3	17.874	–66.976	13.7	2.7	0.13	1.2	1.21	261	B	4	4
10/29/2000	33:37.3	17.978	–66.939	16.6	3.6	0.07	1.07	1.14	196	B	8	2
12/20/2000	46:05.9	17.891	–66.931	20.4	2.2	0.09	1.56	4.02	249	B	4	4
5/28/2001	31:42.6	17.943	–66.943	12.6	2.1	0.11	0.79	1.08	226	B	5	5
6/13/2001	39:06.4	17.93	–66.918	4.6	2.8	0.09	0.61	0.46	231	B	7	7
7/4/2001	12:50.8	17.873	–66.939	13	2.7	0.12	1.35	1.08	255	B	5	5
7/8/2001	33:15.7	17.875	–66.943	12.6	2.9	0.12	1.61	0.92	253	B	6	6
7/31/2001	07:34.5	17.976	–67.009	6.3	2.5	0.09	2.2	2.73	216	B	5	5
8/9/2001	19:36.3	17.951	–66.918	0.8	2.7	0.09	0.76	1.31	222	B	6	4
8/25/2001	33:47.0	17.916	–66.935	13.6	2.7	0.06	1.06	1.09	238	B	4	4
8/27/2001	35:51.0	17.928	–66.948	14.1	2.3	0.19	1.97	3.68	234	C	4	4
10/13/2001	15:32.6	17.909	–66.976	13.6	2.9	0.18	1.84	1.65	247	C	5	4
10/25/2001	30:56.1	17.883	–66.921	7.2	2.5	0.12	1.16	2.15	249	B	6	5
2/3/2002	37:36.6	17.965	–66.924	8.9	2.8	0.12	0.81	1.28	208	B	6	5
7/3/2002	15:00.4	17.948	–66.99	24.8	3.1	0.24	2.14	1.68	219	C	9	6
9/1/2002	14:32.0	17.919	–66.942	12.7	2.2	0.09	0.98	0.7	238	B	6	5
9/1/2002	21:49.6	17.921	–66.945	12.9	2.1	0.1	1.14	0.73	237	B	5	6
9/1/2002	55:38.0	17.928	–66.949	12.5	2.4	0.1	1.26	1.26	234	B	6	5
11/8/2002	56:35.8	17.874	–66.939	12.9	2.5	0.11	1.09	1.25	255	B	6	4
11/8/2002	03:22.4	17.904	–66.951	14.9	2.2	0.09	1.1	2.33	245	B	5	4
11/13/2002	12:36.0	17.89	–66.963	10.7	2.4	0.1	1.35	3.04	252	B	6	4
(E) Region 05												
4/27/1987	56:10.9	18.01	–66.73	1	2.5	0.08	0.68	0.69	182	B	4	3
5/1/1987	24:29.7	17.92	–66.8	5.2	3	0.18	2.23	1.86	208	C	4	3
9/16/1990	20:14.1	18.02	–66.87	15.8	2.3	0.25	1.18	2.82	185	C	4	3
10/29/1990	56:29.4	18.04	–66.88	3.3	3	0.09	1.52	0.68	278	B	4	2
5/1/1991	18:51.9	17.92	–66.8	1.5	2.9	0.28	1.32	1.74	208	C	5	3
6/8/1991	10:30.5	17.91	–66.82	3	2.8	0.12	0.78	0.66	210	B	7	2
9/28/1991	18:35.9	17.93	–66.86	9.7	3.1	0.12	0.7	1.05	215	B	7	3
10/12/1991	40:44.5	17.98	–66.75	9.3	2.7	0.1	0.59	1.05	190	B	8	3
6/5/1992	19:52.9	18.03	–66.84	9.2	2.3	0.07	0.43	0.8	177	B	6	5
10/18/1992	14:44.9	18	–66.79	18.4	2.2	0.08	0.49	0.36	192	B	8	8
3/11/1993	29:47.0	17.96	–66.78	11.4	3	0.08	0.62	0.3	219	B	4	6
5/15/1993	47:20.4	17.94	–66.85	2.2	2.6	0.17	0.73	0.7	204	C	7	6
6/27/1993	38:31.7	18.01	–66.75	12.3	3.3	0.15	0.53	0.7	183	B	10	5
6/27/1993	01:13.0	18.02	–66.74	15	2.6	0.09	0.57	1.75	194	B	4	4
6/28/1993	52:00.9	18	–66.74	14	3	0.12	0.63	0.3	192	B	7	8
7/11/1993	27:16.8	18.03	–66.8	12.9	3.2	0.14	0.57	0.45	177	B	9	7
7/11/1993	51:50.6	18.04	–66.77	17.1	2.6	0.18	1.67	0.72	260	C	5	5
11/20/1993	22:24.4	18.01	–66.76	12.6	3.1	0.16	0.56	0.43	182	C	10	9
11/20/1993	41:05.2	18.02	–66.76	12	3.1	0.12	0.49	0.37	181	B	11	10
1/30/1994	37:51.0	17.97	–66.75	13.2	2.7	0.18	1.56	0.43	201	C	8	7
5/22/1994	08:41.3	18.01	–66.71	8.3	3.2	0.17	0.87	0.7	182	C	9	6
5/28/1994	59:55.1	17.94	–66.84	22.7	2.2	0.05	4.34	2.42	219	B	4	4
11/1/1994	42:33.8	17.91	–66.77	12	2.4	0.06	1.48	0.57	291	B	5	3
3/29/1995	38:50.5	17.94	–66.73	14.8	2.8	0.11	1.28	0.64	208	B	7	5
3/30/1995	45:05.2	17.94	–66.73	13.6	3	0.08	0.83	0.68	208	B	6	2
4/3/1995	31:32.1	17.92	–66.73	13.8	2.8	0.08	0.68	0.41	214	B	8	6
4/7/1995	34:46.3	17.92	–66.73	12.6	3.2	0.09	1.06	0.62	215	B	9	2
4/14/1995	58:59.4	17.97	–66.71	19.4	2.4	0.07	2.13	1.8	248	B	4	3
4/14/1995	56:42.9	17.95	–66.72	18.7	2.1	0.04	1.95	0.86	221	B	4	3
4/26/1995	18:56.0	17.98	–66.74	13.6	2.7	0.19	1.06	1.3	200	C	6	4
5/9/1995	42:49.1	17.95	–66.73	14.6	3	0.03	0.9	0.65	207	B	9	2
5/15/1995	47:53.6	17.95	–66.78	16.1	2	0.1	0.79	2.25	224	B	5	5
8/13/1995	29:33.6	18.04	–66.83	12.6	3.2	0.13	0.51	0.49	173	B	11	4
3/13/1996	41:34.7	18.03	–66.79	16.4	3	0.2	0.92	0.59	183	C	9	8

continued

TABLE 1. HYPOCENTERS USED TO COMPUTE COMPOSITE FOCAL MECHANISMS (*continued*)

Date	Time (GMT)	Lat	Long	Depth (km)	M_{PR}	rms	erh	erz	Gap	Q	Ps	Ss
(E) Region 05												
3/13/1996	43:00.6	18.01	–66.8	17.9	2.3	0.18	0.66	0.84	194	C	8	7
3/13/1996	41:02.8	18.01	–66.77	23.5	2	0.06	0.97	0.81	200	B	6	4
3/13/1996	09:37.0	18.01	–66.79	16.2	2.9	0.13	0.69	0.86	190	B	8	6
3/13/1996	18:39.6	18.01	–66.8	15.2	2.9	0.13	0.62	0.95	196	B	7	7
3/13/1996	29:30.0	17.94	–66.78	14.7	2.9	0.11	2.47	1.73	211	B	5	3
3/13/1996	56:59.0	18	–66.8	16.6	2.5	0.18	1.41	0.93	203	C	7	6
3/13/1996	12:15.8	18.01	–66.8	14.2	2.7	0.12	0.52	0.84	194	B	7	6
3/13/1996	23:34.4	18.02	–66.8	23.1	2.2	0.12	0.9	1.23	193	B	5	3
3/13/1996	24:02.4	18.01	–66.8	15	3.1	0.15	0.61	0.85	190	B	10	8
3/14/1996	51:57.2	18.02	–66.8	16.3	3	0.2	0.86	0.75	186	C	10	8
3/14/1996	44:06.3	18.02	–66.8	13.3	3.4	0.19	0.63	0.55	187	C	10	5
3/14/1996	04:14.5	18.02	–66.8	17.1	2.5	0.16	0.82	0.71	188	C	7	7
3/14/1996	06:57.3	18.01	–66.8	15	2.8	0.12	0.57	0.68	190	B	9	7
3/14/1996	38:48.6	18.01	–66.8	15.3	2.9	0.15	0.63	0.76	188	B	9	8
3/14/1996	16:25.8	18.03	–66.8	15	3.2	0.27	1.04	1.18	185	C	11	6
3/14/1996	57:25.3	18.02	–66.78	22.5	2.3	0.11	0.9	1.06	193	B	6	5
3/16/1996	41:56.1	17.94	–66.71	11.8	2.2	0.09	1.41	0.7	280	B	4	4
3/16/1996	32:03.6	18.02	–66.79	21.7	2.3	0.13	0.63	0.77	192	B	6	7
5/26/1996	01:22.3	18.02	–66.8	13.2	2.7	0.19	0.88	0.59	180	C	8	7
5/18/1997	06:54.6	17.98	–66.71	1	2.2	0.21	1.15	1.02	217	C	5	4
12/21/1997	55:36.9	17.92	–66.87	22	2.3	0.24	1.34	2.39	213	C	7	6
4/21/1998	27:51.3	17.98	–66.85	17	2	0.29	2.12	1.57	243	C	6	5
8/11/1998	22:30.0	18.04	–66.74	10	2.6	0.1	0.62	0.53	175	B	6	8
10/15/1998	19:29.5	17.91	–66.88	3.5	2.7	0.15	0.94	0.79	216	B	6	4
11/4/1999	11:47.0	18.017	–66.764	15.1	2.1	0.11	1.15	2.95	197	B	5	4
12/13/1999	40:34.0	17.964	–66.808	17.4	2.5	0.08	1.67	1.67	218	B	4	4
2/17/2000	44:42.0	18.012	–66.743	18.8	2.3	0.09	4.47	0.67	219	B	4	3
2/22/2000	48:26.1	17.943	–66.759	17.8	2	0.07	1.02	2.64	256	B	4	4
4/6/2000	08:05.9	18.038	–66.777	17	2.2	0.15	0.63	1.97	182	B	7	7
4/26/2000	35:12.7	17.993	–66.844	3.2	3	0.12	0.35	0.53	187	B	9	8
6/7/2000	51:53.9	17.916	–66.859	18.3	2.9	0.08	0.74	1.94	237	B	4	4
6/18/2000	28:47.4	17.949	–66.828	4.9	3.3	0.17	0.72	0.49	200	C	9	7
6/18/2000	07:28.7	17.926	–66.841	16.7	2.7	0.1	0.8	2.32	231	B	5	5
6/19/2000	13:25.5	17.968	–66.792	16.4	2.3	0.13	1.49	3.06	216	B	4	4
9/29/2000	13:19.9	17.911	–66.859	7.2	2	0.03	1.04	1.44	236	B	4	4
10/1/2000	28:36.9	17.916	–66.855	5.1	2	0.06	1.09	0.51	235	B	4	4
11/27/2000	02:23.8	17.943	–66.833	9.5	2.9	0.07	0.55	0.78	240	B	6	6
11/27/2000	05:50.0	17.944	–66.832	9.2	2.7	0.18	1	1.23	254	C	5	4
3/6/2001	11:46.0	18.034	–66.802	12.2	2.2	0.09	1.16	2.64	183	B	4	4
4/24/2001	10:31.2	17.928	–66.832	12.9	2.8	0.1	1.96	0.82	230	B	5	5
4/25/2001	28:17.6	17.91	–66.833	12.8	2	0.1	1.54	0.88	237	B	4	4
7/19/2001	21:17.1	17.978	–66.813	14.8	2.6	0.11	0.73	2.8	211	B	6	6
7/19/2001	22:41.6	17.965	–66.794	21.5	2.4	0.13	1.69	2.47	218	B	5	4
9/13/2001	35:31.5	18.007	–66.815	6.3	2.6	0.13	0.61	0.86	197	B	6	5
11/22/2001	06:47.2	17.954	–66.777	18	2.4	0.03	0.62	1.74	224	B	5	4
12/28/2001	07:24.0	18.03	–66.763	14	2.2	0.09	1.1	0.49	188	B	6	5
2/28/2002	07:34.9	17.937	–66.796	6.5	3	0.14	0.82	0.77	202	B	8	8
4/29/2002	16:44.7	17.951	–66.746	7.8	2.2	0.1	1.47	1.31	222	B	4	5
5/24/2002	05:19.6	18.028	–66.812	13.1	2	0.12	0.98	3	186	B	5	5
5/24/2002	12:13.4	18.04	–66.815	11.5	2.4	0.06	0.5	2.79	178	B	5	4
10/26/2002	52:30.9	17.955	–66.802	9.4	3.6	0.18	0.64	0.66	178	C	15	8
11/5/2002	20:48.0	17.917	–66.876	3.5	2.8	0.13	0.45	0.5	201	B	10	10
11/19/2002	19:27.7	18.039	–66.739	11.1	2.1	0.08	0.58	1.82	184	B	4	4
11/25/2002	31:56.2	17.942	–66.855	7.6	3.1	0.25	0.79	1.36	189	C	13	13
11/28/2002	55:50.0	17.932	–66.86	9.2	2.8	0.21	0.78	0.92	194	C	13	11
(F) Region 06												
4/3/1987	22:28.2	18.16	–67.23	18.2	2.7	0.28	2.41	1.11	246	C	4	3
4/23/1988	54:32.8	18.13	–67.29	6.3	2.8	0.22	1.66	0.79	295	C	5	4
6/12/1988	50:47.4	18.14	–67.26	17.9	2.9	0.16	2.61	1.8	289	C	6	5
11/19/1988	56:51.8	18.13	–67.27	12.2	3.3	0.16	1.55	0.93	263	C	4	4
3/30/1989	20:38.9	18.19	–67.21	11.1	3.5	0.13	2.62	2.01	233	B	6	2

continued

TABLE 1. HYPOCENTERS USED TO COMPUTE COMPOSITE FOCAL MECHANISMS (*continued*)

Date	Time (GMT)	Lat	Long	Depth (km)	M_{PR}	rms	erh	erz	Gap	Q	Ps	Ss
(F) Region 06												
7/7/1991	39:20.8	18.16	–67.21	19.1	2.9	0.12	1.88	1.31	239	B	6	3
12/19/1991	42:55.9	18.16	–67.25	12.7	2.8	0.13	1.11	0.56	318	B	4	4
7/31/1992	05:15.6	18.18	–67.21	10.4	2.6	0.16	1.56	1.2	285	C	4	3
11/1/1992	30:22.5	18.19	–67.2	11.4	2.9	0.3	1.85	1.38	270	C	6	5
11/15/1992	18:27.7	18.18	–67.26	27.4	2.6	0.07	0.99	1.49	285	B	4	4
12/19/1992	18:44.9	18.18	–67.25	26.6	2.5	0.07	0.67	1.48	249	B	5	4
3/14/1994	33:33.9	18.14	–67.26	15.2	3.1	0.1	1.2	0.67	257	B	5	4
4/9/1994	47:07.0	18.15	–67.24	12.3	3.1	0.15	1.53	0.8	251	B	6	4
4/9/1994	38:05.5	18.13	–67.28	13.7	2.9	0.3	1.45	0.8	198	C	10	8
5/5/1994	55:38.9	18.18	–67.23	13	3.1	0.28	2.57	1.56	243	C	5	5
5/13/1994	22:59.2	18.13	–67.26	18.3	3.2	0.12	1.29	0.92	258	B	7	6
8/10/1994	51:19.1	18.15	–67.24	3.4	3.7	0.15	0.79	1.28	168	B	4	5
9/20/1994	44:06.4	18.13	–67.25	27.2	2.2	0.12	1.28	1.04	304	B	4	4
5/31/1995	03:15.2	18.17	–67.21	13.2	2.7	0.1	0.61	0.37	164	B	6	5
5/13/1997	02:56.7	18.14	–67.28	3.2	2.8	0.02	0.58	0.99	198	B	4	2
7/9/1997	04:34.7	18.13	–67.25	21.3	2.1	0.13	0.94	1.74	189	B	4	4
4/17/1998	13:16.3	18.17	–67.25	29.8	2.3	0.21	1.82	1.41	294	C	5	4
6/17/1998	34:51.9	18.13	–67.22	17	2.3	0.25	2.33	0.96	291	C	4	4
12/10/1998	33:28.6	18.14	–67.27	25.2	2.5	0.2	2.65	1.65	302	C	4	3
3/1/1999	02:47.5	18.15	–67.24	23.3	2.8	0.22	1.54	1.94	254	C	5	4
4/25/1999	21:46.9	18.15	–67.2	16.5	2.2	0.23	2.1	1.17	283	C	4	5
5/15/1999	32:47.2	18.12	–67.29	14.1	2.8	0.23	1.52	1.91	275	C	7	6
7/25/1999	11:22.5	18.134	–67.231	22.8	2.2	0.13	1.32	0.93	255	B	5	5
10/26/1999	42:11.8	18.124	–67.26	28.6	2.3	0.17	2.73	1.99	306	C	4	3
6/11/2000	42:13.9	18.131	–67.279	18.4	2.6	0.14	1.12	0.61	198	B	5	5
9/23/2000	43:48.3	18.184	–67.208	23.8	3.5	0.25	1.68	2.85	271	C	7	5
9/25/2000	30:18.1	18.124	–67.291	18.7	3.1	0.11	1.08	0.49	296	B	6	6
12/21/2000	07:33.4	18.115	–67.256	22.3	2.9	0.17	1.59	1.71	302	C	5	5
2/1/2001	27:10.3	18.155	–67.274	17.3	2.7	0.3	2.53	1.34	320	C	5	5
2/6/2001	50:52.0	18.174	–67.22	24.8	3	0.29	3.32	1.6	329	C	4	4
2/9/2001	34:05.2	18.186	–67.293	22.5	3.2	0.09	1.22	2.18	322	B	5	4
2/3/2002	22:10.6	18.205	–67.264	15.7	3	0.2	1.44	0.82	284	C	7	7
4/6/2002	21:14.9	18.142	–67.238	21.5	3	0.2	1.16	1.23	283	C	6	6
4/7/2002	39:58.1	18.162	–67.218	21	2.4	0.1	0.92	0.81	288	B	4	4
6/25/2002	48:29.9	18.14	–67.294	17.6	3.2	0.02	0.79	0.56	198	B	6	2
6/28/2002	09:12.0	18.139	–67.268	16.8	2.9	0.09	0.86	0.57	192	B	5	6
7/8/2002	53:14.6	18.128	–67.284	19.2	2.3	0.09	1.07	0.62	201	B	4	3
7/22/2002	51:51.8	18.16	–67.283	13.7	2.6	0.12	1.92	2.14	192	B	4	4
7/27/2002	17:26.0	18.165	–67.287	27.7	2.4	0.28	2.65	2.23	190	C	5	3
(G) Region 07												
12/27/1991	05:07.5	18.24	–67.05	27.7	2.6	0.11	0.71	0.9	222	B	6	6
8/5/1992	46:17.0	18.23	–67.15	31.1	2.8	0.12	0.91	1.07	205	B	8	6
12/27/1993	29:08.8	18.12	–67.09	29.3	2	0.12	1.28	1.34	185	B	4	4
12/27/1993	42:59.4	18.15	–67.01	25.9	2.9	0.16	0.94	1.26	112	C	7	7
1/31/1994	38:09.9	18.16	–67.04	26.4	2.9	0.25	1.09	1.96	115	C	8	8
2/6/1995	26:06.3	18.21	–67	37.7	2.2	0.04	1.4	0.74	184	B	5	3
12/11/1995	00:35.7	18.24	–67	34	2.4	0.21	4.97	3.56	170	C	4	4
2/29/1996	29:25.8	18.19	–67.08	28.3	4.1	0.22	1.08	1.7	114	C	11	4
2/29/1996	39:29.5	18.17	–67.1	33.9	2.6	0.1	1.12	0.88	244	B	4	5
2/29/1996	42:52.8	18.18	–67.1	32.9	2.2	0.09	1.44	1.69	163	B	5	3
2/29/1996	27:18.4	18.2	–67.08	28.6	2.8	0.18	0.73	0.93	115	C	10	9
3/1/1996	44:51.4	18.14	–67.11	35.4	2.2	0.13	1.15	0.73	135	B	5	5
3/22/1996	45:47.1	18.22	–67.1	30.5	2.2	0.13	1	1	256	B	5	5
4/6/1996	41:39.2	18.24	–67.05	34.7	2.5	0.17	0.81	1.09	123	C	6	8
4/25/1996	18:32.7	18.2	–67.09	27.9	2.7	0.29	1.01	1.51	117	C	10	9
11/10/1996	19:19.0	18.11	–67.05	31.3	2.8	0.09	1.43	2.79	300	B	4	2
12/30/1996	29:30.0	18.2	–67.06	25.3	3.2	0.2	1.04	1.75	179	C	7	6
12/31/1996	03:53.2	18.22	–67.02	38.1	2.2	0.1	1.13	0.93	206	B	4	4
2/26/1997	12:39.7	18.26	–67.13	35.8	2.8	0.15	2.36	2.87	275	B	5	2
5/29/1997	01:59.6	18.22	–67.1	25.2	2.3	0.29	0.92	2.06	120	C	9	9
6/6/1998	08:39.1	18.11	–67.15	30.6	2.4	0.17	1.38	1.31	223	C	4	4

continued

TABLE 1. HYPOCENTERS USED TO COMPUTE COMPOSITE FOCAL MECHANISMS (*continued*)

Date	Time (GMT)	Lat	Long	Depth (km)	M_{PR}	rms	erh	erz	Gap	Q	Ps	Ss
(G) Region 07												
6/7/1998	37:29.3	18.18	–67	35.1	2.3	0.13	1.19	2.63	135	B	4	2
10/8/1999	11:40.7	18.127	–67.097	25.4	2.3	0.14	0.67	1.17	187	B	6	6
10/8/1999	41:55.1	18.164	–67.094	34.7	2	0.06	0.59	0.94	185	B	4	4
10/10/1999	30:06.0	18.116	–67.105	26.6	2.2	0.14	0.96	1.15	192	B	6	6
10/18/1999	32:36.3	18.265	–67.052	28.4	2.7	0.12	1.37	1.54	241	B	5	3
5/21/2000	05:00.9	18.277	–67.102	25.3	3.1	0.24	0.63	1.34	136	C	10	9
7/27/2000	04:01.9	18.136	–67.024	27	2.1	0.14	1.3	1.04	198	B	4	5
9/23/2000	24:33.3	18.151	–67.146	26.6	2.3	0.07	0.75	0.81	253	B	4	4
9/30/2000	07:26.0	18.117	–67.099	36.1	2.2	0.14	2.2	1.18	132	B	4	4
11/13/2000	26:46.5	18.169	–67.137	30.2	2.9	0.24	1.73	2.97	249	C	5	4
2/9/2001	57:33.1	18.101	–67.105	25.3	2.8	0.07	1.11	1.99	278	B	4	4
2/9/2001	54:22.6	18.101	–67.11	28.1	2.3	0.1	1.21	1.26	281	B	4	4
2/11/2001	18:19.9	18.101	–67.111	27.6	2.5	0.12	1.32	1.4	282	B	4	4
2/11/2001	55:52.5	18.113	–67.097	29	2.7	0.07	1.04	1.84	272	B	4	3
3/3/2001	11:02.2	18.284	–67.031	35.4	2.6	0.04	1.13	1.2	240	B	4	4
7/11/2001	03:04.5	18.186	–67.096	25.6	2.5	0.17	0.92	1.43	235	C	6	6
7/11/2001	18:59.2	18.256	–67.143	26.4	2.4	0.03	0.83	1.64	274	B	4	3
(H) Region 08												
5/9/1992	22:56.2	17.84	–66.94	26.1	2.4	0.08	1.2	2.19	265	B	4	3
4/20/1994	18:18.6	17.79	–67.1	38.1	3.4	0.11	0.8	1.66	251	B	11	8
7/2/1999	34:50.9	17.815	–66.899	32.5	2.7	0.1	1.04	1.57	241	B	8	5
1/22/2000	44:35.7	17.862	–67	32.6	2.8	0.04	1.01	1.1	269	B	5	4
3/29/2000	57:12.0	17.847	–67.015	37.9	2.3	0.11	1.89	1.26	273	B	5	5
4/4/2000	11:43.6	17.765	–67.058	28.7	2.5	0.1	1.05	1.38	298	B	5	5
10/25/2000	41:25.9	17.851	–66.909	41.7	3.8	0.11	0.87	1.44	235	B	9	4
10/29/2000	21:22.6	17.862	–67.057	34.1	3.2	0.3	3.47	2.2	269	C	8	5
1/19/2001	31:46.1	17.783	–67.112	34.6	3	0.22	2.03	2.76	286	C	8	8
2/25/2001	44:09.1	17.906	–66.948	29.6	2.1	0.11	2.52	3.11	271	B	4	4
6/5/2001	18:56.0	17.809	–67.055	31.1	2.4	0.04	2.06	1.54	292	B	4	3
7/17/2001	22:05.2	17.792	–67.078	31.7	3	0.19	2.08	2.51	281	C	8	6
8/1/2001	30:44.4	17.771	–67.063	27.8	2.9	0.23	2.31	4.2	279	C	7	6
8/13/2001	29:27.2	17.778	–67.062	29.3	2.7	0.11	1.61	2.34	295	B	6	4
11/14/2001	38:00.4	17.802	–67.007	25	2.6	0.22	2.51	3.74	284	C	5	4
12/15/2001	14:05.7	17.874	–67.03	28.5	3.1	0.15	1.93	1.27	264	B	7	6
12/15/2001	46:54.6	17.841	–67.028	28.9	2.5	0.06	1.19	1.74	280	B	4	4
3/15/2002	18:22.5	17.836	–67.083	34.4	2.8	0.22	2.65	3.12	287	C	4	5
4/25/2002	54:11.6	17.851	–67.034	29.8	2.3	0.11	3	1.62	280	B	4	4
4/26/2002	58:22.0	17.764	–67.084	38.1	2.3	0.12	2.4	3.89	291	B	5	4
4/28/2002	48:01.4	17.834	–67.086	33.5	3	0.25	2.42	2.21	281	C	8	7
4/29/2002	24:56.5	17.829	–67.079	35.7	2.4	0.1	3.03	1.66	295	B	4	3
5/16/2002	28:51.5	17.803	–67.041	28.2	3	0.23	2.06	2.39	247	C	7	6
(I) Region 09												
3/18/1987	49:43.4	17.79	–66.79	3.1	3.4	0.17	2.86	2.79	235	C	6	4
8/7/1987	34:52.7	17.84	–66.8	5.6	2.8	0.2	0.86	1.1	225	C	6	2
9/15/1987	18:37.1	17.84	–66.95	5.9	3.3	0.11	1.69	1.16	262	B	4	3
6/25/1988	58:13.3	17.79	–66.69	16.9	2.7	0.22	2.13	2.9	229	C	6	4
1/3/1989	09:00.9	17.8	–66.72	6.9	3.2	0.09	1.53	1.98	243	B	5	3
3/18/1991	57:45.0	17.82	–66.71	3.4	2.8	0.06	0.86	0.42	224	B	5	4
10/1/1991	12:06.2	17.88	–66.8	1.3	2.9	0.14	0.81	1.33	231	B	8	3
4/10/1993	41:00.9	17.85	–66.9	3.1	2.6	0.12	0.96	0.62	233	B	5	4
10/2/1993	49:46.2	17.83	–66.8	3.6	3.3	0.22	0.88	0.6	229	C	11	10
10/3/1993	39:33.0	17.8	–66.78	17.2	2.1	0.09	0.68	0.91	271	B	6	6
10/4/1993	40:35.8	17.84	–66.79	6.4	3.4	0.21	0.73	0.87	224	C	11	8
10/4/1993	59:27.8	17.86	–66.8	7.1	3.3	0.19	0.64	1.18	220	C	11	8
10/4/1993	38:13.9	17.86	–66.79	5.9	2.3	0.15	0.9	0.75	221	B	6	6
10/4/1993	30:16.2	17.87	–66.8	7	3.1	0.12	0.61	0.82	218	B	8	7
10/4/1993	20:07.5	17.88	–66.8	6.6	2.9	0.23	0.7	0.7	217	C	8	8
12/10/1993	18:53.0	17.84	–66.82	6.4	2.8	0.19	0.69	0.81	241	C	6	5
12/31/1993	24:57.6	17.81	–66.75	13.5	2.9	0.07	0.58	0.4	228	B	7	8
12/31/1993	43:33.8	17.82	–66.75	13.5	3	0.09	0.64	0.45	239	B	8	7

continued

TABLE 1. HYPOCENTERS USED TO COMPUTE COMPOSITE FOCAL MECHANISMS (*continued*)

Date	Time (GMT)	Lat	Long	Depth (km)	M_{PR}	rms	erh	erz	Gap	Q	Ps	Ss
(I) Region 09												
1/1/1994	18:34.5	17.77	–66.7	13.2	3.1	0.14	0.8	0.96	246	B	5	5
1/2/1994	35:00.3	17.86	–66.77	5.4	3	0.22	0.96	0.58	219	C	10	10
1/4/1994	30:37.2	17.87	–66.78	0.1	2.6	0.14	0.62	1.49	234	B	5	4
3/7/1994	52:23.6	17.87	–66.87	21.3	2.3	0.03	0.82	1.53	249	B	4	4
3/26/1994	54:46.6	17.76	–66.94	0.8	2.8	0.08	2.03	2.28	266	B	4	2
5/24/1994	42:08.6	17.84	–66.94	5.4	2.2	0.17	1.76	2.74	241	C	6	4
9/22/1994	50:50.5	17.77	–66.78	5.6	2.2	0.03	1.81	3.02	277	B	4	3
10/5/1994	20:47.4	17.78	–66.77	14.1	2.4	0.11	0.81	2.8	259	B	5	4
10/18/1994	44:38.6	17.82	–66.8	11	2.7	0.07	0.75	0.58	231	B	9	4
11/9/1994	27:28.3	17.83	–66.77	6.3	2.2	0.13	2.55	3.45	264	B	6	4
11/16/1994	21:06.6	17.87	–66.78	8	2.7	0.1	0.55	0.98	217	B	11	4
12/10/1994	09:05.1	17.79	–66.77	11.2	2.2	0.08	1.17	0.97	312	B	4	4
7/3/1995	31:54.1	17.88	–66.73	2.6	2.2	0.14	0.86	1.66	252	B	5	5
9/14/1995	54:20.6	17.78	–66.78	6.3	2.5	0.12	2.52	3.71	275	B	5	5
10/28/1995	41:45.0	17.83	–66.68	10.9	3.8	0.13	0.77	0.93	222	B	10	3
11/23/1995	53:47.0	17.82	–66.67	5.6	2.8	0.13	0.5	0.52	223	B	9	7
3/2/1996	14:00.4	17.84	–66.82	6	3.3	0.14	0.51	0.48	238	B	10	8
3/2/1996	27:12.0	17.78	–66.8	13.6	2.4	0.05	0.7	0.54	274	B	5	5
4/9/1996	14:56.0	17.79	–66.95	10	2.3	0.16	0.93	1.94	262	C	7	5
4/10/1996	22:01.7	17.83	–66.82	5.4	2.7	0.26	1.1	0.75	241	C	8	8
4/10/1996	22:34.6	17.84	–66.81	6.6	2.5	0.12	1.03	2.24	237	B	6	3
1/24/1997	00:07.1	17.79	–66.83	3.9	2.8	0.12	1.38	0.86	248	B	4	3
1/26/1997	57:05.2	17.82	–66.83	8.1	2.7	0.08	0.75	1.57	242	B	6	4
4/5/1997	59:25.6	17.86	–66.89	4.7	3.2	0.13	1.44	0.66	280	B	5	4
4/5/1997	51:06.2	17.76	–66.97	19	2.5	0.23	2.81	1.34	304	C	4	3
5/30/1997	40:50.3	17.78	–66.89	11.1	2.6	0.29	1.46	1.57	253	C	6	4
12/3/1997	41:11.3	17.76	–66.81	17.6	2.5	0.17	1.39	1.12	255	C	6	6
12/10/1997	55:51.0	17.88	–66.74	7.1	2.5	0.13	0.68	0.64	213	B	7	6
12/10/1997	14:50.2	17.88	–66.86	4.8	3	0.19	0.59	0.47	223	C	10	9
12/17/1997	52:59.3	17.88	–66.74	7.5	2.6	0.26	0.92	1.91	204	C	7	7
6/27/1998	02:51.9	17.86	–66.91	21.6	2.5	0.21	1.75	2.95	258	C	5	5
8/8/1998	46:35.4	17.87	–66.85	21.3	2.2	0.13	1.64	3.33	250	B	5	3
10/15/1998	20:20.3	17.85	–66.88	3.7	2	0.11	2.19	1.54	257	B	4	3
8/18/1999	10:48.5	17.796	–66.664	10.4	2.7	0.22	1.23	1.58	227	C	6	4
8/18/1999	23:06.6	17.804	–66.677	11.1	2.9	0.15	0.89	1.11	226	B	6	3
8/18/1999	32:25.1	17.835	–66.677	6.7	3.4	0.16	0.95	2.09	220	C	7	4
8/19/1999	38:02.6	17.82	–66.68	8.7	3.3	0.15	0.83	1.44	223	B	9	6
8/22/1999	34:25.6	17.825	–66.938	18.6	2.4	0.07	1.15	0.78	269	B	5	4
11/13/1999	40:08.8	17.847	–66.998	19.2	2.9	0.1	3.17	2.33	272	B	5	2
11/13/1999	50:31.9	17.876	–66.963	12.2	2.8	0.05	0.68	0.51	257	B	5	6
11/14/1999	31:20.1	17.863	–66.98	13.9	2.8	0.1	1.29	0.81	265	B	6	5
11/24/1999	10:03.7	17.889	–66.967	18.2	2.8	0.09	1.01	1.73	253	B	5	4
4/19/2000	54:05.8	17.836	–66.967	15.4	2.4	0.09	1.22	0.93	285	B	5	5
4/23/2000	35:12.7	17.806	–66.957	22.2	2.5	0.29	2.76	4.93	276	C	4	4
5/21/2000	33:34.4	17.822	–66.76	5.5	2.1	0.06	0.8	1.28	267	B	4	4
5/22/2000	26:54.1	17.865	–66.769	3.4	3	0.19	0.66	0.59	219	C	9	9
8/23/2000	36:08.9	17.858	–66.92	1.3	2.7	0.09	2.24	1.85	278	B	4	4
9/1/2000	24:14.3	17.874	–66.976	13.7	2.7	0.13	1.2	1.21	261	B	4	4
2/2/2001	25:26.3	17.843	–66.872	1.7	2.2	0.19	1.39	1.54	261	C	4	2
4/22/2001	59:01.5	17.869	–66.817	19.8	2.3	0.11	1.94	0.73	252	B	4	4
6/12/2001	30:46.6	17.855	–66.953	13.5	2.6	0.04	1.76	0.81	262	B	4	3
7/4/2001	12:50.8	17.873	–66.939	13	2.7	0.12	1.35	1.08	255	B	5	5
7/8/2001	33:15.7	17.875	–66.943	12.6	2.9	0.12	1.61	0.92	253	B	6	6
10/25/2001	30:56.1	17.883	–66.921	7.2	2.5	0.12	1.16	2.15	249	B	6	5
3/23/2002	20:04.7	17.835	–66.767	18.2	2.2	0.12	2	1.3	264	B	4	3
6/29/2002	53:28.1	17.839	–66.964	15.2	3	0.14	1.63	1.23	255	B	9	3
9/2/2002	53:22.0	17.754	–66.994	12.6	2.6	0.15	1.17	0.87	273	B	6	5
11/5/2002	58:59.8	17.864	–66.94	7.9	2.6	0.1	2.53	3.31	258	B	5	5
11/8/2002	56:35.8	17.874	–66.939	12.9	2.5	0.11	1.09	1.25	255	B	6	4

Note: M_{PR} is the magnitude in the Puerto Rico Seismic Network (PRSN) local scale. Location errors are rms in seconds, erh and erz in km. Gap is the geographical network coverage, Q is the PRSN quality factor. Ps and Ss are the number of wave arrival times used to locate the earthquake.

TABLE 2. GEOMETRY AND SEISMIC MOMENT TENSOR OF COMPOSITE FAULT PLANE SOLUTIONS, LOW HEMISPHERE PROJECTION

(A) Region 01

Plane	Geometry			Slip		P		T		Moment Tensor			Consistency	
	Strike	Dip	Rake	Az	Pl	Az	Pl	Az	Pl				Rds	Gds
1	104	40	–8	110	5	75	37	321	28	–0.14	0.21	0.73	114	68%
2	200	84	–129	14	50						0.43	0.53		
												–0.29		

(B) Region 02

Plane	Geometry			Slip		P		T		Moment Tensor			Consistency	
	Strike	Dip	Rake	Az	Pl	Az	Pl	Az	Pl				Rds	Gds
1	120	75	–20	125	19	76	24	168	3	–0.17	–0.13	0.36	97	67%
2	215	70	–164	29	–15						0.91	0.38		
												–0.74		

(C) Region 03

Plane	Geometry			Slip		P		T		Moment Tensor			Consistency	
	Strike	Dip	Rake	Az	Pl	Az	Pl	Az	Pl				Rds	Gds
1	181	70	169	184	10	46	6	139	21	0.12	–0.34	–0.14	292	61%
2	274	79	20	91	19						0.03	0.92		
												–0.15		

(D) Region 04

Plane	Geometry			Slip		P		T		Moment Tensor			Consistency	
	Strike	Dip	Rake	Az	Pl	Az	Pl	Az	Pl				Rds	Gds
1	20	85	–180	–160	1	244	4	335	4	0.00	0.08	–0.03	73	84%
2	110	90	5	110	5						0.64	0.76		
												–0.64		

(E) Region 05

Plane	Geometry			Slip		P		T		Moment Tensor			Consistency	
	Strike	Dip	Rake	Az	Pl	Az	Pl	Az	Pl				Rds	Gds
1	15	60	–155	181	21	230	37	325	6	–0.37	0.38	–0.32	215	72%
2	271	68	–32	285	30						0.42	0.77		
												–0.05		

(F) Region 06

Plane	Geometry			Slip		P		T		Moment Tensor			Consistency	
	Strike	Dip	Rake	Az	Pl	Az	Pl	Az	Pl				Rds	Gds
1	90	35	–70	156	32	119	73	345	11	–0.88	0.32	0.28	34	68%
2	246	57	–103	00	55						0.88	0.20		
												0.00		

(G) Region 07

Plane	Geometry			Slip		P		T		Moment Tensor			Consistency	
	Strike	Dip	Rake	Az	Pl	Az	Pl	Az	Pl				Rds	Gds
1	120	65	–60	156	51	72	58	188	14	–0.66	–0.38	0.46	39	80%
2	246	38	–137	30	25						0.89	–0.06		
												–0.23		

(H) Region 08

Plane	Geometry			Slip		P		T		Moment Tensor			Consistency	
	Strike	Dip	Rake	Az	Pl	Az	Pl	Az	Pl				Rds	Ags
1	115	40	–100	217	39	262	81	32	5	–0.97	0.10	–0.19	54	83%
2	307	50	–81	25	50						0.71	–0.44		
												0.26		

(I) Region 09

Plane	Geometry			Slip		P		T		Moment Tensor			Consistency	
	Strike	Dip	Rake	Az	Pl	Az	Pl	Az	Pl				Rds	Gds
1	340	85	–170	159	9	204	10	114	4	–0.03	0.14	–0.13	122	77%
2	249	80	–5	250	5						–0.63	0.74		
												0.66		

Note: P and T are the eigenvectors, Az and Pl are the azimuth and plunge for the specific vector, Rds is the number of weighted P-wave polarities used in the computation, and Gds is the percent of goodness of the solution. See Snoke (2003) for details about projection or seismic moment tensor representation.

reverse deformation was detected in Sierra Bermeja, Monte Grande, and Cordillera Central.

Composite focal solutions for the groups with intermediate depth events are representative of the oblique compression due to the convergence of the major plates of North America and the Caribbean. In this case, the southwestern Puerto Rico seismic zone, or in general, the Puerto Rico–Virgin Islands are depth-deforming in response to active ENE-WSW compressional stresses (Fig. 7).

CONCLUSIONS

Earthquake solutions provided by the PRSN catalog show that the southwestern area is one of the most active in the Puerto Rico–Virgin Islands region. The distribution of events shallower than 25 km depth exhibits a clear correlation with the geology of the area. Searching the best locations, we found that the shallow seismic activity is concentrated along topographic features distinguished by the presence of serpentinites. An interpretation of the composite focal mechanisms computed with high quality data indicates two different stress patterns in the southwestern Puerto Rico seismic zone: one related to the shallower extensional NW-SE regime and the other a clear response of the Puerto Rico–Virgin Islands due to the oblique compression of the major plates of North America and the Caribbean.

Although there is no historical evidence of major quakes in the southwestern Puerto Rico seismic zone, moderate activity is normal in this area, as reflected by the number of felt events. Prehistoric Holocene faults trenched by Prentice et al. (2000) indicate that large earthquakes have occurred in southwestern Puerto Rico. The GPS motion estimated for sites in Puerto Rico yields errors that are ≤2.0 mm/yr, indicating that relative motions induced by inner structures of the Puerto Rico–Virgin Islands are possible, especially in the great southern Puerto Rico fault zone, where Jansma et al. (2000) proposed a permissible upper bound motion of 1.5 ± 2.0 mm/yr.

ACKNOWLEDGMENTS

We thank the many people that work in the Puerto Rico Seismic Network (PRSN), including students, technicians, and staff. The authors gratefully acknowledge several anonymous software developers under the GNU family. We wish to acknowledge the importance of the work carried out by all the PRSN providers. We also thank the faculty of the geology department of the University of Puerto Rico at Mayagüez, especially Dr. Jim Joyce. We extend special thanks to Dr. Eugenio Asencio, professor in the geology department at University of Puerto Rico–Mayagüez for his attention, advice, and encouragement.

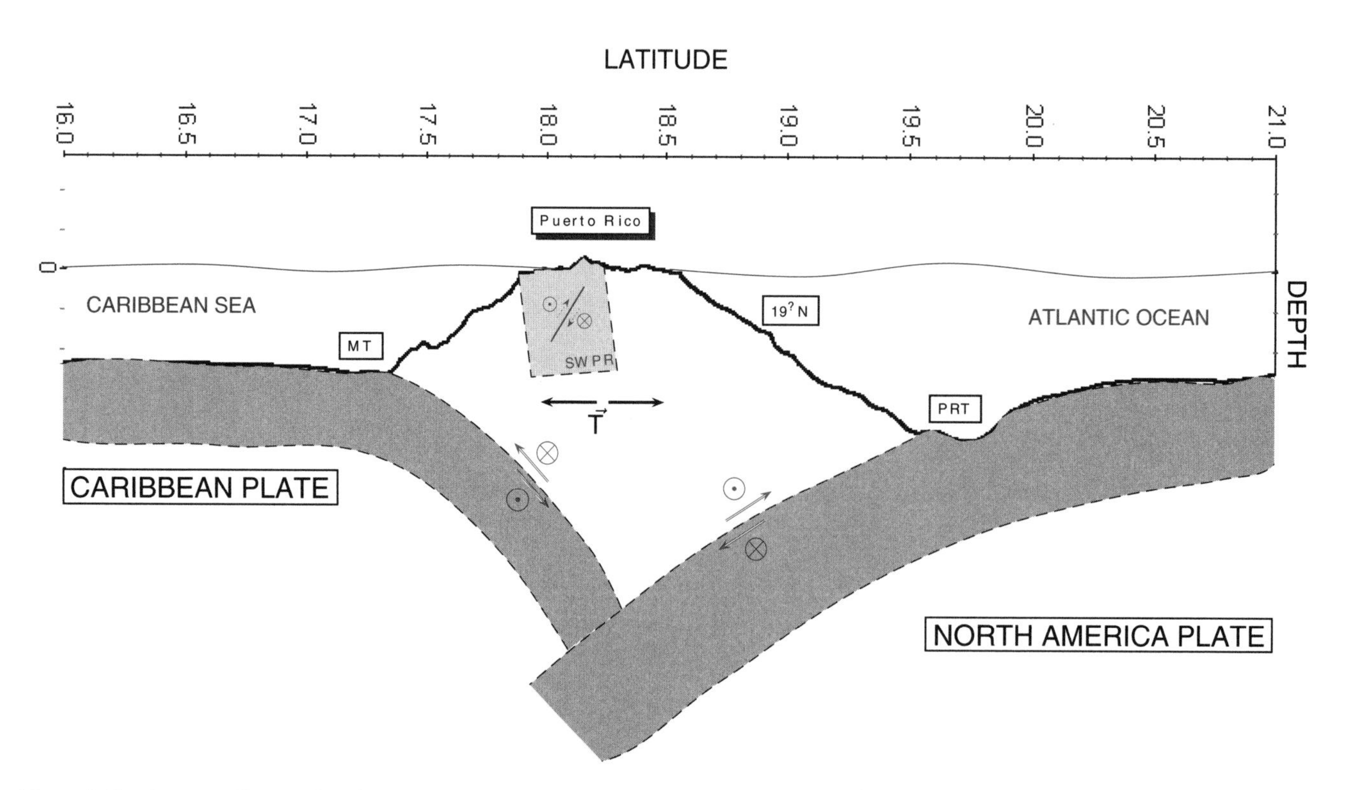

Figure 7. Southwestern Puerto Rico (SWPR) schematic tectonic model projected at long 67°W. The ocean depth has been exaggerated. PRT—Puerto Rico Trench; MT—Muertos trough; T—trough.

REFERENCES CITED

Asencio, E., 1980, Western Puerto Rico seismicity: U.S. Geological Survey Open File Report 80-192, 144 p.

Briggs, R.P., and Akers, J.P., 1965, Hydrogeologic map of Puerto Rico and adjacent islands: U.S. Geological Survey Hydrological Investigations Atlas HA-197, scale 1:240,000.

Dillon, W., ten Brink, U., Frankel, A., Mueller, C., and Rodríguez, R., 1999, Seismic and tsunami hazards in northeast Caribbean addressed at meeting: Eos (Transactions, American Geophysical Union), v. 80, p. 309–310.

Glover, L., and Mattson, P., 1960, Successive thrust and transcurrent faulting during the early Tertiary in south-central PR, *in* Short papers in the geological Sciences: U.S. Geological Survey Professional Paper 400-B, p. 363–365.

Huérfano, V., 1995, Modelo de corteza y mecánica del bloque de Puerto Rico [M.Sc. thesis]: Mayagüez, University of Puerto Rico, 170 p.

Huérfano, V., and Bataille, K., 1995, Crustal structure and stress regime near Puerto Rico: Seismic Bulletin: Preliminary locations of earthquakes recorded near Puerto Rico, Jan.–Dec. 1994: Puerto Rico Seismic Network, p. 15–19.

Jansma, P., Mattioli, G.S., Lopez, A., DeMets, C., Dixon, T.H., Mann, P., and Calais, E., 2000, Neotectonics of Puerto Rico and the Virgin Islands, northeastern Caribbean, from GPS geodesy: Tectonics, v. 19, p. 1021–1037, doi: 10.1029/1999TC001170.

Jolly, W.T., Lidiak, E.G., and Schellekens, J.H., 1998, Volcanism, tectonics, and stratigraphic correlations in Puerto Rico, *in* Lidiak, E.G., and Larue, D.K., eds., Tectonics and geochemistry of the northeastern Caribbean: Geological Society of America Special Paper 322, p. 1–34.

Larue, D.K., Joyce, J. and Ryan, H.F., 1990. Neotectonics of the Puerto Rico trench: Extensional tectonism and forearc subsidence, *in* Larue, D.K., and Draper, G., eds., Transactions 12th Caribbean Geological Conference, St. Croix, Aug. 7–11, 1989: Miami Geological Society, Miami, Florida, p. 231–247.

López, A., Jansma, P., Calais, E., DeMets, C., Dixon, T., Mann, P., and Matiolli, G., 1999, Microplate tectonics along the North American-Caribbean plate boundary: GPS geodetic constraints on rigidity of the Puerto Rico–Northern Virgin Islands (PRVI) block, convergence across the Muertos Trough, and extension in the Mona Canyon: Eos (Transactions, American Geophysical Union), v. 82, p. S77.

Mann, P., Calais, E., Ruegg, J.-C., DeMets, C., Jansma, P., and Mattioli, G.S., 2002, Oblique collision in the northeastern Caribbean from GPS measurements and geological observations: Tectonics, v. 21, no. 6, p. 7-1–7-26, doi: 10.1029/2001TC001304.

Mattson, P.H., 1960, Geology of the Mayagüez area, Puerto Rico: Bulletin of the Geological Society of America, v. 71, p. 319–362.

Pacheco, J.F., and Sykes, L.R., 1992, Seismic moment catalogue of large shallow earthquakes, 1990 to 1989: Bulletin of the Seismological Society of America, v. 82, p. 1306–1349.

Prentice, C., Mann, P., and Burr, G., 2000, Prehistoric earthquakes associated with a late Quaternary fault in the Lajas Valley, southwestern Puerto Rico: Eos (Transactions, American Geophysical Union), v. 181, p. F1182.

Puerto Rico Seismic Network (PRSN), 2003a, Historical Catalogue: http://redsismica.uprm.edu/english/.

Puerto Rico Seismic Network (PRSN), 2003b, Seismic Catalogue: http://temblor.uprm.edu/cgi-bin/new-search.cgi.

Reasenberg, P., and Oppenheimer, D., 1985, FPFIT, FPPLOT and FPPAGE: Fortran computer programs for calculating and displaying earthquakes fault-plane solutions: U.S. Geological Survey Open File Report 85-739, 25 p.

Schellekens, J.H., 1998a, Composition, metamorphic grade, and origin of metabasites in the Bermeja Complex, Puerto Rico: International Geological Reviews, v. 10, p. 722–747.

Schellekens, J.H., 1998b, Geochemical evolution and tectonic history of Puerto Rico, *in* Lidiak, E.G., and Larue, D.K., eds., Tectonics and geochemistry of the northeastern Caribbean: Geological Society of America Special Paper 322, p. 35–66.

Snoke, J.A., 2003, FOCMEC: FOcal MEChanism determinations, *in* Lee, W.H.K., Kanamori, H., Jennings, P.C., and Kisslinger, C., eds., International Handbook of Earthquake and Engineering Seismology: San Diego, Academic Press, Chapter 85.12 (http://www.geol.vt.edu/outreach/vtso/focmec/).

von Hillebrandt, C.G., and Bataille, K., 1995, PRSN formulas for the calculation of magnitude: Puerto Rico Seismic Network Seismic Bulletin: Preliminary location of earthquakes recorded near Puerto Rico, January–December 1993, 36 p.

Manuscript Accepted by the Society 18 August 2004

Geological Society of America
Special Paper 385
2005

Historical earthquakes of the Puerto Rico–Virgin Islands region (1915–1963)

Diane I. Doser*
Christina M. Rodriguez
Claudia Flores
Department of Geological Sciences, University of Texas at El Paso, El Paso, Texas 79968-0555, USA

ABSTRACT

We have collected and modeled the seismograms of historic earthquakes of M >6.0 occurring in the Puerto Rico and Virgin Islands region between 1915 and 1963. Study of offshore events in the north Mona Passage region indicate likely rupture along the North America plate interface in 1915 (M_W = 6.7), 1920 (M_W = 6.5), and 1943 (M_W = 7.8 and 6.0) at depths of 20–30 km. (M_W is moment magnitude.) An event in 1917 (M_W = 6.9) involved strike-slip faulting possibly within the subducting North America plate (36 ± 7 km). The 1918 central Mona Passage earthquake (M_W = 7.2) represents normal-oblique faulting at ~20 km depth. This earthquake generated a tsunami that killed over 100 people on the island of Puerto Rico. The event has a complex source-time function, suggesting rupture along several fault segments. This is consistent with previous tsunami modeling studies. An event in 1916 in southeastern Hispaniola (M_W = 6.8) occurred at a depth of ~16 km, with a reverse faulting mechanism similar to aftershocks of the 1946 great Hispaniola earthquake. Within the Virgin Islands region, we studied three moderate (M ~6–6.5) events. An event in 1930 (M_W = 6.0) appears to involve strike-slip faulting, possibly within the subducting North America plate, while events in 1919 and 1927 (M_W = 6.2 and 5.6) are consistent with rupture along the plate interface. An event in 1939 (M_W = 6.4), located well to the north of the Puerto Rico trench, appears to be a normal faulting event in the outer rise. Slip vectors for earthquakes along the plate interface show northeastward directed slip (~50°) of the Greater Antilles crust relative to the North America plate in the Virgin Island region, north-northeast (~27°) directed slip in the northeast Mona Passage (east of 67.5°W), and east-northeast (~70°) directed slip in the northwest Mona Passage. Normal faulting events in the Greater Antilles crust suggest southeast-northwest oriented extension. Over the past ~85 yr, 87% of the seismic moment release has occurred on structures to the north and northwest of Puerto Rico. This is comparable to estimates of the amount of plate motion that is occurring across these structures from global positioning system and geodesy studies.

Keywords: Puerto Rico–Virgin Islands, historic earthquakes.

*E-mail, Doser: doser@geo.utep.edu. Present addresses: Rodriguez—Exxon-Mobil Corp., Houston, Texas, USA; Flores—University of California at Santa Cruz.

Doser, D.I., Rodriguez, C.M., and Flores, C., 2005, Historical earthquakes of the Puerto Rico–Virgin Islands region, *in* Mann, P., ed., Active tectonics and seismic hazards of Puerto Rico, the Virgin Islands, and offshore areas: Geological Society of America Special Paper 385, p. 103–114. For permission to copy, contact editing@geosociety.org.

INTRODUCTION

The Puerto Rico–Virgin Islands region is located within a zone of complex, oblique subduction (Fig. 1) between the North America and Caribbean plates. East and south of the Puerto Rico–Virgin Islands region, trench normal subduction of North America beneath the Caribbean plate occurs along the Lesser Antilles arc. West of the Puerto Rico–Virgin Islands region, the buoyant Bahamas Platform sits upon the North America plate. The platform resists subduction, causing a transition to transform motion along the northern boundary of the Caribbean plate (e.g., Dolan et al., 1998). Both global positioning system (GPS) results (Jansma et al., 2000) and recent earthquake focal mechanisms (e.g., Deng and Sykes, 1995) indicate left-lateral oblique shortening is occurring north of the Puerto Rico–Virgin Islands region along the Puerto Rico trench (Fig. 1).

Subduction also occurs south of Puerto Rico along the Muertos trough (Fig. 1), with the Caribbean plate thrust beneath the Greater Antilles crust (e.g., Dillon et al., 1994). This leads to active collision between the Caribbean and Atlantic lithospheric slabs beneath eastern Hispaniola and the Mona Passage (Dillon et al., 1994; Dolan et al., 1998). Recent GPS (Jansma et al., 2000) and focal mechanism studies (Huerfano, 1995) suggest there is also a significant component of left-lateral strike-slip motion across the trough. The Muertos trough appears to die out within the southern Virgin Islands (Fig. 1) near 65°W (Masson and Scanlon, 1991).

The Mona Passage (Fig. 1) separates the Puerto Rico–Virgin Islands region from Hispaniola. Structures observed on the seafloor of Mona Passage (Larue and Ryan, 1990; van Gestel et al., 1998) suggest recent, east-west oriented extension within the central Mona Passage. Modeling of tsunami run-ups generated by the 1918 central Mona Passage earthquake (Mercado and McCann, 1998), as well as focal mechanisms of recent earthquakes, indicates normal faulting is presently occurring within the central Mona Passage.

The Anegada Passage forms the southeastern boundary of the Puerto Rico–Virgin Islands region (Fig. 1). It appears to be a zone of left-lateral transtension that transfers displacement from the eastern edge of the Muertos trough to the Puerto Rico trench (Masson and Scanlon, 1991; Jansma et al., 2000).

Thus, the Puerto Rico–Virgin Islands region is sandwiched between four seismogenic zones: the Puerto Rico trench to the north, the Muertos trough to the south, the Mona Passage to the west, and the Anegada Passage to the southeast (Fig. 1). This complex tectonic setting has produced numerous large to great earthquakes, although most have occurred prior to 1963, and thus little is known about the relationship between these events and the regional tectonics.

In order to study the earthquakes of the Puerto Rico–Virgin Islands region, we have divided the region into two areas, based on the clustering of historic and recent events (Fig. 2). The first region (Figs. 4 and 5), located primarily west and northwest of Puerto Rico, is dominated by seismicity associated with the Puerto Rico trench and strike-slip faults within the Greater Antilles crust (Septentrional fault, North Puerto Rico Slope fault, South Puerto Rico Slope fault). The region is located at the eastern edge of the rupture zone of the 1946 great (M_W = 8.0) Hispaniola earthquake (Dolan and Wald, 1998) and includes the source regions of the 1943 (M_W = 7.8) North Mona Passage earthquake (Dolan and Wald, 1998) and the 1787 (M = 8–8¼) Puerto Rico trench earthquake (McCann, 1985) (Fig. 2).

Eastern Puerto Rico and the Virgin Islands form the second study area (Figs. 6 and 7). This region is the site of the 1867 (M_S = 7.5) Anegada Passage earthquake (McCann, 1985) (Fig. 2). Most recent seismicity occurs within the northern portion of the study area. McCann and Sykes (1984) suggest that the seismicity clustered between 64°W and 65°W is caused by the subduction of an aseismic ridge that can be followed to the northwest along strike to the position of the Main Ridge (Fig. 1) within the unsubducted North America plate.

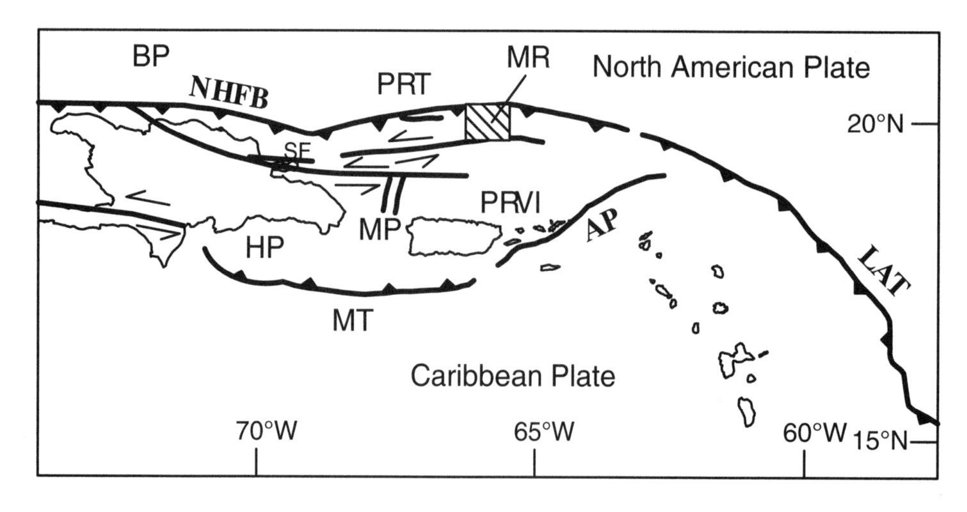

Figure 1. Map of the Puerto Rico–Virgin Islands study area showing major tectonic features modified from Jansma et al. (2000). AP—Anegada Passage; BP—Bahamas Platform; HP—Hispaniola platelet; LAT—Lesser Antilles trench; MP—Mona Passage; MR—Main Ridge; MT—Muertos trough; NHFB—northern Hispaniola fold belt; PRT—Puerto Rico trench; PRVI—Puerto Rico–Virgin Islands block; SF—Septentrional fault

ANALYSIS TECHNIQUES

The earthquakes we have selected for analysis are listed in Table 1. In order to obtain adequate, high quality body-wave information at teleseismic distances, we restricted our study to M >5.9 earthquakes occurring after 1914 and prior to 1964 when the establishment of global seismograph networks allowed for more routine analysis of moderate to large magnitude earthquakes. Within the Puerto Rico–Virgin Islands region, the most recent M >5.9 earthquake occurring prior to 1964 was an aftershock of the 1943 North Mona Passage sequence (Table 1).

Locations for our earthquakes are taken from Russo and Bareford (1993) and Russo (1995, personal commun.). The relocations used a technique similar to that described in Russo and Villaseñor (1995) using the location technique of Wysession et al. (1991) and Russo et al. (1992). Error ellipses were not given for the earthquakes; however, standard deviations in predicted versus observed arrival times for the earthquakes are given in Table 1.

Although we did not have access to the complete phase data set used by Russo and Bareford (1993) and Russo (1995, personal commun.), we did relocate the 1918 central Mona Passage event and the 1920 North Mona Passage event using the iterative bootstrap relocation technique of Petroy and Wiens (1989). Error ellipses for these relocations are shown in Figure 2, and suggest that the uncertainties in the locations of the oldest events (pre-1930, when worldwide seismograph station coverage was very poor) could be as much as 50 km.

Our waveform modeling technique uses the inversion method of Baker and Doser (1988), which inverts body waveforms for focal mechanism, focal depth, and source-time function shape. First motion data, mapped surface faults, focal mechanisms of recent earthquakes and surface wave magnitudes, when available, are used to estimate a priori source parameter values and their associated uncertainties from the

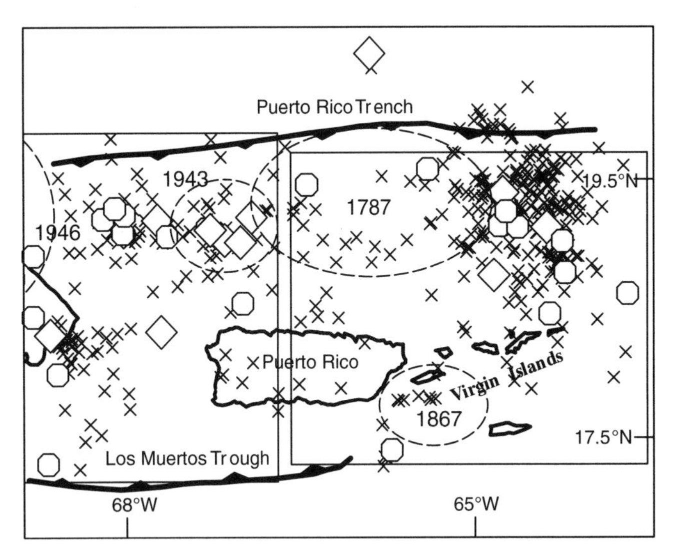

Figure 2. Approximate rupture zones of historic earthquakes (dashed lines) from McCann (1985) and Dolan and Wald (1998), recent seismicity (x's) (NEIC catalog magnitude >3.9 from 1973 to 2001), earthquakes of this study (diamonds) (Table 1), and recent M_W > 4.8 earthquakes (octagons) (Table 2). The boxes indicate the northwestern offshore Puerto Rico and Virgin Islands study areas described in the text and shown in Figures 4 and 6.

TABLE 1. SOURCE PARAMETERS OF HISTORIC EARTHQUAKES

Date (mo/d/yr)	Time	Location* (lat–long)	s.d.* (sec)	Focal mechanism (strike, dip, rake)	Slip[†] vector (deg)	Depth (km)	M_o[§]	M_W[#]	M**
101115	1933	19.04–67.10	4.2	85, 40, 55[††]	50	20[††]	4.5[††]	6.4	6¾
042416	0426	18.26–68.53	3.1	281 ± 38, 50 ± 10, 105 ± 20	9	16 ± 7	19 ± 3	6.8	6¾
072717	0101	19.16–67.66	3.3	65 ± 28, 80 ± 17, −170 ± 13	—	36 ± 7	22 ± 4	6.9	7
101118	1414	18.28–67.62	3.7	207 ± 22, 54 ± 8, −127 ± 28	170	20 ± 7	64 ± 7	7.2	7½
090619	0604	19.36–64.77	3.9	153, 40, 80[††]	65	30[††]	2.4[††]	6.2	6¼
021020	2207	19.07–67.28	3.1	65 ± 40, 30 ± 21, 50 ± 20	20	27 ± 7	5.7 ± 2	6.5	6½
080227	0051	19.10–64.43	2.8	317, 84, 122[††]	47	20[††]	0.5[††]	5.6	6½
062530	1206	18.72–64.87	2.9	167, 87, 168[††]	—	50[††]	1[††]	6.0	6¼
061239	0405	20.41–65.90	2.3	310, 65, −120[††]	—	10[††]	1.5[††]	6.4	6¼
072943	0302	18.99–66.97	2.7	50 ± 21, 30 ± 16, 30 ± 24	23	29 ± 3	529 ± 30	7.8	7¾
				60, 20, 60[§§]		>25[§§]	—	7.9[§§]	
073043	0102	19.17–66.86	2.0	60 ± 40, 30 ± 17, 45 ± 30	18	24 ± 3	1.1 ± 0.2	6.0	6½

*Locations and standard deviations (s.d.) from Russo and Bareford (1993) and R.M. Russo (personal commun., 1995).

†See text for description of how fault plane was selected. Slip vectors were not calculated for events within the subducting North America or Caribbean plates.

§Seismic moment in N-m × 10^{18}.

#Moment magnitude.

**Magnitude from Gutenberg and Richter (1954).

††Results from forward modeling.

§§Results from Dolan and Wald (1998).

TABLE 2. SOURCE PARAMETERS OF RECENT EARTHQUAKES

Date (mo/d/yr)	Time	Location (lat–long)	Focal mechanism (strike, dip, rake)	Slip vector* (deg)	Depth (km)	M_w†	Reference§
052363	0743	19.10–64.78	188, 80, –90	—	50	—	MS
081064	0110	19.03–67.28	168, 82, –100	—	48	5.2 (m_b)	MS
110366	1624	19.17–67.93	94, 39, 9	85	23	5.6	SB
012279	0425	19.10–64.64	75, 14, 28	46	22	5.3	CMT
031579	0658	18.87–68.69	348, 41, 62	—	93	5.0	CMT
110579	0151	17.96–68.46	93, 22, 87	—	78	6.2	CMT
021480	1711	18.43–64.38	138, 17, 82	55	55	4.9	CMT
061182	2157	18.74–64.25	105, 48, 27	85	42	5.1	CMT
093082	1335	18.59–63.76	155, 36, 116	33	37	5.0	CMT
092083	0851	18.39–68.67	50, 43, 64	—	96	5.4	CMT
062685	1710	19.25–64.73	90, 29, 18	73	27	5.9	CMT
072185	1310	19.12–68.11	76, 23, 14	62	23	5.6	CMT
110388	1942	19.04–67.58	63, 38, 0	63	37	5.9	SB
061889	1406	17.28–68.53	34, 64, –22	—	73	5.4	CMT
112392	0631	18.51–66.96	224, 43, –64	110	15	5.3	CMT
080193	1932	17.42–65.71	267, 28, –57	140	15	5.3	CMT
051196	0218	19.54–65.41	143, 39, 134	0	27	5.1	CMT
120898	0232	18.98–64.27	215, 50, –126	125	15	5.4	CMT
121100	1854	19.41–66.40	174, 50, 135	25	15	5.4	CMT
101601	1527	19.31–64.88	40, 18, –24	62	15	5.6	CMT
101701	1129	19.30–64.97	88, 24, 41	48	15	5.8	CMT

*See text for description of how fault plane was selected. Slip vectors were not calculated for events within the subducting North American or Caribbean plates.

†Moment magnitude, except for the 081064 earthquake, where body wave magnitude (m_b) is given.

§MS—Molnar and Sykes (1969); SB—Soto-Cordero and Bataille (1993); CMT—Harvard Centroid Moment Tensor catalog (www.seismology.harvard.edu/CMTsearch.html).

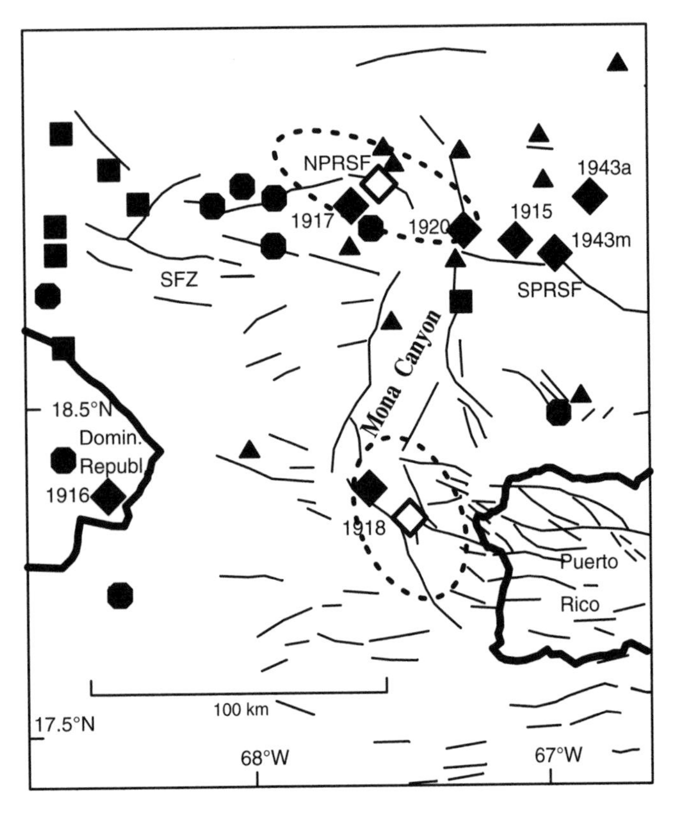

Figure 3. Historic and recent seismicity (M_W >4.8) of the northern offshore Puerto Rico region. Locations of historic events are from Russo and Bareford (1993) and R.M. Russo (1995, personal commun.). Dashed ovals and open symbols show 90% confidence ellipses and locations for the 1918 and 1920 earthquakes, respectively, obtained by using the relocation technique of Petroy and Wiens (1989). See text for details. Triangles are aftershocks of the 1943 North Mona Passage mainshock (labeled 1943m), squares are aftershocks of the 1946 great Hispaniola earthquake, diamonds are historic events of this study, and octagons are recent M_W >4.8 events (Table 2). SFZ—Septentrional fault zone; NPRSF—northern Puerto Rico slope fault zone; SPRSF—southern Puerto Rico slope fault zone.

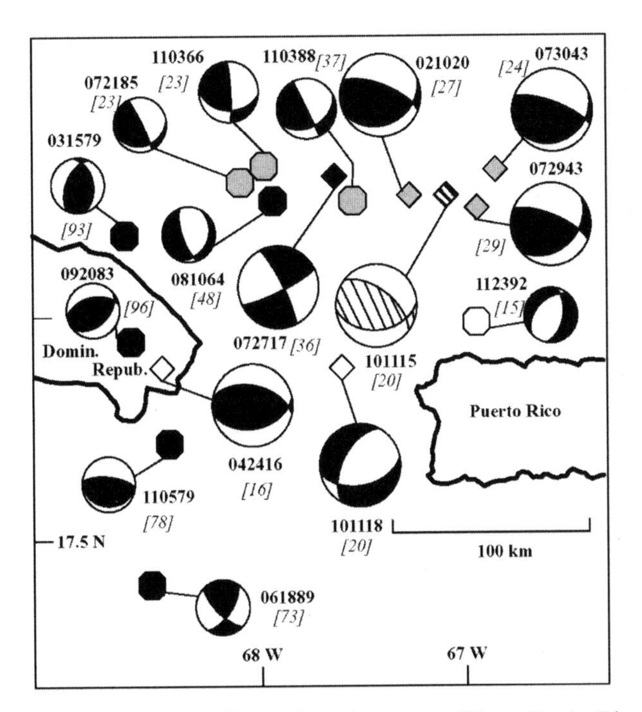

Figure 4. Focal mechanisms of northwestern offshore Puerto Rico earthquakes. Large mechanisms are from this study (Table 1); small mechanisms are for recent M_W >4.8 events (Table 2). Striped mechanisms were obtained from forward modeling studies and are less reliable. The italic numbers in brackets are focal depths. Black symbols are events below the plate interface, gray symbols are events on the plate interface, and open symbols are events above the plate interface. Striped symbols denote suspected plate interface events (based on focal mechanisms) whose depths are poorly constrained. See text for details. A north-south cross section along 67.75°W is shown in Figure 5.

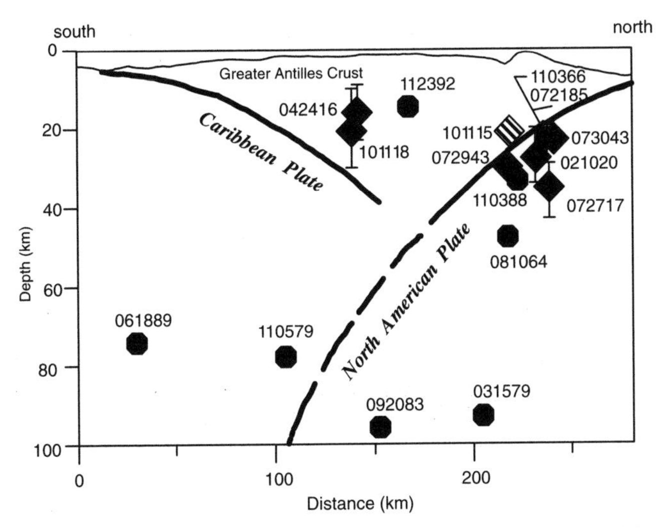

Figure 5. Cross section along 67.75°W showing the locations of the North America and Caribbean plate interfaces, modified from Dolan et al. (1998). Diamonds indicate historic events with error bars shown for results obtained from waveform inversions. Striped symbols denote more poorly constrained depths obtained from forward waveform modeling studies. Note vertical exaggeration.

inversion. Individual seismograms are weighted based on their observed signal to noise ratios and the uncertainties in instrument response parameters. Final covariances in the inversion are calculated using the maximum value of the misfit of the best solution or the initial covariances. These revised covariances result in an estimate of solution uncertainty (Table 1; Appendix, Figs. A1–A11) that is less likely to be biased by overly optimistic estimates of initial data or model quality or by restrictive a priori information (Baker and Doser, 1988). The uncertainties obtained using this inversion technique have been extensively compared to estimates from forward modeling and from other inversion techniques (e.g., Doser et al., 1999; Doser and Webb, 2003) and have generally been found to be more pessimistic than other techniques for describing source parameter uncertainties.

Layered near-source velocity models were built using the crustal velocity models of Huerfano (1995) combined with the structural cross sections of Dolan et al. (1998) to estimate depth to the plate interface. A water layer was also used in our models, since many events are located beneath 4–5 km of water.

Seismograms were hand-digitized. Each seismogram was corrected for mean and trend and resampled at a 0.5 s sampling rate.

Sufficient waveform data (normally 5–6 seismograms from a total of at least 3 stations) to conduct inversions for source parameters were not available for about half of the events. In these cases, as explained in the following sections, we conducted forward modeling studies of possible focal mechanisms for the earthquakes. The results from forward modeling are indicated by the striped focal mechanisms in Figures 4 and 6 and are also given in Table 1. The waveform fits and focal mechanism uncertainties are shown in the Appendix.

RESULTS

Northwestern Offshore Puerto Rico

Waveform analysis was conducted for seven events within the northwestern offshore region. Our analysis included the 1943 North Mona Passage mainshock. Dolan and Wald (1998) had previously studied this earthquake (Table 1), but have not published a complete, detailed account of their waveform modeling analysis procedures. They suggest both the 1943 North Mona Passage and 1946 Hispaniola earthquakes represent rupture along the North America plate interface at strongly coupled asperities associated with underthrusting of the Silver, Navidad, and Mona carbonate banks of the eastern Bahamas platform.

The focal mechanisms of historic and recent earthquakes are shown in Figure 4. Recent mechanisms (smaller size) were obtained from first motion or Centroid Moment Tensor (CMT) studies (see Table 2). The symbol color indicates suspected position relative to the plate interface (Fig. 5). All earthquakes falling within 10 km of the plate interface and having mechanisms consistent with low angle slip along the interface were considered plate interface events. Striped mechanisms indicate poorly constrained mechanisms obtained from forward waveform modeling. Striped symbols indicate events with poorly constrained focal depths obtained from forward modeling.

The earliest event of our study occurred in 1915 and appears to be located west of the 1943 mainshock (Fig. 3). The sparse waveform data from two stations (Fig. A1) did not allow for waveform inversion studies of this event. We used forward modeling to test several possible strike-slip and dip-slip mechanisms (based on mechanisms of surrounding recent and historic events) for this earthquake. The waveforms are most consistent with rupture on a reverse or thrust fault with a focal mechanism similar to the nearby 1920 and 1943 earthquakes. The depth could not be constrained well enough to determine if the earthquake occurred on the plate interface (Fig. 5), but its similarity to other nearby interface events suggests it also may have occurred on the interface.

Inversion of waveform data for the 1916 southeastern Hispaniola earthquake suggests it occurred on a reverse fault within the Greater Antilles crust (Figs. 4, 5, and A2). Its mechanism is similar to some of the aftershocks of the 1946 great Hispaniola earthquake determined by Dolan and Wald (1998). The mechanism indicates continued compression and uplift within eastern Hispaniola due to the collision of the Caribbean and Atlantic slabs beneath this region.

The 1917 North Mona Passage earthquake is located in a region between the 1943 and 1946 aftershock zones (Fig. 3) where bathymetric data reveal a 7-km-deep pull-apart basin (Dolan et al., 1998). Dolan and Wald (1998) suggested the gap between the 1943 and 1946 rupture zones existed either because: (1) a recent earthquake (possibly the 1917 event) had occurred in the region and had relieved the strain along the plate interface, or (2) strain accumulation occurred more slowly along this portion of the plate interface due to a lack of asperities. Although several reverse faulting mechanisms were used as a priori starting models, the waveforms do not appear to be consistent with a reverse mechanism (see modeling results, Fig. A3). The best waveform fit was obtained for strike-slip mechanisms (Table 1; Fig. 4) at a depth of 36 ± 7 km, suggesting faulting within the subducting North America plate (Figs. 4 and 5). Thus, the 1917 earthquake does not appear to have relieved strain along the plate interface.

Of special interest to this study was analysis of the 1918 central Mona Passage earthquake. This earthquake generated a tsunami that struck the western coast of Puerto Rico, resulting in the loss of over 100 lives. Mercado and McCann (1998) modeled tsunami run-up heights for the earthquake, suggesting the earthquake occurred along a series of four normal faults striking 185°–235° with lengths of 4–31 km and displacements of 4 m. Our results are consistent with their multiple fault rupture model. The complexity and duration of the source-time function we obtain (Fig. A4) indicates slip along at least two fault patches, each ~18 km long with ~3 m average slip. The nodal plane striking 210° (Fig. 4) is most consistent with Mercado and McCann's modeling results. Limited first motion data from Vieques, Puerto Rico, and Seminaire Saint Martial, Haiti (Reid and Taber, 1919), are also consistent with this mechanism (Fig. A4).

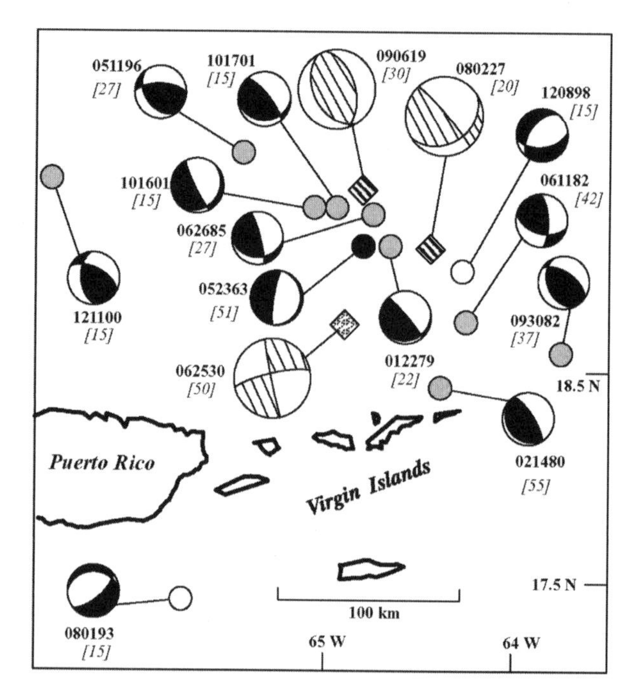

Figure 6. Historic and recent earthquakes of the Virgin Islands region. Large focal mechanisms denote historic earthquakes of this study (Table 1). The striped mechanisms indicate more poorly constrained mechanisms from forward modeling studies. Small focal mechanisms denote recent M_W >4.8 earthquakes (Table 2). Black symbols are events below the plate interface, gray symbols are events on the plate interface, and open symbols are events above the plate interface; the symbol filled with wavy lines denotes an event that is suspected to have occurred within the subducting North America plate. A north-south cross section along 65°W is shown in Figure 7.

Focal mechanisms and depths for the 1920 earthquake and the 1943 North Mona Passage mainshock and its largest aftershock are all consistent with rupture along the plate interface (Figs. 4, 5, A6, A10, and A11). Our results for the 1943 mainshock agree with those of Dolan and Wald (1998), although our best-fit mechanism has a slightly lower rake (Table 1). Dolan and Wald suggest that rupture in the 1943 mainshock propagated in the down-dip direction starting at a depth of ~25 km (comparable to our depth estimate of 29 ± 3 km) and that the lateral extent of the rupture zone was related to the collisional underthrusting of a portion of the Bahamas Bank (Mona Bank). This is consistent with the hypocentral location of the 1943 aftershock we studied (Fig. 5), which indicates the aftershock occurred well up-dip of the mainshock rupture. The hypocenter for the 1920 earthquake, although poorly constrained, also appears to be up-dip of the 1943 mainshock (Fig. 5) and located at the southwestern edge of the 1943 aftershock rupture zone (Fig. 3). Thus, the 1920 event may have relieved stress along the plate interface in this region and arrested continued westward rupture of the plate interface in 1943.

Virgin Islands Region

There were fewer large (M >5.9) earthquakes in the Virgin Island region than the Puerto Rico region during the twentieth century (Fig. 2), although the southwestern Virgin Islands experienced an M ~7.5 earthquake (McCann, 1985) in 1867. We studied three events in this region, as well as an event in 1939 located well to the north of the Puerto Rico trench (Fig. 2). Because these earthquakes were smaller than M = 6.5, very limited seismogram data were available. This required us to conduct forward modeling of the seismograms using a number of starting models for each event that were based on the focal mechanisms and depths of recent earthquakes located within 100 km of their epicenters.

Figure 6 shows focal mechanisms (smaller mechanisms) of recent events (Table 2) in comparison to the historic events. Striped focal mechanisms indicate events with poorly constrained mechanisms obtained from forward modeling studies. Striped symbols indicate events with focal depths poorly constrained by forward modeling, but with mechanisms consistent with rupture along the plate interface. The symbol with wavy lines indicates an event with a focal depth poorly constrained by forward modeling, but with a mechanism that is not consistent with rupture along the plate interface. Note that although the mechanism for earthquake 101601 appears to be normal-oblique (Table 2), only a slight change in the dip (<5°) of the more steeply dipping nodal plane would create a reverse-oblique mechanism similar to the neighboring earthquake on 101701. Thus, we have classified earthquake 101601 as a plate interface event.

A cross section of the region is shown in Figure 7. Again, all earthquakes falling within 10 km of the plate interface and having mechanisms consistent with low angle slip along the interface were considered interface events.

Limited waveform data for the 1919 northern Virgin Islands earthquake (Fig. A5) suggest this event involved reverse/thrust faulting, possibly along the plate interface (Fig. 7). The 1927 earthquake is also consistent with reverse or thrust faulting along the plate interface (Figs. 6, 7, and A7), with a mechanism similar to that of the nearby June 1985 (M_W = 5.9) earthquake.

Waveform modeling suggests a strike-slip mechanism for the 1930 earthquake (Figs. 5 and A8). Reverse and normal faulting mechanisms similar to nearby recent earthquakes did not fit the observed P-wave first motion polarity at La Paz and grossly overestimated P-wave amplitudes at La Paz and DeBilt.

Forward modeling of waveform data for the 1939 earthquake (Fig. A9), located well north of the Puerto Rico trench (Fig. 2), gives a mechanism that suggests normal faulting within the North America plate at a depth of ~10 km (Fig. 7). Thus, this earthquake appears to be an outer rise event.

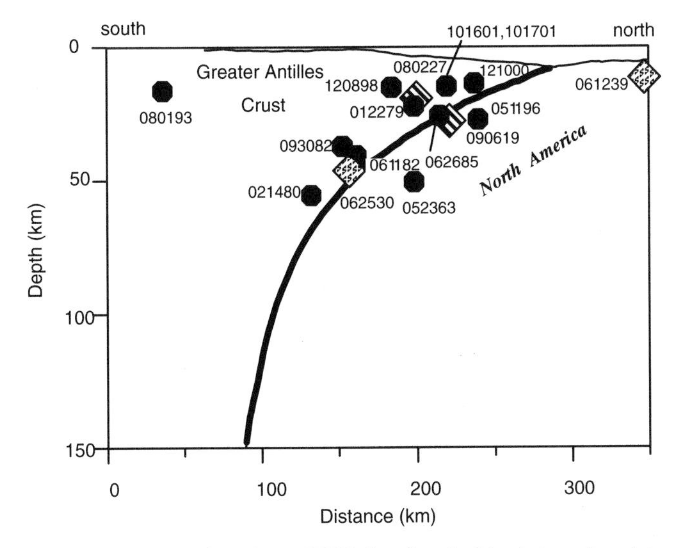

Figure 7. Cross section along 65°W showing the North America plate interface (modified from Dolan et al., 1998). Diamonds are historic events. The focal mechanism for the 1939 earthquake is shown in Figure 8. Symbols are as in Figure 6. Note vertical exaggeration.

SLIP VECTORS AND SEISMIC MOMENT RATES

Slip vectors of events known or suspected to be located above the plate interface (i.e., within the Greater Antilles crust) (open symbols) or on the plate interface (gray or striped symbols) are shown in Figure 8 and given in Tables 1 and 2. Thin arrows indicate events with focal mechanisms that were more poorly resolved (i.e., from forward modeling studies). Striped symbols indicate events with poorly determined focal depths that we suspect occurred on the plate interface. The large arrow in Figure 8 indicates the average direction of motion of the Puerto Rico–Virgin Islands block relative to the North America plate (~17 mm/yr, N68°E) from the GPS studies of Jansma et al. (2000).

For reverse or thrust events of the Puerto Rico region (101115, 042416, 021020, 072943, 073043, 110366, 110388, 072185, 121100), we selected the southward dipping nodal plane as the fault plane, with slip vectors indicating the movements of the south blocks of the faults relative to the north blocks. Note that we grouped the 121100 event with other events in the Puerto Rico region since it is actually closer to the North Mona Passage events than the northern Virgin Island events (Fig. 8). In the case of known or suspected plate interface events, the slip vectors would thus represent the movement of Greater Antilles crust relative to the North America plate.

For normal faulting events on 101118 and 112392, we selected the northeast-southwest striking, northwest dipping nodal planes as the fault planes, with slip vectors indicating the motions of the southeastern fault blocks relative to the northwestern fault blocks. These nodal planes are the most consistent with mapped faults in the region. For the 1918 event, the selected nodal plane is also the nodal plane most consistent with the tsunami modeling studies of Mercado and McCann (1998).

The slip vectors from the reverse or thrust events suggest that in the northeastern Mona Passage motion of the Puerto Rico plate relative to North America is directed toward the north-northeast (an average of 27° for the 5 interplate events located between 66°W and 67.5°W). There is a rotation of slip along the interface to ~70° west of 67.5°W (average of 3 events). This rotation of slip to a most westerly direction west of 67.5°W is consistent with the GPS studies of Mann et al. (2002) that suggest there is a change in the direction and rate of plate motion across the Mona Passage, from a most easterly strike (~80°) and slower rate (4–17 mm/yr) of plate motion within central and eastern Hispaniola to a strike of ~70° and rate of 19–20 mm/yr within the Puerto Rico–Virgin Islands region.

Slip vectors for events within the Greater Antilles crust show southwest to south-southwest directed extension within the central and eastern Mona Passage. The 1916 Hispaniola event shows possible northward directed slip along the thrust fault systems of eastern Hispaniola.

In the Virgin Islands region, we again selected the southward dipping nodal planes as the fault planes for the thrust or reverse events (090619, 080227, 012279, 021480, 061182, 093082, 062685, 051196, 101601, 101701), all of which are known or suspected plate interface events. The average slip vector for these ten events is northeast (50°), rotated ~20° counterclockwise of the motion of the Puerto Rico–Virgin Islands block relative to North America.

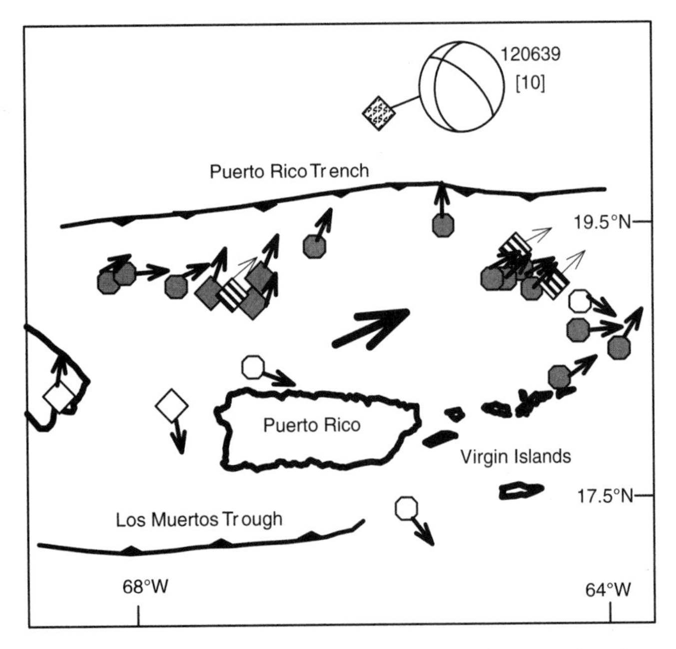

Figure 8. Slip vectors of earthquakes occurring in the upper plate (i.e., Greater Antilles crust) (open symbols) or along the plate interface (i.e., between the Greater Antilles crust and North America plate) (gray symbols). Striped symbols denote events with poorly constrained focal depths that are suspected to have occurred on the plate interface. Slip vectors are given in Tables 1 and 2. Small bold arrows denote events with better determined focal mechanisms. See text for a detailed description of how slip vectors were determined. The large bold arrow is the direction of motion of the Puerto Rico–Virgin Islands block relative to the North America plate (Jansma et al., 2000). The focal mechanism for the 1939 outer rise event (Table 1) is shown at top.

For the 080193 and 120898 normal faulting events, we selected the northwest dipping nodal planes as the fault planes. These gave slip vectors suggesting southeastward directed slip of the southeastern sides of the fault blocks, similar to that observed to the west within the Mona Passage.

The GPS and geodesy studies of Jansma et al. (2000) suggest that ~85% of North America–Caribbean plate motion takes place within the Puerto Rico trench and upon faults located north and northwest of Puerto Rico, with the remainder taken up to the south. This is in good agreement with estimates of seismic moment release over the past ~85 yr, with 87% of seismic moment release (Tables 1 and 2) occurring along structures located north and northwest of Puerto Rico. Another 13% of the seismic moment release took place within the Mona Passage region, where extension rates are estimated to be 2–5 mm/yr (Jansma et al., 2000).

Although offshore faults do not pose as much hazard to Puerto Rico as onshore faults, the effect of rupture directivity should not be overlooked. The 1943 mainshock appears to have ruptured to the west (Dolan and Wald, 1998) and down-dip, causing only intensity V shaking in Puerto Rico (Coffman and von Hake, 1982), and thus may not be representative of the "typical" shaking that could be expected from north Mona Passage events. The 1918 central Mona Passage earthquake, however, with rupture propagation upward to the surface as indicated by the tsunami modeling studies of Mercado and McCann (1998), produced intensity IX effects within Puerto Rico (Coffman and von Hake, 1982).

CONCLUSIONS

Studies of M > 5.9 historic earthquakes of the Puerto Rico–Virgin Islands region indicate that many historic events are consistent with slip along the plate interface, both in the northeastern Mona Passage and the Virgin Islands regions. Crustal deformation has been limited to the 1918 central Mona Passage and the 1916 eastern Hispaniola earthquakes. Focal mechanisms and focal depths for earthquakes in 1917 in the north Mona Passage and 1930 in the Virgin Islands region suggest strike-slip faulting, possibly within the subducting North America plate.

Slip vectors suggest north-northeast to east-northeast directed movement of the Greater Antilles crust relative to North America along the plate interface. The strike of slip appears to change most dramatically across the northern Mona Passage, a result consistent with recent GPS observations of marked differences in the rate and direction of relative plate motion across the Mona Passage.

Seismic moment release within the Virgin Islands region over the past ~85 yr has been nearly two orders of magnitude lower than that of the northwestern offshore Puerto Rico region. This is also consistent with GPS studies suggesting that the structures offshore of northwestern Puerto Rico accommodate ~85% of North America–Caribbean plate motion.

APPENDIX

In the following appendix figures, the observed seismograms are shown on top and the synthetic seismograms on the bottom of each seismogram pair. The labels below seismogram pairs refer to station code and waveform type (z = vertical P-waves, r = radial P-waves, sh = transverse S-waves). The vertical scale shows the amplitudes of the waveforms in centimeters. Seismogram labels with a slash (/) indicate seismograms have been reduced relative to the vertical scale by the given factor. Labels with asterisks indicate that seismograms have been magnified by the given factor relative to the vertical scale. The bold vertical lines above each observed seismogram indicate the start of the phase that was modeled.

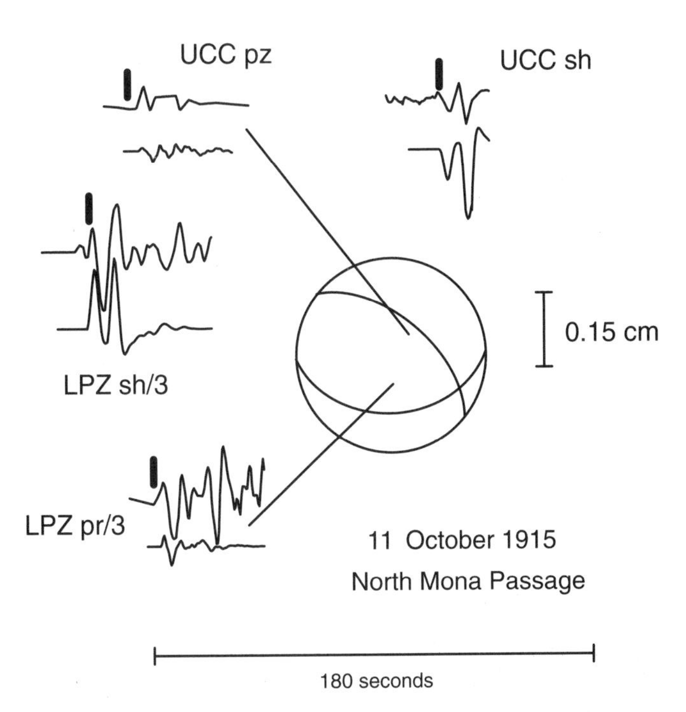

Figure A1. Waveform modeling results obtained from forward modeling of the 1915 North Mona Passage earthquake (see Table 1 and Fig. 4). LPZ—La Paz, Bolivia; UCC—Uccle, Belgium.

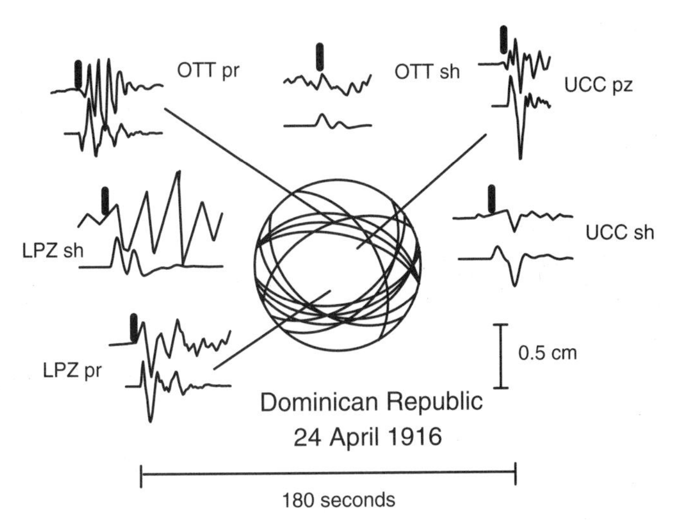

Figure A2. Results of waveform inversion for the 1916 Dominican Republic earthquake. The inversion results are given in Table 1. The range of focal mechanisms shown indicates the solution uncertainty (see Table 1). LPZ—La Paz, Bolivia; OTT—Ottawa, Canada; UCC—Uccle, Belgium.

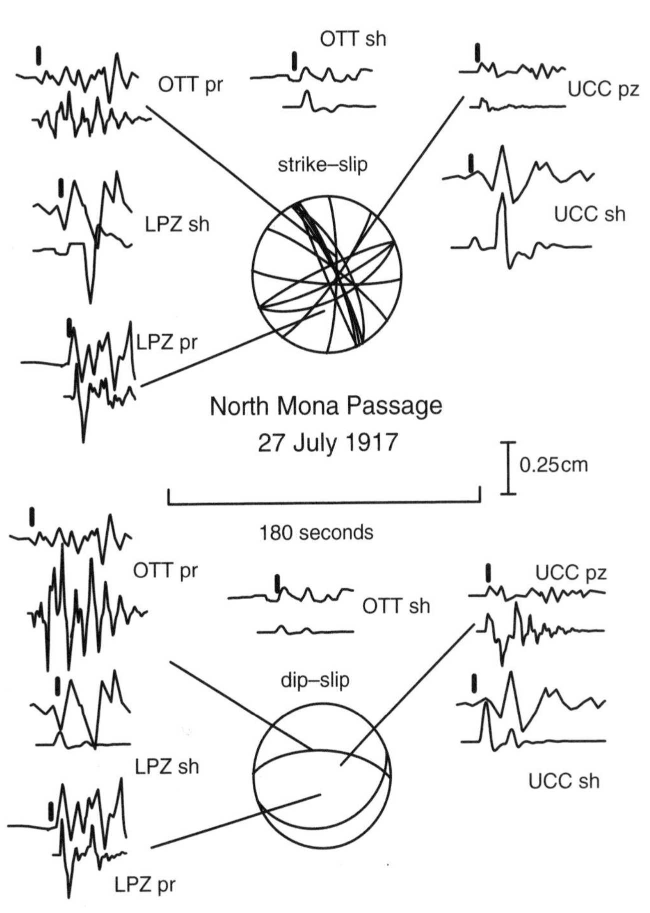

Figure A3. Results of waveform inversion for the 1917 North Mona Passage earthquake. The top synthetic seismograms were obtained for the strike-slip solution shown in Figure 4 and given in Table 1. The range of focal mechanisms shown reflects the uncertainty in the focal mechanism obtained from the inversion. The bottom seismograms were obtained for the dip-slip solution shown. Note that P-wave amplitudes are grossly overestimated at UCC and OTT, S-wave amplitudes are grossly overestimated at UCC, and the polarity of the S-wave at LPZ is mismatched for the dip-slip mechanism. The moment-magnitude of the event would have to be lowered to 6.6 (from 6.9) to obtain an amplitude match to the P-waves observed at UCC and OTT, although Gutenberg and Richter (1954) estimated the event to have a magnitude of 7.0 (Table 1). LPZ—La Paz, Bolivia; OTT—Ottawa, Canada; UCC—Uccle, Belgium.

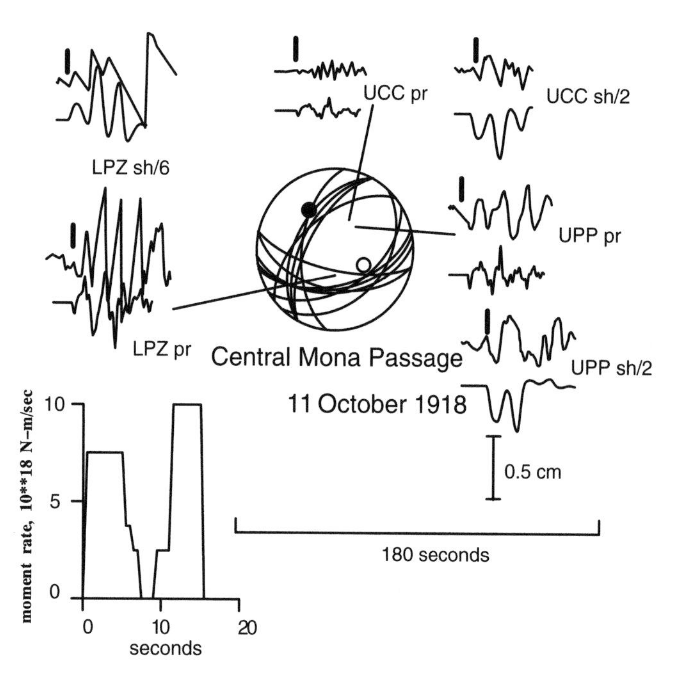

Figure A4. Results of waveform inversion for the 1918 central Mona Passage earthquake. The range of focal mechanisms shown reflects uncertainties obtained in the inversion process (Table 1). The source-time function shown at the lower left indicates the complex nature of the source rupture. The black symbol indicates the compressional first motion observed at Seminaire Saint Martial, Haiti, and the open symbol the dilatational first motion observed at Vieques, Puerto Rico (Reid and Taber, 1919). LPZ—La Paz, Bolivia; UCC—Uccle, Belgium; UPP—Uppsala, Sweden.

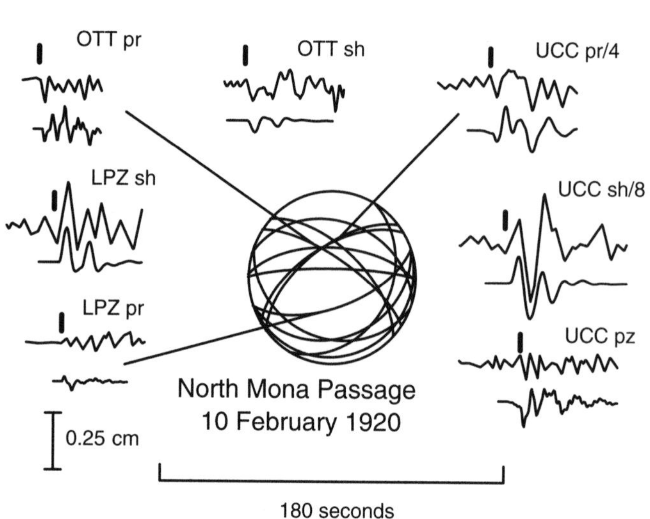

Figure A6. Results from the inversion of waveform data for the 1920 North Mona Passage earthquake. LPZ—La Paz, Bolivia; OTT—Ottawa, Canada; UCC—Uccle, Belgium.

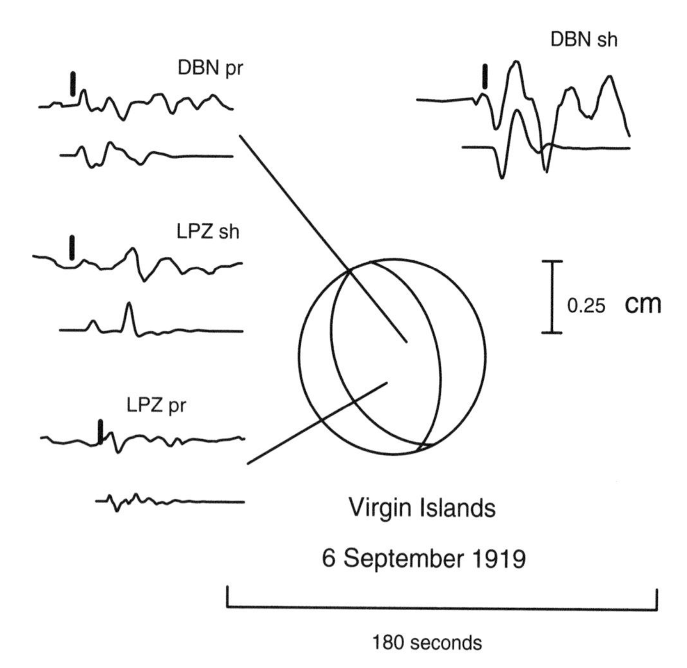

Figure A5. Results of forward modeling of the 1919 Virgin Islands earthquake. DBN—DeBilt, Netherlands; LPZ—La Paz, Bolivia.

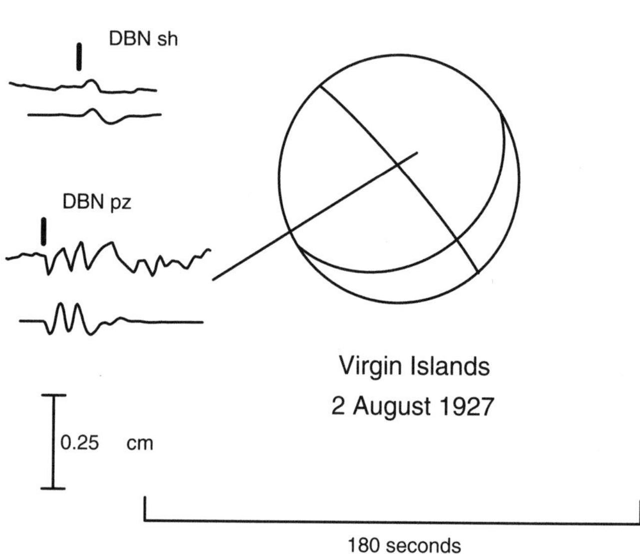

Figure A7. Results obtained from forward modeling of the 1927 Virgin Islands earthquake. DBN—DeBilt, Netherlands.

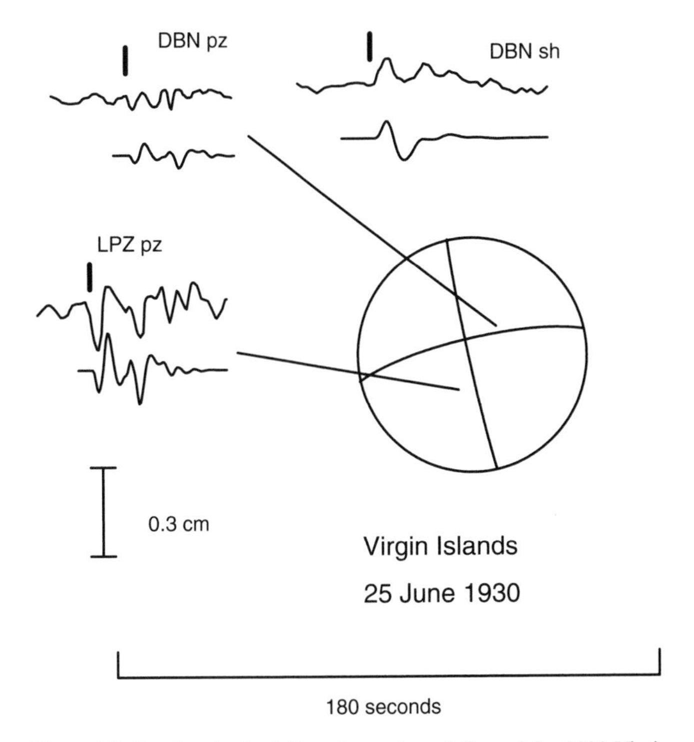

Figure A8. Results obtained from forward modeling of the 1930 Virgin Islands earthquake. DBN—DeBilt, Netherlands; LPZ—La Paz, Bolivia.

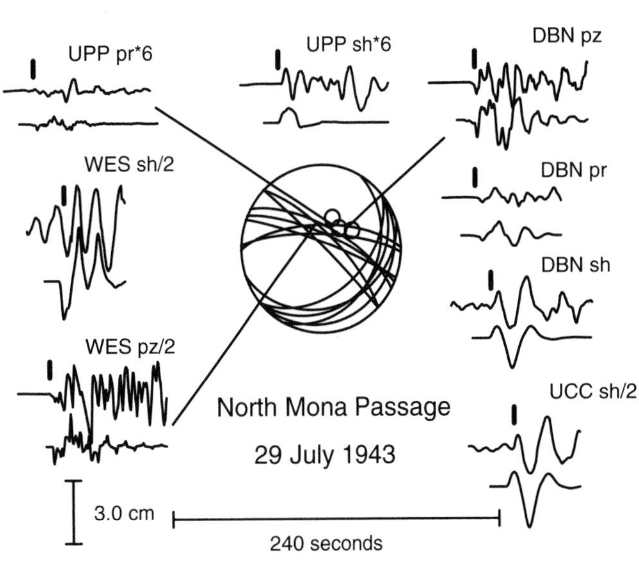

Figure A10. Waveform inversion results for the 1943 North Mona Passage earthquake. Open symbols indicate dilatations obtained from listings of the International Seismological Summary. DBN—DeBilt, Netherlands; UPP—Uppsala, Sweden; WES—Weston, Massachusetts.

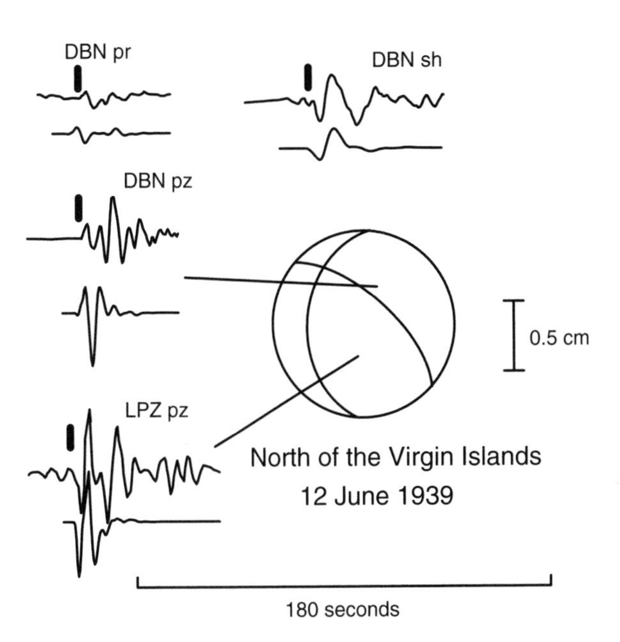

Figure A9. Results obtained from forward modeling of an earthquake located north of the Virgin Islands. This earthquake appears to be an outer rise event (see Fig. 8). DBN—DeBilt, Netherlands; LPZ—La Paz, Bolivia.

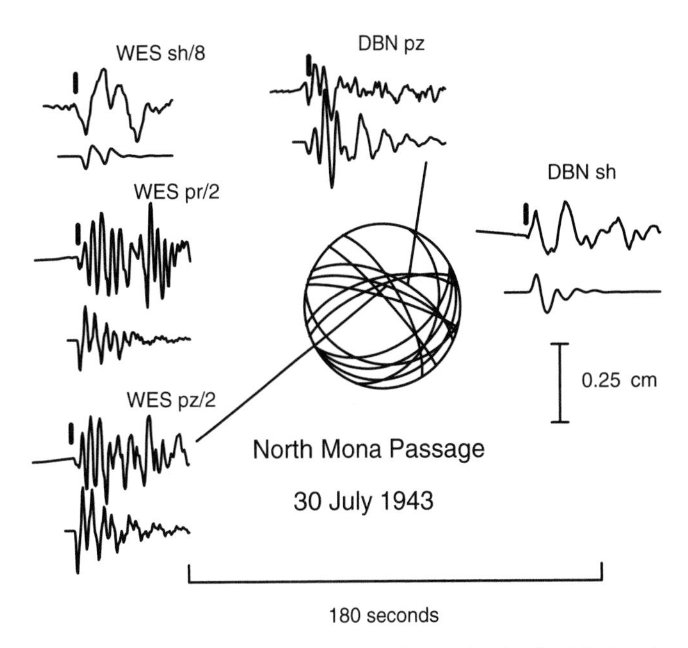

Figure A11. Waveform inversion results for an aftershock of the North Mona Passage earthquake. DBN—DeBilt, Netherlands; WES—Weston, Massachusetts.

ACKNOWLEDGMENTS

We thank seismograph station operators from around the world for sending us copies of the seismograms and instrument response information used in this study. Helpful reviews by J. Dewey, A. Villaseñor, and P. Mann greatly improved the manuscript. We thank C. von Hillebrandt-Andrade for her helpful discussions and comments during the early stages of this research. She pointed us toward many useful references. Discussions and information from P. Jansma and G. Mattioli are appreciated. Funding from the U.S. Geological Survey's Earthquake Hazards Reduction Program (00HQGR0013) is also acknowledged. The views and conclusions contained in this document are those of the authors and should not be interpreted as necessarily representing the official policies, either expressed or implied, of the U.S. Government.

REFERENCES CITED

Baker, M.R., and Doser, D.I., 1988, Joint inversion of regional and teleseismic waveforms: Journal of Geophysical Research, v. 93, p. 2037–2045.

Coffman, J.L., and von Hake, C.A., 1982, Earthquake History of the United States: Boulder, Colorado, U.S. Department of Commerce, National Oceanic and Atmospheric Administration publication 41-1, 258 p.

Deng, J., and Sykes, L.R., 1995, Determination of Euler pole for contemporary relative motion of Caribbean and North America plates using slip vectors of interplate earthquakes: Tectonics, v. 14, p. 39–53, doi: 10.1029/94TC02547.

Dillon, W.P., Edgar, N.T., Scanlon, K.M., and Coleman, D.R., 1994, A review of the tectonic problems of the strike-slip northern boundary of the Caribbean Plate and examination by GLORIA, *in* Gardner, J.V., Field, M.E., and Twichell, D.C., eds., Geology of the United States' seafloor: The view from GLORIA: Cambridge, Cambridge University Press, p. 135–164.

Dolan, J.F., and Wald, D.J., 1998, The 1943–1953 north-central Caribbean earthquakes: active tectonic setting, seismic hazards and implications for Caribbean-North American plate motions, *in* Dolan, J., and Mann, P., eds., Active strike-slip and collisional tectonics of the Northern Caribbean plate boundary zone: Geological Society of America Special Paper 326, p. 143–161.

Dolan, J.F., Mullins, H.T., and Wald, D.J., 1998. Active tectonics of the north-central Caribbean: Oblique collision, strain partitioning, and opposing subducted slabs, *in* Dolan, J., and Mann, P., eds., Active strike-slip and collisional tectonics of the Northern Caribbean plate boundary zone: Geological Society of America Special Paper 326, p. 1–63.

Doser, D.I., and Webb, T.H., 2003, Source parameters of large historical (1917–1961) earthquakes: North Island, New Zealand, v. 152, p. 795–832.

Doser, D.I., Webb, T.H., and Maunder, D.E., 1999, Source parameters of historic (1918–1962) earthquakes, South Island, New Zealand: Geophysical Journal International, v. 139, p. 769–794, doi: 10.1046/j.1365-246x.1999.00986.x.

Gutenberg, B., and Richter, C.H., 1954, Seismicity of the Earth and associated phenomena: New York, Hafner Publishing Company, 245 p.

Huerfano, V., 1995, Crustal structure and stress regime near Puerto Rico [M.S. thesis]: Mayaguëz, University of Puerto Rico.

Jansma, P.E., Mattioli, G.S., Lopez, A., DeMets, C., Dixon, T.H., Mann, P., and Calais, E., 2000, Neotectonics of Puerto Rico and the Virgin Islands, northeastern Caribbean, from GPS geodesy: Tectonics, v. 19, p. 1021–1037, doi: 10.1029/1999TC001170.

Larue, D.K., and Ryan, H.F., 1990, Extensional tectonism in the Mona Passage, Puerto Rico and Hispaniola: A preliminary study: Transactions of the 12th Caribbean Geological Conference, p. 310–313.

Mann, P., Calais, E., Ruegg, J.-C., DeMets, C., Jansma, P.E., and Mattioli, G.S., 2002, Oblique collision in the northeastern Caribbean from GPS measurements and geological observations: Tectonics, v. 21, no. 6, p. 1057, doi: 10.1029/2001TC001304.

Masson, D.G., and Scanlon, K.M., 1991, The neotectonic setting of Puerto Rico: Geological Society of America Bulletin, v. 103, p. 144–154, doi: 10.1130/0016-7606(1991)103<0144:TNSOPR>2.3.CO;2.

McCann, W.R., 1985, On the earthquake hazards of Puerto Rico and the Virgin Islands: Bulletin of the Seismological Society of America, v. 75, p. 251–262.

McCann, W.R., and Sykes, L.R., 1984, Subduction of aseismic ridges beneath the Caribbean plate: implications for the tectonics and seismic potential of the northeastern Caribbean: Journal of Geophysical Research, v. 89, p. 4493–4519.

Mercado, A., and McCann, W., 1998, Numerical simulation of the 1918 Puerto Rico tsunami: Natural Hazards, v. 18, p. 57–76, doi: 10.1023/A:1008091910209.

Molnar, P., and Sykes, L., 1969, Tectonics of the Caribbean and Middle American regions from focal mechanisms and seismicity: Geological Society of America Bulletin, v. 80, p. 1639–1684.

Petroy, D.E., and Wiens, D.A., 1989, Historical seismicity and implications for a diffuse plate convergence in the NE Indian Ocean: Journal of Geophysical Research, v. 94, p. 12,301–12,319.

Reid, H.F., and Taber, S., 1919, The Porto Rico earthquakes of October–November 1918: Bulletin of the Seismological Society of America, v. 9, p. 95–127.

Russo, R.M., and Bareford, C., 1993. Historical seismicity of the Caribbean region, 1933–1963, Caribbean Conference on Volcanism, Seismicity, and Earthquake Engineering, Trinidad, October 1993: University of the West Indies.

Russo, R.M., and Villaseñor, A., 1995, The 1946 Hispaniola earthquakes and the tectonics of the North America–Caribbean plate boundary zone, northeastern Hispaniola: Journal of Geophysical Research, v. 100, p. 6265–6280, doi: 10.1029/94JB02599.

Russo, R.M., Okal, E.A., and Rowley, K.C., 1992, Historical seismicity of the southeastern Caribbean and tectonic implications: Pure and Applied Geophysics, v. 139, p. 87–120.

Soto-Cordero, L., and Bataille, K., 1993, Study of seismicity and stress patterns in the southwestern part of the Puerto Rico trench: Eos (Transactions, American Geophysical Union), v. 74, p. 418.

van Gestel, J.P., Mann, P., Dolan, J., and Grindlay, N., 1998, Structure and tectonics of the upper Cenozoic Puerto Rico–Virgin Islands carbonate platform as determined from seismic reflection studies: Journal of Geophysical Research, v. 103, p. 30,505–30,530, doi: 10.1029/98JB02341.

Wysession, M.E., Okal, E.A., and Miller, K.L., 1991, Intraplate seismicity of the Pacific basin: Pure and Applied Geophysics, v. 135, p. 261–359.

Manuscript Accepted by the Society August 18, 2004

Printed in the USA

Geological Society of America
Special Paper 385
2005

Reconnaissance study of Late Quaternary faulting along Cerro Goden fault zone, western Puerto Rico

Paul Mann*
Institute for Geophysics, Jackson School of Geoscienences, University of Texas at Austin, Austin, Texas 78759, USA

Carol S. Prentice
U.S. Geological Survey, 345 Middlefield Road, MS 977, Menlo Park, California 94025, USA

Jean-Claude Hippolyte
Université de Savoie, UMR-CNRS 5025, LGCA, 73376 Le Bourget de Lac, France

Nancy R. Grindlay
Lewis J. Abrams
Center for Marine Science, University of North Carolina, Wilmington, North Carolina 28409, USA

Daniel Laó-Dávila*
Department of Geology, University of Puerto Rico, Mayaguez, Puerto Rico

ABSTRACT

The Cerro Goden fault zone is associated with a curvilinear, continuous, and prominent topographic lineament in western Puerto Rico. The fault varies in strike from northwest to west. In its westernmost section, the fault is ~500 m south of an abrupt, curvilinear mountain front separating the 270- to 361-m-high La Cadena de San Francisco range from the Rio Añasco alluvial valley. The Quaternary fault of the Añasco Valley is in alignment with the bedrock fault mapped by D. McIntyre (1971) in the Central La Plata quadrangle sheet east of Añasco Valley. Previous workers have postulated that the Cerro Goden fault zone continues southeast from the Añasco Valley and merges with the Great Southern Puerto Rico fault zone of south-central Puerto Rico. West of the Añasco Valley, the fault continues offshore into the Mona Passage (Caribbean Sea) where it is characterized by offsets of seafloor sediments estimated to be of late Quaternary age. Using both 1:18,500 scale air photographs taken in 1936 and 1:40,000 scale photographs taken by the U.S. Department of Agriculture in 1986, we identified geomorphic features suggestive of Quaternary fault movement in the Añasco Valley, including aligned and deflected drainages, apparently offset terrace risers, and mountain-facing scarps. Many of these features suggest right-lateral displacement.

Mapping of Paleogene bedrock units in the uplifted La Cadena range adjacent to the Cerro Goden fault zone reveals the main tectonic events that have culminated

*E-mail, Mann: paulm@ig.utexas.edu. Present address, Lao-Davila: Department of Geology and Planetary Science, University of Pittsburgh, 200 SRCC Building, 4107 O'Hara Street, Pittsburgh, PA 15260-3332

Mann, P., Prentice, C.S., Hippolyte, J.-C., Grindlay, N.R., Abrams, L.J., and Laó-Dávila, D., 2005, Reconnaissance study of Late Quaternary faulting along Cerro Goden fault zone, western Puerto Rico, *in* Mann, P., ed., Active tectonics and seismic hazards of Puerto Rico, the Virgin Islands, and offshore areas: Geological Society of America Special Paper 385, p. 115–138. For permission to copy, contact editing@geosociety.org.

in late Quaternary normal-oblique displacement across the Cerro Goden fault. Cretaceous to Eocene rocks of the La Cadena range exhibit large folds with wavelengths of several kms. The orientation of folds and analysis of fault striations within the folds indicate that the folds formed by northeast-southwest shortening in present-day geographic coordinates. The age of deformation is well constrained as late Eocene–early Oligocene by an angular unconformity separating folded, deep-marine middle Eocene rocks from transgressive, shallow-marine rocks of middle-upper Oligocene age. Rocks of middle Oligocene–early Pliocene age above unconformity are gently folded about the roughly east-west–trending Puerto Rico–Virgin Islands arch, which is well expressed in the geomorphology of western Puerto Rico. Arching appears ongoing because onshore and offshore late Quaternary oblique-slip faults closely parallel the complexly deformed crest of the arch and appear to be related to extensional strains focused in the crest of the arch. We estimate ~4 km of vertical throw on the Cerro Goden fault based on the position of the carbonate cap north of the fault in the La Cadena de San Francisco and its position south of the fault inferred from seismic reflection data in Mayaguez Bay. Based on these observations, our interpretation of the kinematics and history of the Cerro Goden fault zone includes two major phases of motion: (1) Eocene northeast-southwest shortening possibly accompanied by left-lateral shearing as determined by previous workers on the Great Southern Puerto Rico fault zone; and (2) post–early Pliocene regional arching of Puerto Rico accompanied by normal offset and right-lateral shear along faults flanking the crest of the arch. The second phase of deformation accompanied east-west opening of the Mona rift and is inferred to continue to the present day.

Keywords: Puerto Rico, faults, earthquakes, Mona Passage, arch.

INTRODUCTION

Tectonic Setting

The island of Puerto Rico is located within a diffuse and complex plate boundary zone that combines tectonic elements of left-lateral strike-slip faulting, collision, and oblique underthrusting between the North America and Caribbean plates (Dolan et al., 1998; Mann et al., 2002) (Fig. 1). Earthquake data, marine geophysical studies and global positioning system (GPS)–derived velocities suggest that most of Puerto Rico and the Virgin Islands are currently behaving as part of the stable Caribbean plate and are moving in an east-northeast direction (070°) at a rate of 19–20 mm/yr relative to North America (Jansma et al., 2000; Calais et al., 2002; Jansma et al., this volume) (Fig. 1).

Differences in the rate and direction of GPS vectors measured on either side of the Mona Passage between Hispaniola and Puerto Rico indicate that a poorly understood tectonic boundary occupies the shallow marine passage (Fig. 1A). The presence of an active boundary in the Mona Passage is supported by intense shallow seismicity and complexly faulted Neogene strata within the Mona Passage and in western Puerto Rico (van Gestel et al., 1998; Jansma et al., 2000, this volume) (Fig. 1A). Oblique collision between the Bahama Platform and Hispaniola and "pinning" of the Caribbean plate in the Hispaniola region (Dolan et al., 1998; Mann et al., 2002) provide a regional tectonic framework to explain: (1) diffuse shallow crustal seismic activity and deformation of the Mona Passage and western Puerto Rico (Huérfano et al., this volume); (2) Miocene–early Pliocene counterclockwise rotation of Puerto Rico (Reid et al., 1991); and (3) a lack of Neogene convergent structures in Puerto Rico versus widespread late Neogene convergence in Hispaniola (Mann et al., 2002).

Major Plate Boundary Faults near Puerto Rico

An important goal of seismic hazard studies in Puerto Rico and the Virgin Islands is to identify, map, and characterize onshore Holocene faults. On Figure 1A, we compile the results of previous authors to summarize our current understanding of plate-boundary-related, seismogenic faults in the offshore areas of Puerto Rico (Masson and Scanlon, 1991; Grindlay et al., 1997, this volume, Chapters 2 and 7). We describe the currently known offshore structures below to provide an understanding of the regional tectonic environment, which is essential to gaining an understanding of onshore Holocene faults.

Puerto Rico Trench–North Hispaniola Faults

This is a low-angle, submarine thrust fault system along which the North America plate is being subducted southward beneath the Caribbean plate, to depths of several hundred kilometers beneath Puerto Rico (Larue and Ryan, 1998; McCann, 2002; Huérfano et al., this volume) and to an unknown depth beneath Hispaniola (Dolan et al., 1998) (Fig. 1). This thrust system produced several large-magnitude earthquakes in the

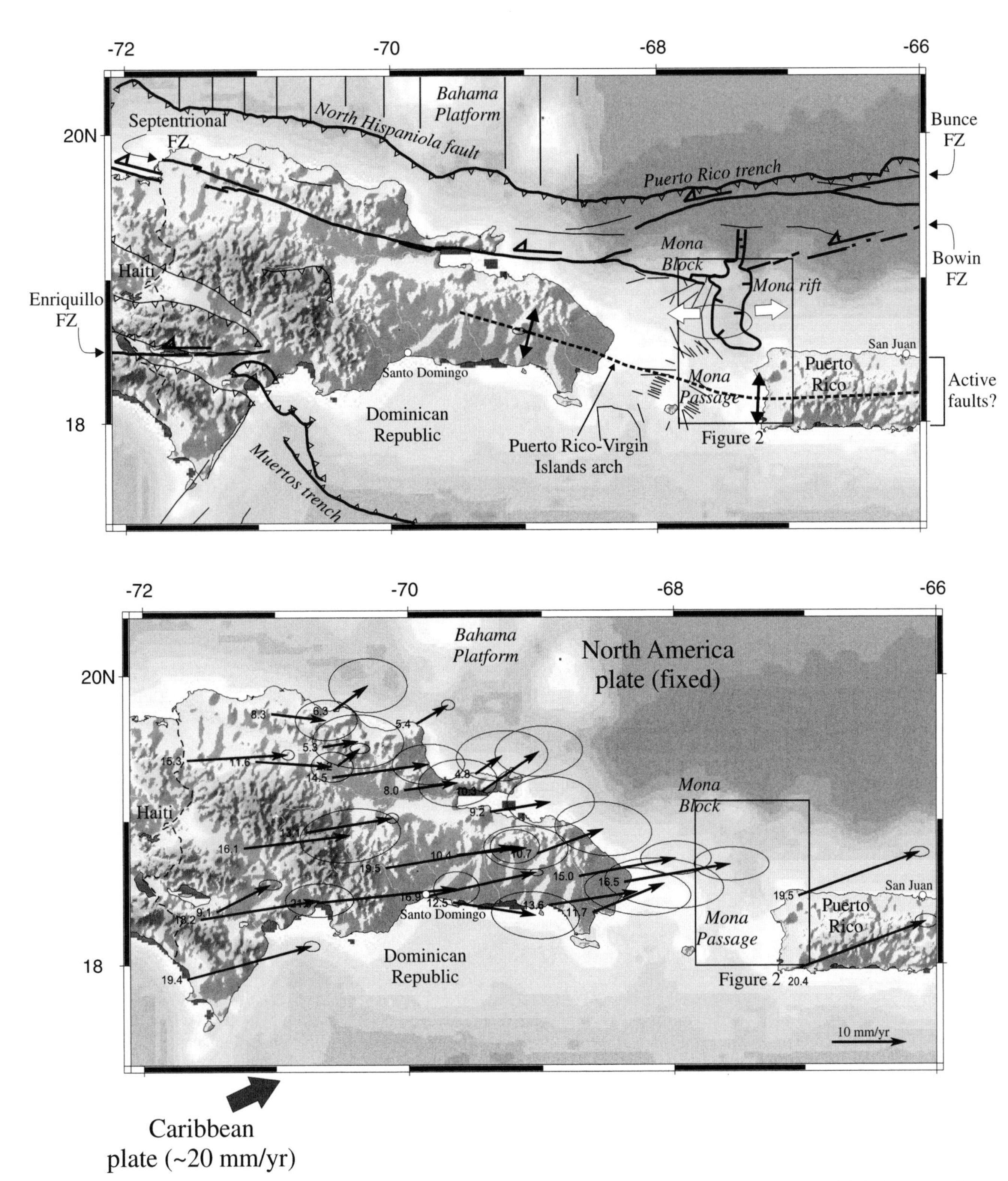

Figure 1. (A) Major faults in Puerto Rico and Hispaniola (Dominican Republic) and box showing more detailed map of faults in western Puerto Rico and the Mona Passage in Figure 2. White arrows indicate inferred direction of opening of the Mona rift based on its bounding faults and onland fault striation studies (Hippolyte et al., this volume). The complex fault pattern shown in the Mona Passage was mapped by Grindlay et al. (1997) using sidescan sonar. Note the axis of the post–early Pliocene Puerto Rico–Virgin Islands arch extending across Puerto Rico and into the eastern Dominican Republic. (B) GPS-derived velocities with respect to the North America plate (modified from Calais et al., 2002). The arrow length is proportional to the displacement rate, indicated in mm/yr next to the vector. The ellipses represent 95% confidence intervals. The large arrow in the lower left shows the predicted Caribbean plate velocity from DeMets et al. (2000). Note change in rate and direction of GPS vectors between western Puerto Rico and Dominican Republic. Modeling of these GPS data suggests ~5 ± 4 mm/yr of east-west extension across the Mona Passage and Mona rift (Jansma et al., 2000).

Hispaniola-Bahama platform collision zone, including several magnitude 7 and 8 events in the period from 1943 to 1953 (Dolan and Wald, 1998) and possibly a magnitude 6.5 event in September 2003 (Mann et al., 2004; Dolan and Bowman, 2004). GPS vectors from northern Hispaniola, south of the Puerto Rico trench–North Hispaniola fault exhibit more northeastwardly directions than vectors from central and eastern Hispaniola; this is consistent with oblique thrusting along the North Hispaniola fault and the tectonic process of strain partitioning in the zone of collision between the Bahama platform and Hispaniola (Dolan et al., 1998; Calais, et al., 2002).

Bunce Fault

This left-lateral strike-slip fault forms a linear valley, subparallel to and south of the Puerto Rico trench for a distance of ~240 km, and accommodates an unknown component of interplate oblique slip (Masson and Scanlon, 1991; Grindlay et al., 1997; Grindlay et al., this volume, Chapter 2) (Fig. 1). No large historical earthquakes are known to have been associated with this fault. The proximity of the Bunce strike-slip fault and the Puerto Rico trench low-angle thrust fault suggests a large degree of strain partitioning between the two parallel faults (ten Brink and Lin, 2004).

Septentrional Fault Zone

This left-lateral strike-slip fault can be traced as a prominent geomorphic feature across the Cibao Valley of the Dominican Republic (Prentice et al., 1993; Mann et al., 1998) and as a submarine feature from the eastern Cibao Valley to the Mona rift in the northern Mona Passage (Dolan et al., 1998). In the central Cibao Valley, this fault experienced at least 4.6 m of oblique normal and left-lateral displacement during a large earthquake ~800 yr ago (Prentice et al., 1993; 2003). Offset stream terraces yield a long-term Holocene slip rate of 6–12 mm/yr on the Septentrional fault, and geodetic models suggest a short-term slip rate of 12.8 ± 2.5 mm/yr (Calais et al., 2002). Geodetic data also indicate that the Septentrional fault forms a boundary between an area to the north moving in a northeast direction relative to the North America plate and an area to the south moving in a more eastwardly direction (Fig. 1). The Septentrional fault is roughly colinear with the Bowin fault to the east of the Mona rift (Fig. 2). The Mona rift north of its intersection with the Septentrional-Bowin fault becomes deeper and exhibits a north-northwest trend, whereas south of the intersection the Mona rift is shallower and exhibits a north-northeast trend. In addition, the Mona rift shows an apparent right-lateral jog where it intersects the Septentrional fault system, which is likely the result of ongoing interaction between the rift and the strike-slip fault (Fig. 1A).

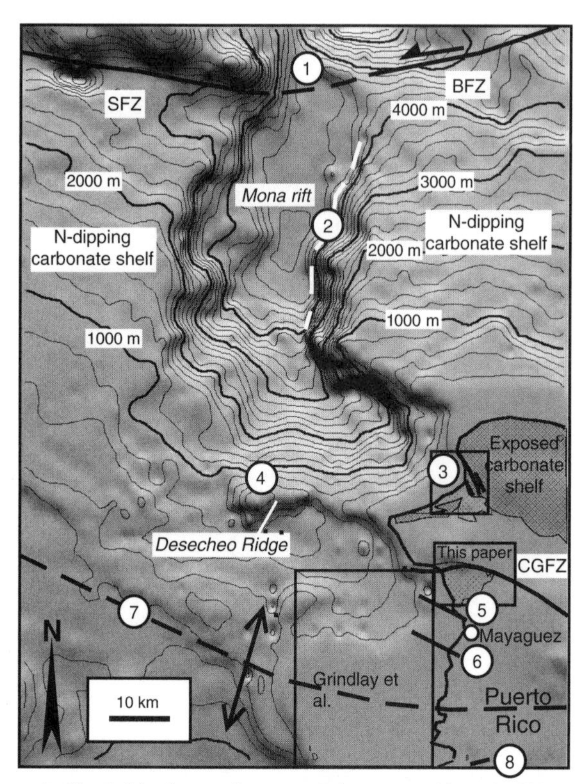

Figure 2. Shaded bathymetric map of the Mona rift and offshore insular shelf of western Puerto Rico with illumination from the north (modified from Hippolyte et al., this volume). Contour interval is 200 m. Seafloor morphology reflects northward-dipping limb of Puerto Rico–Virgin Islands arch and fault-bounded Mona rift. Dashed area numbered one is the proposed link across the Mona rift between the Septentrional fault (SFZ) to the west and the Bowin fault zone (BFZ) to the east. Fault scarps bounding the Mona rift are suspected to be the source of the M7 1918 earthquake that produced widespread damage and a damaging tsunami in western Puerto Rico. Mercado and McCann (1998) proposed that the 1918 earthquake originated along the fault shown on the eastern edge of the rift (dashed line numbered 2) while Doser et al. (this volume) proposed that the epicenter of the event was along the Desecheo ridge forming the southern edge of the rift (area numbered 4). Note that the onland Cerro Goden fault zone (CGFZ) and the offshore Desecheo ridge form parallel, east-west–trending lineaments that define the southern edge of the Mona rift. Boxes indicate our study area along the CGFZ and the study area of Grindlay et al. (this volume, Chapter 7). Faults numbered 5 and 6 were mapped by Grindlay et al. (this volume, Chapter 7) in the coastal area near Mayaguez. Hippolyte et al. (this volume) mapped normal faults with east-west opening numbered 3 that are interpreted to be the southeastward and onshore continuation of faults from the eastern margin of the Mona rift. The Puerto Rico–Virgin Island arch numbered 7 was mapped by Grindlay et al. (this volume, Chapter 7) and forms the southern limit of north-dipping rocks in western Puerto Rico. Fault 8 is the Holocene South Lajas fault mapped by Prentice and Mann (this volume).

Bowin Fault

This youthful fault forms a prominent fault scarp on the seafloor over a distance of ~330 km from the Puerto Rico trench to the Mona rift, as mapped using both seismic reflection and marine magnetic data (Muszala et al., 1999). The fault is inferred to be an active, left-lateral strike-slip fault based on its convergent and divergent stepovers along its trace (Grindlay et al., this volume, Chapter 2). The fault has been interpreted to represent the path of the southern edge of the obliquely subducted Bahama Bank (Mona block) as it slid past the Virgin Islands and Puerto Rico to its present geographic position marked by the morphologic Mona block (Mann et al., 2002; Grindlay et al., this volume, Chapter 2). The fault's strike is subparallel to the 070-trending, GPS-derived North America–Caribbean plate vector measured at sites in Puerto Rico (Jansma et al., 2000). The proximity of this fault to the oversteepened and scalloped shelf margin of northern Puerto Rico makes it particularly hazardous for triggering catastrophic submarine landslides and associated tsunamis that could affect the heavily populated coastline of northern Puerto Rico, including the 1.8 million inhabitants of San Juan (Hearne et al., 2003) (Fig. 1). No known historical earthquakes are associated with this fault.

Mona Rift Fault

Seismic reflection data reveal that this north-south–striking normal fault is very young and is the west-dipping "master normal fault" controlling the formation of the Mona rift (van Gestel et al., 1998). The bathymetric escarpment of the eastern margin of the Mona rift reaches 3 km in relief (Fig. 2). Mercado and McCann (1998) used seismic reflection profiles collected by Western Geophysical in 1972 to infer that the north-south–striking eastern border fault of the Mona rift was the source of the 1918 earthquake and related tsunami. To the east of the Mona rift, the north flank of the Puerto Rico–Virgin Islands arch has a slightly steeper dip than the area to the west of the rift as seen both in the bathymetry shown on Figure 2 and on seismic lines (van Gestel et al., 1998, 1999).

Distributed Faults in the Mona Passage

These roughly southeast-striking normal and oblique-slip faults mapped during a marine survey by Dolan et al. (1998) and by Grindlay et al. (1997; this volume, Chapter 2) occur within a broad area of the Oligocene–early Pliocene carbonate platform and may accommodate diffuse extension and counterclockwise rotation between Puerto Rico and Hispaniola as a result of the Bahama platform–Hispaniola collision to the northeast (Mann et al., 2002) (Fig. 1).

Puerto Rico–Virgin Islands Arch

This large tectonic arch is described in detail by van Gestel et al. (1998) and affects the entire region. Its crest (shown on Fig. 1) defines the topographic and bathymetric divide between the Caribbean Sea and the Atlantic Ocean in eastern Hispaniola, the Mona Passage, Puerto Rico, and the Virgin Islands. Dips on Neogene rocks on the north flank of the arch average 4° and exhibit few faults while those on the south flank dip slightly more steeply (up to 30°) and are more intensively faulted (van Gestel et al., 1998; 1999). Northern Puerto Rico is situated on the northern, north-dipping limb of the arch (Fig. 2).

GEOLOGIC SETTING OF THE CERRO GODEN FAULT ZONE AND OTHER FAULTS IN WESTERN PUERTO RICO

Our studies are aimed at understanding the sense of slip and youngest age of displacement on the east-northeast–trending faults in western Puerto Rico, including the Cerro Goden fault. Geologic mapping of faults cropping out in Eocene and older lithologies has shown that these structures are transpressional, left-lateral faults associated with folding (Glover, 1971; Erikson et al., 1990). However, because these studies only examined older rocks, it should not be assumed that this style of deformation continues into the present. In fact, there are few folds and thrusts present in the Neogene sections of Puerto Rico indicative of continued, large scale sinistral transpression (Mann et al., this volume, Chapter 9). Moreover, ongoing left-lateral slip on these faults would close, not open, the Mona rift (Fig. 2).

The radar image of northwestern Puerto Rico shown in Figure 3 reveals several tectonically and erosionally controlled morphologic features in the vicinity of the Cerro Goden fault zone. This image shows that scarps typically face southward and have been formed in the north-dipping carbonate platform of Oligocene to early Pliocene age. We interpret some scarps, such as the scarp that has developed at the base of the Neogene carbonate cap overlying Eocene and older rocks, to be a product of differential erosion rather than fault activity. Two widespread ancient surfaces are visible on Figure 3. One is a northern surface, formed on top of early Miocene–early Pliocene shallow marine carbonate rocks in post–early Pliocene time (Moussa et al., 1987). Tropical erosion of these carbonate rocks has produced a prominent karst topography visible on the radar image in Figure 3. The second prominent surface is developed on the folded Eocene rocks of the north-tilted La Cadena de San Francisco range (Fig. 3). A possible third surface may be present south of the Cerro Goden fault zone, developed on Eocene and Cretaceous rocks, but is more heavily dissected by erosion and is obscured by Quaternary alluvial deposits of the Rio Añasco. These surfaces now dip 4–10° northwards, reflecting folding associated with the post–early Pliocene Puerto Rico–Virgin Islands arch (Fig. 1). Since the tops of most carbonate platforms have no depositional dip, all of the northward dip is thought to be tectonic in origin (Grindlay et al., this volume, Chapter 2).

The most prominent lineament of western Puerto Rico is the mountain front along the southern edge of the La Cadena de San Francisco Range, which is parallel to and closely associated with the late Quaternary Cerro Goden fault zone (Fig. 3). The Cerro Goden fault zone was originally mapped in Cretaceous and Eocene rocks within the Central La Plata quadrangle to the northeast of Mayaguez by McIntyre (1971).

We selected the Cerro Goden fault zone for a reconnaissance study because of: (1) the curvilinearity and youthful appearance of the mountain front (Fig. 3); (2) its close association with the uplifted La Cadena de San Francisco (Fig. 3); (3) its continuity with submarine faults with suspected Holocene activity (Grindlay et al., this volume, Chapter 7) (Fig. 2); (4) its apparent link to suspected seismogenic faults in the offshore Mona rift (Mercado and McCann, 1998) (Fig. 2); and (5) its possible continuity with the Great Southern Puerto Rico fault zone of the south coast of Puerto Rico (Fig. 1) (Mann et al., this volume, Chapter 9).

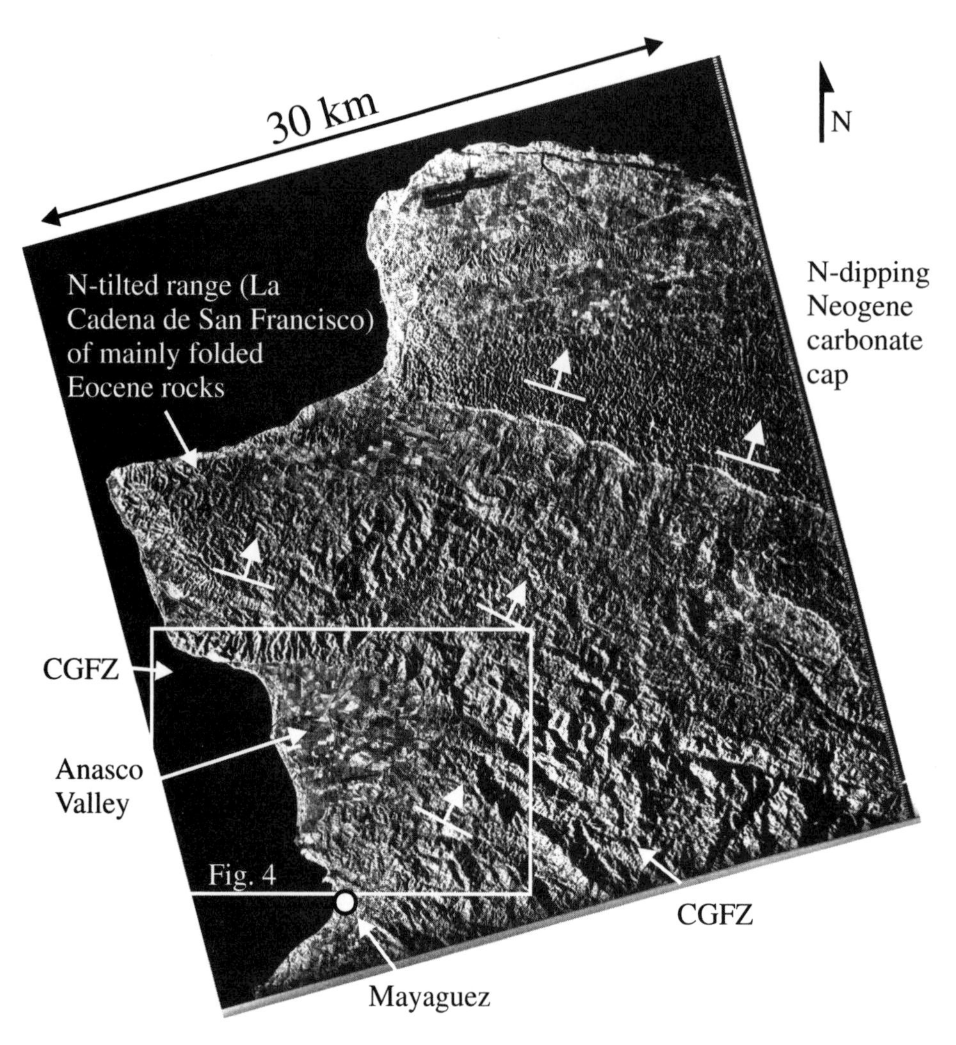

Figure 3. Radar image of western Puerto Rico showing major topographic lineament associated with the Cerro Goden fault zone (CGFZ) along the northern edge of the Añasco Valley. Rocks of the planar Eocene unconformity surface and overlying rocks of Oligocene and younger age have a uniform, 4–10° tilt to the NNE. This reflects the position of northwestern Puerto Rico on the northern flank of the roughly EW-trending Puerto Rico–Virgin Islands arch whose axis is shown on the map in Figure 1.

EVIDENCE FOR LATE QUATERNARY DEFORMATION ALONG THE CERRO GODEN FAULT ZONE

Based on our examination of aerial photographs, topographic maps and satellite imagery, we subdivide the 15-km-long trace of the Cerro Goden fault zone examined for this study indicated in the boxed area of Figure 3 into three distinct morphologic sections shown in greater detail in Figure 4. The west-northwest–trending eastern section mapped by McIntyre (1971) in the western part of the Central La Plata quadrangle occupies the deeply entrenched, straight, northwest-trending valley of the Rio Añasco (Figs. 4 and 5B). The fault separates Cretaceous rocks of the Rio Blanco Formation and the Rio Yauco Formation along this section (Fig. 6). We did not work in this area as part of this study and instead relied on the published work of McIntyre (1971) who describes several bedrock exposures of the fault. Where exposed on the sides of the Añasco River valley the fault consists of a shear zone several meters in width, as reported by McIntyre (1971). The fault continues to the southeast as a lineament roughly parallel to the projected trace of the Great Southern Puerto Rico fault zone (Glover, 1971; Erikson et al., 1990) (Fig. 1). This section of the fault exhibits a 3.15 km, apparent right-lateral offset of the Rio Añasco as indicated in Figure 6.

The west-northwest–trending central section was partially mapped by McIntyre (1971) in the westernmost part of the Central La Plata quadrangle. This section traverses a low range of bedrock hills that form the eastern edge of the alluvial Añasco Valley (Figs. 4 and 5B). As in areas to the east, the fault forms a very sharp and curvilinear boundary between the Eocene Culebrinas Formation to the north and the Cretaceous Yauco Formation to the south (Fig. 6).

The third section is the east-west–trending western section that traverses the Añasco Valley between the bedrock hills and the coast. This section was not mapped by McIntyre (1971) and is associated with the prominent mountain front on the northern edge of the Añasco Valley (Figs. 2, 4, and 5A). The central and western sections of the fault near the Añasco Valley form the primary focus of this study.

Geomorphic Features Indicative of Late Quaternary Fault Displacement

We analyzed two sets of aerial photographs in order to better identify geomorphic features related to Quaternary fault activity, along the central and western sections of the Cerro Goden fault: (1) 1:40,000 scale photographs taken by the U.S. Department of Agriculture in 1986; we used these photographs to make the photomosaic in Figure 4; (2) 1:18,500 scale photographs taken in 1936. The scale and vintage of the latter set of photos make them invaluable for identifying geomorphic features prior to development of this area in the latter half of the twentieth century. We made a photomosaic of the 1936 aerial photographs of the central and western sections of the Cerro Goden fault zone, shown in Figure 6. Comparison of the 1936 mosaic with the 1986 mosaic in Figure 4 shows the intensive development of the area in the intervening 50 years, much of which overlies the projected trace of the Cerro Goden fault zone in the Añasco Valley.

Fault-Line Scarp

We postulate that the Quaternary fault trace of the western and central sections of the Cerro Goden fault lies ~500 m south of the prominent mountain front of the La Cadena de San Francisco range (Fig. 2, 4, and 6) and that the mountain front is therefore a fault-line scarp produced by erosional retreat. This relationship of a prominent mountain-front fault-line scarp located hundreds of meters from the more subtly expressed Quaternary fault is typical of strike-slip faults, for example, the Septentrional fault of Hispaniola (Prentice et al., 1993; Mann et al., 1998; Prentice et al., 2003). We base our interpretation on: (1) the westward projection of the bedrock fault trace into the alluvial area of the Añasco Valley (Fig. 5B); (2) our identification of aligned geomorphic features indicative of Quaternary faulting, and (3) the highly embayed character of the mountain front, which exhibits no evidence for late Quaternary fault activity (Figs. 4, 5A, and 5B).

Using both sets of aerial photos, we identified geomorphic features suggestive of late Quaternary fault movement along the Cerro Goden fault, including aligned and deflected drainages, apparently offset terrace risers, and mountain-facing scarps (Figs. 6, 7, and 8). These features are both to the east and to the west of the low hills forming the eastern edge of the Añasco Valley (Fig. 4). The eastern end of the valley is also the least developed part of the region and therefore it is probably not coincidental that this region is where we found evidence of recent fault movement preserved. We found no geomorphic features associated with Quaternary faulting in the alluvial plain west of Añasco, where cultural impact has been intensive since Spanish colonists began cultivating the area in the early sixteenth century (Fig. 6).

We mapped a west-northwest alignment with right-laterally deflected streams and fluvial terrace risers along the western section of the fault east of Añasco (Fig. 7A). Another right-laterally deflected drainage is located due north of Añasco (white arrows indicate ends of deflected streams in Fig. 6).

Figure 7A shows the contrast between the eroded mountain front that we interpret as a fault-line scarp and the curvilinear alignment of features we interpret as marking the trace of the Quaternary Cerro Goden fault. This alignment of features is along the projection of the bedrock Cerro Goden fault mapped by McIntyre (1971) in the low hills to the east (Figs. 5A, 5B, and 6).

In Figure 7A, the southern banks of two streams each exhibit ~50 m of apparent right-lateral offset. Unfortunately, development in this area has recently obscured these features (Fig. 7B). The photograph in Figure 7B (taken in 2000) shows the embayed mountain front north of the streams, the low bedrock hills rising up to the east of the streams, and the large amount of development that has occurred in the area since 1936.

Part of the Añasco River valley lies eastward of the low bedrock hills and follows the Cerro Goden fault zone for a distance of ~3.15 km. (Fig. 6). We suggest that this west-northwest stretch of the Añasco River may represent a right-lateral offset of the river (Fig. 6). Note the linear character of the bedrock fault mapped by McIntyre (1971). We were unable to locate any exposures of this fault in the low bedrock hills of the central fault section, possibly because the cleared areas of the fault lineament shown on the 1936 air photo in Figure 8A are now densely vegetated. The projected scarp of the Cerro Goden fault traverses the fluvial terrace above the Añasco River shown in Figure 8A. This scarp could be either a fault scarp or a terrace riser, but its location along the projection of the bedrock fault in the low hills of the central fault section suggests the possibility that it is a fault-controlled feature.

In the westernmost area of the Añasco Valley, we did not find any geomorphic features associated with Quaternary faulting along the projected western trace of the fault. Rapid sedimentation in the Añasco River valley and intense cultural overprinting have obscured any Quaternary fault features that may have occurred in this area. We also did not find any evidence for shearing in the well-exposed, south-dipping Eocene rocks along the mountain front, consistent with our interpretation that this mountain front is a fault-line scarp (Fig. 9).

AGE OF DEFORMATION OF CRETACEOUS AND PALEOGENE ROCKS ALONG THE CERRO GODEN FAULT ZONE

Introduction

Structural mapping by Glover (1971) and fault striation studies by Erikson et al. (1990) in southern Puerto Rico have shown that sinistral strike-slip faulting characterized that part of Puerto Rico during the middle Eocene through early Oligocene time and may have extended to deform rocks as young as Miocene in age. In order to better understand the Cerro Goden fault, which is on strike from the faults described by Glover (1971) and Erikson et al. (1990) (Fig. 1), we have compiled geologic data from McIntyre (1971) and Tobisch and Turner (1971), and

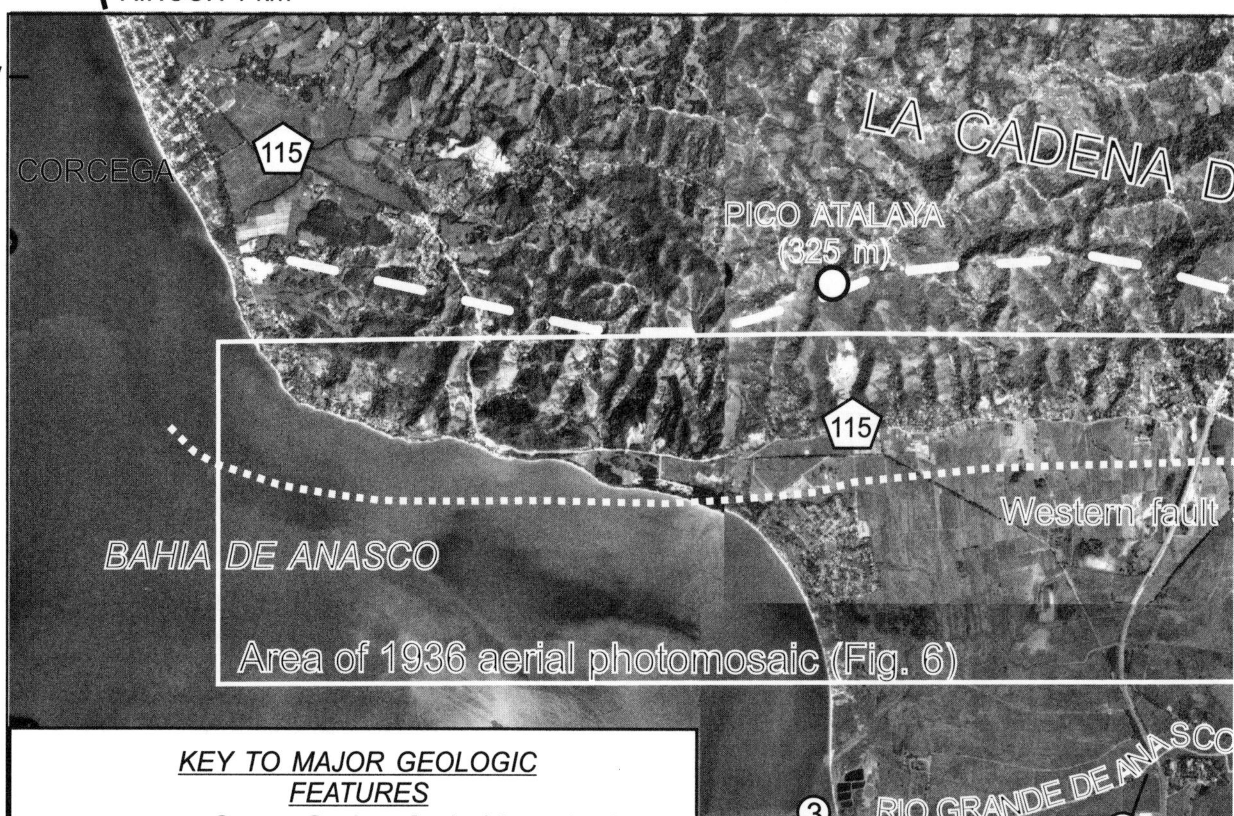

KEY TO MAJOR GEOLOGIC FEATURES

Cerro Goden fault (dotted where inferred)

Crest of topographic ridge parallel to Cerro Goden fault zone

1 km

KEY TO MAJOR CULTURAL FEATURES

2 Highway 2 to San Juan

115 Highway 115 to Rincon

① Industrial park and residential area of Anasco

② Mayaguez regional airport

③ Regional sewage treatment plant

④ Bridges over Rio Anasco

⑤ Northern residential and business area of Mayaguez

Figure 4 (*on this and previous page*). Photomontage made from 1986, 1:40,000-scale air photos taken by the U.S. Department of Agriculture of the southern La Cadena de San Francisco, the Cerro Goden fault zone, and the Añasco Valley. The Cerro Goden fault zone is within 10 km of the city of Mayaguez. Other major cultural and infrastructure features that could be adversely affected by a large earthquake in this region are indicated.

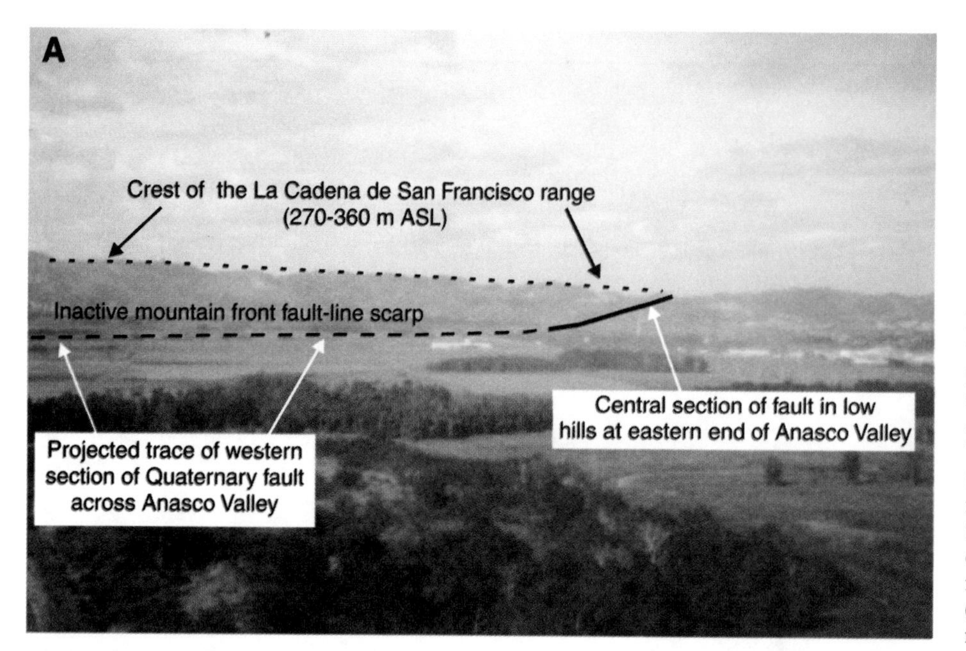

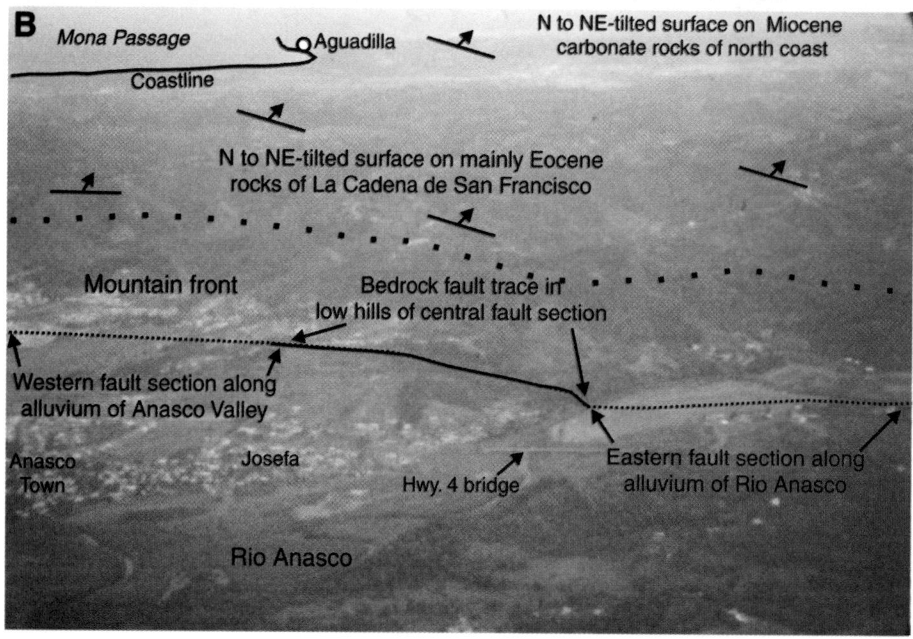

Figure 5. (A) Aerial view looking northward across the Añasco Valley. The projected trace of the Cerro Goden fault is ~500 m south of the mountain front of the La Cadena de San Francisco. This mountain front is inferred to be a fault-line scarp produced by erosional retreat. The top of the fault-line scarp forms a very uniform crest along the entire length of the mountain front that ranges in elevation from 270 to 360 m (cf. Fig. 4). (B) Aerial view looking northward across the section of the Cerro Goden fault zone expressed in bedrock in the low hills at the eastern end of the Añasco Valley. The gently tilted, NNE-tilted erosion surface of the La Cadena de San Francisco range is apparent in this photo. We attribute the northward tilted erosion surface to late Quaternary horizontal and vertical motion on the Cerro Goden fault zone. In the distance, NNE-tilted carbonate rocks of Miocene age are apparent and constrain a post-Miocene age of tilting related to the Puerto Rico–Virgin Islands arch (cf. Fig. 3).

we conducted our own original mapping of the rocks adjacent to the central and western sections of the Cerro Goden fault zone (Figs. 3 and 5). This information is summarized in a series of five cross sections, four of which cross the trace of the Cerro Goden fault zone (lines of section shown on inset of Fig. 10). Another goal of this compilation is to place constraints on the amount of vertical throw on the Cerro Goden fault by correlating units north of the fault in the La Cadena de San Francisco with units south of the fault beneath Añasco Bay.

A section of middle Eocene marine sandstone and shale (Rio Culebrinas Formation) greater than 2800 m thick conformably overlies basalts of Cretaceous-Eocene age in the region of the Cerro Goden fault. To the southeast, marine sedimentary rocks equivalent to the Rio Culebrinas Formation are interbedded with volcanic arc rocks, indicating that arc activity continued into middle Eocene time. The volcanic basement is part of the volcanic arc sequence of Puerto Rico and the overlying Rio Culebrinas Formation formed in an intra-arc basin on top of that arc (Dolan et al., 1991). An angular unconformity is present between the folded Rio Culebrinas Formation and the gently north-tilted carbonate platform units of the Lares and Cibao Formations. The carbonate platform rocks can be projected above the La Cadena de San Francisco using their measured 6–7° northward dips as shown on Figure 10, sections A and E.

Deformation of Rocks Adjacent to the Cerro Goden Fault Zone

Eocene rocks and their underlying basement exhibit large-scale folds and thrust faults, which are either upright folds or slightly overturned to the northeast (Fig. 10) (Tobisch and Turner, 1971). In general, folds are broad, large-wavelength (>4 km) structures except for one fault block adjacent to the westernmost Cerro Goden fault, where much shorter wavelength folds (<1 km) are found (Fig. 10, section B). The largest fold in the area (Cerro Goden anticline mapped by McIntyre, 1971) is spatially associated with the Cerro Goden fault and is shown on the inset map in Figure 10.

In order to better constrain the direction of shortening responsible for the large folds, we have compiled poles to bedding for the U.S. Geological Survey quadrangle maps of McIntyre (1971) and Tobisch and Turner (1971) along with our bedding measurements collected in the western area (Fig. 11A). The plots show a uniform, northeast-to-southwest direction of shortening (Fig. 11B).

Eocene rocks exposed along the fault-line scarp of the Cerro Goden fault zone, such as those shown in Figure 9, are well exposed by quarrying along the steep slopes of the mountain front and by road building along the length of the mountain front. We measured the orientations and sense of slip on many exposed mesofault planes in Eocene rocks of the Culebrinas Formation at six sites shown on the map in Figure 11C in an effort to better understand the state of paleostress responsible for the formation of the fault populations. Analysis of brittle tectonic features is based on the assumption that slickenside lineations on a fault plane indicate the direction and sense of maximum resolved shear stress on that fault plane (Angelier, 1990). Directions of maximum compressive stress derived from mesofaults (Fig. 11D) are similar to the directions derived from the plots of poles to bedding (Fig. 11B) and suggest that both the large folds and mesoscale faults formed during the same northeast to southwest shortening event. We found no evidence for shear planes parallel to the direction of the Cerro Goden faults shown on Figure 6.

The age of shortening is well constrained by the ages of the deformed rocks shown on the cross sections in Figure 10. The angular unconformity between the middle Eocene Rio Culebrinas Formation and the early Oligocene shallow-marine and nearshore rocks of the San Sebastian and Juana Diaz Formations implies a major late Eocene tectonic shortening event that resulted in the uplift of basinal rocks to shallow depths, coincident with the cessation of arc activity (Dolan et al., 1991).

CORRELATION OF OFFSHORE STRATIGRAPHY WITH ONSHORE STRATIGRAPHY

We examined records of marine seismic lines collected in 1972–1973 by the Western Geophysical Company and Fugro (Western Geophysical Company of America and Fugro, Inc., 1973) that were donated in the late 1970s to the Puerto Rico Seismic Network of the University of Puerto Rico, Mayaguez (Fig. 12). These multi-channel seismic lines are valuable because they provide deep penetration into basinal stratigraphy and provide images of fault planes in Añasco Bay where we and Grindlay et al. (this volume, Chapter 7) project the Cerro Goden fault.

Figure 12A shows the locations of two of the seismic lines that provide important constraints on the location and age of the offshore section of the Cerro Goden fault zone. Line XI-108D crosses the fault and reveals a south-dipping fault plane, which we interpret to be the Cerro Goden fault zone (Fig. 12B). Reflectors beneath the upthrown footwall block are chaotic and we interpret these as tilted, folded and faulted middle Eocene rocks of the Rio Culebrinas Formation (Fig. 10). We subdivide rocks of the hanging wall block into a wedge-shaped package of presumably clastic origin and an underlying pre–middle Pliocene carbonate section that exhibits uniform thickness over most of the length of the cross section (Fig. 12B).

Line XI-110D in Figure 12C does not cross the Cerro Goden fault, but lies entirely on the hanging wall block. This seismic line shows an image of the stratigraphy underlying Añasco Bay that is discussed in detail by Grindlay et al. (this volume, Chapter 7). The prominent reflectors are interpreted to represent the top of the north-dipping, pre–middle Pliocene carbonate cap that is exposed onshore north of the Cerro Goden fault (Moussa et al., 1987) (Fig. 3). We interpret the material overlying the carbonate cap as representing middle Pliocene to recent terrigenous, clastic, subhorizontal sediments deposited in a wedge-shaped geometry over the northward-sloping and faulted pre–middle Pliocene carbonate cap south of the Cerro Goden fault.

Seismic line ID-131-D from Western Geophysical is interpreted and discussed by Grindlay et al. (this volume, Chapter 7). In Figure 13, we show the cross section interpreted from this seismic line with no vertical exaggeration and join it with geologic cross section C from Figure 10. On the offshore part of the cross section, pre–middle Pliocene carbonate rocks are interpreted to lie beneath a major angular unconformity that is comparable to the unconformity surface of northern Puerto Rico shown at the northern end of cross sections D and E in Figure 10. In the offshore area, we interpret the overlying rocks, which are deposited in half-grabens and lack convergent structures, as middle Pliocene to recent (Fig. 13). Faults in the offshore area dip to the south, and are associated with growth wedges of sediment indicating that faults remained active during deposition. In this interpretation, the La Cadena de San Francisco range is uplifted within the footwall of the Cerro Goden fault zone. A small outcrop of Eocene basalt near the coast shows that vertical throw of the Cerro Goden fault has elevated Eocene and older arc-related basement rocks to the surface.

We project the base of the Miocene–early Pliocene carbonate platform exposed onshore north of the La Cadena de San Francisco to the Cerro Goden fault zone (Fig. 13). We correlate this horizon with the interpreted base of the carbonate cap

Figure 6. Photomontage of 1936 1:20,000 scale air photos of the Cerro Goden fault zone. These photos precede the surge of residential and industrial development in the Añasco Valley in the latter half of the twentieth century and are useful for locating relatively undisturbed geomorphic features. Black arrow indicates westernmost limit of identifiable geomorphic features that may represent Quaternary displacement across the Cerro Goden fault. We suspect features are obscured by the long history of agriculture in this area, coupled with a high rate of

offshore not imaged on the seismic line but inferred from the known thickness of the carbonate cap in northern Puerto Rico (van Gestel et al., 1998). We estimate a vertical separation of ~4 km on the Cerro Goden fault based on our projection of the onshore carbonate cap as shown on Figure 13. This would yield a very rapid uplift rate of ~1 mm/yr based on the fault offsetting the Oligocene–early Pliocene carbonate platform following the deposition of early Pliocene rocks. If the faulting and fault-related uplift accompanied the formation of the Oligocene–early Pliocene platform, the rate would have been much slower.

DISCUSSION

Figure 14A shows a bathymetric, tectonic, and earthquake epicenter map of the Puerto Rico–Virgin Islands region based on marine geophysical mapping by Grindlay et al. (1997) and van Gestel et al. (1998). Faults in the Puerto Rico trench area trend east-northeast and are interpreted to reflect the path of the subducted Bahama Platform along the northern edge of the Caribbean forearc (Grindlay et al., this volume, Chapter 2). These fault trends, which include strike-slip faults, such as the Bunce and Bowin faults, and low-angle thrusts, like the Puerto Rico trench, parallel the present-day GPS vector between Puerto Rico and North America. GPS results from Jansma et al. (2000) and Jansma and Mattioli (this volume) suggest that Puerto Rico is moving as part of the Caribbean plate. Modeling of GPS data suggests the possibility of a small amount of east-west extension within Puerto Rico (Jansma and Mattioli, this volume).

The eastern margin of the now-subducted Bahama Platform (Mona block) coincides with the eastern edge of the Mona rift, a 25-km-wide graben that has disrupted the Oligocene to early Pliocene carbonate platform (Fig. 14A). An incompletely mapped zone of normal and oblique faults extends to the south and southwest of the Mona rift and deforms the carbonate platform in the central part of the Mona Passage. The carbonate platform is also deformed as part of the Puerto Rico–Virgin Islands arch, the structural crest of which parallels the topographically highest land areas of the Virgin Islands, Puerto Rico, and the Dominican Republic. The northwest to east-west orientation of the Cerro Goden fault (and parallel faults in western Puerto Rico) relative to GPS vectors in Puerto Rico predict right-lateral oblique-slip in the current tectonic regime, consistent with the geologic observations presented in this paper.

The relation of the Cerro Goden fault to current seismicity is poorly understood. Composite focal mechanisms constructed by Huérfano et al. (this volume) indicate left-lateral shear, while we propose right-lateral slip across the Cerro Goden fault. One plausible explanation may be that the Cerro Goden fault is locked and accumulating elastic strain and therefore the composite focal mechanisms may not represent the overall, long-term tectonic

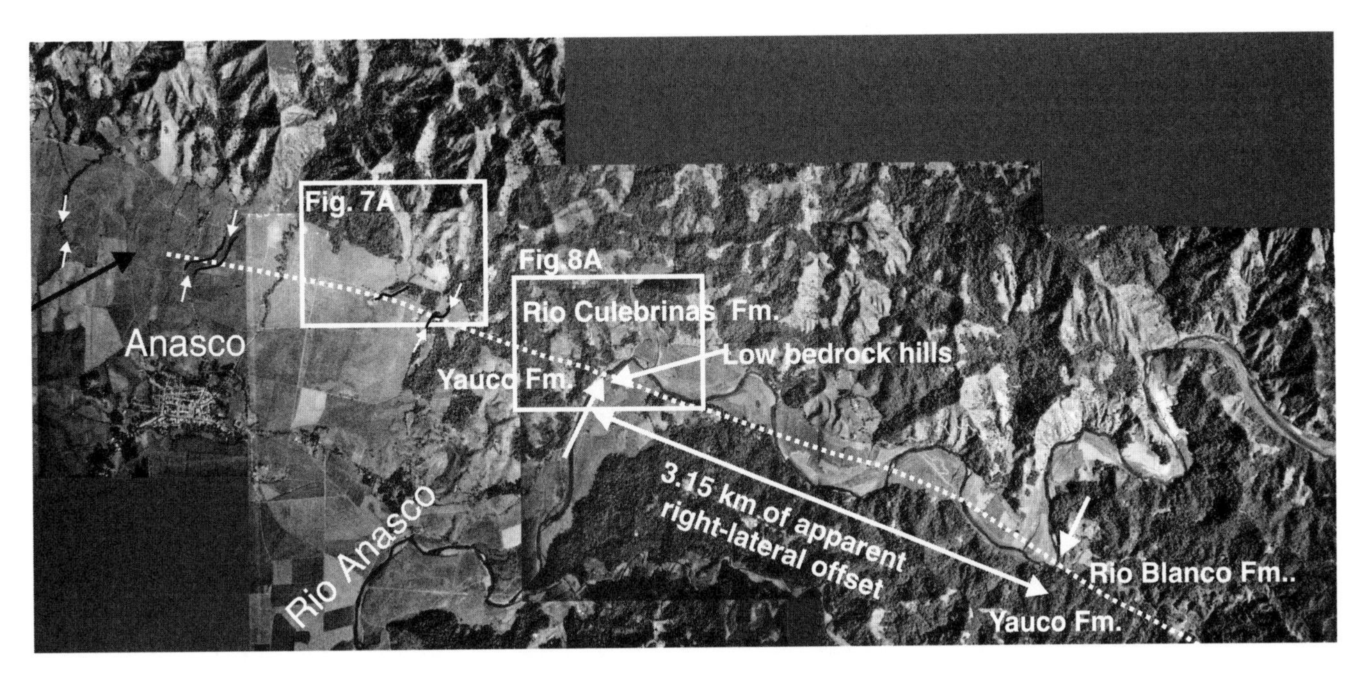

alluvial deposition in the valley. A 3.15 km right-lateral offset of the Rio Añasco can be measured between the white arrows shown. The Cerro Goden fault zone juxtaposes rock types of different lithology and age as indicated: Eocene Rio Culebrinas Formation against late Cretaceous Yauco Formation and Late Cretaceous Rio Blanco Formation against Late Cretaceous Yauco Formation. Boxes show areas of enlarged 1936 photos in Figures 7A and 8A.

style. Huérfano et al. (this volume) takes the alternative view that microearthquakes faithfully express the long-term style of deformation across the Cerro Goden fault. Paleoseismic studies that demonstrate whether or not Holocene displacement has occurred and that demonstrate the sense of displacement across the fault are needed in order to resolve this controversy.

The three main phases of deformation affecting Puerto Rico suggested in this and previous studies (c.f. van Gestel et al. 1998; 1999) are shown schematically in Figure 14C. The oldest event involved Eocene shortening and transpression that produced large-scale folds and thrust faults whose axes trend northwest. This event produced the major angular unconformity separating folded rocks of middle Eocene age and overlying much less folded rocks of Miocene age as seen on the northern ends of cross sections D and E in Figure 10. We propose that the Cerro Goden fault may have nucleated as a left-lateral reverse fault during this phase of deformation.

A second phase of north-south shortening produced the east-west–trending Puerto Rico–Virgin Islands arch in post–early Pliocene time (the youngest unit of the dipping northern limb of the arch is early Pliocene (Moussa et al., 1987). The geomorphology of offshore (Fig. 2) and onshore (Fig. 3) northwestern Puerto Rico reflects its present-day position on the northern limb of this arch (van Gestel et al., 1998). Arching is proposed to account for the large vertical component of deformation on faults confined to the 25-km-wide crestal area of the arch (Fig. 14D) that is also a zone of intense crustal seismicity (Huérfano et al., this volume). We propose that during this phase of deformation, the Cerro Goden fault was reactivated as a dominantly normal fault accommodating extension across the crest of the arch.

The youngest phase of deformation is associated with the present-day east-northeast–directed motion of Puerto Rico away from the collisionally pinned area of Hispaniola (Fig. 14A). Our model suggests that this motion causes opening of the Mona rift and induces right-lateral displacement across the Cerro Goden fault zone. We propose that the process of opening the Mona rift and the more regional arching across Puerto Rico may be occurring concurrently today (Fig. 14C).

Major faults in western Puerto Rico appear to be closely related to the axis of the Puerto Rico–Virgin Islands arch (Fig. 14D). Faults to the north of the axis of the arch dip southward while faults south of the arch dip northward. Together, these two sets of faults form a 25-km-wide, symmetrical pattern about the crest of the arch as shown schematically on Figure 14D. We interpret this relationship between the faults and the axis of the arch to indicate that the major faults of western Puerto Rico are predominantly extensional structures produced in response to the formation of the arch. Extension off the northwest coast of Puerto Rico as indicated by the arrows in Figure 1 produces right-lateral offsets on at least one of these faults, the Cerro Goden.

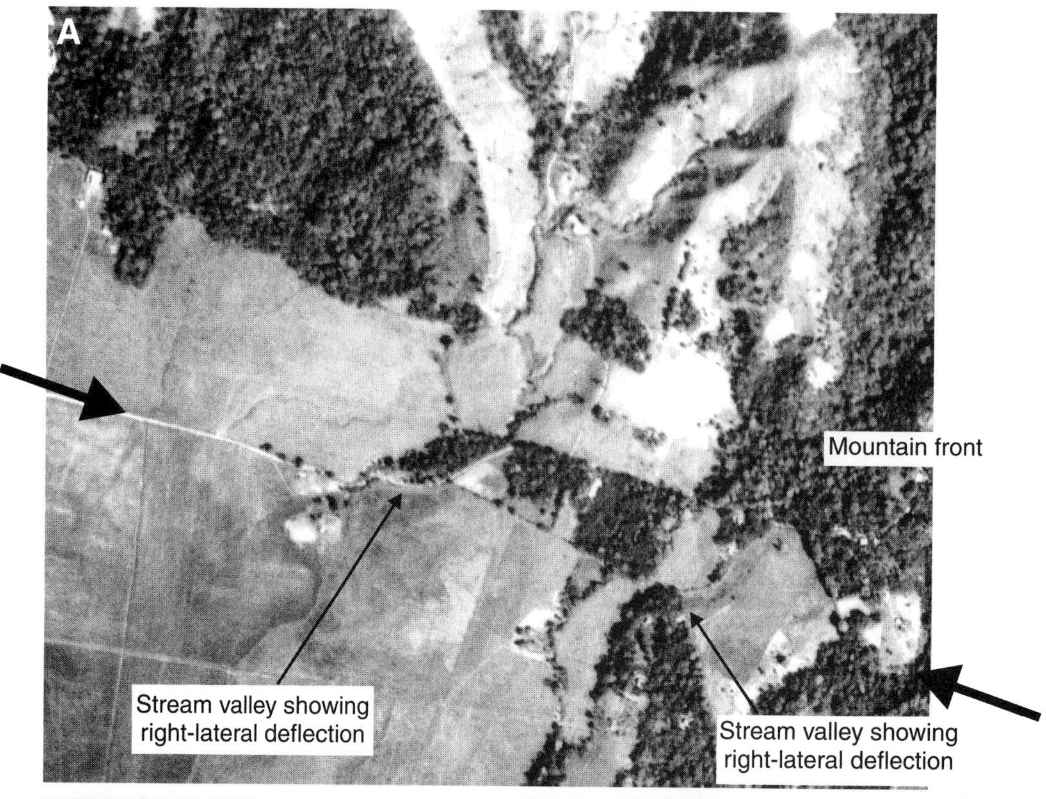

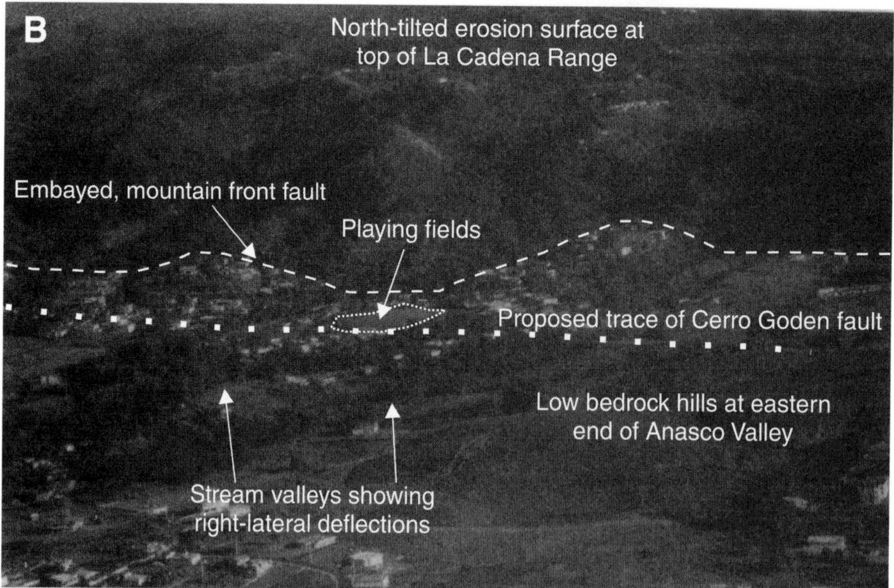

Figure 7. (A) Enlargement of region indicated on Figure 6 showing two right-laterally deflected streams that indicate proposed location of Cerro Goden fault. Black arrows indicate location of fault trace. Arrow to east is along the projection of the bedrock fault trace mapped in the low hills east of the Añasco Valley (cf. Fig. 5B). (B) 2001 aerial view of the same area shown in A. Note extensive residential development along proposed fault trace indicated.

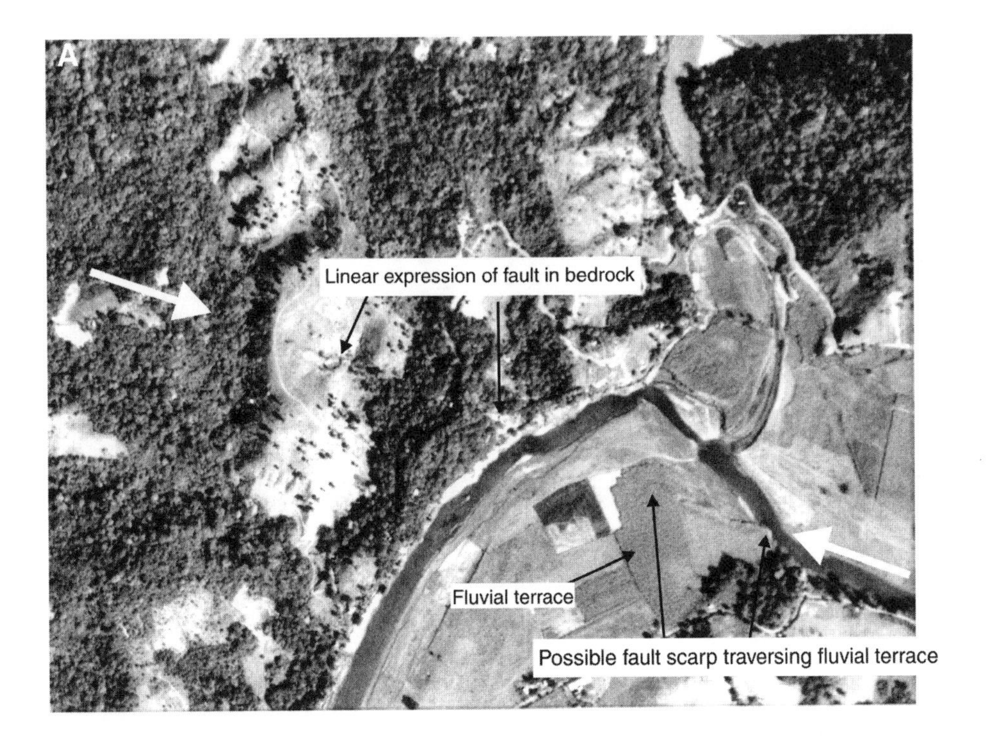

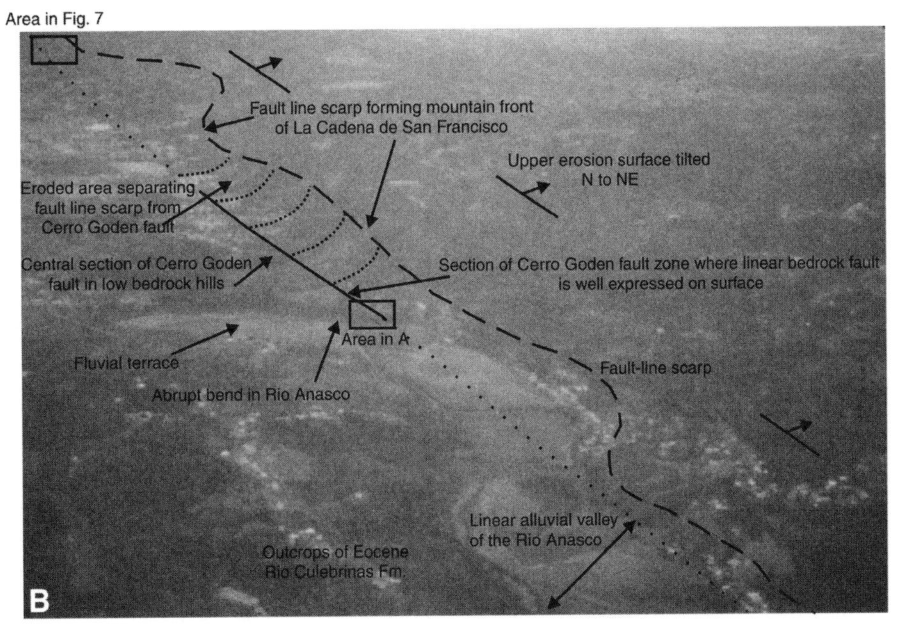

Figure 8. (A) Enlargement of 1936 aerial photograph showing eastern section of Cerro Goden fault (location given in Fig. 6). (B) 2001 aerial view showing eastern section of Cerro Goden fault. Note that the abrupt bend in the Rio Añasco in both photos can be used as a point of reference.

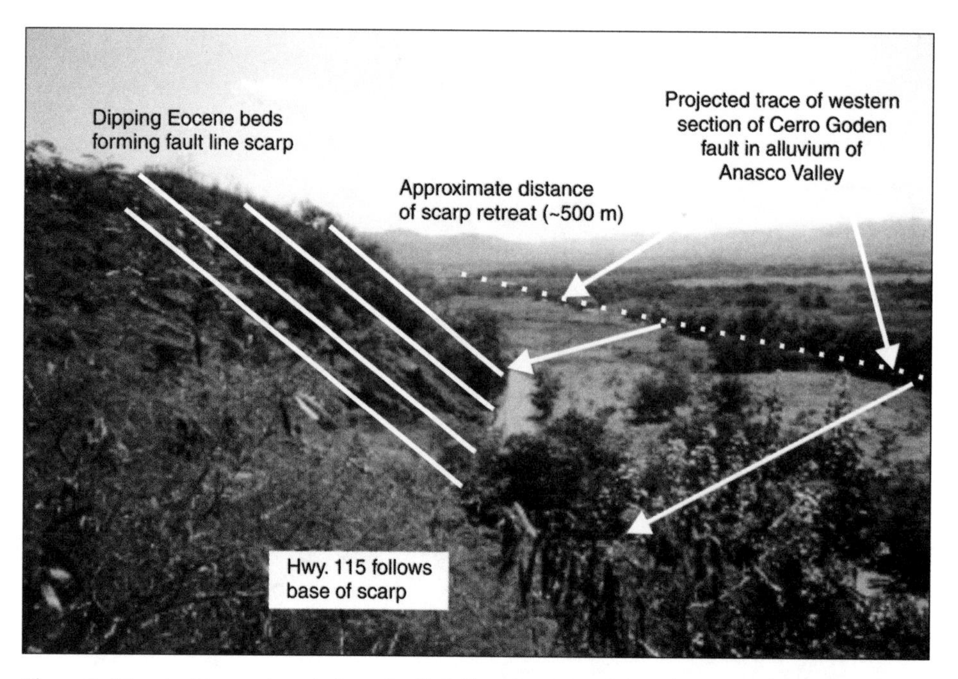

Figure 9. View looking eastward along the fault-line scarp that forms the mountain front of the La Cadena de San Francisco range. The projected trace of the Cerro Goden fault ~500 m south of the mountain front is indicated (white arrows indicate inferred distance of scarp retreat). Dipping Eocene beds indicated by white lines form the fault-line scarp and exhibit no evidence for shearing or other fault-related deformation.

CONCLUSIONS

This reconnaissance study presents some basic observations on the late Quaternary displacement of the Cerro Goden fault, its relation to onland and offshore faults in western Puerto Rico and the Mona Passage, and its tectonic development starting in the Eocene. Geomorphic features along the western and central sections of the Cerro Goden fault zone suggest the possibility of late Quaternary right-lateral fault movement. It is possible that the other parallel faults in western Puerto Rico with known or suspected late Quaternary activity shown on Figure 2 are also right-lateral faults, but more fieldwork is needed to test this hypothesis. Seismic lines show all faults in Añasco Bay dip to the south. Faults exhibit a normal character in young sediments with no signs of inversion or associated folding. Compilation of regional mapping by previous workers, mapping conducted for this study, and our detailed studies of striated fault planes indicate two periods of shortening. Rocks in Puerto Rico were shortened in a northeast-southwest direction by a folding event that occurred between the middle Eocene and early Oligocene. A second deep-seated, shortening event starting in post–early Pliocene time created the Hispaniola–Puerto Rico arch, a major structural feature of Puerto Rico, Hispaniola, and the Virgin Islands (van Gestel et al., 1998). Arching appears responsible for a 25-km-wide zone of extensional deformation in the crest of the arch (Fig. 14D). In this paper, we propose that symmetrical, crestward-dipping faulting includes two faults on the north flank of the arch (Cerro Goden and Mayaguez faults—Grindlay et al. this volume, Chapter 7) and two faults on the south flank (South Lajas fault and Guanajibo River fault). Grindlay et al. (this volume, Chapter 7) discuss the possibility that the fault zones in the crestal area are underlain by serpentinite and that the presence of serpentinite increases local seismicity along these faults (Huérfano et al., this volume).

GPS data suggest that faulting in Puerto Rico is related to east-west extension between Hispaniola, which is colliding with the Bahama platform on the North America plate and Puerto Rico, which is not colliding with the Bahama platform and is therefore free to move at a faster rate to the east-northeast (Mann et al., 2002; Calais et al., 2002) (Fig. 1). GPS models suggest that the Puerto Rico–Virgin Islands area is a rigid block moving with the larger Caribbean plate, so these data do not predict the type of vertical and horizontal fault movements that we describe or infer. Minor amounts of east-west extension are permitted by the GPS data and may affect Puerto Rico (Jansma and Mattioli, this volume).

The relation of the Cerro Goden fault to modern seismicity is poorly understood. Composite focal mechanisms by Huérfano et al. (this volume) indicate left-lateral shear, while data presented in this paper support Quaternary right-lateral displacement across the Cerro Goden fault. We identified several promising sites for paleoseismic studies along the Cerro Goden fault where excavations could determine whether or not the fault has experienced

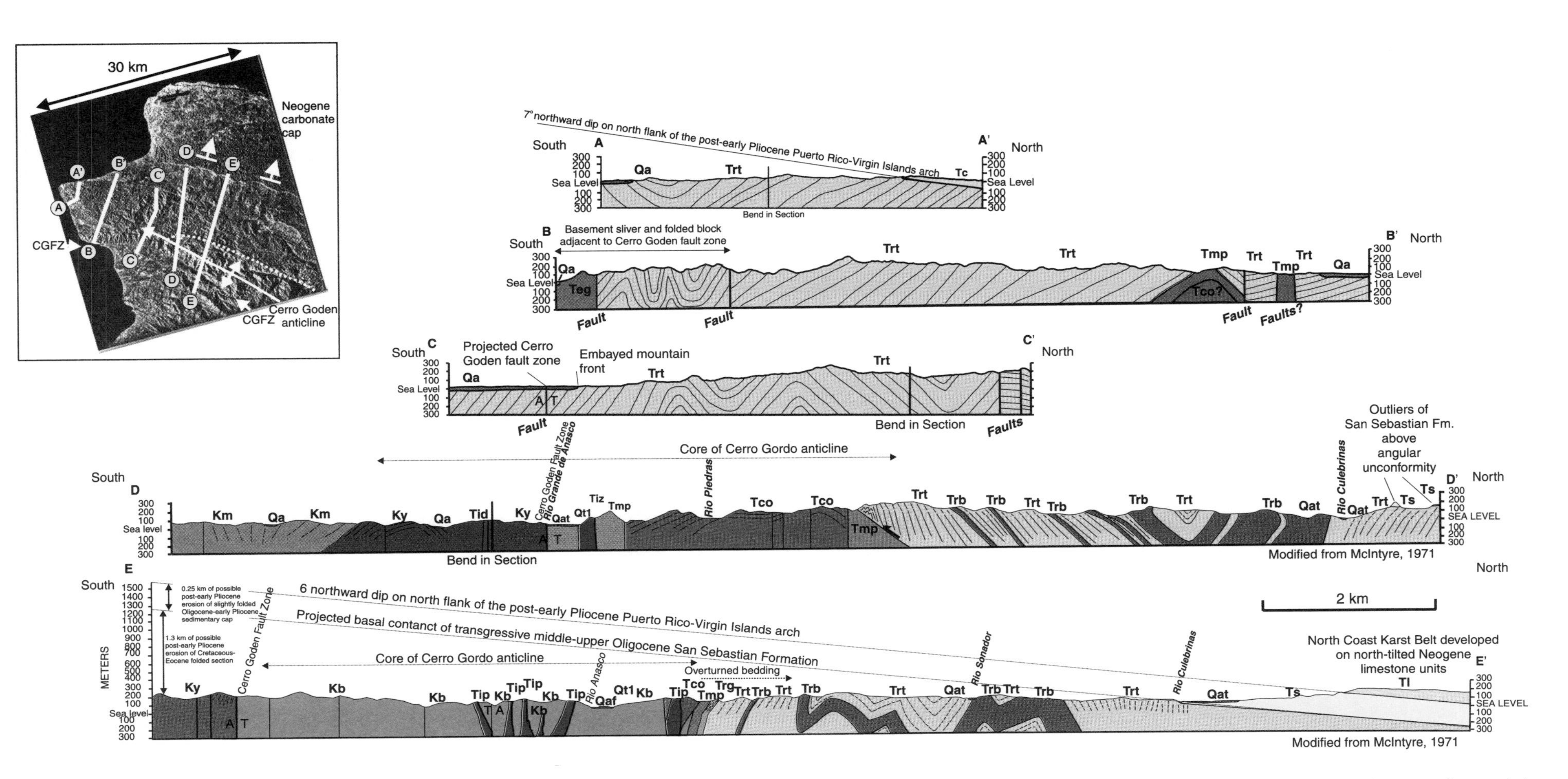

Figure 10. Regional structural cross sections A–E illustrating regional structure and stratigraphic units juxtaposed by the Cerro Goden fault zone (CGFZ). Sections A–C are based on mapping carried out for this study and sections D and E are modified from McIntyre (1971). Inset map shows location of sections relative to the topographic lineament of the Cerro Goden fault zone as seen on the radar image (cf. Fig. 3). Inset map also shows axial trace of Cerro Goden anticline deforming Eocene rocks. Key to stratigraphic units described in detail by McIntyre (1971) and Tobisch and Turner (1971): Ky—Upper Cretaceous Yauco Formation; Kb—Upper Cretaceous Rio Blanco Formation; Km—Upper Cretaceous Maricao Basalt; Tip—Eocene dikes and small stocks; Teg—Eocene Caguabo Formation; Tco—Eocene Concepcion Formation; Tmp—Eocene Mal Paso Formation; Trg—Eocene Guacio Member of the Rio Culebrinas Formation; Trb—Eocene tuff breccia of the Rio Culebrinas Formation; Trt—Eocene volcanic sandstone, siltstone, and mudstone of the Rio Culebrinas Formation; Tid—Eocene pyroxene diorite, hornblende diorite and quartz diorite; Tiz—Eocene aphanitic porphyry with pale pink zeolitized plagioclase phenocrysts; Ts—Oligocene San Sebastian Formation; Tl—Miocene Lares Formation, Tc—Miocene Cibao Formation; Qa—Quaternary alluvium; Qat—Pleistocene and Holocene alluvium of Río Culebrinas; Qt_1—Pleistocene and Holocene alluvium of lower terrace of Río Añasco; Qaf—Pleistocene and Holocene floodplain alluvium of Río Añasco. Arc basement rocks are shown by purple and gray units on the cross sections, while the overlying Rio Culebrinas Formation is shown in brown. Rio Culebrinas Formation and the gently north-tilted carbonate platform units of the Lares and Cibao Formations are shown in yellow and light blue on the sections.

A. map showing locations of USGS quads and cross sections and sections mapped in this study (Rincon Quad)

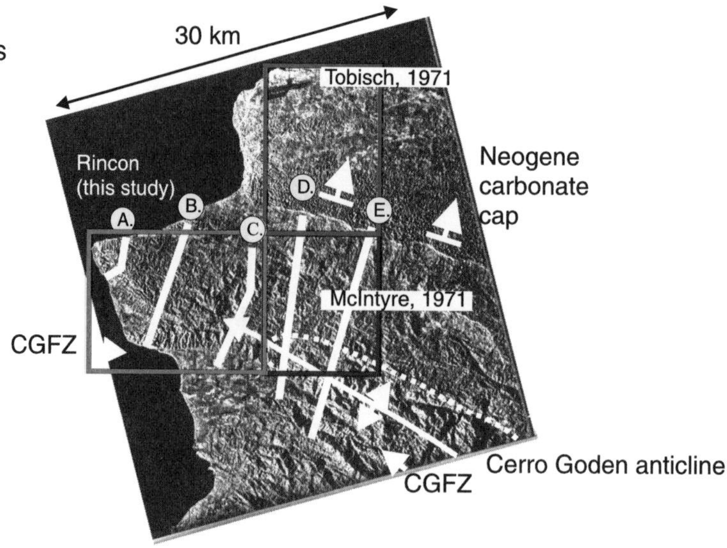

B. Dips of bedding planes in rocks ranging in age from late Cretaceous to early Miocene

Late Oligocene-Early Miocene carbonate cap

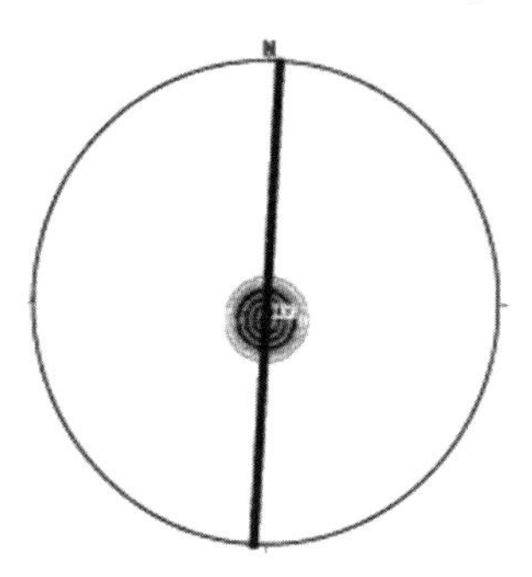

Middle Eocene Rio Culebrinas Fm. Adjacent to CerroGoden fault zone in Rincon Quad

Eocene Concepcion and Mal Paso Formation

Middle Eocene Rio Culebrinas Fm. in Rincon Quad

Middle Eocene Rio Culebrinas Fm. in Rio La Plata Quad

All Cretaceous arc-related formations in Rincon and Rio La Plata Quads

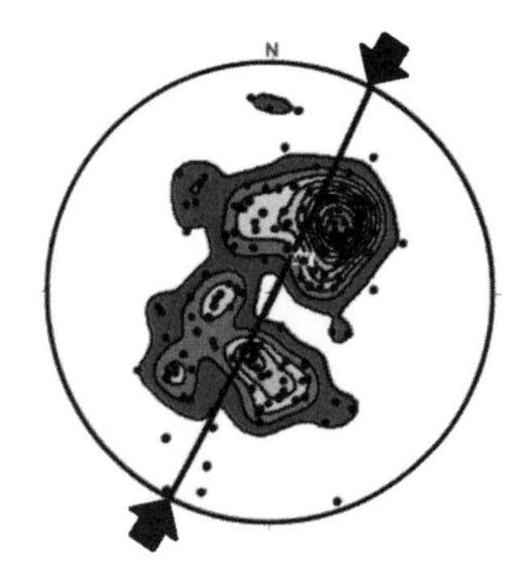

Figure 11 (*on this and following page*). (A) Radar image with superimposed outlines of U.S. Geological Survey geologic quad maps (McIntyre, 1971; Tobisch and Turner, 1971) and the area to the west mapped for this study. (B) Lower hemisphere equal-area stereographic projections showing contoured poles to bedding for formations adjacent to the Cerro Goden fault zone taken from maps by McIntyre (1971), Tobisch and Turner (1971), and the area mapped in this study (Rincon quadrangle). Great circle indicates best fit of maximum density of poles to bedding and black arrows indicate best fit shortening directions. Note that late Oligocene–early Miocene rocks are only gently tilted (6–8°) along the northern limb of the Puerto Rico–Virgin Islands arch.

C. Map showing direction of maximum compressive stress inferred from fault striation studies near the Cerro Goden fault zone

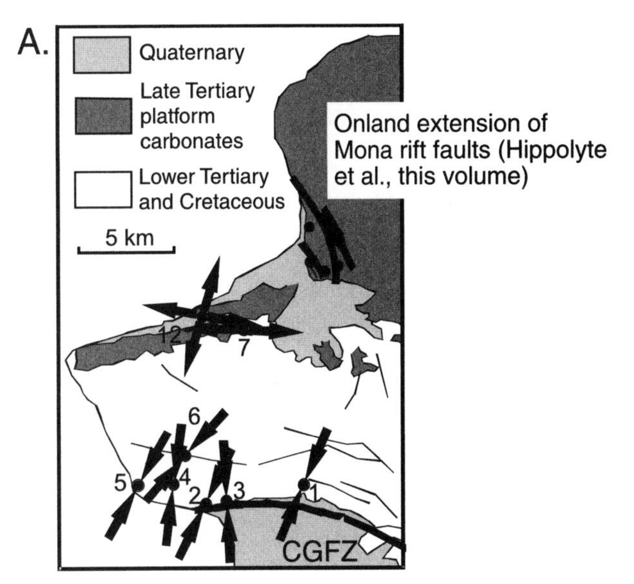

D. Stress directions from striated fault planes in Eocene sedimentary rocks adjacent to the Cerro Goden fault line scarp

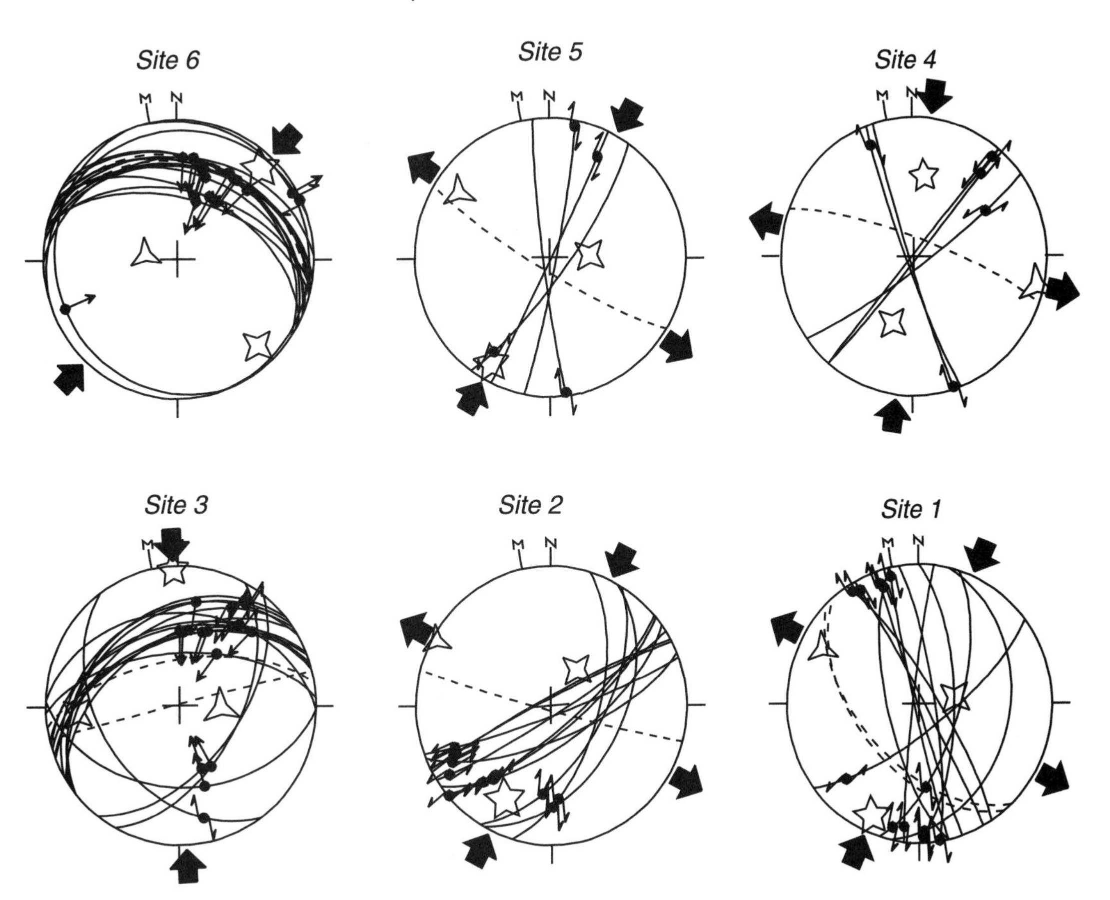

Figure 11 (*continued*). (C) Map showing direction of maximum compressive stress (inward-directed black arrows) inferred from fault striation studies in Eocene rocks exposed along the fault-line scarp of the Cerro Goden fault zone. Directions of maximum compressive stress are similar to shortening directions inferred from bedding plane dips in B. Also shown are directions of maximum extension (outward-directed black arrows) in the area to the north from Hippolyte et al. (this volume). (D) Stress directions derived from striated fault planes in Eocene sedimentary rocks at sites 1–6 (locations shown in C). Five-branch star—σ_1 (maximum principal stress axis); four-branch star—σ_2 (intermediate principal stress axis); three-branch start—σ_3 (minimum principal stress axis); bedding planes are shown as dashed lines.

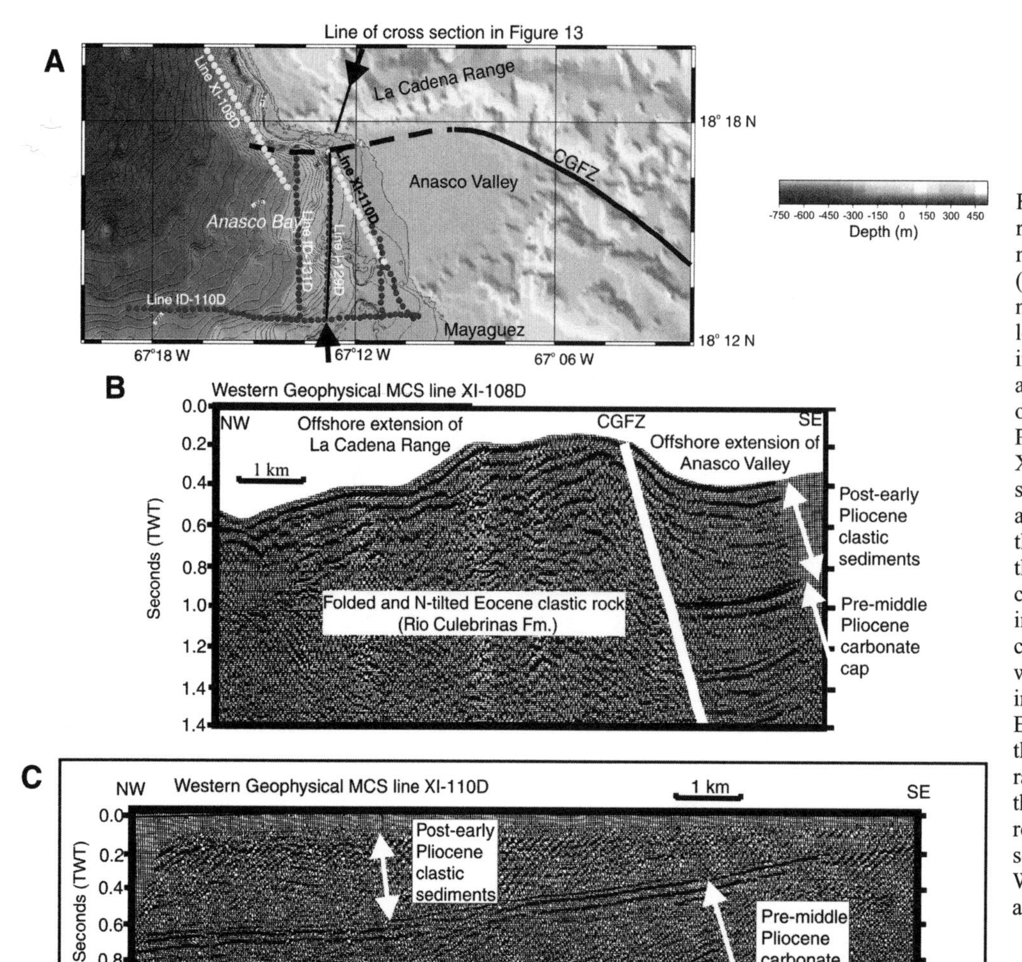

Figure 12. (A) Onland topography and offshore bathymetry near the Cerro Goden fault zone (CGFZ). Dotted lines represent multi-channel seismic lines collected by Western Geophysical in 1972 that were kindly made available to us by the Department of Geology of the University of Puerto Rico–Mayagüez. (B) Line XI-108D crosses the CGFZ and shows a south-dipping fault with an upper wedge of sediments on the downthrown block. We infer that the wedge is post–early Pliocene in age and that the underlying rocks are pre–middle Pliocene in age. Rocks in the footwall which lack coherent reflectors are inferred to be equivalent to folded Eocene rocks exposed onshore in the La Cadena de San Francisco range. (C) Line XI-110D shows the tilted but unfolded nature of rocks underlying Añasco Bay, south of the Cerro Goden fault. We infer the same age relations as shown in B.

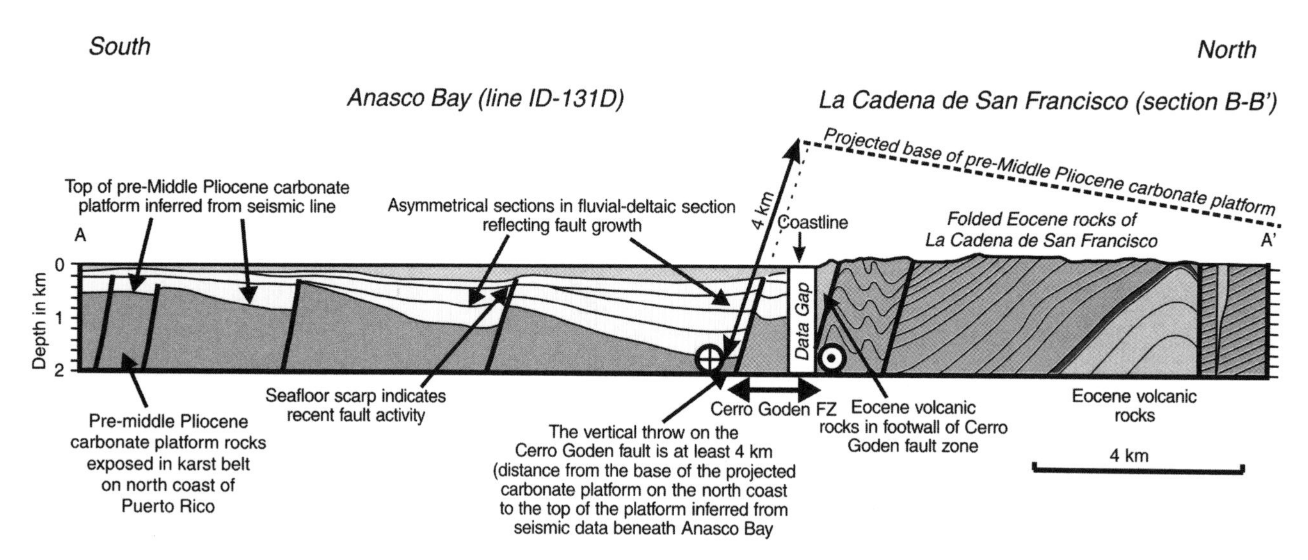

Figure 13. Interpretation of depth-converted Western Geophysical line ID-131-D combined with onshore structural cross section C from Figure 10. Location of line ID-131-D and onland section (cross section B in Fig. 10) are shown on Figure 12. Folded Eocene sedimentary and volcanic rocks are interpreted to form the basement of Añasco Bay beneath a 1–2-km-thick section of Miocene-Recent carbonate and clastic rocks. Note the blockfaulted, tilted nature of the upper units. We infer right-lateral oblique motion along the Cerro Goden fault zone as indicated. We estimate the normal component of throw to be ~4 km as measured from the projected base of the Miocene carbonate platform seen on the sections in Figure 10A and B and the interpreted base of the Miocene carbonate platform beneath Añasco Bay.

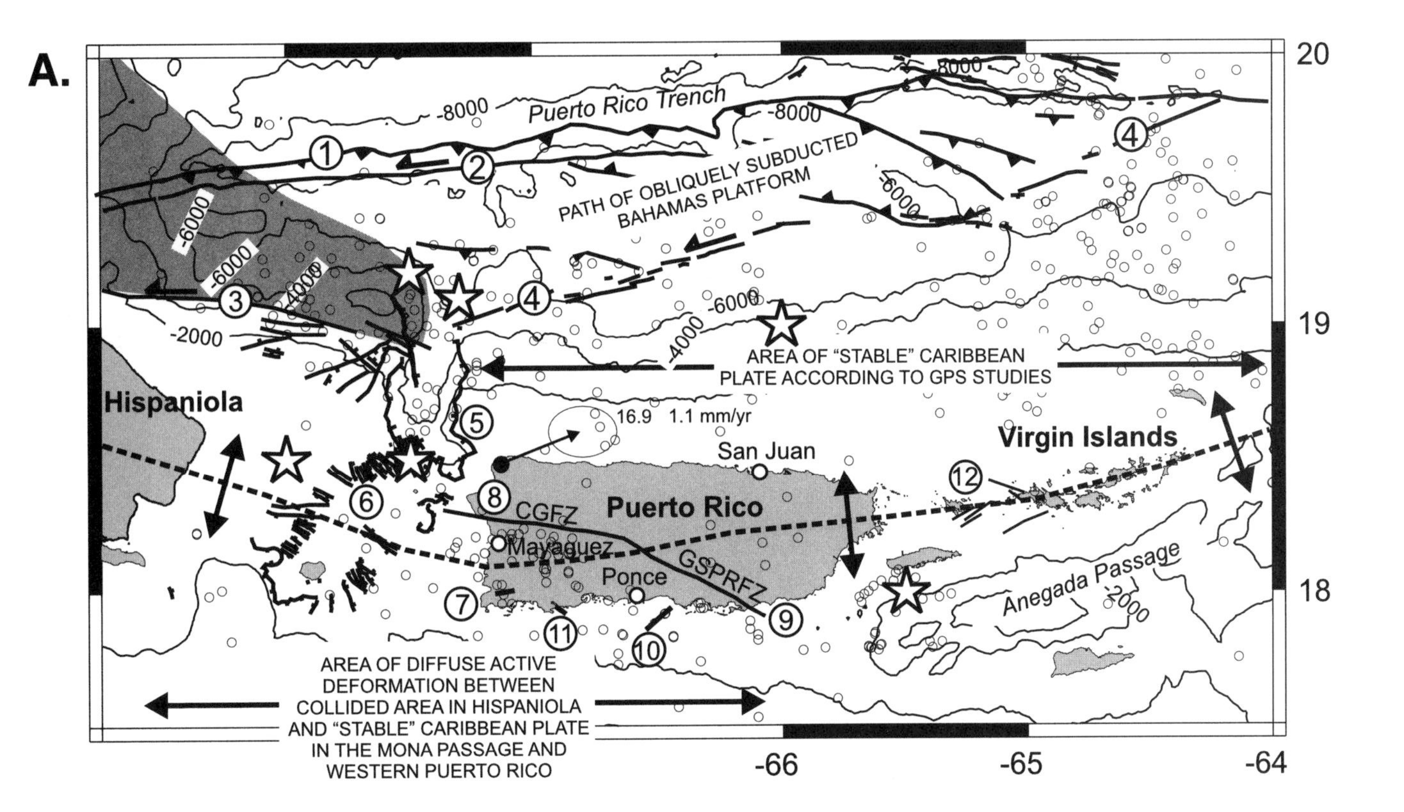

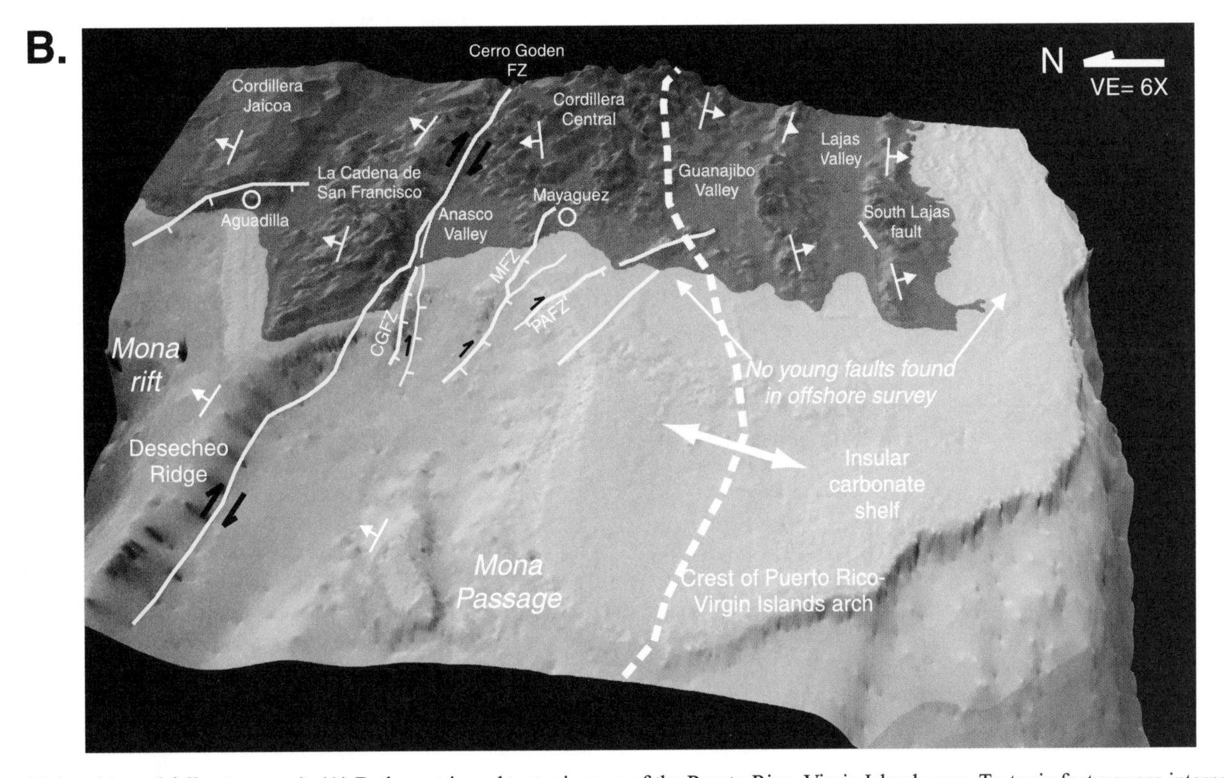

Figure 14 (*on this and following page*). (A) Bathymetric and tectonic map of the Puerto Rico–Virgin Islands area. Tectonic features are interpreted from sidescan imagery, single-channel seismic profiles, and Hydrosweep bathymetry (van Gestel et al., 1998; Grindlay et al., 1997, this volume, Chapters 2 and 7). Epicenters of earthquakes from 1973 to 2001 at depths of <30 km are plotted as open circles. Large magnitude historical earthquakes are shown as stars. Key to faults: 1—Puerto Rico trench fault zone; 2—Bunce fault zone; 3—Septentrional fault zone; 4—Bowin fault zone; 5—eastern boundary fault of Mona rift; 6—faulted area of Mona Passage; 7—South Lajas fault (Prentice and Mann, this volume); 8—Cerro Goden fault zone; 9—Great Southern Puerto Rico fault zone; 10—Caja de Muertos fault zone; 11—Guanica fault zone; 12—seafloor faults in the Virgin Islands. (B) Three-dimensional perspective of western Puerto Rico looking eastward modified from Grindlay et al. (this volume, Chapter 7). Illumination is from the northwest. Offshore faults identified by Grindlay et al. (this volume, Chapter 7) are tentatively traced to faults onshore. The axis of the Puerto Rico–Virgin Islands arch is shown.

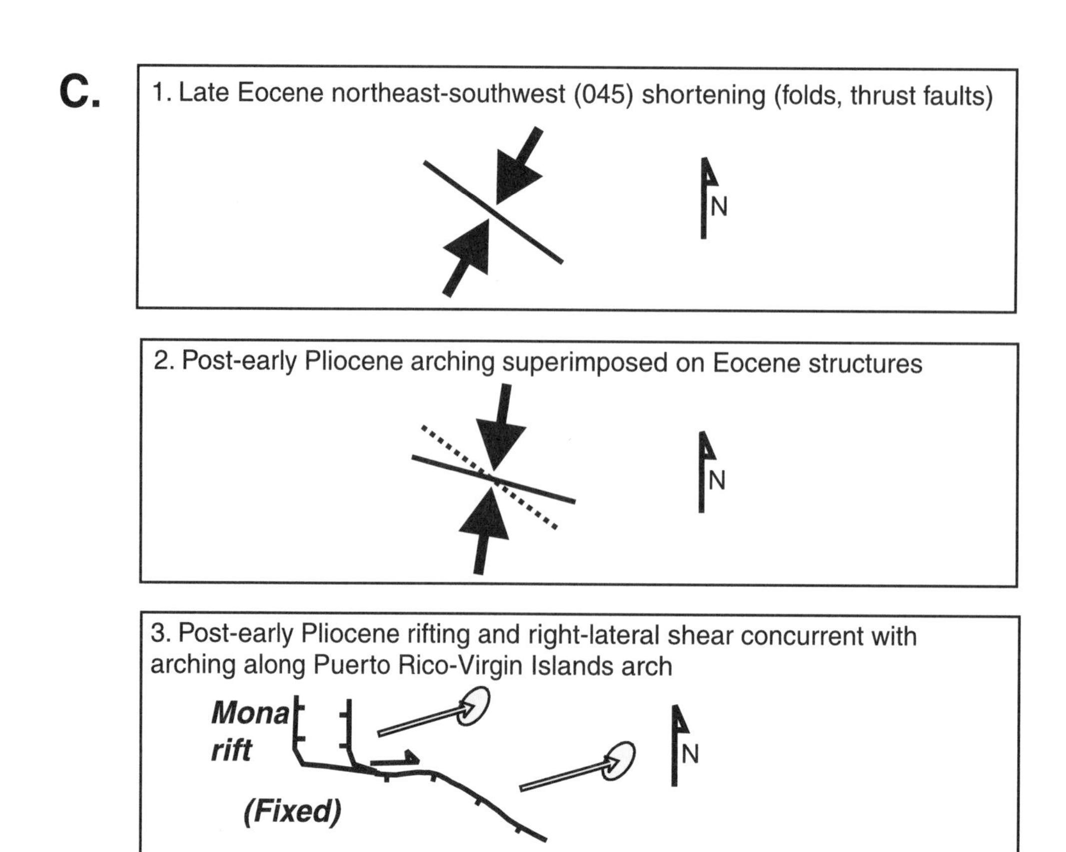

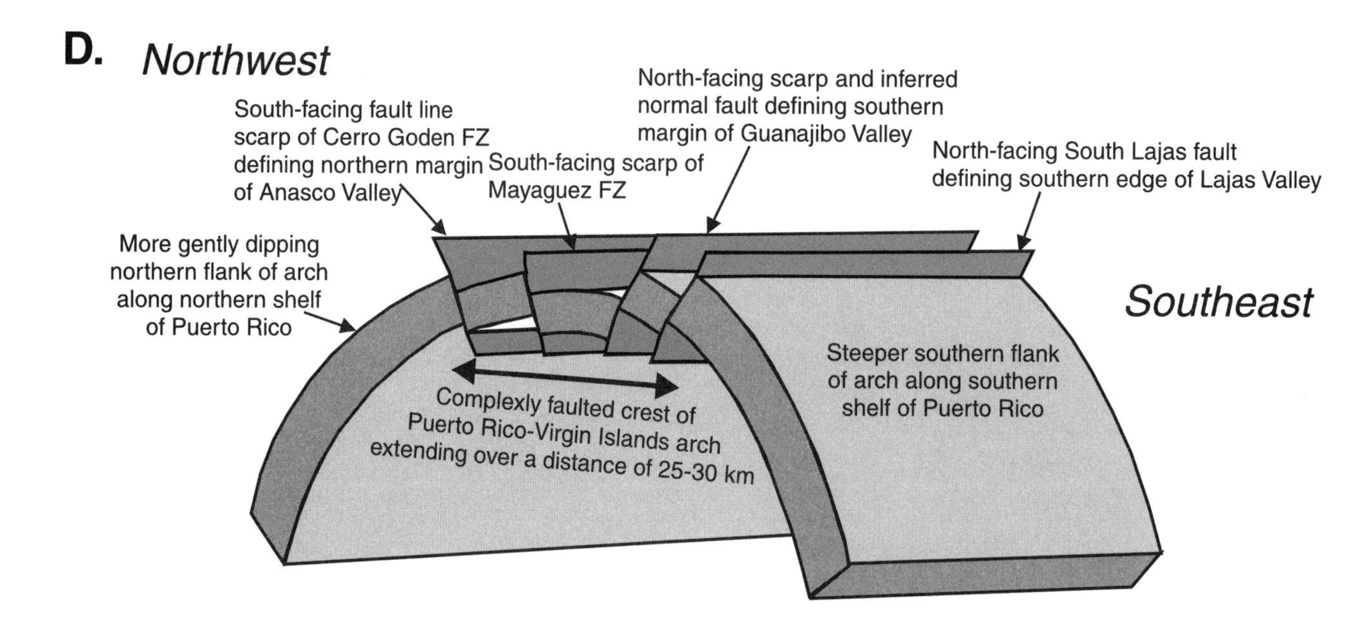

Figure 14 (*continued*). (C) Major tectonic stages in the structural evolution of western Puerto Rico based on results presented in this paper: (1) late Eocene shortening; (2) post–early Pliocene arching; and (3) post–early Pliocene to late Quaternary rifting and right-lateral shear concurrent with arching along the Puerto Rico–Virgin Island arch. (D) Schematic block diagram showing relationship of major faults in western Puerto Rico (including Cerro Goden fault) to axis of post–early Pliocene Puerto Rico–Virgin Islands arch. Major fault scarps appear to be confined to near the crest of the arch where extensional strains are largest. Morphology of western Puerto Rico seen on Figure 14B indicates that arching may continue to the present. Note that vertical throws on mountain fronts in western Puerto Rico indicate a regularity in deformation patterns: faults of north arch are downthrown to the south; faults to the south of arch are downthrown to the north.

late Holocene displacement, and the sense of slip across the fault. This fault is located ~10 km from the city of Mayagüez and its rapidly growing suburbs and should be considered a priority for further paleoseismic and geophysical study.

ACKNOWLEDGMENTS

This work was supported by external grant no. 00HQGR0064 to P. Mann and C. Prentice from the National Earthquake Hazards Reduction Program (NEHRP) of the U.S. Geological Survey. We thank Christa von Hillebrant-Andrade and Victor Huérfano of the Puerto Rico Seismic Network, University of Puerto Rico, for their logistical support of this study and for making available the offshore seismic reflection data in Añasco Bay from Western Geophysical. We also thank Eugenio Asencio, James Joyce, Hans Schellekens, and Hernan Santos of the Department of Geology, University of Puerto Rico, for their continued support. University of Texas at Austin Institute for Geophysics (UTIG) contribution no. 1686. The authors acknowledge the financial support for this publication provided by The University of Texas at Austin's Geology Foundation and Jackson School of Geosciences.

REFERENCES CITED

Angelier, J., 1990, Inversion of field data in fault tectonics to obtain the regional stress: A new rapid direct inversion method by analytical means: Geophysical Journal International, v. 103, p. 363–367.

Calais, E., Mazabraud, Y., Mercier de Lepinay, B., Mann, P., Mattioli, G., and Jansma, P., 2002, 2002, Strain partitioning and fault slip rates in the northeastern Caribbean from GPS measurements: Geophysical Research Letters, v. 29, no. 18, p. 1856, doi: 10.1029/2002GL015397.

DeMets, C., Jansma, P., Mattioli, G., Dixon, T., Farina, F., Bilham, R., Calais, E., and Mann, P., 2000, GPS geodetic constraints on Caribbean–North America plate motion: Geophysical Research Letters, v. 27, p. 437–440, doi: 10.1029/1999GL005436.

Dolan, J., and Bowman, D., 2004, Tectonic and seismologic setting of the 22 September 2003, Puerto Plata, Dominican Republic earthquake: Implications for earthquake hazard in Hispaniola: Seismological Research Letters, v. 75, p. 587-597.

Dolan, J., and Wald, D., 1998, The 1943–1953 north-central Caribbean earthquakes: Active tectonic setting, seismic hazards, and implications for Caribbean–North America plate motions, *in* Dolan, J., and Mann, P., eds., Active Strike-slip and Collisional Tectonics of the Northern Caribbean plate boundary zone: Geological Society of America Special Paper 326, p. 143–169.

Dolan, J., Mann, P., de Zoeten, R., and Heubeck, C., 1991, Sedimentologic, stratigraphic, and tectonic synthesis of Eocene-Miocene sedimentary basins, Hispaniola and Puerto Rico, *in* Mann, P., Draper, G., and Lewis, J., eds., Geologic and Tectonic Development of the North America-Caribbean plate boundary in Hispaniola: Geological Society of America Special Paper 262, p. 217–264.

Dolan, J., Mullins, H., and Wald, D., 1998, Active tectonics of the north-central Caribbean: Oblique collision, strain partitioning, and opposing subducting slabs, *in* Dolan, J., and Mann, P., eds., Active Strike-slip and Collisional Tectonics of the Northern Caribbean plate boundary zone: Geological Society of America Special Paper 326, p. 1–61.

Erikson, J., Pindell, J., and Larue, D., 1990, Mid-Eocene–early Oligocene sinistral transcurrent faulting in Puerto Rico associated with formation of the Northern Caribbean plate boundary zone: Journal of Geology, v. 98, p. 365–384.

Glover, L., III, 1971, Geology of the Coamo area, Puerto Rico, and its relation to the volcanic arc-trench association: U.S. Geological Survey Professional Paper 636, 102 p.

Grindlay, N., Mann, P., and Dolan, J., 1997, Researchers investigate submarine faults north of Puerto Rico: Eos (Transactions, American Geophysical Union), v. 78, p. 401.

Hearne, M., Grindlay, N., and Mann, P., 2003, Landslide deposits, cookie-bites, and crescentic fracturing along the northern Puerto Rico–Virgin Islands margin: Implications for potential tsunamigenesis: Eos (Transactions, American Geophysical Union), v. 84.

Jansma, P., Mattioli, G., Lopez, A., DeMets, C., Dixon, T., Mann, P., and Calais, E., 2000, Neotectonics of Puerto Rico and the Virgin Islands, northeastern Caribbean, from GPS geodesy: Tectonics, v. 19, p. 1021–1037.

Larue, D., and Ryan, H., 1998, Seismic reflection profiles of the Puerto Rico trench: Shortening between the North American and Caribbean plates, *in* Lidiak, E., and Larue, D., eds., Tectonics and Geochemistry of the Northeastern Caribbean: Geological Society of America Special Paper 322, p. 193–210.

Mann, P., Prentice, C., Burr, G., Pena, L., and Taylor, F., 1998, Tectonic geomorphology and paleoseismology of the Septentrional fault system, Dominican Republic, *in* Dolan, J., and Mann, P., eds., Active strike-slip and collisional tectonics of the Northern Caribbean plate boundary zone: Geological Society of America Special Paper 326, p. 63–123.

Mann, P., Calais, E., Ruegg, J. C., DeMets, C., Jansma, P., and Mattioli, G., 2002, Oblique collision in the northeastern Caribbean from GPS measurements and geological observations: Tectonics, v. 21, no. 6, 1057, doi: 10.1029/2001TC001304, 2002.

Mann, P., Calais, E., and Huérfano, V., 2004, Earthquake shakes "big bend" region of North America–Caribbean boundary zone: Eos (Transactions, American Geophysical Union), v. 85, no. 8.

Masson, D., and Scanlon, K., 1991, The neotectonic setting of Puerto Rico: Geological Society of America Bulletin, v. 103, p. 144–154, doi: 10.1130/0016-7606(1991)103<0144:TNSOPR>2.3.CO;2.

McCann, W., 2002, Microearthquake data elucidate details of Caribbean subduction zone: Seismological Research Letters, v. 73, p. 25–32.

McIntyre, D., 1971, Geologic map of the central La Plata quadrangle, Puerto Rico: U.S. Geological Survey Miscellaneous Geologic Investigations Series Map I-660, scale 1:20,000.

Mercado, A., and McCann, W., 1998, Numerical simulation of the 1918 Puerto Rico tsunami: Natural Hazards, v. 18, p. 57–76, doi: 10.1023/A:1008091910209.

Moussa, M., Seiglie, G., Meyerhoff, A., and Taner, I., 1987, The Quebradillas limestone (Miocene-Pliocene), northern Puerto Rico, and tectonics of the northeastern Caribbean: Geological Society of America Bulletin, v. 99, p. 427–439.

Muszala, S., Grindlay, N., and Bird, R., 1999, Three-dimensional Euler deconvolution and tectonic interpretation of marine magnetic anomaly data in the Puerto Rico trench: Journal of Geophysical Research, v. 104, p. 29,175–29,187, doi: 10.1029/1999JB900233.

Prentice, C.S., Mann, P., Taylor, F.W., Burr, G., and Valastro, S., Jr., 1993, Paleoseismicity of the North America–Caribbean plate boundary (Septentrional fault), Dominican Republic: Geology, v. 21, p. 49–52, doi: 10.1130/0091-7613(1993)0212.3.CO;2.

Prentice, C., Mann, P., Pena, L., and Burr, G., 2003, Slip rate and earthquake recurrence along the central Septentrional fault, North American–Caribbean plate boundary, Dominican Republic: Journal of Geophysical Research, v. 108, no. B3, p. 2149, doi: 10.1029/2001JB000442, 2003.

Reid, J., Plumley, P., and Schellekens, H., 1991, Paleomagnetic evidence for Late Miocene counterclockwise rotation of the North Coast carbonate sequence, Puerto Rico: Geophysical Research Letters, v. 18, p. 565–568.

ten Brink, U., and Lin, J., 2004, Stress interaction between subduction earthquakes and forearc strike-slip faults: Modeling and application to the northern Caribbean plate boundary: Journal of Geophysical Research, v. 109, B12310, doi: 10.1029/2004JB003031.

Tobisch, O., and Turner, M., 1971, Geologic map of the San Sebastian quadrangle, Puerto Rico: U.S. Geological Survey Miscellaneous Investigations Series Map I-661, scale 1:20,000.

van Gestel, J.P., Mann, P., Dolan, J., and Grindlay, N., 1998, Structure and tectonics of upper Cenozoic–Puerto Rico–Virgin Islands carbonate platform as determined from seismic reflection studies: Journal of Geophysical Research, v. 103, p. 30,505–30,530, doi: 10.1029/98JB02341.

van Gestel, J.P., Mann, P., Grindlay, N., and Dolan, J., 1999, Three-phase tectonic evolution of the northern margin of Puerto Rico as inferred from an integration of seismic reflection, well, and outcrop data: Marine Geology, v. 161, p. 257–286, doi: 10.1016/S0025-3227(99)00035-3.

Western Geophysical Company of America and Fugro, Inc., 1973, Geological-geophysical reconnaissance of Puerto Rico for siting of nuclear power plants: San Juan, Puerto Rico, The Puerto Rico Water Resources Authority, 127 p.

Manuscript Accepted by the Society 18 August 2004

Geological Society of America
Special Paper 385
2005

Toward an integrated understanding of Holocene fault activity in western Puerto Rico: Constraints from high-resolution seismic and sidescan sonar data

Nancy R. Grindlay*
Lewis J. Abrams
Luke Del Greco
Center for Marine Science and Department of Earth Sciences, University of North Carolina at Wilmington, Wilmington, North Carolina 28409, USA

Paul Mann
Institute for Geophysics, Jackson School of Geosciences, University of Texas at Austin, Austin, Texas 78759-8500, USA

ABSTRACT

It has been postulated that the western boundary of the Puerto Rico–Virgin Islands microplate lies within the Mona Passage and extends onland into southwestern Puerto Rico. This region is seismically active, averaging one event of magnitude 2.0 or larger per day, and over 150 events of magnitude 3.0 or greater occurred during the past five years. Moreover, there have been at least 13 historical events of intensity VI (MM) or greater in the past 500 years. We conducted a high-resolution seismic and sidescan sonar survey of the insular shelf of western and southern Puerto Rico during May 2000 in an effort to identify Holocene faults and to further assess the seismic hazard in the region. We focus on an ~175 km^2 part of the surveyed area offshore of western Puerto Rico, extending from Punta Higuero to Boquerón Bay. This area was targeted as a likely place to image recent faults, because multi-channel seismic profiles offshore western Puerto Rico show numerous WNW-trending normal and strike-slip faults that offset Oligocene-Pliocene age carbonate rocks and underlying Cretaceous basement rocks. Analyses of these data identify three zones of active deformation within the survey area: (1) the Cerro Goden fault zone; (2) the Punta Algarrobo/Mayagüez fault zone that lies offshore the city of Mayagüez; and; (3) the Punta Guanajibo/Punta Arenas fault zone. Two of the offshore fault zones, the Cerro Goden and Punta Algarrobo, show strong correlation with fault zones onland, Cerro Goden and Cordillera, respectively. Many of the mapped faults offshore appear to reactivate older WNW-trending basement structures and show evidence of some component of right-lateral motion that is consistent with geodetic measurements. The offshore deformation zones are also associated with headlands and linear NW-SE magnetization lows (serpentinite dikes?) mapped offshore. Elongate outcrops of serpentinite in western Puerto Rico are colinear with the fault zones we have mapped offshore,

*grindlayn@uncw.edu

Grindlay, N.R., Abrams, L.J., Del Greco, L., and Mann, P., 2005, Toward an integrated understanding of Holocene fault activity in western Puerto Rico: Constraints from high-resolution seismic and sidescan sonar data, *in* Mann, P., ed., Active tectonics and seismic hazards of Puerto Rico, the Virgin Islands, and offshore areas: Geological Society of America Special Paper 385, p. 139–160. For permission to copy, contact editing@geosociety.org.

suggesting that either the presence of serpentinite has localized fault activity or that fault activity has remobilized serpentinite. This offshore study improves assessments of the seismic hazard in Puerto Rico by identifying targets for onshore paleoseismic studies and by better defining the total length of offshore Holocene faults.

Keywords: Holocene faulting, tectonics, Puerto Rico, sidescan sonar, high-resolution seismic reflection.

INTRODUCTION

The onshore and offshore region of western Puerto Rico is one of the most seismically active regions beneath the island of Puerto Rico. During the past five years alone, over 150 earthquakes with a magnitude of 3.0 or greater have been recorded by the local seismic network in the western region of Puerto Rico (Puerto Rico Seismic Network, 2004). There is historical evidence for at least 48 felt seismic events in the Mona Passage and western Puerto Rico between 1524 and 1958, 13 of which had estimated intensities of >VI (Modified Mercalli) (Ascencio, 1980). Most notable is the 1918 (estimated 7.3 Ms) earthquake in the Mona Passage (Figs. 1 and 2) and the resulting tsunami, which together killed at least 116 people and caused four million dollars in damage in northwestern Puerto Rico (Reid and Taber, 1919), at a time when the population of Mayagüez was approximately 17,000 people. Today, Mayagüez is the third most populous city on the island of Puerto Rico with more than 150,000 inhabitants.

Because southwestern Puerto Rico is characterized by "basin-and-range" style topography and frequent shallow (<50 km) seismicity (McCann et al., 1987; Joyce et al., 1987; Ascencio, 1980; Puerto Rico Seismic Network, 2004), this region has been the focus of studies to locate Holocene faults onland. Two seismic reflection lines across the southern margin of the Lajas valley imaged displacements in Quaternary lacustrine sediments and Cretaceous basement rocks (Meltzer and Almy, 2000). Offsets within the sediments and the basement rocks indicate Quaternary faulting, most likely of a transtensional nature (Meltzer and Almy, 2000). Most recently, trenching studies of the South Lajas fault near the town of Boquerón showed evidence of Holocene activity (Fig. 3). Displacements in Holocene alluvium show valley-side-down, normal separation with a component of strike-slip motion (Prentice and Mann, this volume).

Marine geophysical studies of the Mona Passage indicate several sets of youthful, large displacement, normal faults (Gardner et al., 1980; Larue and Ryan, 1990; Grindlay et al. 1997; van Gestel et al., 1998). The Mona rift, located at the northern end of the Mona Passage (Figs. 1 and 2), is a N-S–trending graben with extremely large throws, in many instances over 2 km (Gardner et al., 1980; Larue and Ryan, 1990; Grindlay et al., 1997; van Gestel et al., 1998). Recent tectonic activity in the area is suggested by uneroded rocks on the fault scarps, and by normal faults and vertical fissures which cut tilted surface layers (Gardner et al., 1980). South of the Mona rift, single-channel (SCS) and multichannel seismic (MCS) data show an abundance of normal faults with generally WNW and E-W trends (Fig. 2; Western Geophysical Company and Fugro, Inc., 1973; Larue and Ryan, 1990; Grindlay et al., 1997; van Gestel et al., 1998, 1999).

Because most of the large historical earthquakes have been located far offshore and existing nearshore geophysical data is sparse, an understanding of the distribution and nature of nearshore faults and the seismogenic potential on the insular shelf of western Puerto Rico is lacking. For this reason, we conducted a high-resolution geophysical survey (SCS and sidescan sonar imaging) in May 2000 to characterize the structure of the seafloor and sub-bottom of the shallow insular shelf (Fig. 3). The primary objective of this study was to search for youthful faulting (e.g., seafloor offsets), and to correlate offshore faults with possible Quaternary faults mapped onshore to better define their total length in western Puerto Rico. The delineation of nearshore fault systems provided by this study, combined with their direction and rate of slip from onland surveys, will assist in the improvement of hazard models. Until now, hazard models have relied only on offshore source zones for hazard prediction, because of the lack of evidence for nearshore and onshore faulting.

TECTONIC SETTING OF PUERTO RICO AND THE VIRGIN ISLANDS

The island of Puerto Rico is located within a diffuse and complex plate boundary zone between the North America and Caribbean plates (Fig. 1). Seismicity and marine geophysical studies suggest that Puerto Rico and the Virgin Islands are the emergent part of a microplate that lies within the North America–Caribbean plate boundary zone (Byrne et al., 1985; Masson and Scanlon, 1991; Mann et al., 1995). The Puerto Rico trench and the Muertos trough form the northern and southern boundaries, respectively, of the Puerto Rico–Virgin Islands microplate (Fig. 1). The eastern boundary of the microplate lies within the Anegada Passage. The western boundary of the microplate is poorly defined; its northern portion may be the Mona rift, while its southern portion may lie beneath the Mona Passage or extend into southwestern Puerto Rico.

Phases of Deformation Impacting the Region

It is likely that the style, geometry, and distribution of faults within the Mona Passage and western Puerto Rico are a function of at least three separate and sequential phases of deformation to impact this region. The first phase is an Eocene age transpressive event focused along two major shear zones the Great Northern

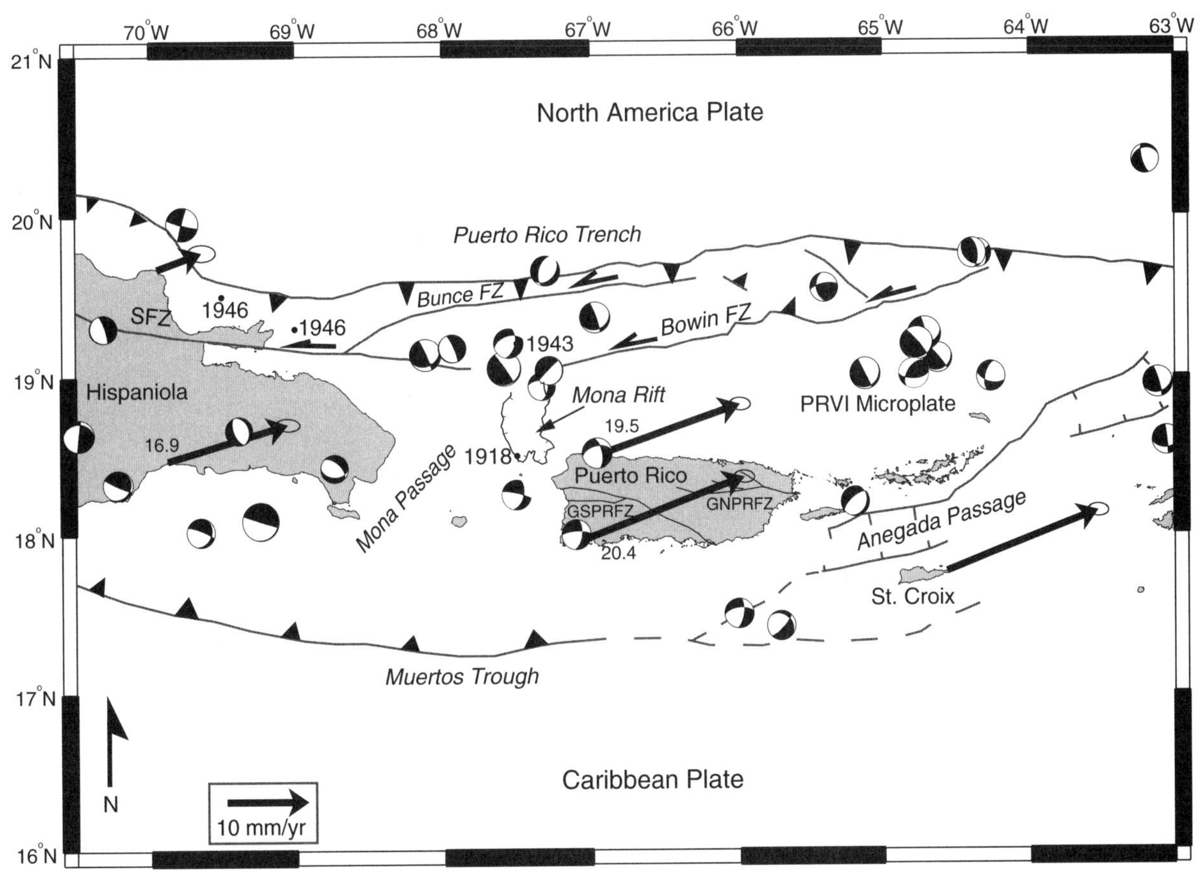

Figure 1. Tectonic setting of the northeastern Caribbean region. Plate vectors relative to a fixed North America plate from Mann et al., 2002. Focal Mechanisms from Harvard CTM database, Molnar and Sykes (1969) and Huérfano (1994). Epicenters of recent large magnitude earthquakes in the region also labeled with year of occurrence (U.S. Geological Survey National Earthquake Information Center [NEIC] PDE database). GNPRFZ—Great Northern Puerto Rico fault zone; GSPRFZ—Great Southern Puerto Rico fault zone; PRVI—Puerto Rico–Virgin Islands; SFZ—Septentrional fault zone.

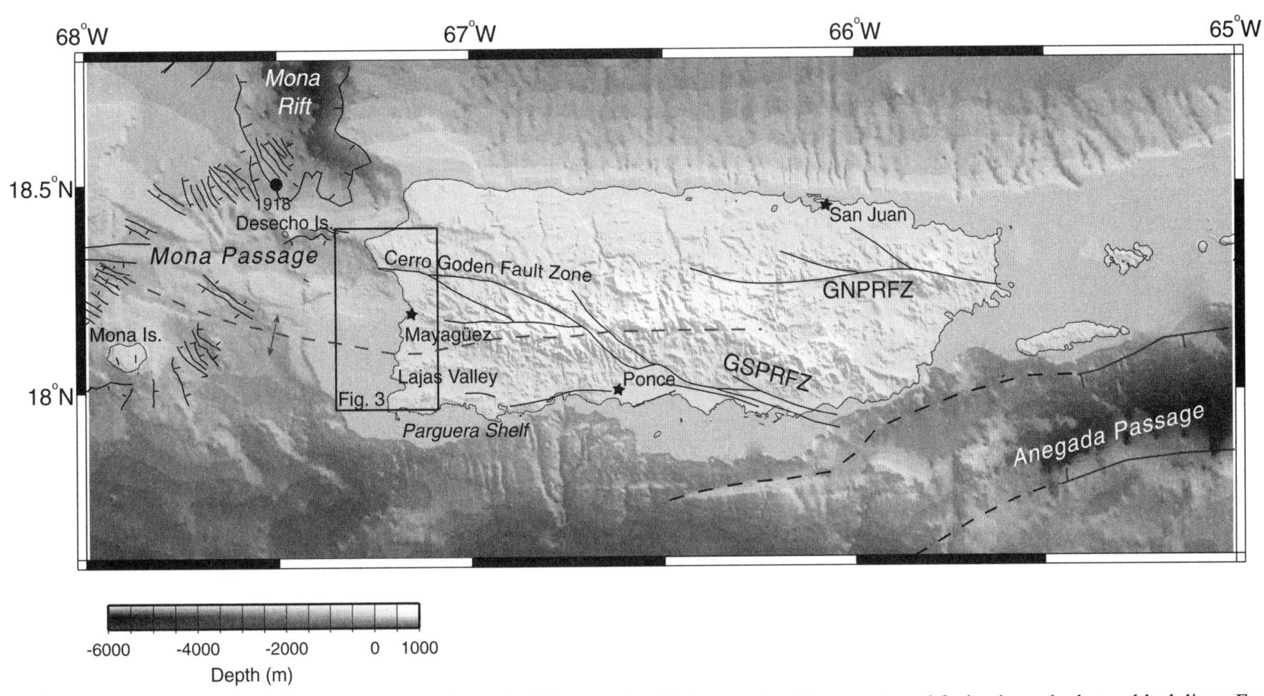

Figure 2. Shaded relief map of the island of Puerto Rico and offshore region. Major structural lineaments and faults shown by heavy black lines. Faults in Mona Passage identified by Grindlay et al. (1997) and van Gestel et al. (1998). Structural lineaments onland from Glover (1971). Digital elevation model (DEM) created from U.S. Geological Survey 3 arc second DEM for Puerto Rico and National Oceanic and Atmospheric Administration (NOAA) hydrographic data. Location of the 1918 earthquake shown by filled circle. Boxed area shows coverage of Figure 3. Inferred axis of E-W–trending arch shown as dashed thick gray line. GNPRFZ—Great Northern Puerto Rico fault zone; GSPRFZ—Great Southern Puerto Rico fault zone.

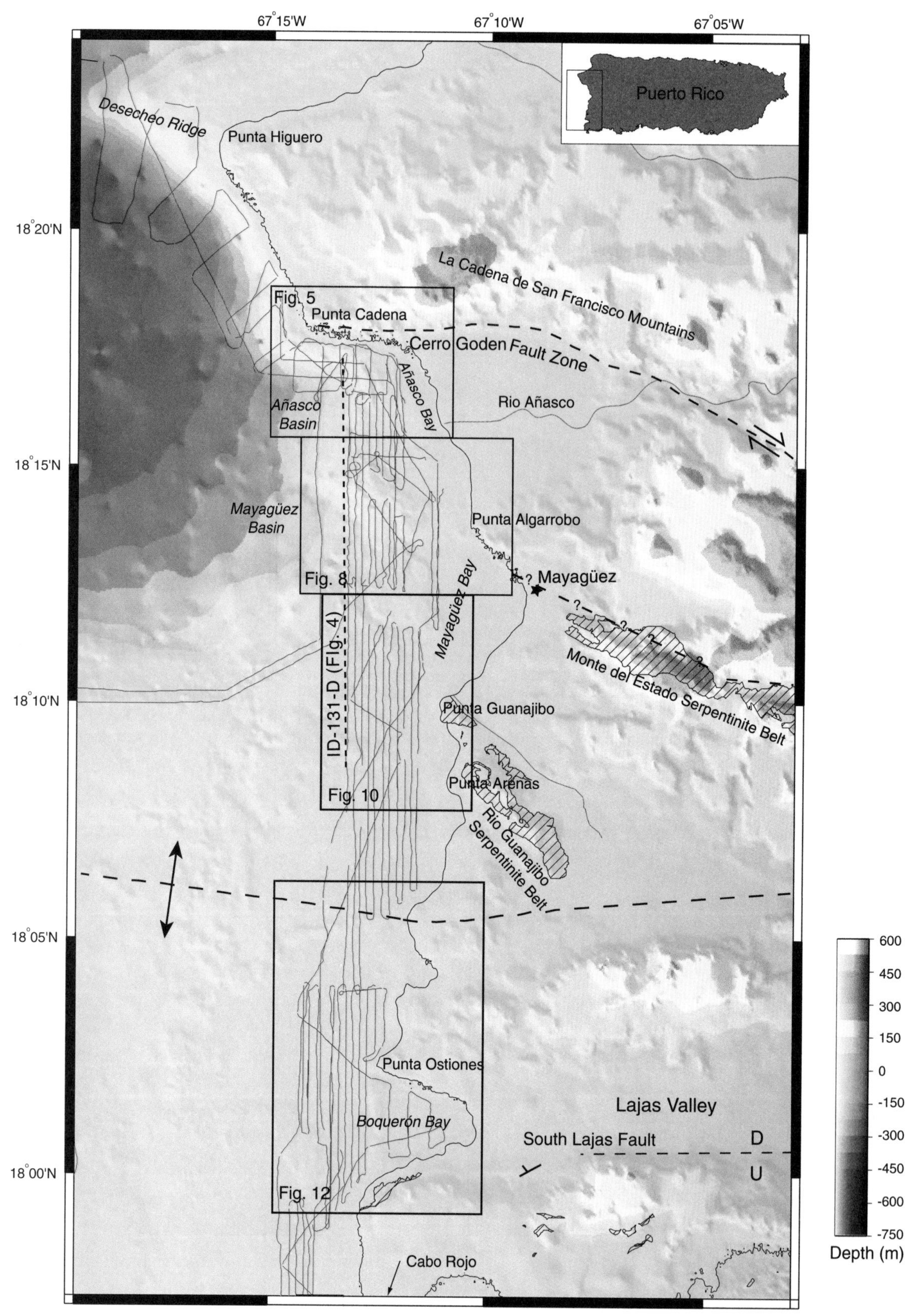

Figure 3. Shaded relief map of western Puerto Rico and offshore region. Ship tracks for May 2000 survey shown in thin red lines. Trackline for a portion of multichannel seismic (MCS) line ID-131-D (Fig. 4) is shown as dashed line. The four boxes represent coverage area of maps shown in Figures 5, 8, 10, and 12. Location of the Cerro Goden fault zone determined from McIntyre (1971) and Mann et al. (this volume, Ch. 6) and location of South Lajas fault from Glover (1971), Meltzer and Almy (2000), and Prentice and Mann (this volume). Dashed line with question marks is the hypothesized eastward extension of the Cordillera fault. Serpentinite bodies mapped onland are shown as areas with striped fill pattern. Inferred axis of E-W–trending anticline formed during early Pliocene N-S–shortening event shown as dashed thick black line.

Puerto Rico fault zone and the Great Southern Puerto Rico fault zone (Figs. 1 and 2; Erikson et al., 1990, 1991).

The second phase of deformation of N-S–shortening occurred in response to post–Eocene-Neogene convergence between the North America and Caribbean plates (Dillon et al., 1996; van Gestel et al., 1998). The episode of deformation resulted in a broad, 120-km-wide arch that has an axis that extends from eastern Hispaniola through the Mona Passage to the western side of the island of Puerto Rico (van Gestel et al., 1998) (Figs. 2 and 3). Localized extension at the crest of the arch presumably caused many small faults with a range of orientations, including many E-W–trending normal faults.

The third and ongoing phase of deformation is characterized by extension in the Mona Passage and possibly western Puerto Rico. The extension is believed to be caused by the differential ENE relative motion of the Puerto Rico–Virgin Islands microplate with respect to eastern Hispaniola (Grindlay et al., 1997; van Gestel et al., 1998; Jansma et al., 2000; Mann et al., 2002). The extension could be accommodated by reactivation of WNW-trending normal faults or shear zones developed during previous deformational phases. Recently published GPS geodetic measurements collected during a 10-year period in the region suggest that the amount of differential motion between Hispaniola and Puerto Rico is ~5 mm/yr (Lopez et al., 1999, Jansma et al., 2000; Mann et al., 2002). Geodetic studies also suggest that Puerto Rico and the Virgin Islands are currently behaving as part of the stable Caribbean plate and are moving in an ENE direction (~070°) at a rate of 19–20 mm/yr relative to North America (Fig. 1; Jansma et al., 2000; Mann et al., 2002).

GEOPHYSICAL DATA COLLECTION AND PROCESSING

We conducted a 10-day marine geophysical survey using the University of Puerto Rico's R/V *Isla Magueyes* in May 2000. We surveyed extensively (728 line km of data) the insular shelf off of western Puerto Rico, from Punta Higuero to Cabo Rojo (Fig. 3). Sidescan sonar and high-resolution sub-bottom profiler data were acquired simultaneously and merged with Differential Global Positioning Satellite (DGPS) navigation (±5 m resolution). Surveying extended very close to shore in water depths as shallow as 2 m, increasing our ability to correlate with known onshore faults. Trackline spacing was ~300 m. The sidescan sonar system generated 300–400-m-wide swaths of 100 kHz and 500 kHz seabed reflectivity data, which were later assembled into mosaic images with 1 m pixel resolution. All of the sonographs have been filtered, slant range corrected, bottom corrected, destriped, and beam-angle corrected before being placed into a georeferenced digital mosaic. The SCS boomer system operated at 280 Joules and 2 shots per second. At the average ship speed of 5 kts, this resulted in a shot spacing of ~1.3 m. The SCS data have been processed as shown in Table 1. Vertical resolution is estimated at 1 m, and penetration ranged from only imaging the seafloor in reefal areas to over 100 milliseconds below the seafloor (msbsf) in areas where relatively thick accumulations of unlithified alluvium were present.

Existing Seismic Data

MCS data were acquired in 1972–1973 by Western Geophysical Company and Fugro for the Puerto Rico Water Resources Authority (Western Geophysical Company of America and Fugro, Inc., 1973). MCS profiles separated by 5–10 km surround the island and were collected in a site assessment for a nuclear power plant. The MCS data were acquired with a 24-channel streamer with 67 m group spacing and were digitized and recorded using a SDS 1010 system. After acquisition, these data were processed using a single deconvolution operator, limited velocity analyses, and stacking. The MCS data set is owned by, and stored at, UPR as large-scale paper and Mylar records. We digitally scanned and examined the limited number of MCS lines that cross or coincide with our shallow-water survey of the inner shelf. In addition, we have used one MCS line that has been reprocessed to include migration (Fig. 4; Detrich, 1995). Initial reports provided by Western Geophysical Company of America and Fugro, Inc. (1973) indicate an abundance of faults that cut deeply into or through the Oligocene-Pliocene carbonate platform off western Puerto Rico. We have used mapped traces of these faults, projected to the surface as originally proposed by Western Geophysical Company of America and Fugro, Inc. (1973), where they could be verified by re-inspection of the original MCS data.

SEISMIC STRATIGRAPHY AND GEOLOGY OFFSHORE WESTERN PUERTO RICO

U.S. Geological Survey marine geologic maps based on grab and dive samples and 3.5 kHz seismic records show surface sediment thickness ranging from <1 to >10 m (Grove, 1983; Schlee et al., 1999) on the western insular shelf of Puerto Rico. Seafloor sediment composition is mixed terrigenous and skeletal sand and mud nearshore, and mostly carbonate mud farther offshore, punctuated by carbonate hardgrounds and shallow reefs (Grove, 1983; Schlee et al., 1999). Presently, the Añasco River deposits siliciclastic material in the nearshore environment, resulting in relatively thick (up to100 m) local accumulations of

TABLE 1. PROCESSING STEPS FOR SINGLE-CHANNEL SEISMIC DATA

Convert files from SEG-Y format
Spherical Divergence Gain (t^2)
Bandpass Filter: 100-150-1000-1100 Hz
Spectral Whitening: 150-1000 Hz
Digitization of the seafloor
Application of seafloor mute
Trace Mixing: 3 traces, not weighted
Application of AGC: 50 ms

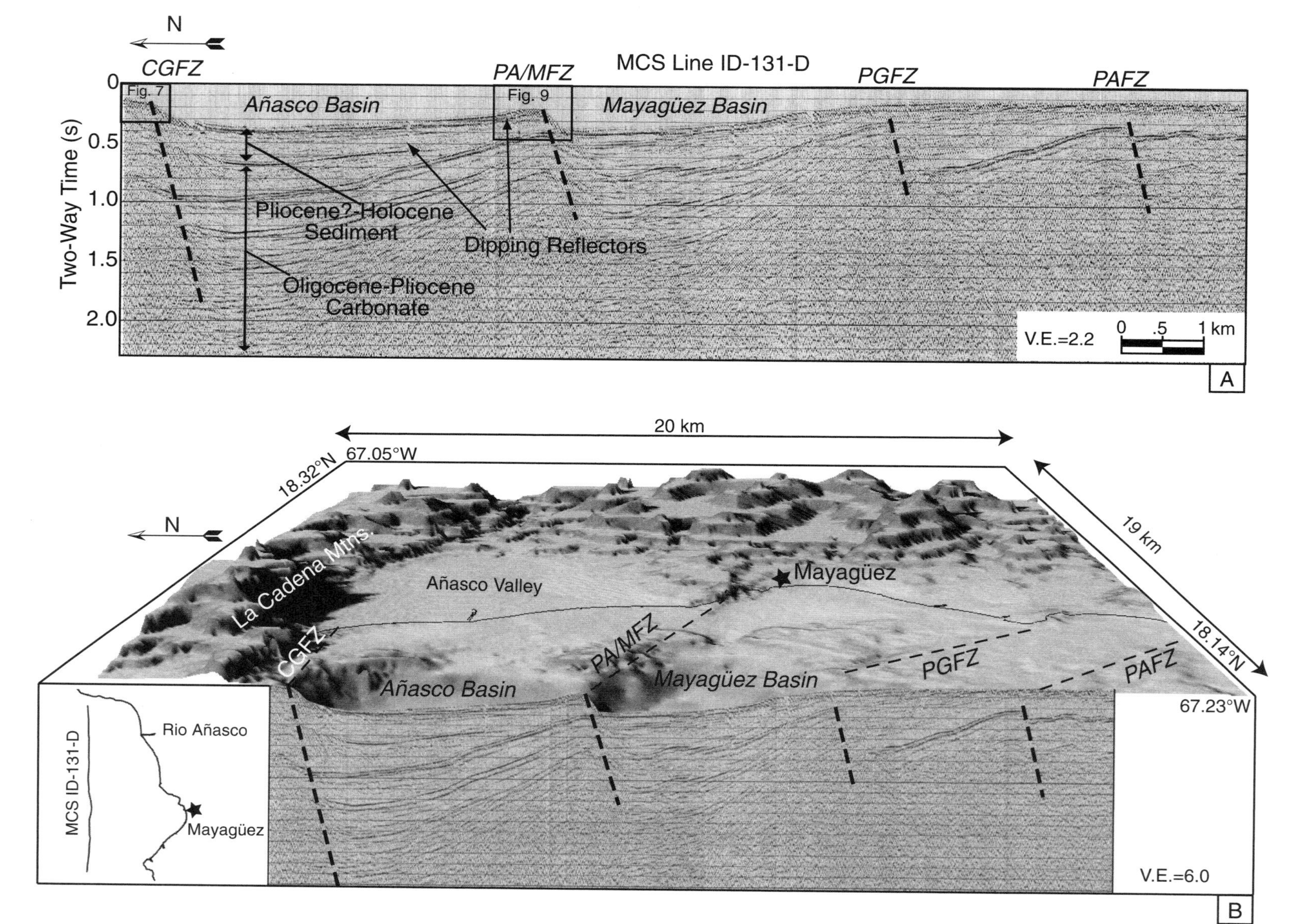

Figure 4. (A and B) Western Geophysical multichannel seismic (MCS) line ID131-D reprocessed including migration from Detrich (1995). This MCS line parallels the west coast of Puerto Rico (see inset for location) and clearly shows north-dipping reflectors truncated by high-angle, normal faults (dashed lines), forming several structural basins. The north-dipping reflectors are interpreted to be the Oligocene-Pliocene carbonate platform that experienced deformation associated with a north-south–shortening episode post–early Pliocene. The profile shown is located along the north-dipping limb of a regional arch developed during Phase 2 (e.g., Mann et al., this volume, Ch. 6, fig. 14D). Boxes show location of seismic profiles shown in Figures 7 and 9 that parallel this MCS line and image similar features. The large normal faults appear coincident with seafloor features identified in this study that define the Cerro Goden fault zone (CGFZ), Punta Algarrobo/Mayagüez fault zone (PA/MFZ), Punta Guanajibo fault zone (PGFZ), and Punta Arenas fault zone (PAFZ). (B) Three-dimensional shaded relief perspective of western Puerto Rico looking eastward combined with MCS line ID131-D. Illumination is from the northwest.

siliciclastic sediments. The age of the unlithified material imaged by SCS data is unknown, but is inferred to be primarily alluvium deposited since the last sea-level lowstand (i.e., Holocene) on an erosional (often hardground) surface.

The insular shelf to slope break within the study area is marked by an almost continuous line of submerged coral reefs that rise to water depths of 15–20 m. The entire insular shelf and upper slope were exposed to subaerial erosion from before 15,000 years until after 10,000 years ago (Morelock et al., 1994). In the southwest along the Parguera shelf (Fig. 2), the Holocene shelf-edge reefs began growing 8,000–9,000 yr ago on the newly submerged Pleistocene surface (Morelock, 1994; Hubbard et al., 1996). These coral died off 6,000 yr ago and the surface was barren until massive coral began to grow a few hundred years ago (Morelock, 1994; Hubbard et al., 1996).

Underlying the Holocene sediments is a thick (up to 1500m) Oligocene–early Pliocene age sequence of shallow-water limestone that lies unconformably over Cretaceous-Eocene volcanics (Moussa et al., 1987). It is assumed that these carbonates are contiguous with the tilted carbonate sequences that extend offshore on the northern and southern insular slopes of the islands (Moussa et al., 1987; van Gestel et al., 1998). The tilted platforms of opposing dip are interpreted to be limbs of a roughly east-west–trending, 120-km-wide arch that resulted from a north-south–shortening event (Phase 2 described in section II) from post Eocene to late Neogene (Dillon et al., 1996; van Gestel et al., 1998). The MCS profile shown in Figure 4 images the northern limb of this arch. Existing MCS profiles offshore western and southern Puerto Rico, collected by Western Geophysical Company and Furgo, show numerous E-W– and WNW-trending normal and strike-slip faults offsetting the carbonate platform strata and the underlying Cretaceous-Eocene volcanic basement (Larue and Ryan, 1990).

The primary limitation of these MCS data for this study is the lack of resolution of structure in the uppermost sediments and the limited coverage in shallow water. The high-resolution SCS profiles from our survey, therefore, complement the existing MCS profiles collected by Western Geophysical Company and Fugro, Inc. SCS profiling of carbonate hardgrounds and reefs resulted in high-amplitude seafloor reflections with very limited sub-seafloor penetration along with significant seafloor multiples obscuring any possible sub-bottom structure. Sediment thicknesses of <1 m, which is often the case, were not resolvable. In other locations, sediment stratigraphy is well imaged above a relatively high-amplitude reflection(s) forming acoustic basement, interpreted to be the top of the carbonate platform strata. Acoustic basement on SCS data often appears as a uniformly dipping reflection or series of reflections beneath onlapping and/or draping, semi-continuous, relatively flat-lying reflections to hummocky clinoforms. In other areas, stratigraphic relationships indicate acoustic basement was affected by cut and fill processes.

For the purposes of this study, youthful faulting is identified by displacements of the seafloor and within the uppermost, unlithified (Holocene?) sediment, lying unconformably over the faulted carbonate platform and/or linear disruptions of the sediment at the seafloor observed on the sidescan sonar mosaics.

OBSERVATIONS AND INTERPRETATIONS

In this section, we present the observations and interpretations of the sidescan sonar and SCS data collected during the May 2000 cruise. Data are presented in four geographic regions (Fig. 3): the North Añasco Bay region, the North Mayagüez Bay region, the South Mayagüez Bay region, and the Boquerón Bay region.

North Añasco Bay Region

The sidescan sonar mosaic of the northern Añasco Bay (Fig. 5) shows a low reflectivity area on the ~3-km-wide insular shelf near the mouth of the Añasco River. The areas of low reflectivity are interpreted to be deposits of fine-grained sediments, mainly mud and sands (Grove, 1983). The areas of high and chaotic reflectivity near the shelf edge are interpreted as unsedimented reefs. The shelf becomes narrower (<1 km) and more reef- and reef debris–dominated to the north, indicated by more and larger areas of high and chaotic reflectivity. The shelf edge parallels the coastline, and south and west of Punta Cadena, the seafloor drops off rapidly to depths in excess of 300 m. Two parallel, curvilinear, high-reflectance features that are separated by 200–300 m are clearly visible in the sidescan sonar mosaic (Figs. 5 and 6). These linear features parallel the coastline as it changes trend from northeast-southwest, north of Punta Cadena to roughly east-west, south of Punta Cadena. Seismic profiles (Fig. 6) show two areas of steeper slope ~10–15 m high, at ~80 m and ~40 m depth, which correspond to the location of the highly reflective linear features seen on the sidescan sonar mosaic. It has been suggested that during the last transgression, sea level rise either stopped, slowed dramatically, or even dropped to form paleo-shores (e.g., mid-Atlantic continental shelf; Emery and Uchupi, 1984). We hypothesize that the two "steps," or terraces, imaged on the slope represent stillstands or slow downs in sea-level rise where reefal growth was able to keep pace with sea-level rise. These submerged terraces are comparable to the submerged reefs found at 20 m water depths at the edge of the insular shelf. Most importantly, offsets in these linear features can be used to pinpoint locations of youthful faults offshore and potentially determine sense of relative motion along the fault trace.

Shore-parallel SCS lines just seaward of the Añasco River reveal the thickest sections (>100 ms) of continuous flat-lying reflections that onlap and bury reef-like pinnacles, sometimes display cut and fill geometries and occasionally are severely attenuated, presumably the result of gas-charged sediments. This seismic facies is interpreted as well-stratified alluvium from the Añasco River. These deposits thin and extend to the northern, east-west–trending shore, where they appear as relatively thick (25 m) accumulations between shallow reef heads. Several of our high-resolution SCS profiles (Fig. 7) clearly show displacements of the seafloor and these alluvial sediments that we interpret to be Holocene age.

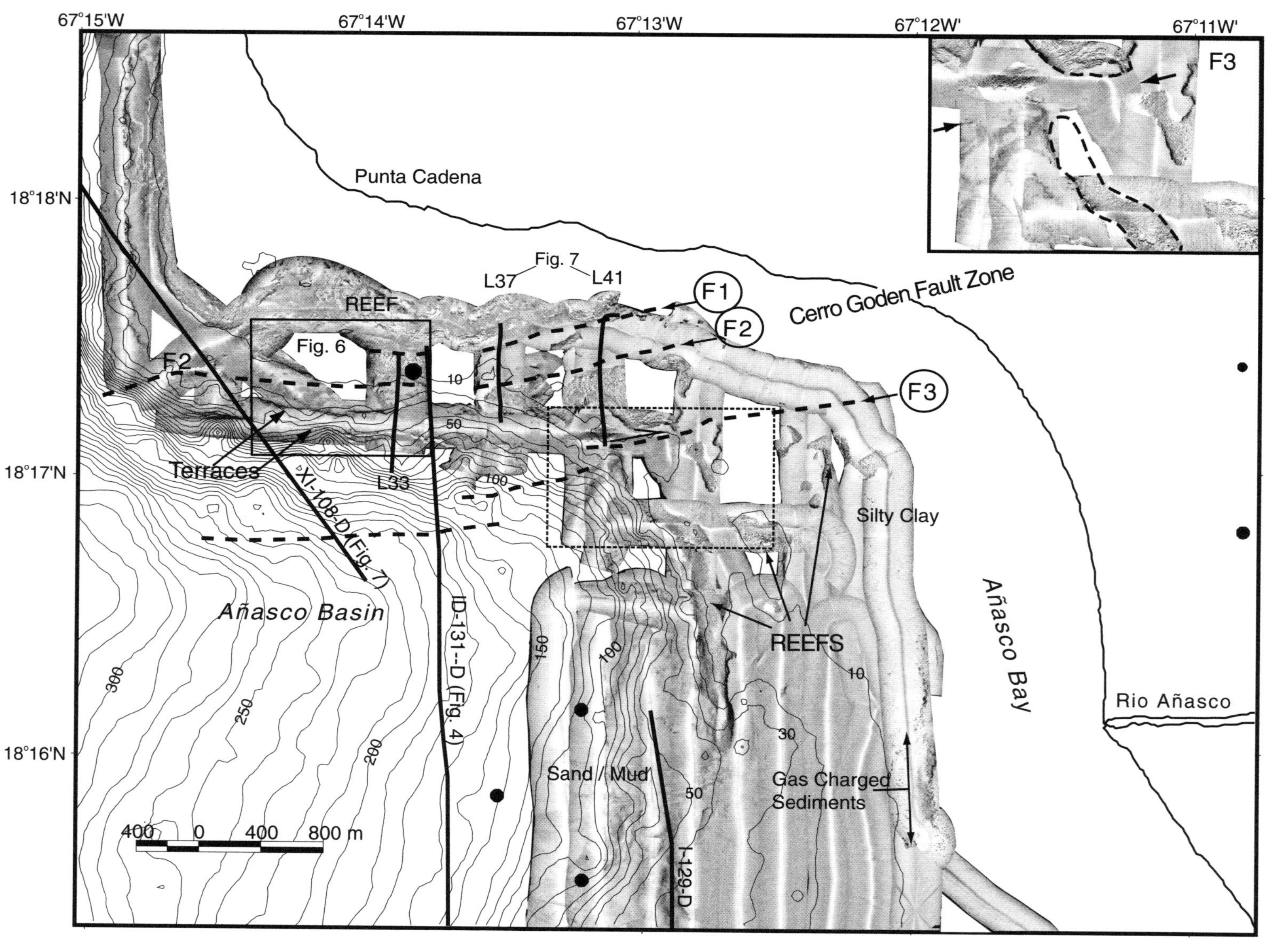

Figure 5. Sidescan sonar mosaic of North Añasco Bay (Fig. 3 shows location). Location of seismic profiles indicated by thick solid lines. Multichannel seismic (MCS) tracks are ID-131-D, XI-108-D and I-129-D; single-channel seismic (SCS) tracks are L37 and L41. MCS profiles ID-131-D and I-129-D extend beyond southern boundary of figure. Dashed lines mark location of linear features visible in sidescan sonar mosaic, seafloor offsets in SCS profiles and offsets of deeper structures in MCS profiles projected to the seafloor. These features are interpreted as offshore extensions of the Cerro Goden fault zone (CGFZ). F1, F2, and F3 mark the location of semi-continuous fault traces within the CGFZ. Earthquake epicenters (U.S. Geological Survey National Earthquake Information Center [NEIC] PDE database 1973–2000) of earthquakes with magnitudes >2.5 shown as filled circles. Bathymetry contour interval = 10 m, and 1 m pixel resolution mosaic. Imagery within boxed area is shown in Figure 6 and dashed box area is enlarged and shown as an inset. Reefs enclosed by dashed lines appear to be offset by ~350–400 m right-laterally (F2). F3 identified on SCS lines is not apparent on MCS lines XI-108-D (Fig. 7) or ID-131-D (Fig. 4).

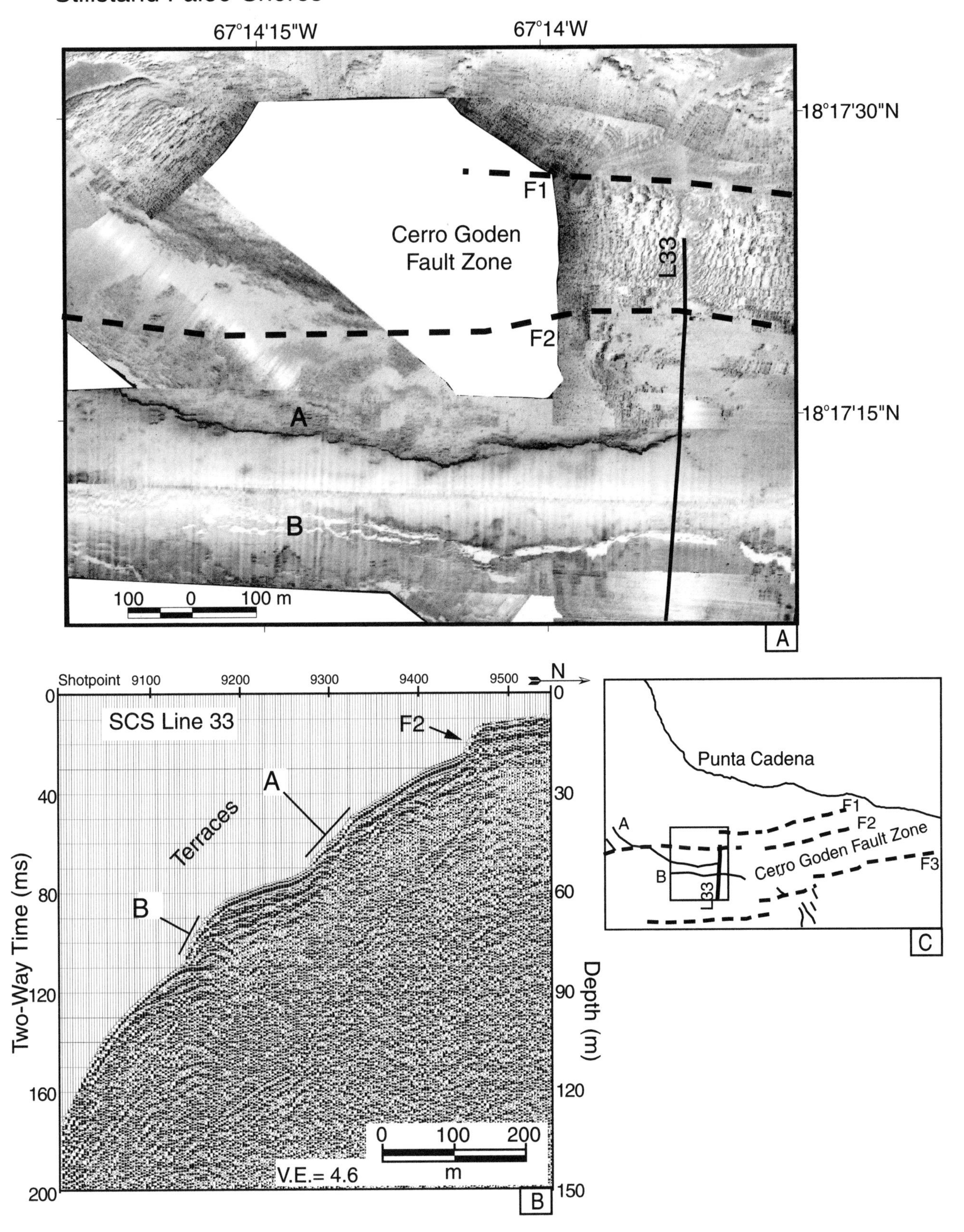

Figure 6. (A) Sidescan sonar enlargement (Fig. 5) showing terraces labeled "A" and "B" and a portion of the Cerro Goden fault zone (CGFZ), including F1 and F2 as in Figure 5. (B) Single-channel seismic (SCS) Line 33 showing the two terraces at ~80 m and ~40 m formed by sea-level stillstands or slow downs during the most recent transgression (depth scale assumes V = 1500 m/s). (C) Map view showing SCS line 33 (solid line), terraces as dotted lines and fault traces making up the CGFZ as dashed lines.

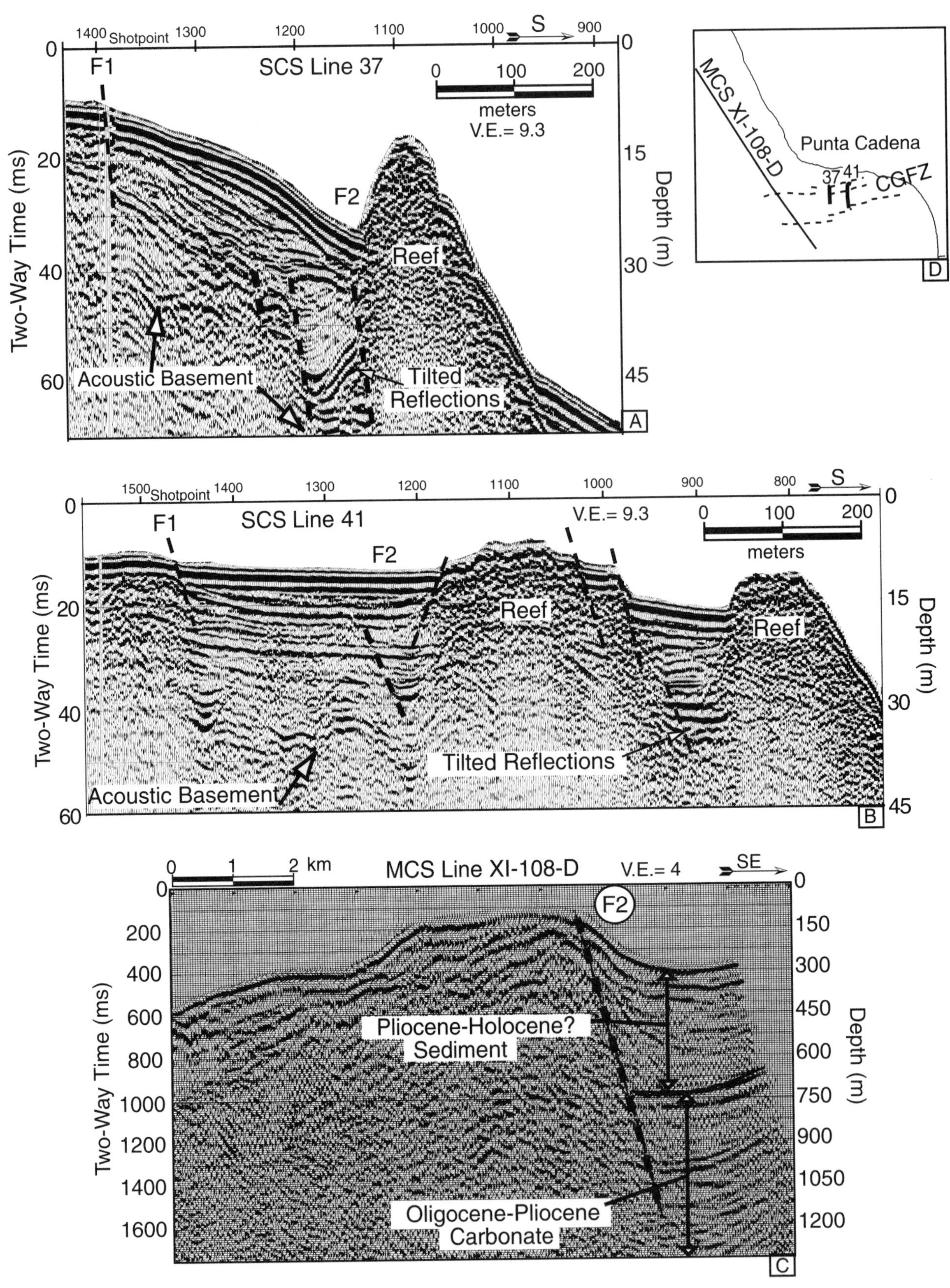

Figure 7. (A and B) Single-channel seismic (SCS) profiles across the Cerro Goden fault zone (CGFZ) (depth scale assumes V = 1500 m/s). Approximately 20–25 m of sediment fills depressions between exposed reef heads. Acoustic basement in these profiles is interpreted as the boundary between alluvium (Holocene?) and well-indurated sediments and sedimentary rocks. Offsets (dashed lines) of seafloor and offsets and tilting of sediments indicate recent fault displacement. (C) Multichannel seismic (MCS) Line XI-108-D shows offset (dashed line) of the Oligocene-Pliocene carbonate platform and overlying section which is assumed to be Pliocene and younger in age. The trace of this normal fault projected to the seafloor aligns with fault (F2). Note that fault F3 is not associated with any deep structure imaged. (D) Map view showing MCS and SCS tracklines (solid lines) and the fault segments within the CGFZ (dashed lines).

On the basis of the SCS, sidescan sonar and bathymetry data, we have identified parallel, semi-continuous fault traces (Figs. 5, 6, and 7) that trend ENE. One of these fault traces (F3, Fig. 5) cuts across the submarine terraces at nearly right angles and appears to offset them in a right-lateral sense. The offset in the reef at 20 m water depth is most clearly distinguishable and is ~400 m (Fig. 5, inset). Fault traces can be traced close to shore, but on SCS lines closest to the shore, no displacements are observed in the well-stratified and relatively thick sediments.

MCS profiles XI-108D (Fig. 7) and ID-131-D (Fig. 4, north end) cross the narrow shelf south of Punta Cadena and reveal a high-angle normal fault offsetting the Oligocene-Pliocene carbonate strata. It is not clear from the MCS profiles if this fault displaces sediments on the seafloor. The map view projection of these deep faults to the seafloor lies within the zone of parallel and semi-continuous fault traces that displace the seafloor and Holocene sediments. This narrow band of faults is interpreted as an offshore portion of the Cerro Goden fault zone (Figs. 3 and 4).

North Mayagüez Bay Region

The sidescan sonar mosaic of the North Mayagüez Bay region (Fig. 8) shows a wider (3–5 km) section of the insular shelf that is dominated by low-reflectivity seafloor that is determined from grab samples to be mainly silty sand and mud (Schlee et al., 1999). Two large areas of chaotic and high reflectivity at the mouth of the bay are interpreted to be reefal areas. The western edges of the reefs form a semicircle that is cut at its midpoint by a narrow NW-SE–trending channel (Fig. 8). The sidescan sonar mosaic shows an abrupt change in reflectivity along the southern edge of the channel that corresponds to small offsets, down to the south, in the SCS data that project northwestward to a very steep, south-facing slope (Fig. 9). West of the shelf edge the seafloor deepens, forming a bowl-shaped basin called the Mayagüez basin (Figs. 4, 8, and 9). In the sidescan sonar imagery, we observed at least two linear and parallel, high-reflectivity features that follow contours along the southern edge of this basin, which we interpret to be submarine terraces comparable to those observed in North Añasco Bay (Fig. 8). These terraces also appear to terminate at the southern edge of the NW-SE–trending channel.

MCS lines running north-south offshore Mayagüez Bay clearly show north-dipping reflectors truncated by high-angle, normal faults forming a significant scarp (Figs. 4 and 9). The north-dipping reflectors are interpreted to be the Oligocene-Pliocene carbonate platform and the scarp separates the clastic sediment depocenters of Mayagüez Bay and Añasco Bay into two sub-basins: the Mayagüez and Añasco basins (Fig. 4). SCS profiles collected in the Mayagüez Bay area parallel to, and overlapping these MCS lines (Fig. 9) show this scarp as a steep WNW-ESE–trending slope, with 50–100 m of relief. Several parallel, high-amplitude reflectors, which dip north at ~6°, terminate at or near the surface beneath a thin section of flat-lying continuous reflections. The dip and normal component of fault motion can be seen on the MCS profiles (Figs. 4 and 9B); however, due to their low resolution, it is not clear whether the seafloor or uppermost sediments are disrupted. The SCS profiles also show the north-dipping reflectors truncated by the south-facing scarp near the seafloor and only a thin (<2 m) veneer of sediment at the ridge (Fig. 9A). The thin Holocene (?) sediments on the top of the ridge are consistent with a recent uplift history, but are too thin to enable offset strata to be seen. Additionally, the lack of piercing points makes determination of the direction and magnitude of strike-slip motion (if any) impossible to determine.

Sidescan sonar imagery shows areas of lower reflectivity "streaking" down the south face of the scarp (filled arrows, Fig. 8) and, along with the SCS data, indicates sediment transport through breaks in the reef. SCS profiles show a thick wedge of sediments accumulated at the base of the slope (Fig. 9). Since no river empties directly into Mayagüez Bay, the source of these sediments is interpreted to be the Añasco River to the north. Offsets of this alluvium and seafloor at the base of the scarp align with displacements and terminations of linear seafloor features observed closer to shore in SCS and sidescan sonar data. The set of offsets that projects from the head of the scarp landward toward Punta Algarrobo and those that project toward Mayagüez Bay are referred to as Punta Algarrobo and Mayagüez fault zones, respectively (Fig. 4).

South Mayagüez Bay Region

The sidescan sonar mosaic of South Mayagüez Bay (Fig. 10), on a relatively wide section of the insular shelf (8–10 km), is dominated by highly reflective seafloor interpreted to be large expanses of coral reefs. The highly reflective area is truncated to the north by a NW-SE–trending lineament that marks an abrupt change to lower reflectivity seafloor (muds and sands). Corresponding to this boundary and visible on adjacent SCS profiles is a narrow (400-m-wide) trough, with a nearly vertical 10 ms offset (down to the south) at the seafloor that trends NW-SE (Figs. 10 and 11). An ~20 ms interval of flat-lying, continuous reflections overlying dipping and diverging reflections terminate abruptly at the north wall of this trough (Fig. 11). This seismic image is interpreted as a relatively thick section (~15 m) of well stratified alluvium (Holocene?) that has filled a fault-bounded depression that has continued to experience a component of vertical motion since infilling began. In addition, contours marking the submerged reef at the shelf edge (20 m depth) appear to be offset ~750 m in the right-lateral sense across the trace of this fault (Fig. 10). Other, smaller vertical offsets of the seafloor and surface sediments are also imaged on parallel and adjacent SCS profiles. The set of parallel and semi-continuous fault traces, trending NW-SE, that project toward Punta Guanajibo, define the offshore section of the Punta Guanajibo fault zone (Figs. 4, 10, and 11).

Offshore from Punta Arenas, the reefal areas are cut by several discontinuous NW-SE–trending lineaments (Punta Are-

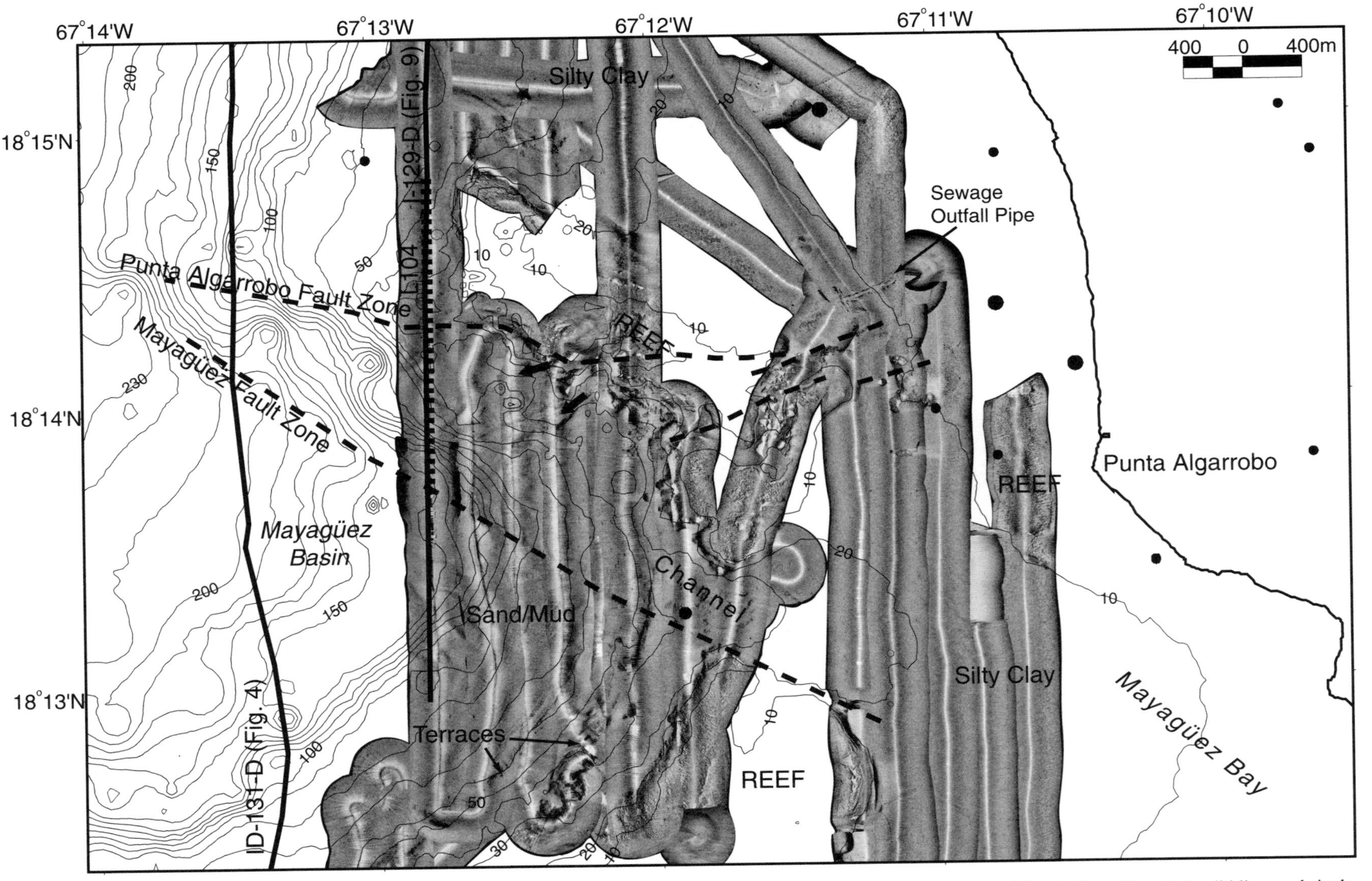

Figure 8. Sidescan sonar mosaic of the North Mayagüez Bay area (Fig. 3 shows location). Location of multichannel seismic (MCS) tracklines indicated by thick solid lines and single-channel seismic (SCS) line 104 shown as dotted line. Dashed lines indicate location of linear features visible in sidescan sonar mosaic, seafloor offsets in SCS profiles and offsets of deeper structures in MCS profiles projected to the seafloor. These features locate the Punta Algarrobo and Mayagüez fault zones (PA/MFZ). Short black arrows indicate location and direction of sediment transport from the Añasco Basin into the Mayagüez Basin. Earthquake epicenters (U.S. Geological Survey National Earthquake Information Center [NEIC] PDE database 1973–2000) of earthquakes with magnitudes >2.5 shown as filled circles. Bathymetry contour interval = 10 m, and 1 m pixel resolution mosaic.

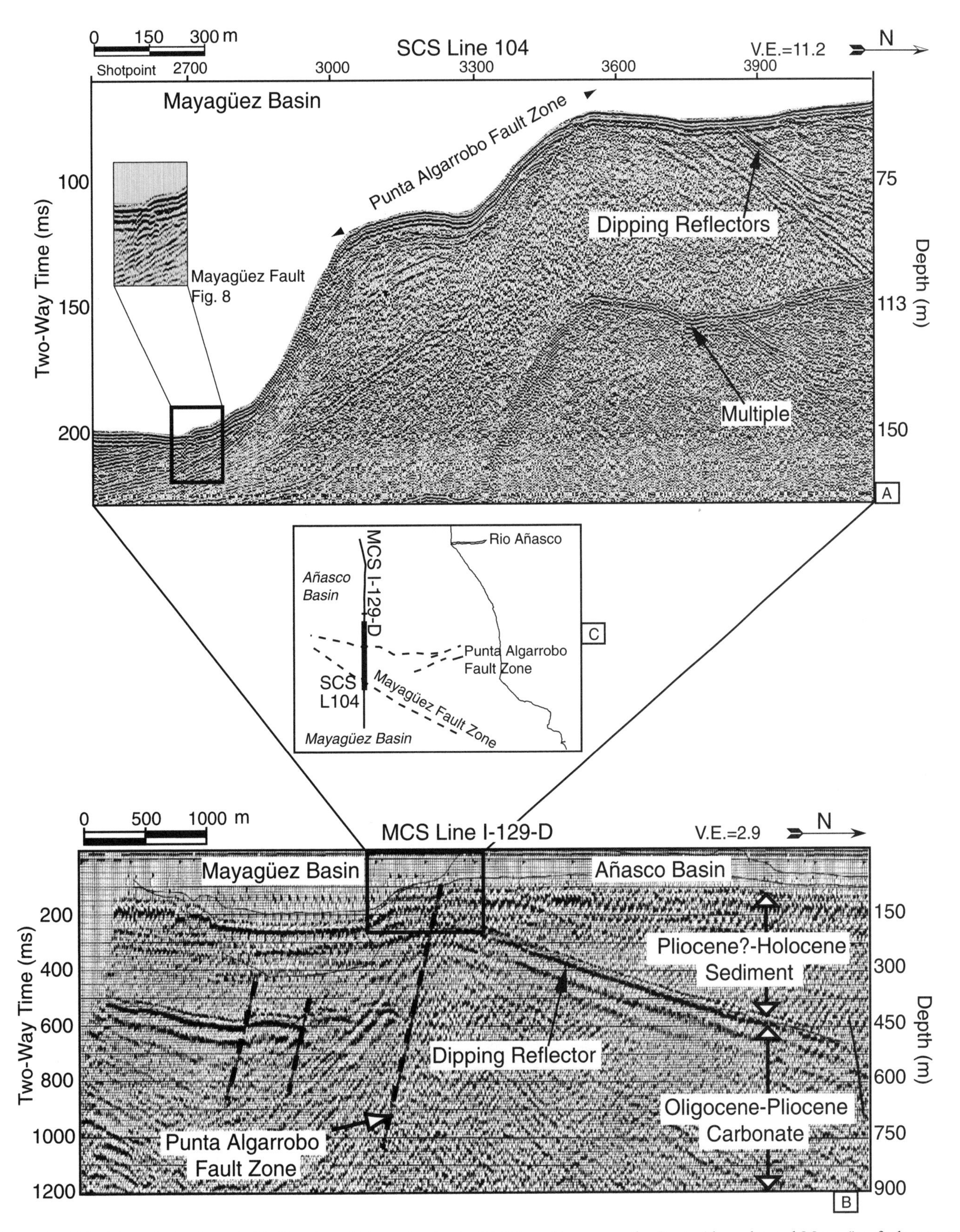

Figure 9. Coincident single-channel (SCS) and multichannel seismic (MCS) profiles across the Punta Algarrobo and Mayagüez fault zones (PA/MFZ) in Mayagüez Bay. (A) SCS line 104 shows that the high-angle normal fault imaged in MCS creates south-facing scarps and truncates north-dipping reflectors, which outcrop at the seafloor or are thinly sedimented. Boxed area is enlarged and reveals an offset in the seafloor and near surface alluvium accumulating in the Mayagüez Basin. (B) MCS line I-129-D clearly shows north-dipping reflectors truncated by high-angle, normal faults (dashed lines). The north-dipping reflectors are interpreted to be the Oligocene-Pliocene carbonate platform (also see Fig. 4). The boxed area shows the area imaged by SCS line 104. (C) Map view showing tracklines of MCS (thin solid line) and SCS (thick solid line) profiles and the fault segments comprising the PA/MFZ (dashed lines).

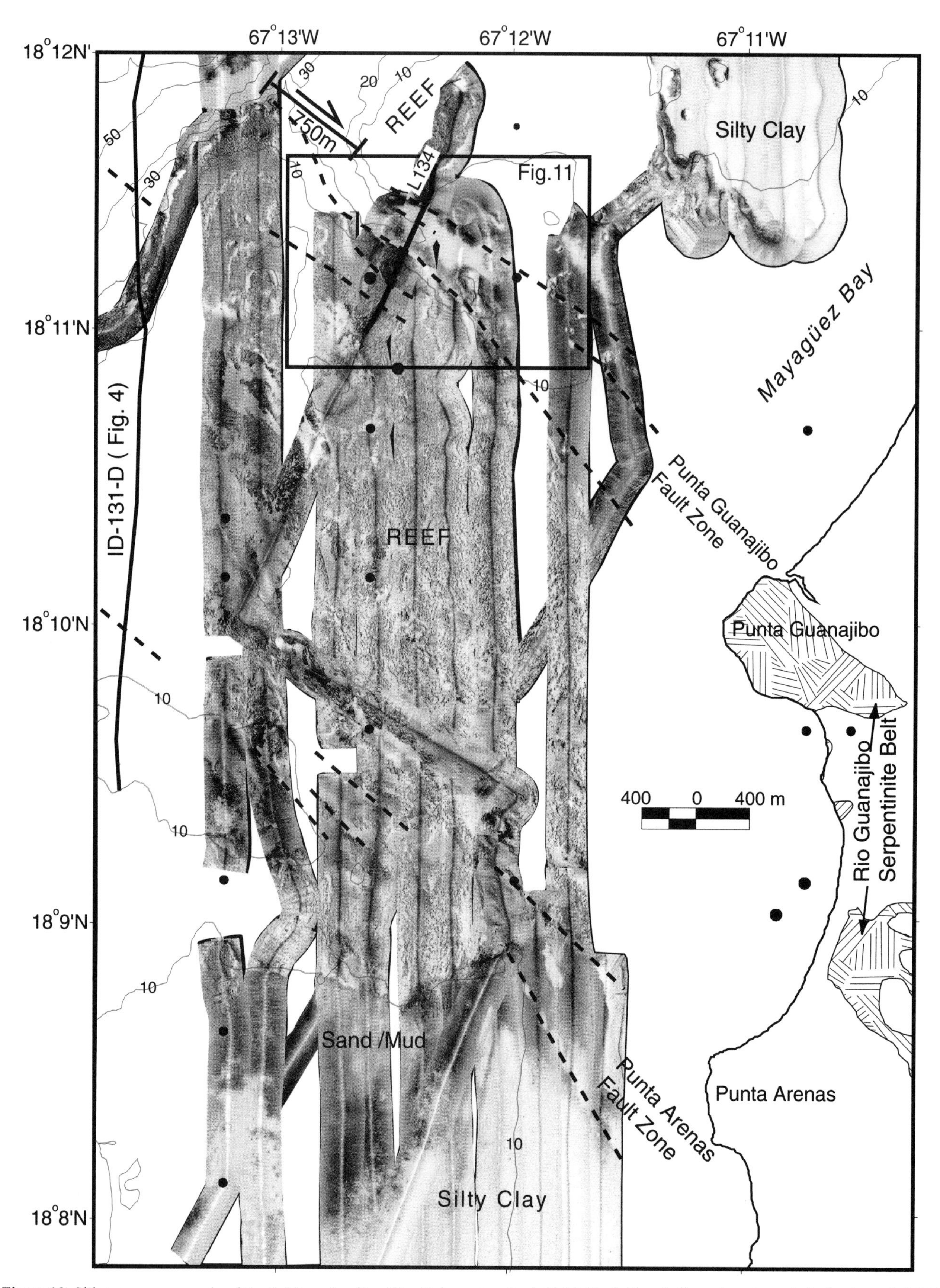

Figure 10. Sidescan sonar mosaic of South Mayagüez Bay (Fig. 3 shows location). Thick black lines indicate locations of single-channel (SCS) and multichannel seismic (MCS) profiles. MCS profile ID131-D extends beyond the northern border of figure (Fig. 4). Dashed lines indicate location of linear features visible in sidescan sonar mosaic, seafloor offsets in SCS profiles and offsets of deeper structures in MCS profiles projected to the seafloor. These features locate the Punta Guanajibo and Punta Arenas fault zones. Filled pattern locates serpentinite. Earthquake epicenters (U.S. Geological Survey National Earthquake Information Center [NEIC] PDE database 1973–2000) of earthquakes with magnitudes >2.5 shown as filled circles. Bathymetry contour interval is 10 m, and 1 m pixel resolution mosaic. Imagery within boxed area is shown in Figure 11.

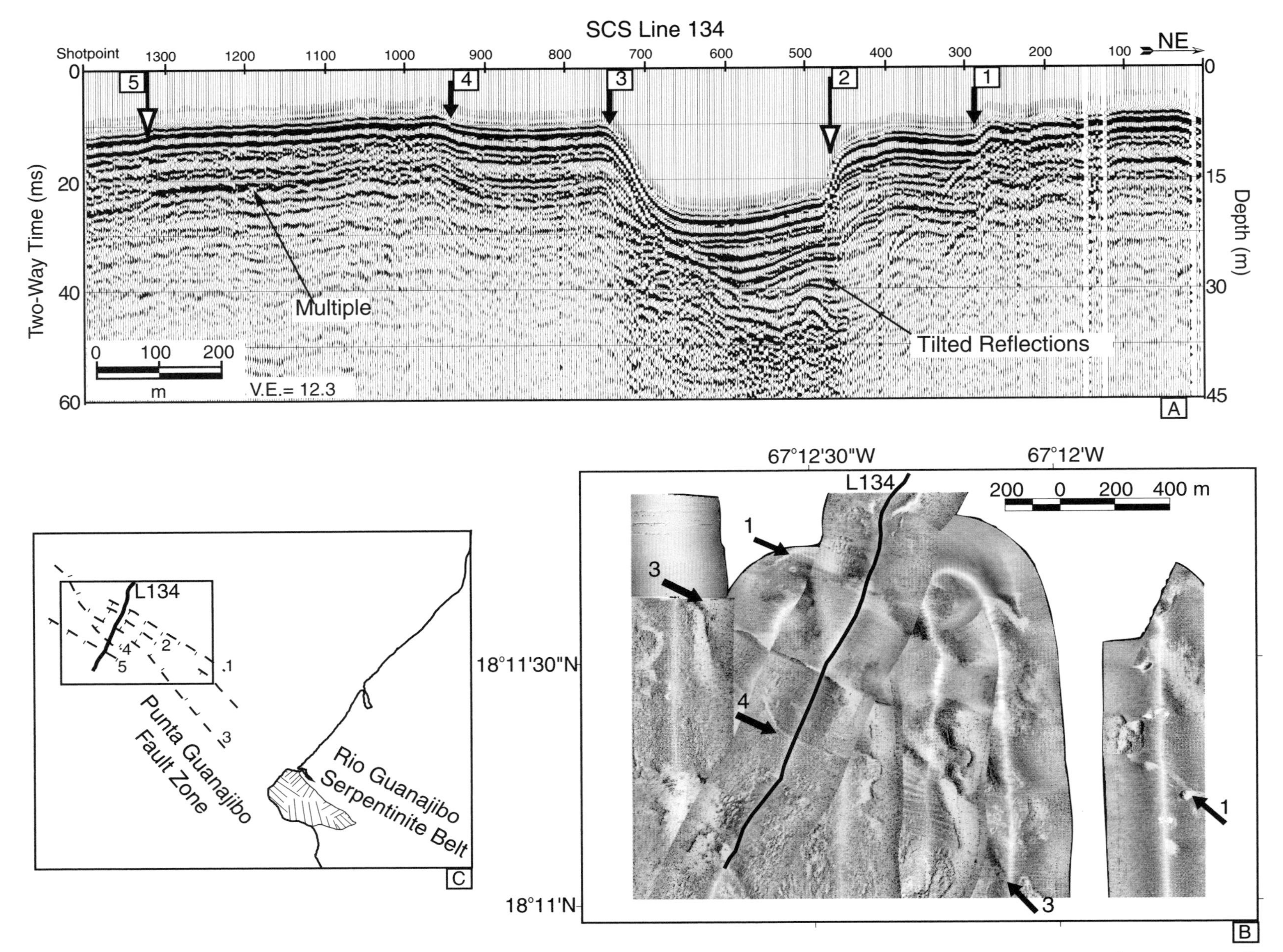

Figure 11. (A) single-channel seismic (SCS) line 134 across the Punata Guanajibo fault zone (PGFZ). Offsets in seafloor and/or surface sediments that are also apparent on adjacent SCS lines are indicated by numbered arrows. Filled arrows number 1, 3, and 4 are offsets that are apparent on both SCS and sidescan sonar records. Well-stratified and tilted sediments fill the fault bounded depression (B) Enlarged area of sidescan sonar mosaic from Figure 10 showing fault lineaments numbered to correspond with SCS. (C) Map view showing SCS trackline (solid lines) and numbered fault segments within the PGFZ (dashed lines). Note that the trend of the PGFZ projects to the Serpentinite belt mapped on shore.

nas fault zone, Fig. 10). The lineaments extend southward into the areas of muds and sands (dark-gray) and silty clays (light gray), but do not correspond to clear offsets of the seafloor in the SCS data. This is not to say that offsets do not exist. Sediment thicknesses in this area are extremely thin (<1 m, below system resolution) and the hardbottom nature of the seafloor resulted in high-amplitude seafloor reflections with very limited sub-seafloor penetration along with significant seafloor multiples obscuring any possible sub-bottom structure.

Boquerón Bay region

Extensive surveying of the area offshore of the Lajas valley did not reveal any evidence of the South Lajas fault surveyed on land by Meltzer and Almy (2000) and Prentice and Mann (this volume). The sidescan sonar mosaic shows areas of very fine silty clays (light gray) close to shore in Boquerón Bay and in the northern half of the survey area (Fig. 12). Seafloor sediments consisting of sands and muds (Schlee et al., 1999), seen as dark gray areas in the sidescan sonar mosaic, dominate the southern portion of the survey area. Within Boquerón Bay and at the mouth of the bay, SCS profiles show south-dipping reflectors, which we interpret to be the Oligocene-Pliocene carbonate platform, overlain by the relatively flat-lying reflections interpreted as Pliocene-Holocene sediments (Fig. 13). The presence of dipping reflectors on adjacent profiles with dissimilar azimuths enabled us to calculate a true dip of 3° to the SSW. The dipping reflectors extend to the seafloor where they appear to be truncated along an erosional surface, on top of which lies a thin covering of sediment (Holocene?). The dip of the reflectors abruptly changes to the north on the north side of an E-W–trending bathymetric high at 18°06′N. We interpret this change in dip to mark the axis of the regional E-W–trending arch (Figs. 3 and 12).

DISCUSSION

Evidence for Holocene Faulting Offshore Western Puerto Rico and Onland Extensions

The systematic sidescan sonar and high-resolution SCS survey of the western insular shelf of Puerto Rico has revealed at least three regions of active deformation.

1. The first region includes the offshore extension of the Cerro Goden fault zone (Figs. 14 and 15). Onland this fault parallels an abrupt, linear mountain front separating the 270–361-m-high La Cadena de San Francisco mountain front from the alluvial Añasco valley (Figs. 3 and 4). Previous workers have postulated that the Cerro Goden fault zone continues to the southeast as the Great Southern Puerto Rico fault zone and to the west, merges with the normal fault forming the eastern wall of the Mona rift (Garrison and Buell, 1971; McCann, 1985) (Figs. 1 and 15). Onland ~500 m south of the prominent La Cadena mountain front, Mann et al. (this volume, Ch. 6) found geomorphic features suggestive of right-lateral Quaternary fault activity, including offset terrace risers and upslope-facing faults. The projection of the mapped onland trace of the Cerro Goden fault zone offshore lies along the northern edge of Añasco Bay and appears as a set of semi-continuous fault traces indicating Holocene activity (Figs. 5, 6, and 7). To the north and west of our survey area, MCS profiles indicate that the fault zone extends to the Desecheo Ridge, the southern limit of the Mona rift (Figs. 3 and 15).

2. The second zone of deformation is the Punta Algarrobo/Mayagüez fault zone (Figs. 14 and 15). The slope and truncated dipping reflectors that indicate the position of the Punta Algarrobo fault zone are not visible on the SCS profiles nearest to shore due to the thickness of the sediment column resulting from rapid accumulation of alluvium, therefore it cannot be directly linked to faults mapped onshore. However, the similarity in trend, location, and type of motion suggest that the fault zone imaged in the northern Mayagüez Bay area is the offshore extension of the Cordillera fault mapped by Moya (1994), who proposed activity within the Quaternary. The offshore fault zone projects into the WNW-ESE–trending Monte del Estado peridotite/serpentinite belt onland (Figs. 3 and 14).

3. The third zone of deformation is the Punta Guanajibo/Punta Arenas fault zone that projects to Punta Guanajibo and Punta Arenas (Figs. 14 and 15). These areas are characterized by the NW-SE–trending Rio Guanajibo serpentinite belt onland (Figs. 3 and 14). In addition, Moya (1994) identified a WNW-trending fault with vertical offsets along the southern flank of the Cordillera de Saban Alta, the seawardmost unit of the serpentinite belt (Figs. 3 and 14). Moya (1994) suggests that this fault shows Quaternary activity.

Both the Punta Algarrobo/Mayagüez and Punta Guanajibo/Punta Arenas fault zones project to, and are colinear with, the trend of onland serpentinite bodies. The serpentinite bodies are associated with magnetization lows that extend offshore, suggesting that the serpentinite also extends offshore (Fig. 14). The faults mapped offshore in this study correspond to the magnetization lows. This relationship suggests that either the presence of serpentinite has localized fault activity or that fault activity has remobilized serpentinite. In either case, the relatively weak and ductile serpentinite may be accommodating part of the strain associated with the regional stress regime by many frequent, small magnitude events, thus accounting for the intense, low magnitude seismic activity that characterizes southwestern Puerto Rico.

The lack of evidence in the offshore for young faulting associated with the South Lajas fault does not necessarily mean there is no Holocene offshore displacement on this structure. This fault has at least a component of strike-slip displacement and is only known to have had two displacement events in the Holocene (Prentice and Mann, this volume). With such a small amount of Holocene displacement, it is possible that it is not resolved even with a high-resolution seismic system, especially if most of the motion is strike-slip. However, the sidescan sonar imagery also failed to detect evidence of the fault, suggesting that if a young fault does exist, recent sedimentation has masked its surfical expression on

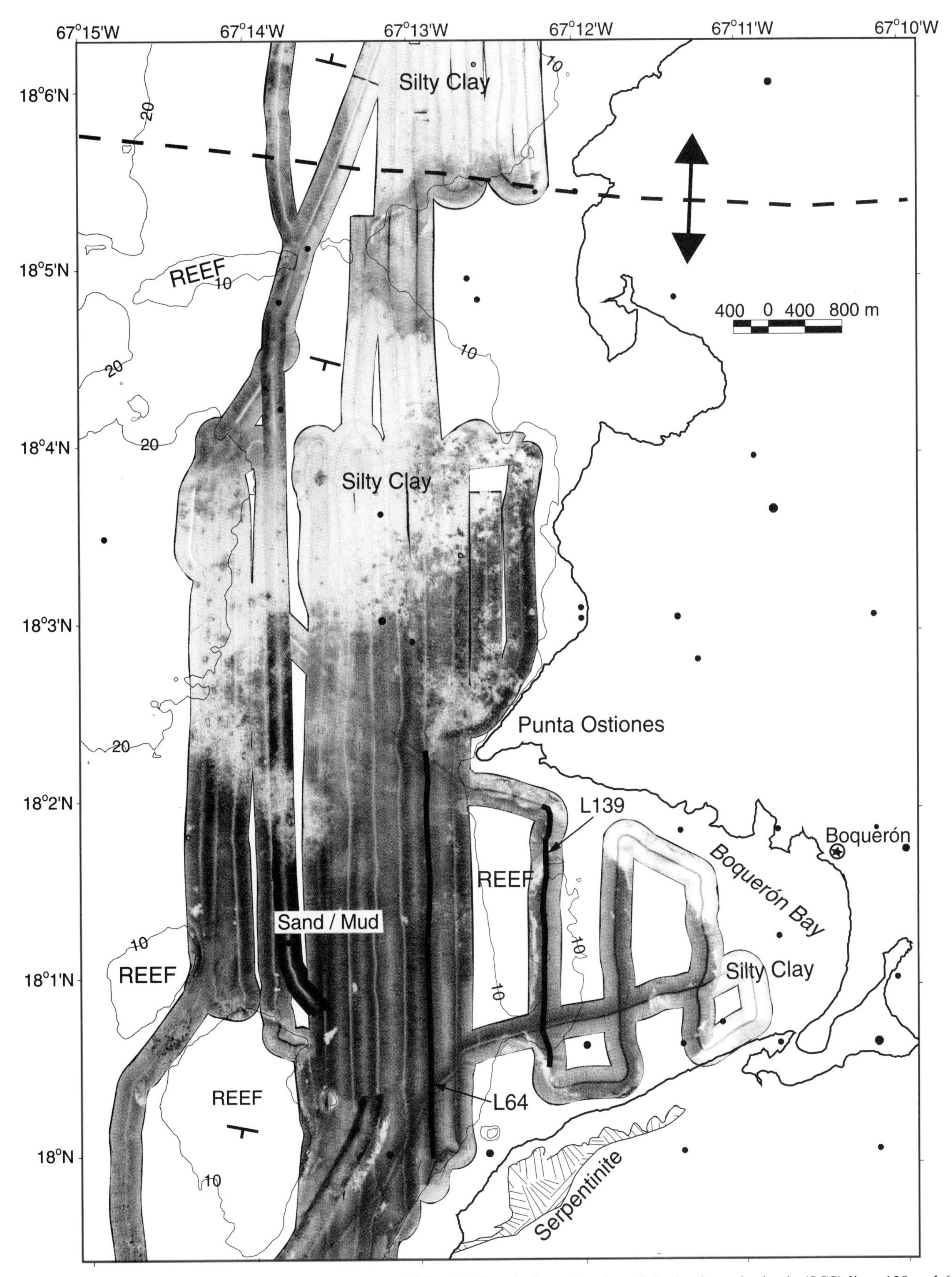

Figure 12. Sidescan sonar mosaic of Boquerón Bay area. Thick solid lines indicate location of single-channel seismic (SCS) lines 139 and 64 shown in Figure 13. Axis of E-W–trending arch shown as dashed thick black line. South of the axis, reflectors in SCS profiles dip SSW; north of the axis reflectors dips NNE. Filled pattern locates serpentinite. Earthquake epicenters (U.S. Geological Survey National Earthquake Information Center [NEIC] PDE database 1973–2000) of earthquakes with magnitudes >2.5 shown as filled circles. Bathymetry contour interval is 10 m, and 1 m pixel resolution mosaic. Sidescan sonar data were not able to resolve any seafloor displacements offshore of the Lajas valley.

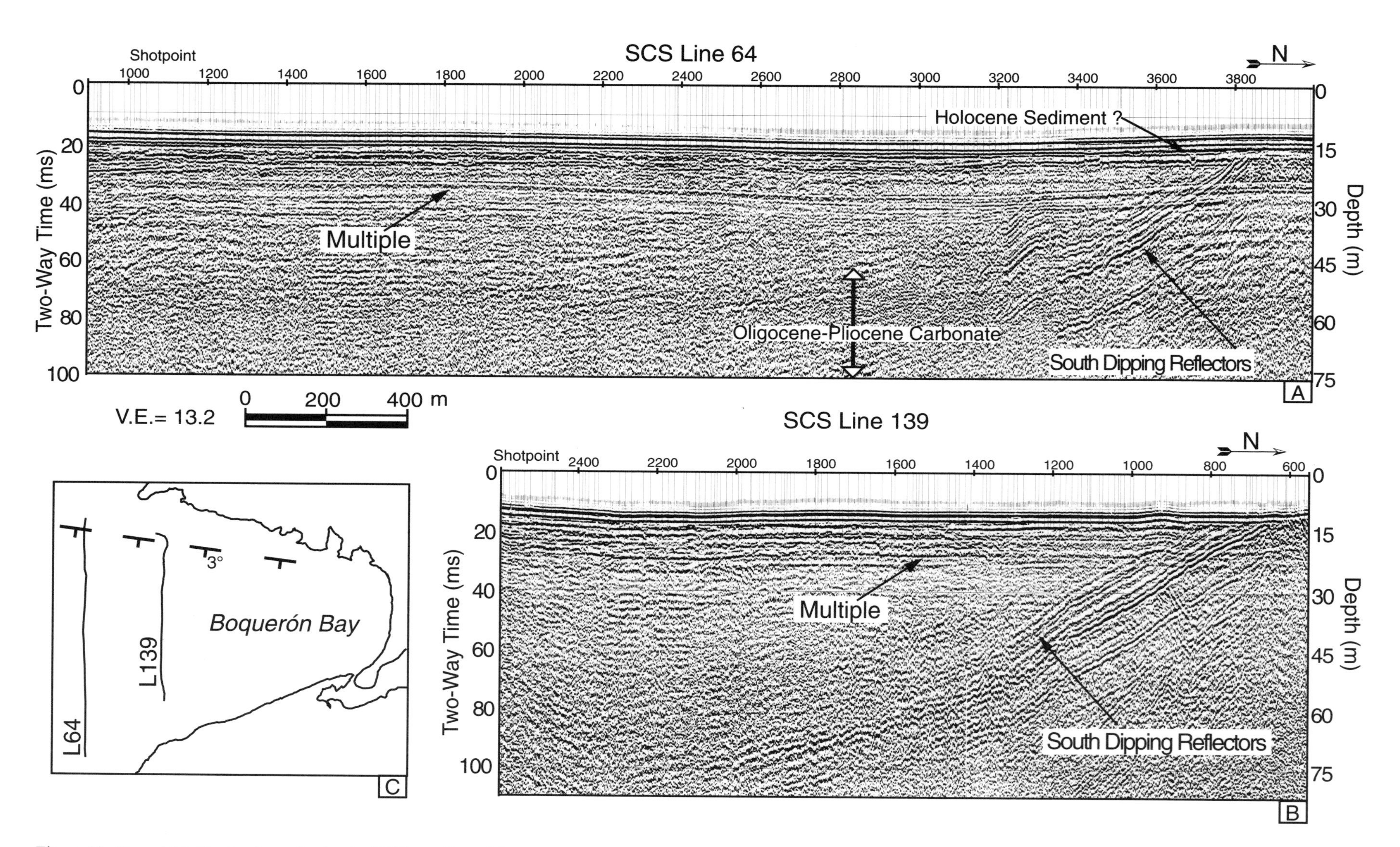

Figure 13. (A and B) Single-channel seismic (SCS) profiles of Boquerón Bay. South-dipping reflections are interpreted as the uppermost portion of the Oligocene-Pliocene carbonate platform. This dip is consistent with structure imaged by offshore multichannel seismic (MCS) profiles. An angular unconformity is apparent between the dipping carbonate strata and the uppermost flat-lying, continuous reflections that are interpreted as Holocene in age. Seafloor or surficial sediment offsets are not observed in any of the SCS profiles offshore of the Lajas valley. (C) Map view showing SCS trackline (solid lines). Series of strike and dip symbols for top of the carbonate platform imaged in SCS and MCS indicating location of dipping reflectors projected to the seafloor.

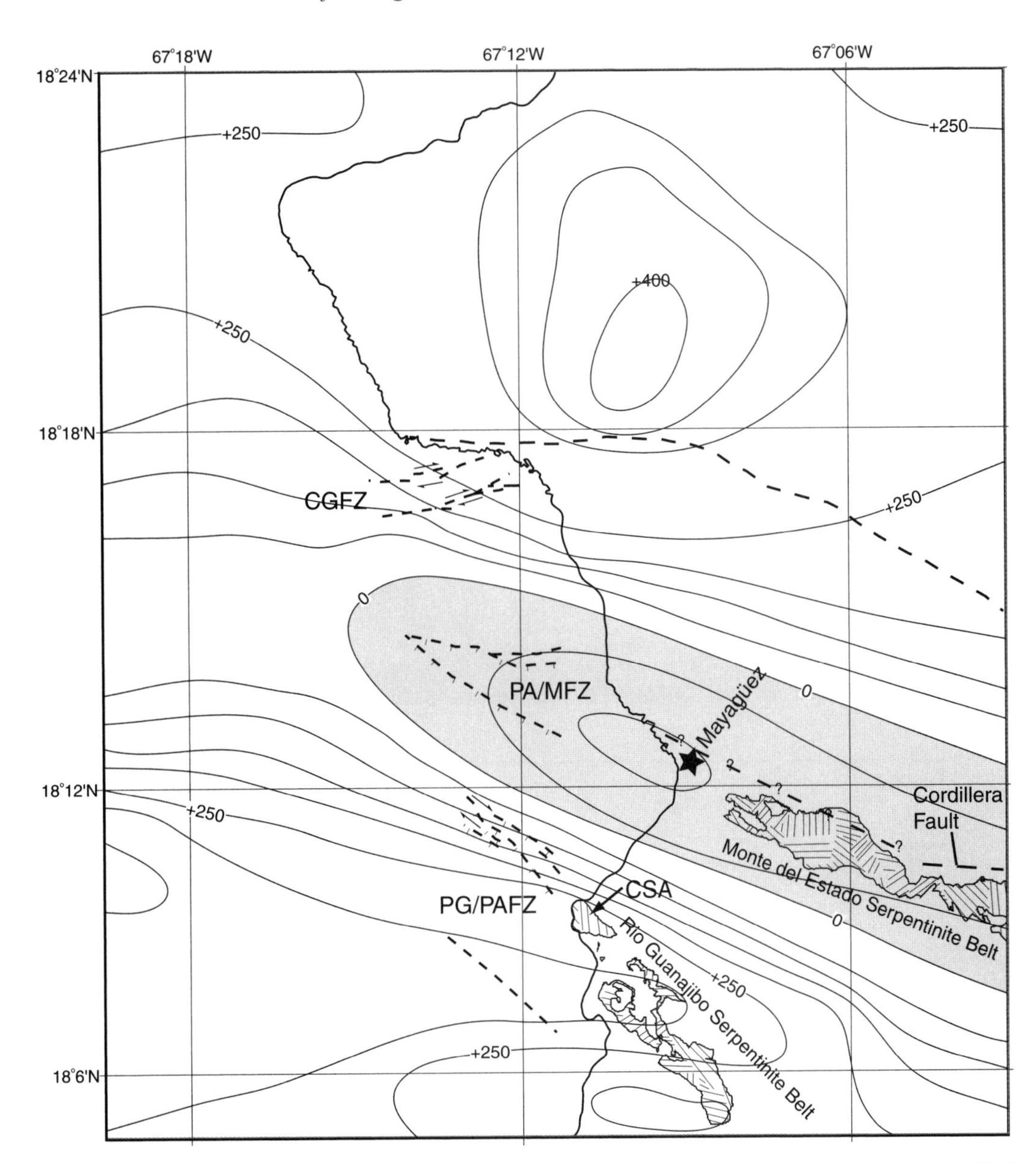

Figure 14. Residual magnetic intensity map for western Puerto Rico (after Bracey, 1968). Contour interval = 50 gammas. Filled pattern locates serpentinite and offshore faults identified by this study and faults with proposed Quaternary activity onland shown by dashed lines. Dashed line with question marks locates the projection of the cordillera fault along the serpentinite belt to the western coast. Note the correspondence of the Punta Algarrobo/Mayagüez fault zone (PA/MFZ), the Monte del Estado peridotite/serpentinite belt onland and shaded zone of low magnetic intensity. CSA—Cordillera de Sabana Alta; PG/PAFZ—Punta Guanajibo and Punta Arenas fault zones.

the seafloor. Alternatively, the South Lajas fault could be one of many short fault segments that, when taken together, form a regional fault system within southern Puerto Rico.

Implications for Neotectonics of Western Puerto Rico and the Mona Passage

On the basis of previous marine geophysical studies as well as the recent onland and marine studies, it is becoming increasingly apparent that the deformational history of Puerto Rico and its vicinity is very complex. One result of our offshore survey and the coordinated onland trenching and mapping studies (Prentice and Mann, this volume; Mann et al., this volume, Ch. 6) is to show the presence of several Quaternary faults intersecting the west coast, and that many of these faults appear to be reactivated, older, WNW-trending basement structures.

The youthful faults we mapped offshore appear to be largely confined to the northern and central areas of the western

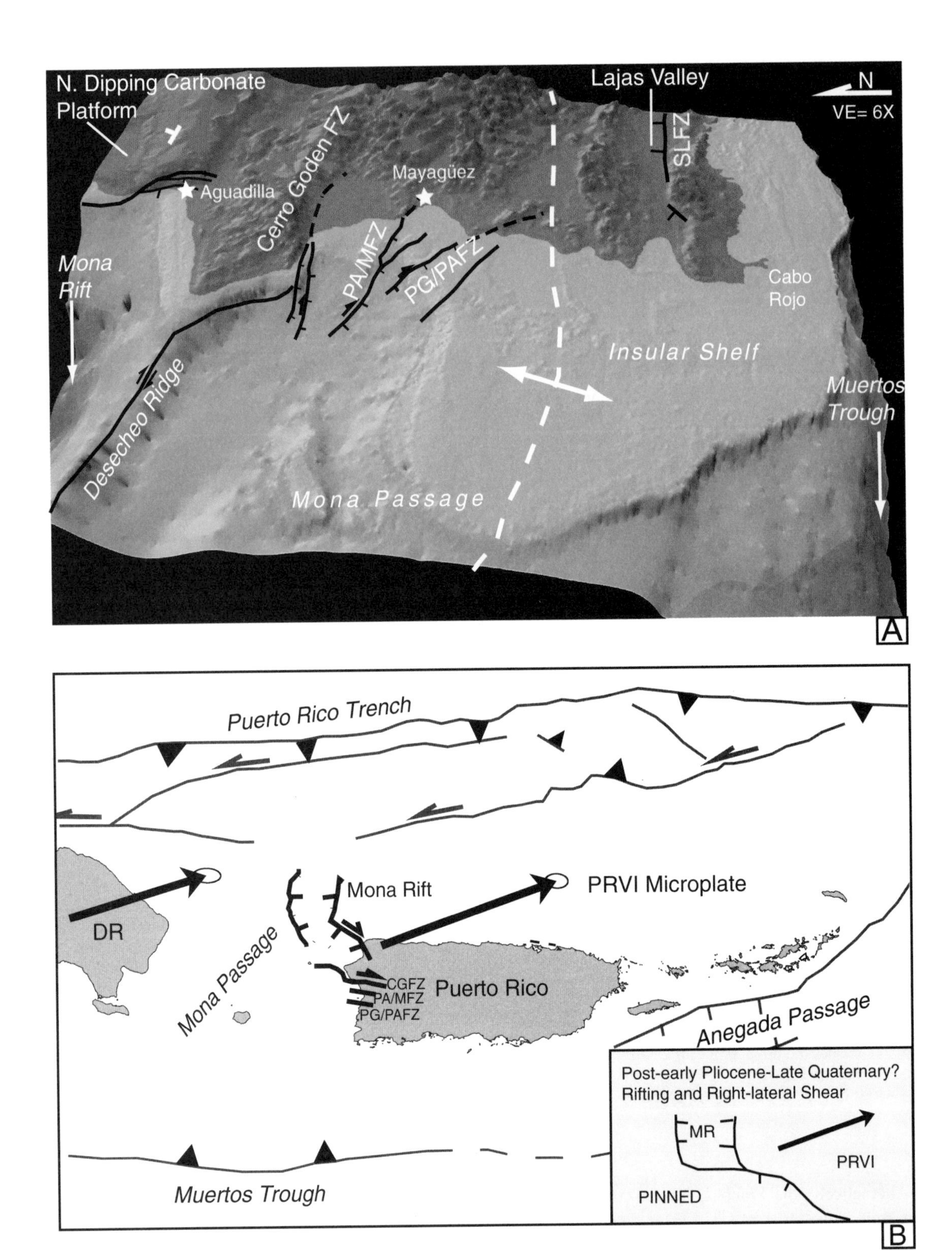

Figure 15. (A) Three-dimensional perspective of western Puerto Rico looking eastward. Illumination is from the northwest. Faults identified during this survey are marked as black lines and are tentatively traced to Quaternary faults onland. Faults onland in the vicinity of Aguadilla from Hippolyte et al. (this volume). Also shown is the axis of the E-W–trending arch. Note the change in insular shelf morphology south to north from broad, flat and undeformed to relatively narrow, deformed and dissected. PA/MFZ—Punta Algarrobo/Mayagüez fault zone; PG/PAFZ—Punta Guanajibo/Punta Arenas fault zone; SLFZ—South Lajas fault zone. (B) Conceptual model for the neotectonics of the Puerto Rico region. Much of the Quaternary faulting is primarily right-lateral transtension. This extension is the result of differential relative motions (shown by the GPS vectors) both in direction and velocity between Hispaniola, Puerto Rico–Virgin Islands (PRVI) microplate and the Caribbean plate.

Puerto Rico insular shelf. The rugged seafloor morphology in this region reflects this wide zone of deformation (Fig. 15). In marked contrast, the southwestern portion of the insular shelf, where we found little or no evidence of youthful faults, is wide, smooth, and flat (Fig. 15). The offshore sections of the Cerro Goden, Punta Algarrobo/Mayagüez, and Punta Guanajibo/Punta Arenas fault zones are all observed to have normal fault geometries, and in the case of the Cerro Goden and Punta Guanajibo fault zones, some evidence suggesting right-lateral strike-slip component is observed. The onshore sections of both of these faults also exhibit right-lateral motion of probable Quaternary age (Mann et al., this volume, Ch. 6; Moya, 1994). If the faults observed offshore are Holocene, then by their coincidence in location and style of motion with older structures (faults in the Oligocene-Pliocene carbonates) they likely represent reactivation or continued motion on those structures (Fig. 4). The overall stress regime suggested by motion on these faults therefore is extension aligned roughly NE-SW. (Fig. 15B). This is consistent with geodetic studies that suggest ENE extension is occurring in the Mona Passage and is largely accommodated by the opening of the Mona rift and associated faults (Jansma et al., 2000; Mann et al., 2002), (Fig. 15B).

CONCLUSIONS

In summary, both the offshore and onland studies have documented for the first time evidence suggesting Quaternary fault activity associated with a broad zone of deformation in the Mona Passage and western Puerto Rico. Three fault zones have been identified offshore western Puerto Rico: the Cerro Goden, the Punta Algarrobo/Mayagüez, and the Punta Guanajibo/Punta Arenas fault zones. These faults appear to coincide with large WNW-trending offsets in the underlying Cretaceous volcanics and Oligocene-Pliocene carbonate sequences, suggesting that many of these faults represent zones of weaknesses that are being reactivated in the most recent phase of NE-SW extensional deformation. Two of the fault zones, the Cerro Goden and Punta Algarrobo, show strong correlation with fault zones onland, Cerro Goden and Cordillera, respectively. Our studies, coupled with the onland studies of Prentice and Mann (this volume), Mann et al. (this volume, Ch. 6), and Hippolyte et al. (this volume) suggest that these faults pose a seismic hazard and must be considered in future seismic hazard models for the island of Puerto Rico, one of the most densely populated regions of the western hemisphere.

ACKNOWLEDGMENTS

We gratefully acknowledge Captain Hector and the crew of the R/V *Isla Magueyes* for their skill and enthusiasm in navigating the shallow reef-infested waters of western and southern Puerto Rico. John "super tech" Murray ensured the smooth operation of the geophysical gear during the cruise. We thank Osku Backstrom and Daniel Lao for watchstanding during the cruise. Christa von Hilldebrant, Victor Huérfano, and other members of the Puerto Rico seismic network and the University of Puerto Rico–Mayagüez Geology Department generously provided time and logistical support for the field work conducted in Puerto Rico. We thank Carol Prentice and Tom Pratt for their helpful reviews of this manuscript. This research was funded by U.S. Geological Survey Grant no. 00HQGR0010. University of Texas at Austin Institute for Geophysics (UTIG) contribution no. 1700.

REFERENCES CITED

Ascencio, E., 1980, Western Puerto Rico seismicity: U.S. Geological Survey Open-File Report 80-192, 135 p.

Bracey, D.R., 1968, Structural implications of magnetic anomalies north of the Bahama-Antilles islands: Geophysics, v. 33, p. 950–961, doi: 10.1190/1.1439989.

Byrne, D.B., Suarez, G., and McCann, W.R., 1985, Muertos Trough subduction—Microplate tectonics in the northern Caribbean: Nature, v. 317, p. 420–421.

Detrich, C., 1995, Characterization of faults located off the western coast of Puerto Rico using seismic reflection profiles [M.S. Thesis]: Lehigh University, 122 p.

Dillon, W.P., Edgar, T., Scanlon, K., and Coleman, D., 1996, A review of the tectonic problems of the strike-slip northern boundary of the Caribbean plate and examination by GLORIA, *in* Gardner, J.V., Field, M.E. and Twichell, D., eds., Geology of the United States' Seafloor; The View from GLORIA: Cambridge University Press, p. 135–164.

Emery, K.O., and Uchupi, E., 1984, The Geology of the Atlantic Ocean: New York, Springer, 1050 p.

Erikson, J., Pindell, J., and Larue, D.K., 1990, Tectonic evolution of the south-central Puerto Rico region: Evidence for transpressional tectonism: Journal of Geology, v. 98, p. 365–386.

Erikson, J., Pindell, J., and Larue, D.K., 1991, Fault zone deformational constraints on Paleogene tectonic evolution in southern Puerto Rico: Geophysical Research Letters, v. 18, p. 569–572.

Gardner, W.O., Glover, and L.K., Hollister, C.D., 1980, Canyons off northwest Puerto Rico: Studies of their origin and maintenance with the nuclear research submarine NR-1: Marine Geology, v. 37, p. 41–70.

Garrison, L.E., and Buell, M.W., 1971, Sea-floor structure of the Eastern Greater Antilles, *in* Symposium on Investigations and Resources of the Caribbean Sea and Adjacent Regions, p. 241–245, UNESCO, Paris, France.

Glover, L., III, 1971, Geology of the Coamo area, Puerto Rico, and its relation to the volcanic arc-trench association: U.S. Geological Survey Professional Paper 636, 102 p.

Grindlay, N.R., Mann, P., and Dolan, J., 1997, Researchers investigate submarine faults north of Puerto Rico: Eos (Transactions, American Geophysical Union), v. 78, p. 404.

Grove, K., 1983, Marine Geologic Map of the Puerto Rico insular shelf, northwestern area—Rio Grande de Añasco to Rio Camuy: U.S. Geological Survey Miscellaneous Investigations Series Map I-1418, scale 1:40,000.

Huérfano, V., 1994, Crustal structure and stress regime near Puerto Rico [M.S. thesis]: Mayagüez, University of Puerto Rico, 89 p.

Hubbard, D.K., Gill, I.P., Ruebenstone, J.L., and Fairbanks, R.G., 1996, Paleoclimate and paleoceanographic controls of shelf-edge reef development in Puerto Rico and the Virgin Islands: Geological Society of America Abstracts with Programs, v. 28, no. 7, p. A-489.

Jansma, P., Mattioli, G., Lopez, A., DeMets, C., Dixon, T., Mann, P., and Calais, E., 2000, Neotectonics of Puerto Rico and the Virgin Islands, northeastern Caribbean, from GPS geodesy: Tectonics, v. 19, p. 1021–1037, doi: 10.1029/1999TC001170.

Joyce, J., McCann, W.R., and Lithgow, C., 1987, Onland active faulting in the Puerto Rico platelet: Eos (Transactions, American Geophysical Union), v. 68, p. 1483.

Larue, D.K., and Ryan, H.F., 1990, Extensional tectonism in the Mona Passage, Puerto Rico and Hispaniola: A preliminary study: Transactions, 12th Caribbean Geological Conference, p. 223–230.

Lopez, A., Jansma, P., Calais, E., DeMets, C., Dixon, T., Mann, P., and Mattioli, G., 1999, Microplate tectonics along the North American–Caribbean plate boundary: GPS geodetic constraints on rigidity of the Puerto Rico-Northern

Virgin Islands (PRVI) block, convergence across the Muertos Trough, and extension in the Mona Canyon: Eos (Transactions, American Geophysical Union), v. 82, p. S77.
Mann, P., Taylor, F.W., Edwards, R.L., and Ku, T.L., 1995, Actively evolving microplate formation by oblique collision and sideways motion along strike-slip faults: An example from the northeastern Caribbean plate margin: Tectonophysics, v. 246, p. 1–69, doi: 10.1016/0040-1951(94)00268-E.
Mann, P., Calais, E., Ruegg, J.-C., DeMets, C., Jansma, P., and Mattioli, G., 2002, Oblique collision in the northeastern Caribbean from GPS measurements and geological observations: Tectonics, v. 21, no. 6, p. 1057, doi: 10.1029/2001TC001304.
Masson, D.G., and Scanlon, K.M., 1991, The neotectonic setting of Puerto Rico: Geological Society of America Bulletin, v. 103, p. 144–154, doi: 10.1130/0016-7606(1991)103<0144:TNSOPR>2.3.CO;2.
McCann, W.R., 1985, On the earthquake hazards of Puerto Rico and the Virgin Islands: Bulletin of the Seismological Society of America, v. 75, p. 251–262.
McCann, W.R., Joyce, J., and Lithgow, C., 1987, The Puerto Rico platelet at the northeastern edge of the Caribbean plate: Eos (Transactions, American Geophysical Union), v. 68, p. 1483.
McIntyre, D., 1971, Geologic map of the central La Plata quadrangle, Puerto Rico: U.S. Geological Survey Miscellaneous Geologic Investigations Series Map I-660, scale 1:20,000.
Meltzer, A., and Almy, C., 2000, Fault structure and earthquake potential Lajas Valley, SW Puerto Rico: Eos (Transactions, American Geophysical Union), v. 81, p. F1181.
Molnar, P., and Sykes, L., 1969, Tectonics of the Caribbean and Middle America regions from focal mechanisms and seismicity: Geological Society of America Bulletin, v. 80, p. 1639–1670.
Morelock, J., Winget, E., and Goenaga, C., 1994, Geologic maps of the Puerto Rico insular Parguera to Guánica: U.S. Geological Survey Miscellaneous Investigations Series Map I-2387, scale 1:40,000.
Moya, J.C., 1994, Geological, geomorphical and geophysical evidence for paleoseismic events in western Puerto Rico: Proceedings of the Workshop on Paleoseismology, U.S. Geological Survey Open File Report 94-568, p. 127.
Moussa, M.T., Seiglie, G.A., Meyerhoff, A.A., and Taner, I., 1987, The Quebradillas Limestone (Miocene-Pliocene), northern Puerto Rico, and tectonics of the northeastern Caribbean margin: Geological Society of America Bulletin, v. 99, p. 427–439.
Puerto Rico Seismic Network, 2004, Annual Seismic Reports http://redsismica.uprm.edu/english/seismicity/reports (accessed October 2004).
Reid, H., and Taber, S., 1919, The Puerto Rico Earthquakes of October-November 1918: Bulletin of the Seismological Society of America, v. 9, p. 95–127.
Schlee, J.S., Rodriguez, R.W., Webb, R.M., and Carlo, M.A., 1999, Marine geologic map of the southwestern insular shelf of Puerto Rico—Mayagüez to Cabo Rojo. U.S. Geological Survey Geologic Investigations Series Map I-2615, scale 1:40,000.
van Gestel, J.-P., Mann, P., Dolan, J., and Grindlay, N.R., 1998, Structure and tectonics of the upper Cenozoic Puerto Rico–Virgin Islands carbonate platform as determined from seismic reflection studies: Journal of Geophysical Research, v. 103, p. 30,505–30,530, doi: 10.1029/98JB02341.
van Gestel, J.-P., Mann, P., Grindlay, N.R., and Dolan, J.F., 1999, Three-phase tectonic evolution of the northern margin of Puerto Rico as inferred from an integration of seismic reflection, well, and outcrop: Marine Geology, v. 161, p. 257–288, doi: 10.1016/S0025-3227(99)00035-3.
Western Geophysical Company of America and Fugro, Inc., 1973, Geological-geophysical reconnaissance of Puerto Rico for citing of nuclear power plants. The Puerto Rico Water Resources Authority, San Juan, P.R., 127 p.

Manuscript Accepted by the Society 18 August 2004

Printed in the USA

Geological Society of America
Special Paper 385
2005

Geologic evidence for the prolongation of active normal faults of the Mona rift into northwestern Puerto Rico

Jean-Claude Hippolyte*
UMR CNRS 5025 Laboratoire de Géodynamique des Chaînes Alpines, Université de Savoie—Campus Scientifique Technolac, 73376 Le Bourget du Lac Cedex, France

Paul Mann
Institute for Geophysics, Jackson School of Geosciences, University of Texas at Austin, 4412 Spicewood Springs Road, Bldg. 600, Austin, Texas 78759 USA

Nancy R. Grindlay
Center for Marine Science, University of North Carolina, Wilmington, North Carolina 28409, USA

ABSTRACT

Topography, bathymetry, regional structural observations, and fault slip measurements support the idea that the Mona rift is an active, offshore extensional structure separating a colliding area (eastern Hispaniola) from a subducting area (northwestern Puerto Rico). Near the city of Aguadilla in northwestern Puerto Rico, paleostress reconstruction through fault slip analysis demonstrates that the Mona rift is opening in an E-W direction. This fault slip analysis also indicates that the opening is oblique in the southern part of the Mona rift. We propose that oblique rifting results from accommodation of E-W extension by oblique right-lateral reactivation of previously mapped, NW-trending Eocene basement convergent structures (Aguadilla faults, Cerro Goden fault). The evolution of the stress field during the Miocene and the present E-W opening of the Mona rift support the assumption that the Miocene 25° counterclockwise rotation of Puerto Rico has stopped and that this island is presently moving to the east relative to the colliding Hispaniola.

Keywords: active faults, paleostress, Caribbean, Puerto Rico, Mona rift.

INTRODUCTION

Plate Boundary Zone

The island of Puerto Rico is located within a ~250-km-wide deformation zone formed at the northern boundary between the Caribbean and North America plates (Fig. 1). Eastward motion of the Caribbean plate at a rate of ~2 cm/yr relative to the North America plate determined by recent GPS measurements (DeMets et al., 2000) indicates that this active deformation within this plate boundary zone is largely controlled by left-lateral strike-slip faulting. Because Puerto Rico and eastern Hispaniola are underthrust in the north by North American lithosphere and in the south by Caribbean lithosphere, it is likely that Puerto Rico behaves independently from the two larger plates (Fig. 1).

*Jean-Claude.Hippolyte@univ-savoie.fr

Hippolyte, J.-C., Mann, P., and Grindlay, N.R., 2005, Geologic evidence for the prolongation of active normal faults of the Mona rift into northwestern Puerto Rico, *in* Mann, P., ed., Active tectonics and seismic hazards of Puerto Rico, the Virgin Islands, and offshore areas: Geological Society of America Special Paper 385, p. 161–171. For permission to copy, contact editing@geosociety.org.

Microplate Tectonics

On the basis of seismotectonic and structural data, three microplates have been identified and named within this diffuse boundary zone. From west to east, the microplates include Gonave, Hispaniola, and Puerto Rico–Virgin Islands (e.g., Byrne et al., 1985; Masson and Scanlon, 1991; Mann et al., 1995). Of particular interest to this study is the nature of the active tectonic transition between Hispaniola, where collisional and strike-slip deformation is occurring (e.g., Mann et al., 1995), and Puerto Rico–Virgin Islands, where subduction occurs (McCann, 2002).

One regional model for deformation is that collision of Hispaniola with the Bahamas bank pins or impedes the eastward motion of the Hispaniola segment of the plate boundary zone. Tectonic pinning or collision results in localized extension in the marine straits (Mona Passage) between Hispaniola and Puerto Rico (Vogt et al., 1976). This idea is consistent with GPS measurements that indicate 5 ± 4 mm/yr of opening between the two microplates (Jansma et al., 2000). These authors suggest that the approximately E-W–induced extension is accommodated by the Mona rift (Fig. 1).

Mechanisms for Opening the Mona rift

The kinematic of opening for the Mona rift is not well constrained. Two earthquake focal plane mechanisms near the Mona rift indicate SE-trending and ESE-trending T-axes or opening directions (Fig. 1). These trends contrast with the shape of the rift that corresponds to a deep canyon that trends N-S in the north and SE in the south (Fig. 2). Moreover, the main neotectonic event shown by the geology of Puerto Rico is the arching of a Miocene carbonate platform suggestive of NS shortening (van Gestel et al., 1998).

As seen on the bathymetric map in Figure 2, the Mona rift changes trend abruptly from NS to NW-SE. Northwestern Puerto Rico is the onshore area closest to the Mona rift and can yield information that is not possible to obtain offshore. Here, we describe recent onshore faulting in this region that could then be used in turn to predict the locations, trends, opening directions, and earthquake hazards of offshore faults. We present field-based fault observations of the fault kinematics of recent, onland faults belonging to the Mona rift fault system; the results of the first fault slip measurements from Neogene carbonate rocks in Puerto Rico; and an analysis of the fault slip measurements in terms of stresses using the Direct Inversion Method of Angelier (1990). This method allows us to calculate an average reduced stress tensor [orientation of the stress axes $\sigma1$, $\sigma2$, and $\sigma3$ and the ratio $\Phi = (\sigma2 - \sigma3)/(\sigma1 - \sigma3)$] with a homogenous fault slip data set. Within a heterogeneous fault slip data set, and taking into account field observations, it is possible to separate different data subsets corresponding to successive fault generations. It is therefore possible to reconstruct the tectonic and paleostress evolution of this area from Eocene to recent times.

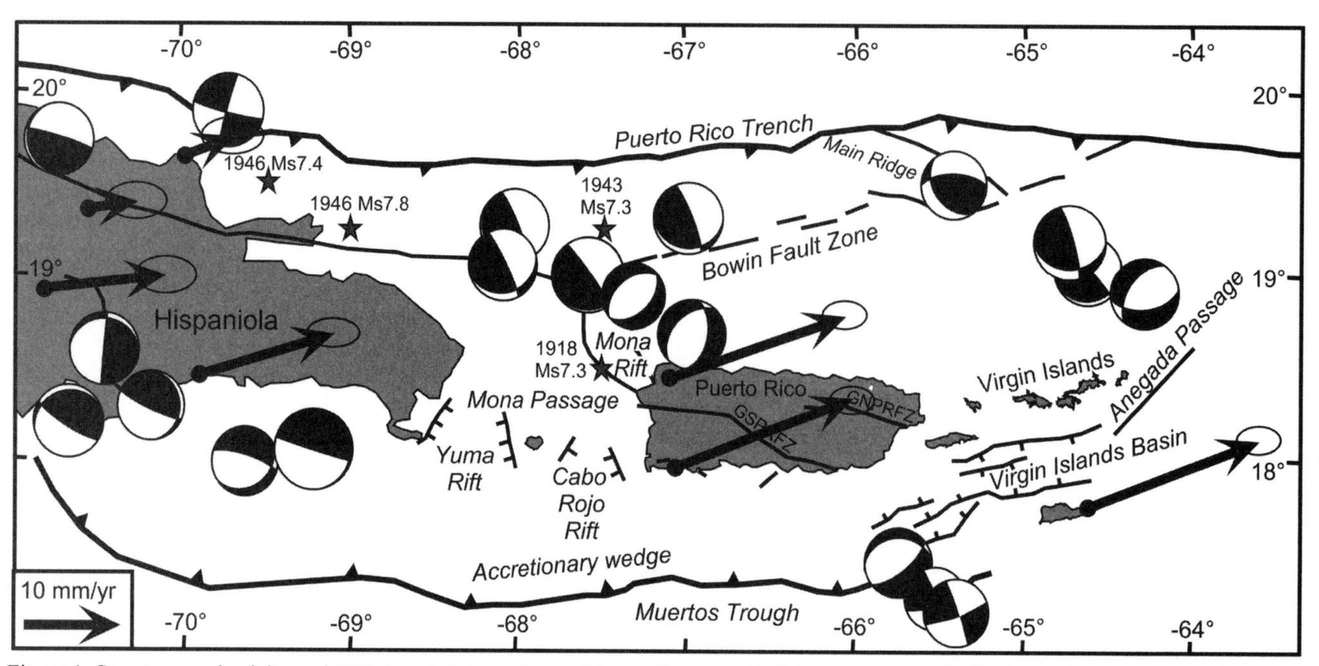

Figure 1. Structures, seismicity and GPS-based plate motions of the northeastern Caribbean plate margin. Faults are from Masson and Scanlon (1991) and van Gestel, et al. (1998) Focal plane mechanisms for depth < 35 km are from the Harvard Centroid Moment Tensor catalogue and from Huérfano (1995) (south of the Muertos trough). Historical seismicity is from Pacheco and Sykes (1992). GPS velocities relative to North America are from Jansma et al. (2000) and Mann et al. (2002).

IMPORTANCE OF THE MONA PASSAGE FOR UNDERSTANDING THE GEODYNAMICS OF NORTHWESTERN CARIBBEAN

Structural Setting of Puerto Rico

The Puerto Rico–Virgin Islands microplate is limited to the north and south by subduction zones, and to the east and the west by zones of probable extension or transtension (Fig. 1). To the north, the microplate is underthrusted by North America lithosphere along the EW-striking Puerto Rico trench that reaches a water depth of ~8 km (Fig. 1). To the south, the Puerto Rico–Virgin Islands microplate is bounded by the Muertos trough, an EW-striking, 5-km-deep bathymetric feature, where northward subduction of Caribbean lithosphere is occurring (Ladd et al., 1977; Byrne et al., 1985). At least 40 km of underthrusting of the Caribbean plate occurs south of eastern Hispaniola (Ladd et al., 1977) . This value decreases toward the east, and underthrusting is ultimately replaced by extension in the Virgin Island basin complex (Masson and Scanlon, 1991). This basin, along with the Anegada Passage, probably form a transtensional zone that may define the eastern boundary of the Puerto Rico–Virgin Islands microplate (Jansma et al., 2000).

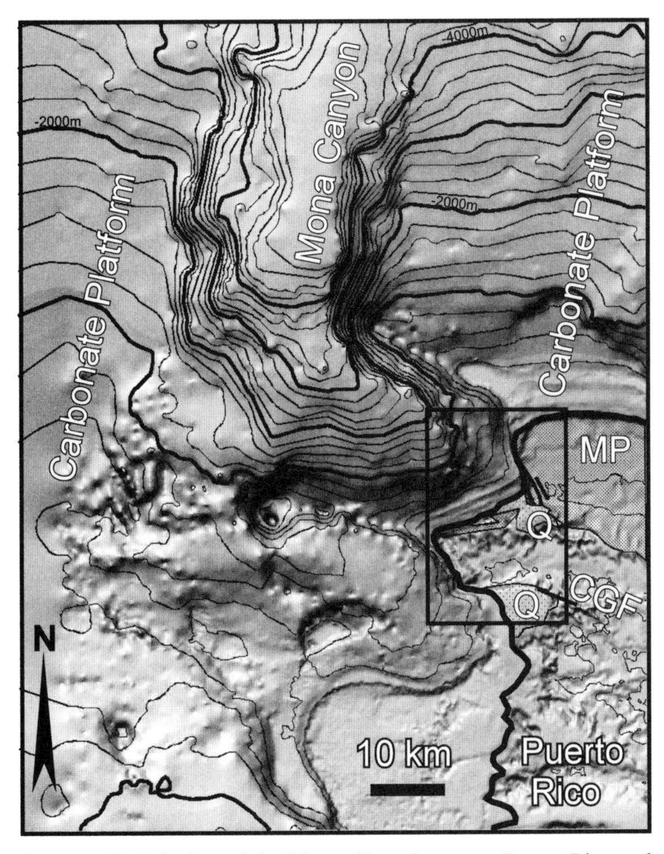

Figure 2. Shaded view of the Mona rift and western Puerto Rico and location of the study area (Fig. 3B). Illumination is from SSE. Contour inferred is 200 m. MP—exposed Miocene–early Pliocene platform carbonates; Q—Quaternary deposits; CGF—Cerro Goden active fault (Mann et al., this volume, Chapter 6). The Mona rift trends N-S in the north, then turns to the southeast toward the northwestern corner of Puerto Rico. Note NNW-trending faults cutting the Neogene carbonates in northwest Puerto Rico that are close to the offshore steep eastern wall of the rift.

The western boundary of the Puerto Rico–Virgin Islands microplate in the Mona Passage is mainly characterized by NW-trending normal faults (e.g., Case and Holcombe, 1980; Masson and Scanlon, 1991; Grindlay et al., 1997) and by three (Yuma, Cabo Rojo and Mona; Fig. 1) roughly N-S–trending deep canyons interpreted as late Neogene rift structures (e.g., Masson and Scanlon, 1991; Grindlay et al., 1997; van Gestel et al., 1998). Of these rifts, by far the most morphologically prominent is the Mona rift (Fig. 2). It is likely that the box-shaped profile and the steep walls of the Mona rift are controlled by N-S and NW-SE–oriented faults. However, only a few other N-S–trending faults have been recognized in the Mona Passage (Manson and Scanlon, 1991; Grindlay et al., 1997; van Gestel et al., 1998)(Fig. 1).

There are no drill cores to constrain the age of these onland faults despite their apparent activity in late Neogene times. Because the offshore faults can be seen cutting seismic lines through the entire Neogene carbonate section, the onland faults are also inferred to be Pliocene-Quaternary in age (e.g., van Gestel et al., 1998).

Present Deformation

The historical seismicity in the Mona Passage (McCann, 2002; Doser et al., this volume) and the steep walls of the Mona rift (Fig. 2) support the interpretation that this structure remains active today. In 1918, a significant (Ms ~ 7.3) earthquake occurred offshore of northwest Puerto Rico (Pacheco and Sykes, 1992) and was followed by a 4–6-m-high tsunami in western Puerto Rico (Fig. 1). The epicenter of this earthquake is thought to have been located northwest of Puerto Rico and the generation of a tsunami is consistent with the hypothesis that normal faults are active in the Mona rift (Fig. 1). Accordingly, it was a rupture along four segments of a N-S–trending normal fault in the Mona rift that was used in numerical simulation of the tsunami run-up in western Puerto Rico (Mercado and McCann, 1998).

On the basis of GPS measurements and two-dimensional elastic modeling of the elastic strain accumulation in Hispaniola, Jansma et al. (2000) inferred an eastward motion of Puerto Rico relative to central Hispaniola of 5 ± 4 mm/yr. Assuming all motion is accommodated across the Mona rift (whose total tectonic extension is ~6 km; van Gestel et al., 1998), and assuming that the rate of opening is constant, these authors estimate that minimum and maximum ages of the Mona rift are <1 m.y. and 6 m.y., respectively.

Present-day east-west extension in the Mona Passage apparently disagrees with geodynamic models that predict a counterclockwise rotation of the Puerto Rico–Virgin Islands microplate within a broad left-lateral shear zone (Fink and Harrison, 1972; Masson and Scanlon, 1991; Huérfano, 1995). Such rotation explains extensional structures of the Anegada Passage, explains the variation of shortening in the Muertos trough, and predicts extension to the northwest in the Puerto Rico trench (Schell and Tarr, 1978; Masson and Scanlon, 1991). Paleomagnetic data from Cretaceous and Eocene rocks exposed in Puerto Rico effectively record >45° of counterclockwise rotation since the Eocene and 24° since the Miocene (Fink and Harrison, 1972; Van Fossen et al., 1989; Reid et al., 1991).

It is possible that this rotation has ceased. Because the paleomagnetic declination of Pliocene carbonates from northern Puerto Rico was not statistically different from those expected (1.7 ± 9.9°), the rotations observed in Miocene rocks were assumed by Reid et al. (1991) to have occurred between ~11 and 4.5 Ma. Despite the low number of sites measured in the eastern part of the Puerto Rico–Virgin Islands block, GPS data seem to support the conclusion that the counterclockwise rotation has ceased (Jansma et al., 2000). Offshore seismic profiles were interpreted to document a change from extension to shortening across the Puerto Rico trench (Larue and Ryan, 1998).

Another late Neogene geodynamic change has been postulated for the Anegada passage with extensional movements changing to dextral transtensional in the Pliocene (Jany et al., 1987, 1990; Mauffret and Jany, 1990) (Fig. 1). Our analysis of the fracturing of northwestern Puerto Rico (in the prolongation of the Mona rift) will allow us to evalutate these models and to determine at what intervals major geodynamic changes occurred from the Eocene to Recent.

EOCENE TO RECENT EVOLUTION OF ROCK FRACTURING IN NORTHWEST PUERTO RICO

Puerto Rico consists of strongly folded and faulted early Cretaceous-Eocene volcanic arc rocks unconformably overlain along the north and south coasts by weakly deformed Oligocene-Pliocene siliciclastic rocks and platform carbonates (Fig. 3).

Structure of Cretaceous-Eocene Arc Basement Rocks

The Cretaceous to Eocene volcanic, plutonic, and metasedimentary volcanic arc units of Puerto Rico were deformed during late Eocene to earliest Oligocene (Dolan et al., 1991). Faults trend mainly E-W to NW-SE (Monroe, 1969; Bawiec, 2001) (Fig. 3).

Two main fault zones are usually distinguished: the ~8-km-wide Great Southern Puerto Rico Fault Zone and the diffuse Great Northern Puerto Rico Fault Zone in northeastern Puerto Rico (Fig. 3A). Both faults may have undergone sinistral strike-slip motions in Eocene times in agreement with the general left-lateral oblique collision zone of the Greater Antilles (Pindell and Barrett, 1990; Gordon et al., 1997).

Great Southern Puerto Rico Fault Zone

In the southeastern extremity of the Great Southern Puerto Rico Fault Zone, Erikson et al. (1990) found fault striations in mainly Eocene rocks compatible with E-W compression and in agreement with the left-lateral deformational model (Fig. 3A). Erikson et al. (1990) also found reverse faults yielding NNE-trending compression (Fig. 3A), which is in agreement with the presence of northeast-directed thrust structures. These thrust structures are compatible with the NNW-SSE to NNE-SSW trends of compression characterized by Larue et al. (1998) in north-central Puerto Rico (Fig. 3A). Along the Cerro Goden fault zone (Fig. 3B), Mann et al. (this volume, Chapter 6) found evidence for recent right-lateral displacement of geomorphic features.

In Eocene rocks of northwestern Puerto Rico near the Cerro Goden fault trend, we measured fault planes with striations and clear sense indicators including stratigraphic offsets, calcite steps, and Riedel fractures. Steeply dipping bedding planes strike E-W to NW-SE (cf. the dashed lines in on stereo diagrams in Fig. 3) in agreement with many Eocene and older fold axes trends measured in Puerto Rico (e.g., Erikson et al., 1990).

Paleostress computation using all strike-slip and reverse faults measured in the sites 1–6 from Eocene basement rocks near the Cerro Goden fault yield NNE-SSW compressional trends (Fig. 3; Table 1). From the symmetry of fault planes with bedding planes and the plunge of the paleostress axes, we can infer that faulting occurred before, during, and after the folding (e.g., Hippolyte and Sandulescu, 1996). At sites with vertical bedding planes in Eocene rocks (sites 5, 4, 3, and 2; Fig. 3), we could not distinguish whether some observed normal faults are strike-slip faults that have been tilted 90° or recent normal faults reflecting E-W extension. However, at site 6, compressional faults have been tilted to the north and some reverse faults are now horizontal or north-dipping normal faults. We conclude, from this observation and from the systematic perpendicularity of the compressional trend with the strike of the bedding planes, that NNE-trending compression is the main event responsible for the ESE-trending folding in western Puerto Rico (Fig. 3A).

DEFORMATION OF THE NEOGENE CARBONATE PLATFORM OF PUERTO RICO

In northern Puerto Rico, the pre-Oligocene, faulted basement is unconformably overlain by a tabular 4° to 8° northward-dipping sequence of mainly shallow-water limestone. This tilted carbonate platform extends offshore to 4000 m below sea level at a latitude of ~19°N (Moussa et al., 1987; van Gestel et al., 1998)(Fig. 2). Together with the carbonates cropping out on the southern side of the island, this tilted platform is interpreted as part of an ~120-km-wide arch formed from post-Eocene to late Neogene times and resulting from north-south shortening (Dillon et al., 1994; van Gestel et al., 1998).

Offshore of northern Puerto Rico, this platform is remarkably free of any faults or folds except a few southwest-dipping recent normal faults near the Mona rift (van Gestel et al., 1998). Onshore, Monroe (1980) mapped two minor faults in north-central Puerto Rico and four SE-striking faults in the northwestern corner of the island (Fig. 3B). In the northwestern corner of the

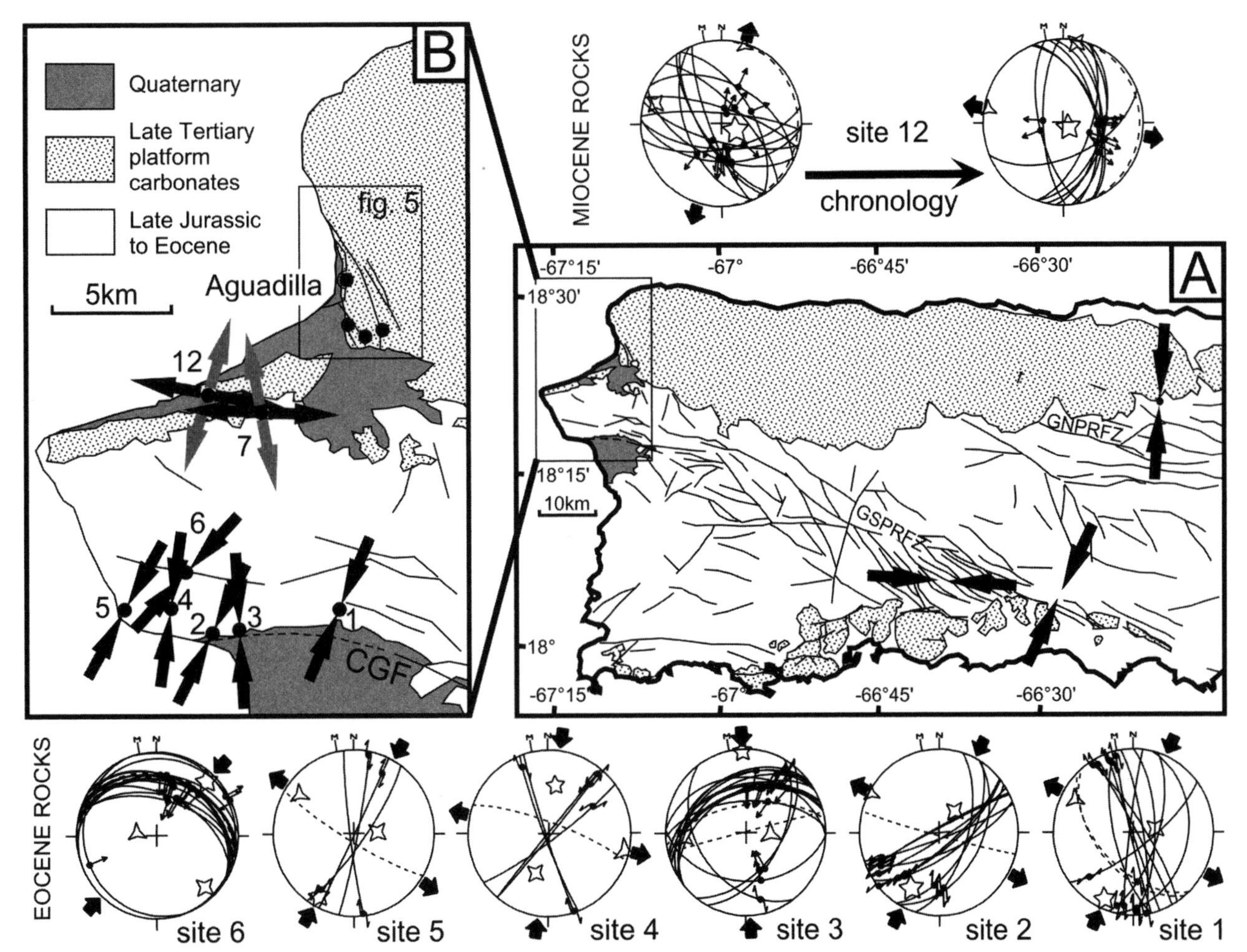

Figure 3. Location of the studied area and Schmidt's lower hemisphere projections of faults measured in the Miocene platform carbonates and in the folded Eocene basement. (A) Structural sketch of western Puerto Rico (modified from Bawiec, 2001) with directions of compression determined by Erikson et al. (1990) and Larue et al. (1998) from faults slip measured in Cretaceous–Lower Tertiary formations. GSPRFZ—Great Southern Puerto Rico Fault Zone; GNPRFZ—Great Northern Puerto Rico Fault Zone. (B) Structural sketch of the studied area (modified from Monroe, 1980) with directions of compression (σ_1) and extension (σ_3) derived from this study. CGF—Cerro-Goden fault. For each site the projections show the measured striated fault planes and the computed stress axes. Five-branch star—σ_1 (maximum principal stress axis); four-branch star—σ_2 (intermediate principal stress axis); three-branch star—σ_3 (minimum principal stress axis); bedding planes as broken lines.

TABLE 1 PALEOSTRESS TENSORS COMPUTED FROM FAULT-SLIP DATA*

Site no.	Formation	Age of rocks	Stress regime	No. of faults	σ_1		σ_2		σ_3		Φ	ANG	RUP
					Trend	Plunge	Trend	Plunge	Trend	Plunge			
1	Culebrinas	Eocene	S	12	204	13	81	67	299	18	0.47	11	33
2	Culebrinas	Eocene	S	12	207	26	37	64	299	4	0.23	13	40
3	Culebrinas	Eocene	C	16	357	3	266	25	94	65	0.68	12	28
4	Culebrinas	Eocene	S	5	8	42	199	48	103	6	0.71	8	25
5	Culebrinas	Eocene	S	4	210	14	87	66	305	20	0.43	24	55
6	Culebrinas	Eocene	C	16	43	12	137	15	277	70	0.32	14	43
7	Cibao	Early Miocene	E	16	134	77	258	7	349	11	0.23	9	31
7	Cibao	Early Miocene	E	14	232	78	180	7	92	9	0.28	8	29
8	Cibao	Early Miocene	E	27	60	79	172	4	263	10	0.45	9	32
9	Cibao	Early Miocene	E	7	82	82	179	1	269	8	0.45	8	23
10	Aguada	Early-middle Miocene	E	15	303	84	178	4	87	5	0.22	15	37
11	Cibao	Early Miocene	E	12	185	79	19	10	288	2	0.47	4	19
11	Cibao	Early Miocene	S	8	209	2	322	85	119	5	0.43	11	37
12	Cibao	Early Miocene	E	11	130	83	10	4	280	6	0.32	9	25
12	Cibao	Early Miocene	E	14	115	74	286	15	17	2	0.26	11	38

*Sites of Figures 3, 4, 5, and 6. Stress regimes: C—compressional; S—strike-slip; E—extensional; σ_1, σ_2, σ_3—maximum, intermediate, and minimum principal stress axis, respectively; trend—trend, north to east; plunge—plunge (°) of the stress axes; $\Phi = (\sigma_2 - \sigma_3)/(\sigma_1 - \sigma_3)$; ANG—average angle between computed shear stress and observed slickenside lineation (°). RUP—ratio upsilon: quality estimator ($0 \leq$ RUP ≤ 200) taking into account the relative magnitude of the shear stress on fault planes (cf. Angelier, 1990).

island, we measured 125 faults with striations and a clear sense of motion that were present in Miocene carbonates. Sites 7 and 12 (Fig. 3B) in the Cibao Formation revealed only normal faults (Figs. 3 and 4). On the basis of field observations and on paleostress analyses, we distinguish two fault generations. First, we recognize a N-S extension ($\sigma 3$ trending NNW to NNE). At site 12 (Fig. 3), this extension corresponds to prelithification faulting; at site 7, this extension corresponds to syndepositional and sealed faults with metric-scale offsets (Fig. 4). Second, we recognize an extension with $\sigma 3$ trending E-W to ESE (Figs. 3 and 4) that corresponds to mainly N-S–trending faults cross-cutting and offsetting the older syndepositional faults (Fig. 4).

This analysis shows that the stress field has evolved from Eocene to Recent times. We conclude that the NNE-trending compression (Fig. 3), which could have been responsible for the arching of the platform (van Gestel et al., 1998), characterizes only the Eocene-Oligocene collisional event because (1) the large-scale folds and thrusts affecting pre-Oligocene arc basement do not cut the overlying carbonate platform; and (2) we did not find evidence for compressional faulting in the Neogene cover. In contrast, we find that from early Miocene to recent times, the stress field is extensional.

RECENT FAULTING OF AGUADILLA AREA IN NORTHWESTERN PUERTO RICO

The regularity of the north-dipping monocline of platform carbonates is interrupted at the northwestern corner of Puerto Rico near the probable southwestern prolongation of the Mona Canyon (Fig. 2). Near Aguadilla, four SE-trending faults displace Neogene platform carbonates down to the east from 0 to 25 m (Monroe, 1969) (Figs. 3B and 5). However, the main structure deforming the 4°-north-dipping monocline is a down to the west (coast) monoclinal flexure in bedding of more than 125 m (Figs. 5 and 6).

At sites 8, 9, and 10 in Figure 5, we have measured west-dipping bedding planes (cf. dashed lines in the diagrams of Figures 5 and 6). Their unusual 10–20° westward dips result from a combination of the following: (1) the general 4° northward dip; (2) the effect of the regional flexure to the WSW (3–4°); and (3) local tilts resulting from slip on SSE-trending faults (Fig. 5). The regional flexure in bedding trends approximately NNW-SSE and is parallel to the faults mapped by Monroe (1969) (Figs. 5 and 6). To the west of one of these faults, we observe on a digital elevation model (DEM) north-trending geomorphic lineaments clearly visible despite the karst topography (Fig. 6). They form a horsetail splay suggesting a right lateral component on the main fault trending NW-SE (Figs. 5 and 6).

The fault zone in a quarry at site 8 is represented by decametric fault planes that show only normal slip and indicate E-W extension (Fig. 5, upper left). At site 11, we observed another large fault as mapped by Monroe (1969) (Fig. 5). There, multiple normal faults over a few meters correspond to ESE-trending extension (Fig. 5, upper right) and decimetric secondary faults correspond to a strike-slip state of stress that also has an ESE-trending $\sigma 3$ axis (Fig. 5, lower right). These two stress states are related by a permutation, or switch, in the $\sigma 1$ and $\sigma 2$ axes (Angelier and Bergerat, 1983). Such stress permutations commonly occur during a same tectonic event (Hippolyte et al., 1992). At site 11, the ESE-WNW trend of $\sigma 3$ implies a dextral component on the normal movement of the main fault (Fig. 5). Considering its NW-SE trend, the main fault is probably a pre-Oligocene basement fracture. We interpret this stress permutation as resulting from the propagation of this fault through the upper-lying carbonate and the accommodation of this oblique movement by pure normal and pure strike-slip faults.

Site 10 is in the Aguada limestone (Fig. 6). In contrast with the Cibao Formation, which contains clay and marl levels, the upperlying Aguada limestone is strongly karstified. With such karst erosion, the striated surfaces should have been partly or totally dissolved. However, at karst outcrops near Aguadilla, we observed unweathered planar faults with fresh striations and styloliths (Fig. 6). The fault planes cross-cut the karst-eroded limestone, which explains the large dispersion in the fault attitudes in the diagram of Figure 6. Despite the irregular shape of the fault surfaces, all faults are compatible with an E-W extension similar to those reconstructed in the other sites of Aguadilla area (Fig. 5). This faulting, which clearly postdates the karst, is therefore very recent because the arching and uplift of this area of the island postdates the early Pliocene Camuy limestones (Moussa et al., 1987).

Similar to other sites of the Aguadilla flexure, fracturing of site 9 reveals only extension in an E-W orientation (Fig. 5). Considering that the main faults of Aguadilla area are oblique to this trend of extension (Fig. 5), but parallel to pre-Oligocene arc basement faults (Fig. 3A), we interpret these faults as typical NW-trending basement faults that have been reactivated with a normal-dextral motion. Reactivation generated local stress permutations in order to accommodate oblique slip in the previously unfractured carbonate cover. The existence of the flexure lowering the carbonate platform to the WSW also supports this interpretation of the influence of the prefractured arc basement. This flexure is interpreted as being due to a west-dipping basement fault inherited from the Eocene deformation (Fig. 3A), and reactivated as a normal fault with some superficial, conjugate, and east-dipping faults (Fig. 6).

The horse-tail splay of the main fault (Fig. 5) and its curving to the west indicate that it is the termination of a west-dipping normal dextral fault that continues offshore toward the Mona rift (Fig. 2). We conclude that field data confirm that faults of the Mona rift extends onshore onto Puerto Rico near Aguadilla. The recent age of the fault termination, which is supported both by its morphology and the presence of postkarst striations, is in agreement with an ongoing opening of the Mona rift. Our fault slip measurements indicate that its opening direction is E-W and that the SE trend of the southern part of this rift, offshore and onshore, results from the reactivation of SE-trending basement faults of Eocene age (Fig. 2).

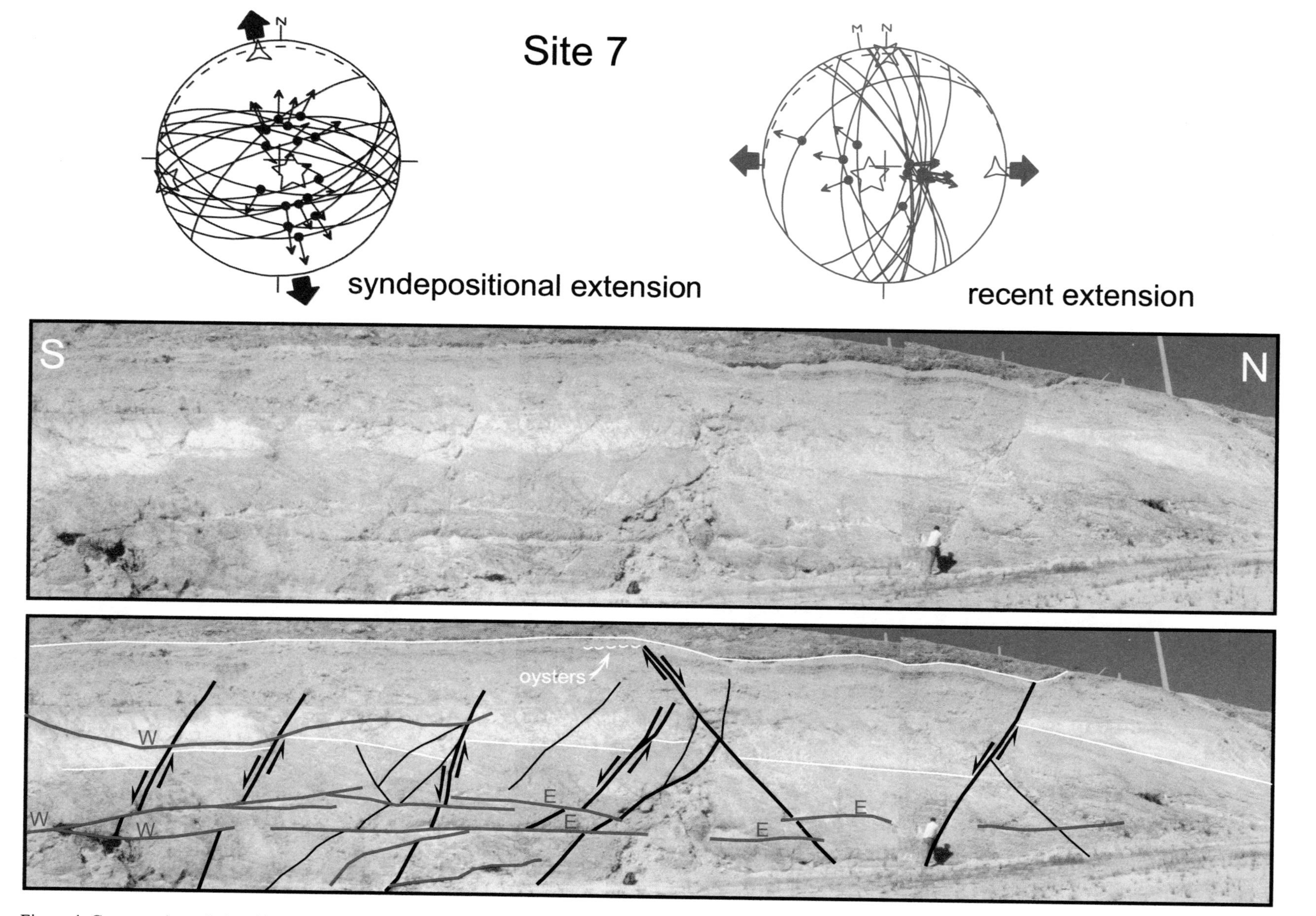

Figure 4. Cross-cutting relationships of normal faults at site 7. White lines show the base of stratigraphic layers. E-W normal faults (in black) corresponding to a NNW-SSE extensional event, are syndepositional. Their activity during the deposition of the Cibao early Miocene carbonates is attested by variations in thickness of the carbonate layers and by the upper layer that seals the faulted blocks. The presence of growth positive oysters at an uplifted fault-block crest confirms the syndepositional faulting interpretation. N-S normal faults (in gray), corresponding to a more recent E-W extensional event, offset the syndepositional faults. E and W indicate the dip directions of the fault planes.

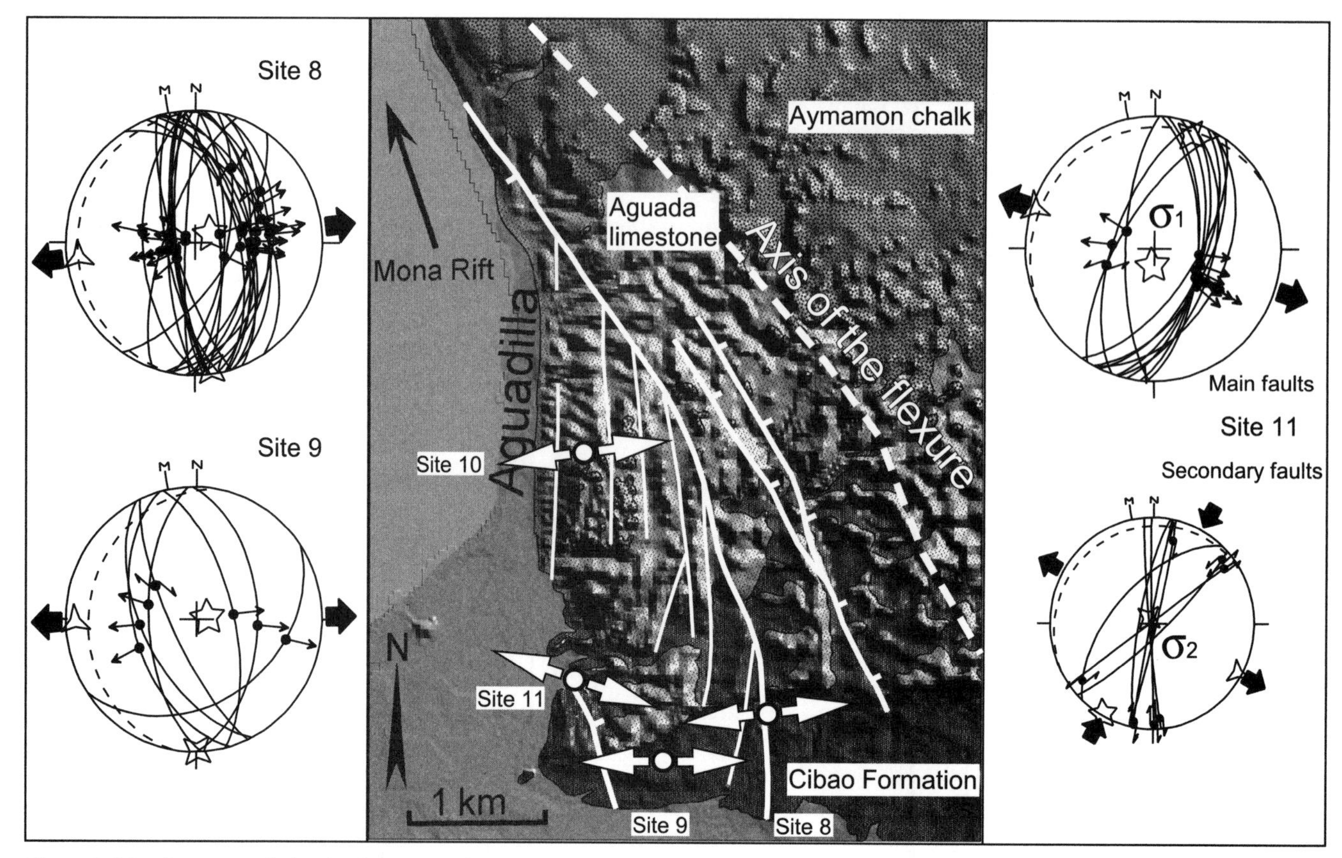

Figure 5. Morphostructural sketch of the Aguadilla area (location in Fig. 3B) and results of stress inversion from faults slip measured in the Miocene limestones. The geological map (Monroe, 1969) is draped on a USGS 30 m Digital Elevation Model with illumination from the NE. Stratigraphic contours and NW-trending faults are from Monroe (1969). The Cibao Formation is Early Miocene, the Aguada limestone is early-middle Miocene and the Aymamon chalk is middle-late Miocene (Seiglie and Moussa, 1984). Note the N-S morphological lineaments of the karst eroded Aguada limestone that branch on a NW-SE fault mapped by Monroe (1969). Lineaments are interpreted as part of a horse tail splay at the termination of the eastern fault boundary of the Mona rift.

DISCUSSION: MODEL FOR THE PRESENT DEFORMATION OF THE PUERTO RICO AREA

The E-W opening of the Mona rift and our preliminary result in southern Puerto Rico (i.e., SE-trending sinistral transtension [Fig. 7] postdating NNE-trending dextral transtension) are not compatible with a present counterclockwise rotation of the Puerto Rico–Virgin Islands microplate. As shown above, the stress field has evolved since Eocene times, and the present geodynamic setting is probably as recent as Late Pliocene–Quaternary. We can reconstruct the Cenozoic history of the island in three tectonic stages.

The orogenic event that affected the northern Caribbean island arc in Paleogene times was characterized in western Puerto Rico by a NNE direction of compression (Fig. 3). Note that this direction was NE-SW before the Miocene-Pliocene counterclockwise rotation of the island. From late Oligocene to recent times, our fault analysis indicates only extension. During deposition of the Neogene carbonate platform and in particular during the early Miocene (Fig. 5), extension is trending N-S. Despite the large (>1000 m) variations of thickness of the carbonate platform normal, faults are remarkably rare in the offshore area north of Puerto Rico (Masson and Scanlon, 1991; van Gestel et al., 1998), indicating that this stress field did not generate large structures. The N-S strike of σ3 suggests that this extensional stress field corresponds to a diffuse postorogenic extensional collapse induced by the presence in front of Puerto Rico of the "free face" of the subducting North America plate. As a model of counterclockwise rotation of Puerto Rico predicts possible extension in the Puerto Rico trench (Masson and Scanlon, 1991), we cannot exclude that this N-S extension is a result of the initiation of the rotation, which mainly occurred between 10 and 4 m.y.

For the late Pliocene-Quaternary period, we did not find any evidence of compression that would have generated an arching of the platform (van Gestel et al., 1998). This conclusion seems valid for most of the island according to another paleostress study

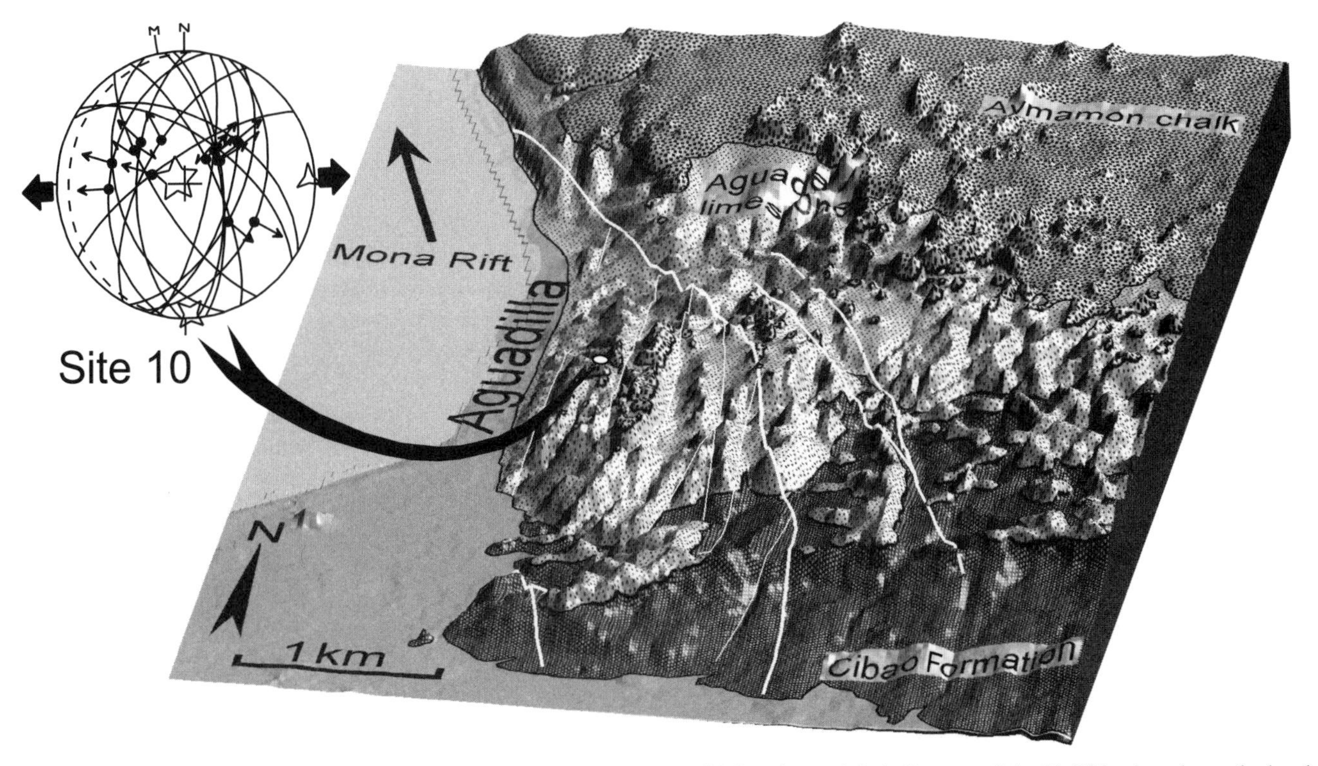

Figure 6. Three-dimensional view of Figure 5 (vertical exaggeration: 3:1) with location and fault diagram of site 10. This view shows the local flexure toward the SW of the general north-dipping platform carbonates (cf. the dip of the contact between the Cibao and the Aguada formations). Note the karst topography in the Aymamon and Aguada carbonates. The faults of site 10 are measured in the highly karst-eroded Aguada limestone. The striations postdate the karst erosion as shown by the irregularity of the striated surfaces crossing the weathered, vuggy limestone. This is illustrated by the various fault attitudes in the Schmidt diagram.

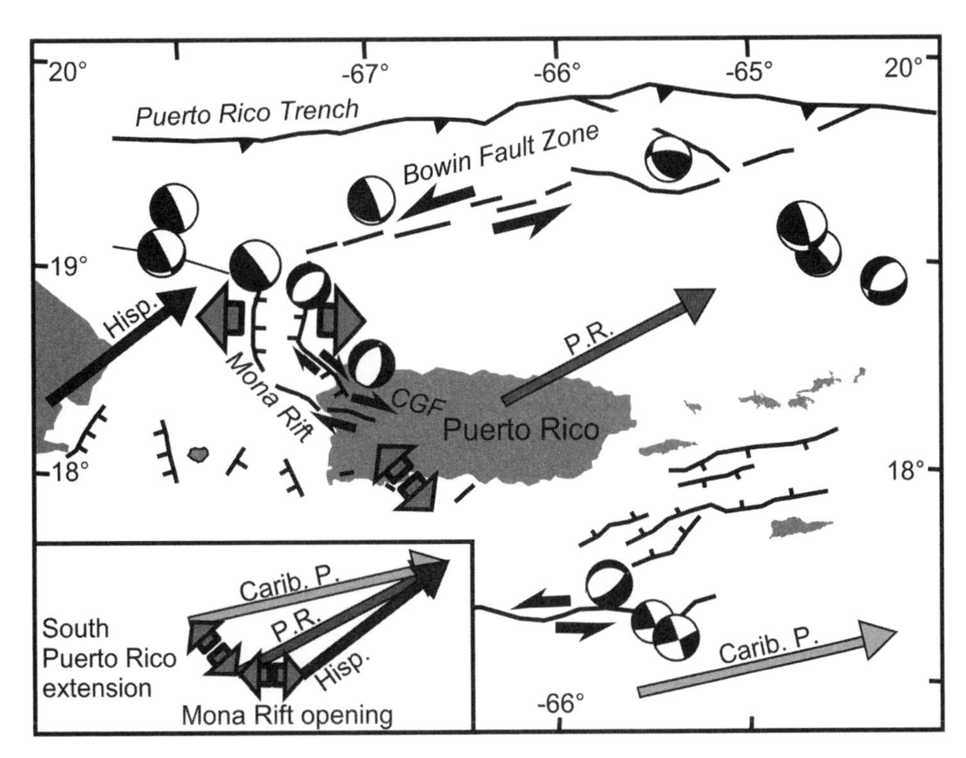

Figure 7. Conceptual model for the Pliocene-Quaternary geodynamics of Puerto Rico. Mona rift opens E-W. The south of Puerto Rico is deformed by SE-trending extension (according to the focal planes mechanisms and to our preliminary results of fault slip measurements), which indicate sinistral transtension on this E-W margin. (Inset) We suppose that the origin and trends of these openings (large arrows) lies in small differences in orientation and velocities of the movements of the Hispaniola, Puerto Rico–Virgin Islands, and Caribbean plates (thin arrows).

in the Neogene carbonates of southern Puerto Rico (Mann et al., this volume, Chapter 9). This suggests that the 4° Pliocene tilt of the carbonate platform north of Puerto Rico (Moussa et al., 1987; van Gestel et al., 1998) may result from tectonic erosion (e.g., Birch, 1986; Grindlay et al., this volume, Chapter 2) rather than from compression. In contrast, we found recent normal faulting at the onshore extension of the Mona rift that supports is present-day activity onland in Puerto Rico. This extensional deformation contrasts with the compressional deformation of a carbonate platform of similar age in northern Hispaniola (e.g., Calais and Mercier de Lepinay, 1991) and supports the idea that the Mona rift is an incipient plate boundary forming in response to collisional processes (e.g., Mann et al., 2002).

The average trend of σ3 is N92°E which implies a right-lateral component on the displacement of SE-trending normal faults in the southern Mona rift (Fig. 2). Moreover, this direction suggests that the mainly dextral late Holocene ESE-trending Cerro-Goden fault (Figs. 2 and 7; Mann et al., this volume, Chapter 6) is also part of the fault termination of the Mona rift (Fig. 7) and could be interpreted as a transfer fault. On Figure 2, one can note that the steepest wall of the Mona rift is its eastern side, where there are the largest normal displacements (van Gestel et al., 1998; Mann et al., 2002). In the southern part of the Mona rift, the main scarp trends NW-SE and is in agreement with the structural grain of the pre-Oligocene basement. The occurrence of recent deformation at the prolongation of the northeastern scarp (Fig. 5), on the Cerro Goden fault at the prolongation of another SW-dipping scarp (Fig. 2), and between these two zones (Fig. 3B; sites 7 and 12) indicates that all the northwestern side of the island is an area of diffuse normal oblique faulting representing the southeastern termination of the Mona rift (Fig. 2 and 7).

The data presented here support the hypothesis that differences in velocities between Hispaniola and Puerto Rico leads to extension in the Mona Passage (Vogt et al., 1976). This model is now confirmed by GPS measurements that indicate slower velocities in Hispaniola than in Puerto Rico relative to stable North America (Jansma et al., 2000)(Fig. 1). However, in the Hispaniola convergent zone, GPS velocities may mostly represent elastic strain accumulation on seismic faults. To determine the real movement of eastern Hispaniola, one must integrate the coseismic slips of every seismic fault, which at this moment is only known for the Septentrional fault (9 ± 3 mm/yr, Prentice et al., 2003).

Jansma et al. (2000) tried to deduce the actual motion of Hispaniola from two-dimensional elastic modeling and proposed that Hispaniola is moving 5 ± 4 mm/yr slower than Puerto Rico. They concluded that the direction of the divergent motion between Hispaniola and Puerto Rico is roughly east-west with a possible northward component predicting minor left-lateral slip along the N-S–trending faults of the Mona rift.

Our observation of recent normal faulting is in agreement with this conclusion of differential velocities between Hispaniola and Puerto Rico, but the computed direction of σ3 (N92°E) does not seem to confirm the northward component of motion predicted from this two-dimensional elastic strain modeling. Note that this average trend of σ3 in northwestern Puerto Rico is between the SE trend of the T-axes of the two extensional focal planes mechanisms of the Mona rift area and the ENE trends of the GPS vectors (Fig. 1). Even if this orientation was obtained at a rift termination where oblique slips on large faults can generate stress perturbations and rotate the σ3 axis, its perpendicularity with the main northern part of the rift suggests that the possible stress rotation is small and that this trend must be close to the opening direction of the Mona rift (Fig. 7).

To generate an E-W opening, we suspect that the real motion of Hispaniola relative to North America is not more easterly than the motion of Puerto Rico (Jansma et al., 2000), but more northerly (Fig. 7). This inference is in agreement with three-dimensional modeling of elastic strain accumulation resulting from the motion of the Caribbean plate on four faults in Hispaniola (Mann et al., 2002). These GPS results and modeling point out that a significant amount of fault reverse motion must exist. If coseismic NE-SW reverse movements are added to the elastic strain accumulation on mostly strike-slip faults (Fig. 1), the long-term movement of Hispaniola is more northerly than the GPS vectors indicate (Fig. 1). In the same way, the SE-trending sinistral transtension on the southern margin of Puerto Rico, as indicated by our preliminary fault slip measurements in South Puerto Rico (Mann et al., this volume, Chapter 9), and focal plane mechanisms (Fig. 1), might result from a slower and more northerly long-term motion of Puerto Rico than the Lesser Antilles (relative to fixed North America) (Fig. 7, inset). Despite the short time of observations and the absence of large earthquakes, GPS measurements of Jansma et al. (2000) agree with a difference in velocities between Puerto Rico and the Caribbean Plate, supporting our model (Fig. 7, inset).

ACKNOWLEDGMENTS

This work was partly supported by the University of Savoy, France, and by external grant no. DOHQGR0064 to P. Mann and C. Prentice from the National Earthquake Hazards Reduction Program (NEHRP) of the U.S. Geological Survey. We thank Grenville Draper and Mark Gordon for their constructive reviews. University of Texas at Austin Institute for Geophysics (UTIG) contribution no. 1701.

REFERENCES CITED

Angelier, J., 1990, Inversion of field data in fault tectonics to obtain the regional stress—III. A new rapid direct inversion method by analytical means: Geophysical Journal International, v. 103, p. 363–376.

Angelier, J., and Bergerat, F., 1983, Systèmes de contrainte et extension intracontinentale: Bulletin des Centres de Recherches Exploration-Production Elf-Aquitaine, v. 7, p. 137–147.

Bawiec, W.J., 2001, Geologic terranes of Puerto Rico, *in* Bawiec, W., ed., Geology, geochemistry, geophysics, mineral occurrences and mineral resource assessment for the Commonwealth of Puerto Rico: U.S. Geological Survey Open-File Report 98-38, p. 59–65.

Birch, F.S., 1986, Isostatic, thermal, and flexural models of the subsidence of the north coast of Puerto Rico: Geology, v. 14, p. 427–429.

Byrne, D.B., Suarez, G., and McCann, W.R., 1985, Muertos trough subduction—Microplate tectonics in the northern Caribbean: Nature, v. 317, p. 420–421.

Calais, E., and Mercier de Lepinay, B., 1991, From transpression to transtension along the northern Caribbean plate boundary: Implications for the recent motion of the Caribbean plate: Tectonophysics, v. 186, p. 329–350, doi: 10.1016/0040-1951(91)90367-2.
Case, J.E., and Holcombe, T.L., 1980, Geologic-tectonic map of the Caribbean region: U.S. Geological Survey Miscellaneous Investigation Series Map I-1100, scale 1:2 500 000.
DeMets, C., Jansma, P., Mattioli, G., Dixon, T., Farina, F., Bilham, R., Calais, E., and Mann, P., 2000, GPS geodetic constraints on Caribbean–North America plate motion: Geophysical Research Letters, v. 27, p. 437–440, doi: 10.1029/1999GL005436.
Dillon, W.P., Terence Edgar, N., Scanlon, K.M., and Coleman, D.F., 1994, A review of the tectonic problems of the strike-slip northern boundary of the Caribbean plate and examination by GLORIA, *in* Gardner, J.V., et al., eds., Geology of the United States' seafloor: The view from GLORIA: Cambridge, UK, Cambridge University Press, p. 135–164.
Dolan, J.P., Mann, P., de Zoeten, R., Heubeck, C., Shimora, J., and Monechi, S., 1991, Sedimentologic, stratigraphic, and tectonic synthesis of Eocene-Miocene sedimentary basins, Hispagnola and Puerto Rico, *in* Mann, P., et al., eds., Geologic and tectonic development of the North America–Caribbean Plate boundary in Hispaniola: Geological Society of America Special Paper 262, p. 1–28.
Erikson, J.P.E., Pindell, J.L., and Larue, D.K., 1990, Mid-Eocene–Early Oligocene sinistral trancurrent faulting in Puerto Rico associated with formation of the northern Caribbean Plate Boundary Zone: Journal of Geology, v. 98, p. 365–384.
Fink, L.K., and Harrison, C.G.A., 1972, Palaeomagnetic investigations of selected lava units on Puerto Rico. Proceedings of the 6th Caribbean Geological Conference, p.379.
Gordon, M.B., Mann, P., Caceres, D., and Flores, R., 1997, Cenozoic tectonic history of the North America–Caribbean plate boundary zone in western Cuba: Journal of Geophysical Research, v. 102, no. B5, p. 10,055–10,082, doi: 10.1029/96JB03177.
Grindlay, N.R., Mann, P., and Dolan, J.F., 1997, Researchers investigate submarine faults north of Puerto Rico: Eos (Transactions, American Geophysical Union), v. 78, p. 404.
Hippolyte, J.-C., and Sandulescu, M., 1996, Paleostress characterization of the "Wallachian phase" in its type area (southeastern Carpathians, Romania): Tectonophysics, v. 263, p. 235–248, doi: 10.1016/S0040-1951(96)00041-8.
Hippolyte, J.-C., Angelier, J., and Roure, F., 1992, Les permutations de contraintes dans un orogène: Exemple des terrains quaternaires du sud de l'Apennines: Comptes Rendu de l'Académie des Sciences de Paris, t. 315, ser. 2, p. 89–95.
Huérfano, V., 1995, Crustal structure and stress regime near Puerto Rico [M.S. thesis]: University of Puerto Rico, Mayagüez.
Jansma, P.E., Mattioli, G.S., Lopez, A., DeMets, C., Dixon, T.H., Mann, P., and Calais, E., 2000, Neotectonics of Puerto Rico and the Virgin Islands, northeastern Caribbean, from GPS geodesy: Tectonics, v. 19, no. 6, p. 1021–1037, doi: 10.1029/1999TC001170.
Jany, I., Mauffret, A., Bouysse, P., Mascle, A., Mercier de Lépinay, B., Renard, V., and Stephan, J.-F., 1987, Relevé bathymétrique Seabeam et tectonique en décrochement au sud des îles Vierges (Nord-Est Caraïbes): Comptes Rendu de l'Académie des Sciences de Paris, v. 304, p. 527–532.
Jany, I., Scanlon, K.M., and Mauffret, A., 1990, Geological interpretation of combined seabeam, Gloria and Seismic data from Anegada Passage (Virgin Islands, North Caribbean): Marine Geophysical Researches, v. 12, p. 173–196.
Ladd, J.W., Worzel, J.L., and Watkins, J.S., 1977, Multifold seismic reflection records from the Northern Venezuela Basin and the north slope of Muertos trench, *in* Talwani, M., and Pitman, W.C., eds., Island arcs, deep-sea trenches and back-arc basins: Washington, D.C., American Geophysical Union, p. 41–56.
Larue, D.K., and Ryan, H.F., 1998, Seismic reflection profiles of the Puerto Rico trench: Shortening between the North American and Caribbean plates, *in* Lidiak, E.G., and Larue, D.K., eds., Tectonics and geochemistry of the Northeastern Caribbean: Geological Society of America Special Paper 322, p. 193–210.
Larue, D.K., Torrini, R., Jr., Smith, A.L., and Joyce, J., 1998, North Coast Tertiary basin of Puerto Rico: From arc basin to carbonate platform to arc-massif slope: Geological Society of America Special Paper 322, p. 155–176.
Mann, P., Taylor, F.W., Edwards, R.L., and Ku, T.L., 1995, Actively evolving microplate formation by oblique collision and sideways motions along strike-slip faults: An example from the northeastern Caribbean plate margin: Tectonophysics, v. 246, p. 1–69, doi: 10.1016/0040-1951(94)00268-E.
Mann, P., Calais, E., Ruegg, J.C., DeMets, C., Jansma, P.E., and Mattioli, G., 2002, Oblique collision in the northeastern Caribbean from GPS measurements and geological observations: Tectonics, v. 21, no. 6, 1057, doi: 10.1029/2001TC001304.
Masson, D.G., and Scanlon, K.M., 1991, Neotectonic setting of Puerto Rico: Geological Society of America Bulletin, v. 103, no. 1, p. 144–154, doi: 10.1130/0016-7606(1991)1032.3.CO;2.
Mauffret, A. and Jany, I., 1990, Collision et tectonique d'expulsion le long de la frontière Nord-Caraïbe. Oceanologica Acta, v. 10, p. 97–116.
McCann, W., 2002, Microearthquake data elucidate details of Caribbean subduction zone: Seismological research Letters, v. 73, no. 1, p. 25–32.
Mercado, A., and McCann, W., 1998, Numerical simulation of the 1918 Puerto Rico tsunami: Natural Hazards, v. 18, p. 57–76, doi: 10.1023/A:1008091910209.
Monroe, W.H., 1969, Geologic map of the Aguadilla Quadrangle, Puerto Rico: U.S. Geological Survey Miscellaneous Investigations Maps I-569, scale 1:20 000.
Monroe, W.H., 1980, Geology of the Middle Tertiary Formations of Puerto Rico: Geological Survey Professional Paper 953, 1 map, 93 p.
Moussa, M.T., Seiglie, G.A., Meyerhoff, A.A., and Taner, I., 1987, The Quebradillas Limestone (Miocene-Pliocene), northern Puerto Rico, and tectonics of the northeastern Caribbean margin: Geological Society of America Bulletin, v. 99, p. 427–439.
Pacheco, J.F., and Sykes, L.R., 1992, Seismic moment catalog of large shallow earthquakes, 1900 to 1989: Bulletin of the Seismological Society of America, v. 82, p. 1306–1349.
Pindell, J.L., and Barrett, S.F., 1990, Geological evolution of the Caribbean region: A plate-tectonic perspective, *in* Dengo, G., and Case, J.E., eds., The geology of North America, vol. H, The Caribbean region: Boulder, Colorado, Geological Society of America, p. 405–432.
Prentice, C.S., Mann, P., Peña, L.R., and Burr, G., 2003, Slip rate and earthquake recurrence along the central Septentrional fault, North American–Caribbean plate boundary, Dominican Republic: Journal of Geophysical Research, v. 108, no. B3, 2149, doi: 10.1029/2001JB000442.
Reid, J.A., Plumley, P.W., and Schellekens, J.H., 1991, Paleomagnetic evidence for Late Miocene counterclockwise rotation of north coast carbonate sequence, Puerto Rico: Geophysical Research Letters, v. 18, p. 565–568.
Schell, B.A., and Tarr, A.C., 1978, Plate tectonics of the north eastern Caribbean Sea region: Geologie en Mijnbouw, v. 57, p. 319–324.
Seiglie, G.A., and Moussa, M.T., 1984, Late Oligocene–Pliocene transgressive-regressive cycles of sedimentation in northwestern Puerto Rico, *in* Schlee, J.S., ed., Interregional unconformities and hydrocarbon accumulation: American Association of Petroleum Geologists Memoir 36, p. 89–95.
Van Fossen, M.C., Channel, J.T., and Schellekens, J.H., 1989, Paleomagnetic evidence for Tertiary anticlockwise rotation in southwest Puerto Rico: Geophysical Research Letters, v. 16, p. 819–822.
van Gestel, J.P., Mann, P., Dolan, J.F., and Grindlay, N.R., 1998, Structure and tectonics of the upper Cenozoic Puerto Rico–Virgin Islands carbonate platform as determined from seismic reflection studies: Journal of Geophysical Research, v. 103, p. 30,505–30,530, doi: 10.1029/98JB02341.
Vogt, P.R., Lowrie, A., Bracey, D.R., and Hey, R.N., 1976, Subduction of aseismic ridges: Effects on shape, seismicity, and other characteristics of consuming plate boundaries: Geological Society of America Special Paper 172, p. 1–59.

Manuscript Accepted by the Society August 18, 2004

Geological Society of America
Special Paper 385
2005

Neotectonics of southern Puerto Rico and its offshore margin

Paul Mann*
Institute for Geophysics, Jackson School of Geosciences, University of Texas at Austin, 4412 Spicewood Springs Road, Building 600, Austin, Texas 78759, USA

Jean-Claude Hippolyte
Laboratoire de Geodynamique des Chaines Alpines, Université de Savoie, UMR-CNRS 5065, Campus Scientifique Technolac, 73376 Le Bourget de Lac Cedex, France

Nancy R. Grindlay
Lewis J. Abrams
Center for Marine Science, University of North Carolina Wilmington, Wilmington, North Carolina 28409, USA

ABSTRACT

Puerto Rico is located within a zone of tectonic transition between mainly east-west, North America–Caribbean strike-slip motion to the west in Hispaniola and east-north-east–oriented underthrusting to the east beneath the Lesser Antilles island arc. Various models and tectonic mechanisms have been proposed for the Neogene to present-day deformation of southern Puerto Rico, its island margin, and the Muertos trench by previous workers that include normal, thrust, and strike-slip faulting accompanied by large-scale rotations. In this study, we present the results of a regional study integrating onland mapping of striated fault surfaces in rocks ranging in age from Oligocene to possibly as young as earliest Pliocene, and offshore mapping of faults deforming the uppermost sediments beneath the seafloor.

The tectonic geomorphology and distribution of late Quaternary marine terraces and beach ridges in south-central Puerto Rico suggest either stability or slow late Quaternary uplift along the south-central part of the coast. In contrast, the coastline of southwestern Puerto Rico exhibits no late Quaternary coastal sediments and a pattern of long-term drowning of coastal features. Fault striation studies of three formations composing the Puerto Rico–Virgin Islands carbonate platform of south-central Puerto Rico (Juana Diaz Formation basal clastic unit, Juana Diaz Formation upper carbonate unit, Ponce Formation) indicate two distinct extensional phases affecting the youngest formation (Ponce Formation of middle Miocene–early Pliocene age). The first event, a north-northeast–directed extensional event is accommodated by normal faults striking mainly to the west-northwest. A second, southeast-directed extensional event crosscut and reactivated faults formed during the first event and produced at least one northeast-trending Quaternary rift bounded by northeast-striking normal faults (Ponce basin).

Offshore seismic profiling by previous workers and reported in this study support the presence of late Holocene seafloor-rupturing, northeast-striking normal faults that

*E-mail: paulm@ig.utexas.edu.

Mann, P., Hippolyte, J.-C., Grindlay, N.R., and Abrams, L.J., 2005, Neotectonics of southern Puerto Rico and its offshore margin, *in* Mann, P., ed., Active tectonics and seismic hazards of Puerto Rico, the Virgin Islands, and offshore areas: Geological Society of America Special Paper 385, p. 173–214. For permission to copy, contact editing@geosociety.org.

accommodate southeast extension of the southern margin of Puerto Rico. The post–early Pliocene extension direction is roughly perpendicular to the east-northeast–trending sections of the stable or slowly uplifting coastline along much of southern Puerto Rico. In addition to northeast-striking normal faults, offshore profiles confirm the presence of late Holocene, seafloor-rupturing left lateral strike-slip faults along the offshore extension of the Great Southern Puerto Rico fault zone. Where the Great Southern Puerto Rico fault zone curves to the northeast, the fault becomes less strike-slip and more normal in character and produces greater extensional and tilting effects in the linked Whiting half-graben.

A neotectonic model for southern Puerto Rico to explain both directions of extension known from fault striation studies and the present tectonic geomorphology of the preserved Puerto Rico–Virgin Islands carbonate platform in south-central Puerto Rico involves late Miocene–early Pliocene oblique collision of the Bahama Platform with Hispaniola to the northwest of Puerto Rico and counterclockwise rotation and extension of the area of southern Puerto Rico. A later crosscutting extensional event during the post–early Pliocene involves left-lateral transtension of the southern margin of Puerto Rico with most strike-slip motion concentrated along the Great Southern Puerto Rico fault zone.

Keywords: Puerto Rico, active tectonics, faults, geomorphology, fault striations, seismic profiles.

INTRODUCTION

The island of Puerto Rico is roughly rectangular in shape, ~175 km long east-west, and 65 km long north-south with a land area of ~8500 km^2 (Fig. 1). The island is located within a complex, 150-km-wide, seismogenic plate boundary zone separating the North America and Caribbean plates (Fig. 1). To the north of Puerto Rico is the 8.5-km-deep Puerto Rico trench, a site of highly oblique, southward underthrusting of Atlantic Ocean floor of the North America plate beneath Puerto Rico to depths greater than 100 km (Dolan et al., 1998; Larue et al., 1998; McCann, 2002; Huerfano et al., this volume) (Fig. 1). To the south of Puerto Rico is the 5-km-deep Muertos trench, a site of oblique northward underthrusting of Caribbean ocean floor beneath Puerto Rico to depths of at least 50 km (Byrne et al., 1985; Dillon et al., 1994). A system of east-west–trending strike-slip faults lie parallel to the coast of northern Puerto Rico trench as described by Grindlay et al. (this volume, Chapter 2) (Fig. 1). The overall motion between the two plates at the longitude of Puerto Rico is left-lateral strike-slip as shown by recent global positioning system (GPS) studies (DeMets et al., 2000; Jansma et al., 2000; Mann et al., 2002; Jansma and Mattioli, this volume). The tectonic component of North America–Caribbean shortening increases to the west of Puerto Rico on the island of Hispaniola as a result of its late Miocene to recent collision with the southeastern edge of the Bahama carbonate platform (Mann et al., 2002; Calais et al., 2002).

Objectives and Data Sources for This Paper

Recent work on the neotectonics of Puerto Rico has focused on mapping active structures along the north or west coasts and shelf areas of Puerto Rico as reflected by most of the papers in this special paper (e.g., Grindlay et al., Chapters 2 and 7; Hippolyte et al., Prentice and Mann). In order to provide a more regional and balanced view of Puerto Rican neotectonics, this paper focuses on the neotectonics of the southern margin of Puerto Rico and its offshore shelf. Here, we integrate several different data sets collected over the past 40 years by previous workers, including geologic results from quadrangle mapping done in the 1960s through 1980s, at a scale of 1:20,000 by various geologists of the U.S. Geological Survey and offshore sparker profiling collected across the insular shelf of southern Puerto Rico by the U.S. Geological Survey in the late 1960s (Garrison, 1969).

These previously published data are combined with results presented in this paper for the first time and include:

- Offshore single-channel seismic profiles collected by some of us (PM, LA, NG) on the RV *Isla Magueyes* in 2000. These data were collected as a one day reconnaissance study of the southern coast of Puerto Rico. A much more complete seismic and sidescan sonar data set was collected over a period of 9 days from the west coast of Puerto Rico (Grindlay et al., this volume, Chapter 7).
- Offshore multi-channel seismic profiles collected by Western Geophysical and Fugro in 1969 that were donated to the Department of Geology of the University of Puerto Rico. This is an extensive survey that was carried out in relation to the construction of proposed nuclear reactors in Puerto Rico and was focused on the western and northern margins of the island. Only a few select reflection lines from this important data set have been previously interpreted and published (Larue and Ryan, 1998; Larue et al., 1998; Grindlay et al., this volume, Chapter 7).

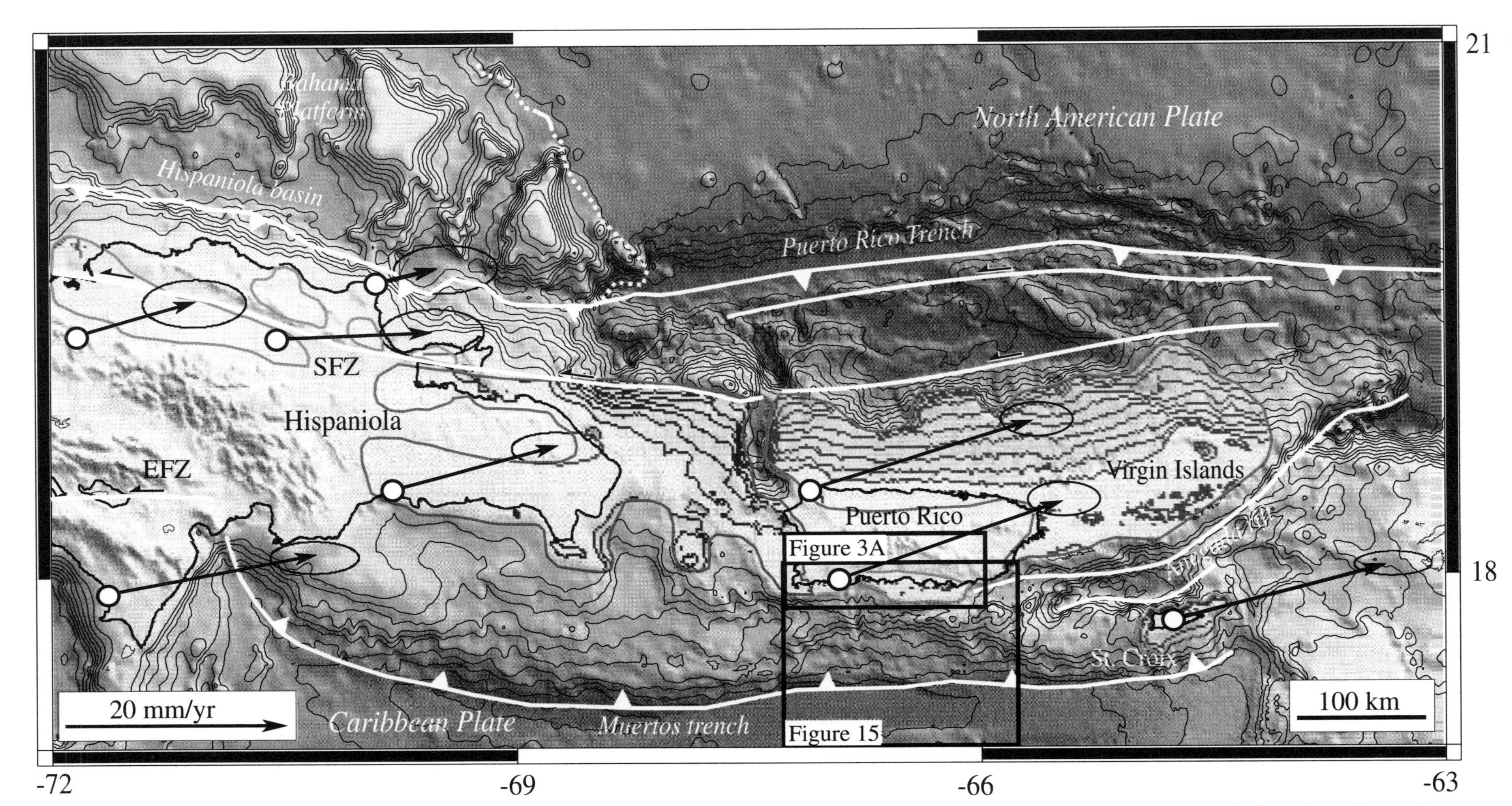

Figure 1. Bathymetric-topographic map of the northeastern Caribbean–North America plate boundary modified from Grindlay et al. (1997) and Grindlay et al. (this volume, Chapter 2). Arrows indicate GPS vectors relative to a fixed North America plate from Jansma et al. (2000; Jansma and Mattioli, this volume). The arrow length of GPS vectors is proportional to the displacement rate. The ellipses represent 95% confidence intervals. Yellowish areas surrounded by a brown contour show extent of Oligocene-Pliocene carbonate platform rocks in the Virgin Islands, St. Croix, Puerto Rico, and Hispaniola (Dominican Republic and Haiti). Boxed area for Figure 3A shows location of more detailed map of onland study area of southern Puerto Rico; second boxed area for Figure 15 shows more detailed map of adjacent offshore area mainly studied with existing seismic profiles. EFZ—Enriquillo fault zone. SFZ—Septentrional fault zone.

- Onland fault striation measurements collected by some of us from the Oligocene–early Pliocene stratigraphic section of southern Puerto Rico (JCH, PM) in 2001. These measurements were carried out to better understand the kinematics of onland faults and the Neogene history of major faults in both northwestern (Hippolyte et al., this volume) and southern Puerto Rico.

PREVIOUS MODELS FOR THE NEOTECTONICS OF SOUTHERN PUERTO RICO

In Figure 2, we have subdivided previous neotectonic models for southern Puerto Rico into three major groups. The first group predicts that the island of Puerto Rico was part of a broad left-lateral shear zone or "plate boundary zone" accommodating diffuse, east-west–directed, North America–Caribbean relative plate motion (Mann and Burke, 1984) (Fig. 2A). All major faults in the area, including those forming the Anegada Passage, were inferred to be left-lateral faults created by interplate strike-slip motion and accommodating various amounts of the total, interplate motion between the Caribbean and North America now known from GPS studies to total ~21 mm/yr (DeMets et al., 2000; Jansma and Mattioli, this volume). Similar models proposing broad zones of distributed, east-west–trending left-lateral shear were proposed during the early 1980s for the land areas west of Puerto Rico including Hispaniola (Mann et al., 1984), Jamaica, and northern Central America (Burke et al., 1980; Mann et al., 1985).

A second group of models were based on the interpretation that faults lying parallel to the submarine Anegada Passage separating Puerto Rico from St. Croix are active right-lateral strike-slip faults (Jany et al., 1990; Masson and Scanlon, 1991). This interpretation is based on the right-stepping geometry of inferred pull-apart basins in the fault-bounded Anegada Passage (Virgin Islands and Sombrero basins shown in Fig. 2B). Jany et al. (1990) and Masson and Scanlon (1991) used regional sidescan sonar and seismic profiles to constrain the fault and basin geometries.

Different workers proposed different tectonic mechanisms to explain the origin of right-lateral shear and right-stepping pull-apart basins along the Anegada Passage. Stephan et al. (1986) propose that east-west shortening of east-west–trending land area or "frame" of Hispaniola and Puerto Rico creates bending, right-lateral shear, and local areas of extension in the manner shown in the inset of Figure 2B. Jany et al. (1990) and Mauffret and Jany (1990) proposed that the block bounded on the south by faults of the Anegada Passage and on the north by faults of the Puerto Rico trench was "escaping" to the east-northeast in the direction of subducting Atlantic oceanic crust. Finally, Masson and Scanlon (1991) used new regional sidescan data to propose that large-scale, counterclockwise rotation of a rigid Puerto Rico–Virgin Islands block or microplate produced transtensional right-lateral shear along the Anegada Passage, left-lateral shear along the Puerto Rico trench, and shortening across the western Muertos trench (Fig. 2B).

A third and final group of models predicted southeast-northwest opening across the Anegada Passage that results in a broad zone of rifting that includes southeastern Puerto Rico, St. Croix, and the northern Lesser Antilles (Speed and Larue, 1991; Feuillet et al., 2002)(Fig. 2C). Speed and Larue (1991) propose that the plate boundary-normal component of extension was produced by subduction rollback, or, in their words, "the northward pullaway of North America from its south-dipping slab beneath Puerto Rico." Another linked prediction of this model is that the Puerto Rico trench formed during this same phase of rollback-induced extension—rather than by oblique underthrusting (Speed and Larue, 1991). Feuillet et al. (2002) propose that strain partitioning related to oblique underthrusting of North America produces a plate boundary–normal component of extension across wide areas of the Caribbean–North America plate boundary (Fig. 2C). This plate boundary–normal component of extension is the tectonic mechanism for creating the Anegada rift system and a parallel group of northeast-trending rifts found in the northern Lesser Antilles arc and Anegada Passage.

To better evaluate all three sets of models, we begin by reviewing the overall geologic setting of Puerto Rico and then focus on the tectonic geomorphology and structure of the area of southern Puerto Rico and its offshore area. The objective of this paper is to better understand the late Neogene tectonic style of deformation in southern Puerto Rico by integrating information on its onland and offshore geomorphology, structure, and stratigraphy (Fig. 1). An improved understanding of the neotectonic setting of Puerto Rico along with the locations and paleoseismicity of active faults in Puerto Rico and its offshore areas is a necessary step to provide an accurate assessment of earthquake and related tsunami hazards.

GEOLOGIC SETTING OF NEOGENE FAULTS IN SOUTHERN PUERTO RICO

Cretaceous-Eocene Island Arc History of Puerto Rico

The metamorphic, plutonic, volcanic, and sedimentary basement rocks of Puerto Rico formed in the northeastern segment of the Caribbean island arc, a lower Cretaceous to Holocene, continuous island arc chain extending from western Cuba to the north coast of South America (Burke, 1988; Lidiak and Larue, 1998) (Fig. 3A). Land areas of Puerto Rico and its northern island shelf are underlain by the volcanic arc and its plutonic basement (Moussa et al., 1987; Larue et al., 1998; van Gestel et al., 1998; 1999). Regional magnetic data and dredging show that the northern island slope and northern edge of the Puerto Rico trench are underlain by an elongate, Cretaceous blueschist belt running roughly parallel to the trend of the island (Perfit et al., 1980; Muszala et al., 1999).

The northern segment of the Caribbean island arc became inactive during a Paleocene-Oligocene collision event with a now-subducted feature with a crustal thickness greater than normal oceanic crust that lay to the south and southeast of the present-day

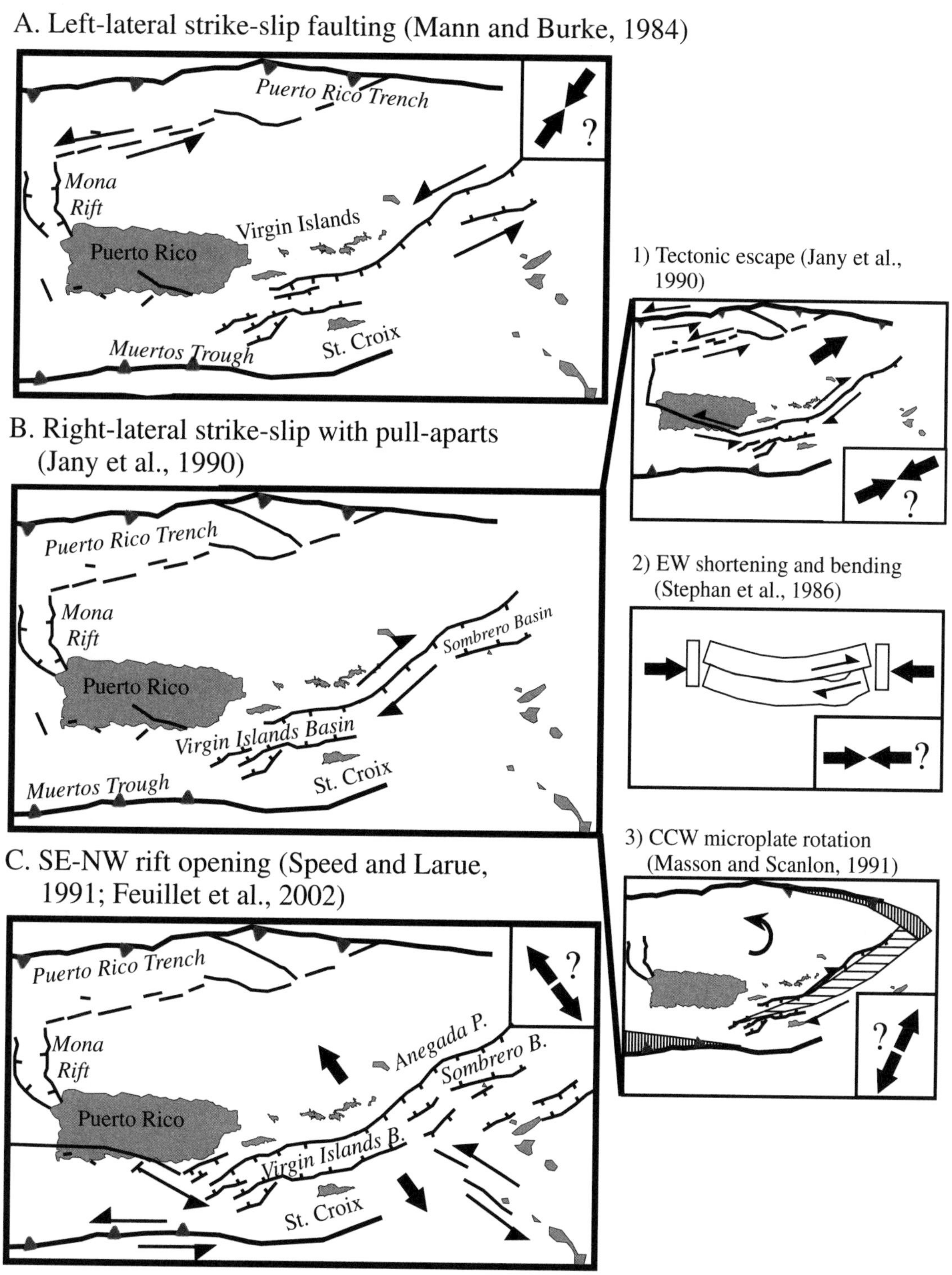

A. Left-lateral strike-slip faulting (Mann and Burke, 1984)

B. Right-lateral strike-slip with pull-aparts (Jany et al., 1990)

C. SE-NW rift opening (Speed and Larue, 1991; Feuillet et al., 2002)

1) Tectonic escape (Jany et al., 1990)

2) EW shortening and bending (Stephan et al., 1986)

3) CCW microplate rotation (Masson and Scanlon, 1991)

Figure 2. Summary of previously proposed tectonic models for opening of the Anegada rift system (Virgin Islands and Sombrero basins). Note how land areas of eastern Puerto Rico and the Virgin Islands form the northern uplifted shoulder of the submarine rift basin and the island of St. Croix (U.S. Virgin Islands) forms its southern uplifted shoulder. Black arrows in inset drawings show the inferred directions of maximum compression and extension implied by these models. The wide variety of models is related to the lack of large earthquakes in recent years and corresponding earthquake focal mechanism studies, the slow (<5 mm/yr) rate of opening based on GPS studies done to date (Jansma et al., 2000; Jansma and Mattioli, this volume), and, finally, the lack of direct studies of striated fault planes affecting Neogene rocks. (A) Left-lateral strike-slip models (e.g., Mann and Burke, 1984) are related to the sub-parallelism of the Anegada rift to known strike-slip faults in the Puerto Rico trench north of Puerto Rico. (B) Right-lateral strike-slip models of Jany et al. (1990) are based on the inferred pull-apart geometry and origin of the submarine Virgin Islands and St. Croix basins within the Anegada rift shown on Figure 1. Three tectonic mechanisms for active right-lateral deformation along the Anegada rift system have been proposed and are shown in the insets to the right: (1) tectonic escape of the Puerto Rico block to the northeast (Jany et al., 1990); (2) east-west shortening (Stephan et al., 1986); and (3) a counterclockwise-rotating Puerto Rico microplate (Masson and Scanlon, 1991). (C) Northwest-southeast rift opening models by Speed and Larue (1991) and Feuillet et al. (2002) in the central Lesser Antilles arc that propose extension parallel to the trend of the Caribbean volcanic arc.

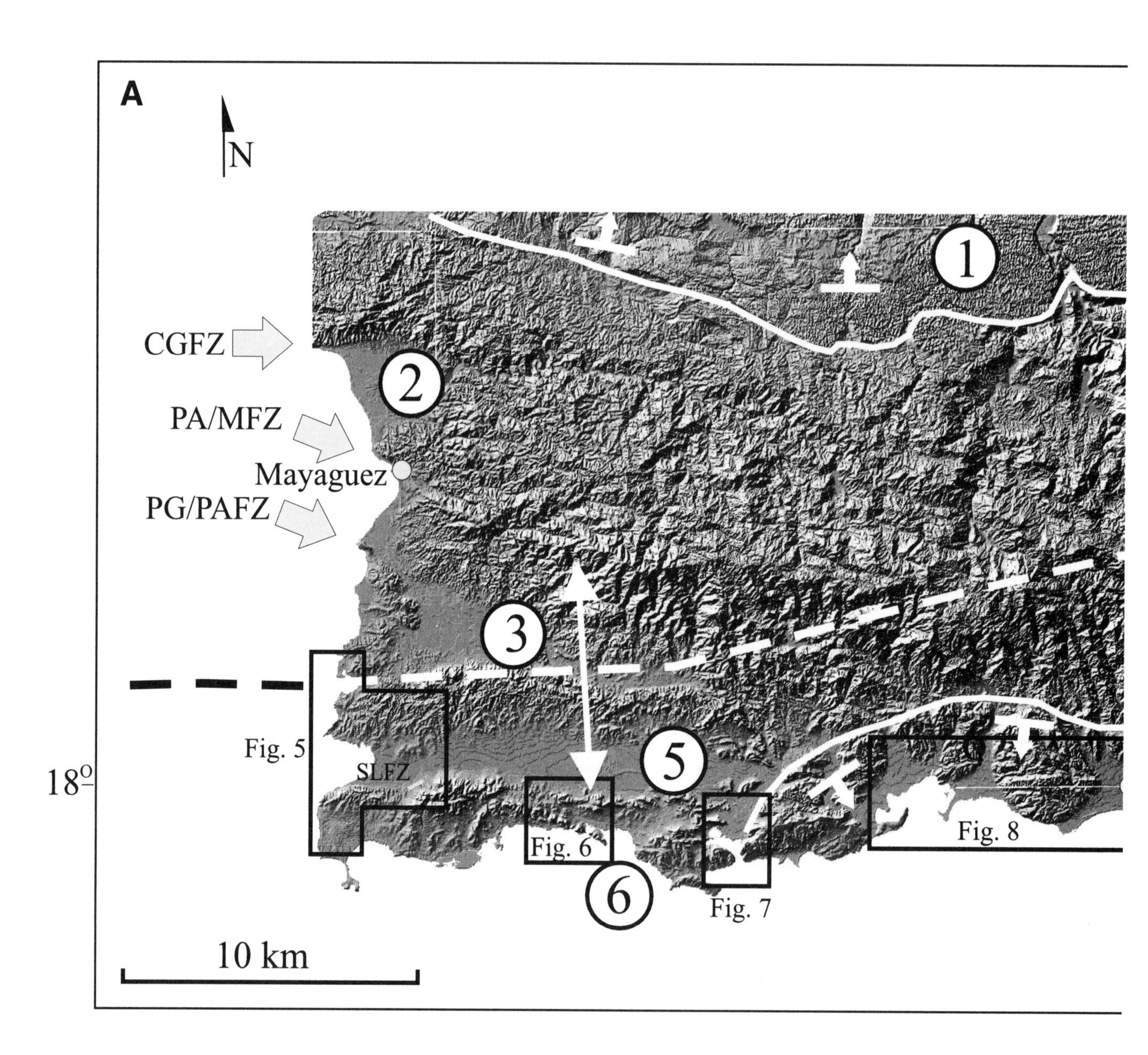
A
N
CGFZ
PA/MFZ
Mayaguez
PG/PAFZ
1
2
3
5
6
Fig. 5
SLFZ
Fig. 6
Fig. 7
Fig. 8
18°
10 km

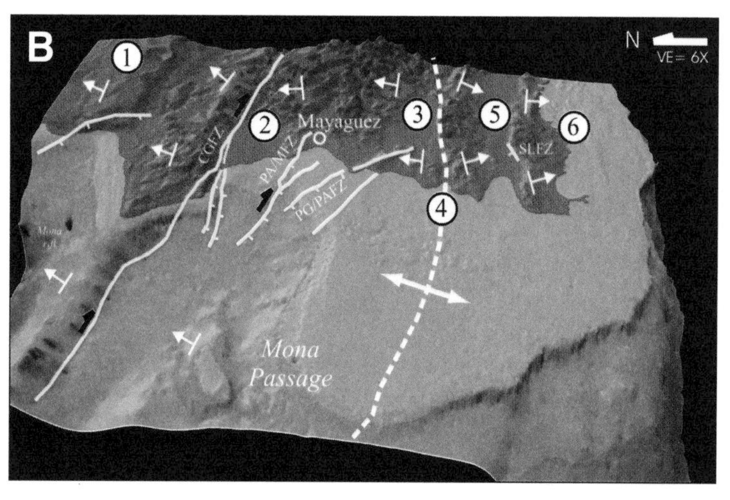
B
N
VE= 6X
1
2
3
4
5
6
Mayaguez
CGFZ
PA/MFZ
PG/PAFZ
SLFZ
Mona
Passage

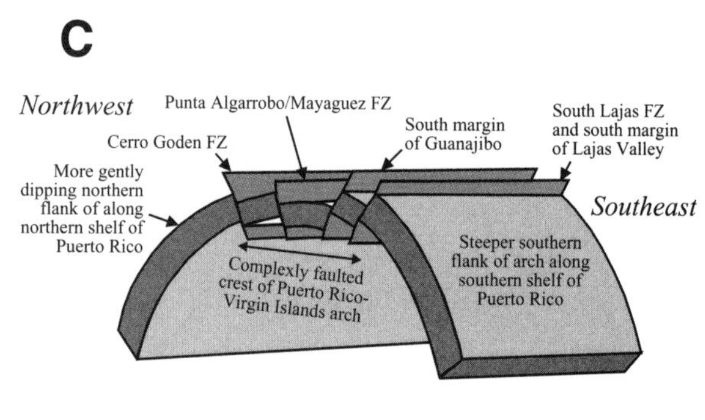
C
Northwest
Southeast
Punta Algarrobo/Mayaguez FZ
Cerro Goden FZ
South margin of Guanajibo
South Lajas FZ and south margin of Lajas Valley
More gently dipping northern flank of along northern shelf of Puerto Rico
Complexly faulted crest of Puerto Rico-Virgin Islands arch
Steeper southern flank of arch along southern shelf of Puerto Rico

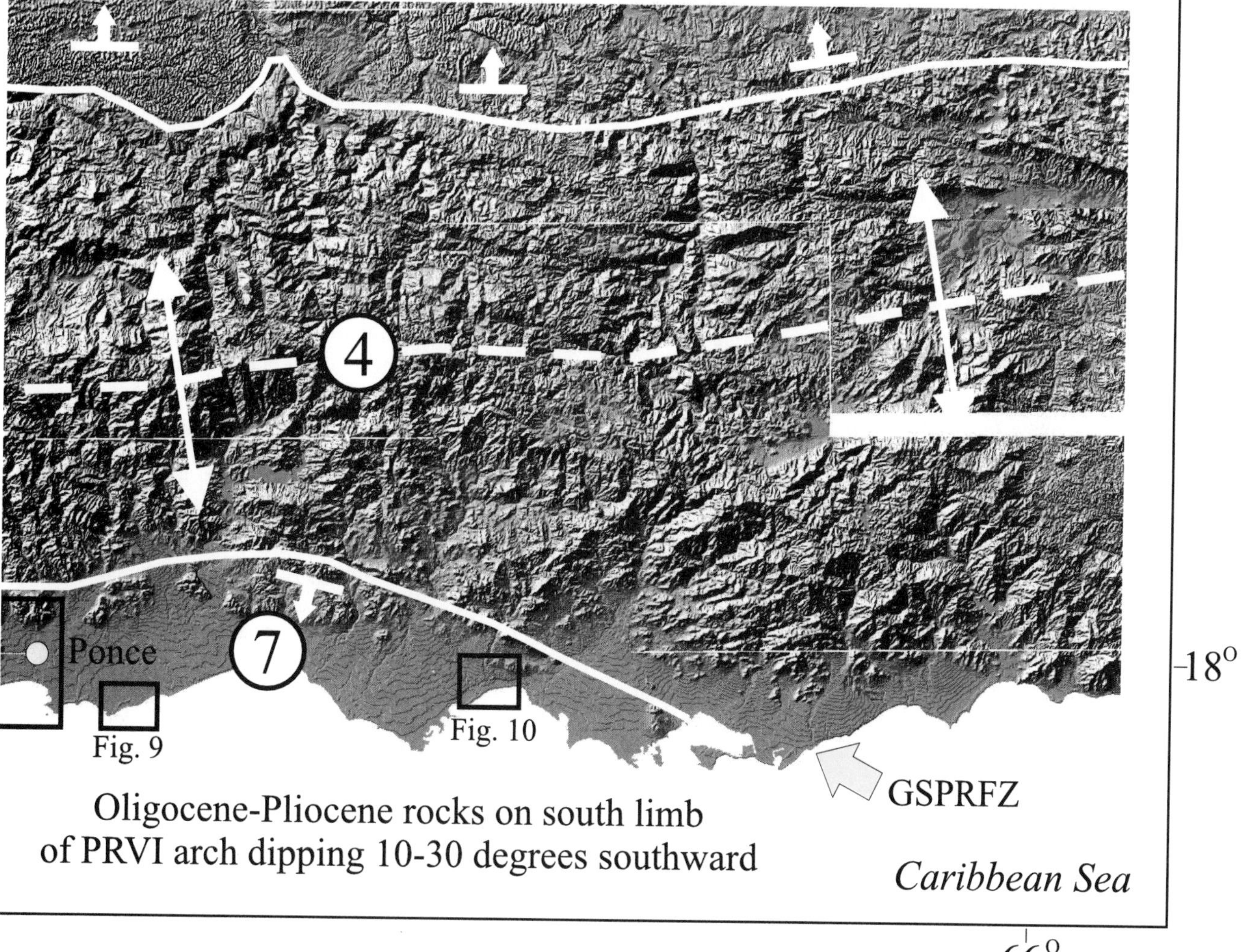

Figure 3. (A) Digital elevation model (DEM) for Puerto Rico showing the trace of the Puerto Rico–Virgin Islands (PRVI) arch parallel to the main topographic divide of the island and the outcrops of Oligocene-Pliocene sedimentary rocks on the northern and southern flanks of the arch. DEM created from U.S. Geological Survey 3 arc second DEM for Puerto Rico (30 m resolution). Yellow arrows indicate faults with known or suspected Holocene movement. CGFZ—Cerro Goden fault zone; PA/MFZ—Punta Algarrobo/Mayaguez fault zone; PG/PAFZ—Punta Guanajibo/Punta Arenas fault zone; SLFZ—South Lajas fault zone; GSPRFZ—Great Southern Puerto Rico fault zone. Numbered features: 1—north-dipping Oligocene-Pliocene sedimentary rocks on northern flank of Puerto Rico–Virgin Islands arch; 2—Anasco Valley; 3—Guanajibo Valley; 4—crest of Puerto Rico–Virgin Islands arch and approximate topographic divide of island; 5—Lajas Valley; 6—drowned coastline of southwestern Puerto Rico; 7—south-dipping Oligocene-Pliocene sedimentary rocks on southern flank of Puerto Rico–Virgin Islands arch. Boxed areas represent detailed maps shown in Figures 5–10. (B) Three-dimensional perspective on the bathymetry and topography of western Puerto Rico looking eastward (modified from Grindlay et al., this volume, Chapter 7). Illumination is from the northwest. The crest of the Puerto Rico–Virgin Islands arch is shown by heavy, dashed white line and numbered areas correspond to same features shown in map view in A. Faults with known or suspected Holocene activity shown on the map in A are indicated by yellow lines. (C) Schematic block diagram showing relationship of major faults in western Puerto Rico (including suspected and known Holocene faults of western Puerto Rico shown in A and B) to axis of post–early Pliocene Puerto Rico–Virgin Islands arch. Most prominent fault scarps with known or suspected Holocene movement appear to be confined to near the crest of the arch where extensional strains are largest. Note that vertical throws on mountain fronts in western Puerto Rico indicate a regularity in the deformation pattern: faults north of the crest of the arch are downthrown to the south; faults south of the crest of the arch are downthrown to the north.

Bahama carbonate platform (Fig. 1). The age of this Paleogene collisional event was diachronous from west to east and occurred in late Paleocene–earliest Eocene in western Cuba (Gordon et al., 1997), early to middle Eocene time in central Cuba (Hempton and Barros, 1993), middle Eocene in Hispaniola (Mann et al., 1991), and late Eocene in Puerto Rico (Dolan et al., 1991; Mann et al., this volume, Chapter 6). In all these areas, fold-thrust belts and strike-slip faults recorded this collisional phase; arc activity terminated along this northern segment of the Caribbean arc and the main locus of arc activity shifted to the Lesser Antilles volcanic arc that remains active today (Lidiak and Larue, 1998) (Fig. 4).

Eocene Deformation in Puerto Rico

Following collision of the Puerto Rico arc and cessation of arc volcanism, most deformation in Puerto Rico was focused on the high-angle Great Northern and Great Southern Puerto Rico fault zones (Fig. 3A). The Great Southern Puerto Rico fault zone preserves a younger, 110-km-long, 5–10 km wide strip of middle Eocene marine sedimentary and volcanic rocks (Cerrillos belt). These submarine rocks of mainly marine turbidite facies and localized volcanic centers were deposited during the last phase of the middle Eocene Caribbean volcanic arc (Glover, 1971; Dolan et al., 1991).

Rocks of the Cerrillos belt and adjacent arc rocks exhibit widespread effects of convergent deformation that also affected the underlying volcanic arc. Most thrusts in the Cerrillos belt dip ~30° to the southwest and folds are asymmetrical with the axial plane inclined toward the southwest (Glover, 1971). Detachment planes at shallow depth are indicated by small-scale isoclinal folds in thinly bedded units. Folds wider than 1 km trend N70–80W parallel to the dominant direction of pre-Oligocene folds in Puerto Rico. Smaller folds are isoclinal and do not parallel the regional structure as closely as the larger folds.

The Great Northern Puerto Rico fault zone is a west-northwest to west-northwest–striking system of high-angle faults deforming rocks of early Cretaceous to Eocene in age and extending across the northeast sector of the island. According to Briggs et al. (1970) and Monroe (1980), all strands of the Great Northern fault zone are unconformably overlain by rocks of the Oligocene–early Pliocene carbonate shelf of northern Puerto Rico and are therefore older than Oligocene in age. Briggs et al. (1970) state that the total, pre-Oligocene left-lateral offset of all strands of the fault zone is greater than 60 km, although some individual strands of the fault zone show minor right-lateral offsets.

Glover (1971) and Erikson et al. (1990) infer that the main period of left-lateral and thrust motion along the Great Southern Puerto Rico fault zone is confined to Eocene and older times because several strands of both fault systems are unconformably overlain by the Oligocene–early Pliocene carbonate cap of northern and southern Puerto Rico (Figs. 3A and 4). Faults and folds of the late Eocene to earliest Oligocene shortening event in Puerto Rico do not affect overlying and much less deformed Oligocene shallow marine and nearshore rocks of the San Sebastian Formation of early Oligocene age on the north coast of Puerto Rico or rocks of the early Oligocene Juana Diaz Formation on the south coast of Puerto Rico (Fig. 4). Rocks of the San Sebastian Formation unconformably overlie the middle Eocene Cerrillos belt rocks as described at several localities in northwestern Puerto Rico (McIntyre, 1971; Tobisch and Turner, 1971; Mann et al., this volume, Chapter 6). Similarly, rocks of the Juana Diaz Formation unconformably overlie folded and thrusted middle Eocene rocks of the Great Southern Puerto Rico fault zone of south-central Puerto Rico (Glover, 1971) (Fig. 4).

Puerto Rico–Virgin Islands Carbonate Platform

The widespread angular unconformity indicates that the late Eocene shortening event resulted in abrupt uplift of the deep marine sedimentary rocks of the Cerrillos belt to shallow depths. Oligocene siliciclastic rocks of the San Sebastian Formation in northern Puerto Rico (Monroe, 1980; van Gestel et al., 1999) and the Juana Diaz Formation in southern Puerto Rico were deposited in shallow-water to continental environments. In southern Puerto Rico, clastic rocks of the Juana Diaz Formation lie in angular unconformity with all older rocks ranging in age from Cretaceous to Eocene, but at most places the Juana Diaz Formation is in fault contact with these older rocks and the unconformity cannot be seen (Monroe, 1973; 1980). These clastic rocks form the base for the Puerto Rico–Virgin Islands carbonate platform (van Gestel et al., 1998, 1999) (Fig. 4). For reasons that are not clear, Oligocene clastic facies persisted longer and are slightly thicker in southern Puerto Rico than in northern Puerto Rico (Monroe, 1980) (Fig. 4). Clastic rocks of the Juana Diaz Formation of southern Puerto Rico pass upwards into Oligocene chalks, chalky limestone, and calcareous sandstone informally called the carbonate part of the formation by Monroe (1980).

The Puerto Rico–Virgin Islands carbonate platform covers an area of ~18,000 km^2 and extends from the eastern Dominican Republic on the island of Hispaniola, west of Puerto Rico, to the Virgin Islands, east of Puerto Rico (van Gestel et al., 1998) (Fig. 1). The continuity and general similarity of shallow-water carbonate facies across the platform in all of these areas indicates a remarkable tectonic stability for the region for a period of 35 m.y. (early Oligocene–early Pliocene) (Fig. 4). Detailed studies of the platform rocks in northern Puerto Rico by Monroe (1980) and Moussa et al. (1987) and by Frost et al. (1983) in southern Puerto Rico show that deposition of the platform occurred near sea level with minor periods of early Pliocene subsidence during the youngest period of platform development in northwestern Puerto Rico (Moussa et al., 1987). The thickness of the platform in northern and southern Puerto Rico ranges between 500 and 1000 m thick (Fig. 4). Breaks in deposition do not appear to coincide on the two sides of the island.

The less steeply dipping, less faulted northern part of the Puerto Rico–Virgin Islands carbonate platform is characterized by extensive karst topography that is not observed in the more heavily faulted and steeper dipping southern area (Figs. 3A, 3B, and 3C).

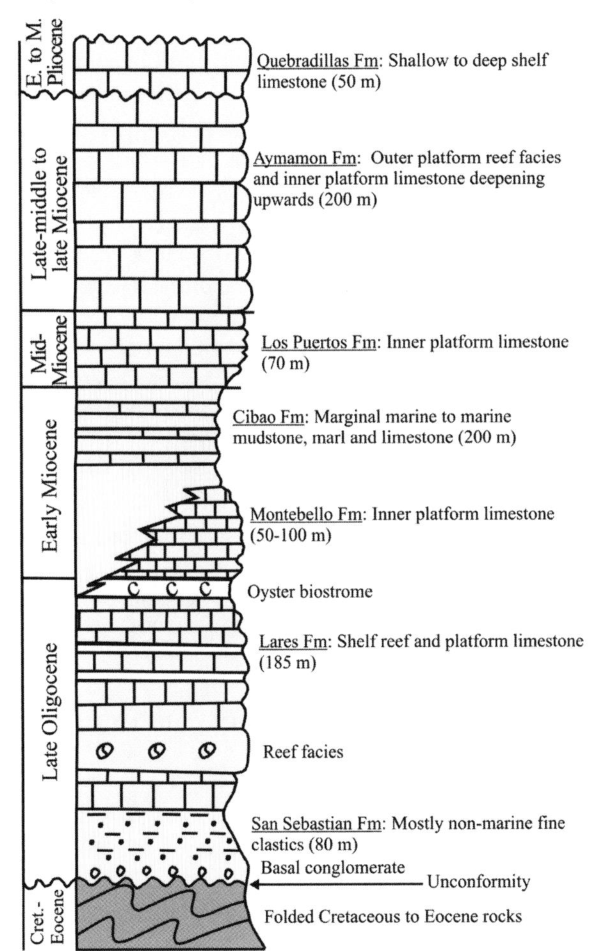

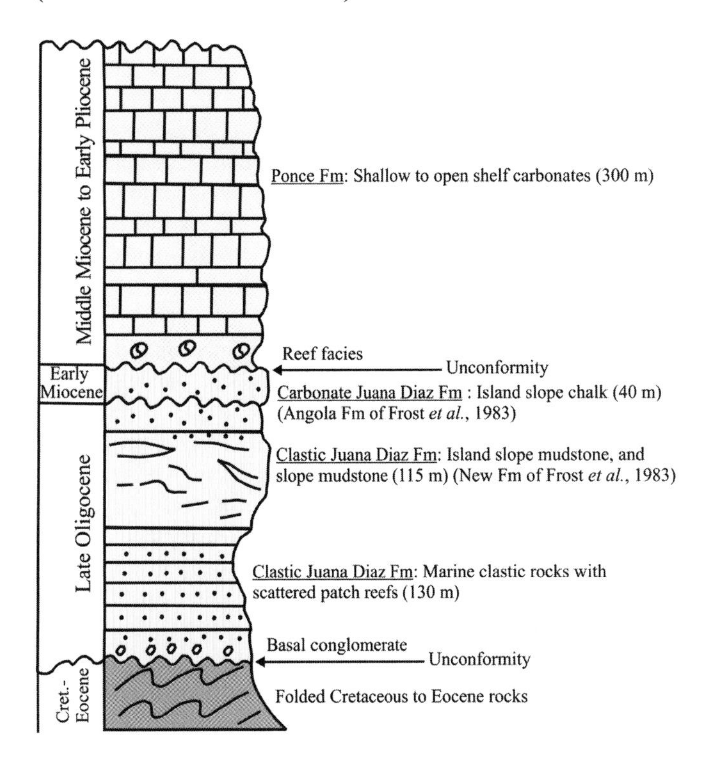

Modified from Frost *et al.*, (1983)

Figure 4. Comparison of Oligocene to Pliocene stratigraphic sequence from north coast of Puerto Rico (Lares area) and southwestern coast (Guanica-Penuelas area) modified from Frost et al. (1983). Locations of the northern and southern outcrop areas on the flanks of the Puerto Rico-Virgin Islands arch are shown on the map in Figure 3.

Although the carbonate sections on the north and south coasts are of similar age and of generally similar composition, exact correlations of individual stratigraphic units has not been possible between the two areas because of differences in their lithology and paleontology (Monroe, 1973, 1980; Frost et al., 1983) (Fig. 4).

Post-Eocene Deformation in Puerto Rico

Field evidence for post-Eocene displacement affecting rocks of the Puerto Rico–Virgin Islands carbonate platform or Quaternary units along the Great Northern and Southern Puerto Rico fault zones is sparse. Geomatrix Consultants (1988), however, reports several east-west–trending zones of inferred late Quaternary deformation along the Great Southern Puerto Rico fault zone in south-central Puerto Rico (Fig. 3A). The distal end of the Great Southern Puerto Rico fault zone adjacent to the coastal plain area consists of the parallel Rio Jueyes and Esmeralda fault strands. Both fault strands are covered by tilted but largely unfaulted post-Eocene clastic and carbonate strata (Glover and Mattson, 1960; Glover, 1971; Geomatrix Consultants, 1988). Glover (1971) uses unpublished gravity data from the Puerto Rico coastal plain to propose that the Rio Jueyes fault zone extends to the southeast beneath the sediments of the Quaternary alluvial plain along with another fault trace to the south that is informally named the Salinas fault by Geomatrix Consultants (1988). Garrison (1969) and McCann (1985) propose that the southeastward extension of the Great Southern Puerto Rico fault zone immediately offshore of south-central Puerto Rico deforms recent sediments imaged on seismic profiles, although there has been no direct dating of the seafloor sediments disturbed by the faulting.

Post–Early Pliocene Formation of the Puerto Rico–Virgin Islands Arch

The uniform dip of the Puerto Rico–Virgin Islands carbonate platform by 6–11° off the north coast of Puerto Rico and steeper, more variable dips in the range of 10–30° off the south coast of Puerto Rico define the Puerto Rico–Virgin Islands arch whose axis is indicated on the map in Figure 3A and is shown in perspective view in Figure 3B. The elongate arc core of the island rises 1.3 km above sea level in the central range, or Cordillera Central, of Puerto Rico (Fig. 3A).

Glover (1971) named this immense feature the "Puerto Rico geanticline" and postulated that deformation of the feature began in the middle Eocene–early Oligocene and continues to the present. Glover (1971) presented no evidence for his interpretation for the age of arching. The crest of the arch nearly coincides with the present-day drainage divide of Puerto Rico, which also lies nearer the south coast of the island (Glover, 1971) (Fig. 3A).

In the north, the platform rocks of the gentler-dipping northern limb of the arch are deformed by a few small anticlines and faults having displacements of 35 m or less. In the northwestern corner of the island, fault displacements are much more significant due to their proximity to the actively opening seismogenic Mona rift (Monroe, 1973; 1980; Hippolyte et al., this volume) and right-lateral Cerro Goden fault zone (Mann et al., this volume, Chapter 6) (Fig. 1). Rocks of the more steeply dipping southern limb of the arch are broken by many faults, some with displacements up to several hundred meters (Monroe, 1973; 1980).

Seismic profiling has defined the crest of the arch shown on Figure 3A beneath the coastal shelf of western Puerto Rico (Grindlay et al., this volume, Chapter 7) and beneath the coastal shelf separating eastern Puerto Rico and the Virgin Islands (van Gestel et al., 1998). Remnants of the Puerto Rico–Virgin Islands carbonate platform are not exposed on the western and eastern ends of Puerto Rico because the outcrop pattern of Oligocene–early Pliocene rocks on the south flank of the arch forms an arcuate pattern as seen on the map in Figure 3A. The arcuate pattern for the north flank of the arch is particularly evident when the thicknesses of the offshore carbonate units are displayed together with its equivalent onland outcrop in map view (van Gestel et al., 1999).

Joyce et al. (1987) propose that the Mona Passage extension may control the "basin and range topography" of western Puerto Rico (Fig. 3). Grindlay et al. (this volume, Chapter 2) and Mann et al. (this volume, Chapter 6) have identified offshore and onland faults concentrated in the basement core of the arch and propose that the localization of these faults may reflect the higher strains in the crest area of the arch (Figs. 3B and 3C). Grindlay et al. (this volume, Chapter 7) show a close spatial correlation between active faulting and the locations of serpentinite belts and propose that there may also be a lithologic control on faulting.

In general, major normal and oblique-normal faults dip to the south on the northern flank of the Puerto Rico–Virgin Islands arch and to the north on the southern flank as seen on the topographic perspective diagram in Figure 3C. These onland fault scarps control flat-bottomed, alluvial basins of likely half-graben origin along both the western and southeastern coasts of the island (Mann et al., this volume, Chapter 6) (Fig. 3A).

The age of the arch is assumed to be post–early Pliocene, the age of the highest tabular, stratigraphic unit affected by arching on the north coast of Puerto Rico (Quebradillas Formation; Moussa et al., 1987) (Fig. 4). At all previous times during the Oligocene–late Miocene the tabular carbonate units of northern and southern Puerto Rico remained near sea level and were apparently unaffected by arching (van Gestel et al., 1998; 1999). The prominent geomorphology of fault scarps and basins of western Puerto Rico indicates that arching continues to the present-day (Mann et al., this volume, Chapter 6) (Fig. 3C).

The tectonic origin of the Puerto Rico–Virgin Islands arch is controversial. Proposed tectonic mechanisms include: (1) counterclockwise rotation of the Puerto Rico block (Jany et al., 1990) (Fig. 2B); (2) northwest to southeast rifting related to subduction rollback with the formation of rollover anticlines (Speed and Larue, 1991, Fig. 2C); (3) interaction of the Caribbean and North America subducted slabs at depth beneath Puerto Rico (Dolan et al., 1998; van Gestel et al., 1998); and (4) late Neogene collisional interaction between the easternmost tip of Bahama carbonate platform and Puerto Rico (Mann et al., 2002; Grindlay et al., this volume, Chapter 2). Whatever its tectonic origin, the Puerto Rico–Virgin Islands arch appears to be a younger deformational event that is superimposed on the older, convergent structures formed during the late Eocene tectonic shortening event described by Glover (1971), Dolan et al. (1991), and Mann et al. (this volume, Chapter 6).

TECTONIC GEOMORPHOLOGY OF SOUTHERN PUERTO RICO

Overview and Methods

We select several key areas along the coast of southern Puerto Rico in Figures 5–10 to illustrate its along-strike changes in tectonic geomorphology using aerial photographs and published U.S. Geological Survey 1:20,000-scale geologic maps available in GIS format (Bawiec, 2001). In most areas, we utilize photomosaics composed of 1:40,000-scale photographs taken in 1988. In some cases, we use 1:18,700-scale 1936–1937 photographs that are especially helpful for establishing the geomorphology of the south coast area prior to the rapid phase of development in agriculture and urbanization that affected the region in the latter half of the twentieth century.

Western Lajas Valley and Environs

Setting

The Lajas Valley and its offshore extension into Boqueron Bay forms the most southerly and longest of the three, east-west–trending alluvial basins of western Puerto Rico that are prominent on the regional DEM in Figures 3A, 3B, and 3C. The other two east-west–trending valleys are the Anasco Valley in the north that is discussed by Mann et al. (this volume, Chapter 6) and Grindlay et al. (this volume, Chapter 7) and the Guanajibo Valley in the central part of western Puerto Rico that is discussed by Grindlay et al. (this volume, Chapter 7) (Fig. 3A).

The Lajas valley is 30 km long and ranges from 1.5 to 4.5 km in width (Fig. 5). The valley is open at both ends and drains to Boqueron Bay in the west and Guanica Bay to the east via a manmade canal system. The elevation of the valley floor varies from sea level at Boqueron Bay to 15 m above sea level (ASL) along the east-west drainage divide located 10 km east of Boqueron Bay. The mountain ranges bounding the valley rise up to elevations of 200 m ASL in the Sierra de Guanajibo north of the valley and 300 m ASL in the Sierra de Bermeja south of the valley (Fig. 5).

Streams flowing out of these ranges into the valley are small and ephemeral. Following heavy rains, flow from these streams disappears into the alluvial deposits of the valley (Anderson, 1977). Surface-water bodies within the Lajas Valley include Laguna Cartagena and a large mangrove swamp east of Boqueron Bay (Fig. 5). A larger natural lake in the western part of the valley (Lake Guanica) was drained in 1955 for agricultural purposes by a canal system leading to Guanica Bay on the southern coast of Puerto Rico (Anderson, 1977).

Tectonic Geomorphology and Faulting

The Lajas Valley has long been suspected as a possible area for active faulting because of its prominent east-west topographic expression, linear mountain fronts, and closed topographic depressions occupied by lakes or dried lakes within the valley (Almy, 1969; McCann et al., 1987; Joyce et al., 1987) (Fig. 5).

Quadrangle mapping by Volckmann (1984a, 1984b) along with subsequent, detailed mapping studies of early Tertiary and Mesozoic rocks summarized by Jolly et al. (1998) reveal that the dominant fault trend in the older rocks exposed in ranges flanking the valley is to the northeast and oblique to the east-west topographic trend of the Lajas Valley (Fig. 5B). Northeast-trending fault-bounded belts in both the Sierra de Bermeja south of the valley and the Sierra de Guanajibo north of the valley include abundant belts of serpentinized periodotite of Mesozoic age (Fig. 5B). No Quaternary faults are shown on 1:25,000-scale geologic quadrangle maps by Volckmann (1984a, 1984b) of the area of the Lajas Valley and its bounding mountain ranges that is shown in the map area of Figure 5. An extensive field mapping effort by Geomatrix Consultants (1988) also does not reveal any evidence for late Quaternary faults either within or along the edges of the Lajas Valley.

High-resolution onland seismic reflection profiles described by Meltzer et al. (1995) and Meltzer and Almy (2000) across the southern margin of the valley revealed east-west–trending faulted offsets of reflectors that are correlated to Quaternary lacustrine sediments of Laguna Cartagena and its larger, predecessor lake which once filled the closed topographic depression of the present-day, central valley floor. These authors interpret the subsurface offsets along the southern edge of the Lajas Valley as late Quaternary faults of transtensional origin.

A grid of high-resolution offshore seismic profiles and accompanying sidescan sonar mosaic described by Grindlay et al. (this volume, Chapter 7) from Boqueron Bay did not reveal the seaward projection of any of the active, east-west–trending faults proposed by Meltzer et al. (1995) and Meltzer and Almy (2000) along the southern edge of the bay (Fig. 5B). Single-channel seismic lines did reveal basinal reflectors dipping 3° to the south, which Grindlay et al. (this volume, Chapter 7) interpret as the Oligocene-Pliocene Puerto Rico–Virgin Islands carbonate platform tilted on the southern limb of the Puerto Rico–Virgin Islands arch (Fig. 3). These dipping reflectors are overlain by relatively flat-lying reflectors interpreted to be the Pliocene-Holocene sedimentary section known from water wells drilled on the floor of the Lajas Valley (Anderson, 1977).

Grindlay et al. (this volume, Chapter 7) interpret the southward dip of basinal reflectors as evidence for a large normal fault on the south side of the valley that was not imaged by their survey. This half-graben interpretation would also explain the overall topography of the valley with a much steeper gradient in topography on the southern edge of the valley than on the northern flank (Fig. 5).

Prentice and Mann (this volume) identify the South Lajas fault zone, a 1-km-long, 1.5–3.0-m-high scarp cutting an alluvial fan on the south side of the Lajas Valley (Fig. 5). The fault has a northeasterly trend that is oblique to the overall east-west trend of the Lajas Valley and does not project parallel to the east-west-trending, southern edge of the Lajas Valley or in the direction of Boqueron Bay (Grindlay et al., this volume, Chapter 7). Trenching by Prentice and Mann (this volume) revealed two fault zones, ~1 m apart, that disrupt Quaternary alluvial fan deposits, radiocarbon dated at ~5000 yr B.P. The presence of two colluvial wedges indicates the occurrence of at least two earthquakes during the past 7000 yr. Relations indicate normal faulting, valley-side down with a component of strike-slip motion. On the basis of a single trench exposure, it could not be determined whether the horizontal component of displacement was right-lateral or left-lateral.

Interpretation of Tectonic Geomorphology and Faulting

The northeast-striking South Lajas fault and the northeast-striking faults affecting pre-Oligocene basement rocks, including elongate belts of serpentine, are subparallel and may indicate preferential Holocene reactivation of these older fault systems and accompanying uplift of their intervening basement blocks (Fig. 5). For example, the scarp of the South Lajas fault is 3 km to the west and subparallel to a major range-bounding fault that defines the western edge of the northeast-trending exposure of pre-Oligocene rocks of the Sierra de Bermeja (Volckmann, 1984a) (Fig. 5B). Volckmann (1984a) notes that the fault is a high-angle, northeast-trending fault marked by a zone of intense shearing ~60 m wide. The northwest side of the fault has been downdropped relative to the southeast side as Tertiary rocks dip toward and appear to have been faulted against older Mesozoic rocks. To the west of this fault is a topographic saddle separating the pre-Oligocene rocks of the Sierra de Bermeja with an extensive area of Oligocene and Miocene carbonate rocks. One possible interpretation is that the carbonate rocks are preserved

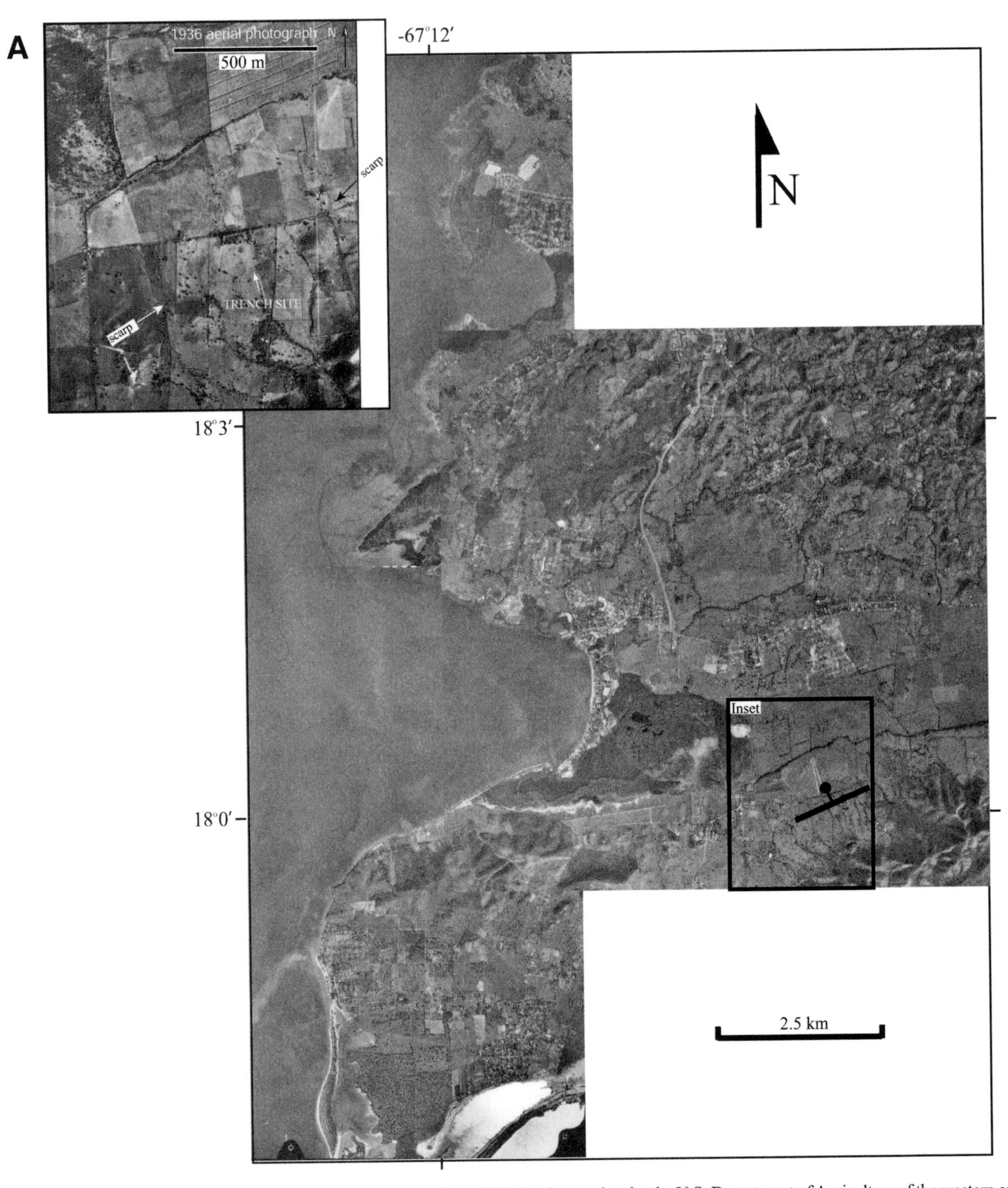

Figure 5 (*on this and following page*). (A) Photomontage made from 1986 air photos taken by the U.S. Department of Agriculture of the western end of the Sierra de Guanajibo, the western end of the Lajas Valley and Boqueron Bay, and the Sierra de Bermeja, southwestern Puerto Rico (original scale: 1:40,000). Location of map shown on regional DEM in Figure 3A. Inset map is an enlargement of a 1:18,700 scale, 1936 air photograph showing the scarp of the South Lajas fault, the only documented Holocene fault in Puerto Rico (Prentice and Mann, this volume). Box on the larger mosaic shows the location of the air photo and South Lajas fault scarp. (B) Simplified geologic map of the same area in A based on 1:25,000-scale geologic map information provided in GIS format by the U.S. Geological Survey (Bawiec, 2001). Note the presence of three trends: (1) a northeast trend defined by the strike of pre-Quaternary bedrock units and older faults affecting these units; (2) an east-west trend defined by the Lajas Valley and flanking east-west–trending topographic ranges; and (3) a northeast trend defined by the Holocene South Lajas fault zone. Southward dips of Oligocene and Miocene rocks reflect their position on south limb of Puerto Rico–Virgin Islands arch (cf. Fig. 3A). The 5-km-long Holocene scarp of the South Lajas fault is indicated. Qa—Quaternary alluvium; Qb—Quaternary beach deposits; Qs—Quaternary swamp deposits; QTs—Quaternary?-late Tertiary? Deposits; Tpo—Miocene-Pliocene Ponce Formation; Tgua—upper Miocene Guanajibo Formation; Km—Maestrichtian-Campanian Melones Formation; Kcot—Maestrichtian-Campanian Cotui Formation; Kp—Maestrichtian-Santonian Parguera Formation; Ks—Maestrichtian-Turonian Sabana Grande Formation; Tky—upper Cretaceous-lower Tertiary Yauco Formation; Kpob—upper Cretaceous two-pyroxene olivine basalt.

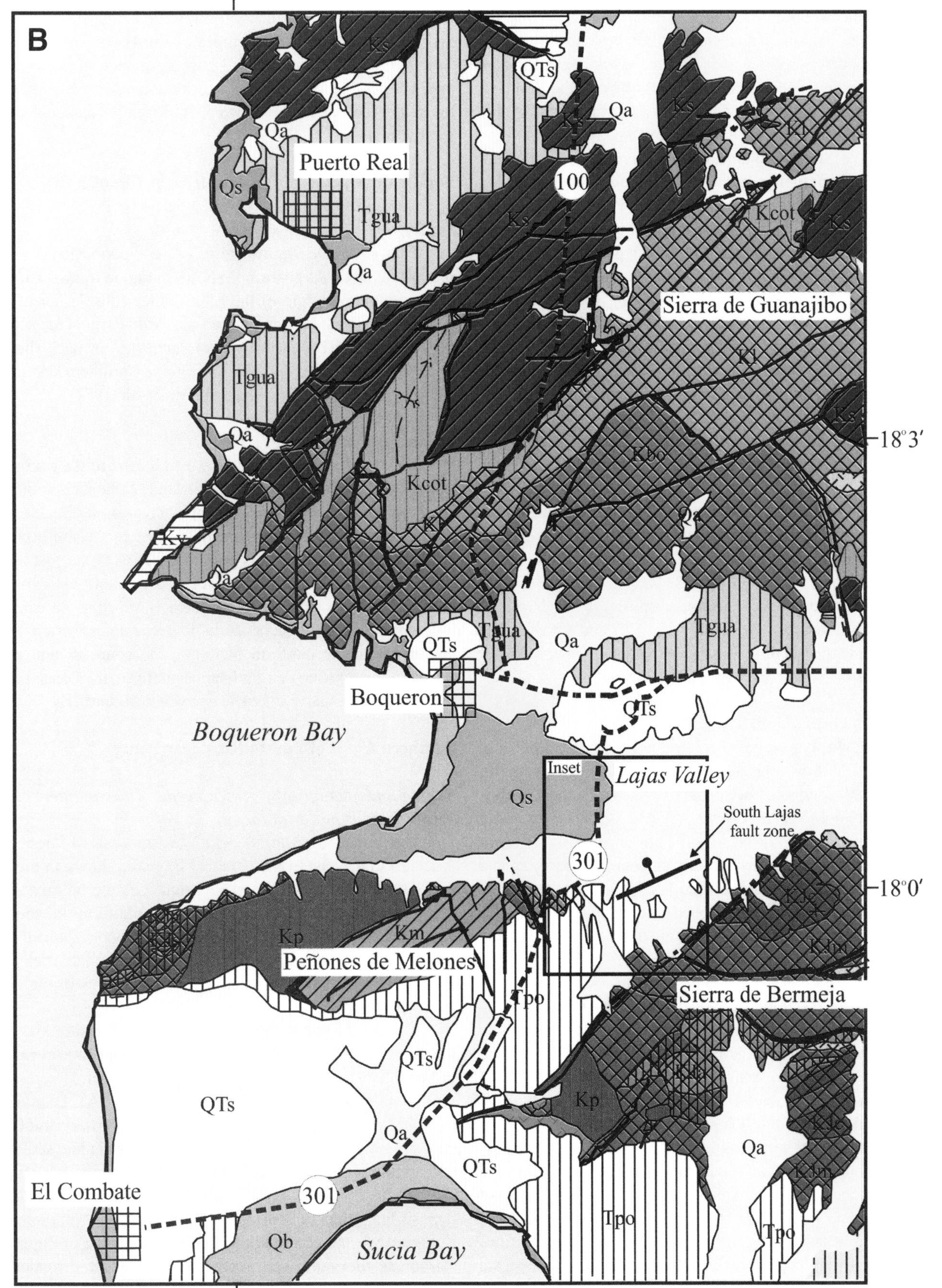
-67°12′
B
Puerto Real
Sierra de Guanajibo
18°3′
Boqueron
Boqueron Bay
Inset
Lajas Valley
South Lajas fault zone
18°0′
Peñones de Melones
Sierra de Bermeja
El Combate
Sucia Bay
100
301
301
Ks
QTs
Qa
Qs
Tgua
Kl
Kcot
Kbo
TKy
Kp
Km
Tpo
Qb

on the downthrown side of a large normal fault bounding the basement uplift. An alternative view by Prentice and Mann (this volume) is that the South Lajas fault forms a segment of a more east-west–trending fault system that aligns with the southern topographic edge of the valley and the linear former shorelines of lakes that currently or previously occupied the lake floor.

Grindlay et al. (this volume, Chapter 7) has noted the spatial correlation and parallelism between possible northwest-striking Holocene faults mapped offshore of western Puerto Rico and the northwest trend of two onland serpentinite belts (Rio Guanajibo and Monte del Estado). The relationship suggests that either the presence of the serpentinite has localized the inferred Holocene fault zones or the inferred Holocene faulting has remobilized the serpentinite. In the case of the south Lajas Valley and Sierra de Bermeja, the subparallelism between the South Lajas fault of documented Holocene age and the basement trends in the uplifted Sierra de Bermeja suggests a similar interpretation.

Southwestern Coast of Puerto Rico near La Paguera

Setting

This part of the southwestern coast is 5 km south of the southern margin of the Lajas Valley (Fig. 6). The topographic crest of the narrow range of low hills separating the Lajas Valley from the Caribbean Sea is ~150 m ASL. The La Paguera shelf has been extensively studied using seismic profiles, dredging, and bottom observations by Saunders (1973) and Morelock et al. (1994) and consists of a broad, unfaulted carbonate platform characterized by small coral cays and shallow shoals.

Tectonic Geomorphology

The southwest coast near La Paguera is of interest because of evidence for drowning and lack of late Quaternary uplift. Unlike the shelf south of Ponce to the east, the La Paguera shelf is relatively flat, and structureless exhibiting no evidence for active faulting. Drowning is indicated by the presence of older Mesozoic rocks in the surf zone without any evidence of overlying late Neogene carbonate rocks or uplifted late Quaternary coral reef limestone. Older rocks are deformed into a large, west-northwesterly trending syncline formed during the Eocene convergent event (Volckmann, 1984a, 1984b) (Fig. 6).

Interpretation of Tectonic Geomorphology and Faulting

The drowned coastline is consistent with a lack of tectonic activity, faulting, or late Quaternary uplift in the coastal area. The only indication of tectonic activity is the South Lajas fault located 7 km to the north in the interior part of the Lajas Valley (Fig. 3A). Grindlay et al. (this volume, Chapter 7) conducted a single-channel seismic and sonar survey in the area of Cabo Rojo and Sucia Bay to the south of the Lajas area shown in Figure 5 and to the southwest of the La Paguera area shown in Figure 6 and did not find any evidence for seafloor faulting.

The map of the La Paguera area in Figure 7B derived from the quadrangle mapping of Volckmann (1984a, 1984b) shows the presence of the angular unconformity separating steeply dipping and folded Cretaceous units in the northeastern part of the map from the more gently dipping Ponce Formation of late Neogene age. The gentle dip of the Ponce formation reflects its position on the southern limb of the Puerto Rico–Virgin Islands arch shown in Figure 3B.

Southwestern Coast of Puerto Rico at Guanica Bay

Setting

This part of the southwestern coast of Puerto Rico is 5 km to the east of the La Parguera area shown in Figure 6 and 3 km south of the southern margin of the Lajas Valley (Fig. 7). There is no topographic range separating the Lajas Valley from Guanica Bay because the deep bay probably represents the drowned valley of a river that once flowed southward from the Cordillera Central and Lajas Valley to the Caribbean Sea (Anderson, 1977).

Tectonic geomorphology

Guanica Bay represents the western limit of the continuous outcrop of the Puerto Rico–Virgin Islands carbonate platform of southern Puerto Rico (Fig. 3A). For that reason, we interpret it as the western edge of a broad zone of late Neogene stability or slow uplift that extends from this point for 30 km to the east toward Ponce (Fig. 3A). West of Guanica Bay are mainly Mesozoic and early Tertiary rocks; east of Guanica Bay begin the extensive outcrops of the Oligocene–early Pliocene Puerto Rico–Virgin Islands carbonate platform including the youngest unit of the southern Puerto Rico–Virgin Islands platform, the Ponce Formation of Miocene–early Pliocene age (Figs. 4B and 7B).

Southern Coast of Puerto Rico near Ponce

Setting and Stratigraphy of Oligocene-Miocene Puerto Rico–Virgin Islands platform rocks

This semiarid part of the south-central coast of Puerto Rico is 10 km to the east of the Guanica Bay area shown in Figure 7 and near the center of the most extensive outcrop area of the Puerto Rico–Virgin Islands carbonate platform in southern Puerto Rico (Monroe, 1980) (Fig. 3A). Topographically, the area consists of large, domal outcrops of carbonate rocks generally capped by more erosionally resistant units of the Ponce Formation.

The city of Ponce marks a geomorphic transition between domal, carbonate hills to the west and a broad coastal plain to the east that extends west of the Great Southern Puerto Rico fault zone in southeastern Puerto Rico (Fig. 3A). Domal hills of Neogene carbonate rocks and older early Tertiary and Cretaceous rocks protrude locally through the coastal plain sediments and are thought to underlie both the alluvial plain and offshore shelf (Glover, 1971). River valleys separate domal areas and deposit fan deltas in coastal areas as seen on the photomosaic in Figure 8A and compilation of USGS mapping in Figure 8B. Carbonate rocks are well exposed in quarries and roadcuts in

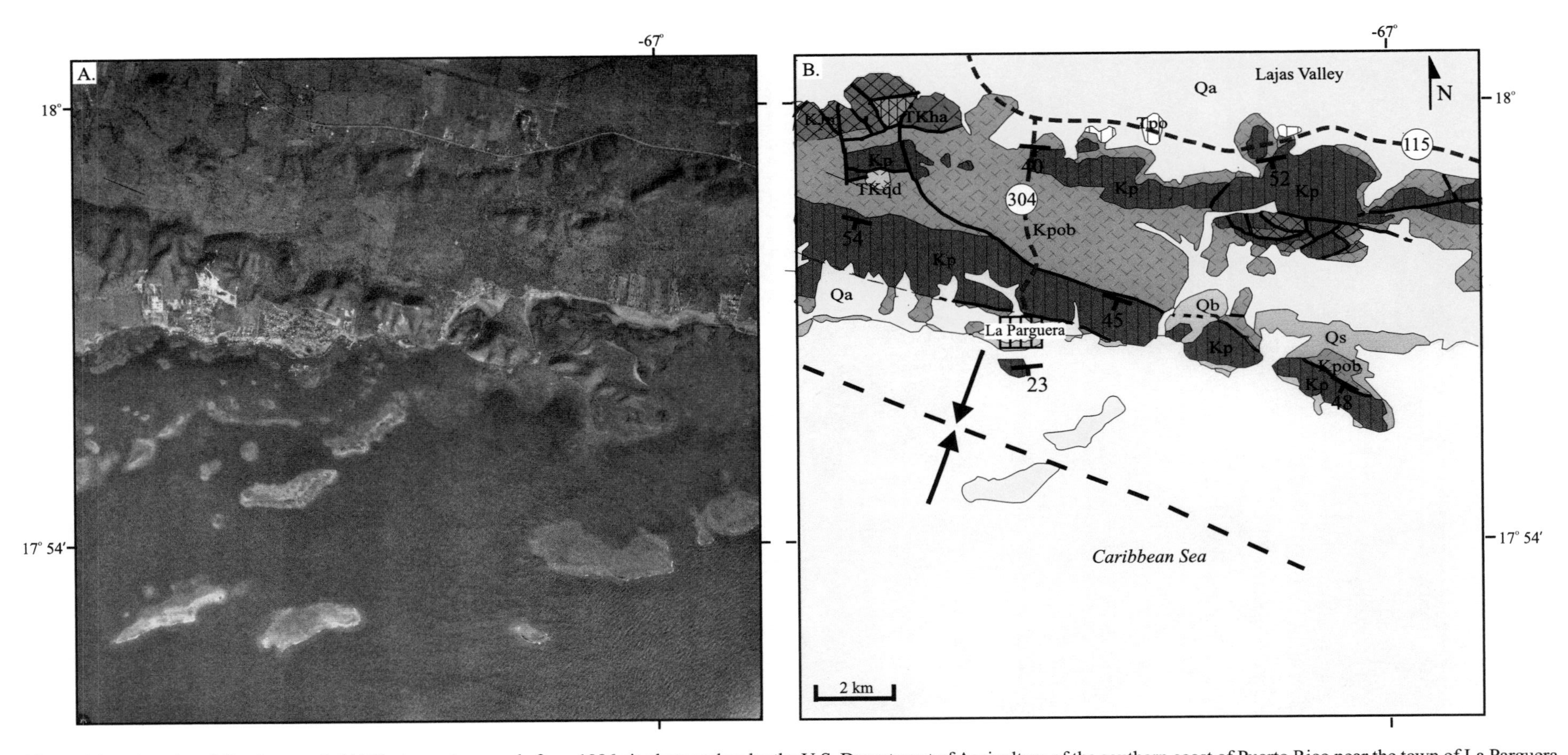

Figure 6 (*continued on following page*). (A) Photomontage made from 1986 air photos taken by the U.S. Department of Agriculture of the southern coast of Puerto Rico near the town of La Parguera (original scale: 1:40,000). Location of map shown on regional DEM in Figure 3A. (B) Simplified geologic map of the same area in A based on 1:25,000-scale geologic mapping (Bawiec, 2001). Note the presence of a west-northwest trend defined by the strike of folded and faulted pre-Oligocene bedrock units. Unconformably overlying Oligocene-Miocene sedimentary rocks dip southward and reflect their position on the south limb of the Puerto Rico–Virgin Islands arch. The embayed coast formed by submergence of the pre-Oligocene fold belt that is subparallel to the coast. Qb—Quaternary beach deposits; Qs—Quaternary swamp deposits; Tkha—Cretaceous?-Tertiary? hydrothermally-altered rock; Kpob—upper Cretaceous two-pyroxene olivine basalt; Kp—Maestrichtian-Santonian Parguera Formation; Ks—Maestrichtian-Turonian Sabana Grande Formation.

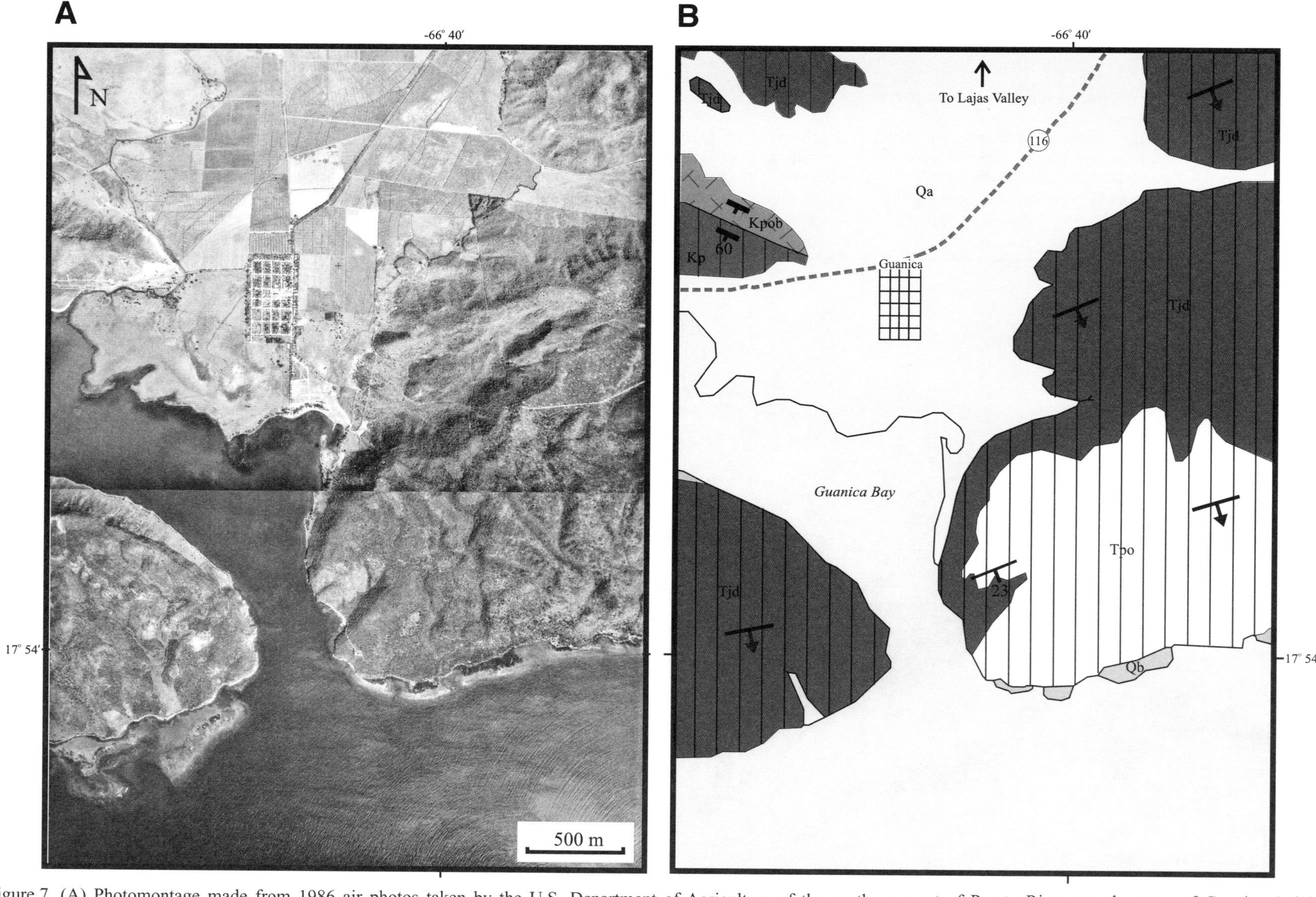

Figure 7. (A) Photomontage made from 1986 air photos taken by the U.S. Department of Agriculture of the southern coast of Puerto Rico near the town of Guanica (original scale: 1:40,000). Location of map shown on regional DEM in Figure 3A. (B) Simplified geologic map of the same area in A based on 1:25,000-scale geologic mapping (Bawiec, 2001). Note the presence of an east-northeast trend defined by the strike of folded pre-Oligocene bedrock units and older faults affecting these units. The coast exhibits a drowned morphology with Guanica Bay a likely drowned river valley. Note the moderate southward dips of Oligocene and Miocene carbonate units forming the southern flank of the Puerto Rico–Virgin Islands arch (cf. map in Fig. 3A). Guanica Bay represents the approximate eastern limit of Oligocene-Pliocene carbonate units exposed in southern Puerto Rico. Qa—Quaternary alluvium; Tpo—Miocene-Pliocene Ponce Formation; Tjd—Oligocene-Miocene Juana Diaz Formation; Kpo—Maestrichtian-Santonian Pozas Formation; Kp—Maestrichtian-Santonian Parguera Formation; Kpob—upper Cretaceous two-pyroxene olivine basalt.

the domal areas around the city of Ponce where limestone quarrying supports a large cement industry. Older limestone quarries are now being used as landfills by Ponce, Juana Diaz, and other major cities in the area.

The southern coastal plain of Puerto Rico is made up of several large and coalesced alluvial fans originating in the Cordillera Central (Figs. 3A and 8B). The mountain front bordering these fans is highly irregular and contains several large embayments where the major rivers and their fans exit the mountains (Geomatrix Consultants, 1988). Alluvial fan environments merge in a southward and downdip direction with late Quaternary swamp and beach deposits in the coastal area.

Tectonic Geomorphology of the Coast and Offshore Islands

The arcuate distribution of domal-shaped carbonate units at the coast indicate the possibility of active emergence of the southern limb of the Puerto Rico–Virgin Islands carbonate platform as suggested by large-scale structure of the arch shown on Figure 3C. Previous workers have identified marine terraces along the southern coast between Ponce and Guayama that suggest that this part of the coast is possibly slowly emergent. Terraces include a 1.5–2.0-m-high marine terrace deposit on Isla Caja de Muertos on the shallow shelf south of Ponce (Fig. 3). The marine terrace contains fragments of shells and coral heads (Lobeck, 1922). Two areas of beach ridges up to several meters above sea level are present along the coast near Ponce (Figs. 8A, inset, and 9), and one 3-m-high, wave-cut terrace at an elevation of 3 m ASL is found near Salinas (Prentice et al., 2004) (Fig. 10).

The steep and uneroded morphology of the soft shoreline material in the beach ridges and the terrace near Salinas suggests that these features are young, perhaps Holocene in age (Prentice et al., 2004). The low elevations (<3 m ASL) of the marine terrace on Caja de Muertos, the beach ridges near Ponce, and the marine terrace near Salinas make it unlikely that these features formed 120,000 yr B.P. when global sea level was significantly higher (+6 m) than it is today. Their low elevations are more consistent with higher stands of Holocene global sea level or perhaps tectonic uplift in late Holocene time. Despite the presence of these terraces, Kaye (1959) and Monroe (1968) conclude that there has been no significant late Pleistocene crustal movement in southern Puerto Rico. In contrast, Horsfield (1975) concludes that slow rates of late Quaternary uplift have affected southern Puerto Rico although he had no direct radiometric data.

The scalloped "zediform" shape of the modern coastline of southern Puerto Rico (Taylor and Stone, 1996; Otvos, 2000) shown on Figure 3A is closely controlled by the position of the major alluvial fans, the southeast-prevailing trade wind and wave direction (with consequent east-to-west coast-parallel transport of sand), and, possibly, by localized tectonic uplift along northeast-trending fault systems.

Tectonic Geomorphology of the Coastal Plain

In the coastal plain area, Geomatrix Consultants (1988) provide maps of well-preserved fluvial terrace sequences that are incised into the coalesced alluvial fans. Prominent fluvial terraces are found on several major north-south–trending rivers draining the Cordillera Central northeast of Ponce including the Rio Portugues, Descalabrado, Coamo, and Jueyes. Typically, these terraces exhibit original depositional morphology including meander scars, mid-channel bars, and scroll-bar topography. Detailed mapping using aerial photograph study combined with field observations by Geomatrix Consultants (1988) was not able to establish any fault lineaments affecting these terrace sequences.

The area of Oligocene-Pliocene carbonate rocks of southern Puerto Rico shown on Figure 3A is deformed by a system of widely spaced faults with common orientations of east-west or north-south (Glover, 1971; Krushensky and Monroe, 1975; 1978; Monroe, 1980) (Fig. 8B). These high-angle faults are generally discontinuous and cannot be traced for distances greater than 10 km (Monroe, 1980). In outcrop the faults are high-angle and exhibit normal throws to the north or south. As can be seen on the photomontage in Figure 8A, the faults form prominent lineaments and fault valleys in the Neogene carbonate lithologies. Monroe (1980) proposes that the east-west striking faults are displaced by north-south–striking faults. Later work by Geomatrix Consultants (1988), however, indicates that one east-west–striking fault (San Marcos fault) displaces a north-south–trending one. Glover (1971) and Erikson et al. (1990) describe a subsurface fault known from drill data in the Ponce area (Fig. 8A). The fault is northeast-trending with its downthrown side to the southeast.

One of the most detailed studies of a possible late Quaternary fault in southern Puerto Rico was carried out by Geomatrix Consultants (1988) on the San Marcos fault because the fault is located at a critical site 3 km south of the Portugues dam northeast of Ponce. The fault was originally mapped by Krushensky and Monroe (1975, 1978) as a 3.8-km-long normal fault with down-to-the-south displacement and occurring predominantly within Oligocene age lithologies of the Juana Diaz Formation. The eastern end of the fault is buried by alluvium deposited by the Rio Portugues. The western end of the fault is showed by Krushensky and Monroe (1978) as being truncated by an inferred north-south trending fault beneath alluvium of the Rio Canas Valley. Fieldwork by Geomatrix Consultants (1988) showed that the San Marcos fault is not truncated, is ~11 km long, and is onlapped by late Quaternary fluvial terraces of the Rio Portugues that are not affected by late Quaternary faulting.

Geomatrix Consultants (1988) also investigated in detail fault that is exposed on a small hill (Cerro del Muerto) on the southern coastal plain. The hill exposes a down-to-the-south normal fault with 35 m of post–middle Oligocene vertical displacement. The fault was projected to the southeast by Geomatrix Consultants (1988) and is named the Salinas fault. Aerial photograph study indicates the presence of numerous, generally east-west–trending lineaments on the coalesced alluvial fan complex between Rio Coamo and Rio Jueyes. However, no late Quaternary deformation could be established in the field for these lineaments.

A

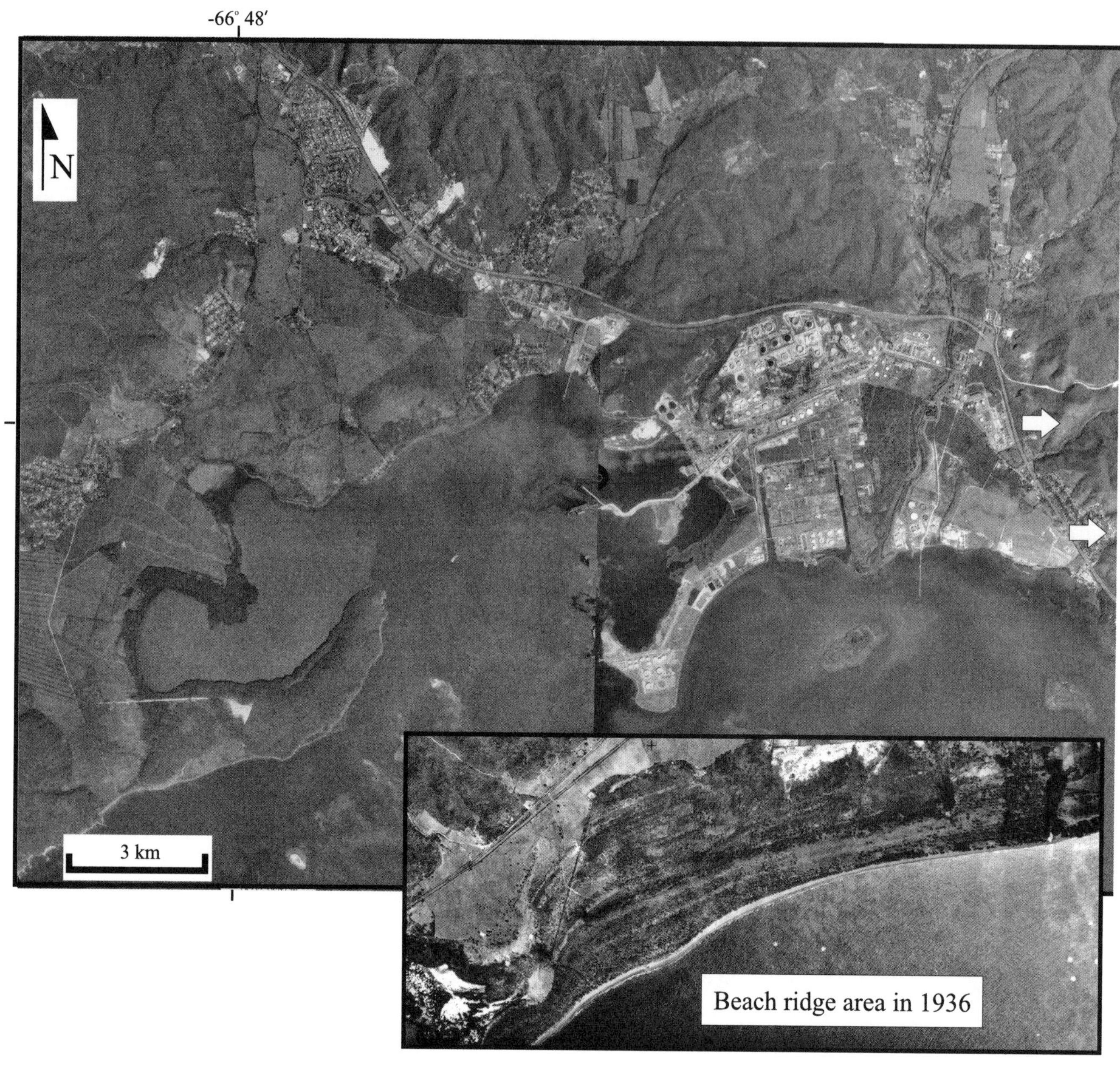

Figure 8 (*on this and following two pages*). (A) Photomontage made from 1986 air photos taken by the U.S. Department of Agriculture of the southern coast of Puerto Rico near the city of Ponce (original scale: 1:40,000). Location of map shown on regional DEM in Figure 3A. Inset map is an enlargement of a 1:18,700-scale, 1936 air photograph showing beach ridges in a coastal area southwest of Ponce that was developed by the time the 1986 photographs were taken. Box on the larger mosaic shows the location of the 1936 air photo. White arrows show lineaments in Mio-Pliocene carbonate bedrock.

B

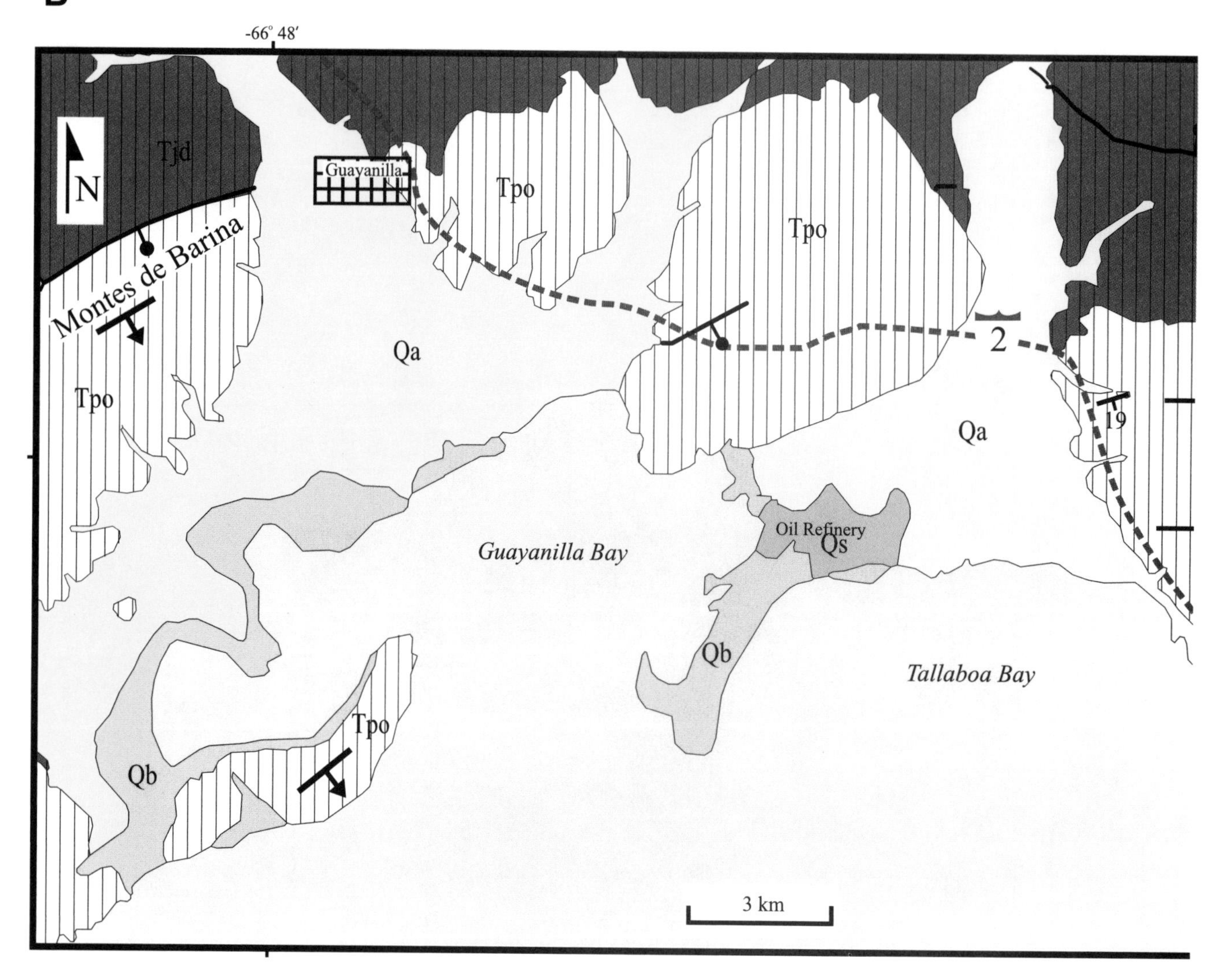
-66° 48′
N
Tjd
Guayanilla
Tpo
Tpo
Montes de Barina
Qa
2
Tpo
19
Qa
Oil Refinery
Qs
Guayanilla Bay
Qb
Tallaboa Bay
Tpo
Qb
3 km

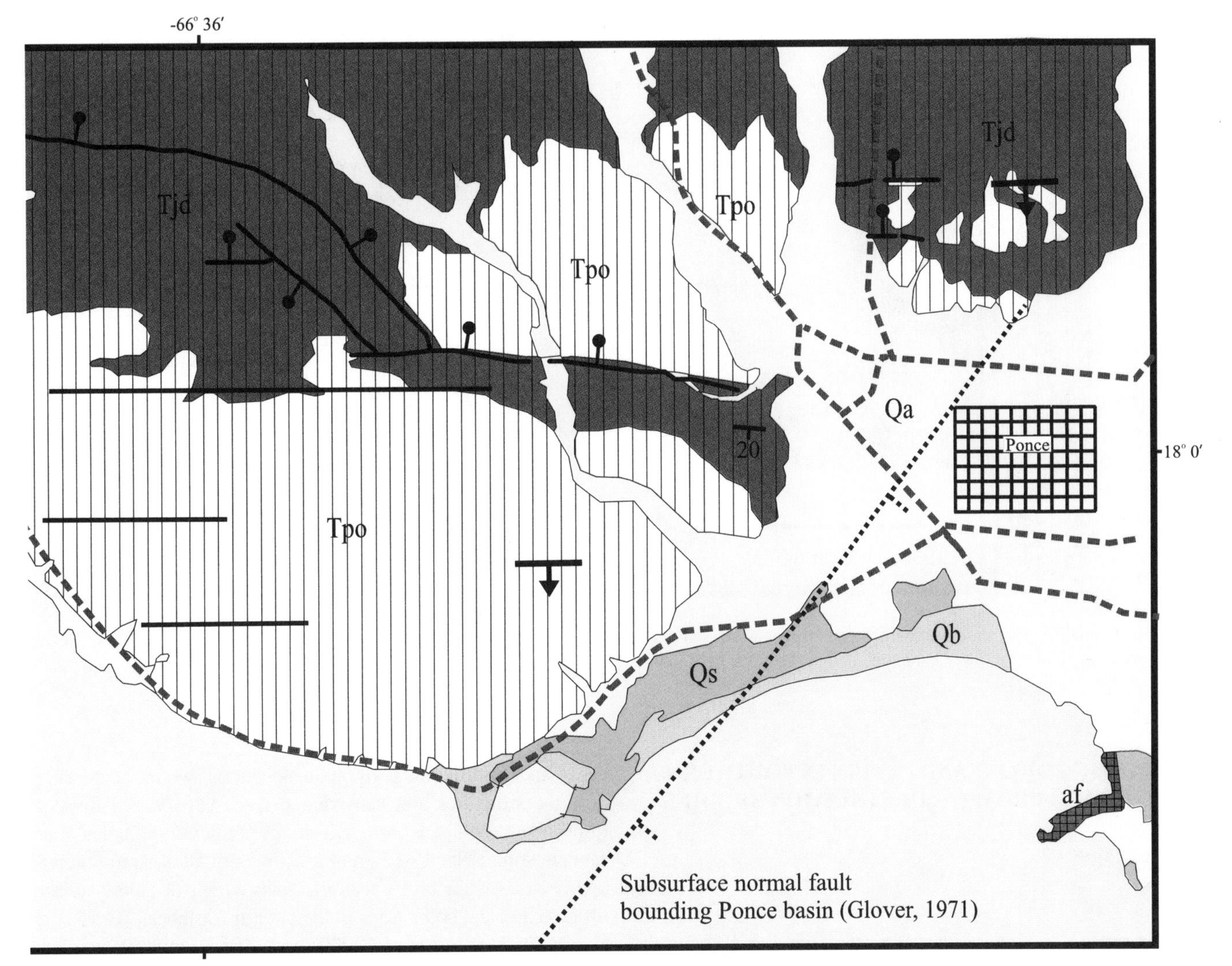

Figure 8 (*continued*). (B) Simplified geologic map of the same area in A based on 1:25,000-scale geologic mapping (Bawiec, 2001). Note the presence of northeast trends defining the outcrop limits of Oligocene-Pliocene carbonate units and east-west faults trends within the carbonate units. Note the moderate southward dips of Oligocene and Miocene carbonate units forming the southern flank of the Puerto Rico–Virgin Islands arch (cf. map in Fig. 3A). This area is in the approximate center of Oligocene-Pliocene carbonate units exposed in southern Puerto Rico. Dashed line is coastal road. af—artificial harbor fill; Qa—Quaternary alluvium; Qb—Quaternary beach deposits; Qs—Quaternary swamp deposits; Tpo—Miocene-Pliocene Ponce Formation; Tjd—Oligocene-Miocene Juana Diaz Formation.

Figure 9. 1936 aerial photograph showing beach ridges southeast of Ponce (location shown on map in Figure 3A; original scale: 1:20,000). Beach ridges are undated but are likely Holocene in age (Prentice et al., 2004).

KINEMATICS OF ONLAND FAULTS IN SOUTHERN PUERTO RICO FROM FAULT STRIATION STUDIES

Methodology

Utility

We applied methods of brittle tectonic analysis (Hancock, 1985; Angelier, 1990) to fresh exposures of striated, mesoscopic fault planes generally found in limestone rock quarries, roadcuts and excavations for the sites of new houses and buildings (Fig. 11). Our goal was to collect quantitative mesoscopic fault data across southern Puerto Rico and to use these data to reconstruct the Miocene-Recent plate boundary tectonic forces that deformed these rocks and influence and perhaps control the complex pattern of intense crustal seismicity (McCann, 2002; Huerfano et al., this volume) and regional arching (Fig. 3).

By systematically looking at striated faults in rock units of different ages (generally in the range of Oligocene to early Pliocene) and in different geographic locations, we can determine how the structures and causative stresses have varied through time and can compare our results to the predictions of conceptual models summarized in Figure 2. The "structural-stratigraphic" approach used for brittle tectonic analysis is particularly useful for arcuate plate boundaries in the circum-Caribbean (Gordon et al., 1997; Pubellier et al., 2000) and in the Alpine-Pannonian-Apennine system (Hippolyte et al., 1993; 1994; 1999), where the regional stress field can change abruptly in response to rapidly changing plate boundary conditions. To our knowledge, only one previous study (Erikson et al., 1990) has applied brittle tectonic analysis to faults in Puerto Rico, but it was in Cretaceous-Eocene rocks that mainly record intense convergent deformation predating the deposition of the Oligocene-Miocene carbonates.

Theoretical Basis

Brittle tectonic analysis is based on the assumption that slickenside lineations on a fault plane indicate the direction and

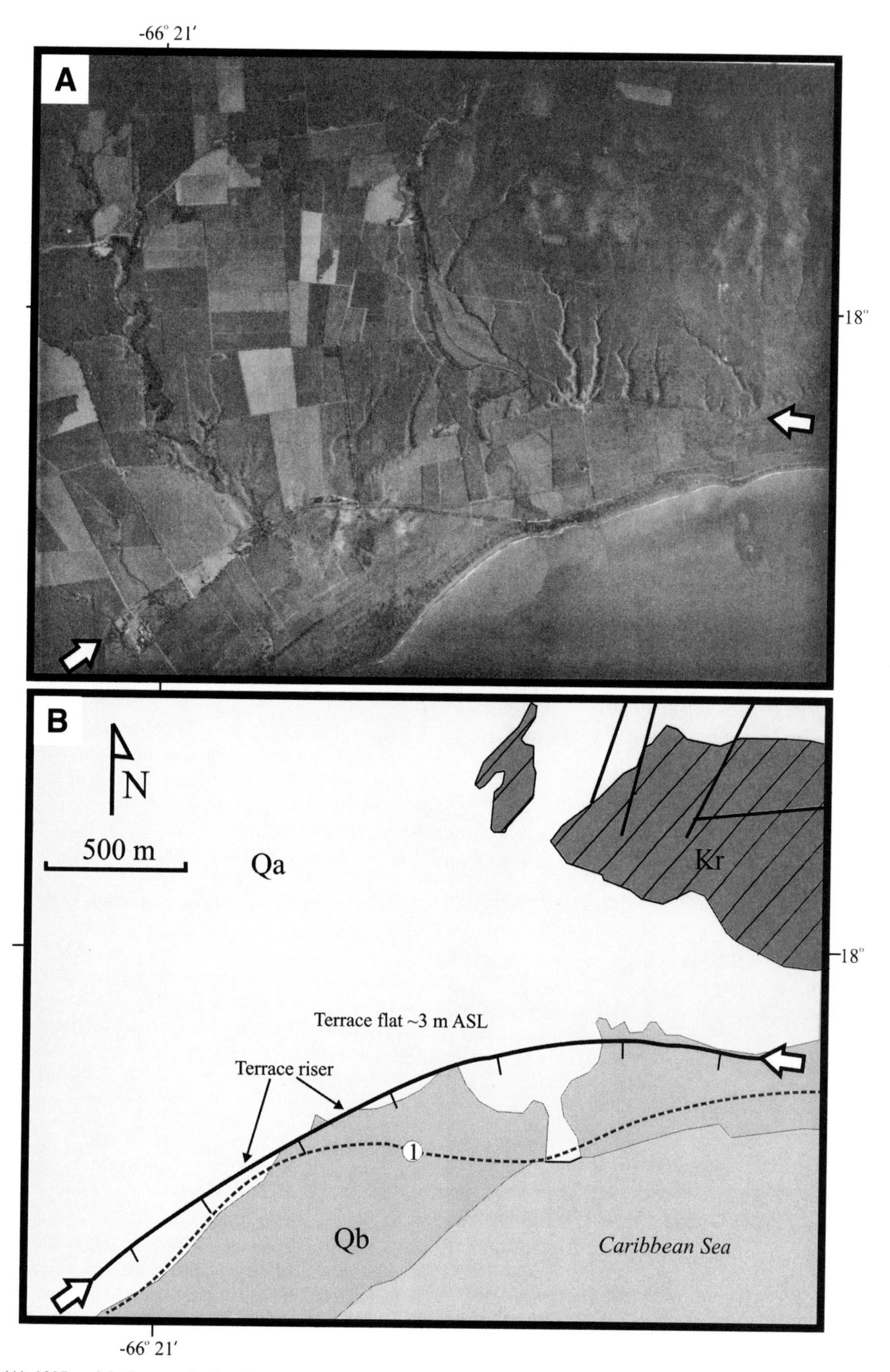

Figure 10. (A) 1937 aerial photograph showing young, stranded shoreline near Salinas (location shown on map in Figure 3A; original scale: 1:20,000). (B) Simplified geologic map of the same area in A based on 1:25,000-scale geologic mapping (Bawiec, 2001). This area is near the eastern limit of Oligocene-Pliocene carbonate units exposed in southern Puerto Rico and 3 km south of the Great Southern Puerto Rico fault zone (cf. map in Fig. 3A). Qa—Quaternary alluvium; Qb—Quaternary beach deposits; Kr—Cretaceous basalt.

sense of maximum resolved shear stress on that fault plane. The method assumes that the theoretical stress-shear relationships proposed by Wallace (1951) and Bott (1959) are valid for natural faults. If one knows the stress state, one can determine the shear stress orientation on any fault plane. By measuring the slip directions and hence the shear stress orientations on fault planes of various orientations, one can solve the inverse problem and determine the average stress state.

Field Methods

Hancock (1985) and Petit (1987) review standard field methods employed by us in Puerto Rico for collecting mesoscale fault data including determination of sense of slip from striated fault planes. Because carbonate lithologies form many of the faulted outcrops we examined, we were able to use accretion steps or stylolitic peaks as reliable sense of slip indicators (Hancock, 1985). The sense of slip observed on fault planes was frequently confirmed by stratigraphic offsets. During field work, we identify different tectonic phases responsible for homogeneous fault populations and conjugate fault sets. Normal and strike-slip faults are separated at this stage and oblique slip faults are plotted with the most compatible fault population (Angelier, 1994). The sequence of tectonic events is determined by comparison of fault sets to the stratigraphic age of rocks they affect and by examining overprinting of one fault set upon another. Quarry and roadcuts provided many excellent exposures of faults including the one shown in Figure 11 near the western edge of Ponce. These freshly excavated faults provided unequivocal evidence for the sense of displacement.

Figure 11. View of large, striated normal fault plane exposed in Miocene limestone of the Ponce Formation at site 3 west of the city of Ponce (cf. Figure 13 for location). Arrows indicate direction of movement of downthrown (southeast) block in the foreground. Post-Miocene to post-Pliocene striated fault planes collected at this and other localities shown in Figures 13 and 14 indicate that dip-slip and oblique-slip normal faulting is the predominate Neogene deformational style in sedimentary rocks along the south coast of Puerto Rico.

Computer Analysis

Angelier (1990, 1994) and Gordon et al. (1997) review the computer-based methods used by us for analyzing fault data and by Hippolyte et al. (this volume). In the case of a single phase of deformation the stress inversion is straightforward. In the case of polyphase deformation, a tectonic phase must be assigned to each fault according to field observations (conjugate sets, cross-cutting relationships, slickenside chronologies) prior to inverting the fault data for stress directions and test their compatibility. When the analysis is complete, the stress directions and fault data are plotted using a standard set of symbols. Data output includes the following quantities listed on Table 1 for each of the sample localities:

- Axes of principal stresses σ_1, σ_2, and σ_3 ($\sigma_1 \geq \sigma_2 \geq \sigma_3$), compression positive.
- $\Phi = (\sigma_2 - \sigma_3)/(\sigma_1 - \sigma_3)$
- ANG is the angle between the observed fault slip direction (slickenside lineation) and the maximum resolved shear stress on the plane (as calculated from the σ_1, σ_2, σ_3, and Φ value given above).
- RUP ("ratio upsilon") is defined by Angelier (1990).

Data Quality

The values of Φ, ANG, and RUP provide a basis for assessing the quality of data and confirms or rejects the fault population as a single tectonic phase. Ideal values of Φ for high-quality data are ~0.5. ANG and RUP are two different methods of showing the deviation between the observed fault slip and the computed shear stress along the fault (Angelier, 1990).

Distribution and Ages of Oligocene-Miocene Stratigraphic Units

The Ponce Formation overlies older, mixed carbonate and clastic units of the Oligocene-Miocene Juana Diaz Formation of Glover (1971) and Monroe (1980). More detailed stratigraphic and biostratigraphic work by Frost et al. (1983) further subdivided the upper part of the Juana Diaz Formation into the New Formation, Angola Formation, and an unnamed reef facies as shown on Figure 4. For the purpose of the fault measurements made on these rocks, we follow the earlier stratigraphic usage of Krushensky and Monroe (1975, 1978, 1979) and Monroe (1980) that simply divides the Oligocene-Pliocene carbonate sections into three distinctive units: clastic Oligocene Juana Diaz Formation, carbonate Oligocene Juana Diaz Formation, and Miocene–early Pliocene Ponce Formation (cf. Monroe, 1980, for a detailed account of the evolution of the stratigraphic terminology of these units). We could readily identify these units in the field using the maps and descriptions of Monroe (1980) whereas the different classification and distribution of units proposed by Frost et al. (1983) was not always apparent in the field. These units are shown on the maps in Figures 13 and 14.

TABLE 1. MICROTECTONIC DATA SUMMARY FOR THE SOUTHERN MARGIN OF PUERTO RICO

Site	Site Name	Formation	Formation age	Stress regime*	No. of faults	σ_1		σ_2		σ_3		Φ	ANG	RUP
						Trend	Plunge	Trend	Plunge	Trend	Plunge			
1	1 km south of Yauco	Juana Diaz clastic beds	Oligocene-Miocene	E	33	295	78	102	12	197	3	0.34	13	40
1	1 km south of Yauco	Juana Diaz clastic beds	Oligocene-Miocene	E	27	280	80	50	6	141	8	0.43	16	39
2	Punta Cuchara	Ponce	Miocene-Pliocene	E	25	347	87	236	1	146	3	0.5	12	35
3	4 km west of Ponce	Juana Diaz Limestone Mb. and Ponce Fm.	Oligocene-Pliocene	E	29	25	77	282	3	192	12	0.26	11	35
3	4 km west of Ponce	Juana Diaz Limestone Mb. and Ponce Fm.	Oligocene-Pliocene	E	9	247	74	47	15	138	5	0.55	8	31
4	Ponce main quarry	Juana Diaz Limestone Mb.	Oligocene-Miocene	E	14	354	78	118	7	209	10	0.24	15	40
6	Puerto de Guayanilla	Ponce	Miocene-Pliocene	E	8	279	76	46	9	138	11	0.41	14	30
7	3 km south of Yauco	Juana Diaz Limestone Mb.	Oligocene-Miocene	S	21	73	3	179	79	343	11	0.47	8	25
8	3 km east of Guanica	Juana Diaz clastic beds	Oligocene-Miocene	E	6	69	73	238	17	329	3	0.42	10	27
9	Guayanilla	Juana Diaz Limestone Mb. and Ponce Fm.	Oligocene-Pliocene	E	6	344	80	137	9	227	5	0.43	15	29
9	Guayanilla	Juana Diaz Limestone Mb. and Ponce Fm.	Oligocene-Pliocene	S	5	70	8	252	82	160	0	0.34	7	20
10	Palomas Landfill	Juana Diaz clastic beds	Oligocene-Miocene	E	22	32	87	282	1	192	3	0.38	11	30
11	Valle de Lajas	Juana Diaz Limestone member	Oligocene-Miocene	E	8	241	81	107	6	16	6	0.32	5	14
11	Valle de Lajas	Juana Diaz Limestone member	Oligocene-Miocene	E	5	32	79	210	11	300	0	0.65	3	30
12	4 km east of Ponce	Juana Diaz Limestone member	Oligocene-Miocene	E	17	114	80	271	9	2	4	0.3	13	30
13	2 km east of Ponce	Ponce	Miocene-Pliocene	E	9	116	63	282	26	14	6	0.54	10	41
14	2 km east of Juana Diaz	Juana Diaz clastic beds	Oligocene-Miocene	E	5	82	89	272	1	182	0	0.5	13	29
15	3 km west of Yauco	Juana Diaz clastic beds	Oligocene-Miocene	E	15	287	60	94	29	187	6	0.44	6	23
15	3 km west of Yauco	Juana Diaz clastic beds	Oligocene-Miocene	E	11	214	81	47	9	316	2	0.43	9	29
15	3 km west of Yauco	Juana Diaz clastic beds	Oligocene-Miocene	S	15	248	5	95	84	338	3	0.66	13	36

*E—extensional; S—strike-slip

Figure 12. View of crosscutting fault relationships in Miocene limestone of the Ponce Formation at site 3 west of the city of Ponce (cf. Figure 13 for location). Faults of the southeast extensional event crosscut older faults formed by the north-northeast–directed extensional event.

A slight angular unconformity separates the uppermost carbonate rocks of the Oligocene Juana Diaz Formation from the lowermost rocks of the early Miocene–early Pliocene Ponce Formation (Monroe, 1980). The younger age of the shallow-water Ponce Formation is not well determined because of its shallow carbonate shelf depositional environment which excluded age-diagnostic planktonic foraminifera and calcareous nannoplankton (Frost et al., 1983). Van den Bold (1969) identifies 13 species of ostracodes from localities in what is apparently the upper part of the Ponce Formation in southern Puerto Rico. He correlates the ostracode assemblages with mollusk and planktonic foraminiferal zones that mark the latest Miocene–earliest Pliocene (Moussa and Seiglie, 1975, p. 168). Frost et al. (1983) conclude that "the Ponce Formation spans virtually all of the Middle Miocene and may extend into the earliest Pliocene."

Results

Results of the fault study are summarized on the maps in Figure 13 and 14 and in Table 1. Two phases of late Neogene faulting are identified in southern Puerto Rico. A north-northeast–directed extension was identified as a regional event affecting Oligocene carbonate and clastic Juana Diaz Formation and Miocene–early Pliocene Ponce Formation (Fig. 4). Variability in the direction of extension is generally less than 20°. The north-northeast extensional event was recognized across most of southern Puerto Rico although most observations are limited to the extensive limestone quarry exposures in the Ponce area (Fig. 13).

The north-northeast extensional event may be responsible for: (1) the formation of the east-west–trending fault zone bounding the northern limit of the Oligocene-Pliocene carbonate formations (Monroe, 1980; Figs. 13 and 14); (2) east-west–trending topographic features such as the Lajas Valley of southwestern Puerto Rico; and (3) east-west–trending sections of the coastline of southern Puerto Rico (Fig. 5). The precise age of the Lajas Valley rift is required to test this hypothesis for its origin.

A younger extensional event oriented in a northwest-southeast direction is found associated with minor strike-slip movements (Fig. 14). At site 3 near Ponce, faults of this younger event crosscut faults formed by the north-northeast–directed extensional event and are therefore younger in age (Fig. 12). Also at site 3, southeast-striking faults formed during the north-northeast extensional event were reactivated during southeastward extension (cf. diagram of site 3 in Fig. 14). At site 15, the relative chronology of striations also confirms that southeastward extension is younger than east-northeastward extension (Fig. 14).

Southeastward extension is roughly perpendicular to east-northeast–trending sections of the zediform, or Z-shaped, coastline (Bloom, 1986) between La Parguera and the area east of Ponce (Fig. 14). The two extensional events are found in stratigraphic sections of all middle-late Tertiary carbonate rocks examined. The youngest carbonate rocks of southern Puerto Rico, the Ponce formation of middle Miocene–early Pliocene (Frost et al., 1983) are faulted by these two extensional events at site 3 (Fig. 12). Pending more precise biostratigralphic dating of the Ponce Formation, both faulting events are post–early Pliocene in age.

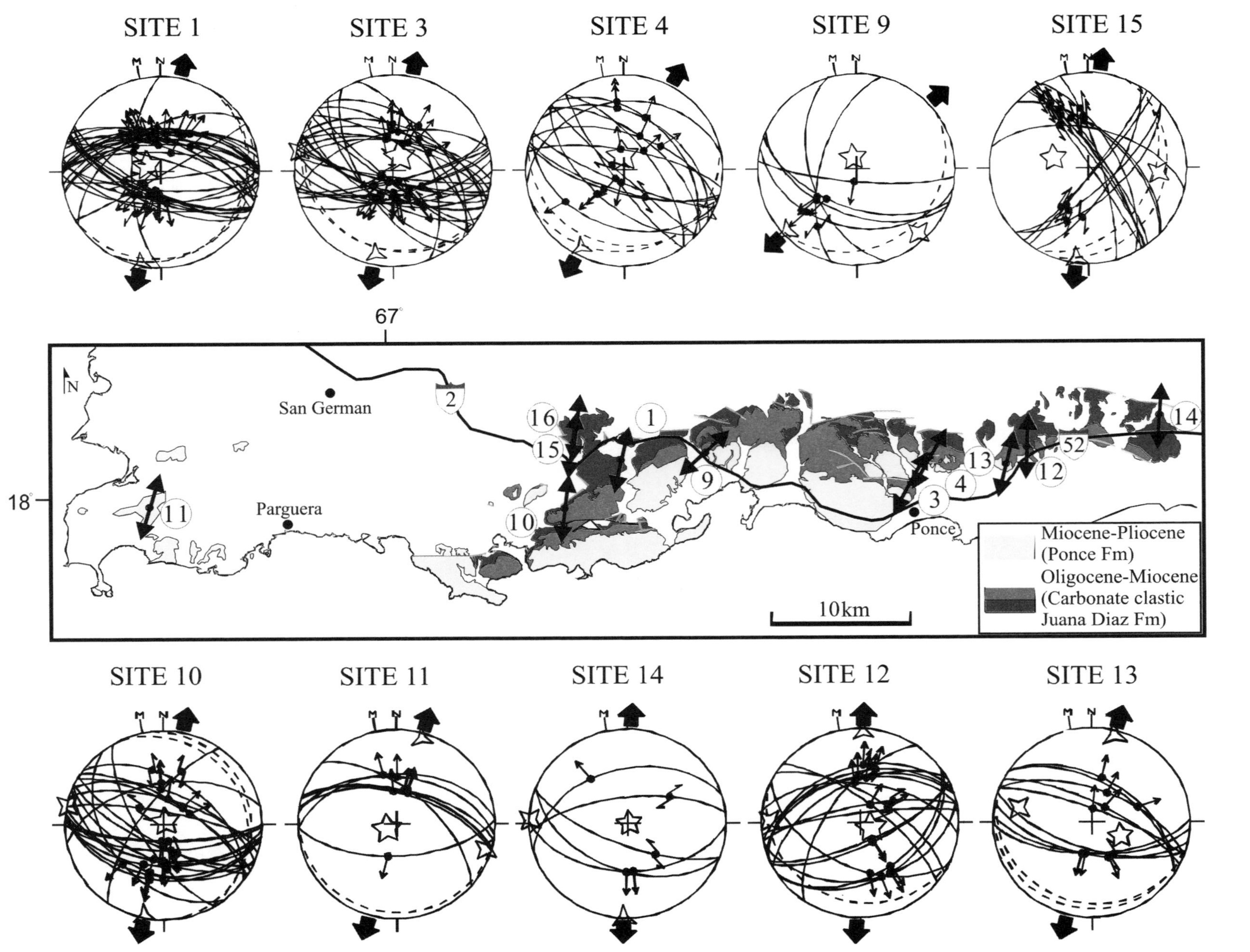

Figure 13. Summary of fault striation results from 10 localities in Miocene-Pliocene sedimentary rocks of southern Puerto Rico constraining a post–early Pliocene phase of north-north-east–trending right-lateral transtension. This tectonic phase may be responsible for east-west–trending "basin and range" landforms of southwestern Puerto Rico (cf. Figure 3A, B).

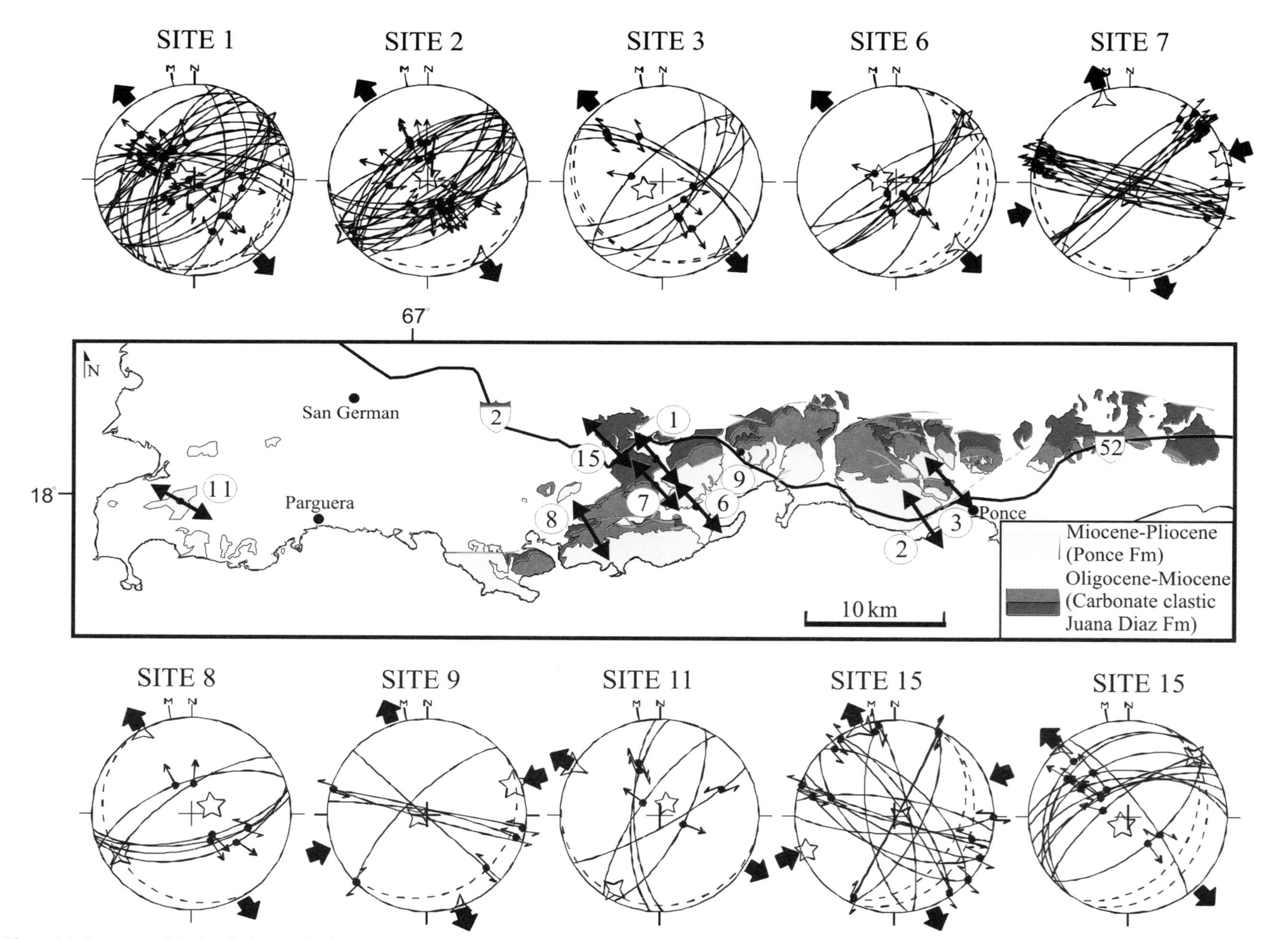

Figure 14. Summary of fault striation results from 10 localities in Miocene-Pliocene sedimentary rocks of southern Puerto Rico constraining a post–early Pliocene phase of southeast-trending extension that crosscuts and reactivates normal faults formed during the earlier phase of north-northeast–trending extension (cf. Fig. 13). This tectonic phase may be recorded in northeast-trending embayments, elongate islands, and submarine morphology of the southwestern shelf and slope of Puerto Rico and in the formation of the Anegada rift southeast of Puerto Rico (cf. Figs. 3A, B).

OFFSHORE LATE NEOGENE STRUCTURES OF SOUTHERN PUERTO RICO

Major Tectonic Features and Faults Offshore of Southern Puerto Rico

The offshore geology of Puerto Rico is key to understanding the larger scale relationships between the onland tectonic geomorphology and the two extensional events of Miocene-Quaternary age identified by onland fault striation studies. On Figure 15, we summarize the main offshore structures in a zone bounded by the Muertos trench to the south and the southern coast of Puerto Rico to the north.

Muertos Trench (Number 1)

This east-west–trending trench along which the Venezuelan basin subducts beneath Puerto Rico and Hispaniola takes its name from the largest island off the south coast of Puerto Rico, Caja de Muertos, which is 25 m ASL and is cored by an outcrop of Eocene volcanic rocks similar to those found in southern Puerto Rico (Glover, 1971) (Fig. 15). The Muertos trench is filled by an unfolded section of Neogene sediments (number 2) (Case and Holcombe, 1980) (Fig. 15). Case and Holcombe (1980), Jany et al. (1990) and Masson and Scanlon (1991) review information showing active deformation along the axis of the trench.

Muertos Accretionary Prism (Number 3)

This east-west fold-thrust belt has formed by the subduction-accretion of Cretaceous-Neogene marine sediments against Puerto Rico (Fig. 15). According to Jany et al. (1990) and Masson and Scanlon (1991), the accretionary prism of the Muertos trench disappears at longitude 65°W and by this point all convergence between the Puerto Rico block and the Caribbean plate has been transferred to strike-slip or oblique-slip faults in the Anegada Passage.

Investigator Fault and Bank (Number 4)

This 75-km-long fault forms a prominent east-west–trending submarine valley that is flanked to the north by the slope of the carbonate platform of southern Puerto Rico and to the south by the elongate and submerged Investigator Bank (Fig. 15). The sense and activity of the Investigator fault has not been studied, although it is possible that this fault is an active left-lateral fault that accommodates left-lateral slip related to oblique subduction at the Muertos trench. The fault terminates in the vicinity of the Grappler bank and Whiting basin.

Faults on the Muertos Shelf (Numbers 7, 8, 9, and 11)

The Muertos shelf forms a broad, southward-convex, lobate-shaped carbonate platform that extends 55 km along the south-central coast of Puerto Rico. Water depths range from ~18 m on the northern inner shelf and 40 m near the shelf edge (Beach and Trumbull, 1981; Trias, 1991) (Fig. 15). The carbonate platform extends to a shelf edge 15–18 km south of Puerto Rico marked by a single or double rim of coral reefs deposited at times in the early Pleistocene when sea level was lower (Garrison, 1969; Beach and Trumbull, 1981). The depth of these ridges is ~20 m below sea level, and they are inferred to be drowned barrier reefs of late Quaternary age. Observations of both the outer drowned reefs and smaller patch reefs on the shelf show the presence of few corals. Instead, these shelf and slope buildups are composed mainly of encrustations of coralline algae and overgrowths of gorgonians and sponges (Frost et al., 1983). MacIntyre (1972) proposes that the rapid Holocene rise of sea level is the probable cause for death of these and similar reefs in the Caribbean rather than other causes like changes in the supply of nutrients and changes in water temperature.

The inner part of the platform near the south coast of Puerto Rico is characterized by small, low-lying, and uninhabited islets and cays formed by coral reef rock and calcareous sand. The most prominent island, Caja de Muertos ("Coffin Island" due to its longitudinal topographic profile that resembles a coffin), exhibits a distinctive north-east trend (Fig. 15). Unnumbered faults on the compilation map shown on Figure 15 were taken from the Case and Holcombe (1980) compilation of faults in the area.

Caja de Muertos Fault Zone and Adjacent Ponce Basin (Number 7)

This fault, bounding the northeast-trending, elongate submarine ridge on which Caja de Muertos island is located, was first identified in a seismic reflection survey during hydrocarbon exploration by the United Geophysical Company (Denning, 1955). The fault is reflected in the present bathymetry where the edge of the uplifted footwall block to the southeast forms the southeastern boundary of the basin. The Caja de Muertos fault is described by Kaye (1959) and Glover (1971) using bathymetric data. Denning (1955) concludes from seismic reflection data that the Caja de Muertos fault was downthrown to the northeast by as much as 1.1 km near Caja de Muertos Island, but that the downward sense of throw dissipates in a scissor-like manner to zero by the time the fault crosses the shoreline southeast of Ponce at Punta Petrona (Fig. 15).

Garrison (1969) and Beach and Trumbull (1981) used single-channel seismic reflection profiles to map the Caja de Muertos fault south of Ponce. These data showed that the fault formed a structural edge of the Ponce basin, a northeastern-trending bathymetric and sedimentary rift south of Ponce. The water depth of the Ponce basin varies from 12 m south of the coast near Ponce to more than 40 m near the shelf edge. The basin is dotted by local bathymetric highs most likely related to local coral patch reefs. Terrigenous mud is accumulating on the floor of the Ponce basin in comparison to skeletal carbonate sand and gravel on the adjacent platform areas (Beach, 1976). Glover (1971) was not able to show any onland continuation of the Caja de Muertos fault in the Ponce area.

It is interesting to note that the beach ridge area southeast of Ponce that is shown in Figure 9 is located on the east side of Punta Petrona and may reflect recent uplift on this fault zone

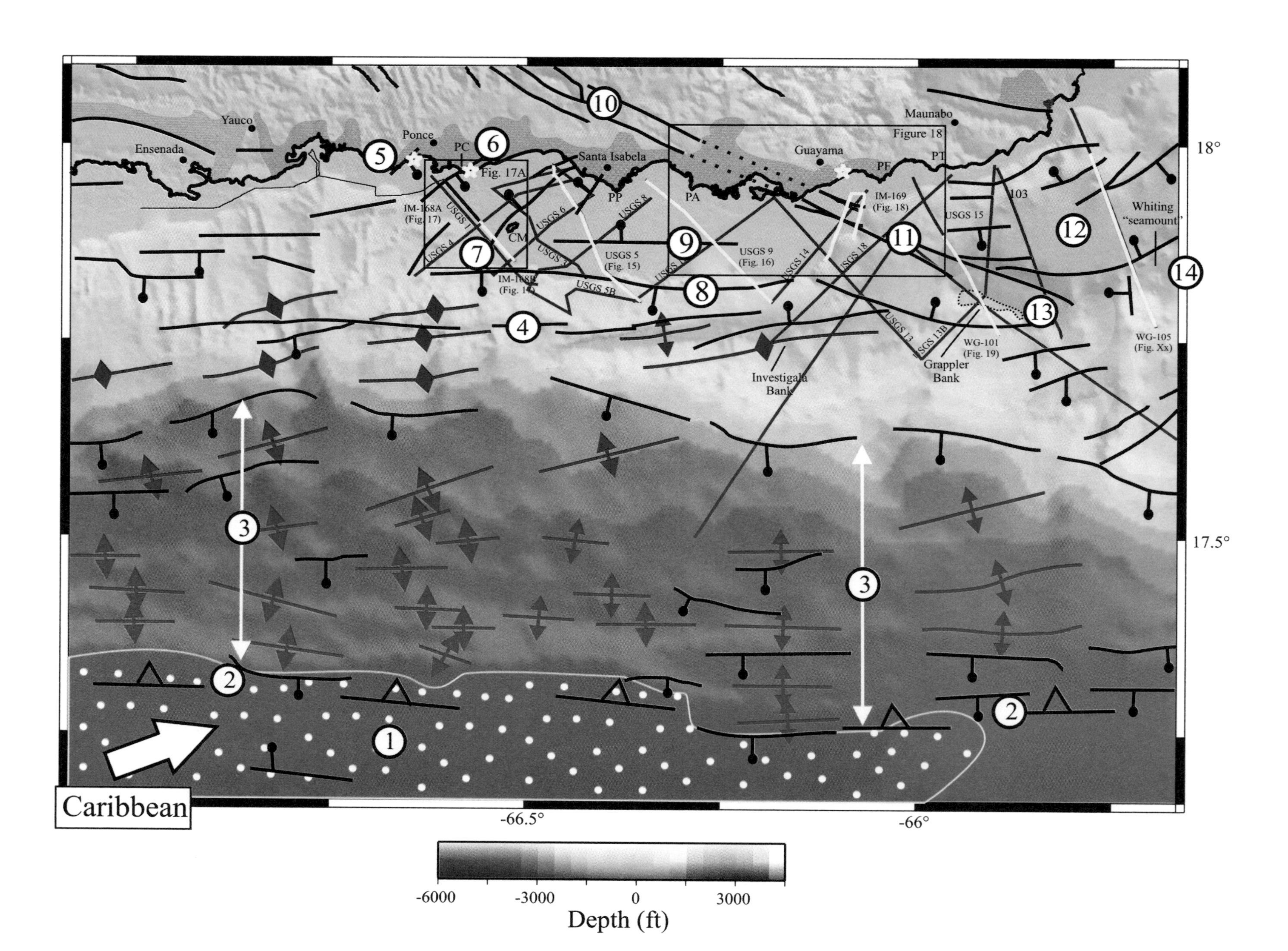
Yauco
Ensenada
Ponce
PC
Fig. 17A
Santa Isabela
PP
PA
Guayama
PF
PT
Maunabo
Figure 18
IM-168A
(Fig. 17)
IM-168B
(Fig. 14)
CM
USGS 1
USGS 4
USGS 6
USGS 3
USGS 8
USGS 5
(Fig. 15)
USGS 5B
USGS 9
(Fig. 16)
USGS 14
USGS 18
IM-169
(Fig. 18)
USGS 15
103
USGS 13
USGS 13B
WG-101
(Fig. 19)
Grappler
Bank
Investigala
Bank
Whiting
"seamount"
WG-105
(Fig. Xx)
1
2
3
4
5
6
7
8
9
10
11
12
13
14
18°
17.5°
-66.5°
-66°
Caribbean
-6000
-3000
0
3000
Depth (ft)

(beach ridge areas are shown as yellow stars on the map in Fig. 15). The western edge of the Ponce basin is defined by the Bajo Tasmanian fault and adjacent carbonate platform (Garrison, 1969).

Bajo Tasmanian Fault (number 6)

Seismic profiling by Garrison (1969) and Beach and Trumbull (1981) show that this fault has a down-to-the-southeast displacement and forms the northern edge of the Ponce basin. The onland projection of the fault at Punta Cabullon was not detected by Glover (1971). It is interesting to note that the beach ridge area southwest of Ponce that is shown in the inset of Figure 8A is located on the east side of Punta Cabullon and may reflect recent uplift on this fault zone (beach ridge areas are shown as yellow stars on the map in Fig. 15).

Great Southern Puerto Rico Fault Zone (Numbers 10–11)

According to Garrison (1969), USGS profile 14 showed that this fault extends in a southeastward direction and crosses the coastline near Bahia de Jobos. He interpreted the profile as showing that the offshore extension of the Great Southern fault zone is "several hundred meters wide and involves multiple fractures." Vertical movement was inferred to be down to the southwest. The amount of vertical displacement was not possible to estimate. Garrison (1969) correlates this fault to the onland Great Southern Puerto Rico fault zone described in detail by Glover (1971). Griscom and Geddes (1966) identify a linear discontinuity in the total magnetic intensity contours in the area that they relate to the presence of the Great Southern Puerto Rico fault zone in the coastal plain of southeastern Puerto Rico.

Whiting Half-Graben (Number 12)

This deep basin forms the southeastern margin of Puerto Rico. It is bounded on the south by the flat-topped Grappler bank (Fig. 14) and to the southeast by the Whiting bank (Fig. 15). The Whiting basin occurs near the intersection of two linear fault trends: the Investigator fault (no. 4) and the Great Southern Puerto Rico fault zone (no. 10–11).

New Observations on Offshore Faults of Southern Puerto Rico

USGS Lines across the Muertos Shelf

In 1969, the U.S. Geological Survey and the Puerto Rico Economic Development Program collected a loose grid of intersecting and continuous single-channel seismic profiles crossing the Muertos shelf in northeast and southwest directions (lines indicated by USGS and line number in Fig. 15). The sound source used was a 13,000 J stored-energy sparker. These data were published in a brief, 9-page U.S. Geological Survey Open File Report (Garrison, 1969).

In Figures 16 and 17, we reproduce two of these lines to illustrate the presence of two fault trends affecting the uppermost part of the sedimentary section of the Muertos shelf. The Central fault of Garrison (1969) is an east-west–trending fault on the shelf that consists of three individual faults downthrown to the north (USGS line 5 in Figure 16 and USGS line 9 in Fig. 17). Other unnamed faults with downthrow to the north are present on the northern end of line 15. A fault with downthrow to the south is inferred to form the shelf-slope break at the southern end of lines 5 and 9. The northward dip of these faults is consistent with the northward dip of the major scarp-forming faults of southern Puerto Rico as seen on the perspective view of the Puerto Rico–Virgin Islands arch in Figure 3C.

Single-Channel Seismic Data Collected by RV Isla Magueyes *in 2000*

We obtained a single track along the southern coast of Puerto Rico during a one-day survey of that area. Since we only obtained a single track from this area, we do not display the narrow swath of sidescan data and instead only display the single-channel seismic data that is more useful for identifying faults.

The seismic profiles shown were processed with the following parameters: spherical divergence correction, spectral whitening, time variant band-pass filtered at 20–2400 Hz, and trace mixing. A water bottom mute and a 10 msec automatic gain control (AGC) have also been applied. Single-channel seismic profiling of carbonate hard grounds that characterize much of the southern shelf of Puerto Rico resulted in high-amplitude seafloor reflections with limited sub-seafloor penetration and obscuring of sub-seafloor structure by multiples.

←

Figure 15. Digital elevation model (DEM) and structural map of southern Puerto Rico, the Puerto Rico island shelf and slope, and the Muertos trench. Bathymetry is from NOAA hydrographic data and contoured in feet. Numbered features: 1—Quaternary turbidites (dotted area) filling axis of Muertos trench; 2—active deformation front of Muertos trench; thrust faults (barbs mark overthrust block); 3—width of fold-thrust belt of Muertos accretionary prism; 4—linear expression of Investigator fault with possible strike-slip movement; 5—Ponce normal fault inferred by Glover (1971) using subsurface data (ticks mark downthrown block); 6—Bajo Tasmanian normal fault of Garrison (1969); 7—Caja de Muertos (CM) normal fault of Garrison (1969) and this paper; 8—unnamed fault along island slope from Case and Holcombe (1980); 9—Central fault zone of Garrison (1969); 10—left-lateral, transpressional Great Southern Puerto Rico fault zone of Glover (1971) and Erikson et al. (1990); 11—submarine extension of Great Southern Puerto Rico fault zone (Esmeralda and Rio Jueyes fault zones); 12—Whiting half-graben; 13—flat-topped Whiting bank (footwall uplift); 14—Whiting "seamount," or footwall uplift. Single- and multi-channel seismic reflection lines used in this paper to illustrate these features are marked in yellow. Red lines with double-headed arrows are anticlines mapped by Case and Holcombe (1980) using seismic reflection data. Red lines with diamonds are elevated fault blocks also mapped by Case and Holcombe (1980). Abbreviations of major "points" or headlands along the south coast of Puerto Rico: PC—Punta Cabullones; PP—Punta Petrona; PA—Punta Arenas; PF—Punta Figueras; and PT—Punta Tuna. Large white arrow indicates direction major plate motion.

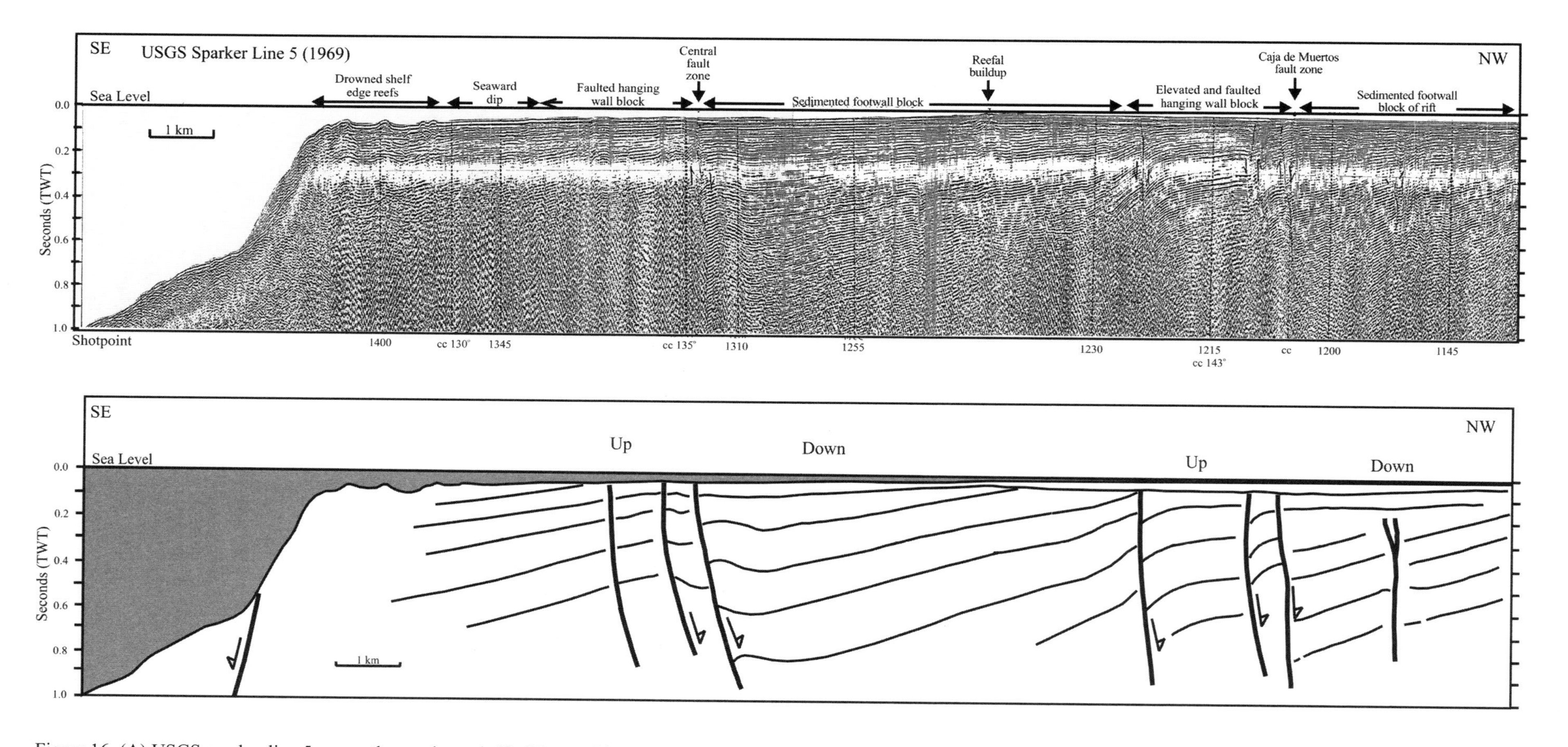

Figure 16. (A) USGS sparker line 5 across the southern shelf of Puerto Rico recorded to a depth of 1 second modified and reinterpreted from Garrison (1969). White band between 0.2 and 0.3 seconds is related to poor reproduction of original analog data record. (B) Structural summary showing overall southward dip of reflectors and overall northward dip of normal faults. Note that this direction of stratal and fault dip is similar to that proposed for westernmost Puerto Rico (cf. Figure 3A, B, C).

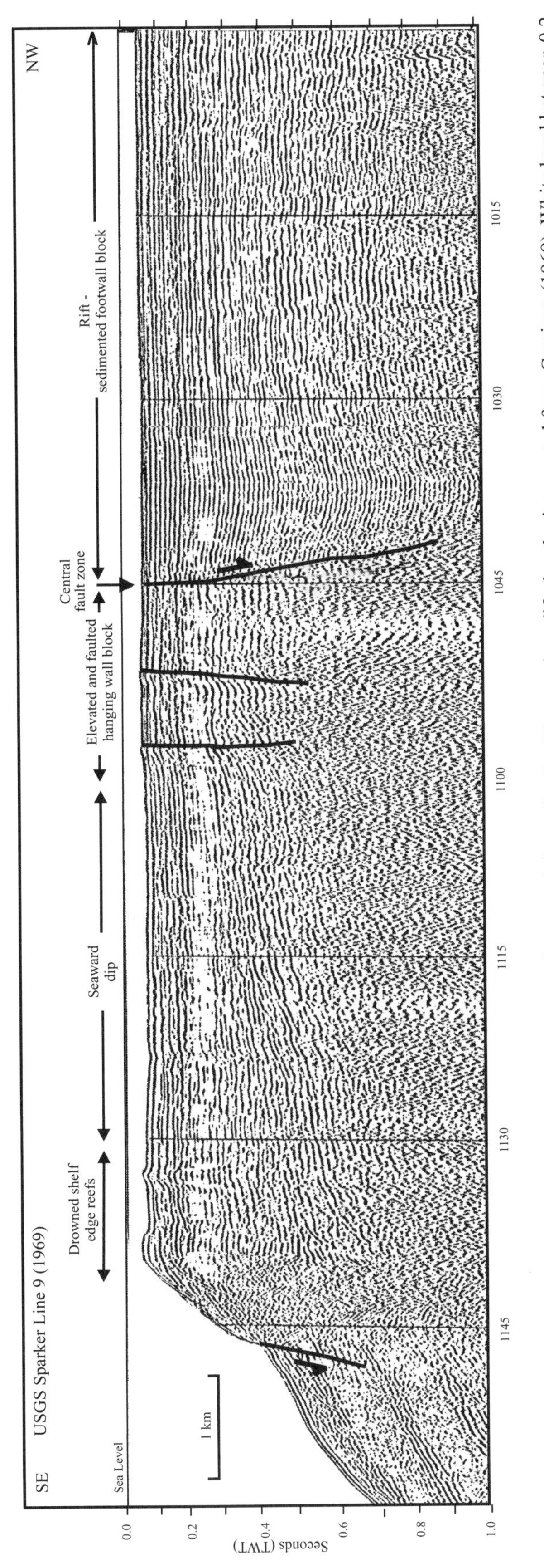

Figure 17. USGS sparker line 9 across the southern shelf of Puerto Rico recorded to a depth of 1 second modified and reinterpreted from Garrison (1969). White band between 0.2 and 0.3 seconds is related to poor reproduction of original analog data record.

Our offshore survey in May 2000, identified three areas of late Quaternary seafloor faulting shown on Figure 15. The Bajo Tasmanian and Caja de Muertos faults on the shelf south of Ponce had been previously recognized through USGS seismic surveys and were named and described by Garrison (1969).

The Bajo Tasmanian and Caja de Muertos faults together isolate the large, offshore Ponce basin of Garrison (1969) and Glover (1971). This offshore basin has a northeast trend and is similar to the onland Ponce basin described by Glover (1971) using drill data (Fig. 15). The island of Caja de Muertos occupies the crest of a narrow, emergent horst block adjacent to southeastern edge of the Ponce basin. Line 168 shown in Figure 18 shows the recent scarp on the seafloor that bounds the ridge upon which the island is located and forms ~20 m of seafloor relief. According to Glover (1971) and Erikson et al. (1990), unpublished borehole data obtained by these authors in the area west of Ponce indicates the presence of an unnamed fault-bounding the subaerial, Quaternary, sediment-filled basin upon which the city of Ponce is built. The location of this fault is shown on Figure 8B.

A second area of late Quaternary seafloor faulting was observed along the Esmeralda and Rio Jueyes faults, the offshore projections of the onland strands of the same name of the Great Southern fault zone that was mapped by Glover (1971). The Esmeralda fault strikes at almost right angles to the northeast-striking shelf faults south of Ponce (Fig. 15). On these lines, both zones of offshore faulting of the seafloor correspond well to the linear projections of the Esmerala and Rio Jueyes faults as proposed by Glover (1971). We see no evidence for a third strand of the fault zone (Salinas fault) that was proposed by Geomatrix Consultants (1988) in the area south of the Esmeralda fault zone.

Multi-Channel Seismic Data from the Whiting Half-Graben

We interpret existing 24-channel seismic profiles collected by Western Geophysical Company of America and Fugro, Inc. (1973), and provided to us for the purpose of this study by the Puerto Rico Seismic Network of the University of Puerto Rico at Mayaguez. These data are useful for constraining faulting at depth but are less useful for resolving recent deformation at the seafloor.

The two lines cross the faults bounding the Whiting half-graben (Fig. 15). Line R-101 crosses the southwestern edge of the basin including the offshore projection of the Great Southern fault zone and the Grappler bank. Line R-105 is more to the east and crosses the central part of the basin and the adjacent Whiting "seamount" or steeply tilted submarine ridge. Both lines reveal the half-graben structure of the Whiting basin that includes: a tilted hanging wall block, an asymmetric wedge indicative of syntectonic filing of the half-graben, and a tilted footwall block. Tilting of the footwall and hanging wall blocks is less on line R-101 probably because the southeast-trending fault on the south side of the basin in this area is a dominantly left-lateral strike-slip that corresponds to the offshore projection of the Great Southern Puerto Rico fault zone (Fig. 15). Tilting of the hanging and

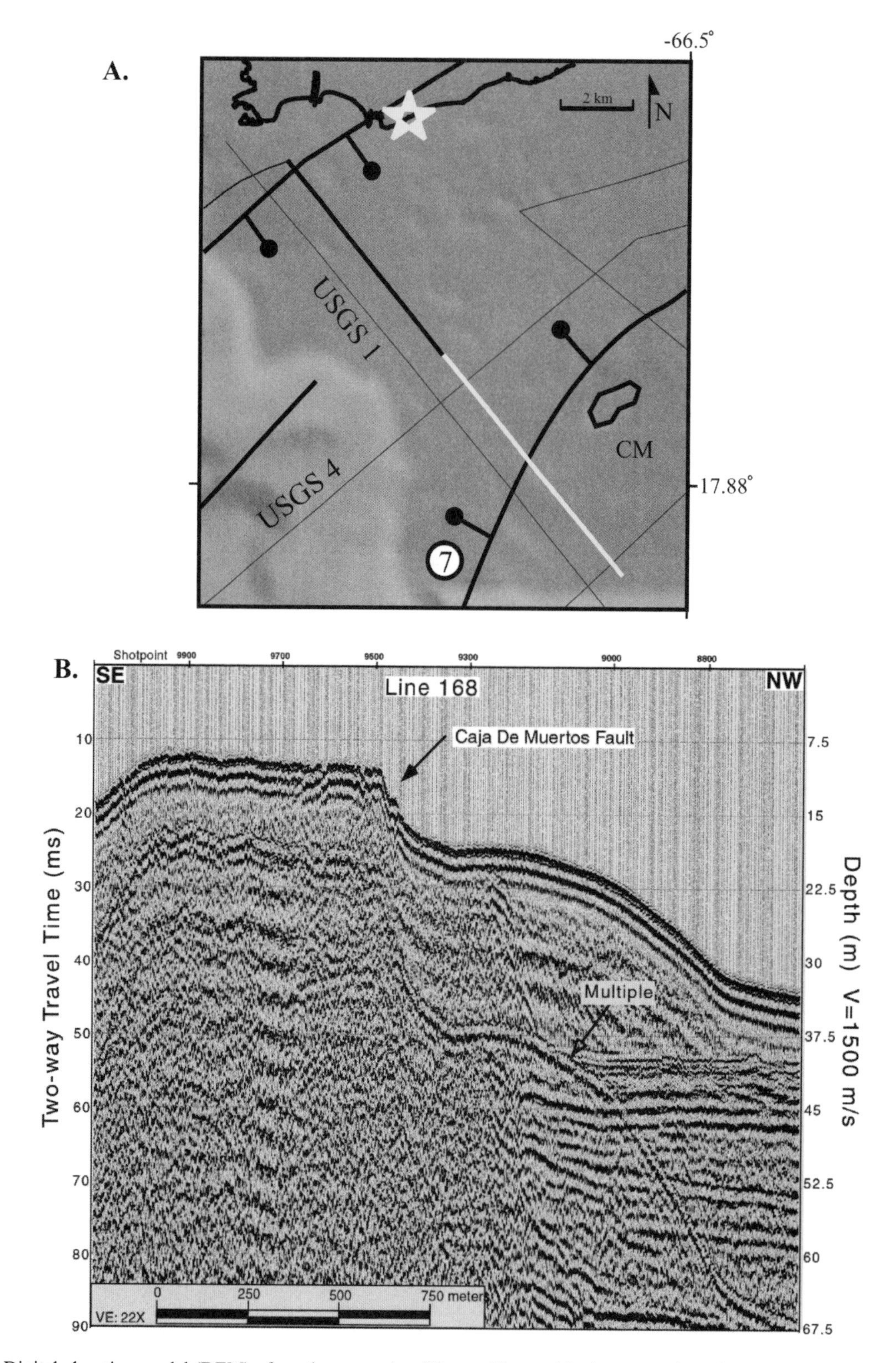

Figure 18. (A) Digital elevation model (DEM) of southern margin of Puerto Rico and bathymetry of southern shelf from NOAA in the area of the Caja de Muertos (CM) normal fault. (B). Single-channel seismic line 168 across the Caja de Muertos normal fault 0.5 km south of Caja de Muertos Island. Scarp of the fault and its down-to-the-northwest sense of throw are evident as earlier noted by Garrison (1969) using sparker lines in the same orientation.

footwall blocks of the basin becomes much more pronounced on line R-105 because the fault controlling the basin trends more to the northeast and therefore has a much larger normal component.

DISCUSSION: NEOGENE TECTONIC SETTING OF SOUTHERN PUERTO RICO

The following discussion summarizes the main observations made using onland tectonic geomorphology, fault striation studies, and offshore seismic profiles and compares these to the model predictions summarized in Figure 2.

Main Observations from Tectonic Geomorphology of Southern Puerto Rico

Observations presented support the idea of stability or extremely slow late Quaternary uplift in the area of the widest outcrop area of the Puerto Rico–Virgin Islands carbonate platform in southern Puerto Rico (Fig. 3A). In this area, late Neogene carbonate rocks are found near the coastline and are associated with limited marine terrace deposits of possible Holocene age (Prentice et al., 2004). To the west of the main outcrop area near La Paguera, the coastline is drowned and indicates either active drowned or a prior episode of drowning followed by a present-day period of stability. To the east of the main outcrop area, the coast is characterized by a broad coastal plain indicative of stability or subsidence (Fig. 3A).

Beach ridges of possible Holocene age are found at two areas along the south coast of Puerto Rico. Both areas are associated with the onland projection of northeast-trending normal faults observed on offshore seismic reflection lines as shown on Figure 15. More work is needed to verify whether these ridges are Holocene in age and have been uplifted.

Main Observations from Fault Striation Studies of Southern Puerto Rico

Results of the fault study are summarized on the maps in Figure 13 and 14 and in Table 1. Two phases of late Neogene extensional faulting are identified in southern Puerto Rico. A north-northeast–directed extension was identified as a regional event affecting Oligocene-Miocene carbonate units of southern Puerto Rico including the youngest unit (Ponce Formation). The north-northeast extensional event was recognized across most of southern Puerto Rico although the most observations are limited to the area of the best rock exposures in the Ponce area (Fig. 13).

The north-northeast extensional event may be responsible for the formation of east-west–trending topographic features like the Lajas Valley of southwestern Puerto Rico and east-west–trending sections of the coastline (Fig. 5), although the age of initiation of this presumed Lajas half-graben structure is not known.

A younger extensional event oriented in a northwest-southeast direction also deforms the youngest carbonate unit (Ponce formation of Miocene–early Pliocene age). Faults of this event crosscut faults of the north-northeast–directed extensional event and are therefore younger in age (post–early Pliocene to Quaternary). The southeast extension direction is roughly perpendicular to east-northeast–trending sections of the zediform coastline between La Parguera and the area east of Ponce (Fig. 14). It is not clear whether this episode of tectonic extension shaped the coastline or whether the coastline is simply the result of non-tectonic forces including the prevailing southeast trade winds and longshore drift of sand.

Main Observations from Offshore Seismic Profiles

Seafloor faulting of presumed Holocene age was identified at two localities using higher resolution data collected on the RV *Isla Magueyes* in 2000. The Caja de Muertos normal fault, previously recognized by Garrison (1969), bounds an elongate ridge upon which the island of Caja de Muertos is located (Figs. 15 and 18). This fault together with the Bajo Tasmanian fault to the northwest bound a graben that extends onshore in the Ponce area and has been named the Ponce basin (Fig. 15).

Seismic profiling in 2000 also confirmed active deformation of the seafloor along the southeastern extension of the Esmeralda and Rio Jueyes fault zones (Fig. 19) as previously proposed by Garrison (1969) using an earlier set of seismic profiles. These earlier profiles partially reproduced on Figures 16 and 17 show both northeast- and east-west–trending faults deforming the upper sediments of the offshore Caja de Muertos carbonate shelf (Fig. 15).

The set of northeast-trending faults is inferred to have been generated by the tectonic event responsible for striated faults of this trend onland in Puerto Rico (Fig. 14, Table 1). The age of this event is assumed to be Pliocene or younger as the youngest ages from the youngest carbonate unit (Ponce Formation) is early Pliocene (Frost et al., 1983). Southeast extension would induce normal faulting on faults trending to the northeast such as the Cajo de Muertos and Bajo Tasmanian faults and left-lateral strike-slip faults on faults trending west-northwest like the Great Southern Puerto Rico fault. Where the Great Southern Puerto Rico fault zone curves to the northeast, it becomes less strike-slip and more normal in character and produces greater extensional and tilting effects in the Whiting half-graben (Figs. 20 and 21).

Neotectonic Model

These results shown above for southern Puerto Rico are most consistent with the neotectonic model proposed by Speed and Larue (1991) and Feuillet et al. (2002) showing northwest-southeast opening of the Anegada rift system (Fig. 2C). In Figures 22A and 22B, we summarize some slight variations of this northwest-southeast extension model, including the earlier development of the fault pattern that was initiated by an oblique collision between the southeastern Bahama carbonate platform and Puerto Rico (Mann et al., 2002).

Figure 22A shows a bathymetric, tectonic, and earthquake epicenter map of the Puerto Rico and Virgin Islands area based on

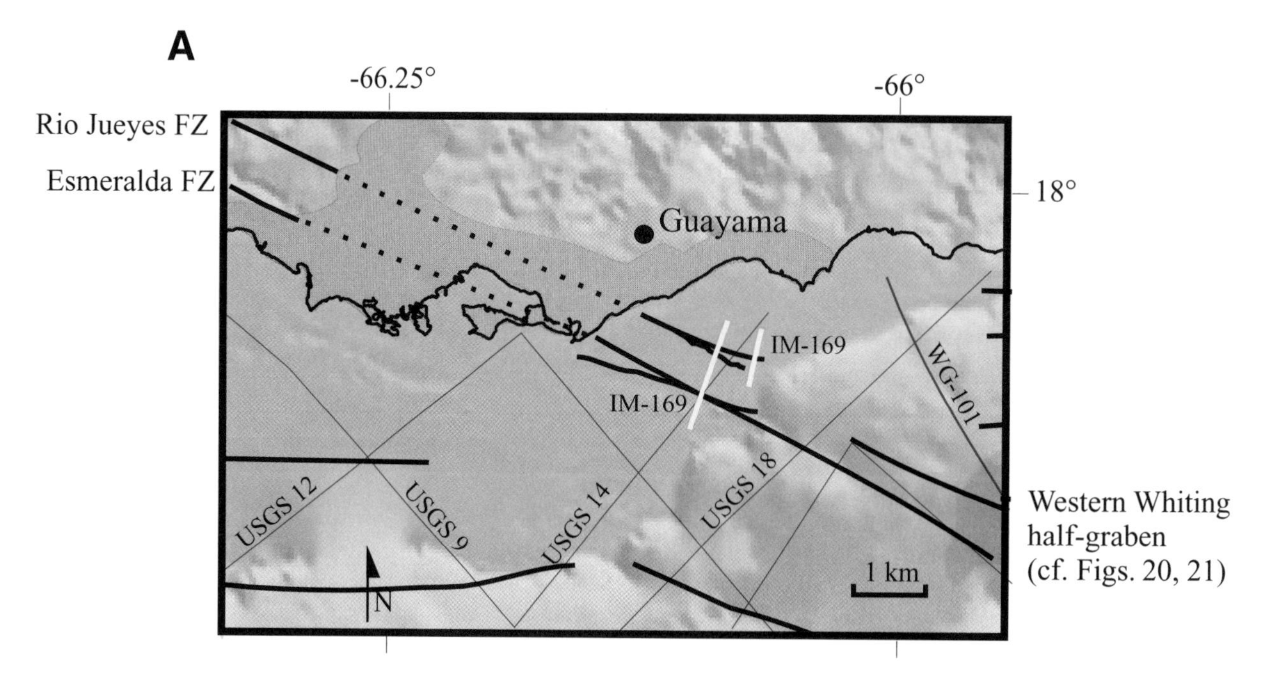

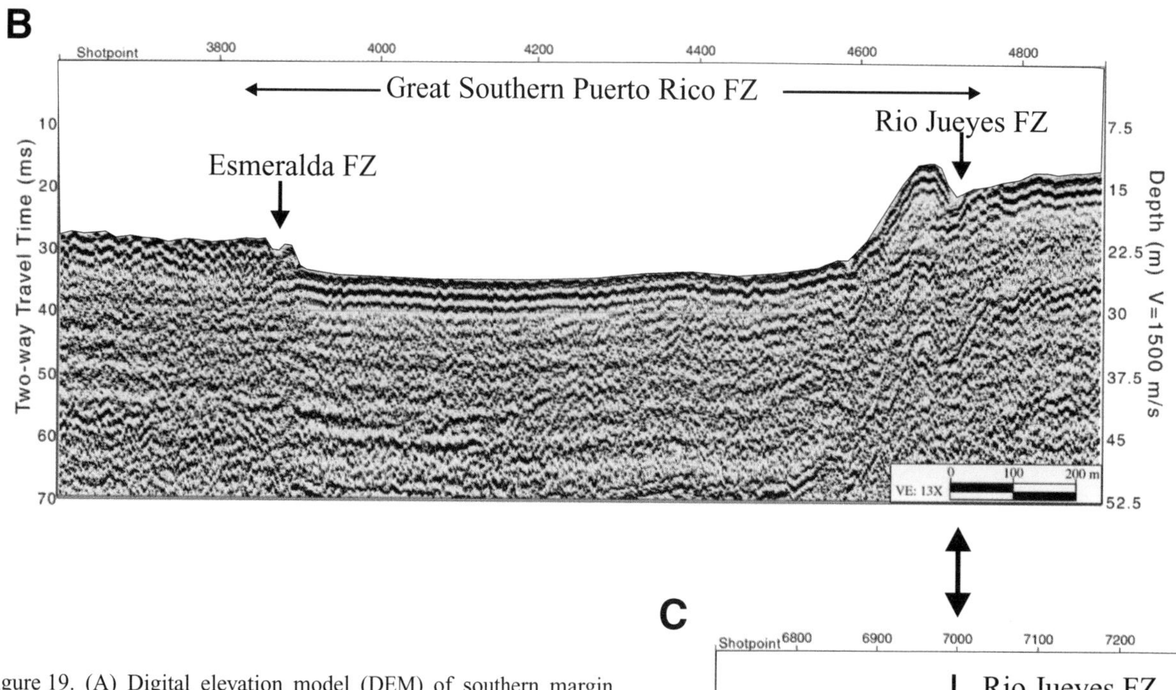

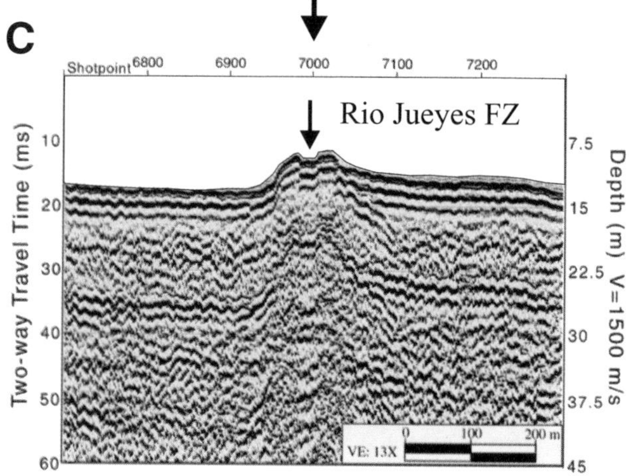

Figure 19. (A) Digital elevation model (DEM) of southern margin of Puerto Rico and bathymetry of southern shelf from NOAA in the area of the Great Southern Puerto Rico left-lateral strike-slip fault. (B) Single-channel seismic line IM-169 collected by RV *Isla Magueyes* in 2000 across the fault zone south of Guayama. The southern edge of the fault zone is defined by the submarine projection of the Esmeralda fault as projected across the coastal plain area by Glover (1971). The northern edge is defined by the submarine projection of the Rio Jueyes fault also projected by Glover (1971). The area between the two faults is marked by a shallow undisturbed depression several meters deeper than the areas north and south of its bounding faults.

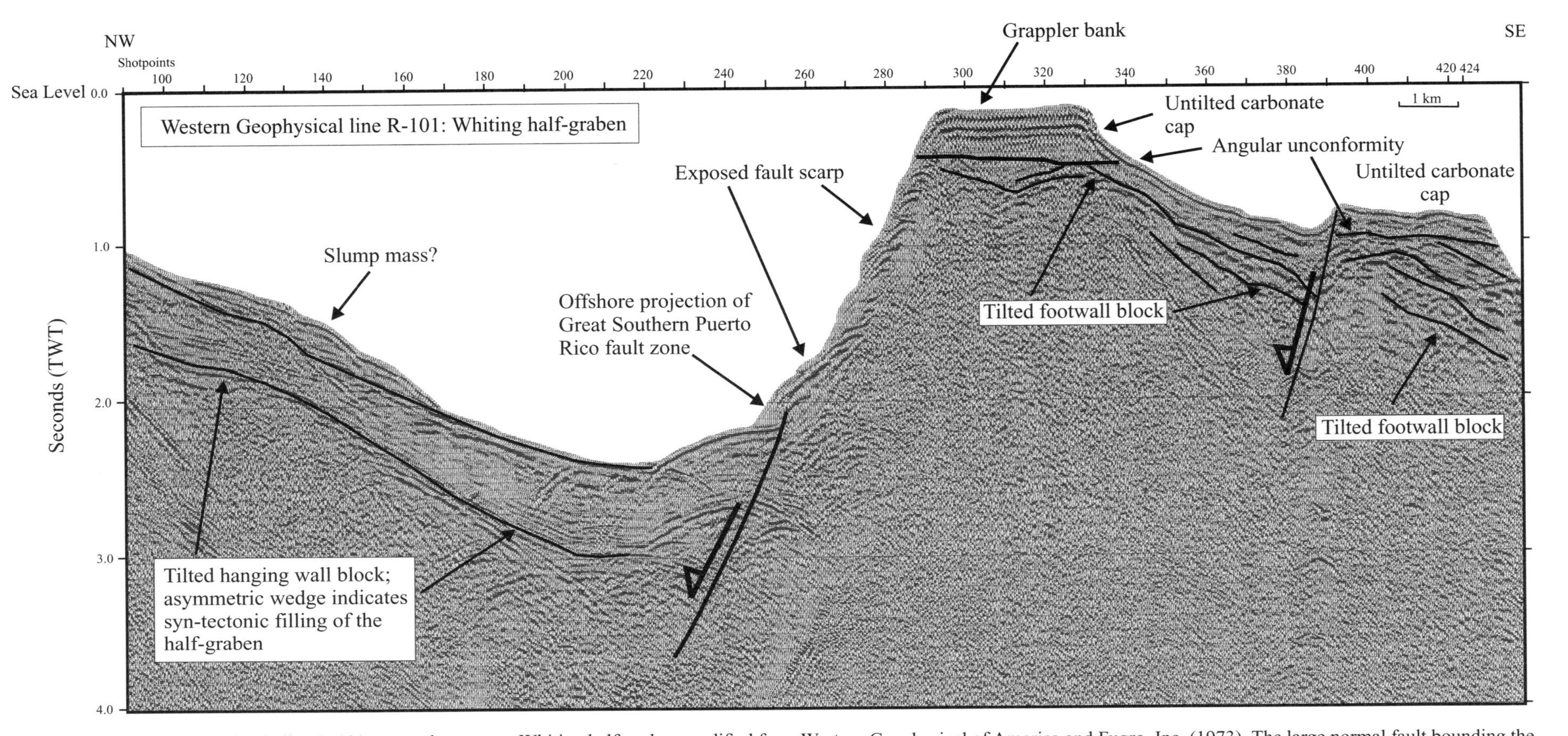

Figure 20. Multi-channel seismic line R-101 across the western Whiting half-graben modified from Western Geophysical of America and Fugro, Inc. (1973). The large normal fault bounding the flat-topped Whiting bank is the southeastward continuation of the Great Southern Puerto Rico fault zone shown in Figure 19 and seen in map view in Figure 15. At this locality the fault probably maintains a significant left-lateral strike-slip component. Whiting bank is a footwall uplift on the fault. The flat carbonate cap of the bank indicates that tilting is minimal.

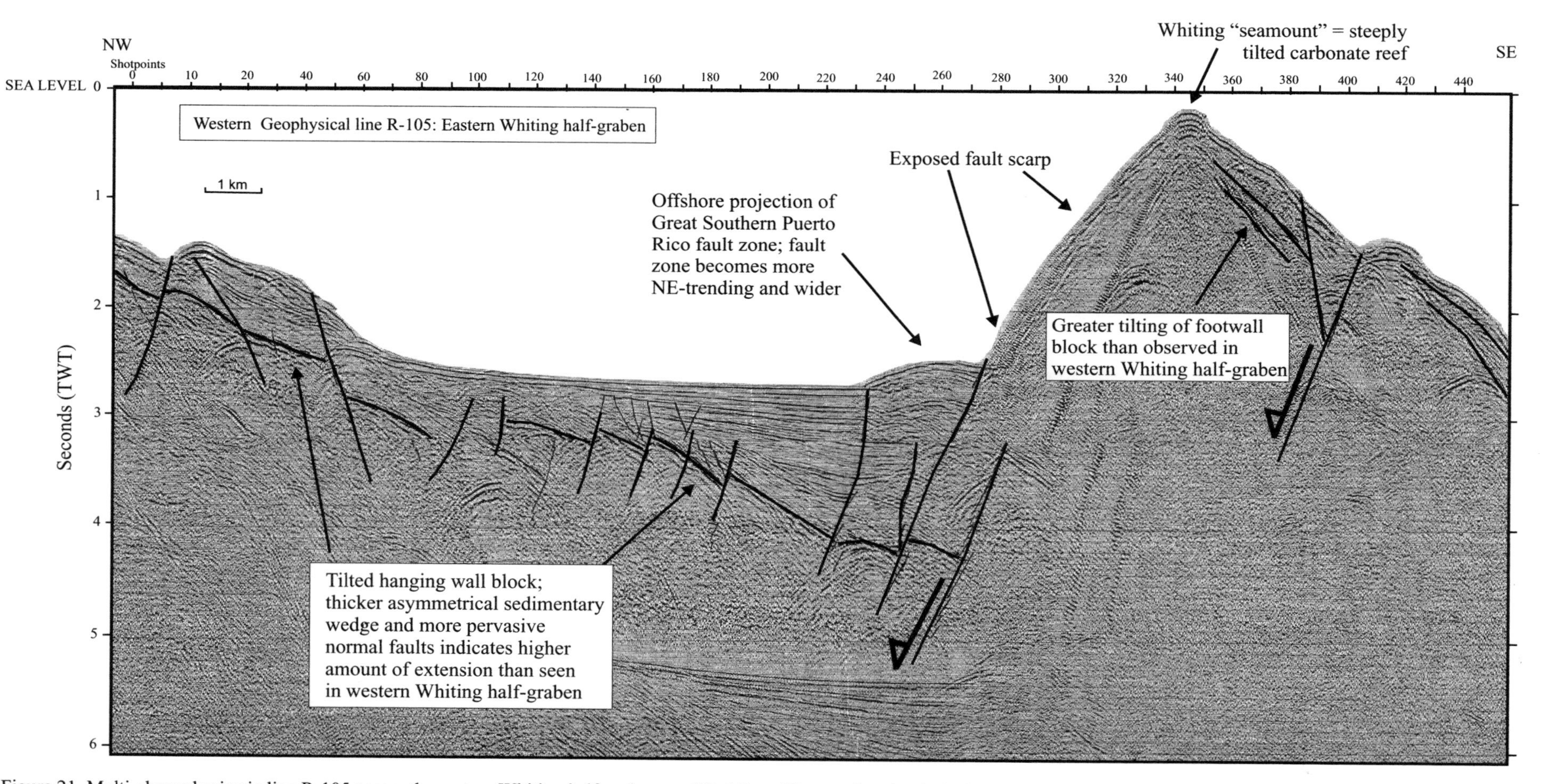

Figure 21. Multi-channel seismic line R-105 across the eastern Whiting half-graben modified from Western Geophysical of America and Fugro, Inc. (1973). The large normal fault bounding the flat-topped Whiting "seamount" is the southeastward continuation of the Great Southern Puerto Rico fault zone shown in Figure 19 and seen in map view in Figure 15. At this locality the fault assumed a greater amount of northeast-southwest extension due to the changing orientation of the fault (cf. Fig. 15). Whiting "seamount" is a footwall uplift on the fault. The absence of the flat carbonate cap seen to the west and the steeper basement dips indicates that the amount of extension has increased in the eastward direction.

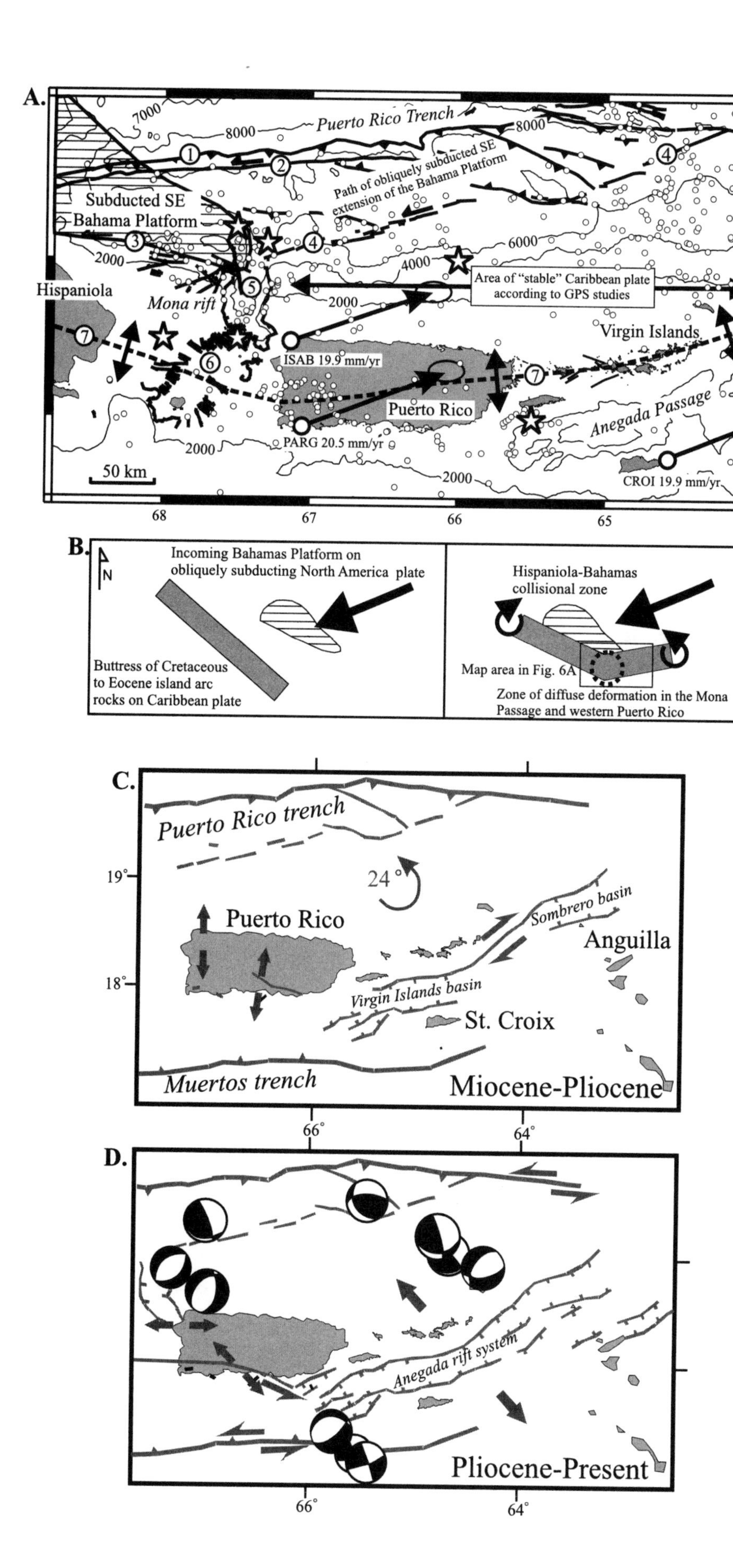

Figure 22. (A) Bathymetric and tectonic map of the eastern Hispaniola–Puerto Rico–Virgin Islands area modified from Grindlay et al. (1997; this volume, Chapter 7) and van Gestel et al. (1998) showing faulting and earthquakes near the eastern edge of the Hispaniola-Bahama collision zone. Motions of GPS sites at St. Croix (CROI), Parguera (PARG) and Isabella (ISAB) are indistinguishable from the larger Caribbean plate. Open circles are earthquake epicenters from 1973 to 1999 occurring at depths <30 km. Stars represent large magnitude historic earthquakes. Numbered features: 1—Puerto Rico trench; 2—Bunce fault; 3—Septentrional fault; 4—Bowin fault; 5—area of normal faults bounding the Mona rift; 6—area of diffuse normal and oblique-slip faults in the Mona Passage; 7—Puerto Rico–Virgin Islands arch. (B) Schematic diagram illustrating regional effect of oblique Bahama collision on the tectonics of the area of the Mona Passage and western Puerto Rico. Hispaniola has a post–middle Miocene history of crustal convergence related to the Bahama collision while the same period in Puerto Rico is dominated by extension probably related to a 25° counterclockwise rotation that accompanied the collision and indentation of the buttress of ancient arc rocks in Hispaniola and Puerto Rico. (C) Proposed regional tectonic interpretation of first phase of Miocene-Pliocene, north-south extension during 25° of counterclockwise rotation of the Puerto Rico block and right-lateral shear along the Anegada Passage (Virgin Islands and Sombrero basins) during collision between Hispaniola and the Bahama carbonate platform as shown schematically in Figure 22B. (D) Proposed regional tectonic interpretation of second phase of Pliocene-Present northeast-southwest extension southern Puerto Rico and opening across the Anegada rift system. Harvard earthquake focal mechanisms show that present-day deformation is dominated by normal and strike-slip faulting.

geophysical mapping by Grindlay et al. (1997), van Gestel et al. (1998, 1999), and Grindlay et al. (this volume, Chapter 2). Faults in the Puerto Rico trench area trend east-northeast and reflect the path of the subducted southeast extension of the Bahama Platform along the southern edge of the Caribbean forearc of Puerto Rico. These fault trends that include both strike-slip faults like the Bunce and Bowin faults (Fig. 1) and low-angle thrust faults like the Puerto Rico trench fault, parallel the present-day GPS velocity of Puerto Rico relative to North America. The end of the now-subducted Bahama platform (Mona block) now coincides with the eastern edge of the Mona rift, a 25-km-wide full graben that has disrupted the Oligocene to early Pliocene Puerto Rico–Virgin Islands carbonate platform. An incompletely mapped zone of northwest-striking normal and oblique-slip normal faults extends to the south and southwest of the Mona rift and deforms the carbonate platform in the central part of the Mona Passage. The Puerto Rico–Virgin Islands carbonate platform is also deformed into the Puerto Rico–Virgin Islands arch whose structural crest parallels the topographically highest land areas of the Virgin Islands, Puerto Rico, and eastern Hispaniola (Fig. 22A).

Our simplified interpretation of this complex pattern of deformation follows that of Mann et al. (2002) and Grindlay et al. (this volume, Chapter 2) and is shown schematically in Figure 22B. As the incoming Bahama platform has obliquely subducted beneath the edge of the Caribbean plate, the subducting platform has indented a linear belt or buttress of underlying Cretaceous-Eocene island arc rocks on the Caribbean plate in Hispaniola and the western part of the Mona Passage. Complex rifting in the Mona Passage reflects rifting and rotation as the uncollided areas to the east rotate in a counterclockwise direction. Paleomagnetic studies of the Puerto Rico–Virgin Islands carbonate platform in northern Puerto Rico confirm 25° of counterclockwise rotation of the island in late Miocene–Pliocene time (Reid et al., 1991). The collided area of the Caribbean plate to the west of the Mona Passage undergoes widespread Neogene shortening (Pubellier et al., 2000) while the uncollided area to the east of the Mona Passage in Puerto Rico and the Virgin Islands is characterized mainly by counterclockwise rotation about a hinge point in the Mona Passage, broad arching of the Puerto Rico–Virgin Islands arch, normal faulting related to separation from the collided area of Hispaniola, and strike-slip faulting.

The early phase of north-northeast extension affecting southern Puerto Rico is inferred to represent extension accompanying this earlier counterclockwise phase of rotation as shown in Figure 22C. This phase of extension is followed by a Quaternary phase of northwest-southeast extension that accompanied opening of the Anegada Passage as shown in Figure 22D. Earthquake focal mechanisms are complex but some are compatible with northeast-trending normal faults.

ACKNOWLEDGMENTS

The offshore surveying component of this study was supported by external National Earthquake Hazards Reduction Program (NEHRP) grant 00HQGR0032 to N. Grindlay and L. Abrams, and grant 00HQGR0033 to P. Mann. The onland fault mapping component of the project by Mann and Hippolyte was funded by external NEHRP grant 00HQGR0064 to P. Mann and grant 00HQGR0065 to C. Prentice (U.S. Geological Survey, Menlo Park). We thank Christa von Hillebrant-Andrade and Victor Huerfano of the Puerto Rico Seismic Network, University of Puerto Rico, for their logistical support and for archiving and making available the Western Geophysical offshore seismic reflection data from the southern margin of Puerto Rico. We also thank Jim Joyce, Hans Schellekens, and Hernan Santos of the Department of Geology, University of Puerto Rico at Mayaguez, and Charles Almy, Guilford College, for many useful discussions on the geology of this area. Special thanks to Uri ten Brink (USGS–Woods Hole) for making available archived paper copies of the USGS single-channel data collected by Garrison (1969) and to Rob Rogers (UTIG) for providing the DEM shown in Figure 3. University of Texas at Austin Institute for Geophysics (UTIG) contribution no. 1745. The authors acknowledge the financial support for this publication provided by The University of Texas at Austin's Geology Foundation and Jackson School of Geosciences.

REFERENCES CITED

Almy, C., Jr., 1969, Sedimentation and tectonism in the Upper Cretaceous Puerto Rican portion of the Caribbean island arc: Transactions of the Gulf Coast Association of Geological Sciences, v. 19, p. 269–279.

Anderson, H., 1977, Ground water in the Lajas Valley: U.S. Geological Survey Water-Resources Investigation Report 68-76, 45 p.

Angelier, J., 1990, Inversion of field data in fault tectonics to obtain the regional stress: A new rapid direct inversion method by analytical means: Geophysical Journal International, v. 103, p. 363–367.

Angelier, J., 1994, Fault slip analysis and paleostress reconstruction, *in* Hancock, P., ed., Continental Deformation: Pergamon Press, p. 53–100.

Bawiec, W., ed., 2001, Geology, geochemistry, geophysics, mineral occurrences, and mineral resource assessment for the Commonwealth of Puerto Rico: U.S. Geological Survey Open File Report 98-38 (CD-ROM).

Beach, D., 1976, Sedimentation on the western Isla Caja de Muertos insular shelf, Puerto Rico [unpublished M.S. thesis]: Mayaguez, University of Puerto Rico, 100 p.

Beach, D., and Trumbull, J., 1981, Marine geologic map of the Puerto Rico insular shelf, Isla Caja de Muertos area: U.S. Geological Survey Miscellaneous Investigations Map I-1265.

Bloom, A.L., 1986, Coastal landforms, *in* Short, N.M., and Blair, R.W., Jr., eds., Geomorphology from space: Washington, D.C., NASA Special Publication SP-486, p. 353–406.

Bott, M., 1959, The mechanics of oblique-slip faulting: Geological Magazine, v. 96, p. 109–117.

Briggs, R., Gelabert, P., Jordan, D., and Aguilar, E., Alonso, R., and Valentin, R., 1970, Engineering geology in Puerto Rico: Annual Meeting, Association of Engineering Geologists, Field trip number 6, p. 2–7.

Burke, K., 1988, Tectonic evolution of the Caribbean: Annual Review of Earth and Planetary Sciences, v. 16, p. 201–230.

Burke, K., Grippi, J., Şengör, C., 1980, Neogene structures in Jamaica and the tectonic style of the northern Caribbean Plate boundary: Journal of Geology, v. 88, p. 375–386.

Byrne, D., Suarez, G., and McCann, W., 1985, Muertos trough subduction—Microplate tectonics in the northern Caribbean?: Nature, v. 317, p. 420–421.

Calais, E., Mazabraud, Y., Mercier de Lepinay, B., Mann, P., Mattioli, G., and Jansma, P., 2002, Strain partitioning and fault slip rates in the northeastern Caribbean from GPS measurements: Geophysical Research Letters, v. 29, no. 18, 1856, doi:10.1029/2002GL015397, 2002.

Case, J.E., and Holcombe, T.L., 1980, Geologic-tectonic map of the Caribbean region: Reston, Virginia, USA, U.S. Geological Survey Miscellaneous Investigations Series Map I-1100, scale: 1:2,500,000.

DeMets, C., Jansma, P., Mattioli, G., Dixon, T., Farina, F., Bilham, R., Calais, E., and Mann, P., 2000, GPS geodetic constraints on Caribbean–North America plate motion: Geophysical Research Letters, v. 27, p. 437–440, doi: 10.1029/1999GL005436.

Denning, W., 1955, Preliminary results of geophysical exploration for oil and gas on the south coast of Puerto Rico: Commonwealth of Puerto Rico Economic Development Administration Division, Mineralogy and Geology, Bulletin 2, 17 p.

Dillon, W.P., Edgar, N.T., Scanlon, K.M., and Parson, L.M., 1994, A review of the tectonic problems of the strike-slip northern boundary of the Caribbean plate examination by GLORIA, in Gardner, J., Field, M., and Twichell, D., eds., Geology of the United States Seafloor: The View from Gloria: United Kingdom, Cambridge University Press, p. 135–164.

Dolan, J., Mann, P., de Zoeten, R., and Heubeck, C., 1991, Sedimentologic, stratigraphic, and tectonic synthesis of Eocene-Miocene sedimentary basins, Hispaniola and Puerto Rico, *in* Mann, P., Draper, G., and Lewis, J., eds., Geologic and Tectonic Development of the North America-Caribbean plate boundary in Hispaniola: Geological Society of America Special Paper 262, p. 217–264.

Dolan, J., Mullins, H., and Wald, D., 1998, Active tectonics of the north-central Caribbean: Oblique collision, strain partitioning, and opposing subducting slabs, *in* Dolan, J., and Mann, P., eds., Active Strike-slip and Collisional Tectonics of the Northern Caribbean plate boundary zone: Geological Society of America Special Paper 326, p. 1–61.

Erikson, J., Pindell, J., and Larue, D., 1990, Mid-Eocene–early Oligocene sinistral transcurrent faulting in Puerto Rico associated with formation of the Northern Caribbean plate boundary zone: Journal of Geology, v. 98, p. 365–368.

Frost, S.H., Harbour, J.L., Realini, M.J., and Harris, P.M., 1983, Oligocene Reef Tract development southwestern Puerto Rico Part 1, report: Miami, Florida, University of Miami, 144 p.

Feuillet, N., Manighetti, I., Tapponnier, P., Jacques, E., 2002, Arc parallel extension and localization of volcanic complexes in Guadeloupe, Lesser Antilles: Journal of Geophysical Research, B, Solid Earth and Planets, v. 107, (B12), 2331, doi: 10.1029/2001JB000308.

Garrison, L., 1969, Structural geology of the Muertos insular shelf, Puerto Rico: U.S. Geological Survey Open-File Report no. 1270, 9 p.

Geomatrix Consultants, 1988, Geological-seismological evaluation to assess potential earthquake ground motions for the Portugues dam, Puerto Rico: unpublished report submitted to the Department of the Army, Jacksonville Corps of Engineers, Jacksonville, Florida, 89 p.

Glover, L., III, 1971, Geology of the Coamo area, Puerto Rico, and its relation to the volcanic arc-trench association: U.S. Geological Survey Professional Paper 636, 102 p.

Glover, L., III, and Mattson, P., 1960, Successive thrust and transcurrent faulting during the early Tertiary in south-central Puerto Rico, *in* Short Papers in the Geological Sciences, U.S. Geological Survey Professional Paper 400-B, p. 363–365.

Gordon, M., Mann, P., Caceres, D., and Flores, R., 1997, Cenozoic tectonic history of the North America–Caribbean plate boundary zone in western Cuba: Journal of Geophysical Research, v. 102, p. 10,055–10,082, doi: 10.1029/96JB03177.

Grindlay, N., Mann, P., and Dolan, J., 1997, Researchers investigate submarine faults north of Puerto Rico: Eos (Transactions, American Geophysical Union), v. 78, p. 401.

Griscom, A., and Geddes, W., 1966, Island-arc structure interpreted from aeromagnetic data near Puerto Rico and the Virgin Islands: Geological Society of America Bulletin, v. 66, p. 153–162.

Hancock, P., 1985, Brittle microtectonics: Principles and practice: Journal of Structural Geology, v. 7, p. 437–457, doi: 10.1016/0191-8141(85)90048-3.

Hempton, M., and Barros, A., 1993, Mesozoic stratigraphy of Cuba: Depositional architecture of a southeast-facing continental margin, *in* Pindell, J., and Perkins, R., eds., Mesozoic and Early Cenozoic Development of the Gulf of Mexico and Caribbean Region: A Context for Hydrocarbon Exploration: Gulf Coast Section, Society of Economic Paleontologists and Mineralogists, Houston, Texas, p. 193–209.

Hippolyte, J.C., Angelier, J., Bergerat, F., and Guieu, G., 1993, Tectonic-stratigraphic record of paleostress time changes in the Oligocene basins of the Provence, southern France: Tectonophysics, v. 226, p. 15–35, doi: 10.1016/0040-1951(93)90108-V.

Hippolyte, J.C., Angelier, J., Roure, F., and Casero, P., 1994, Piggyback basin development and thrust belt evolution: Structural and paleostress analyses of Plio-Quaternary basins in the southern Apennines: Journal of Structural Geology, v. 16, p. 159–173, doi: 10.1016/0191-8141(94)90102-3.

Hippolyte, J.C., Badescu, D., and Constantin, P., 1999, Evolution of the transport direction of the Carpathian belt during its collision with the east European platform: Tectonics, v. 18, p. 1120–1138, doi: 10.1029/1999TC900027.

Horsfield, W.T., 1975, Quaternary vertical movements in the Greater Antilles: Geological Society of America Bulletin, v. 86, p. 933–938.

Jansma, P., Mattioli, G., Lopez, A., DeMets, C., Dixon, T., Mann, P., and Calais, E., 2000, Neotectonics of Puerto Rico and the Virgin Islands, northeastern Caribbean, from GPS geodesy: Tectonics, v. 19, p. 1021–1037, doi: 10.1029/1999TC001170.

Jany, I., Scanlon, K.M., Mauffret, A., 1990, Geological interpretation of combined Seabeam, GLORIA and seismic data from Anegada Passage (Virgin Islands, North Caribbean): Marine Geophysical Researches, v. 12, no. 3, p. 173–196.

Jolly, W., Lidiak, E., Schellekens, J., and Santos, H., 1998, Volcanism, tectonics, and stratigraphic correlations in Puerto Rico, *in* Lidiak, E., and Larue, D., eds., Tectonics and Geochemistry of the Northeastern Caribbean: Geological Society of America Special Paper 322, p. 1–34.

Joyce, J., McCann, W., and Lithgow, C., 1987, Onland active faulting in the Puerto Rico platelet: Eos (Transactions, American Geophysical Union), v. 68, p. 1483.

Kaye, C., 1959, Shoreline features and Quaternary shoreline changes, Puerto Rico: U.S. Geological Survey Professional Paper 317-B, p. 49–140.

Krushensky, R., and Monroe, W., 1975, Geologic map of the Ponce quadrangle, Puerto Rico: U.S. Geological Survey Miscellaneous Investigations Map I-1042, scale 1:20,000.

Krushensky, R., and Monroe, W., 1978, Geologic map of the Penuelas and Punta Cuchara quadrangles, Puerto Rico: U.S. Geological Survey Miscellaneous Investigations Map I-1042, scale 1:20,000.

Krushensky, R.D., Monroe, W.H., 1979, Geologic map of the Yauco and Punta Verraco quadrangles, Puerto Rico: Reston, Virginia, USA, U.S. Geological Survey Miscellaneous Investigations Series Map I-1147, scale 1:20,000.

Larue, D., and Ryan, H., 1998, Seismic reflection profiles of the Puerto Rico trench: Shortening between the North American and Caribbean plates, *in* Lidiak, E., and Larue, D., eds., Tectonics and Geochemistry of the Northeastern Caribbean: Geological Society of America Special Paper 322, p. 193–210.

Larue, D., Torrini, R., Jr., Smith, A., and Joyce, J., 1998, North Coast Tertiary basin of Puerto Rico: From arc basin to carbonate platform to arc-massif slope, *in* Lidiak, E., and Larue, D., eds., Tectonics and Geochemistry of the Northeastern Caribbean: Geological Society of America Special Paper 322, p. 155–176.

Lidiak, E.G., Larue, D.K., editors, 1998, Tectonics and geochemistry of the northeastern Caribbean: Geological Society of America Special Paper 322, 210 p..

Lobeck, A.K., 1922, The physiography of Porto Rico: Scientific Survey of Porto Rico and the Virgin Islands, v. 1, no. Part 4, p. 301–379.

Mann, P., and Burke, K., 1984, Neotectonics of the Caribbean: Reviews of Geophysics and Space Physics, v. 22, p. 309–362.

Mann, P., Burke, K., and Matumoto, T., 1984, Neotectonics of Hispaniola: Plate motion, sedimentation, and seismicity at a restraining bend: Earth and Planetary Science Letters, v. 70, p. 311–324.

Mann, P., Draper, G., and Burke, K., 1985, Neotectonics of a strike-slip restraining bend system, Jamaica, *in* Biddle, K.T., and Christie-Blick, N., eds., Strike-slip Deformation, Basin Formation, and Sedimentation: Society of Economic Paleontologists and Mineralogists Special Publication no. 37, p. 211–226.

Mann, P., Draper, G., and Lewis, J., 1991, Overview of the geologic and tectonic development of Hispaniola, *in* Mann, P., Draper, G., and Lewis, J.F., eds., Geologic and Tectonic Development of the North America–Caribbean Plate Boundary in Hispaniola: Geological Society of America Special Paper 262, p. 1–28.

Mann, P., Calais, E., Ruegg, J. C., DeMets, C., Jansma, P., and Mattioli, G., 2002, Oblique collision in the northeastern Caribbean from GPS measurements and geological observations: Tectonics, v. 21, no. 6, 1057, doi: 10.1029/2001TC001304, 2002.

Masson, D., and Scanlon, K., 1991, The neotectonic setting of Puerto Rico: Geological Society of America Bulletin, v. 103, p. 144–154, doi: 10.1130/0016-7606(1991)103<0144:TNSOPR>2.3.CO;2.

Mauffret, A., and Jany, I., 1990, Collision et tectonique d'expulsion le long de la frontiere Nord-Caraibe: Oceanologica Acta, v. 10, p. 97–116.

McCann, W.R., 1985, On the earthquake hazards of Puerto Rico and the Virgin Islands: Bulletin of the Seismological Society of America, v. 75, p. 251–262.

McCann, W., Joyce, J., and Lithgow, C., 1987, The Puerto Rico platelet at the northeastern edge of the Caribbean plate: Eos (Transactions, American Geophysical Union), v. 68, p. 1483.

McCann, W., 2002, Microearthquake data elucidate details of Caribbean subduction zone: Seismological Research Letters, v. 73, p. 25–32.
McIntyre, D., 1971, Geologic map of the central La Plata quadrangle, Puerto Rico: U.S. Geological Survey Miscellaneous Geologic Investigations Series Map I-660, scale 1:20,000.
MacIntyre, I.G., 1972, Submerged reefs of the eastern Caribbean: Bulletin of the American Association of Petroleum Geologists, v. 56, p. 720–738.
Meltzer, A., and Almy, C., 2000, Fault structure and earthquake potential Lajas Valley, southwest Puerto Rico: Eos (Transactions, American Geophysical Union), v. 81, Suppl., p. F1181.
Meltzer, A.S., Schoemann, M.L., Dietrich, C., Almy, C.C., Schellekens, H., 1995, Characterization of faulting; Southwest Puerto Rico: Geological Society of America Abstracts with Programs, v. 27, no. 6, p. 227–228.
Mercado, A., 1994, Digitization of National Ocean Survey hydrographic survey "smooth" sheets for Puerto Rico and the U.S. Virgin Islands, Sea Grant College Program, 116 p.
Monroe, W.H., 1968, The age of the Puerto Rico Trench: Geological Society of America Bulletin, v. 79, p. 487–494.
Monroe, W., 1973, Stratigraphy and petroleum possibilities of middle Tertiary rocks in Puerto Rico: Bulletin of the American Association of Petroleum Geologists, v. 57, p. 1086–1099.
Monroe, W., 1980, Geology of the middle Tertiary formations of Puerto Rico: U.S. Geological Survey Professional Paper 953, 93 p.
Morelock, J., Winget, A., and Goenaga, C., 1994, Geologic maps of southwestern Puerto Rico: Parguera to Guanica insular shelf: U.S. Geological Survey Miscellaneous Investigations Series Map I-2387, 1:40,000 scale.
Moussa, M.T., and Seiglie, G.A., 1975, Stratigraphy and petroleum possibilities of middle Tertiary rocks in Puerto Rico: Discussion: American Association of Petroleum Geologists Bulletin, v. 59, no. 1, p. 163–168.
Moussa, M., Seiglie, G., Meyerhoff, A., and Taner, I., 1987, The Quebradillas limestone (Miocene-Pliocene), northern Puerto Rico, and tectonics of the northeastern Caribbean: Geological Society of America Bulletin, v. 99, p. 427–439.
Muszala, S., Grindlay, N., and Bird, R., 1999, Three-dimensional Euler deconvolution and tectonic interpretation of marine magnetic anomaly data in the Puerto Rico trench: Journal of Geophysical Research, v. 104, p. 29,175–29,187, doi: 10.1029/1999JB900233.
Otvos, E., 2000, Beach ridges—Definitions and significance: Geomorphology, v. 32, p. 83–108, doi: 10.1016/S0169-555X(99)00075-6.
Perfit, M., Heezen, B., Rawson, M., and Donnelly, T., 1980, Chemistry, origin and tectonic significance of metamorphic rocks from the Puerto Rico trench: Marine Geology, v. 34, p. 125–156, doi: 10.1016/0025-3227(80)90069-9.
Petit, J., 1987, Criteria for the sense of movement on fault surfaces in brittle rocks: Journal of Structural Geology, v. 9, p. 597–608, doi: 10.1016/0191-8141(87)90145-3.
Prentice, C.S., McGeehin, J., Simmons, K.R., Muhs, D.R., Roig, C., Joyce, J., and Taggart, B., 2004, Holocene marine terraces in Puerto Rico: evidence for tectonic uplift?: Geological Society of America Abstracts with Programs, v. 36, no. 5, p. 288.
Pubellier, M., Mauffret, A., Leroy, S., Vila, J., and Amilcar, H., 2000, Plate boundary readjustment in oblique convergence: Example of the Neogene of Hispaniola: Greater Antilles: Tectonics, v. 19, p. 630–648.
Reid, J., Plumley, P., and Schellekens, H., 1991, Paleomagnetic evidence for Late Miocene counterclockwise rotation of the North Coast carbonate sequence, Puerto Rico: Geophysical Research Letters, v. 18, p. 565–568.
Saunders, C., 1973, Carbonate sedimentation on the inner shelf, Isla Magueyes, Puerto Rico [unpublished M.S. thesis]: Mayaguez, University of Puerto Rico, 52 p.
Schoemann, M., and Meltzer, A., 1995, High-resolution seismic reflection profiling and its application to the imaging of neotectonic deformation in the Lajas Valley, Puerto Rico: Eos, (Transactions, American Geophysical Union), v. 76, p. F391.
Speed, R.C., Larue, D.K., 1991, Extension and transtension in the plate boundary zone of the northeastern Caribbean: Geophysical Research Letters, v. 18, p. 573–576.
Stephan, J.F., Blanchet, R., and Mercier de Lepinay, B., 1986, Northern and southern Caribbean festoons (Panama, Colombia–Venezuela, and Hispaniola–Puerto Rico), interpreted as pseudo subductions induced by east-west shortening of the peri-Caribbean continental frame, *in* Wezel, F.-C., ed., The Origin of Arcs: New York, Elsevier, p. 401–422.
Taylor, M., and Stone, G., 1996, Beach-ridges: A review: Journal of Coastal Research, v. 12, p. 612–621.
Tobisch, O., and Turner, M., 1971, Geologic map of the San Sebastian quadrangle, Puerto Rico: U.S. Geological Survey Miscellaneous Investigations Series Map I-661, scale 1:20,000.
Trias, J., 1991, Marine geologic map of the Puerto Rico insular shelf—Guanica to Ponce area: U.S. Geological Survey Miscellaneous Investigations Series Map I-2263, 1:40,000 scale.
van den Bold, W.A., 1969, Neogene Ostracoda from southern Puerto Rico: Caribbean Journal of Science, v. 9, no. 3–4, p. 117–125.
van Gestel, J.P., Mann, P., Dolan, J., and Grindlay, N., 1998, Structure and tectonics of upper Cenozoic–Puerto Rico–Virgin Islands carbonate platform as determined from seismic reflection studies: Journal of Geophysical Research, v. 103, p. 30,505–30,530, doi: 10.1029/98JB02341.
van Gestel, J.P., Mann, P., Grindlay, N., and Dolan, J., 1999, Three-phase tectonic evolution of the northern margin of Puerto Rico as inferred from an integration of seismic reflection, well, and outcrop data: Marine Geology, v. 161, p. 257–286, doi: 10.1016/S0025-3227(99)00035-3.
Volckmann, R., 1984a, Geologic map of the Cabo Rojo and Parguera quadrangles, southwest Puerto Rico: Rico: Reston, Virginia, USA, U.S. Geological Survey Miscellaneous Investigations Series, Map I-1157, scale 1:20,000.
Volckmann, R.P., 1984b, Geologic map of the Puerto Real Quadrangle, Southwest Puerto Rico: Reston, Virginia, USA, U.S. Geological Survey Miscellaneous Investigations Series Map I-1559, scale 1:20,000.
Wallace, R., 1951, Geometry of shearing stress and relation to faulting: Journal of Geology, v. 59, p. 118–130.
Western Geophysical Company of America and Fugro, Inc., 1973, Geological-geophysical reconnaissance of Puerto Rico for siting of nuclear power plants: The Puerto Rico Water Resources Authority, San Juan, Puerto Rico, 127 p.

Manuscript Accepted by the Society 18 August 2004

Printed in the USA

Geological Society of America
Special Paper 385
2005

Paleoseismic study of the South Lajas fault: First documentation of an onshore Holocene fault in Puerto Rico

Carol S. Prentice
U.S. Geological Survey, 345 Middlefield Road, MS 977, Menlo Park, California 94025, USA

Paul Mann
Institute for Geophysics, 4412 Spicewood Springs Road, University of Texas, Austin, Texas 78713, USA

ABSTRACT

The island of Puerto Rico is located within the complex boundary between the North America and Caribbean plates. The relative motion along this boundary is dominantly left-lateral strike slip, but compression and extension are locally significant. Although tectonic models proposed for the region suggest the presence of onshore active faults in Puerto Rico, no faults with Holocene displacement have been documented on the island before this study. Current seismic hazard assessments primarily consider only the impact of distant, offshore seismic sources because onshore fault hazard is unknown. Our mapping and trenching studies demonstrate Holocene surface rupture on a previously undocumented fault in southwestern Puerto Rico. We excavated a trench across a scarp near the southern edge of the Lajas Valley that exposed a narrow fault zone disrupting alluvial deposits. Structural relations indicate valley-side-down fault slip, with a component of strike-slip motion. Radiocarbon analyses of organic material collected from the sediments suggest that the most recent surface rupture occurred during the past 5000 yr, but no minimum age has yet been established. This fault may be part of a larger fault zone that extends from the western end of the Lajas Valley toward Ponce, the second largest city in Puerto Rico.

Keywords: seismic hazard, Puerto Rico, active fault, paleoseismology.

INTRODUCTION AND TECTONIC SETTING

Puerto Rico is located within a broad and complex plate boundary between the North America and Caribbean plates (Fig. 1). The plate boundary in this region is transitional between subduction, dominant to the east, and strike slip, dominant to the west (Mann et al., 1998; DeMets et al., 2000; Bilich et al., 2001; McCann, 2002; Prentice et al., 2003). Recent geodetic studies (cf. Jansma et al., 2000) show that most of the plate motion between the North America and Caribbean plates occurs offshore north of Puerto Rico. However, Puerto Rico lies within an ~350-km-wide deformation zone between the Puerto Rico trench on the north and the Muertos trench on the south. Puerto Rico is also bounded on the east and west by faults in the Anagada passage and Mona passage, respectively (Masson and Scanlon, 1991).

Strike-slip, reverse, and normal faulting accommodate plate motion in the Puerto Rico–Virgin Islands area (McCann and Pannington, 1990). The major structures associated with the plate boundary are submerged and inaccessible for quantitative field study. Therefore, recent neotectonic models rely on teleseismic studies (Deng and Sykes, 1995; Dolan and Wald, 1998), GPS-based geodetic studies (e.g., Jansma et al., 2000), and marine

Prentice, C.S., and Mann, P., 2005, Paleoseismic study of the South Lajas fault: First documentation of an onshore Holocene fault in Puerto Rico, *in* Mann, P., ed., Active tectonics and seismic hazards of Puerto Rico, the Virgin Islands, and offshore areas: Geological Society of America Special Paper 385, p. 215–222. For permission to copy, contact editing@geosociety.org.

geophysical surveys (Dillon et al., 1996; Grindlay et al., 1997; van Gestel et al., 1998; Dolan et al., 1998). These studies suggest that the island of Puerto Rico is moving away from the island of Hispaniola at a rate of ~5 mm/yr (Jansma, et al., 2000). The extensional boundary between Hispaniola and Puerto Rico is potentially as wide as 150 km, but most data sets, including recent geodetic results, indicate that the boundary is primarily within westernmost Puerto Rico and the Mona Passage. The deepest part of the Mona Passage is the Mona rift, a graben bounded by late Quaternary faults. Deformation in this area, including normal faulting in the Mona Passage and western Puerto Rico, appears to be driven by the oblique collision of the Bahama carbonate platform with Hispaniola, west of Puerto Rico (Vogt et al., 1976; Mann et al., 1995; Dolan et al., 1998). This oblique collision has pinned or impeded the eastward motion of Hispaniola relative to North America and has produced a complex zone of rotational and extensional deformation in the Mona Passage and western Puerto Rico. Marine geophysical studies in the Mona Passage map several sets of large displacement, normal, and oblique-slip faults (e.g., Dolan et al., 1998; van Gestel et al., 1998). These faults are consistent with GPS results indicating that northwestern Puerto Rico is moving faster than is Hispaniola, relative to the North America plate (Jansma et al., 2000). The faster relative motion of Puerto Rico is consistent with the model that Hispaniola is pinned by the Bahama collision, producing opening of the Mona rift and right-lateral strike-slip motion along W-NW-striking faults in Puerto Rico (Dolan et al., 1998).

Jansma et al. (2000) use GPS observations to show that little (<3 mm/yr), if any, plate-boundary motion is accommodated within Puerto Rico onshore. However, shallow onshore seismicity, especially in the southwestern part of the island, suggests upper-plate deformation (Asencio, 1980). Abundant shallow

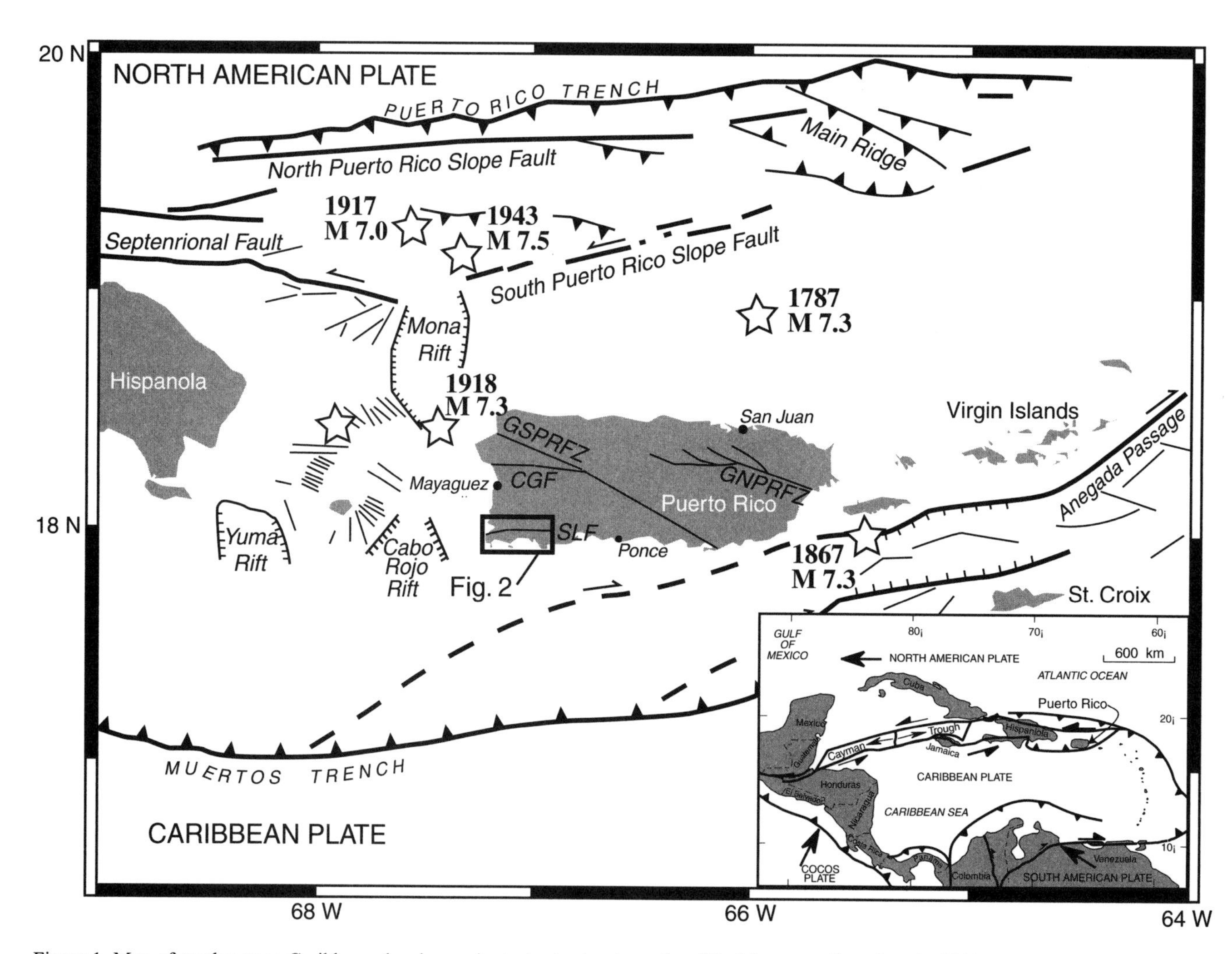

Figure 1. Map of northeastern Caribbean showing major tectonic structures (modified from van Gestel et al., 1998; Jansma et al., 2000). Stars represent approximate locations of epicenters of large-magnitude historical earthquakes. Rectangle indicates the Lajas Valley study area, shown in Figure 2. GNPRFZ—great northern Puerto Rico fault zone; GSPRFZ—great southern Puerto Rico fault zone; CGF—Cerro Goden fault; SLF—South Lajas fault, the subject of this study. (Inset) Tectonic setting of the Caribbean region, modified from Mann et al. (1990).

seismicity and the location of Puerto Rico between the Puerto Rico and Muertos trenches imply the possibility of Holocene faults onshore that could constitute an important seismic hazard to the island. Current seismic hazard assessments focus on the impact of the distant seismic sources offshore and on smoothed seismicity because no Holocene faults have previously been documented onshore. Our new mapping and paleoseismic studies demonstrate that Holocene fault rupture has occurred on at least one onshore fault in southwestern Puerto Rico, herein referred to as the South Lajas fault.

LAJAS VALLEY

The Lajas Valley of southwestern Puerto Rico is an E-W trending, 30-km-long linear depression bounded by hills on its northern and southern edges (Fig. 2). This geomorphology and the closed drainage of the valley suggest fault control, and therefore we chose this region to focus our search for evidence of Holocene faulting. Other workers have recognized possible fault control for the Lajas Valley on the basis of geomorphology (Joyce et al., 1987) and on seismic reflection profiles (Meltzer and Almy, 2000). However, before our study, no faults with Holocene displacement had been documented.

Geologic maps of the Lajas Valley area include 1:20,000-scale quadrangle maps of the San German and Puerto Real quadrangles (Volckmann, 1984a, 1984b). Neither map shows faults bounding the Lajas Valley, with the exception of a 5-km-long E-W– to NE-trending bedrock fault separating Mesozoic units in the far western part of the valley, south of Bahía de Boquerón. This fault is shown as being truncated to the east by a N-NW–trending bedrock fault and covered by alluvium to the west. These maps show no faults displacing Quaternary alluvium in the Lajas Valley. Earlier mapping by Mattson (1960) shows a fault in nearly the same position in the western Lajas Valley as on Volkman's maps, but Mattson extends the fault for a few hundred meters under the alluvium to the east, not to the west, and queries it under the alluvium. Briggs and Akers (1965) show a queried fault under alluvium in the western Lajas Valley that extends from the southern edge of the Bahia de Boquerón 5 km to the northeast. Almy (1969) inferred a buried, pre-Oligocene strike-slip fault under the entire length of the Lajas Valley. Subsequent publications depicting the geology of Puerto Rico variously show no faults in the Lajas Valley, a short fault segment in the far west of the valley, or a fault through the entire length of the valley. None addresses the age or style of faulting, or gives any fault mapped in the Lajas Valley a name.

We analyzed 1936 aerial photography of the Lajas Valley and identifed a 1-km-long NE-trending scarp crossing an alluvial fan on the southern side of the western Lajas Valley (Fig. 3) in the approximate position of the queried fault beneath alluvium shown by Briggs and Akers (1965). The scarp is ~1.5 m high at the excavation site (Fig. 4), and decreases in height along strike to the west.

Figure 2. Landsat Thematic Mapper image of southwestern Puerto Rico showing E-W–trending Lajas Valley, location of fault scarp, and inferred fault east of the excavation site. Box shows area of Figure 3. Image processed by Michael Rymer, U.S. Geological Survey.

Figure 3. Aerial photograph of southwestern Lajas Valley showing scarp crossing alluvial fan surface and location of excavation site. Black dot with arrow indicates view direction of photograph in Figure 4. Aerial photograph taken February 13, 1936 (number K-4-320), available from Puerto Rico Department of Transportation, San Juan.

Figure 4. Photograph, looking southward toward excavation site, showing scarp crossing alluvial fan surface. Arrow indicates trend of scarp crossing alluvial fan surface. See Figure 3 for location.

Approximately 0.5 km southwest of the excavation site, the scarp disappears, apparently buried by very young alluvial fan deposits. We are uncertain where or whether the fault continues west of this point. Marine geophysical studies in Bahía de Boquerón do not find evidence of recent faulting under the bay (Grindlay et al., 2000). This indicates that the fault is either deeply buried by recent sedimentation in the bay, ends east of the bay, or is too close to the shoreline to be imaged by this marine survey, or that the fault follows a more southerly course along the south side of the Peñones de Melones (Fig. 3). Further investigation is needed to determine which, if any, of these possibilities is correct.

To the east of the excavation site, the scarp increases in height along strike. Approximately 0.5 km northeast of the excavation site, the scarp is ~3 m high. Farther northeast, the scarp diminishes in height and disappears, and is apparently buried beneath alluvium shed from the Sierra Bermeja. Laguna Cartegena, a shallow lake that was partially drained as part of a large irrigation and drainage project completed in 1955 (Anderson, 1977), occupies a closed depression at the base of the Sierra Bermeja ~4 km to the northeast along the southern margin of the Lajas Valley (Fig. 2). We suggest that the southern shoreline of this lake is controlled by the South Lajas fault and that this location may be an ideal excavation site for future study of this fault. Similarly, Laguna de Guanica, also drained in 1955, is along strike of this inferred fault 16 km east of Laguna Cartagena, and its former northern shore is very linear (Fig. 2). The north shore of Laguna de Guanica may also be controlled by the South Lajas fault, providing another possible future target for paleoseismic studies. We have, at this time, been unable to trace the fault east of the Lajas Valley, and we note that additional study is needed to determine whether or not the fault continues eastward toward the city of Ponce.

EXCAVATION

We excavated a 1.5-m-deep trench across the scarp on the southern side of the Lajas Valley near the town of Boquerón at the site shown in Figures 2, 3, and 4. This site is located on an alluvial fan derived from the Sierra Bermeja, the mountains that bound the south side of the valley. Our excavation exposed a sequence of alluvial sediments (Table 1) that are disrupted by a subvertical fault zone (Fig. 5). The fault zone terminates upward ~0.5 m below the ground surface. This fault, labeled F1 on Figure 5, extends through depositional unit 30 and into the lower part of unit 20, but does not break unit 10.

Four samples of organic material collected from the alluvial fan sediments yielded Holocene radiocarbon ages (Table 2). We observed no evidence of bioturbation, such as krotovina or active burrows, in these alluvial sediments, and we therefore assume that the charcoal samples were deposited in place at the time the

TABLE 1. DESCRIPTIONS OF ALLUVIAL SEDIMETARY UNITS EXPOSED IN EXCAVATION ACROSS SOUTH LAJAS FAULT, SOUTHWESTERN PUERTO RICO

Unit number	Description	Comments
10	Massive, loose, sandy silt; few scattered pebbles. Lower contact gradational with underlying undisturbed unit 20.	Plow zone and modern soil.
20	Loose, gray, silty, sandy, pebble gravel. Lower contact with underlying unit 30 sharp and distinct.	Sample 1, which yielded a modern radiocarbon age collected from this unit. There is a significant hiatus between this and the underlying unit 30.
30	Very hard, massive, red silty clay with few scattered pebbles. Lower contact with underlying unit 40 gradational	Samples 3 and 5 collected from this unit. Sample 3: 4840–5040 cal yr B.P.; sample 5: 5650–5990 cal yr. B.P. Unit interpreted to represent a well developed B horizon of paleosol.
40	Hard, massive, sandy, pebbly clay. Abundant calcium carbonate nodules. Bottoms of clasts coated with calcium carbonate. Lower contact with underlying unit 50/55 sharp and distinct.	Aluvial fan deposits
50	Sandy clay containing calcium carbonate nodules. Contacts sharp and distinct.	Aluvial fan deposits
55	Sandy clay containing calcium carbonate nodules. Contacts sharp and distinct.	Aluvial fan deposits
60	Loose, reddish, pebble gravel. Contacts sharp and distinct.	Aluvial fan deposits
65	Hard, massive, silty, tan clay, containing calcium carbonate nodules. Contacts sharp and distinct.	Sample 2 collected from this unit: 7280–7550 cal yr B.P.
70	Hard, massive, tan clayey gravel. Bottoms of pebbles coated with calcium carbonate. Contains calcium carbonate nodules.	Aluvial fan deposits

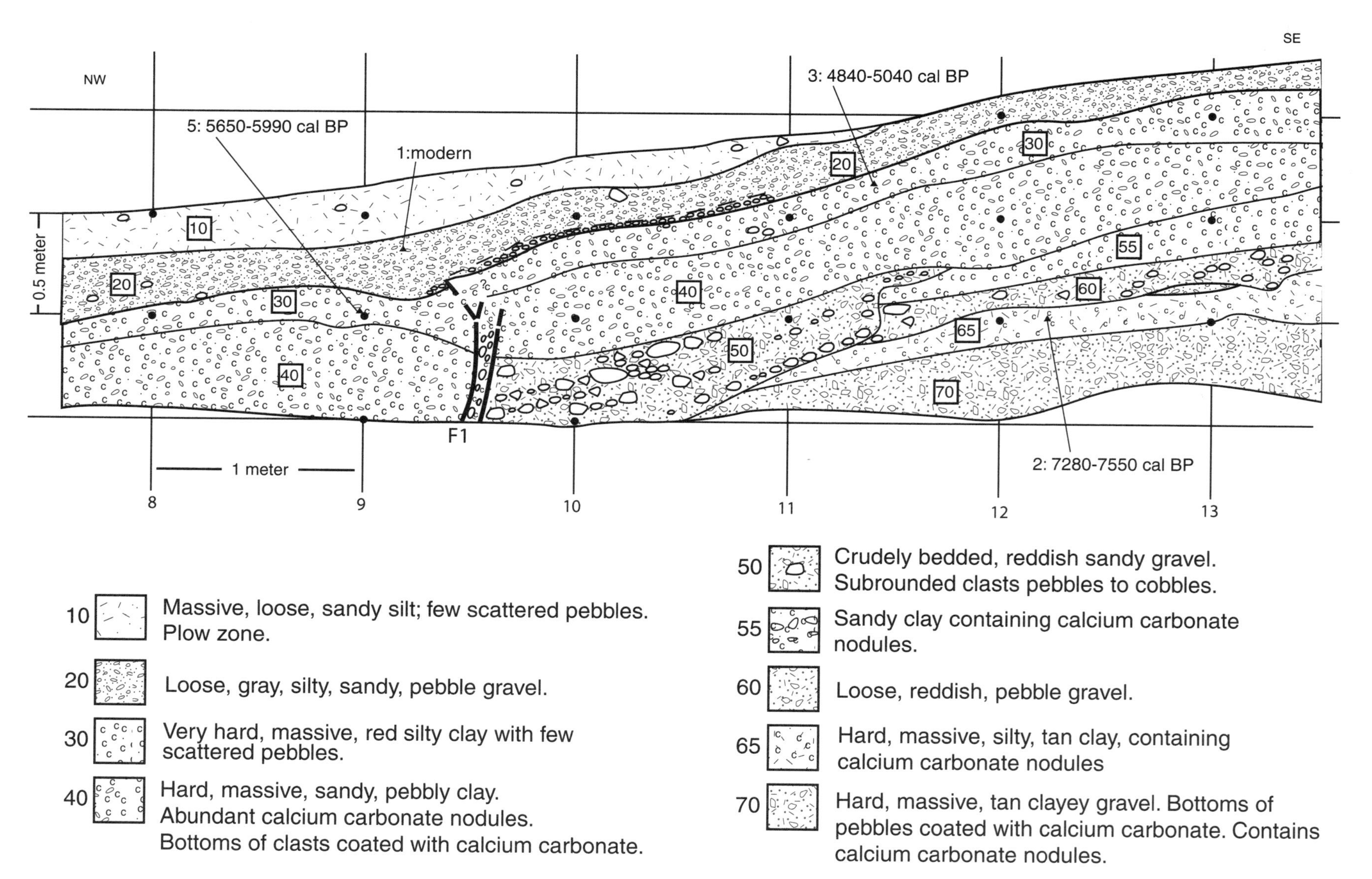

Figure 5. Log of excavation across fault scarp shown in Figure 4. Black dots represent 50 × 100–cm grid used to log the excavation. Fault zone F1 offsets unit 20, but does not offset unit 10, indicating a surface-rupturing earthquake occurred after unit 20 was deposited, and before deposition of unit 10. Radiocarbon dates indicate this surface rupture occurred within the past 5000 yr.

sediments were deposited. The ages of the samples are then at least slightly older than the age of deposition. The ages of these samples suggest that the most recent surface rupture occurred during the past 5000 yr (Fig. 5), based on radiocarbon analysis of sample 3, which yielded a 2σ calibrated age range of 4840–5040 yr B.P. The most recent event occurred before the deposition of sample 1, which yielded a modern age. Therefore, the most recent earthquake recorded in the excavation occurred after 5040 Cal yr B.P., but its minimum age is unconstrained. However, no large earthquake is reported in this region in the 500-yr-long historical record of Puerto Rico (McCann, 1985), suggesting that the most recent event occurred more than 500 yr ago.

Structural and stratigraphic relations in the excavation indicate a component of valley-side (north side)-down displacement as well as a component of strike-slip displacement. Strike-slip displacement is suggested by the thickness changes of units 30 and 40 across F1 (Fig. 5). Relations exposed in this excavation do not allow us to determine whether the strike-slip displacement is right or left lateral. We believe that additional excavations at this site would provide data to determine the sense of lateral motion.

DISCUSSION

The results of this study demonstrate Holocene surface rupture on a previously unrecognized onshore fault in Puerto Rico. Further study is required to document total length, style of deformation, and rate of fault slip. The length of the fault and the amount of slip per event need to be determined in order to estimate the magnitude of past earthquakes and the potential magnitude of future earthquakes. Nonetheless, our preliminary study clearly demonstrates that this fault is capable of producing earthquakes large enough to cause surface rupture. Most earthquakes that rupture to the surface are greater than M6, and are typically in the range of M7 or greater. An earthquake of M7 on the South Lajas fault could cause damage in the city of Ponce as well as in the towns within the Lajas Valley.

An important result of our study is to highlight the inadequacy of GPS methods to provide sufficient data to completely characterize Holocene faulting. Jansma et al. (2000) conclude that no onshore faults are present in Puerto Rico. Our study shows that this conclusion is incorrect. GPS data alone are not capable of providing a complete understanding of seismic hazard. Appropriate geologic studies are essential, as is well demonstrated by this example.

Additional work may identify other onshore seismic sources and is critical to a complete evaluation of seismic hazard for the island. We consider the Cerro Goden fault, near the city of Mayagüez, as another probable onshore late Quaternary structure (Mann et al., this volume Ch. 6). Puerto Rico is densely populated, with ~3.8 million people inhabiting the island. Thus, a moderate to large onshore earthquake could have a significant impact. Additional work aimed at identifying onshore Quaternary faults and quantifying their recent behavior is essential to more adequately characterize the seismic hazard for the island.

ACKNOWLEDGMENTS

We thank Carlos Pacheco for allowing us to dig the excavation on his property, and we thank all the property owners in Puerto Rico who allowed us access to their land during the course of this study. We thank Raquel Robledo of the USDA for assistance in locating aerial photography. Thanks to Christa von Hilebrandt of Red Sísmica at the University of Puerto Rico, Mayagüez, for assistance and support, and the Geology Department of the University of Puerto Rico, Mayagüez, especially Pamela Jansma, James Joyce, Glen Mattioli, and Robert Ripperdan. Thanks also to Melita Cutcher for field assistance. We are grateful for reviews of this paper by Eugene Schweig and

TABLE 2. CALCULATED DATES FROM ^{14}C ANALYSIS OF CHARCOAL, SOUTH LAJAS FAULT, PUERTO RICO

No.	Sample	$\delta^{13}C$ (‰)	^{14}C age* (yr B.P.)	Calibrated yr B.P.† (2σ)	Probability distribution	
1	AA-36450	–21.8	Post-bomb	0		
2	AA36451	–22.9	6495 ± 60	7270–7550	7554–7541	0.024
					7560–7495	0.019
					7492–7305	0.911
					7303–7287	0.029
					7281–7272	0.016
3	AA-36452	–23.5	4360 ± 45	4840–5040	5044–4839	1.000
5	AA-36454	–22.0	5100 ± 80	5650–5990	6164–6160	0.002
					5992–5652	0.998

*Conventional radiocarbon ages reported by University of Arizona. Calculations assume a Libby half-life (5568 yr). Uncertainties are 1 standard deviation counting errors.

†Dendrochronologically calibrated, calendar age ranges from CALIB Rev. 4.3, Method B, 2 standard deviation uncertainty, rounded to nearest decade (Stuiver et al., 1998; Stuiver and Reimer, 1993).

Patricia McCrory. This work was supported by external grant no. 00HQGR0064 to P. Mann and C. Prentice from the National Earthquake Hazards Reduction Program of the U.S. Geological Survey. University of Texas at Austin Institute for Geophysics (UTIG) contribution no. 1743.

REFERENCES CITED

Almy, C.C., Jr., 1969, Sedimentation and tectonism in the Upper Cretaceous Puerto Rican portion of the Caribbean island arc: Transactions of the Gulf Coast Association of Geological Sciences, v. 19, p. 269–279.

Anderson, H.R., 1977, Ground water in the Lajas Valley, Puerto Rico, U.S. Geological Survey Water-Resources Investigations 76-68, 52 p.

Asencio, E., 1980, Western Puerto Rico seismicity: U.S. Geological Survey Open-File Report 80-192, 135 p.

Bilich, A., Frohlich, C., and Mann, P., 2001, Global seismicity characteristics of subduction-to-strike-slip transitions: Journal of Geophysical Research, v. 106, p. 19,443–19,452, doi: 10.1029/2000JB900309.

Briggs, R.P., and Akers, J.P., 1965, Hydrogeologic map of Puerto Rico and adjacent islands: U.S. Geological Survey Hydrologic Investigations Atlas HA-197.

DeMets, C., Jansma, P., Mattioli, G., Dixon, T., Farina, F., Bilham, R., Calais, E., and Mann, P., 2000, GPS geodetic constraints on Caribbean–North America plate motion: Geophysical Research Letters, v. 27, p. 437–440, doi: 10.1029/1999GL005436.

Deng, J., and Sykes, L., 1995, Determination of Euler poles for Caribbean–North American plate using slip vectors of interplate earthquakes: Tectonics, v. 14, p. 39–53, doi: 10.1029/94TC02547.

Dillon, W.P., Edgar, T., Sclanon, K., and Coleman, D., 1996, A review of the tectonic problems of the strike-slip northern boundary of the Caribbean plate and examination by GLORIA, *in* Gardner, J., et al., eds., Geology of the United States seafloor: The view from GLORIA: Cambridge, UK, Cambridge University Press, p. 135–164.

Dolan, J.F., Mullins, H.T., and Wald, D.J., 1998, Active tectonics of the north-central Caribbean: Oblique collision, strain partitioning, and opposing subducted slabs, *in* Dolan, J.F., and Mann, P., eds., Active strike-slip and collisional tectonics of the Northern Caribbean Plate Boundary Zone: Geological Society of America Special Paper 326, p. 1–61.

Dolan, J.F., and Wald, D.J., 1998, The 1943–1953 north-central Caribbean earthquakes: Active tectonic setting, seismic hazards, and implications for Caribbean–North America plate motions, *in* Dolan, J.F., and Mann, P., eds., Active strike-slip and collisional tectonics of the Northern Caribbean Plate Boundary Zone: Geological Society of America Special Paper 326, p. 143–169.

Grindlay, N., Mann, P., and Dolan, J., 1997, Researchers investigate submarine faults north of Puerto Rico: Eos (Transactions, American Geophysical Union), v. 78, p. 404.

Grindlay, N., Abrams, L., Mann, P., and Del Greco, L., 2000, A high-resolution sidescan and seismic survey reveals evidence of late Holocene fault activity offshore western and southern Puerto Rico: Eos (Transactions, American Geophysical Union), v. 81, p. F1181.

Jansma, P., Lopez, A., Mattioli, G., DeMets, C., Dixon, T., Mann, P., and Calais, E., 2000, Neotectonics of Puerto Rico and the Virgin Islands, northeastern Caribbean, from GPS geodesy: Tectonics, v. 19, p. 1021–1037, doi: 10.1029/1999TC001170.

Joyce, J., McCann, W.R., and Lithgow, C., 1987, Onland active faulting in the Puerto Rico platelet: Eos (Transactions, American Geophysical Union), v. 68, p. 1483.

Mann, P., Schubert, C., and Burke, K., 1990, Review of Caribbean neotectonics, *in* Dengo, G., and Case, J.E., eds., The geology of North America, vol. H, The Caribbean region: Boulder, Colorado, Geological Society of America, p. 307–338.

Mann, P., Prentice, C.S., Burr, G., Peña, L.R., and Taylor, F.W., 1998, Tectonic geomorphology and paleoseismology of the Septentrional fault system, Dominican Republic, *in* Dolan, J.F., and Mann, P., eds., Active strike-slip and collisional tectonics of the Northern Caribbean Plate Boundary Zone: Geological Society of America Special Paper 326, p. 63–123.

Mann, P., Taylor, F.W., Edwards, R.L., and Ku, R., 1995, Actively evolving microplate formation by oblique collision and sideways motion along strike-slip faults: An example from the northeastern Caribbean plate margin: Tectonophysics, v. 246, p. 1–69, doi: 10.1016/0040-1951(94)00268-E.

Masson, D.G., and Scanlon, K.M., 1991, The neotectonic setting of Puerto Rico: Geological Society of America Bulletin, v. 103, p. 144–154, doi: 10.1130/0016-7606(1991)103<0144:TNSOPR>2.3.CO;2.

Mattson, P.H., 1960, Geology of the Mayagüez Area, Puerto Rico: Geological Society of America Bulletin, v. 71, p. 319–362.

McCann, W.R., 2002, Microearthquake data elucidates Caribbean subduction zone: Seismological Research Letters, v. 73, p. 25–32.

McCann, W.R., and Pannington, W.D., 1990, Seismicity, large earthquakes, and the margin of the Caribbean plate, *in* Dengo, G., and Case, J.E., eds., The geology of North America, vol. H, The Caribbean region: Boulder, Colorado, Geological Society of America, p. 291–306.

McCann, W.R., 1985, On the earthquake hazards of Puerto Rico and the Virgin Islands: Seismological Society of America Bulletin, v. 75, p. 251–262.

Meltzer, A., and Almy, C., 2000, Fault structure and earthquake potential, Lajas Valley, SW Puerto Rico: Eos (Transactions, American Geophysical Union), v. 81, p. F1181.

Prentice, C.S., Mann, P., Peña, L.R., and Burr, G., 2003, Slip rate and earthquake recurrence along the central Septentrional fault, North American–Caribbean plate boundary, Dominican Republic: Journal of Geophysical Research, v. 108 (B3), doi: 10.1029/2001JB000442.

Stuiver, M., and Reimer, P.J., 1993, Extended ^{14}C data base and revised CALIB 3.0 ^{14}C age calibration program: Radiocarbon, v. 35, p. 215–230.

Stuiver, M., Reimer, P.J., Bard, E., Beck, J.W., Burr, G.S., Hughen, K.A., Kromer, B., McCormac, F.G., v.d. Plicht, J., and Spurk, M., 1998, INTCAL98 radiocarbon age calibration, 24,000—cal BP: Radiocarbon, v. 40, p. 1041–1083.

van Gestel, J.-P., Mann, P., Dolan, J., and Grindlay, N.R., 1998, Structure and tectonics of the upper Cenozoic Puerto Rico–Virgin Islands carbonate platform as determined from seismic reflection studies: Journal of Geophysical Research, v. 103, p. 30,505–30,530, doi: 10.1029/98JB02341.

Vogt, P.R., Lowrie, A., Bracey, D.R., and Hey, R.N., 1976, Subduction of aseismic ridges: effects on shape, seismicity and other characteristics of consuming plate boundaries: Geological Society of America Special Paper 172, p. 1–59.

Volckmann, R.P., 1984a, Geologic map of the San German Quadrangle, southwest Puerto Rico: U.S. Geological Survey Map I-1558, scale 1:20,000.

Volckmann, R.P., 1984b, Geologic map of the Puerto Real Quadrangle, southwest Puerto Rico: U.S. Geological Survey Map I-1559, scale 1:20,000.

Manuscript Accepted by the Society August 18, 2004

Printed in the USA

Geological Society of America
Special Paper 385
2005

A seismic source model for Puerto Rico, for use in probabilistic ground motion hazard analyses

Roland C. LaForge
U.S. Bureau of Reclamation, Seismotectonics and Geophysics Group, Box 25007, D-8330, Denver Federal Center, Denver, Colorado 80225, USA

William R. McCann
Earth Scientific Consultants, Westminster, Colorado 80021, USA

ABSTRACT

Here we present a seismic source model for Puerto Rico, designed for use in probabilistic ground motion hazard analyses. The model consists of characterizations of known on- and offshore faults and their estimated geometries, the magnitudes of maximum earthquakes, one or more recurrence models for each fault, and activity rates for randomly occurring upper-crustal seismicity beneath the island. Slip rates for faults in the Mona and Anegada Passages were estimated by the allocation of GPS-based regional horizontal geodetic vectors onto appropriate groupings of faults. Rates for other sources were based on GPS vectors or empirical fits to Puerto Rico Seismic Network (PRSN) data.

Hazard curves for peak horizontal acceleration were computed by using the model for the four corners of the island. The hazard appears to be greatest in the western part of the island, which is closer to the more active sections of the two subduction zones and faults of the Mona Passage extensional regime. To the east the hazard is less because of presumed lower coupling between the North America and Puerto Rico–Virgin Islands microplate, and because of dispersion of relative motion at the Muertos subduction zone onto faults in the Anegada Passage and the Investigator faults south of the island. Recommendations for further work include further development of attenuation parameters, refinement of the local seismicity catalog, and further investigations into faults on and near the island.

Keywords: Puerto Rico; probabilistic seismic hazard analysis; seismic source model.

1. INTRODUCTION

Puerto Rico, a densely populated island of ~4 million inhabitants, lies in the northeast corner of the zone of convergence between the Caribbean and North America plates (Fig. 1). Puerto Rico and the Virgin Islands lie on the Puerto Rico–Virgin Islands platform, a shallow, partially subaerial submarine bank at the easternmost end of the Greater Antilles island chain. These islands lie near the northeastern corner of the Caribbean Plate. Slow subduction of the North America plate beneath the leading edge of the Caribbean plate dominates the tectonic environment of the northeastern Caribbean (Sykes et al., 1982; DeMets et al., 1990). To the north and east lies the west-southwesterly moving North America plate, locally

LaForge, R.C., and McCann, W.R., 2005, A seismic source model for Puerto Rico, for use in probabilistic ground motion hazard analyses, *in* Mann, P., ed., Active tectonics and seismic hazards of Puerto Rico, the Virgin Islands, and offshore areas: Geological Society of America Special Paper 385, p. 223–248. For permission to copy, contact editing@geosociety.org.

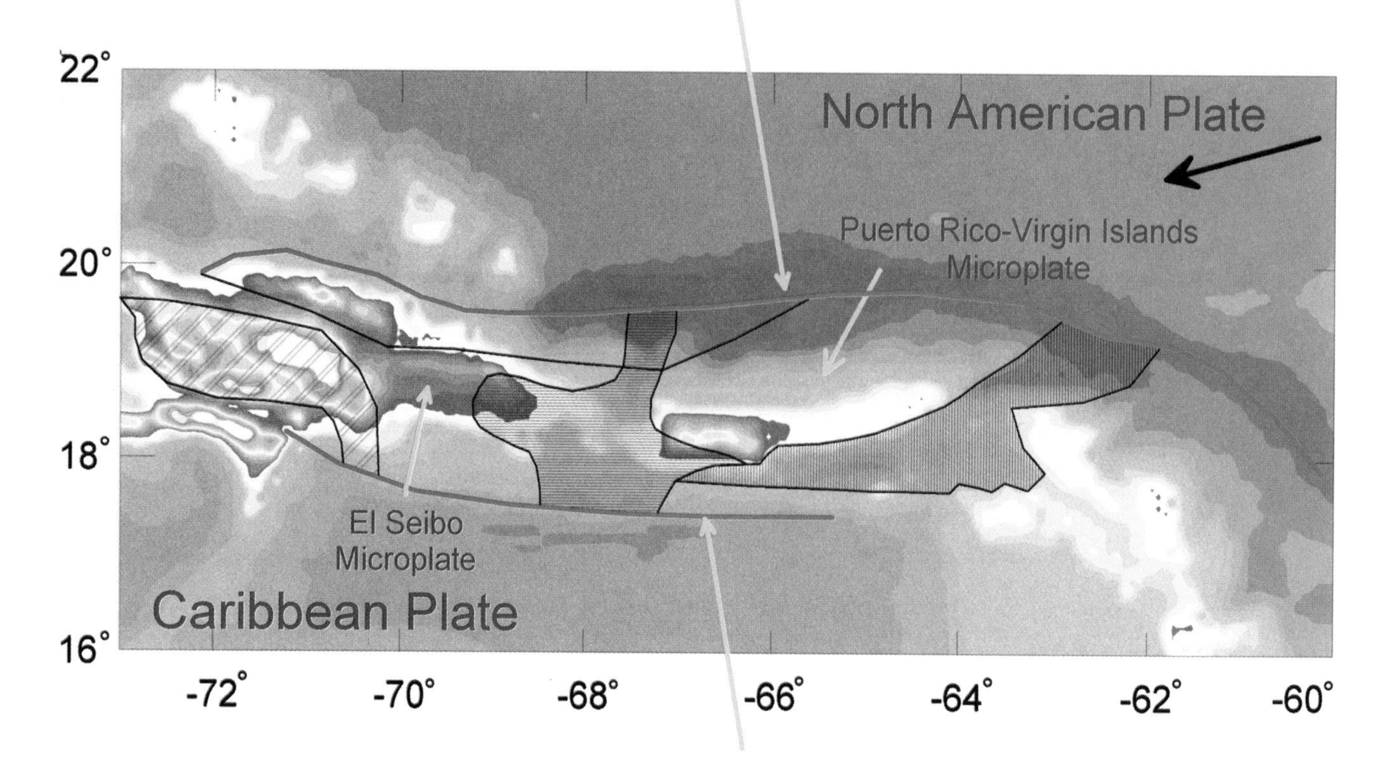

Figure 1. Regional map showing major tectonic elements. Puerto Rico–Lesser Antilles subduction zone is trace of south-dipping seismic sources, including shallow plate interface and deeper intraslab events. Los Muertos trough is trace of northerly dipping Caribbean Plate and associated seismic zones. Mona Passage extensional zone is region with horizontal lines and Anegada trough extensional zone is region with vertically oriented lines. Diagonally hatched region is collisional zone at west end of El Seibo microplate in Hispaniola. Arrow in upper right corner is motion vector for North America plate with respect to a fixed Caribbean plate (Jansma et al., 2000).

represented by the floor of the Atlantic Ocean and the Bahama platform. To the south and west lie the various basins of the Caribbean Sea, within the rigid Caribbean plate. The Puerto Rico–Virgin Islands platform is part of an arc massif along the northeast fringe of the Caribbean plate. It straddles and is cut by major tectonic and seismically active features that form the plate boundary zone between the major plates.

The Puerto Rico trench and the Muertos trough bound the Puerto Rico microplate to the north and south, respectively. Motion along these features reflects oblique convergent slip between the major plates and the Puerto Rico microplate. To the west the microplate ends in a wide zone of NE-SW directed transtension in western Puerto Rico, Mona Passage, and the eastern Dominican Republic (van Gestel et al., 1998; McCann, 1999; Mann et al., this volume). To the east, its margin lies at the extensional Anegada trough (Lithgow et al., 1987). Deformation in the Mona Passage and Anegada trough occurs along the edges of microplates within the plate boundary zone (Byrne et al., 1985) and serves to transfer slip from the Muertos trough to the Puerto Rico trench. Two active faults we assume to be related to the Mona Passage deformation have been identified on land in western Puerto Rico (Mann and Prentice, 2001; Prentice et al., 2003).

Because the North America plate motion is nearly parallel to the Puerto Rico trench (Calais et al., 2002), a number of transtensional features have developed in the accretionary wedge above the downgoing North America plate. In our study area, the most important of these are the Septentrional fault and the Bowin fault zone (Fig. 2). In central Hispaniola, the Septentrional fault exhibits 6–12 mm/yr of slip and accounts for a significant proportion of the relative plate motion (Prentice et al., 2003). To the east, however, this rate decreases by an undetermined amount because of the presence of similar features that serve the same function. The Bowin fault zone lies on the east side of Mona Canyon,

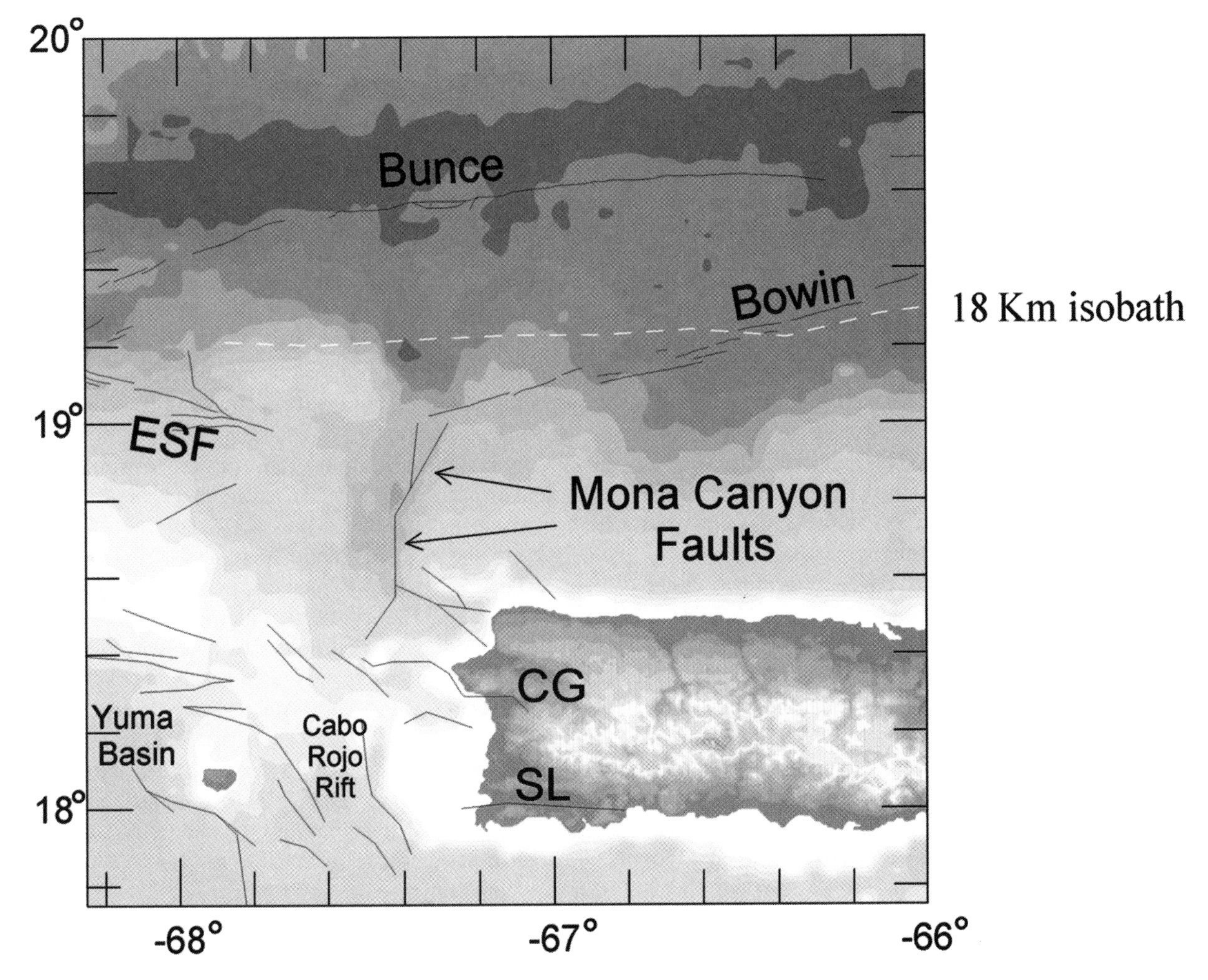

Figure 2. Surface traces of faults in the Mona Passage area as modeled for the probabilistic analysis. CG—Cerro Goden fault; SL—South Lajas fault; ESF—eastern Septentrional fault. Bunce and Bowin faults, Yuma Basin, and Cabo Rojo Rift are shown. Depth to North America plate shown as 18-km isobath.

adjacent to the east end of the Septentrional fault, in unknown relation to it and with an unknown slip rate.

In addition to these seismogenic features, ground-shaking hazards are also presented by upper crustal randomly occurring earthquakes not associated with known structures. Given the list of known seismic sources and a history of damaging events (McCann, 2002), earthquakes present a serious threat to the engineered infrastructure and public safety of the island.

In this paper, we construct a seismic source model for use in probabilistic ground motion calculations and calculate hazard curves for peak horizontal acceleration for the four geographic corners of the island. Results from these sites are meant to show the effect of proximity to the various sources and their relative contributions. The construction of a hazard map, which involves computing the hazard for a dense grid of sites and contouring the results, was beyond the scope of this work. The model consists of a catalog of known faults, their estimated geometries, magnitudes of maximum events, recurrence parameters, and activity rates, as well as activity rates for randomly occurring upper-crustal seismicity on the island. The major source regions are as follows:

1. The Puerto Rico subduction zone, which accommodates highly oblique convergence between the oceanic North America plate and the Puerto Rico–Virgin Islands microplate. This source is further subdivided into a shallow zone, the source of large thrust events along the plate interface, and a deeper "intraslab" zone, where earthquakes occur within the downgoing North America plate as it moves westward within the upper mantle.

2. The Mona Passage region, which lies between Puerto Rico and the island of Hispaniola. This region of active deformation is responding to the eastward movement of Puerto Rico relative to Hispaniola (Jansma et al., 2000; Calais et al., 2002; Mann et al., 2002). This region is marked by canyons and basins on the north and south edges, and numerous faults in between (McCann, 1998; van Gestel et al., 1998).

3. The eastern end of the Septentrional fault, which accommodates a predominantly strike-slip component of the nearly east-west North America–Puerto Rico–Virgin Islands plate motion (Dolan et al., 1998). The Bowin fault zone begins ~50 km east of the Septentrional fault and trends ~100 km in a northeasterly direction.

4. The Muertos subduction zone lies parallel to and south of the south coast of Puerto Rico and accommodates closure between the Caribbean plate and the Puerto Rico–Virgin Islands microplate. The Muertos subduction zone gradually dies out to the east, as the plate motion is apparently transferred to faults to the north (the Investigator faults) and faults in the Anegada trough.

5. The Anegada trough, a region of deep basins formed during a previous episode of northwest directed extension, but still active as an east-west extensional regime today (Jany et al., 1990).

6. Upper crustal seismicity beneath the island of Puerto Rico (McCann, 2002), unassociated with known faults, characterized as an areal zone of randomly occurring earthquakes.

Two Quaternary faults on land have been identified on the basis of geomorphic and paleoseismic studies (Prentice et al., 2003; Mann, this volume), and submarine faults were identified through analysis of ship-based seismic reflection and bathymetric data (Mercado and McCann, 1998). Slip rates on many of the faults were derived from recently published GPS data (Jansma et al., 2000; Calais et al., 2002). For the Mona Passage, Muertos trough, and Anegada trough zones, slip was apportioned onto each fault on the basis of its relative seismic moment contribution, then corrected for the inferred three-dimensional (3-D) geometry of the faults. The only data available for corroboration of the calculated seismicity rates are the historic seismicity record and paleoseismic information for the South Lajas fault in southwestern Puerto Rico. Slip rates for the eastern Septentrional and Bowin faults relied on bathymetric evidence and historic seismicity. Rates for the areal zone and North America plate intraslab seismicity were calculated from a portion of the historic seismicity catalog.

This study represents an attempt to construct a comprehensive, integrated 3-D seismic source model based on current knowledge and to compute representative mean probabilistic results. As might be expected, the quality of information is better in some regions than others, and therefore the level of uncertainty in the results can be expected to be quite variable. Slip rates are characterized as a single best-guess value, maximum magnitudes are presented as distributions, and alternative recurrence models are treated in a logic tree format. Although previous uncertainties were incorporated where deemed appropriate, a quantitative analysis of the uncertainties in the input parameters and their effect on posterior distributions is beyond the scope of this paper.

Our knowledge of the details of tectonic processes and structural dynamics of the Puerto Rico region has increased dramatically in the past decade. Much has been learned, but the nature—and quantitative details—of many processes remain poorly constrained. This should be kept in mind when interpreting these results. One goal of this work, however, is to establish a framework into which future advances and refinements can be incorporated. Another goal, a side benefit of probabilistic analyses, is to identify which areas of research could have the most impact in reducing the uncertainties in our estimates of the hazard posed by ground shaking from earthquakes.

2. SEISMIC SOURCES

2.1. Puerto Rico Subduction Zone

The Puerto Rico trench lies ~135 km north of the northern shore of Puerto Rico and marks the initial descent of the North America plate beneath the Puerto Rico microplate (Fig. 1). The trench is the westerly extension of the bathymetric low to the east of the Lesser Antilles. There it trends northerly or northwesterly, but gradually changes trend and deepens as it approaches Puerto Rico, where its axis reaches depths of 8 km or more. The trench continues to the west of Puerto Rico, shallowing some 4 km between Puerto Rico and Hispaniola, being filled by the buoyant Bahama Bank. The Atlantic seafloor continues south of the trench, forming the lower portion of the interplate zone of contact, the upper portion being the Puerto Rico microplate. Seismic activity indicates that the North America slab descends beneath the islands to depths of 125–150 km.

Recent GPS studies indicate a relative convergence between the North America and the Puerto Rico microplate of 16.9 ± 1 mm/yr in a WSW direction (Jansma et al., 2000; Calais et al., 2002), on the basis of data collected from 1994 to 2000. This is intermediate between values of 11 mm/yr (DeMets et al., 1990) and 37 mm/yr (Sykes et al., 1982), which were previously estimated on the bases of global plate motion models and the length of the downgoing slab, respectively.

Seismicity related to subducted slabs is of two general types: interface earthquakes are the result of the direct contact between the downgoing and overriding plates, and intraslab earthquakes are the result of internal deformation in the downgoing plate at greater depths. Global surveys show that interface earthquakes begin at ~20-km depth (Byrne et al., 1988) and end at ~40-km depth (Tichelaar and Ruff, 1993). However, dense metamorphic rocks have been found just south of the trench (Larue and Ryan, 1998), suggesting to us strong coupling at relatively shallow depths. The upper limit of the seismogenic surface was therefore assumed to be at a depth of 10 km below sea level. The location of the 40-km depth level of the Atlantic plate was identified by microearthquake activity (McCann, 2002), and the dip of the slab is assumed to be planar from the trench to this location. Assuming the 130-km depth contour of the subducted slab to lie beneath the south end of the island (McCann and Sykes, 1984; McCann, 2002) and 40-km depth to mark the downdip limit of the plate interface, the geometric model shown in Figures 3 and 4 was derived. The shallow subduction zone dips 21°, and the deep zone 41°. This is a reasonable approximation to the continually steepening dip seen in the cross sections of microseismicity (McCann, 2002).

2.1.1. Shallow Interface

The Puerto Rico trench is a linear, continuous feature between Mona Canyon and the Main Ridge, a high-topography feature on the North America plate. McCann (1985) suggested that a magnitude 8–8.25 interface event occurred on this segment in 1787, rupturing from roughly Mona Canyon on the west to the Main Ridge on the east (all magnitudes in this paper are assumed to be moment magnitude—M_W—unless otherwise noted). In 1943, a magnitude 7.8 event ruptured an ~80-km-wide section of the subduction zone across Mona Canyon, and on the basis of a focal mechanism, it was judged to have occurred on the shallow interface (Dolan and Wald, 1998). Because the Bahama platform impinges on the Caribbean plate at Mona Canyon but not to the east, the percentage of aseismic slip is likely to be greater to the east than in the 1943 zone. We have therefore divided the shallow interface into a western zone with relatively strong coupling and an eastern zone with weaker coupling.

Estimates of the percentage of seismic coupling were derived from comparisons to other island arcs, as described in Scholz and Campos (1995). For the eastern zone, the seismic coupling coefficient was estimated to have a best estimate value of 20%. Multiplying these values by the 16.9 mm/yr value of Jansma et al. (2000) results in corresponding seismic slip rate of 3.4 mm/yr. For the western zone, a best estimate of 80% was used. Here, we first subtract 2 mm/yr because of partitioning of the North America–Puerto Rico–Virgin Islands motion onto the East Septentrional fault (section 2.5). This results in a slip rate of 11.9 mm/yr for the western shallow interface. This characterization assumes that future behavior of the shallow Puerto Rico subduction zone will consist of repeats of the 1787 and 1943 events.

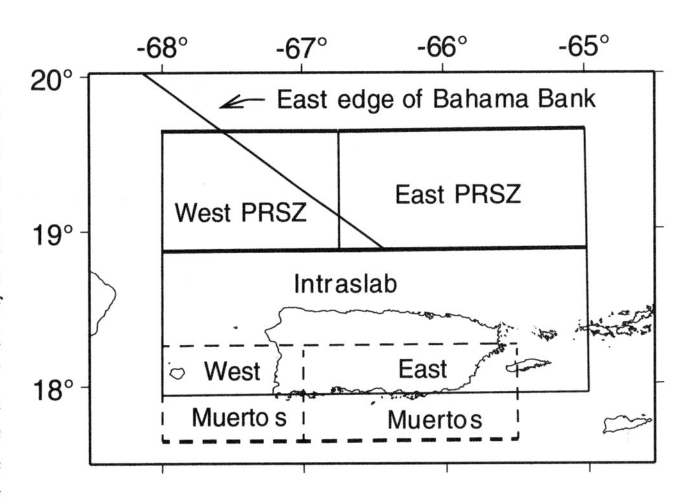

Figure 3. Map projection of modeled Puerto Rico subduction zone, intraslab, and Muertos subduction zone (dashed lines). Heavy line signifies updip boundary. Eastern boundary of Bahama Bank shown. PRSZ—Puerto Rico subduction zone.

It is assumed here that the Main Ridge to the east and the presence of the Bahama Bank and other seafloor relief in the shallow subduction zone to the west will act as rupture containment bounds for future large events in the eastern shallow Puerto Rico subduction zone. On the basis of the bathymetry of the

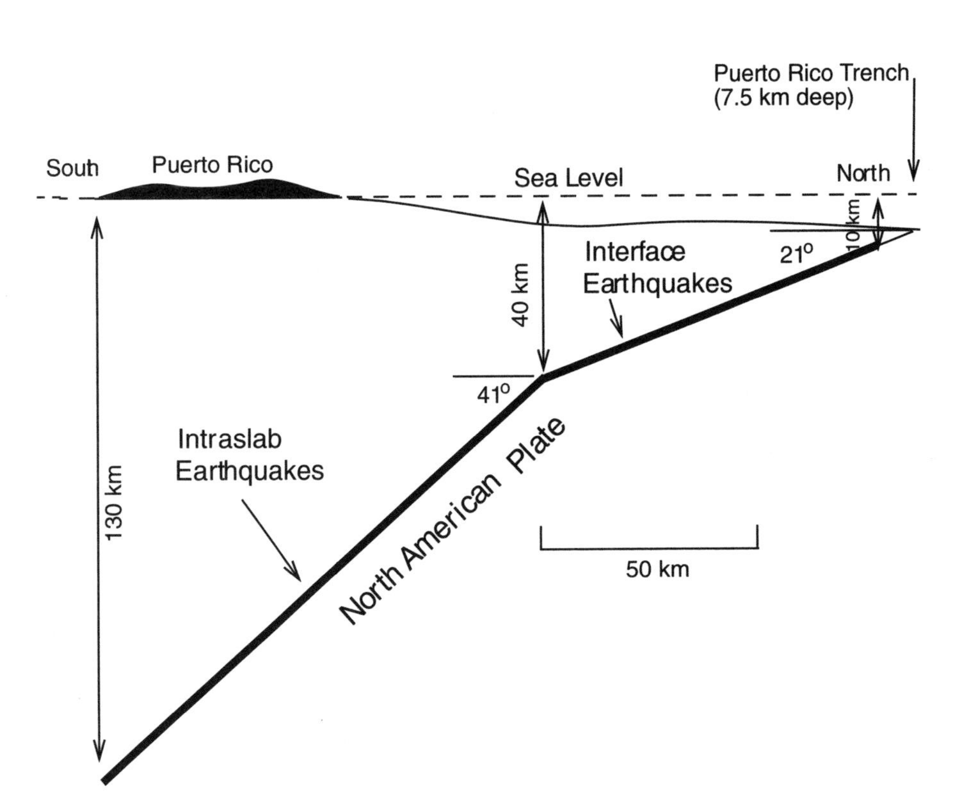

Figure 4. North-south cross section of North America plate as modeled. Island topography and bathymetry are schematic. Configuration based on McCann and Sykes (1984) and McCann (2002).

trench, minimum and maximum rupture lengths for the eastern zone were estimated to be 117 and 290 km, respectively. The minimum estimate assumes the rupture bounds to be the western edge of the Main Ridge and at the Mona Canyon. The maximum estimate assumes the bounds to be the center of the Main Ridge and at the Mona Canyon (McCann and Sykes, 1984). Given a rupture width of 84 km (Fig. 4) and the magnitude-fault area relation of Geomatrix (1993) for global shallow thrust earthquakes in subduction zones, the mean maximum magnitudes range between 7.9 and 8.2. We therefore impose an upper magnitude limit of 8.2 for this segment.

Although we can mark the eastern end of possible western interface ruptures at ~67°W on the basis of impingement of the Bahama Bank, judging where the western limit to a large rupture might be is problematical. Although the 1943 event apparently ended at ~67.75°W, we know of no strong argument that would preclude the extension of a future rupture further to the west. Although it is possible that Mona Canyon may be a significant enough feature to inhibit rupture propagation on the shallow interface beneath it, this remains a speculative idea at present. We therefore impose the same upper magnitude limit as on the eastern segment.

2.1.2. Deep Puerto Rico Subduction Zone

The deep portion of the Puerto Rico subduction zone is modeled as the down-dip continuation of the shallow zone (Figs. 3 and 4). Earthquakes in this zone result from internal stresses within the slab, caused by such factors as slab push or pull, bending moments, and density contrasts with the surrounding upper mantle. Similar to the shallow interface, a planar surface is used to represent a continually dipping structure.

In contrast to the shallow plate interface, activity rates of the deep portion of the Puerto Rico subduction zone are significantly lower, and no geomorphic features can be observed or investigated through geophysical methods. We are therefore forced to exclusively use the historic seismicity record to estimate seismicity rates. On the basis of arguments presented in section 2.6, we use the period August 1975 through December 1978, and take the region 65.5°W–68°W, the region south of 19°N, and events deeper than 40 km. This selection may include deeper events occurring in the north-dipping Caribbean plate as well. As in section 2.6, we assume the catalog to be complete at the magnitude 3.0 level. The search yielded 22 events of magnitude 3 and 6 events of magnitude 4—a small number from which to extrapolate occurrence rates of much larger earthquakes. On the basis of the two magnitude ranges (3 and 4), the b value in the Gutenberg-Richter relation $\log(N) = a - b(\mathrm{M})$ (where N is the number of earthquakes at and above magnitude M) is 0.56. This in turn implies that a magnitude 7 event should occur every 15 yr, which is obviously not happening. Estimates of b values in regions globally usually center around 1.0.

To estimate recurrence times for the larger intraslab events, we simply assume a b value of 1.0 and set the a value in the recurrence equation based on 28 events of magnitude 3 and greater over a period of 41 months. As in LaForge and Hawkins (1999), we assume magnitude 7.5 to be the upper magnitude limit for intraslab events. This results in return periods of 84 yr for magnitude 6.0 and above, 850 yr for 7.0 and above, and 2700 yr for magnitude 7.5. Compared with the historic record, these appear reasonable. The Instituto Panamericano de Geografía y Historia (IPGH) catalog for the Caribbean region (Shepherd et al., 1994) lists two magnitude 6 events that occurred between 64°W and 67°W, deeper than 50 km, from 1900 to 2001. One nineteenth-century earthquake in the historic record could have been a large intraslab event: in 1844, moderate shaking and low-level damage was reported uniformly throughout the central-eastern part of the island. No aftershocks were noted. The event was not reported to have been felt in Hispaniola, but was claimed to have been felt in St. Thomas and Guadeloupe (McCann, 2003).

2.2. Mona Passage Faults

The Mona Passage is underlain by the submerged Puerto Rico–Virgin Islands platform between the islands of Puerto Rico and Hispaniola (Fig. 2). In north-south cross section, the platform has an archlike structure, being quite shallow in the central part (three islands are present in the Mona Passage) and deepening to the north and south (van Gestel et al., 1998). GPS measurements indicate 5 mm/yr (with 3 mm/yr standard error) of extension between the two islands (Jansma et al., 2000; Calais et al., 2002). It is postulated that extension in the Mona Passage is due to stronger coupling of the subducted Bahama Bank in the shallow North America–Puerto Rico–Virgin Islands subduction zone, which inhibits eastward relative movement of Hispaniola while Puerto Rico moves faster to the east, unimpeded (McCann et al., 1987; McCann, 1999; Dolan et al., 1998; Mann et al., 2002).

Active structures clearly indicative of crustal extension exist in the seafloor bathymetry. Mona Canyon is a north-south–oriented canyon cut into the northeastern part of the platform, ~50 km northwest of the northwest corner of the island. It is ~50 km long and shows ~3.5 km of vertical relief (Fig. 2). Abundant seismicity is observed in the vicinity of the canyon, although these events are not located well enough to draw correlations to particular faults associated with the canyon. An earthquake of magnitude 7.3 (Doser et al., this volume) occurred near the canyon in 1918 and was responsible for a tsunami that caused 116 deaths and substantial monetary losses. Soon after the event, Reid and Taber (1919) postulated that the earthquake was associated with the canyon. Tsunami modeling by Mercado and McCann (1998) successfully matched the observed effects on land with the rupture of the long normal fault bounding its east side (Fig. 2).

On the south side of the island platform, two similar features are cut into the island platform. The Yuma Basin, to the southwest, is wider and covers a greater area than Mona Canyon, but is not as deep (Fig. 2). The Cabo Rojo rift lies in the southeast part of the platform (Fig. 2), and appears to be a smaller version of the Yuma Basin (van Gestel et al., 1998). These features undoubtedly perform a role similar to that of Mona Canyon in the extension of the passage. Faults in south-central Mona Passage appear to be associated with these features. Faults on the west side of Yuma Basin probably exist but have not been identified because of the lack of oceanographic surveys in that part of the passage.

McCann (1998) identified 84 active normal faults in the Mona Passage from seismic reflection records. The recency of activity was deduced from the youthful appearance of the scarps, and the continuity of some of the faults with Quaternary faults on land. Although a few are correlated with Mona Canyon (also called Mona Rift), most consist of short features trending generally NW-SE, with one trending N-S in the south-central part of the passage. On the basis of the faulting pattern, the azimuth of regional extension was estimated to be approximately N45°E. Of these, 26 that were long enough to generate M_W 6.5 or larger events (using the fault length-magnitude relations of Wells and Coppersmith [1994]) were included in the seismic source model (Fig. 2). Although McCann (1998) was unable to observe fault dips directly, a mean estimate of 63° was made on the basis of theoretical considerations.

In addition to these, two faults found on the island of Puerto Rico were included in the model. The Cerro Goden fault consists of both onland and offshore segments (Mercado and McCann, 1998; Grindlay et al., this volume, Chapter 7). The onshore segment forms the southern range front of the La Cadena range in west-central Puerto Rico and shows evidence of Holocene right-lateral and normal movement (Mann and Prentice, 2001; Mann et al., this volume). The right-lateral movement is consistent with northeast oriented extension. The north-dipping South Lajas fault trends obliquely to the east-west geomorphic southern edge of the Lajas Valley (Meltzer, 1997). A single trench site yielded evidence for two surface faulting events in the past 7000 yr (Mann and Prentice, 2001; Prentice, 2003). The trench information showed normal movement of ~1 m and also lateral movement of unknown direction and magnitude.

Although the identification of the major faults involved in the Mona Passage extensional regime is an important step in describing the hazard from this region, for probabilistic analyses, it is necessary to estimate slip rates on the faults. With the possible exception of the fault on the east side of the Mona Canyon (Fig. 2), direct evidence for fault displacement is not available, and has not been investigated, for the Mona Passage faults. We have therefore developed a method by which a regional horizontal extension azimuth and displacement rate is apportioned onto the faults.

The method is applied by assigning to each fault a magnitude based on surface length, using the relations of Wells and Coppersmith (1994). The magnitude is then converted to seismic moment as described in Hanks and Kanamori (1979). The proportion that each fault contributes to the total seismic moment is then calculated. That proportion of the regional slip vector is then assigned to the fault's slip vector. Because of the 3-D nature of the slip vector, however, the summed horizontal slip components over the region will be less than the regional value. The correction factor is determined and applied uniformly to each fault. The result is a slip vector and rate for each fault apportioned by its presumed relative seismic moment capability, which correctly accounts for the strike and dip of each fault and its angular relationship to the regional stress field, and whose summed horizontal slip rate components in the direction of regional extension match the geodetically observed value (Calais et al., 2002; LaForge and McCann, 2003).

The regional extension was assumed to be 5 mm/yr, as reported from the latest GPS analysis (Jansma et al., 2000; Calais et al., 2002), directed along an azimuth of N45°E, as estimated by McCann (1998). Because of the lack of seismic reflection data in the southwest part of the passage, unidentified faults associated with the Yuma Basin (particularly on the west side of the basin) were assumed to account for 20% of the total regional slip magnitude, resulting in a value of 4 mm/yr distributed over the faults shown in Figure 2. All faults were assumed to have a dip of 63°. Table A1 gives the slip rate estimated for each fault using this methodology and the return period for the estimated maximum fault magnitude using the fault area-moment relation of Aki and Richards (1980) (this paper, Equation 2) and fault area-magnitude relations (for "all" fault types) of Wells and Coppersmith (1994). Magnitudes were estimated on the basis of fault length. Fault areas are also shown in Table A1. The allocated slip rates are on the order of tenths or hundredths of millimeters per year, and return periods range from ~2000–10,000 yr.

This method is admittedly crude in that it makes use of poorly constrained data and produces parameter values for which we have little corroborative information. However, some rough comparisons can be drawn from Table A1. If we take the number of submarine faults capable of magnitude 7 events and sum their activity rates, we find that the return period for a magnitude 7 event is ~500 yr. Because we have observed one such event in the 500-yr observation period (the 1918 Mona Canyon event), this does not appear unreasonable. We also have an observation of two surface faulting events on the South Lajas fault in 7000 yr (Prentice et al., 2003). The value of 3400 yr for this fault (MP-LV in Table A1) compares favorably to this observation.

The Cerro Goden fault is a complex fault with several changes in trend in onland as well as offshore segments. The fault trends E-W, where it goes offshore and soon changes strike to NW-SE to parallel the coast for several kilometers before it changes strike again to E-W, veering away from the west coast of Puerto Rico (Fig. 2). The NW-trending segment, as all others, experiences oblique-normal slip, and this vertical motion helps form the troughs in the Mona Passage and probably influences that portion of the coast of Puerto Rico that lies on the footwall of the fault. Taggart (1992) studied late Quaternary terraces along the coast of northwest Puerto Rico and observed regional uplift of the northwest corner of the island at a rate of 0.034 mm/yr for the past few million years. Because outcrops of these terraces are not continuous, correlations between them to determine a regional uplift are tenuous. Nevertheless, it is interesting to note that Taggart observes uplift along that portion of coast just landward of the Cerro Goden fault. Jackson and McKenzie (1983) calculated that the ratio of footwall uplift to hanging wall subsidence is ~10%. Applying this observation to the Cerro

Goden fault means that it should experience ~0.3 mm/yr vertical motion to cause the 0.034 mm/yr uplift observed by Taggart. We estimate, through apportionment of measured GPS displacements, 0.65 mm/yr total slip rate in reasonable agreement with the uplift observed along the coast. This results in a return period of 2200 yr for a magnitude 7.1 event (MP-CG in Table A1).

2.3. Muertos Subduction Zone–Investigator Faults

The surface trace of the Muertos subduction zone lies ~80 km off the south coast of the island and accommodates closure between the Caribbean plate and Puerto Rico–Virgin Islands microplate. The most recent GPS estimate of Caribbean–Puerto Rico–Virgin Islands plate motion (fixed Caribbean plate) gives a mean value of 2.4 mm/yr in a direction of N79°W (Jansma et al., 2000). Uncertainties on this vector are large—1.4 mm/yr standard error for magnitude and 26° for azimuth. Four of the seven stations used to estimate this vector show relative azimuths of east-west or northwest-southeast. Extension directions derived from focal mechanisms of several Anegada trough earthquakes mimic this range of azimuths (McCann and Lithgow-Bertelloni, 2002).

Somewhere between 66°W and 64°W, the bathymetric expression of the Muertos trough dies out as the relative plate motion changes to the transtensional regime of the Anegada trough to the east (Masson and Scanlon, 1991) (Fig. 1). The apportionment of the Caribbean–Puerto Rico–Virgin Islands vector onto Anegada trough faults is described in section 2.4.

Two additional features significant in Caribbean–Puerto Rico–Virgin Islands interaction are the two segments of the Investigator faults, seen in the bathymetry. As seen in Figure 5, these are expressed as linear E-W–trending benchlike features in the slope from the island platform down to the Muertos trough. There appear to be two distinct segments, separated at ~66.4°W. The western segment, termed here the West Investigator fault, is ~53 km long. The eastern segment, termed the East Investigator fault, is ~46 km long and is marked by a wider bench (actually a trough) that increases in width to the east and merges into the western end of the Whiting Basin.

We interpret the tectonic role of these features as accommodating some of the change in tectonic style from compression at the Muertos trough to extension in the Anegada trough. As slip on the Muertos subduction zone decreases to the east, it must increase on Anegada trough structures. The fact that the East Investigator fault trough widens as it approaches the Whiting Basin suggests an increasing magnitude of both slip and extension from west to east, manifested as probably predominant strike-slip on the West Investigator fault, strike-slip with an increased normal component on the East Investigator fault, to large normal-oblique slip on faults bounding the Whiting Basin. It is postulated here that the Investigator faults mark the transfer of some of the Muertos subduction zone motion to faults of the Anegada trough.

Any estimate of slip rates on the Muertos subduction zone and Investigator faults must be considered to have a high degree of uncertainty. The Puerto Rico–Virgin Islands–Caribbean plate motion is not well constrained, but it can be said to trend roughly E-W and have a probable rate of at most a few millimeters a year. This is in apparent contradiction to GPS measurements referencing station GORD (at 18.5°N, 64.5°W, on stable Puerto Rico–Virgin Islands plate) to the St. Croix station (CRO1), which show no relative movement within the inversion error (Jansma et al., 2000). Although the major basins of the Anegada Passage slowed their rate of opening ca. 5 Ma (Reid et al., 1991), abundant seismicity in the trough, including a magnitude 7.3 tsunamigenic event in 1867, attests to tectonic activity in the region. Simple plate tectonic arguments demonstrate that the Puerto Rico–Virgin Islands–Caribbean vector applies to the Anegada trough as well.

The degree of partitioning between the Muertos subduction zone and the Investigator faults can only be estimated. All we really know of the Investigator faults is their existence and their likely role in the current tectonic setting. Microearthquakes have been associated with the Muertos subduction zone south of Puerto Rico, but no moderate to large events have been recorded or exist in the historic record.

In estimating the relative slip partitions on the Muertos subduction zone and Investigator faults, we postulate that at some point near western Puerto Rico (at or west of the west end of the West Investigator fault) slip on the Muertos subduction zone starts to decrease, and reaches zero where bathymetric expression of the Muertos trough ceases to exist, somewhere between 65°W and 66°W. We assume a simple linear decrease to the east, although the decrease could easily have a nonlinear shape. Three scenarios were considered: (1) the decrease starting at 67°W and extending to 65°W; (2) starting at 67°W and extending to 66°W; and (3) starting at 67.5°W and extending to 65.5°W. The plate motion was assumed to be 2.4 mm/yr in the direction of N79°E (Jansma et al., 2000). The midpoint of each fault was taken as its application point, and the plate motion apportioned accordingly. Earthquake magnitudes and return periods were based on the fault area-magnitude relations of Wells and Coppersmith (1994) and Equation 1. Magnitudes based on fault length were also from Wells and Coppersmith (1994) for the Investigator faults, and from Geomatrix (1993) for Muertos subduction zone events.

Table 1 shows the results of this analysis. For the West Investigator fault, slip rates range from 0.4 to 1.0 mm/yr, with corresponding return periods between 1800 and 4100 yr for a M_W 7.1 event. For the East Investigator fault, slip rates range from 0.8 to 1.4 mm/yr, with corresponding return periods between 930 and 1700 yr for a M_W 7.0 event. The third scenario gives generally higher slip rates because transfer off the Muertos subduction zone begins furthest west. The highest slip rate for the East Investigator fault is from the second scenario, which gives ~70% transfer close to the midpoint of the fault.

Considering the uncertainties in the magnitude of the Puerto Rico–Virgin Islands–Caribbean plate motion would broaden the uncertainties on these values. For the West Investigator fault, application of one standard error to the above analysis results in ranges of 0.2–1.5 mm/yr (1200–8800 yr) and 0.4–2.4 mm/yr (600–4400 yr) for the East Investigator fault.

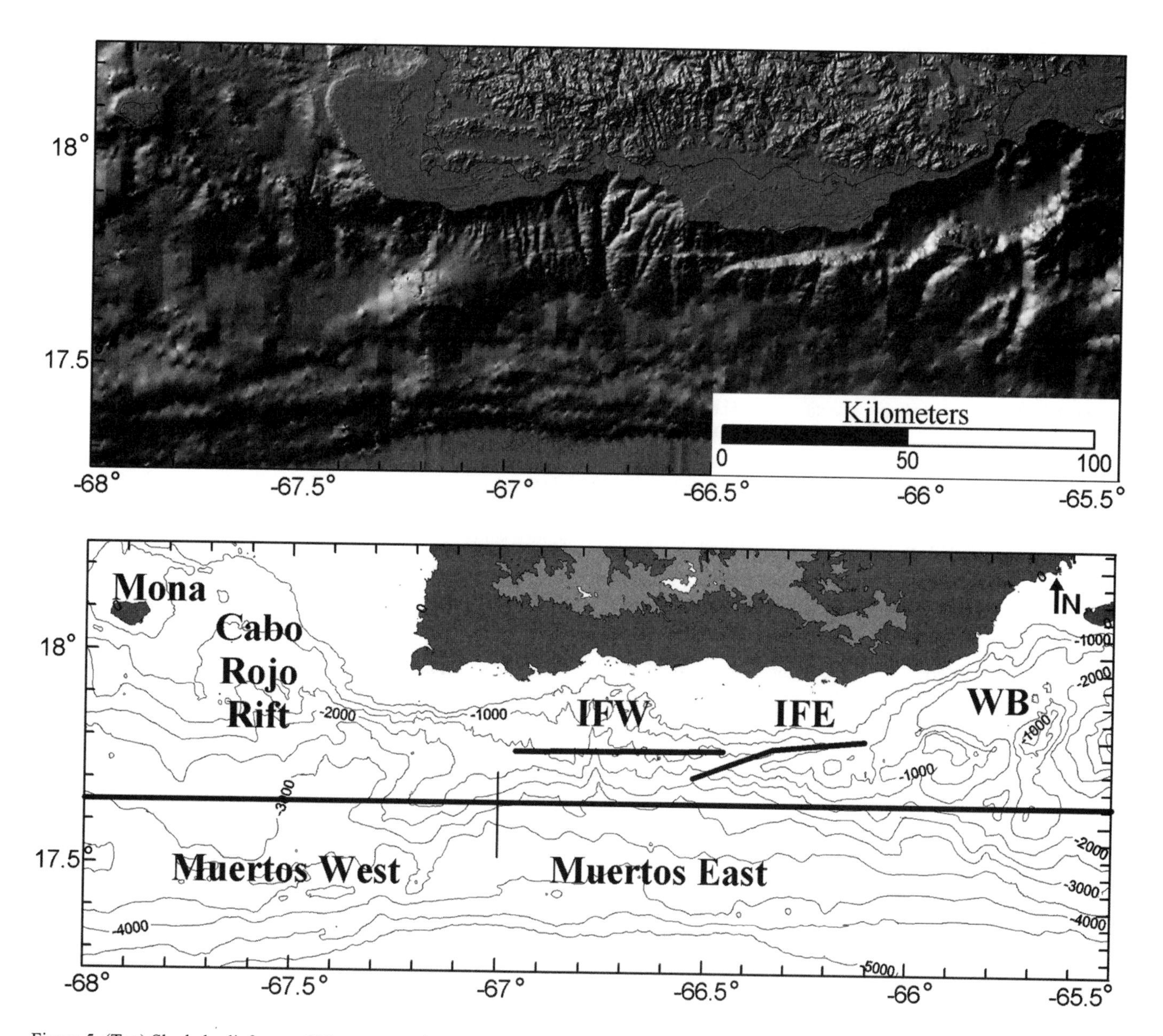

Figure 5. (Top) Shaded relief map of Muertos trough north slope and southern part of Puerto Rico. Illumination is from the northwest. Note the clear expression of the two segments of the Investigator fault, just south of the south coast of Puerto Rico. (Bottom) Contoured relief and location of modeled west and east Investigator fault segments (IFW and IFE) and Muertos subduction zone. WB—Whiting Basin.

TABLE 1. MAXIMUM MAGNITUDES AND POSTULATED SLIP RATES FOR INVESTIGATOR FAULTS

Fault	Magnitude	67°W–65°W		67°W–66°W		67.5°W–65.5°W	
		Slip rate (mm/yr)	Return period (yr)	Slip rate (mm/yr)	Return period (yr)	Slip rate (mm/yr)	Return period (yr)
West investigator	7.1	0.43	4100	0.84	2100	1.00	1800
East investigator	7.0	0.84	1700	1.56	930	1.40	1000

The final values used in the model are shown in Table A2. For the West Investigator fault, an average slip rate of 0.76 mm/yr was used; for East Investigator, 1.27 mm/yr. This results in return periods of 2200 and 1200 yr, respectively. These rates are significant for faults about which so little is known, and which lie near the south coast of the island.

Given this slip partitioning onto the Investigator faults, we have also divided the Muertos subduction zone into two segments: a western segment with no partitioning, and an eastern one with diminishing slip that eventually reaches zero at 65.5°W (Figs. 3 and 5). The western segment was originally assigned the full 2.4 mm/yr rate, whereas the eastern segment was given half that rate, 1.2 mm/yr (these rates were subsequently reduced for the probabilistic ground motion analysis in light of the historic seismicity record, as described in section 3.3). The western segment is continuous from 67°W to ~70°W, a length of ~300 km. Given the fault length-magnitude relations of Geomatrix (1995), a magnitude of 8.2 is possible. The eastern segment, with a length of 250 km, would be also be capable of a magnitude 8.2 event. However, given the decreasing slip rate from west to east and the probable discontinuous nature of the plate interface, we have assigned the eastern segment a maximum magnitude of 7.8.

A dip angle of 17° was estimated for the Muertos subduction zone on the basis of the hypocentral location of a M_S 6.7 thrust event that occurred in the western Muertos subduction zone in 1984 (Byrne et al., 1985). A focal mechanism was computed for this event, but the shallow-dipping plane was not well constrained. Earthquakes on the Muertos subduction zone were assumed to occur between 20 and 40 km depth. The modeled fault plane is shown in map view in Figure 3.

2.4. Anegada Passage

The Anegada Passage is underlain by a late Neogene complex of extensional basins and intervening ridges in the northeastern Caribbean. It cuts the older Antillean arc platform, from the Puerto Rico–Lesser Antilles trench in the northeast to the Muertos trough in the southwest. It is an ENE-WSW–trending extensional zone ~50 km wide, which separates the Puerto Rico–Virgin Islands microplate from the Caribbean plate to the south (Jany et al., 1990). Several deep basins, including the Virgin Islands and Whiting Basins, were formed between 11 and 4.5 Ma as the Puerto Rico–Virgin Islands microplate underwent ~20° of counterclockwise rotation. This rotation is postulated to be due to the impingement of the Bahama Bank on the northwest corner of Puerto Rico (Fink and Harrison, 1972; Van Fossen et al., 1989; Reid et al., 1991). Although no rotation has been noted during the last few million years, active deformation in the Anegada Passage basins is evidenced by abundant seismicity, including a tsunamigenic event with an estimated magnitude of 7.3 that occurred in 1867 (McCann, 1994, McCann and Lithgow-Bertelloni, 2002).

As discussed in section 2.3, the GPS vector between station GORD and CRO1, which straddles the Anegada Passage, shows negligible movement (<1 mm/yr) across the passage. However, the standard error on this vector is large, >2 mm/yr. It is clear that the GPS data are not able to give meaningful slip rates across the Anegada Passage at this time. We therefore assume the 2.4 mm/yr Puerto Rico–Virgin Islands–Caribbean rate that was used for the Muertos-Investigator faults. This is higher than the 1.47 mm/yr rate estimated by McCann (1994) on the basis of deformation over a period of ~17 m.y. Focal mechanisms for several small events off the southeast corner of the island show normal faulting due to east-west extension, but with a wide variation in the orientation of the tension axis (McCann and Lithgow-Bertelloni, 2002).

The major Anegada Passage faults were identified from seismic reflection records by means of a procedure similar to that used to identify faults in the Mona Passage (McCann, 1998). These are shown in map view in Figure 6. The faults have been categorized into three genetic groups. The first group, group A, consists of faults bounding the northern edges of the Virgin Islands and Whiting Basins. Group B consists of the faults bounding the southern edges of these basins (there is no such identified fault along the south edge of Whiting Basin). The third group, group C, consists of shorter faults cutting the island platform west of St. Croix.

The genetic groups reflect the likely role that the faults in each group plays in the overall deformation of the region. Fault groups A and B played major roles during the Late Neogene development of the Anegada trough (McCann and Lithgow-Bertelloni, 2002). Fault group A experienced primarily sinistral slip, whereas group B experienced normal faulting. Although relative amounts of extensional and sinistral slip vary rapidly along the trough, slip on this pair of faults, when summed together, yields the vast majority of Late Neogene transtensional deformation in the western part of the Anegada trough. Group C normal faults served to pass slip from the easterly dying Muertos trough to the easterly more important transtensional faults (groups A and B) in the main part of the trough. Today, east-west extension appears to be more regionally constant. Groups A and B still play an important role, but the role of group C may be diminished by the presence of the Investigator fault because it also serves to transfer slip from the thrust zone to the trough via strain partitioning. Any remaining slip on the Muertos thrust zone would then transfer through the group C faults to the group A and B faults.

Because of the genetic difference between the groups of faults, it was decided to assign 45% of the regional slip vector to group A, 45% to group B, and 10% to group C. Thus, groups A and B are each assigned 1.08 mm/yr, and group C is assigned 0.24 mm/yr. Because of the proximity of the two northwest group A faults to the coastline (AP-1 and AP-2 in Table A3), they were included despite having fault-length-based magnitudes less than 6.5.

Slip was allocated among the faults by means of the same methodology as for the Mona Passage faults. Table A3 shows the results. The two large bounding faults of the Virgin Islands Basin (AP-1 and AP-4) were allocated ~0.8 mm/yr each. These are capable of generating earthquakes in the magnitude 7.0–7.5 range, and one was the probable source of the 1867 magnitude 7.3 event. The return periods of the two faults imply that magnitude 7.0–7.5 events should occur on one or the other about every 1500 yr.

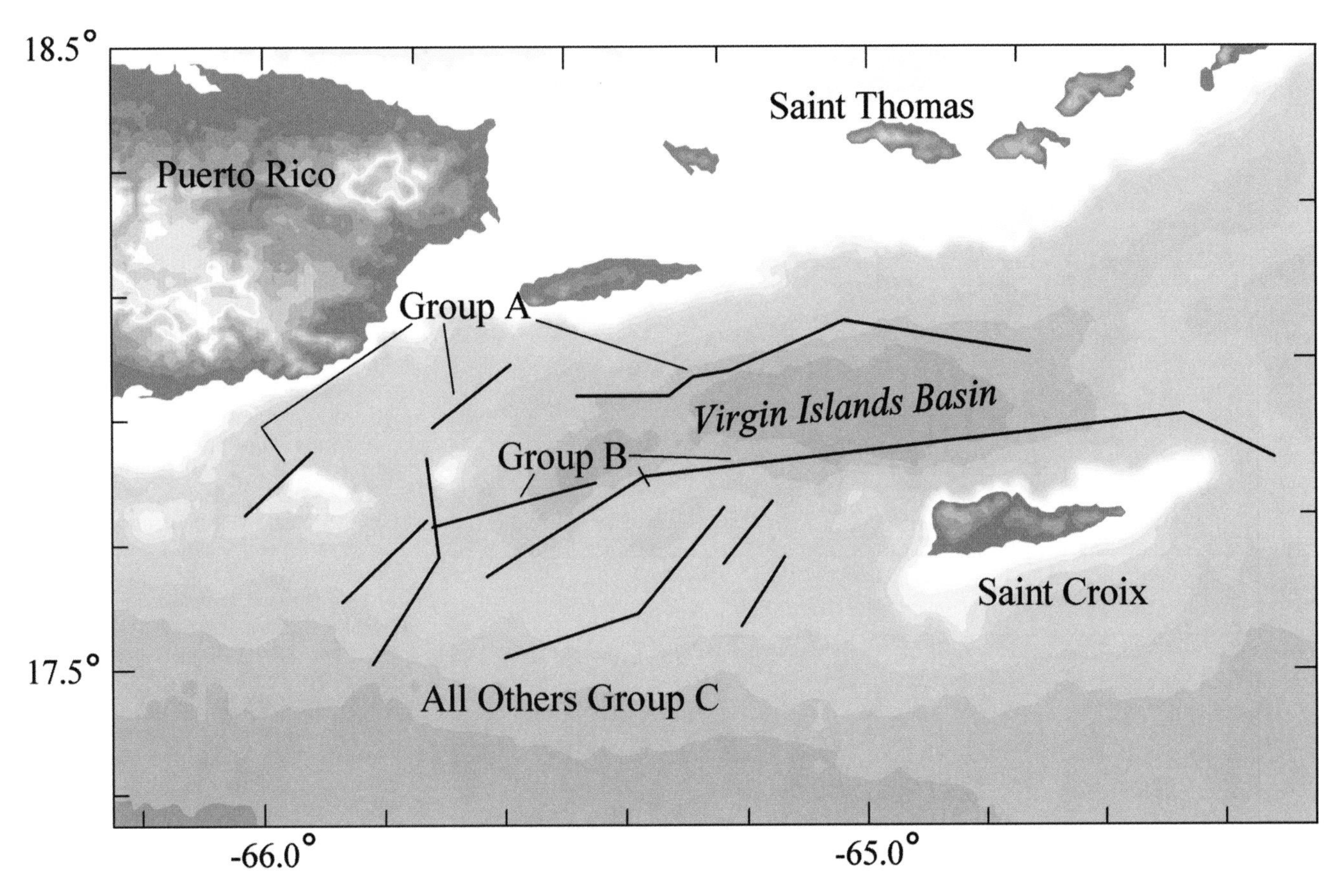

Figure 6. Contour map of Anegada trough relief. Faults in the region, as used in the probabilistic analysis, were grouped according to their genetic differences. See text for details.

2.5. East Septentrional–North Puerto Rico Slope Faults

The eastern Septentrional fault (the eastern section of the Septentrional fault) is a left lateral fault zone that parallels the Puerto Rico trench between Mona Canyon and western Hispaniola (Fig. 2). Its role in the larger tectonic setting is to accommodate an increasingly greater proportion (from east to west) of the motion between the North America plate and the various microplates between the North America and Caribbean plates. The volume of material above the subducted North America plate, between the Septentrional fault and eastern Septentrional fault and the trench, is termed Septentrional Terrane. In western Hispaniola, the subduction zone dies out and the boundary (there called the Oriente fault zone) becomes purely strike-slip in nature. The eastern Septentrional fault, and probably the Bunce and Bowin fault zones (Fig. 2), thus mark the beginning of the partitioning of North America relative plate motion between oblique underthrusting in the Puerto Rico subduction zone and left-lateral strike-slip. In Hispaniola, the fault appears to have a Holocene slip rate of 6–12 mm/yr, and trenching studies at one site revealed one surface faulting event ~1000 yr ago (Mann et al., 1998; Prentice et al., 2003). West of Puerto Rico, the fault intersects Mona Canyon at ~19°N, 67.5°W (Figs. 2 and 7).

The Bowin fault is an apparent continuation of the eastern Septentrional fault to the east of Mona Canyon (Grindlay et al., this volume, Chapter 7). Here the fault changes azimuth from E-W to ENE-WSW. Seafloor bathymetry and seismic reflection data show the Bowin fault to consist of a set of poorly aligned, discontinuous segments (Grindlay et al., 1997; Fig. 2). Although the most likely source for the 1787 earthquake was the Puerto Rico subduction zone, Dolan and Wald (1998) point out the possibility that it may have occurred on the Bowin fault.

Another left lateral fault zone, the Bunce fault, has been identified parallel to, and just several kilometers south of, the Puerto Rico trench (Dolan et al., 1998). This zone undoubtedly plays a part in taking up some of the North America–Puerto Rico–Virgin Islands plate motion as left-lateral strike-slip. However, because the depth to the downgoing slab here cannot be more than a few kilometers and rocks here are not likely to be very competent, the Bunce fault is not considered here to be a seismogenic source. This judgment is supported by Dolan and Wald (1998).

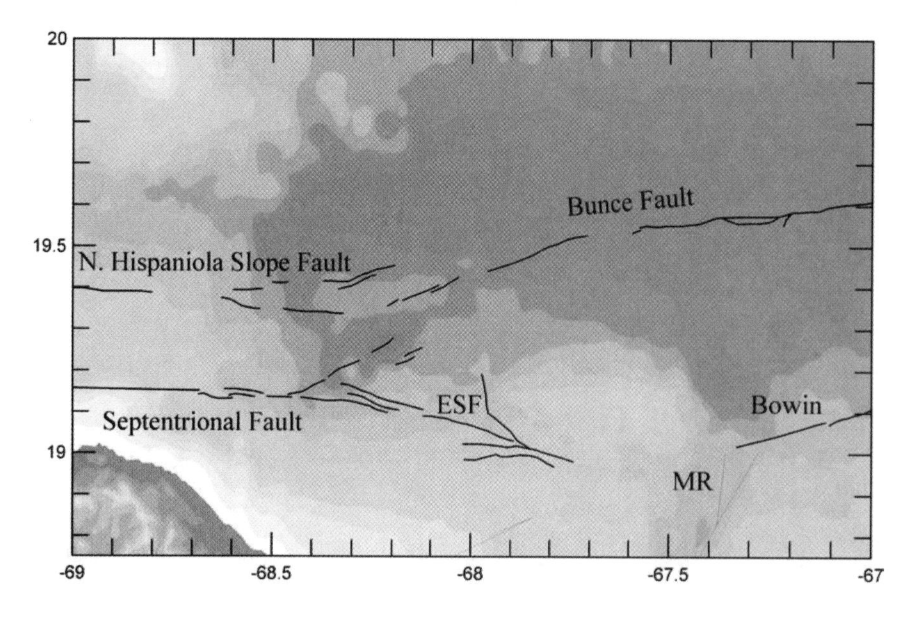

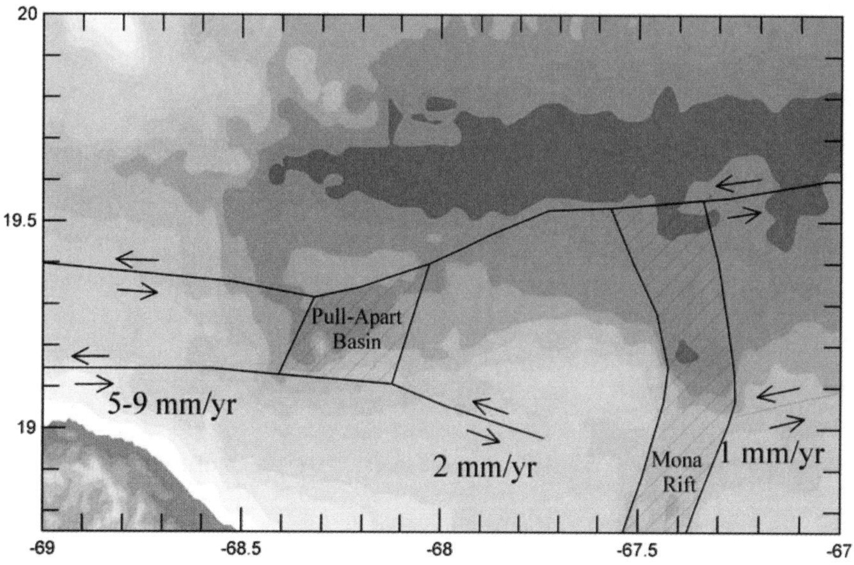

Figure 7. (A) Major faults on the inner wall of the Puerto Rico trench and accretionary wedge. Most faults represent the sinistral strike-slip component of slip in a strain-partitioned, obliquely convergent margin. The Bunce and North Hispaniola faults appear to be two segments of one fault system. Faults are from Dolan et al. (1998) and Grindlay et al. (1997). Septentrional fault and East Septentrional fault are shown. The eastern Septentrional fault appears as a probable distinct fault system. A group of normal faults connects the Septentrional and Hispaniola fault systems along a 7-km deep pull-apart basin. ESF—East Septentrional fault; MR—Mona Rift. (B) Cartoon of simplified faults and location of other major tectonic features. Slip along the Septentrional fault is transferred to the normal faults in the pull-apart basin, then to the plate interface and possibly the Bunce fault. A similar transfer of slip may occur at Mona Rift, although recent data suggest that the eastern Septentrional fault does not reach the western edge of that rift (ten Brink, 2003).

From seafloor expression, the Septentrional fault appears to be the dominant structure taking up left lateral motion in this region. However, the existence of strike-slip and other transtensional features similar to the Bowin fault between Mona Canyon and eastern Hispaniola indicate that slip on the eastern Septentrional fault is likely to decrease significantly from central Hispaniola to Mona Canyon and further east. One hundred kilometers west of Mona Canyon (~68.5°E), the northeastern Septentrional fault splays off to the northeast from the main Septentrional fault (Fig. 7). At this location, the Septentrional fault–eastern Septentrional fault trend is marked by a 20° change in strike and a pull-apart basin. The northeastern Septentrional fault trends northeast for 50 km, where it meets the Northeast Hispaniola fault zone and the possible extension of the Bunce fault. A 7-km-deep pull-apart basin exists at this location, indicating significant recent extension (Dolan et al., 1998). The seafloor bathymetry indicates that although strands of the eastern Septentrional fault appear to be the dominant feature west of Mona Canyon, transtensional features are found near the eastern Septentrional fault in a zone several kilometers wide. This argues for distributed deformation throughout the Septentrional Terrane, with the distributed nature increasing from west to east.

On the basis of the above considerations, we propose that east of 68.5°E, the eastern Septentrional fault will rupture independently of the Septentrional fault. East of this point, it appears that the longest continuous segment of the eastern Septentrional fault is ~75 km. By using the fault length-magnitude relations of Wells and Coppersmith (1994), we deduce that this indicates a potential magnitude of 7.3. Given the apparent 5-km offset in the fault trace at 67.8°E (Fig. 2), this may be a conservative assessment.

Unfortunately, there are no detailed slip rate data for the eastern Septentrional fault, Northeast Hispaniola fault zone, or Bowin fault, and we are forced to make educated guesses as to the eastern Septentrional fault slip rate. One observation is that there have been no felt events originating on the eastern Septentrional fault during the historic record of 500 yr. If felt reports require a magnitude 7.3 event, we can combine the equation

$$\dot{Mo} = \mu A\dot{s}, \quad (1)$$

where Mo is moment rate, μ is shear modulus, A is fault area, and s is slip rate (Aki and Richards, 1980), with the moment-magnitude conversion of Hanks and Kanamori (1979) and the Poisson probability formula to compute the probability of *not* seeing such an event, given various possible slip rates. Performing the calculations, we find that for a slip rate of 5 mm/yr, there is a 19% probability of not seeing such an event; for 3 mm/yr, a 37% probability, and for 2 mm/yr, a 51% probability. Because 2 mm/yr is close to a 0.50 probability, we consider this value a reasonable median estimate for the eastern Septentrional fault slip rate. The historic record does in fact contain an event of M_S 7.0 (McCann and Sykes, 1984) in this region. For an observation period of 300 yr, this implies a 67% probability of not seeing an event this size for a 2 mm/yr slip rate. However, this event could have occurred on one of the Mona Canyon faults or in the Puerto Rico subduction zone.

Slip rate data also do not exist for the Bowin fault, and the fault zone is too far away from the island for microseismicity to be correlated to the fault zone. An important clue as to slip on the Bowin fault lies in observations of the bathymetric configuration of Mona Canyon (Fig. 2). South of 19.0°N, the floor of the canyon trends NNE and is ~15 km wide. North of 19.0°N, where the eastern Septentrional fault and Bowin fault intersect it, the canyon floor drops 2.5 km, narrows to a width of ~8 km, and appears to change trend NNW. There is clearly a change in the canyon's expression at this latitude, possibly related to the intersection of the two faults, the thinning of the island platform along its northern boundary that changes its deformation response, or a tectonic rotation. Another interpretation is that the feature north of 19.0°N is there by coincidence and genetically unrelated to Mona Canyon. Thus, the Bowin fault, its relationship to the eastern Septentrional fault and Mona Canyon faults, and its role in regional tectonic processes remain enigmatic.

The above observations point to a model in which partitioning of North America–Puerto Rico–Virgin Islands plate motion onto the Septentrional fault begins north of Puerto Rico, where it is low in magnitude and highly distributed in nature. Between there and western Hispaniola, the proportion of motion applied to the Septentrional fault increases, and it becomes more "focused" onto the main Septentrional fault strand as related transtensional features become less important.

For the purposes of this model, we adopt a mean value of 2 mm/yr for the eastern Septentrional fault. Seismicity in this region is diffuse but abundant (Fig. 8). Although the diffuseness may be due to poor location capabilities of the PRSN this far from the island, complex tectonic activity should be expected in this zone of converging first-order structural features.

We adopt a value of 1 mm/yr for the Bowin fault, despite its discontinuous nature and apparent lack of effect on the trend of Mona Canyon. The possibility exists that this feature is inactive or not capable of generating damaging earthquakes. Where the fault zone intersects Mona Canyon, the depth to the plate interface is ~30 km (McCann, 2002). Over an extent of 90 km to the ENE, this depth decreases to 10 km. We estimate this to constitute the maximum probable continuous rupture length for this feature. By using the Wells and Coppersmith (1994) relations, this indicates a possible magnitude of 7.3.

2.6. Random Earthquakes

In addition to hazards posed by identified active faults seen on land and on the seafloor, damaging earthquakes can also occur on unidentified or buried faults in the upper crust beneath the island. We assume that such earthquakes can have magnitudes as high as 7.0. Events this large unassociated with mapped faults have occurred globally, and it is quite possible that because of high erosion rates on the island and the lack of detailed field investigations, not all Quaternary surface faults have been discovered. A lower limit of 5.0 is imposed on the assumption that events smaller than this will not cause significant damage. In this section, we compute recurrence parameters for such activity, on the basis of observed seismicity.

The PRSN has been operating more or less continuously since 1975, with a complement of about a dozen stations. However, there have been a number of changes in station configuration, frequency response of the recording system, and analysis techniques. Of particular relevance to this study, magnitude determinations throughout much of the network history cannot be considered accurate or consistent (McCann, 1994). On the basis of technical considerations, only the August 1975 through December 1978 network data were used to estimate recurrence parameters. The network configuration at this time consisted of 12 stations that were spaced uniformly throughout the island.

A map of all events from this time span is shown in Figure 9. The boundaries about the island are meant to provide consistent location accuracy and magnitude completeness for the data set. In order to leave out events associated with the subducted North America plate, depths were limited to less than 45 km. The pattern shows relatively dense activity in the western and eastern parts of the island, with the central part (between 66.25°W and 67°W) relatively quieter. This is similar to the longer-term pattern seen in Figure 8. Two procedures were necessary before performing recurrence calculations. One was to attempt to eliminate as many quarry explosions as possible, and the other was to eliminate "dependent" events (largely

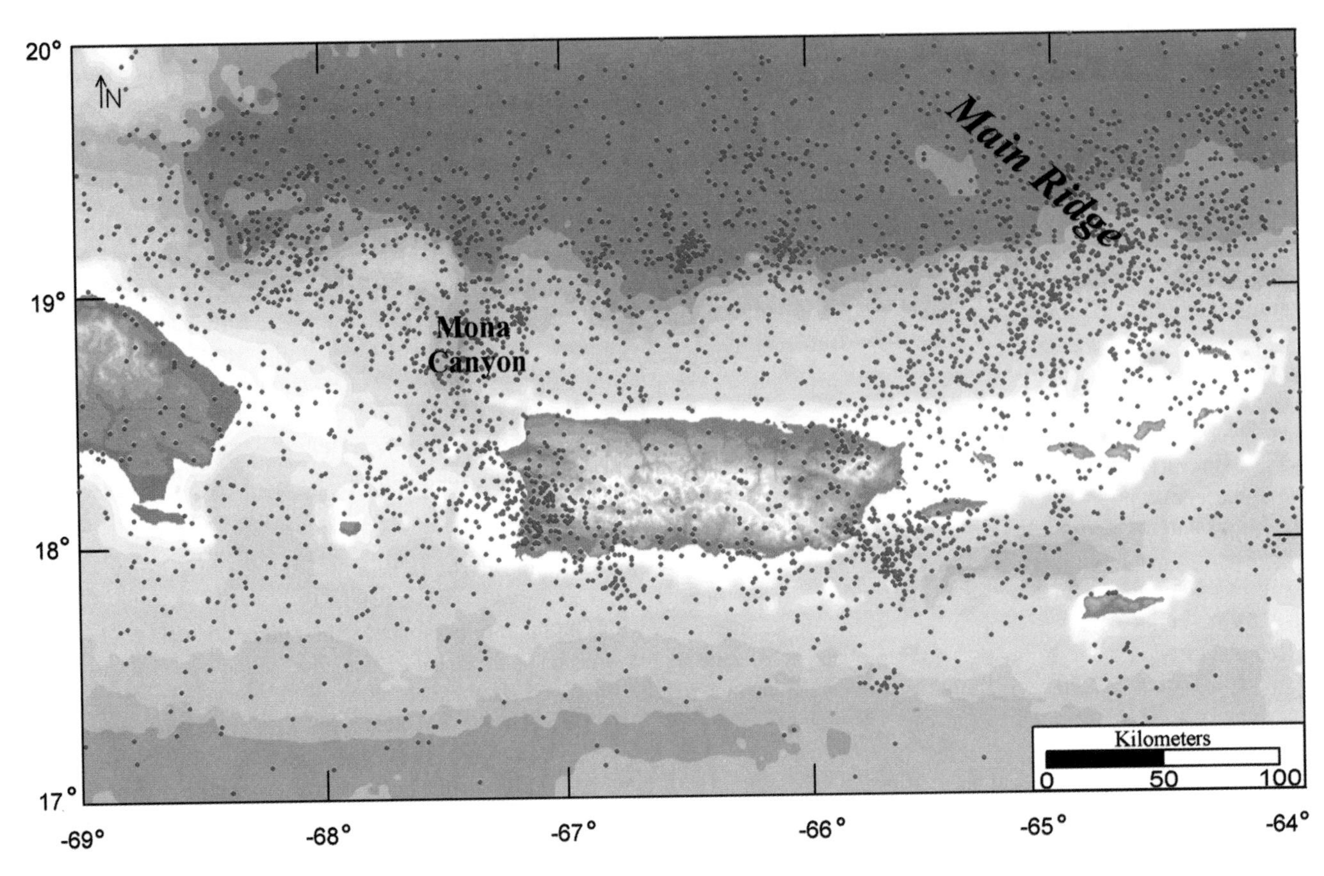

Figure 8. Map of seismicity recorded by the local network from 1975 to 2001. Events shallower than 40 km and of magnitude 3.0 or larger are shown. Note high level of activity in SW Puerto Rico.

aftershocks and swarm events) so that the data set approximates a Poisson distribution.

The elimination of quarry blasts was performed by the identification of five rectangular zones surrounding known quarries that were active at this time. Events within the quarrying areas were then deleted if they occurred between 12:00 noon and 6:00 p.m. local time (the most likely times for blasting), and if their computed depths were less than 10 km. Five catalog entries in the magnitude 2 range were thus eliminated. This procedure is admittedly crude, and it is probable that some man-made explosions remain in the data set.

"Declustering" was performed on this data set by using a modified version of the Reasenberg (1985) algorithm. The technique identifies "dependent" events by using a computed potential aftershock radius based on source theory and assumed Omori aftershock time-decay parameters for time-after-mainshock criteria. Details can be found in Reasenberg (1985) and Savage and dePolo (1993).

For magnitude <5, the 3.5-yr completeness period was used. For magnitude 5 events, completeness was assumed since 1950, based on Sykes and Ewing (1965), and for magnitude 6.0–6.5, events back to 1930 were used, based on a study of strong earthquakes by Gutenberg and Richter (1954). Detection of magnitude 6.5–7.0 events was assumed to be complete back to 1760. Plotting of preliminary recurrence curves showed that events with magnitude <3 were incompletely recorded. Therefore the range was set to magnitude 3.0–7.0. The largest event in the data set was a (PRSN) magnitude 5.2 event that occurred beneath the northwest corner of the island in February 1977.

The earthquakes used to calculate recurrence are shown in Figure 10. It is apparent from this figure that seismicity is greater in the western and eastern parts of the island than in the central part. However, the data were found to be too sparse to increase the number of spatial source zones while maintaining statistically stable results.

An incremental recurrence curve (Fig. 11) was fit to the observations by the maximum-likelihood technique (Weichert, 1980). Errors on the maximum-likelihood fit were computed after Bollinger et al. (1989). The curve shows a b value of 1.3, with events of magnitude ≥5 having a return period of ~17 yr. Uncertainties for the larger events are great; for example, for magnitude 6.5–7.0, the maximum-likelihood return period is 2000 yr, but with 95% confidence bounds of 400 and 10,000 yr.

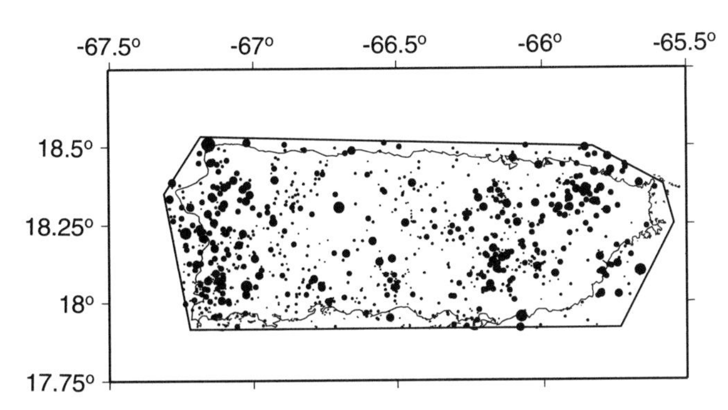

Figure 9. Map of shallow (depth <45 km) island seismicity, August 1975–December 1978. Smallest symbol is magnitude 0, largest is magnitude 5.

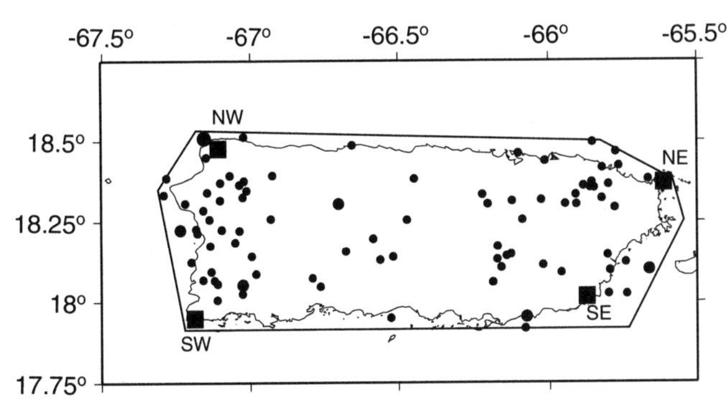

Figure 10. Map of earthquakes used in recurrence calculations. Smallest symbol is magnitude 3, largest is magnitude 5. The four sites for which hazard curves for peak horizontal acceleration were computed are shown.

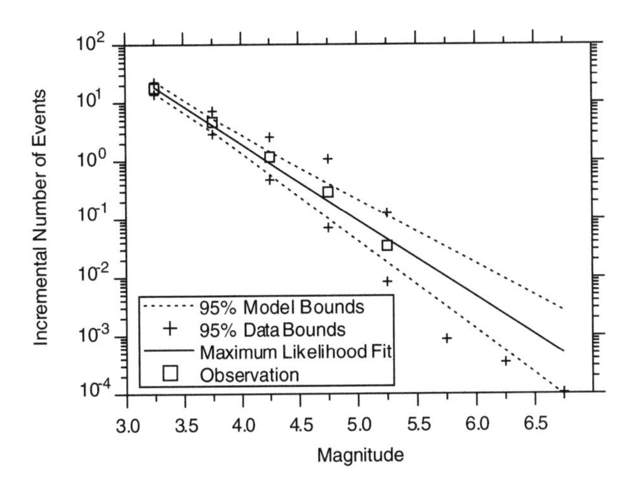

Figure 11. Maximum likelihood incremental recurrence curve for random seismicity. Maximum likelihood fit, and 95% confidence bounds on data and model are shown.

3. PROBABILISTIC SEISMIC HAZARD ANALYSIS

3.1. Introduction

Since its introduction by Cornell (1968), probabilistic seismic hazard analysis has become the method usually preferred for quantifying the ground motion hazard from earthquakes. The hazard is stated in the form of an annual probability of exceedance of a particular ground motion level, at a particular spectral period, according to the equation

$$P(v > V) = \sum_{i=1}^{n} \int_{Mo\,(r=0)}^{Mu} \int^{\infty} f_i(m) f_i(r) P(v > V | m, r)\, dr\, dm, \qquad (2)$$

where the annual probability of ground motion v exceeding some amplitude value V is equal to the probability that earthquakes with magnitude and distance distributions $f(m)$ and $f(r)$, integrated over Mo to Mu and all distances, respectively, cause exceedance of amplitude V. The calculation is performed for the N relevant sources and summed.

In practical terms, $f(m)$ is the recurrence relation, which shows the magnitude distribution and rates of earthquakes emanating from the source, $f(r)$ describes the location of the earthquake source, and $P(v>V/m,r)$ is the attenuation function, which describes the probability of exceeding a ground motion level given a magnitude and distance. For a given spectral response period, a hazard curve can be constructed, with ($\log_{10}$) annual probability of exceedance on the y-axis and ground motion level on the x-axis.

Earthquake recurrence models used here are Poissonian; that is, it is assumed that the occurrence of an earthquake in space and time has no effect on the occurrence of another. With this assumption and relatively small annual probabilities of exceedance, return period can be defined as the reciprocal of annual probability of exceedance.

For this study we compute hazard curves for peak horizontal acceleration for the four corners of the island. Because the island is surrounded by seismic sources of differing characteristics, this gives a glimpse of the relative importance of each source at extreme locations on the island. For simplicity's sake these are labeled NW, SW, NE, and SE. In reality they correspond to Punta Agujereada (NW), Punta Aguila (SW), Punta Gorda (NE), and Punta Yeguas (near the town of Yabucoa) (SE). The locations are shown on Figure 10.

3.2. Attenuation Functions

Two sets of attenuation functions were used in this study. The primary set was used in the earlier stages of this study, and consists of relations developed from California earthquakes for upper crustal seismicity and from worldwide events for subduction zone earthquakes. The second set are the recently developed relations developed for Puerto Rico by Motazedian and Atkinson (this volume). We compare results by using this relation to those obtained from the primary set for all sites on the island.

At the time this work began, no attenuation functions had been developed for the crustal conditions of Puerto Rico (or elsewhere in the Caribbean). We were therefore forced to use functions developed for other tectonic environments. For subduction zone earthquakes, we used the "interface" relations of Youngs et al. (1997), which are based on shallow interface earthquakes worldwide. We also used their "intraslab" relations for intraslab earthquakes for the deeper Puerto Rico subduction zone intraslab events. For all other sources, seismic waves pass through a heterogeneous upper crust of varying post-Cretaceous arc massif material. On Puerto Rico, crustal properties are different in the southwestern and central/eastern parts of the island. These two regions are separated by the Great Southern Puerto Rico fault zone, a thrust/left-lateral feature that was active in the Tertiary (Erickson et al., 1991). In the southwestern region, rocks tend to contain more serpentinite, and gravity anomalies suggest that the crust is denser and/or thinner; to the north and east, it is thicker and/or lighter and dotted by batholiths. No volcanism or hot springs are found on the island. Anderson and Larue (1991) measured a low crustal heat flow, consistent with the lack of hot springs and suggestive of a "cold" subduction zone environment. For this study, we use the "rock" function of Abrahamson and Silva (1997), a recent relation based largely on California records, to describe ground motions from crustal earthquakes. The appropriateness of this relation is discussed in section 4.7.

The relations of Motazedian and Atkinson (this volume) were developed from seismograms recorded by the PRSN, with magnitudes ranging from 3 to 5.5. Because of the lack of records at all magnitudes and distance ranges, the natural database was augmented by synthetic records incorporating the attenuation parameters regressed from the observed data. Their relation reflects a National Earthquake Hazards Reduction Program (NEHRP) "C" site condition, and is thus comparable to the other relations used in this study. Because of the preliminary nature of the study, a rigorous analysis of the uncertainties associated with the ground motions was not conducted. However, it appears that a $\log_{10}$ value of 0.28 (standard deviation of the $\log_{10}$ amplitude value) for all magnitudes, distances, and periods appears to be a reasonable estimate (Motazedian, 2002, personal commun.), and was used in our calculations.

3.3. Recurrence Models

The choice of earthquake recurrence model, which describes the range of magnitudes and activity rate interrelationships, can have a significant impact on probabilistic results (e.g., Youngs and Coppersmith, 1985). For this study we consider two, the Gutenberg-Richter relation (see section 2.6), where activity rates are distributed exponentially with respect to magnitude, and the maximum-moment model (e.g., Wesnousky, 1986), which assumes that seismic slip is restricted to a narrow range of large earthquakes. The exponential model uses the maximum value listed in Tables A1 to A3, a lower bound of 6.0 or 6.5, and an assumed *b* value of 1.0.

For the Puerto Rico subduction zone, both recurrence models were tested for the east and west zones. For the west zone, the exponential model was constrained to magnitudes between 6.5 and 8.2 and a slip rate of 11.9 mm/yr (section 2.1.1). The distribution was transformed to activity rates as described in Anderson (1979) and Youngs and Coppersmith (1985). For the west zone, the model predicts that earthquakes with magnitude 6.5 and greater should be occurring about every 34 yr, with a magnitude 8 event about every 1400 yr. The historic record indicates that there have been four events of magnitude $M_W \geq 6.5$ since 1915 in this part of the shallow interface (Doser et al., this volume; Shepherd et al., 1994). This rate (four events in 87 yr) is consistent with the exponential model.

In contrast, the eastern shallow interface has probably experienced one magnitude ~8 event (in 1787), with only one with magnitude higher than 6.0 west of 64°W in the past 85 yr (Doser et al., this volume; their Fig. 6). The exponential model, with an assumed slip rate of 3.4 mm/yr and a *b* value of 1.0, predicts an earthquake of magnitude $M_W \geq 6.5$ in 80 yr, $M_W \geq 7.5$ in 800 yr, and $M_W \geq 8.0$ in 3000 yr. The occurrence of one M_W 6.2 event in 1919 during the past 87 yr (Doser et al., this volume) is consistent with the $M_W \geq 6.5$ rate. However, the implication from this characterization is that the 1787 earthquake was a rare event if its magnitude was ~8.

The maximum-moment model was applied to each Puerto Rico subduction zone by assuming that magnitudes could range between 7.0 and 8.2, with equal likelihood. In other words, slip can be distributed as several magnitude 7.0 events occurring relatively frequently, one magnitude 8.2 event occurring less frequently, or corresponding magnitudes and occurrence rates in between. This mimics behavior seen in subduction zones worldwide (Thatcher, 1990). In the western Puerto Rico subduction zone, the assumed slip rate translates to return periods of 11 and 350 yr for magnitude 7.0 and 8.0 events, respectively. Although the M_W 8.0 return period may be plausible, the rate for M_W 7.0 events is not, and events smaller than magnitude 8.0 clearly have occurred. Therefore, only the exponential model was used in the western Puerto Rico subduction zone.

In the eastern Puerto Rico subduction zone, the maximum-moment model predicts return periods of 25 yr if only magnitude 7 events are occurring, 150 yr for magnitude 7.5 events, and 800 yr for magnitude 8 events. Because we feel it necessary to consider the possibility of return periods less than 3000 yr for 1787-size events, we weight the maximum-moment model, with magnitudes ranging from 7.5 to 8.2, equally with the exponential model for the eastern Puerto Rico subduction zone. In support of this, we note that the larger-magnitude historic seismicity here is concentrated around 65°W longitude (where the Main Ridge intersects the trench; see Fig. 8), and the Puerto Rico subduction zone between here and the western Puerto Rico subduction zone has been noticeably quiescent for larger events (Doser et al., this volume).

Similar analyses were applied to the Muertos subduction zone. Here, our proposed slip rates are even lower, increasing return periods and further decreasing the utility of the historic record. The record indicates that there have been no events of magnitude 6 or greater in the past 87 yr, and none with $M_W \geq 6.5$ in the past 500 yr.

The western Muertos zone (Fig. 3) was originally assumed to have a slip rate of 2.4 mm/yr on the basis of poorly constrained GPS data (Jansma et al., 2000), and the eastern Muertos zone half this amount, 1.2 mm/yr. Applying the maximum-moment model means a return period of 34 yr for M_W 7.0 events and 200 yr for M_W 7.5 events. If we calculate the Poisson probability of not seeing these events in the 500 yr observation period, the results are 4×10^{-7} for M_W 7 and 0.073 for M_W 7.5. By using half the assumed slip rate, results are in probabilities of 0.017 for M_W 7 and 0.27 for M_W 7.5. The exponential model for $M_W \geq 6.5$ and a *b* value of 1.0 in both zones gives a nonobservance probability of 0.0013 for the original slip rates, 0.04 for 50% of these, and 0.20 for 25%. Considering the uncertainties involved (*b* value, recurrence model, GPS resolution) we use the maximum-moment model for magnitudes 7.5–8.2 in the western zone and 7.5–7.8 in the eastern zone, and the exponential model for $M_W \geq 6.5$ in each zone, both at half the originally assumed rates. This implies 1.2 mm/yr in the western zone and 0.6 mm/yr in the eastern zone. Data on fault area, and maximum magnitudes and their return periods are shown in Table A2.

For Puerto Rico subduction zone intraslab seismicity, activity rates were estimated from small events from the PRSN catalog and the exponential model. We therefore feel compelled to use the same exponential model for the larger events. This model predicts a return period of 84 yr for magnitude $M_W \geq 6.0$. In the instrumental record (Shepherd et al., 1994), there have been two events of magnitude 6 located between 64°W and 67°W and deeper than 50 km since 1900. The historic data are thus consistent with the exponential model. One earlier event in the historic record may be a deep intraslab earthquake (section 2.1.2). Magnitude bounds of 6.0 and 7.5 were imposed; earthquakes with magnitude <6.0 are deep enough below the island (Fig. 4) to be judged incapable of causing significant damage, and 7.5 appears to be a reasonable bound on such events occurring worldwide. The return period of 84 yr of $M_W \geq 6.0$ events was achieved by assuming a slip rate of 0.16 mm/yr over the slab area shown in Figures 3 and 4. Although it is known that intraslab earthquakes do not occur in this manner, the desired rates are arrived at.

Both exponential and maximum-moment models were applied to the Mona Passage faults. In using the exponential model, we assumed a lower magnitude bound of 5.0, an upper bound of the value shown in Table A1, and a *b* value of 1.0. Summed over the 28 faults, recurrence statistics show a return period of 5 yr for a magnitude ≥5.0 event anywhere in the system. Seismic activity is clearly not this high in the Mona Passage; in the past 50 yr of reasonable teleseismic monitoring, the largest event in the passage was magnitude 6.0 in 1965 (IPGH catalog of Shepherd et al., 1994). Only three events of magnitude 5.0 or above were noted during this period, two in the mid-1960s and both associated with the Mona Canyon, and the other in 1983. This event was associated with either the Mona Canyon or the East Septentrional fault. When one considers a longer period of time—say, the historic record of good reporting since the 1750s—there are only two large earthquakes that have occurred in the Mona Passage: a possible one in 1865, and the 1918 M_S 7.3 Mona Canyon event.

To test the maximum-moment model, it was assumed that all slip results in magnitudes described by a symmetric triangular distribution, with the peak at the value listed in Table A1, and endpoints at ± 0.25 from this value. This resulted in return periods that are more consistent with the observed seismicity. For example, the exponential model predicts a return period of ~8000 yr for magnitude 7 events on the Lajas Valley fault; the maximum-moment model predicts 3400 yr, which is more consistent with the geologic data. The exponential model predicts a recurrence interval of ~5500 yr for a magnitude 7.3 event on the largest Mona Canyon fault (southerly canyon fault, Figure 2; fault MP-16 in Table A1); the maximum-moment model predicts ~1900 yr. The fact that we have experienced one such event in the 500-yr historic record argues for the shorter return period being more probable. Summing the annual frequencies of the Table A1 magnitudes, we find that there should be one such event about every 200 yr. This appears to be more consistent with the historic seismicity record than results from the exponential model. We therefore use the maximum-moment model exclusively for recurrence of earthquakes on faults in the Mona Passage.

Similar calculations and observations give ambiguous results for the Anegada trough faults. Use of the exponential model implies that magnitude $M_W \geq 6.0$ events should occur in the region every 140 yr. Although no events exist in the historic record of this size in this region, our detection capability of such events extends back only ~100 yr. The exponential model predicts maximum events on the two large bounding faults of the Virgin Islands Basin (Fig. 6) about every 2700 yr, whereas the maximum-moment model predicts one every 820 yr. Because the observations do not provide a basis for choosing one model over the other, they were weighted equally.

For the Investigator faults, our estimated slip rates are too low to be able to discriminate between the two recurrence models because of the lack of large events on it in the historic record. The exponential model, for events of magnitude ≥6.0, predicts a 425-yr return period for the West Investigator fault and 250 yr for the East Investigator fault. The maximum-moment model predicts a 2200-yr return period for magnitude 7.1 events on the West fault, and 650 yr for magnitude 7.0 events on the East fault. Results are comparable for the East Septentrional and Bowin fault. We therefore use both models, weighted equally, for these sources. For the exponential model, we applied a lower bound of magnitude 6.0, and upper bounds of 7.3 for the East Septentrional and Bowin fault, 7.1 for the West Investigator fault, and 7.0 for the East Investigator fault (Table 1). The maximum-moment magnitude distribution was applied in the same manner as for the Mona and Anegada Passage faults.

3.4. Probabilistic Seismic Hazard Analysis Computations

Fault sources were modeled as dipping, planar, sometimes multisegmented structures. Dips for nonsubduction zone faults were assumed to be 63°, and lower depth bounds were set at 20 km, except for some of the shorter faults which were set at 15 km to maintain a reasonable aspect ratio (length/width ≥1). The "shortest distance" between the rupture surface and the site as required for the attenuation functions was determined by distributing magnitude-appropriate square rupture areas on the fault. The Wells and Coppersmith (1994) magnitude-rupture area relations were used for the crustal faults, and the Geomatrix (1993) relations were used for subduction zone sources. The pattern of rupture areas was shifted along dip and strike as needed to model equally probable earthquake locations on the fault, and stability of the hazard curves. The areal source was modeled as uniformly spaced points at 5-km intervals. The results were unaffected by closer grid spacings. Hypocentral depths were modeled as a triangular distribution with a peak at 7 km and lower bound at 20 km.

3.5. Results

The probabilistic ground motion results using the first set of attenuation relations are shown in Figures 12–15. The results for each site consist of hazard curves for peak horizontal acceleration

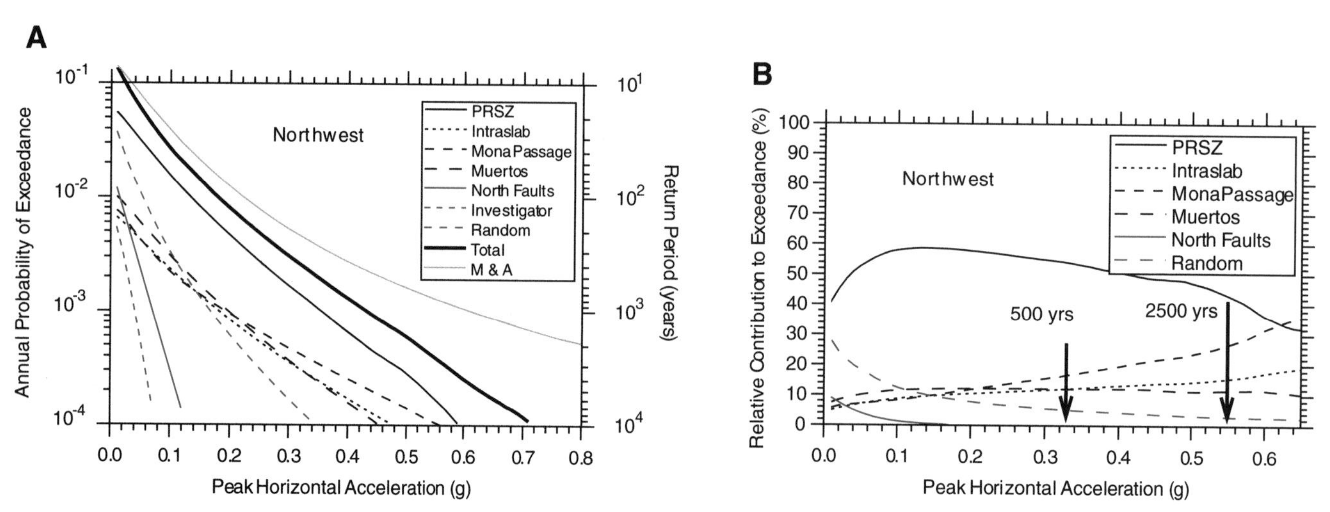

Figure 12. (A) Northwest site peak horizontal acceleration (PHA) hazard curve for major source categories, and total. (B) Relative contributions to PHA exceedance level, with return periods of 500 and 2500 yr. PRSZ—Puerto Rico subduction zone. M&A—Motazedian and Atkinson (this volume).

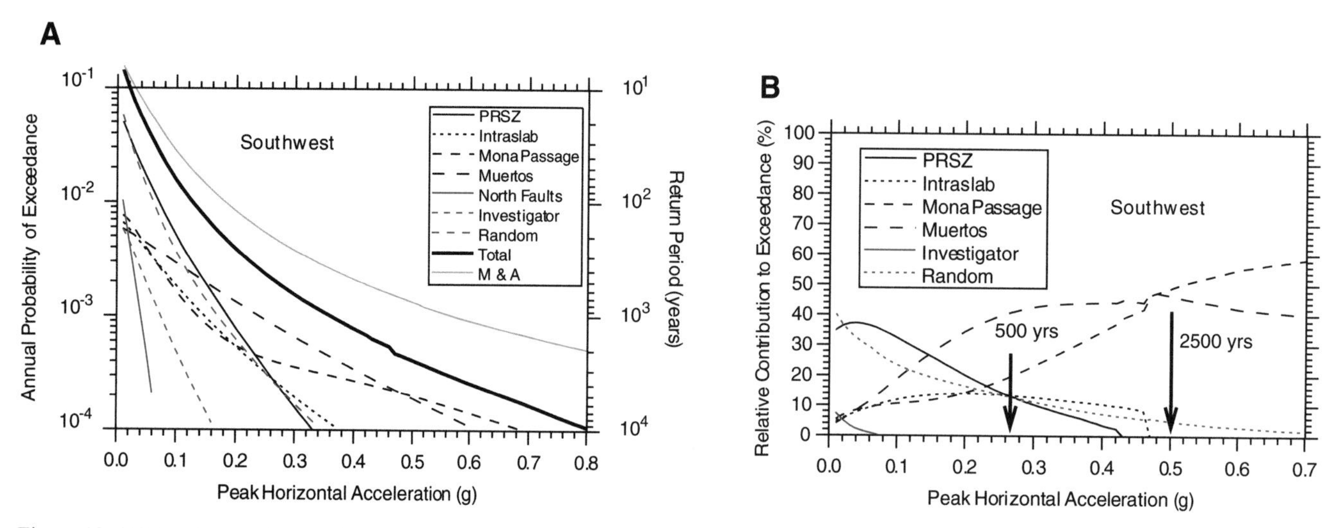

Figure 13. (A) Southwest site peak horizontal acceleration (PHA) hazard curve for major source categories, and total. (B) Relative contributions to PHA exceedance level, with return periods of 500 and 2500 yr. PRSZ—Puerto Rico subduction zone. M&A—Motazedian and Atkinson (this volume).

and a plot showing the relative contributions of each source over the peak acceleration range. The values for return periods of 500 and 2500 yr are shown. The faults have been classified by major source category. Puerto Rico subduction zone refers to both east and west sections of the shallow zone; Muertos refers to both east and west sections of the Muertos subduction zone; and North Faults refers to the East Septentrional and Bowin faults. Categories that produced insignificant results were not included.

For the northwest site (Fig. 12) the major contributor is the shallow Puerto Rico subduction zone. This can be explained by its proximity to the more active western part of the interface. Figure 12B shows it to constitute 30%–60% of the hazard at all ground motion levels. Random seismicity contributes 20%–30% at return periods of less than ~50 yr. The Mona Passage, intraslab, and Muertos sources contribute about equally; at 2500 yr, their combined contribution is ~40%.

The southwest site (Fig. 13) shows a decreased significance of the Puerto Rico subduction zone because of its farther distance and greater importance of Muertos subduction zone and Mona Passage faults. Though not shown, the Mona Passage source at this site is dominated by the Lajas Valley and Cerro Goden land-based faults. At 200 yr, these faults, the Puerto Rico subduction zone, Mona Passage, and Muertos subduction zone contribute about equally. At higher return periods, the land-based faults and Muertos subduction zone increasingly dominate because of their high magnitudes and proximity.

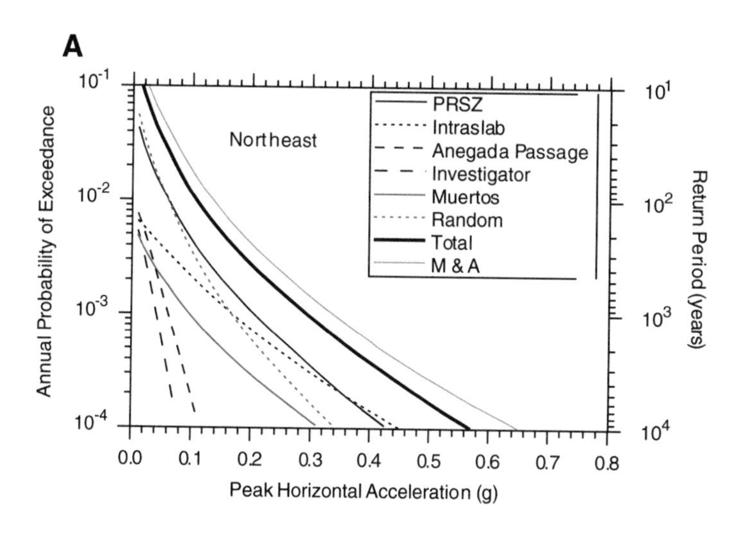

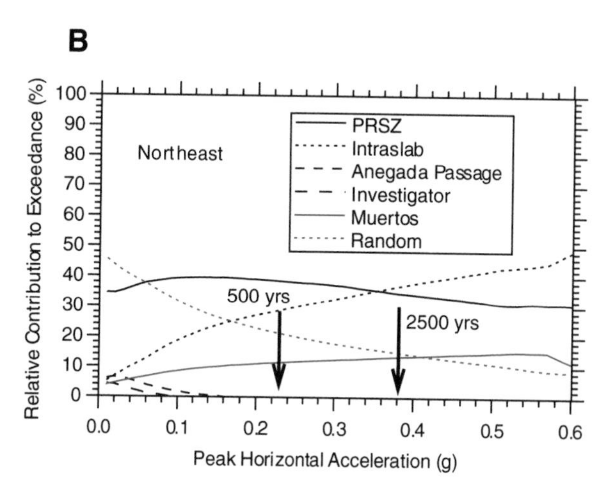

Figure 14. (A) Northeast site peak horizontal acceleration (PHA) hazard curve for major source categories, and total. (B) Relative contributions to PHA exceedance level, with return periods of 500 and 2500 yr. PRSZ—Puerto Rico subduction zone. M&A—Motazedian and Atkinson (this volume).

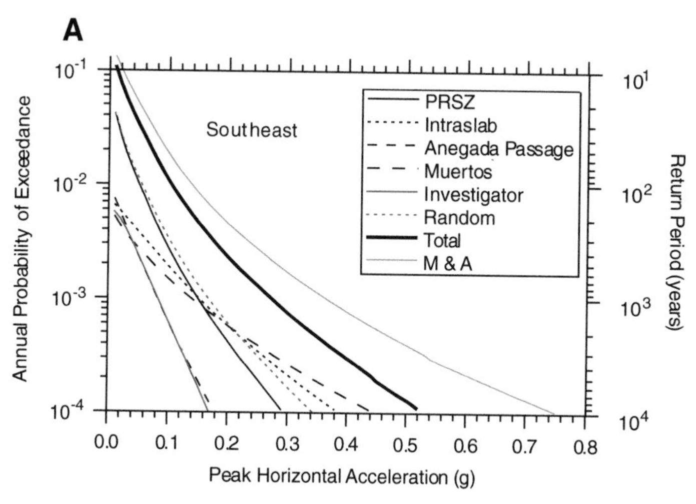

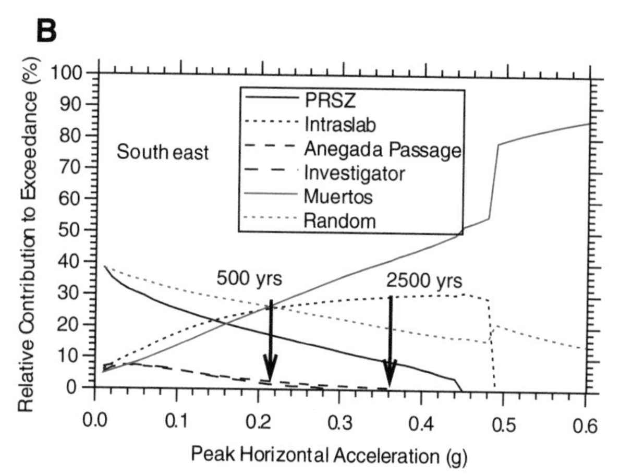

Figure 15. (A) Southeast site peak horizontal acceleration (PHA) hazard curve for major source categories, and total. (B) Relative contributions to PHA exceedance level, with return periods of 500 and 2500 yr. PRSZ—Puerto Rico subduction zone. M&A—Motazedian and Atkinson (this volume).

The northeast site (Fig. 14) shows a lower overall hazard than the sites on the west side of the island because of its distance from the Mona Passage and the more active sections of the Puerto Rico subduction zone and Muertos subduction zone. Although the hazard is dominated by the less-active eastern section of the Puerto Rico subduction zone, the hazard from intraslab events is about equal in importance at a 2500-yr return period.

The southeast site (Fig. 15) shows the lowest overall hazard of the four sites. It is farthest away from the most active source (the western Puerto Rico subduction zone), the most active (western) part of the Muertos subduction zone, and the land-based Mona Passage faults. At 500 yr, the Muertos subduction zone, intraslab events, and random seismicity contribute equally. At longer return periods, the Muertos subduction zone dominates. Surprisingly, the Anegada trough faults do not appear to be significant contributors. The nearest large fault is ~40 km away, and activity rates on them, by our characterization, are relatively low.

In summary, the probabilistic analysis reflects the first-order effects of the salient elements of the model. The hazard from peak horizontal acceleration appears to be greatest near the primary plate boundary source, the Puerto Rico subduction zone. The southwestern site, however, because of the added presence of the land-based faults and the Muertos subduction zone, appears to have a comparable total hazard level. The land-based faults, despite their relatively low slip rates, are significant when close to nearby sites, especially at longer return periods. The eastern sites are further away from these sources and therefore experience lower ground motion levels for a given return period. Random seismicity becomes more important here. At the southeast site, faults in the Anegada trough become important at very long return periods. Of interest is the influence of the North faults (East Septentrional and Bowin) and Investigator faults. These appear to have little importance for the sites analyzed here in relation to the larger sources.

4. DISCUSSION

This study represents an attempt to integrate all current knowledge regarding the location of seismic sources, their activity rates, maximum events, and recurrence characteristics; regional tectonic models, paleoseismic information for land-based faults, and regional space-based horizontal crustal deformation information into a self-consistent seismic source model for use in probabilistic ground motion analyses. In this section, we discuss some of the limitations and sources of uncertainty in the model. Although a detailed sensitivity analysis is beyond the scope of this study, we discuss in general terms how these uncertainties might affect the probabilistic results. We also suggest what further investigations would be appropriate for the purpose of reducing the level of uncertainty in the ground motions predicted from the model.

4.1. Puerto Rico Subduction Zone

The Puerto Rico subduction zone is the dominant contributor for all sites except the southeast. The GPS signal for North America–Puerto Rico–Virgin Islands plate motion is quite strong and stable (Jansma et al., 2000), and unlikely to vary significantly in the future. The most significant source of uncertainty is the level of seismic coupling, which determines what percentage of the motion is released as earthquakes. We have used comparisons to other subduction zones for our estimates; further studies of this type would help in refining the estimate. We have also compared the historic seismicity record to the GPS-coupling-factor-based estimates. A more complete knowledge of preinstrumental earthquakes would be useful, as well as provide a better picture of seismic activity on the Puerto Rico subduction zone from smaller events recorded by the PRSN. Despite its distance from the island, a 3-D velocity model would significantly improve location accuracy of Puerto Rico subduction zone earthquakes. More accurate magnitude determinations based on instrument calibration would allow the calculation of recurrence statistics, which could be compared with our GPS-based estimate.

Intraslab seismicity may be an important factor in the eastern part of the island, especially the southeast, where the Puerto Rico subduction zone influence is diminished. The rate and recurrence model characterization would benefit from the improvements in the PRSN data suggested above. Also, the deeper North American slab is different from those of most other subduction zones in that it has been moving horizontally, rather than vertically, in the upper mantle for ~15 m.y. Gravitational body forces between the slab and surrounding material are therefore more likely to have reached equilibrium, and comparisons to other slabs that are descending vertically may not be appropriate.

4.2. Mona Passage

Characterization of the submarine Mona Passage faults has been derived from ship-based seismic reflection data (Mercado and McCann, 1998). We have assigned slip rates to these faults on the basis of GPS information and the assumptions described in section 2.2. The method of assigning slip rates on the basis of horizontal GPS data is experimental and would benefit from testing in areas where GPS signals are strong and stable, and slip rates for a complete set of faults subjected to a uniform regional stress are known from geologic methods. This would help answer the question of whether a moment-based scheme, as used here, or some other scaling method is appropriate. We have also assumed no aseismic slip on these faults; an assumption of some aseismic slip would lower the hazard estimates presented here. For the method as presented here, the results are sensitive to the assumed dip of the faults. This is especially true as the dips approach vertical, because an unmeasured vertical component of slip must be taken into account. For example, by using a 75° dip for the Mona Passage faults instead of 63°, the slip rate on the Lajas Valley fault increases by 35%. The assumed direction of assumed horizontal stress also affects the results, although to a lesser extent (LaForge and McCann, 2003). The magnitude of horizontal extension scales directly with apportioned slip rate; therefore uncertainties in the GPS-derived magnitude directly affect calculated slip rates. For the most recent GPS estimate of 5 ± 3 mm/yr, this means a 60% variation at the one standard error level. Reductions in the GPS estimate errors, which would entail continued monitoring from more stations, would help in more accurately characterizing activity on faults in the Mona Passage and western Puerto Rico.

The east-bounding fault of Mona Canyon has been identified as the source of a M_S 7.3 event in 1918, whose tsunami and ground shaking caused 116 deaths and $25,000,000 in damage (Mercado and McCann, 1998). Additional research into this feature would refine slip rates assigned to it and further clarify its role in the larger region of Hispaniola–Puerto Rico extension.

Further geologic studies to search for Quaternary faults in addition to the South Lajas and Cerro Goden are important. Despite relatively low slip rates, the potential proximity of such faults to population centers and critical structures means that the consequences of strong earthquakes on them will be severe. Because the relative contribution of these faults to the total hazard increases with return period, they become increasingly important for critical structures, which are usually designed to withstand relatively low probability (and therefore high amplitude) ground motions. The paleoseismic data for the South Lajas fault (Prentice, 2003) have provided an important comparison to slip rates estimated from the regional slip-allocation method. More detailed information about the known land-based faults, and identification and investigations into possible others, are important not only from the hazard characterization standpoint, but in understanding how they relate to regional tectonic models.

Independent evidence for the orientation of horizontal stress comes from focal mechanisms of teleseismically recorded historic and recent shallow platform events that are clearly not associated with the plate interface. Doser et al. (this volume; their Fig. 4) show two such events; the magnitude 7.3 1918 Mona Canyon event indicates normal faulting on NE-SW–oriented planes, and a magnitude 5.3 1992 event with a similar mechanism that occurred ~20 km north of the northwest corner of the island. Jansma et al. (2000; this volume) show an event with a similar mechanism ~60 km northwest of this one. Tension axes for the two events in Doser et al. are oriented near horizontal, with azimuths of 142° and 116°, respectively. This is in contradiction to the NE-SW regional extension suggested by McCann (1998) on the basis of the orientation of faults in the Mona Passage. It also is inconsistent with inferred right-lateral Holocene offset on the Cerro Goden fault (Mann and Prentice, 2001; Mann et al., this volume, Chapter 6). Construction of focal mechanisms from the PRSN data in western Puerto Rico might resolve this discrepancy.

4.3. Muertos Subduction Zone–Investigator Faults

Our results indicate that the Muertos subduction zone becomes a significant contributor in the southwest part of the island at return periods of ~100 yr and greater (Fig. 12B). Because the relative plate motion at the Muertos subduction zone is about an order of magnitude less than that of the Puerto Rico subduction zone, its nature and behavior are more difficult to characterize. Also, seismologic observations are notably lacking, although this in itself supports a relatively low activity rate. As with the Mona Passage faults, uncertainties in the GPS extension magnitude vary directly with the Muertos subduction zone slip rate estimates. Specifically, the plate motion magnitude uncertainty detailed in Jansma et al. (2000; this volume) translates to the 60% variation in slip rates at the level of one standard error. Reduction of the GPS-based motion error estimate would improve rate estimates for the Muertos subduction zone.

Although the sites analyzed here are far enough from the Investigator faults so that their influence appears insignificant, for sites in the south-central part of the island, their activity rates and potential magnitudes indicate that they would assume an importance similar to that of the Lajas Valley fault for sites in the southwest part of the island. Very little is known of these faults, aside from their existence, general bathymetric expression, and presumed role in the kinematics of Caribbean–Puerto Rico–Virgin Islands plate interaction. Considering their proximity to the south coast of the island and current uncertainty regarding their activity rates, research on these faults should be considered important.

4.4. Anegada Trough

Our results indicate that the Anegada trough faults assume an importance only for the southeast part of the island, and only at return periods of a few thousand years and greater. These results, however, are subject to all the uncertainties inherent in the GPS-derived rate estimates, as discussed in the previous section, and also the frequency-dependent considerations of the ground motions discussed in section 4.7 and by Motazedian and Atkinson (this volume).

4.5. East Septentrional–Bowin Faults

These two faults, as modeled, have an effect only at the northwest site and appear to be insignificant compared with the Puerto Rico subduction zone. However, it is necessary to point out that they are among the least known components of the source model presented here, and their activity rates rely on little more than poorly constrained inferences.

4.6. Random Seismicity

The characterization of randomly occurring events indicates that it is an important source for the southwest part of the island at return periods of less than ~100 yr, and for the eastern part of the island at all return periods. Suggestions regarding the analysis of PRSN data are emphasized here. Use of the entire 25-yr catalog, instead of a 3.5-yr period, would greatly improve the stability of calculated upper crustal seismicity rates beneath the island. In addition, use of more data would allow the island to be zoned according to relative activity levels. Specifically, it is clear from the larger data set (Fig. 8) that activity is greater in the southwest part of the island. Because of the small sample size of our data set, we were unable to account for this observation in the model.

4.7. Ground Motion Attenuation Relations

The attenuation function used for nonsubduction zone sources is a significant source of uncertainty in the results presented here. For the shallow plate interface subduction zone sources (Puerto Rico subduction zone and Muertos subduction zone), we have used the "interface" relation of Youngs et al. (1997) and their "intraslab" relation for the deeper Puerto Rico subduction zone. Because these relations are based exclusively on shallow interface and intraslab records, we feel that they are appropriate relations for these sources. For the remaining sources, however, we have relied on a function based largely on crustal attenuation characteristics in western California. Differences, and their consequences, between there and the arc massif environment of Puerto Rico are twofold. First, western California contains an actively deforming strike-slip plate boundary; Puerto Rico is near such a boundary but does not contain it. The southwest part of Puerto Rico, from land-based fault existence and seismicity levels, appears to be actively deforming, but at a rate much lower than that of western California. This suggests that the upper crust of Puerto Rico is less disrupted than that of western California and thus able to transmit seismic energy more efficiently. The attenuation of ground motion is therefore likely less than assumed in this analysis. As a test, for the random source an eastern North America attenuation function (Toro et al., 1997) was substituted for the Abrahamson and Silva (1997) function at the northwest site. The resulting ground motions were a factor of 2 higher for the 500-yr return period and greater than that at longer return periods. Second, Puerto Rico is being subjected to extensional stresses; in western California, they are compressional. Earthquakes with thrust mechanisms have been found to generate higher ground motions than those with strike-slip or normal mechanisms (e.g., Spudich et al., 1999; Sadigh et al., 1997).

An important step in characterizing ground motion hazard on the island is an attenuation function developed by Motazedian and Atkinson (this volume). In that study, PRSN records were calibrated and used to estimate attenuation from all sources about the island. Because of the restricted magnitude and distance range of the earthquakes used, synthetic finite-rupture accelerograms were generated to fill in the gaps, particularly for larger magnitudes and closer distances. Different directivity scenarios were also taken into account.

An important parameter in such relations is the sigma value, or the standard deviation of the lognormal ground motion distribution for a given magnitude and distance. In probabilistic analyses such as this, the distribution is integrated, and thus this uncertainty estimate has a significant effect, especially at longer return periods. Such estimates, when based on purely empirical records, are usually magnitude and response-period dependent (e.g., Abrahamson and Silva, 1997). When developing an attenuation relation from largely synthetic records, such an estimate is difficult because it would be necessary to incorporate all estimated uncertainties in all parameters used to generate the records. However, on the basis of the scatter in the PRSN data used to develop this relation, a rough estimate of 0.28 ($\log_{10}$ units), for all magnitudes and periods, was made (Motazedian, personal commun., 2002). The results show a significant increase in hazard. At the northwest site, at a return period of 500 yr, the peak horizontal acceleration hazard is higher by 30%, and at 1000 yr, it is higher by 40%. Considering the comparisons to eastern North America attenuation discussed above and made by Motazedian and Atkinson (this volume), this may be quite reasonable.

4.8. Other Considerations

Probabilistic results were generated for peak horizontal acceleration only. Because one of the goals of this study was to examine the relative source contributions at different sites, we point out that the computed hazard for longer period ground motions will give different results. For example, if an engineered structure is more responsive to long period vibrations and long motion duration (e.g., a long bridge or embankment dam), a source which generates large magnitude events will have a greater effect (all other conditions being equal) than random seismicity, which is assumed to produce magnitudes of 6.5 and less.

For all fault sources but the Puerto Rico subduction zone and Muertos subduction zone, we have assumed that all crustal strain is expressed as earthquakes. If aseismic slip is occurring, this would have the effect of lowering the activity rates and resulting hazard curve levels.

5. CONCLUSIONS

This study represents an attempt to construct a seismic source model for Puerto Rico for use in probabilistic ground motion calculations. We have used all known faults, characterized them as 3-D planar features, and assigned slip rates on the basis of GPS-based regional extensional rates (Jansma et al., 2000; Calais et al., 2002). Activity rates were compared with historic seismicity and paleoseismic data as an independent check on magnitude distribution and appropriate recurrence model. Random seismicity and intraslab seismicity rates associated with the downgoing North America plate were characterized from a subset of microearthquake data from the PRSN.

Probabilistic hazard curves for peak horizontal acceleration were computed for the four corners of the island (Fig. 10). The results show the hazard to be greatest at the northwest site, near the more active segment of the shallow Puerto Rico subduction zone and faults of the Mona Passage extensional zone. Hazards in the southwest are less because of increased distance from the Puerto Rico subduction zone but increase to the northwest site levels at long return periods because of the presence of the Lajas Valley fault and the more active western part of the Muertos subduction zone. Hazards in the eastern part of the island are significantly less. The northeast is dominated by the Puerto Rico subduction zone and intraslab events, as we have assigned the eastern Puerto Rico subduction zone a lower rate because of lower assumed coupling of the plate interface. The western Puerto Rico subduction zone is assumed to have greater coupling and higher activity rate, because of subduction of the buoyant Bahama Bank. In the southeast, the hazard is lowest of the four sites, with no source clearly dominating.

The information and techniques used to construct the model and perform the probabilistic calculations contain a wide range of assumptions and uncertainties. In this study, we present mean results and discuss uncertainties qualitatively. The model should be considered a work in progress, subject to change because of future discoveries and refinements in its component parts. On the basis of our findings, we propose several research directions that would help constrain and reduce the uncertainties of the source parameters presented here.

1. For many sites on the island, the choice of attenuation function is a significant source of uncertainty. Although western California attenuation relations were used for seismic sources in this paper, preliminary results that use a new relation developed for the island (Motazedian and Atkinson, this volume) indicate ground motion amplitude values could be on the order of 30% higher than those presented here.

2. The PRSN catalog should be refined by developing a 3-D velocity model to obtain improved locations, and by calibrating the network in order to obtain more accurate and consistent magnitude determinations. A moment magnitude catalog should be created for twentieth-century teleseismic events.

3. For several sources, we used regional GPS measurements and apportioned slip onto faults by using relative seismic moment potential and an assumed fault dip. Although the results seem relatively insensitive to the GPS vector azimuth, they are quite sensitive to assumed fault dip (LaForge and McCann, 2003). In addition, any pure strike-slip faults are unlikely to be identifiable from the seismic reflection records. Although the technique appears to give reasonable results for this application, it should be tested in regions where GPS measurements are reliable and slip rates have been measured independently.

4. Further land-based geologic and geomorphic studies should be performed to identify additional possibly active faults, estimate their slip rates, and learn more about young faults and how they fit into regional tectonic models.

5. More refined GPS measurements would help resolve the Mona Passage stress direction discrepancy discussed section 4.2.

6. For sites on the south coast of the island, the Investigator faults may constitute a significant hazard. Further research into these faults is recommended.

7. The potential for large-magnitude events within the deeper North America plate is unknown and may not be comparable to settings where arc-normal subduction dominates. Research into deep seismicity in arc-parallel subduction environments would help resolve this issue.

APPENDIX A. PARAMETERS FOR FAULTS USED IN SOURCE MODEL

TABLE A1. PARAMETERS FOR MONA PASSAGE FAULTS

Fault	Length (km)	Magnitude	Seismic moment proportion	Fault area (km^2)	Slip rate (mm/yr)	Return period (yr)
MP-1	25	6.7	0.025	582	0.18	4700
MP-2	28	6.8	0.029	632	0.21	4400
MP-3	22	6.6	0.018	549	0.13	5200
MP-4	44	7.0	0.063	971	0.45	2800
MP-5	27	6.7	0.028	615	0.20	4400
MP-6	18	6.5	0.013	301	0.09	9400
MP-7	56	7.1	0.097	1156	0.70	2400
MP-8	18	6.5	0.013	324	0.09	8700
MP-9	35	6.9	0.043	761	0.31	3700
MP-10	52	7.1	0.086	1089	0.62	2500
MP-11	22	6.6	0.019	490	0.14	5500
MP-12	20	6.6	0.015	312	0.11	9300
MP-13	16	6.5	0.011	294	0.08	8900
MP-14	18	6.5	0.014	330	0.10	8200
MP-15	42	7.0	0.059	972	0.42	2800
MP-16	66	7.2	0.129	1425	0.92	1900
MP-17	22	6.6	0.019	486	0.14	5600
MP-18	28	6.8	0.029	658	0.21	4100
MP-19	18	6.6	0.014	413	0.10	6800
MP-20	23	6.7	0.021	525	0.15	5400
MP-21	21	6.5	0.017	471	0.12	5800
MP-22	21	6.6	0.017	465	0.12	6100
MP-23	22	6.6	0.019	498	0.14	5600
MP-24	18	6.5	0.013	395	0.09	7100
MP-25	19	6.6	0.015	422	0.11	6400
MP-26	18	6.5	0.013	386	0.09	7000
MP-CG	54	7.1	0.091	1287	0.65	2200
MP-LV	47	7.0	0.071	827	0.51	3400

TABLE A2. PARAMETERS FOR MUERTOS SUBDUCTION ZONE–INVESTIGATOR FAULTS

Fault	Length (km)	Magnitude	Fault area (km^2)	Slip rate (mm/yr)	Return period (yr)
Muertos—West	300	8.2	7236	1.2*	3000
Muertos—East	250	7.8	18090	0.6*	1900
Investigator—West	53	7.1	1063	0.76	2200
Investigator—East	46	7.0	917	1.48	1200

*1/2 assumed tectonic rate.

TABLE A3. PARAMETERS ANEGADA PASSAGE FAULTS

Fault	Length (km)	Magnitude	Seismic moment proportion	Fault area (km^2)	Slip rate (mm/yr)	Return period (yr)
AP-1	84	7.3	0.413	1833	0.86	2500
AP-2	15	6.5	0.021	397	0.04	11,400
AP-3	13	6.4	0.016	362	0.03	12,300
AP-4	117	7.5	0.411	2616	0.83	1200
AP-5	30	6.8	0.039	665	0.08	12,800
AP-6	9	6.2	0.003	321	0.01	32,600
AP-7	10	7.2	0.004	314	0.009	34,000
AP-8	42	7.0	0.045	1160	0.113	8800
AP-9	30	6.8	0.025	736	0.062	14,000
AP-10	22	6.6	0.015	821	0.037	12,600
AP-11	17	6.5	0.009	465	0.022	23,000

ACKNOWLEDGMENTS

We extend our gratitude to Gail Atkinson, Chuck Mueller, Art Frankel, and Paul Mann for careful and thoughtful reviews. This work was supported by the Puerto Rico Electric Power Authority and the U.S. Bureau of Reclamation. Some figures were made with Generic Mapping Tools (Wessel and Smith, 1998).

REFERENCES CITED

Abrahamson, N.A., and Silva, W.J., 1997, Empirical response spectral attenuation relations for shallow crustal earthquakes: Seismological Research Letters, v. 68, p. 94–128.

Aki, K., and Richards, P., 1980, Quantitative seismology: Theory and methods: New York, W.H. Freeman, 557 p.

Anderson, J.G., 1979, Estimating the seismicity from geological structure for seismic-risk studies: Bulletin of the Seismological Society of America, v. 69, p. 135–159.

Anderson, R.N., and Larue, D.K., 1991, Wellbore heat flow from the Toa Baja Scientific Drill hole, Puerto Rico: Geophysical Research Letters, v. 128, p. 537–540.

Bollinger, G.A., Davison, F.C., Sibol, M.S., and Birch, J.B., 1989, Magnitude recurrence relations for the southeastern United States and its subdivisions: Journal of Geophysical Research, v. 94, p. 2857–2873.

Byrne, D., Suarez, G., and McCann, W.R., 1985, Muertos trough subduction-microplate tectonics in the northern Caribbean: Nature, v. 317, p. 420–421.

Byrne, D.E., Davis, D.M., and Sykes, L.R., 1988, Loci and maximum size of thrust earthquakes and the mechanics of the shallow region of subduction zones: Tectonics, v. 7, p. 833–857.

Calais, E., Mazabraud, Y., Mercier de Lepinay, B., Mann, P., Mattioli, G., and Jansma, P., 2002, Strain partitioning and fault slip rates in the northeastern Caribbean from GPS measurements: Geophysical Research Letters, v. 29, no. 18, p. 3-1–3-4.

Cornell, C.A., 1968, Engineering seismic risk analysis: Bulletin of the Seismological Society of America, v. 58, p. 1583–1606.

DeMets, C., Gordon, R., Argus, D., and Stein, S., 1990, Current plate motions: Geophysical Journal International, v. 101, p. 425–478.

Dolan, J.F., and Wald, D., 1998, The 1943–1953 north-central Caribbean earthquakes: Active tectonic setting, seismic hazards, and implications for Caribbean–North America plate motions, *in* Dolan, J., and Mann, P. eds., Active strike-slip and collisional tectonics of the Northern Caribbean Plate Boundary Zone: Boulder, Colorado, Geological Society of America Special Paper 326, p. 143–169.

Dolan, J.F., Mullins, H., and Wald, D., 1998, Active tectonics of the north-central Caribbean: Oblique collision, strain partitioning, and opposing subducted slabs, *in* Dolan, J. and Mann, P., eds., Active strike-slip and Collisional Tectonics of the Northern Caribbean Plate Boundary Zone: Boulder, Colorado, Geological Society of America Special Paper 326, p. 1–61.

Erickson, J.P., Pindell, J.L., and Larue, D.K., 1991, Fault zone deformational constraints on Paleogene tectonic evolution in southern Puerto Rico: Geophysical Research Letters, v. 18, p. 569–572.

Fink, L.K., and Harrison, C.G.A., 1972, Paleomagnetic investigations of selected lava units on Puerto Rico: Transactions of the Caribbean Geological Conference, v. 6, 379.

Geomatrix, 1993, Seismic margin earthquake for the Trojan site: Final unpublished report prepared for Portland General Electric Trojan Nuclear Plant, Rainier, Oregon: San Francisco, California, Geomatrix Consultants.

Geomatrix, 1995, Seismic design mapping, State of Oregon: Consultant's report for Oregon Department of Transportation: San Francisco, California, Geomatrix Consultants.

Grindlay, N., Mann, P., and Dolan, J., 1997, Researchers investigate submarine faults north of Puerto Rico: Eos (Transactions, American Geophysical Union), v. 78, p. 404.

Gutenberg, B., and Richter, C.F., 1954, Seismicity of the Earth: Princeton, New Jersey, Princeton University Press, 310 p.

Hanks, T.C., and Kanamori, H., 1979, A moment magnitude scale: Journal of Geophysical Research, v. 84, no. B5, p. 2348–2350.

Jansma, P., Mattioli, G.S., Lopez, A., DeMets, C., Dixon, T.H., Mann, P., and Calais, E., 2000, Neotectonics of Puerto Rico and the Virgin Islands, northeastern Caribbean, from GPS Geodesy: Tectonics, v. 19, p. 1021–1037, doi: 10.1029/1999TC001170.

Jany, I., Scanlon, K., and Mauffret, A., 1990, Geological interpretation of combined Seabeam, GLORIA, and seismic data from Anegada Passage (Virgin Islands, North Caribbean): Marine Geophysical Research, v. 12, p. 173–196.

LaForge, R., and Hawkins, F.F., 1999, Seismic hazard and ground motion analysis for Carraizo Dam, Puerto Rico: U.S. Bureau of Reclamation Seismotectonic Report 99-2, 29 p.

LaForge, R., and McCann, W.R., 2003, Using GPS data to assign slip rates to a fault set: Mona Passage, Puerto Rico [abs.]: Seismological Research Letters, v. 74, no. 2, p. 200.

Lithgow, C.W., McCann, W., and Joyce, J., 1987, Extensional tectonics at the eastern edge of the Puerto Rico Platelet: Eos (Transactions, American Geophysical Union), v. 68, p. 466–489.

Larue, D.K., and Ryan, H.F., 1998, Seismic reflection profiles of the Puerto Rico trench: Shortening between the North American Caribbean plates, *in* Lidiak, E.G., and Larue, D.K., eds., Tectonics and geochemistry of the northeastern Caribbean: Geological Society of America Special Paper 322, p. 193–210.

Mann, P., and Prentice, C.S., 2001, Collaborative research UTIG and USGS, Towards an integrated understanding of late Holocene fault activity in western Puerto Rico: Onland scarp mapping and fault trenching: U.S. Geological Survey, Final Project Report, 13 p.

Mann, P., Prentice, C.S., Burr, G., Peña, L.R., and Taylor, F.W., 1998, Tectonic geomorphology and paleoseismology of the Septentrional fault system, Dominican Republic, *in* Dolan, J. and Mann, P. eds., Active strike-slip and collisional tectonics of the Northern Caribbean Plate Boundary Zone: Geological Society of America Special Paper 326, p. 63–123.

Mann, P., Calais, E., Ruegg, J.C., De Mets, C., Jansma, P.E., and Mattioli, G.S., 2002, Oblique collision in the northeastern Caribbean from GPS measurements and geological observations: Tectonics, v. 21, no. 6, p. 1057, doi: 10.1029/2001TC001304.

Masson, D.G., and Scanlon, K.M., 1991, The neotectonic setting of Puerto Rico: Geological Society of America Bulletin, v. 103, p. 144–154, doi: 10.1130/0016-7606(1991)103<0144:TNSOPR>2.3.CO;2.

McCann, W.R., 1985, On the earthquake hazard of Puerto Rico and the Virgin Islands: Bulletin of the Seismological Society of America, v. 75, p. 251–262.

McCann, W.R., 1994, Seismic hazard map for Puerto Rico, 1994, prepared for Seismic Safety Commission of Puerto Rico: Westminster, Colorado, Earth Scientific Consultants, 60 p.

McCann, W.R., 1998, Tsunami hazard of western Puerto Rico from local sources: Characteristics of tsunamigenic faults: Westminster, Colorado, Earth Scientific Consultants, 13 p.

McCann, W., 1999, Characterization of important, potentially tsunamigenic faults near the Dominican Republic: Sea grant internal report: Mayagüez, Puerto Rico, University of Puerto Rico, 21 p.

McCann, W., 2002, Microearthquake data elucidates details of Caribbean subduction zone: Seismological Research Letters, v. 73, p. 25–32.

McCann, W.R., 2003, Catalog of felt earthquakes for Puerto Rico and neighboring islands 1492–1899 with additional information for some 20th century earthquakes: Seismological Research Letters, v. 74, p. 199.

McCann, W.R., and Sykes, L., 1984, Subduction of aseismic ridges beneath the Caribbean plate: Implications for the tectonics and seismic potential of the northeastern Caribbean: Journal of Geophysical Research, v. 89, p. 4493–4519.

McCann, W.R., and Lithgow-Bertelloni, C., 2002, Origin and neotectonics of the Anegada trough, Northeastern Caribbean, submitted to Tectonophysics.

McCann, W., Joyce, J., and Lithgow, C., 1987, The Puerto Rico Platelet at the northeastern edge of the Puerto Rico Platelet: Eos (Transactions, American Geophysical Union), v. 68, p. 1483.

Mercado, A., and McCann, W., 1998, Numerical simulation of the 1918 Puerto Rico tsunami: Natural Hazards, v. 18, p. 57–76, doi: 10.1023/A:1008091910209.

Meltzer, A.S., 1997, Fault structure and earthquake potential, Lajas Valley, SW Puerto Rico: U.S. Geological Survey External Grant Report, 16 p.

Prentice, C.S., 2003, Paleoseismic study of the South Lajas fault, Lajas Valley, southwestern Puerto Rico [abs.]: Seismological Research Letters, v. 74, no. 2, p. 200.

Reasenberg, P., 1985, Second-order moment of central California seismicity, 1969–1982: Journal of Geophysical Research, v. 90, p. 5479–5495.

Reid, H., and Taber, S., 1919, The Porto Rico earthquake of 1918, with descriptions of earlier earthquakes (Report of the Earthquake Investigation Commission): Washington, D.C., House of Representatives Document 269, 74 p.

Reid, J., Plumley, P., and Schellekens, J., 1991, Paleomagnetic evidence for late Miocene counterclockwise rotation of north coast carbonate sequence, Puerto Rico: Geophysical Research Letters, v. 18, p. 319–324.

Sadigh, K., Chang, D.-Y., Egan, J.A., Makdisi, F., and Youngs, R.R., 1997, Attenuation relationships for shallow crustal earthquakes based on California strong motion data: Seismological Research letters, v. 68, p. 190–199.

Savage, M.K., and dePolo, D.M., 1993, Foreshock probabilities in the western Great Basin–eastern Sierra Nevada: Bulletin of the Seismological Society of America, v. 83, p. 1910–1939.

Scholz, C.H., and Campos, J., 1995, On the mechanism of seismic decoupling and back arc spreading at subduction zones: Journal of Geophysical Research, v. 100, p. 22,103–22,117, doi: 10.1029/95JB01869.

Shepherd, J.B., Tanner, J.G., and Lynch, L.L., 1994, A revised earthquake catalog for the eastern Caribbean 1530–1992: Proceedings, steering committee meeting: Melbourne, Florida, Latin American and Caribbean Seismic Hazard Project, April 1992, p. 95–158.

Spudich, P., Joyner, W.B., Lindh, A.G., Boore, D.M., Margaris, B.M., and Fletcher, J.B., 1999, A revised ground motion prediction relation for use in extensional tectonic regimes: Bulletin of the Seismological Society of America, v. 89, p. 1156–1170.

Sykes, L., and Ewing, M., 1965, The seismicity of the Caribbean region: Journal of Geophysical Research, v. 70, no. 5065, p. 5–74.

Sykes, L., McCann, W., and Kafka, A., 1982, Motion of Caribbean plate during last seven million years and implications for earlier Cenozoic movements: Journal of Geophysical Research, v. 87, p. 10,656–10,676.

Taggart, B.E., 1992, The tectonic and eustatic significance of the correlation of radiometrically dated Late Quaternary marine terraces on northwestern Puerto Rico and Isla de Mona, Puerto Rico [Ph.D. thesis]: Puerto Rico, University of Puerto Rico, Mayaguez, 255 p.

ten Brink, U.S., 2003, High-resolution bathymetric map of the Puerto Rico trench: Implications for earthquake and tsunami hazards: Seismological Research Letters, v. 74, no. 2, p. 230.

Thatcher, W., 1990, Order and diversity in the modes of circum-Pacific earthquake recurrence: Journal of Geophysical Research, v. 95, no. B3, p. 2609–2623.

Tichelaar, B.W., and Ruff, L.J., 1993, Depth of seismic coupling along subdution zones: Journal of Geophysical Research, v. 98, no. B2, p. 2017–2037.

Toro, G.R., Abrahamson, N.A., and Schneider, J., 1997, Model of strong ground motions from earthquakes in central and eastern North America: Best estimates and uncertainties: Seismological Research Letters, v. 68, p. 41–58.

Van Fossen, M.C., Channel, J., and Schellekens, H., 1989, Paleomagnetic evidence for Tertiary anticlockwise rotation on southwest Puerto Rico: Geophysical Research Letters, v. 16, p. 819–822.

van Gestel, J., Mann, P., Dolan, J.F., and Grindlay, N.R., 1998, Structure and tectonics of the upper Cenozoic Puerto Rico–Virgin Islands carbonate platform as determined from seismic reflection studies: Journal of Geophysical Research, v. 103, no. B-12, p. 30,505–30,530.

Weichert, D., 1980, Estimation of the earthquake recurrence parameters for unequal observation periods for different magnitudes: Bulletin of the Seismological Society of America, v. 70, p. 1337–1347.

Wells, D.L., and Coppersmith, K.J., 1994, New empirical relationships among magnitude, rupture length, rupture width, rupture area, and surface displacement: Bulletin of the Seismological Society of America, v. 84, p. 974–1003.

Wessel, P., and Smith, W.H.F., 1998, New, improved version of Generic Mapping Tools released: Eos (Transactions, American Geophysical Union), v. 79, p. 559.

Wesnousky, S.G., 1986, Earthquakes, Quaternary faults, and seismic hazard in California: Journal of Geophysical Research, v. 91, no. B-12, p. 12587–12631.

Youngs, R.R., and Coppersmith, K.J., 1985, Implications of fault slip rates and earthquake recurrence models to probabilistic seismic hazard estimates: Bulletin of the Seismological Society of America, v. 75, p. 939–965.

Youngs, R.R., Chiou, S.-J., Silva, W.J., and Humphrey, J.R., 1997, Strong ground motion attenuation relationships for subduction zone earthquakes: Seismological Research Letters, v. 68, p. 58–74.

Manuscript Accepted by the Society August 18, 2004

Printed in the USA

Geological Society of America
Special Paper 385
2005

Liquefaction susceptibility zonation map of San Juan, Puerto Rico

James V. Hengesh*
Jeffrey L. Bachhuber*
William Lettis & Associates, 1777 Botelho Road, Walnut Creek, California 94596, USA

ABSTRACT

This paper discusses development of a 1:20,000 scale liquefaction susceptibility map for the San Juan Quadrangle of Puerto Rico. San Juan, with an estimated population of 434,000, is the capital of Puerto Rico, and the most important U.S. commercial and industrial center in the Caribbean region. Puerto Rico is located in a seismically active region characterized by the convergence and lateral translation of the North America and Caribbean plates. Large earthquakes in 1670, 1787, 1867, and 1918 caused significant damage to major parts of the island, including the San Juan area. Paleoliquefaction features, possibly caused by at least three different earthquakes since A.D. 1300, have been found in Holocene floodplain sediments at several sites in western Puerto Rico (Tuttle et al., this volume). The historic earthquakes and paleoliquefaction features demonstrate that the opportunity exists for future liquefaction events to occur in Puerto Rico.

The liquefaction susceptibility map was developed through a five step process including (1) preparation of a detailed Quaternary geology map delineating deposit age, depositional environment, and texture; (2) evaluation of Quaternary deposit thickness and depth to groundwater; (3) initial evaluation of relative liquefaction susceptibility (decision tree); (4) liquefaction triggering evaluation using geotechnical borehole data and the Seed and Idriss (1971b) "simplified procedure"; and, (5) identification of units of similar susceptibility and the definition of liquefaction susceptibility zones. The map depicts five liquefaction hazard zones for the greater San Juan metropolitan area that range from very low hazard to very high hazard. Areas of very high hazard occur along the edges of Bahia de San Juan and Laguna San Jose. Extensive swamp deposits and artificial fill over swamp deposits containing loose sandy soils occur in these areas. Areas of high hazard occur along the beach and areas in the vicinity of the airport. Most of the area to the south of Bahia de San Juan and Laguna San Jose lies within moderate, low, or very low hazard zones.

Keywords: liquefaction, susceptibility map, zonation, San Juan.

*hengesh@lettis.com; bachhuber@lettis.com

Hengesh, J.V., and Bachhuber, J.L., 2005, Liquefaction susceptibility zonation map of San Juan, Puerto Rico, *in* Mann, P., ed., Active tectonics and seismic hazards of Puerto Rico, the Virgin Islands, and offshore areas: Geological Society of America Special Paper 385, p. 249–262. For permission to copy, contact editing@geosociety.org.

INTRODUCTION

This paper discusses development of a liquefaction susceptibility zonation map (1:20,000 scale) for the San Juan Quadrangle of Puerto Rico. The project was completed under a U.S. Geological Survey (USGS) National Earthquake Hazard Reduction Program (NEHRP) research grant. In addition to this liquefaction hazard map, the study developed a companion ground shaking amplification susceptibility map.

San Juan, with an estimated population of 434,000 (U.S. Census Bureau, 2001), is the capital of Puerto Rico and the most important U.S. commercial and industrial center in the Caribbean region. San Juan has expanded across the northern coastal plain (Fig. 1) that includes broad areas of Quaternary deposits occupying deep alluvial valleys, flood plains, beaches, swamps, and lagoons. Extensive development of this region over the past 50 years has led to artificial filling and urbanization of large tracts of land underlain by soft, young deposits. Quaternary sediments underlying the coastal plain contain fine-grained deposits of sand, silt, clay, and peat, artificial fill over clay, and saturated granular sediments that are susceptible to liquefaction. The maps developed during this project will assist Puerto Rico in its efforts to characterize and mitigate seismic hazards, reduce risks from these hazards, and plan future development. The maps have been incorporated into the Puerto Rican Planning Board GIS database for governmental use, and are available to the public through this organization.

REGIONAL SEISMIC SETTING

Puerto Rico lies along the Northern Caribbean Plate Boundary Zone, a seismically active region characterized by convergence and lateral translation of the North America and Caribbean plates (Mann et al., 2002; McCann, 1985, 2002). The major tectonic elements of the region include the Puerto Rico and Muertos trough subduction zones (located north and south of the island, respectively) the Anegada and Mona Passages (located east and west of the island, respectively), and segments of the Great Southern Puerto Rico fault zone that cross the island from northwest to southeast (Masson and Scanlon, 1991). Geodetic data for the northeastern Caribbean region indicate an ~20 mm/yr shortening rate across the Northern Caribbean Plate Boundary Zone (Mann

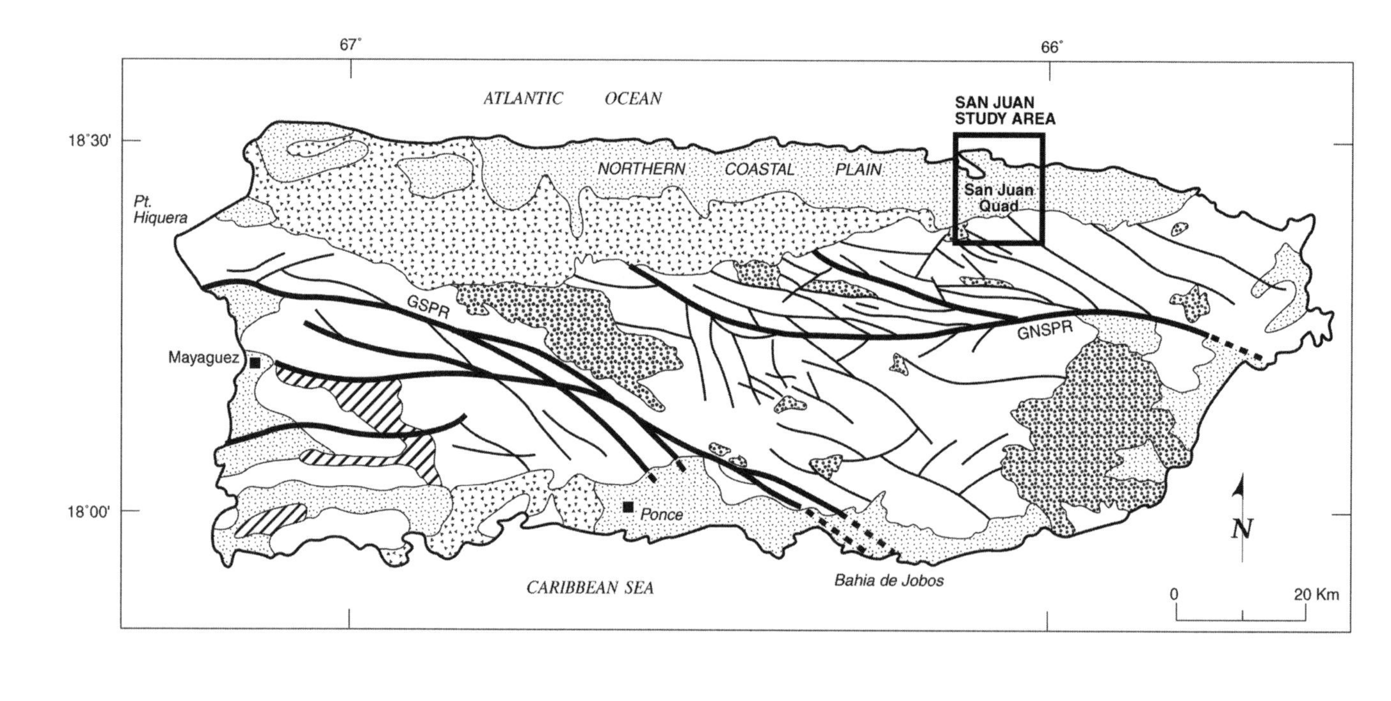

Figure 1. San Juan liquefaction hazard study location and generalized geologic map of Puerto Rico (geology after Briggs, 1964).

et al., 2002). Several large-magnitude historical earthquakes demonstrate the seismic potential of the region. Major earthquakes were reported in 1670, 1787, 1867, and 1918 (McCann 1985, 2002, Doser et al., this volume). The May 2, 1787, earthquake on the Puerto Rico subduction zone produced Modified Mercalli Intensities (MMI) of VII to VIII, and caused significant damage to the northern part of the island (McCann et al., 2002). The November 18, 1867, event in the Anegada Passage produced Rossi-Forel (RF) intensity IX in the Virgin Islands, and MMI intensity VIII in eastern Puerto Rico (Reid and Taber, 1919). The October 18, 1918, M7.5 event in Mona Passage produced MMI intensity IX in western Puerto Rico and was the most destructive earthquake recorded on the island (Reid and Taber, 1919; Doser et al., this volume).

Probabilistic seismic hazard analyses (PSHA) have been conducted for Puerto Rico to estimate the probabilities that certain levels of strong ground shaking might occur during future time intervals (Crouse and Hengesh, 2003; Mueller et al., 2003). The results of the PSHAs for the San Juan area indicate that ground motions with a 10% probability of exceedance in 50 yr (~475 yr return period) are ~0.25 g, and ground motions with a 2% probability of exceedance in 50 yr (2475 yr return period) are ~0.45 g. These two ground motion estimates are well above the triggering threshold values for liquefaction of deposits in the San Juan area, as discussed in later sections of this paper. The historical seismicity and seismotectonic setting of Puerto Rico demonstrate that the opportunity exists to cause liquefaction in susceptible sediments.

GEOLOGICAL CONDITIONS IN THE SAN JUAN AREA, PUERTO RICO

The urbanized northern coastal plain of Puerto Rico (Fig. 1) is underlain by Tertiary limestone and areally extensive Quaternary sediments deposited by fluvial, marine, and eolian processes. Most bedrock units are mantled by thick residual clay soils or deeply weathered saprolite. Quaternary sediments include sands, clayey sands, sand and gravel, soft organic clay, silty clay, peat, and calcareous mud that accumulated in streams, beaches, lagoons, estuaries, and swamps (Fig. 2; Pease and Monroe, 1977). Eolian sands also are deposited along the coastline and increase the sand component of the swamp and estuarine deposits near the beach. Much of the built-up areas of the San Juan Quadrangle along lagoon or coastal margins were formed by artificial filling that locally included sluicing of hydraulic fill (Monroe, 1980; Ellis, 1976). The quality and texture of the fill vary greatly across the study area. In general, older fill along the margins of lagoons and bays is sandy hydraulic fill that was sluiced or dumped into place to form broad tracts of reclaimed land. The filling was typically accomplished using sandy material dredged from the lagoons and was placed mainly before the early 1970s, and in some cases in the late 1800s and early 1900s (Ellis, 1976). Later fill and fill placed in inland areas typically consist of mixtures of clay and sand and locally are engineered and densely compacted. These later-vintage fills therefore are less susceptible to liquefaction.

Based on the geological mapping of Pease and Monroe (1977), aerial photograph interpretation, and field reconnaissance, the Quaternary geologic history of the San Juan metropolitan area involves extensive erosion of material from the highlands and deposition of alluvium and alluvial fan complexes along the stream systems and slopes. These deposits interfinger with coastal lagoonal deposits and beach-eolian sands on the coastal plain. A combination of sea-level change and regional tectonic uplift (Horsfield, 1975; Monroe, 1968; Taggert and Joyce, 1989) caused Holocene river channels to incise deeply into broad Pleistocene flood plains. An example of this process is a large Pleistocene alluvial fan complex in the Hato Rey district that has been incised by Rio Piedras (Fig. 2). Holocene alluvium of Rio Piedras was deposited within terraces and floodplains incised ~5–10 m into the former fan surface. Holocene deposits generally appear confined to the incised channel systems, beach, estuary, and lagoonal environments.

In addition to regional tectonic uplift, Quaternary sea-level fluctuations have changed stream base levels and have influenced the development of stream systems and deposition of sediment along the northern coastal plain. Coastal streams were incised and graded to lower base levels during low stands of sea level. Paleovalleys that formed during sea level low stands were drowned and filled as sea level rose to its present elevation through the Holocene. Former beach sands were blown into dunes and sand sheets that migrated inland and stabilized as eolian sand sheets against hillsides and within topographic depressions. The bay and estuary deposits of Bahia de San Juan and Laguna San Jose, and alluvial deposits shed from the coastal mountains, now blanket much of the former low-lying landscape (Fig. 2). Swamps and mangroves that formed at lower stands of sea level were drowned and are now are marked by buried, preserved peat layers within the bay and estuary deposits.

FACTORS INFLUENCING LIQUEFACTION SUSCEPTIBILITY

Liquefaction susceptibility is controlled by a number of factors including grain size distribution (texture), soil density, depth to groundwater, and the liquefaction triggering threshold with respect to the level of ground shaking that might be anticipated for a study region (Youd et al., 2001; Seed et al., 2002). The texture and soil density are strongly controlled by depositional environment and deposit age, which allows application of Quaternary geological mapping principles to define the distribution and areal extent of potential liquefaction susceptibility map units (Youd and Perkins, 1978). Liquefaction is restricted to areas with a narrow range of geologic and hydrologic characteristics that can be identified and mapped based on established Quaternary mapping techniques (Youd and Perkins, 1978).

The depositional environment controls the texture, sorting, and packing of sediments (Youd and Perkins, 1978). For

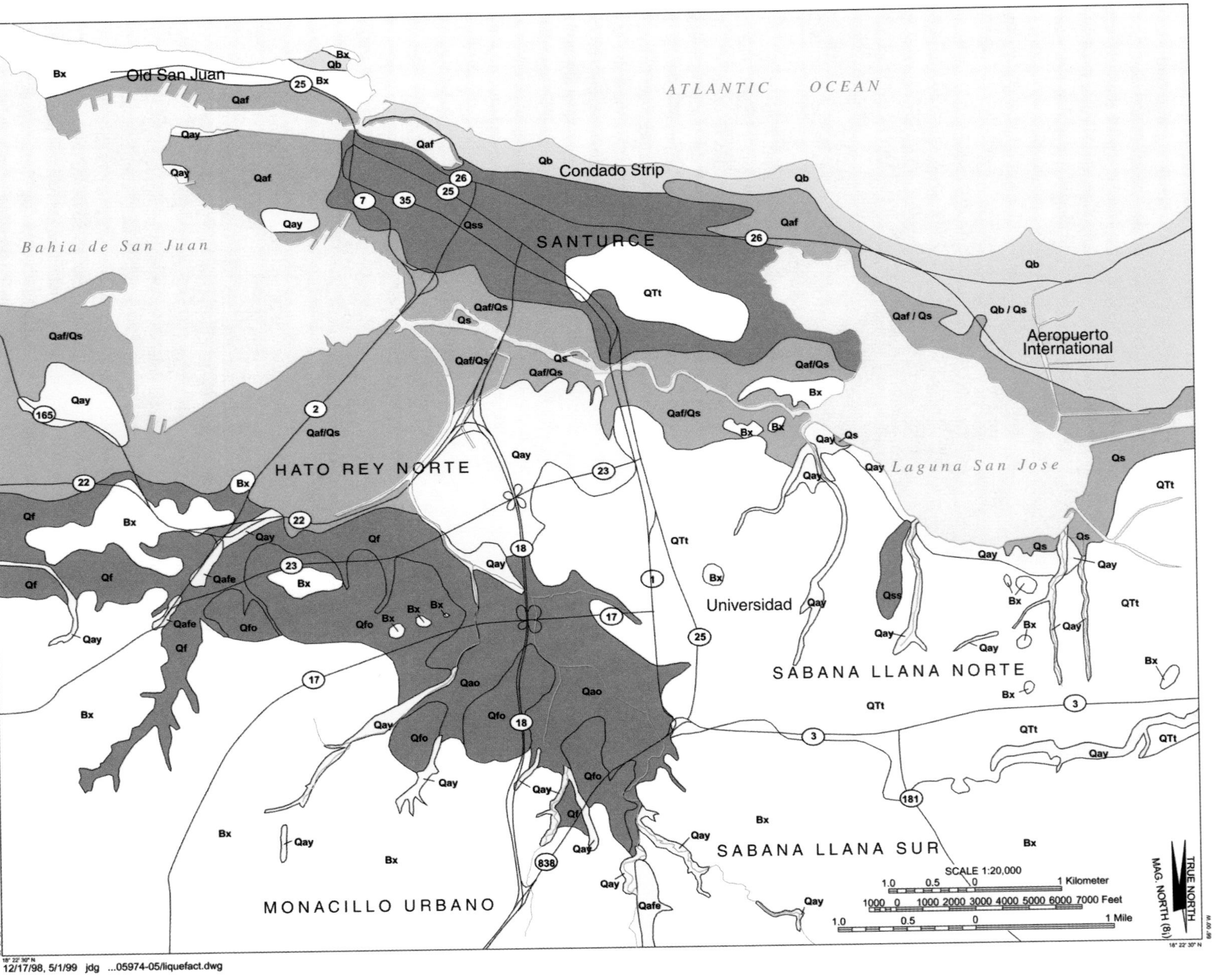

Old San Juan
ATLANTIC OCEAN
Condado Strip
SANTURCE
Bahia de San Juan
Aeropuerto International
HATO REY NORTE
Laguna San Jose
Universidad
SABANA LLANA NORTE
SABANA LLANA SUR
MONACILLO URBANO
Bx
Qb
Qaf
Qay
Qss
Qaf/Qs
Qb / Qs
Qaf / Qs
Qs
QTt
Qf
Qfo
Qao
Qafe
SCALE 1:20,000
1.0 0.5 0 1 Kilometer
1000 0 1000 2000 3000 4000 5000 6000 7000 Feet
1.0 0.5 0 1 Mile
TRUE NORTH
MAG. NORTH (8¡)
12/17/98, 5/1/99 jdg ...05974-05/liquefact.dwg

Explanation

Explanation for liquefaction susceptibility

Geologic Unit	Description	Estimated Percent Liquefiable Texture	Estimated Liquefaction Triggering Acceleration		Typical Groundwater Depth	
			M_W 6.5	M_W 8.0		
Qaf / Qs	Artificial fill over swamps	<50%	0.1g	<0.1g-0.25g	<5'	VERY HIGH - LOW See Note 2
Qs	Holocene swamp deposits	50%	0.15g	0.1g	<5'	VERY HIGH
Qb	Holocene beach deposits	80%	0.2g	0.15g	<10'	MEDIUM - HIGH
Qaf	Artificial fill	<50%	0.2g	0.15g	10-20'	MEDIUM
Qay	Holocene alluvium	40%	0.2g	0.15g	<10'	MEDIUM
Qafe	Artificial road embankment fill	<50%	0.25g	0.20g	5-20'	MEDIUM
Qf	Late Pleistocene to early Holocene fan deposits	<10%	>0.3g	0.25g	5-20'	LOW
Qfo	Mid-Pleistocene to Pliocene fan deposits	<10%	>0.3g	>0.30g	5-20'	LOW
Qao	Late Pleistocene-Pliocene Alluvium	<30%	>0.3g	>0.25g	5-20'	LOW
Qss	Late Pleistocene(?) dune sands	80%	>0.3g	>0.25g	10-30'	LOW - MEDIUM
QTt	Pleistocene alluvium	<10%	>0.3g	>0.3g	10-30'	LOW - VERY LOW
Bx	Bedrock	0%	NA	NA	10-30'	VERY LOW

Symbols

Geologic contact (location approximate)

Highways, Major Roadways

Shorelines and Streams

NOTES

1. **Triggering accelerations shown are for Peak Ground Acceleration (PGA). Estimated triggering levels are based on evaluation of borehole Standard Penetration Test data using Seed Simplified Approach (Seed and others, 1971). The triggering values should be viewed as estimates based on limited borehole data, and should be verified by site specific studies in high or very high zones. Susceptibility ratings reflect both percentage of liquefiable texture and density.**
2. **Susceptibility is dependent on density of fill, which is quite variable. For initial screening, area should be considered to have a high susceptibility until proven otherwise by additional information.**

Figure 2 (*on this and previous page*). Map showing Quaternary geology units and relative liquefaction susceptibility for the San Juan Quadrangle study area.

example, high-energy environments, such as beaches and fast-flowing rivers, preferentially sort grains, and result in a coarse-grained, densely packed deposit. Low energy environments, such as lagoons and estuaries, form predominantly fine-grained, loosely packed deposits. Deposits with higher relative densities and more stable soil structure have a lower susceptibility to liquefaction. The amount of clay in a deposit dramatically affects its susceptibility to liquefaction. In a general sense, cohesive soils that contain more than 10%–30% of plastic clay may be considered nonliquefiable (Seed et al., 2002). For this reason, Quaternary geological mapping can be applied to delineate map units formed in depositional environments characterized by high clay content or to those map units where aging effects have increased the clay content of the deposit (Birkeland, 1984). With increasing age, the relative density of a deposit may increase as particles gradually align and consolidate together and the deposit undergoes long-term drainage. The soil structure also may become more stable with age through particle reorientation or cementation. Additionally, over time, sand and silt grains are reduced in size and undergo mineralogical decomposition to clay by mechanical and chemical weathering (e.g., Mitchell, 1999). These mechanical, mineralogical, and cementation changes in the deposit are permanent effects that become "locked" into the sediment and dramatically reduce their susceptibility to liquefaction. In some cases, older soils have undergone burial and compaction, followed by a period of uplift and erosional unloading. The process of burial and exhumation causes overconsolidation of the soils, which also greatly reduces their susceptibility to liquefaction (Youd et al., 2003).

Quaternary geological and geomorphological mapping provides an important first step in assessing liquefaction susceptibility by identifying geologic deposits whose age and textural characteristics are most susceptible to liquefaction. Most liquefaction occurs in areas of poorly engineered hydraulic fills and in late Holocene fluvial deposits (e.g., Pyke, 2003). Therefore, our Quaternary and geomorphologic mapping for the San Juan Quadrangle concentrated on identification and microzonation (where possible) of artificial fill and latest Holocene deposits. We also estimated the relative age of other Quaternary deposits to form an age-based susceptibility ranking system shown on Figure 3. We use the age of deposits as

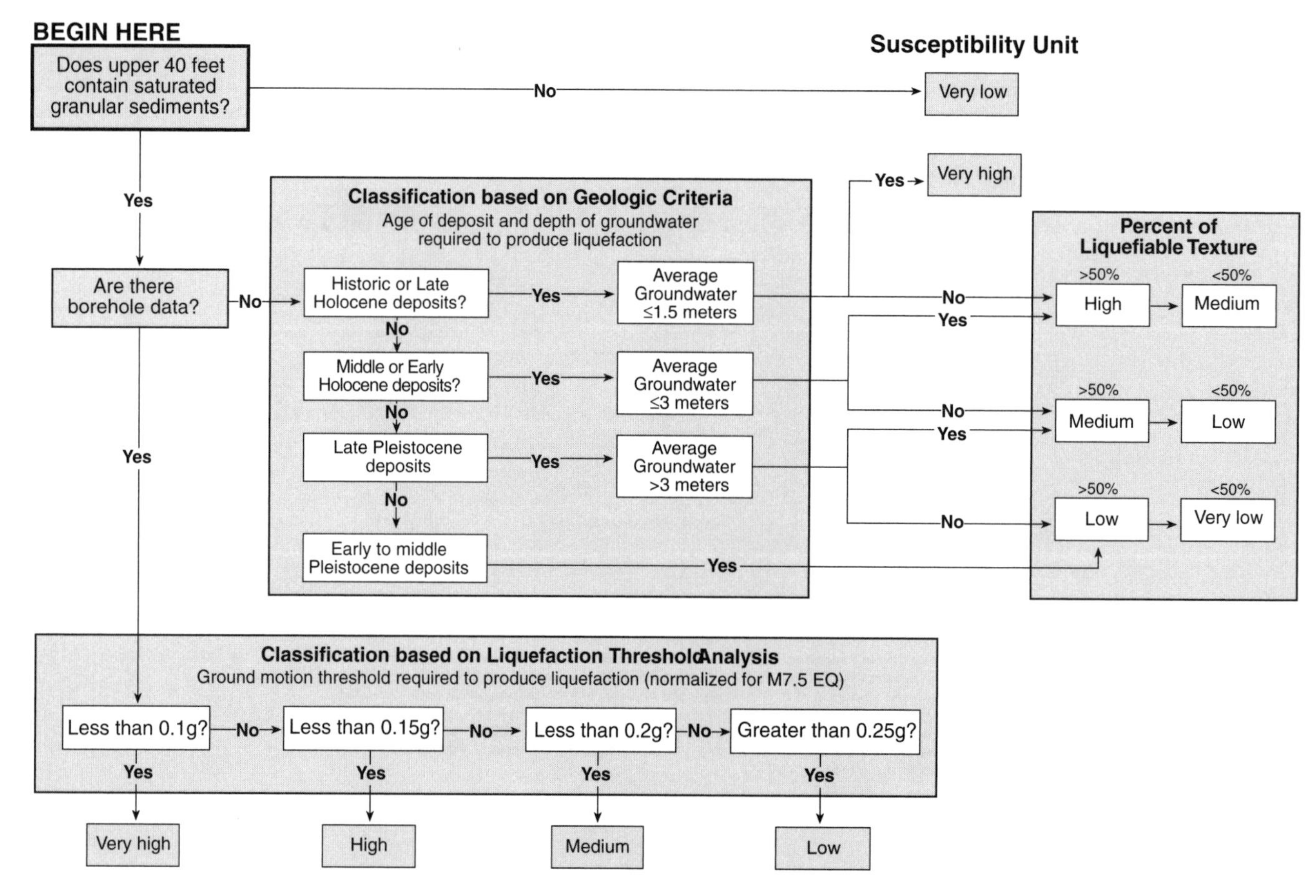

Figure 3. Decision tree used to develop relative ranking of liquefaction susceptibility units.

one criterion to establish the relative liquefaction susceptibility of geologic units (Youd and Perkins, 1978; Youd, 1991; Tinsley et al., 1985; Bachhuber et al., 1994; Hitchcock et al., 1999). Of critical importance is the differentiation of geologic deposits of Holocene and Pleistocene age. In general, historic occurrences of severe liquefaction were largely or wholly restricted to hydraulic fills and natural alluvial deposits of late Holocene age (e.g., Pyke, 2003).

Depth to groundwater is a significant factor governing liquefaction susceptibility. Saturation reduces the normal effective stress acting on loose, sandy sediments. This condition, particularly in the upper 20 m of the ground surface, increases the likelihood of liquefaction and resulting ground failure (Youd and Perkins, 1978). Because groundwater levels may vary due to seasonal variations and historic groundwater use, we used the highest reasonable water levels recorded in borings and wells for the liquefaction susceptibility analysis.

The level of strong ground shaking anticipated for a study region is an important consideration for specifying liquefaction susceptibility classes. If the shaking triggering threshold of a deposit exceeds the maximum anticipated ground motions for a low seismicity region, this deposit may be assigned a low susceptibility classification. However, if the same triggering threshold for a deposit with similar geotechnical properties is exceeded in a more seismically active region, then this deposit should be assigned a higher susceptibility rating. In preparing the maps of the San Juan Quadrangle, we have evaluated the liquefaction triggering thresholds for each of six geological map units in light of the regional seismic context, and compared these triggering thresholds to three levels of ground shaking: 0.1 g, 0.2 g, and 0.3 g.

DEVELOPMENT OF LIQUEFACTION SUSCEPTIBILITY MAPS

The liquefaction susceptibility mapping process involves five steps: (1) creating detailed Quaternary geologic maps delineating deposits of various age, depositional environment, and texture; (2) evaluating Quaternary deposit thickness and depth to groundwater; (3) performing an initial evaluation of relative liquefaction susceptibility using a decision tree process; (4) evaluating liquefaction triggering thresholds using geotechnical borehole data and the Seed and Idriss (1971b) "simplified procedure"; and, (5) identifying units of similar susceptibility and grouping them to form liquefaction susceptibility zones. Figure 4 illustrates the integration of the three primary map layers (topographic map, Quaternary geology map, depth to groundwater map) and geologic and geotechnical analyses to form the final susceptibility map shown on Figure 2.

The distribution of Quaternary map units was refined and augmented with geotechnical borehole data. Because the lithologic and engineering properties of sediments often vary significantly both laterally and vertically, it is necessary to interpret the available surface and subsurface data to explain these variations and extrapolate borehole data to areas within similar map units that lack borehole data. The surficial geologic mapping provides the means to improve correlations among subsurface data and increase confidence in the distribution of susceptibility units.

LIQUEFACTION SUSCEPTIBILITY ANALYSIS

Quaternary Geologic Mapping

The Quaternary geologic map was developed by compiling surficial geological and soils data, analyzing subsurface boring logs, interpreting aerial photographs, and conducting field verification of unit boundaries and descriptions. Data sources for our map analysis included

- USGS geological map of the San Juan Quadrangle (scale 1:20,000) (Pease and Monroe, 1977);
- Geology of the Middle Tertiary Formations of Puerto Rico (Monroe, 1980);
- USDA Soil Conservation Service, Soil Survey of the Arecibo Area (Acevido, 1982);
- Generalized Earthquake Induced Geologic Hazard Map for the San Juan metropolitan area at a scale of 1:40,000 (Molinelli-Freytes, 1985); and
- 224 unpublished logs of geotechnical boreholes and water wells from the Puerto Rico Department of Transportation and Highways, Tren Urbano Project, and USGS files.

The published USGS quadrangle map (Pease and Monroe, 1977) differentiates stream channel, beach, swamp, dune, fluvial and marine terrace, alluvial fan, and artificial fill deposits. Approximate geologic ages or relative ages are assigned to the units and include information on general deposit thicknesses.

Quaternary map units for the San Juan Quadrangle (Pease and Monroe, 1977) were reassessed to confirm unit boundaries and textural characteristics, age, and environment of deposition. Our reevaluation consisted of (1) comparing geologic unit boundaries with stereographic interpretation of USGS (1962) aerial photographs; (2) field reconnaissance of map units; and (3) analysis of borehole stratigraphy. Quaternary deposits and geomorphic surfaces were evaluated on the basis of several stratigraphic, geomorphic, and pedologic criteria, including (1) topographic position in a sequence of inset deposits or surfaces; (2) relative degree of surface modification (e.g., erosional dissection); (3) relative degree of soil-profile development and other surface weathering phenomena; (4) superposition of deposits separated by erosional unconformities and/or buried soils; (5) relative ages of individual deposits; and (6) physical continuity and lateral correlation with other stratigraphic units. The resulting Quaternary geologic map developed for this study and used for our liquefaction susceptibility analysis is shown on Figure 2. Minor modifications to the map unit boundaries of Pease and Monroe (1977) were made, including differentiation of older and younger alluvial deposits, differentiation of older and younger fan deposits, and adjustment of some contact locations. The geologic data were compiled at a scale of 1:20,000.

Evaluation of Quaternary Deposit Thickness and Depth to Groundwater

Subsurface data from 224 exploratory boring logs and well logs were used to assess the texture and relative density of deposits, thickness of Holocene and Pleistocene sediments, and depth to groundwater. Data from the boring and well logs were used to define the unconformity that marks the Holocene-Pleistocene boundary and to estimate the thickness of potentially liquefiable Holocene sediments. In the San Juan area, the base of Holocene deposits typically is an irregular erosional surface recognized in geotechnical borings by a marked increase in SPT blow counts and in well logs by a lithologic change. Older deposits are generally finer grained than the overlying Holocene sediments and represent a different paleo-landform or sedimentary environment than the modern setting. For example, coastal swamp deposits adjacent to Bahia de San Juan are within a modern broad topographic basin eroded into the distal parts of older coalescing alluvial fans and alluvial terraces. The swamp environment is conducive to deposition of loose, sandy and silty layers, and the composite thickness of the swamp deposits ranges between ~5 and 15 m. A buried peat soil often marks the base of the Holocene deposits, but the thickness and depth to the peat layer varies from boring to boring. The underlying Pleistocene fan and alluvial deposits are more dense and typically do not contain loose sediments susceptible to liquefaction.

Groundwater levels used in our liquefaction susceptibility analysis were assessed from geotechnical borings, well logs, and the topographic position of deposits relative to perennial streams and sea level. The depth to groundwater varied between map units, primarily due to differences in the elevations of landforms formed by the deposits and proximity to streams or water bodies. For example, Holocene swamp deposits (Qs) occur at or below mean sea level, and older Tertiary or Quaternary Terrace alluvium (QTt) lies several meters above sea level. However, groundwater typically exists at depths less than ~3–7 m across the entire study area and is <1.5 m deep under the coastal and lowland swamp areas. For the purposes of our liquefaction susceptibility analysis, we conservatively assumed that groundwater can rise to <1.5 m across the entire study area.

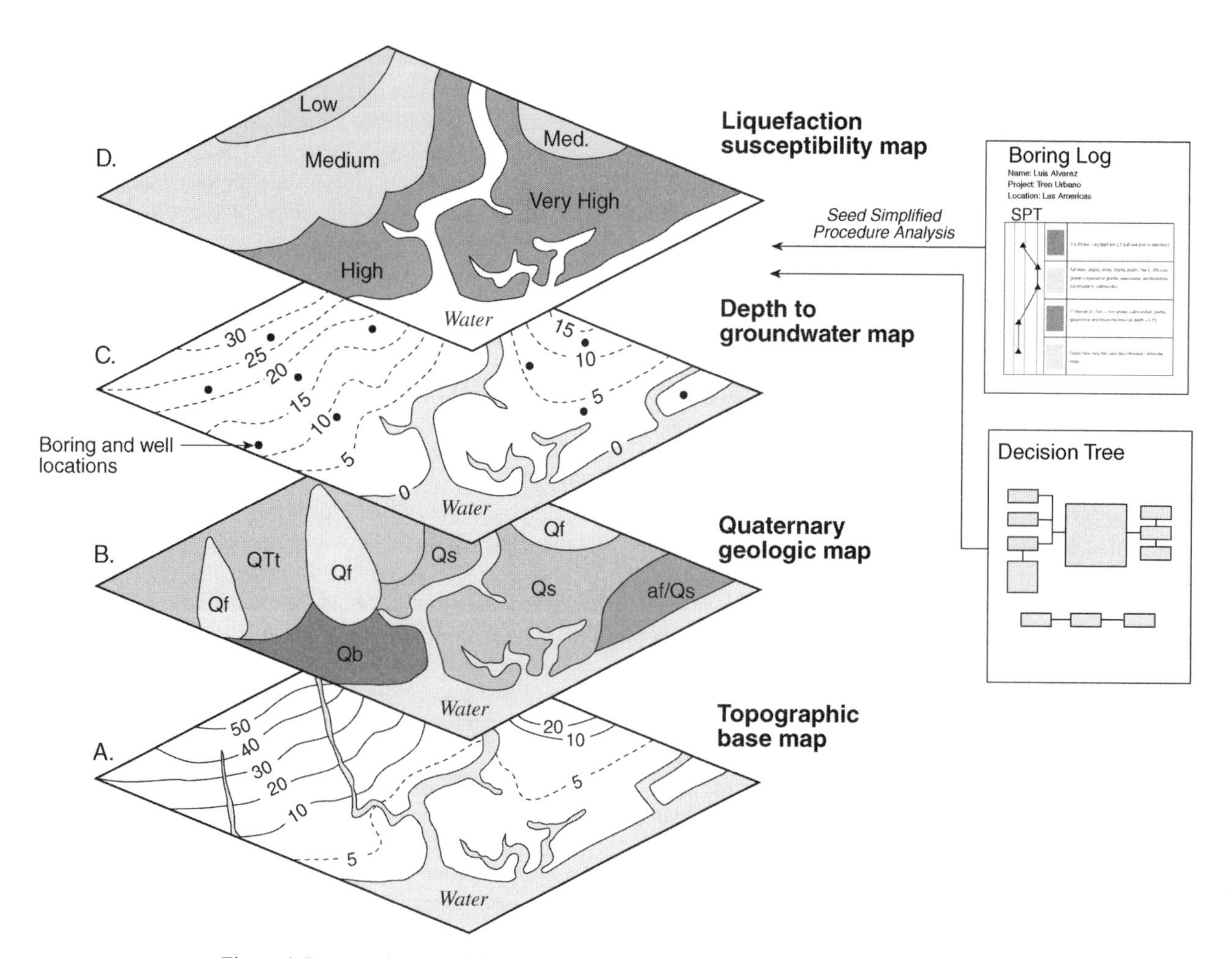

Figure 4. Integrated map and data layers used to create the liquefaction susceptibility map.

Preliminary Liquefaction Susceptibility Evaluation with a Decision Tree

The liquefaction susceptibility of each map unit was initially estimated using the decision tree shown on Figure 3. The decision tree was developed specifically for the San Juan study area and incorporates important local geologic and groundwater occurrence factors that influence liquefaction susceptibility. This decision tree was modified from a similar one adopted by the California Geological Survey (Hitchcock et al., 1999). The results from the decision tree analyses were used to assign a susceptibility classification to each map unit that lacked sufficient subsurface data to quantitatively assess liquefaction triggering thresholds. The decision tree assigns the very high susceptibility classification to late Holocene or modern deposits and the high susceptibility classification to middle and early Holocene deposits. Older deposits are assigned to susceptibility classes that range from very low to medium, depending on the percent liquefiable texture, groundwater conditions, and SPT simplified procedure (Seed and Idriss, 1971b) triggering levels. The percent liquefiable texture within a geologic unit was used to differentiate the susceptibility of more clay-rich deposits from deposits that lack clay and contain significant percentages of sand or silt. The percent liquefiable texture for each geologic unit was estimated on the basis of literature-reported characteristics (e.g., Pease and Monroe, 1977), review of geotechnical boring and well logs, and field examination of deposits in stream banks and road cuts.

The decision tree also was used to define susceptibility classes where sufficient subsurface data were available to characterize liquefaction triggering threshold values. As described next, the liquefaction threshold analysis was based on the simplified procedure (Seed and Idriss, 1971b; Youd et al., 2001).

Evaluation of Borehole Data

Geotechnical borehole data were compiled to document subsurface stratigraphy, texture, soil density, and groundwater conditions for input to quantitative analysis of liquefaction susceptibility. A total of 141 geotechnical borings and 83 water well logs were evaluated. In general, the geotechnical borings include detailed soil descriptions and information on soil density, whereas the well logs generally contain nontechnical stratigraphic descriptions and are mainly useful in determining the depth to bedrock and major geologic stratigraphic units.

The geotechnical borings include SPT data for soils at various intervals throughout the boreholes and also typically indicate depth to groundwater. The SPT is a standard method to evaluate the geotechnical properties of a soil deposit. The SPT data are obtained by driving a hollow sampler of standardized dimensions into the bottom of a borehole with standardized hammering equipment and recording the hammer blows ("*N* count") required to drive the sampler in 6-inch increments for a total drive length of 18 inches. The resulting SPT *N* count is recorded by totaling the blows required to drive the sampler the last 12 inches of penetration. The SPTs in the San Juan Quadrangle were obtained by local drilling companies using standardized 140-pound hammers with 30-inch drop heights. Typically, cathead or wireline hammer triggering equipment was used.

The simplified procedure was used to quantitatively assess the susceptibility of a deposit to liquefaction given various levels of ground shaking (Seed and Idriss, 1971b; Seed et al., 1983; Seed and Harder, 1990). The simplified procedure is an empirically based method for silty and sandy sediments used to compute the peak ground acceleration (PGA) that could trigger liquefaction (triggering threshold) in susceptible sediments and considers the percent fines content, groundwater conditions, overburden loads, SPT-based density correlations, and earthquake loading conditions (cyclic stress ratio, CSR). We note that the San Juan liquefaction susceptibility study was performed between 1996 and 1998 and incorporated the various simplified procedure correlations and methodologies that were standard practice at that time. The simplified procedure has since been modified (e.g., Youd et al., 2001; Idriss and Boulanger, 2004), but the more recent modifications do not affect the relative liquefaction susceptibility classifications established for our study.

The liquefaction triggering thresholds were estimated for three scenario earthquakes that are considered likely in the region. These included M_W 8.0 and M_W 7.5 events located along the Puerto Rico subduction zone, and a M_W 6.5 event located onshore in proximity to San Juan. In the analyses, we also specified three levels of PGA that might be produced from these scenario earthquakes. The ground motions considered are 0.1 g, 0.2 g, and 0.3 g. These ground motion values were based on the tectonic environment of the San Juan region and are considered realistic values with a moderate to high probability of occurrence. In addition to the evaluation of different PGA levels, the effects of different strong ground shaking durations from the range of scenario earthquake magnitudes were analyzed using magnitude correction factors (Seed et al., 1983; Seed and Harder, 1990). For equivalent PGA levels, the increased duration from larger magnitude earthquakes causes an increased potential for liquefaction.

The liquefaction susceptibility analysis for each map unit was performed by calculating whether or not a deposit from the uppermost 20 m of sediment would liquefy for the three scenario ground motions produced by the three distinct scenario earthquakes. We compiled geotechnical data for each geologic map unit into separate Excel worksheets and performed the simplified procedure analyses specifically for each map unit in conjunction with the decision tree evaluation (Fig. 3). This allowed comparison of the liquefaction susceptibility and PGA triggering threshold of distinct map units to establish the relative susceptibility ranking (Fig. 3).

Figure 5 shows summary liquefaction susceptibility plots from the simplified procedure analysis for six map units based on the M_W7.5 scenario earthquake. The plotted SPT data are for deposits described as "silt" or "sand" on the geotechnical boring for the Holocene swamp (Qs), Holocene beach (Qb), Holocene

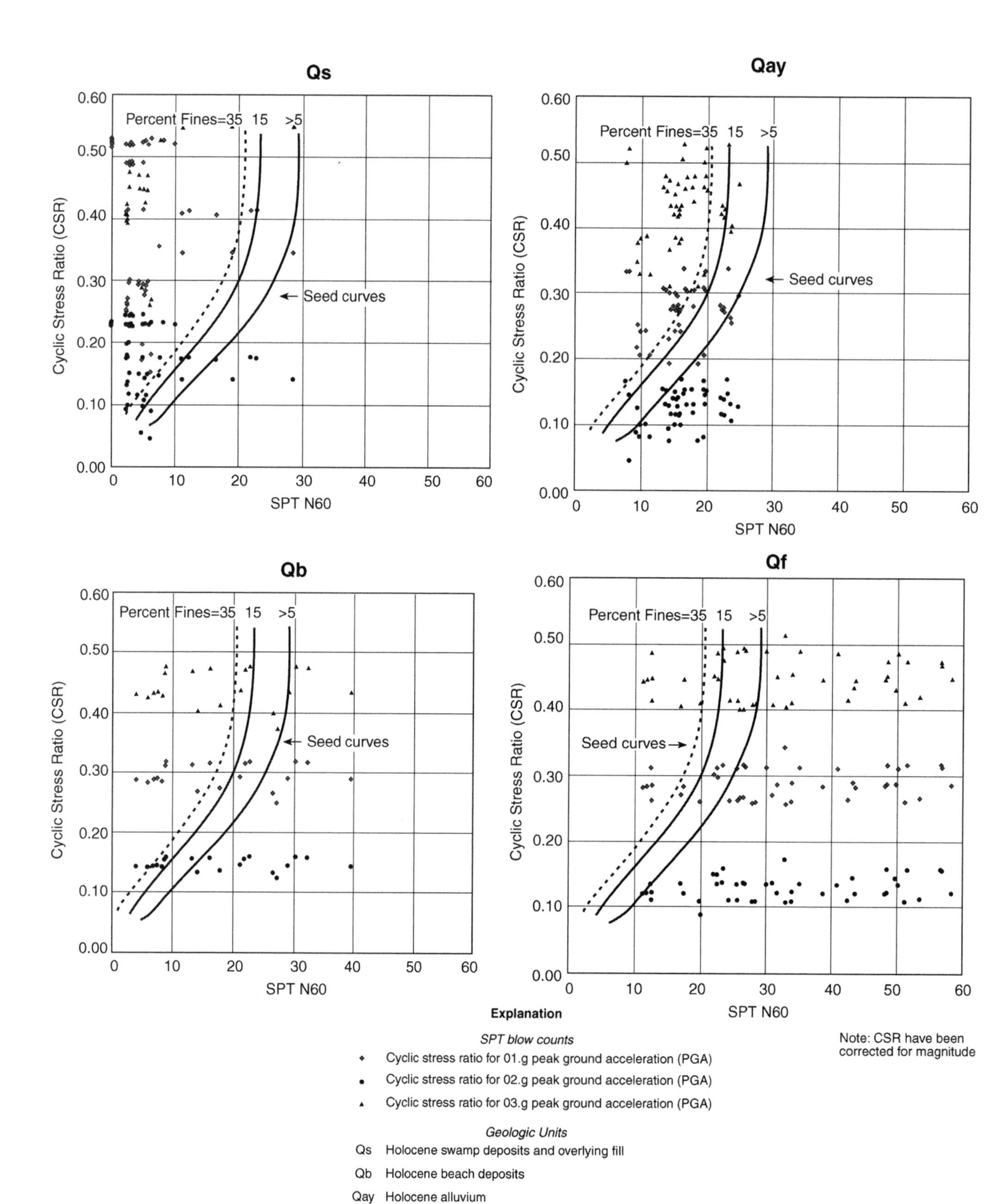

Figure 5 (*on this and following page*). Summary plots of Standard Penetration Test (SPT) blow count data versus cyclic stress ratio (CSR) for various geologic units within the San Juan study area for a M_W 7.5 earthquake. Potentially liquefiable sediments are indicated by blow count points that plot to the left of the reference curves (modified from Seed and Idriss, 1971b).

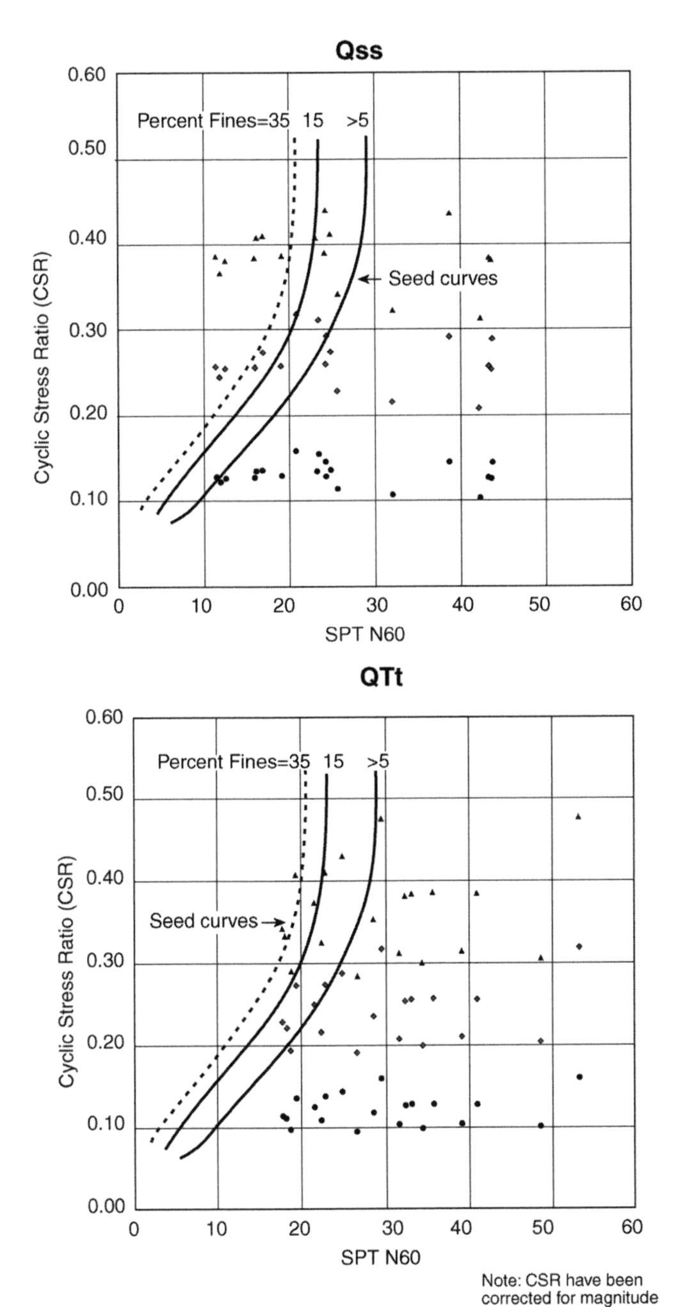

Figure 5 (*continued*).

alluvium (Qay), early Holocene–late Pleistocene silica sand (Qss), Pleistocene alluvial fan and valley fill (Qf), and Pleistocene-Pliocene alluvium (QTt). The summary plots show the amount and variability of SPT data from each map unit. The SPT data are plotted in three populations that represent the 0.1 g, 0.2 g, and 0.3 g input ground acceleration. The plots also show three curves ("Seed curves") that are used to evaluate whether a sample would liquefy or not based on the percent fines (silt clay) fraction of the soil. Points that plot to the left of a seed curve have exceeded the liquefaction triggering threshold and are potentially liquefiable, and points to the right are below the liquefaction triggering threshold and are not expected to liquefy.

The three populations represented on the plots (e.g., 0.1 g, 0.2 g, and 0.3 g) allow a quantitative evaluation of the ground acceleration triggering level. By visually inspecting the SPT blow count clusters, we determine the PGA level where the mean value of the cluster plots to the left of the seed curve. This PGA value is then selected as the general PGA triggering level for that map unit. Additionally, visual comparison of the Figure 5 plots allows evaluation of the relative liquefaction susceptibility of the various map units. For example, the plot for young alluvium (Qay) on Figure 5 indicates that for 0.1 g the majority of the data points fall to the right of the liquefaction boundary curves and therefore this level of PGA would not trigger liquefaction in the deposit. However, for ground motions of 0.2 g or greater, most data points fall to the left of the boundary curves, and therefore the deposit could be triggered into liquefaction. The estimated triggering PGA values for each map unit were used to augment the liquefaction susceptibility classifications (e.g., high, medium, and low) presented on the decision tree (Fig. 3) and on the liquefaction susceptibility map (Fig. 2).

Results—Liquefaction Susceptibility Map

The liquefaction susceptibility map shown on Figure 2 illustrates the general susceptibility of the San Juan area to liquefaction. This map was prepared by integrating the various map and data layers as shown on Figure 4, the susceptibility ranking decision tree shown on Figure 3, and the simplified procedure analyses described above and shown on Figure 5. The susceptibility analysis considered M_W 6.5, M_W 7.5, and M_W 8.0 scenario earthquakes and estimated PGA values of 0.1 g, 0.2 g, and 0.3 g. Five relative susceptibility zones were established: very high, high, medium, low, and very low. Each of these zones has different estimated PGA liquefaction triggering levels (Figs. 2 and 3).

In general, zones ranked as very high and high susceptibility could produce widespread and severe liquefaction under low to moderate to strong earthquake shaking and possible minor to moderate liquefaction under moderate levels of earthquake shaking. Zones ranked as low and very low susceptibility likely would either experience no significant liquefaction or isolated and relatively minor liquefaction even under very strong earthquake shaking. The medium susceptibility zones likely would experience isolated and restricted zones of severe to moderate liquefaction under strong earthquake shaking and only very minor and sparse liquefaction under moderate levels of earthquake shaking.

Very High Susceptibility Zones

The very high susceptibility zones comprise modern swamp deposits (Qs) and areas of artificial fill over swamp deposits (Qaf/Qs) in the reclaimed areas around Bahia de San Juan and Laguna San Jose (Fig. 2). Silt and sand lenses in these deposits have estimated PGA liquefaction triggering thresholds of 0.03–0.1 g and have an estimated percent liquefiable texture of ~50%. Groundwater in these deposits is generally encountered within ~1.5 m of the ground surface. These deposits occur in lagoons or under filled land along lagoon margins and generally consist of soft, unconsolidated, saturated lagoonal sediment, including silt, clay, peat, and discontinuous sand lenses. Borings show that the silt and sand composition of lagoonal deposits is higher near the mouths of streams that discharge into lagoons and where sand is blown inland from beaches and sand spits along lagoon coastal margins. However, silt and sand lenses can occur throughout the swamp deposits. Artificial fill overlying swamp deposits along former lagoon margins (Qaf/Qs) is commonly saturated and varies locally in composition and density. The liquefaction susceptibility of artificial fill ranges from low (coarse, rocky clay fill) to very high (fine sand and silt hydraulic fill) and conservatively should be considered to be very high where fill overlies swamp deposits until proven otherwise by site-specific studies. Liquefied silt and sand lenses in swamp deposits underlying fill also can cause excessive settlement, lateral spreading, oscillation, and fissuring of non-liquefied overlying fill.

Although the Qs and Qaf/Qs deposits have the highest susceptibility to liquefaction in the region, the surface manifestations of liquefaction within the swamp deposits may vary due to the presence of lenticular sand bodies, areas of clayey or fine silt, and/or peat deposits. Surface expression of liquefaction is likely to vary considerably within these units. We conservatively estimate that liquefaction in the most susceptible lagoonal units may be expressed over 25%–50% of the surface area. Liquefaction occurrence and effects likely will be concentrated, and of greatest magnitude, within 100 m of the coastal margin and 50 m of stream channels or free faces.

High Susceptibility Zone

The high susceptibility zone consists of Holocene beach deposits (Qb), which are comprised primarily of sand and silty sand (estimated sand and silt percentages of ~85%), and occur in low areas with a relatively high groundwater table, generally less than ~3 m deep. These units are assigned a medium to high liquefaction susceptibility rating with a PGA triggering threshold of 0.15–0.2 g. We conservatively estimate that liquefaction may occur over 25% of this map unit.

Medium Susceptibility Zone

The medium susceptibility zone consists of Holocene alluvium (Qay), such as channel sediments or low floodplain and terrace deposits along active streams or lagoonal margins, and engineered fill in inland areas (Qafe). The Holocene alluvium typically consists of interbedded clay, silt, and sand with some gravel lenses, with an estimated PGA triggering threshold of 0.15–0.2 g. The percent liquefiable texture ranges between ~40% and 50%, and groundwater levels are ~1.5–7 m deep. Portions of the beach deposits (Qb) also fall within the medium susceptibility zone, although these have conservatively been assigned to the high-susceptibility zone, as described above.

We estimate that liquefaction in the medium susceptibility zones will be localized, probably not exceed 10% of the mapped unit, and be concentrated adjacent to stream channels.

Low Susceptibility Zone

The low susceptibility zone includes Pleistocene-aged alluvium and alluvial fan deposits (Qf, Qfo, Qao), and the Santurce Sands (Qss). Analysis of borehole data suggests that units within this zone have PGA liquefaction triggering thresholds of 0.25–0.3 g and groundwater levels of ~2–7 m depth. These deposits exhibit percent liquefiable textures of ~10%–30%, with the exception of the Santurce Sands that have a percent liquefiable texture of ~80%. The Santurce Sands exhibit a high degree of packing and possible weak ferruginous cementation that should preclude development of extensive liquefaction. If liquefaction occurs in the low susceptibility zone, it likely would be of very limited areal extent (less than ~5% of map area). We would not expect significant settlement or lateral spreads in this unit.

Very Low Susceptibility Zone

The very low susceptibility zone is comprised of Pleistocene alluvium (QTt) that occurs in fans and terraces along the southern edge of the San Juan basin and mapped bedrock areas. These materials exhibit dense grain packing and a low (QTt) to high (bedrock) degree of lithification. Groundwater typically is between 3 and 10 m deep. Analysis of borehole data indicates a PGA liquefaction triggering threshold of over 0.3 g for unlithified sandy sediments within these deposits; however, we believe that aging effects and overconsolidation of these deposits would prevent development of liquefaction.

DISCUSSION

Our regional susceptibility ranking for the San Juan Quadrangle can be compared against relative susceptibility zonation maps used for other geographic areas by comparing the estimated threshold PGA triggering levels. Additionally, the unit-specific PGA triggering thresholds can be used to estimate which map areas could undergo liquefaction from various scenario earthquakes by overprinting the estimated PGA isoseismal contours from a specific earthquake scenario over the liquefaction susceptibility map and shading those areas where the PGA exceeds the liquefaction triggering threshold. However, liquefaction occurrence would not be uniform within the shaded area, and liquefaction would be concentrated and most severe within the zones with the lowest triggering values, rather than in zones where the estimated scenario earthquake PGA is the same as, or slightly greater than, the triggering levels. For the purposes of general planning, the potential area of liquefaction within each triggered zone can be estimated by using the percent area estimates by zone (described above) and by shading the areas within the very high and high susceptibility zones that are within 100 m of shorelines, stream channels, and drainage channels (free face zones). These areas should be the most susceptible to liquefaction effects.

Most liquefaction-induced damage is caused by differential ground settlement, surface cracking, loss of bearing capacity, and lateral spread movements (Seed and Idriss, 1971a). The occurrence of liquefaction does not always cause these effects, particularly if the liquefaction occurs on level ground and the liquefied layer is deeply buried. For example, liquefaction-induced ground failure during the 1999 Kocaeli earthquake in Turkey and 1995 Kobe earthquake in Japan typically were restricted to areas where liquefied layers are thick and occur at shallow depth, and within ~100 m of shorelines and free faces such as stream banks and coastlines (e.g., Bardet and Seed, 2001). Microzonation of these areas and quantification of liquefaction-induced effects requires dense borehole data and specific engineering analyses that are beyond the scope of this regional susceptibility mapping study. However, general estimates of potential settlement and lateral spread movement can be made for the various susceptibility map zones using the scenario-based PGA triggering thresholds, estimated thickness of deposits, and percentage of liquefiable texture listed in Figure 3. Quantitative estimates of the potential settlements and lateral spreads can be developed with empirically based methods developed by Ishihara and Yoshimine (1992) to predict magnitude of settlement, and by Bartlett and Youd (1995) to predict amount of lateral spread movement.

CONCLUSIONS

Liquefaction susceptibility maps were prepared for the San Juan Quadrangle, Puerto Rico, through integration of Quaternary geological data with subsurface data from geotechnical borings and well logs. The hazard characteristics of each significant map unit were established so that geological map units of similar hazard rating could be combined to form relative liquefaction susceptibility hazard zones.

The resulting liquefaction susceptibility map (Fig. 2) shows the relative hazard across the greater San Juan metropolitan area. The liquefaction susceptibility map shows that the areas of very high hazard occur along the edges of Bahia de San Juan and Laguna San Jose. Extensive swamp deposits and artificial fill over swamp deposits occur in these areas. These deposits contain layers of young, saturated, and cohesionless silt and sand. Damage to structures or lifeline facilities from liquefaction in these areas may be extensive. The high liquefaction hazard zone includes the beach and areas in the vicinity of the airport. Most of the area to the south of Bahia de San Juan and Laguna San Jose lie within medium, low, or very low hazard zones.

In general, the areas of very high hazard for both liquefaction and ground shaking amplification lie around the margins of the Bahia de San Juan and the Laguna San Jose. This finding is consistent with previous hazard maps of the area (Molinelli-Freytes, 1985). However, the quantitative basis of this study has led to significant reduction in the areas previously considered to be of high to very high hazard for both liquefaction susceptibility and ground shaking amplification.

Special consideration of the effects of liquefaction and ground shaking amplification should be given during the site selection, design, and construction of structures and underground improvements in the high and very high hazard zones. The presence of these zones should not prevent development, but rather should be used as an initial indication of subsurface conditions that may warrant site-specific investigation. Development within these types of hazard zones is becoming routine engineering practice and the hazards associated with liquefaction can generally be mitigated or avoided.

ACKNOWLEDGMENTS

This research was supported by the U.S. Geological Survey, Department of the Interior, under U.S. Geological Survey award number 1434-HQ-96-GR-02765. The views and conclusions in this document are those of the authors and should not be interpreted as necessarily representing the official policies, either expressed or implied, of the U.S. Government. Numerous people and agencies helped with data compilation for the study, including Mr. Richard Webb of the Bayamon Office, U.S. Geological Survey; Dr. Juan Carlos Moya; Director Ariel Perez and (past director) Jose Hernandez of the Autoridad de Carreteras y Transportación, Puerto Rico; Mr. Mariano Vargas of the Civil Defense Office, Puerto Rico; and the late Dr. Jose Capacetti. Data integration into the Puerto Rican Office of Governor Planning Board GIS database was facilitated by analyst Rebecca de la Cruz Perez. Dr. Paul Mann (University of Texas Institute for Geophysics), Mr. Keith Knudsen (California Geological Survey), and Dr. Clark Fenton (URS Corporation) provided very helpful and constructive review comments for this paper, and Dr. Stephen Thompson performed a final review. Mr. Patrick Drouin assisted with data compilation and analyses, and Mr. Jason Holmberg developed the paper illustrations.

REFERENCES CITED

Acevido, G., 1982, Soil survey of Arecibo area of northern Puerto Rico: U.S. Department of Agriculture, Soil Conservation Service, 88 p.

Bachhuber, J.L., Thompson, S.C., Noller, J.S., and Lettis, W.L., 1994, Liquefaction susceptibility mapping of the San Francisco Bay Area, California [abs.]: Geological Society of America Annual Meeting Abstracts with Programs, v. 26, no. 7, p. A252.

Bardet, J.P. and Seed, R.B., chapter coordinators, 2001, Soil liquefaction, landslides, and subsidences, *in* Kocaeli, Turkey, earthquake of August 17, 1999, reconnaissance report: Earthquake Engineering Research Institute Earthquake Spectra, Supplement A to Volume 16, p. 141–162.

Bartlett, S.F., and Youd, T.L., 1995, Empirical prediction of liquefaction-induced lateral spreads: Journal of Geotechnical Engineering, American Society of Civil Engineers, v. 121, no. 4, p. 316–329.

Birkeland, P.W., 1984, Soils and Geomorphology: New York, Oxford University Press, 372 p.

Briggs, R.P., Gelabert, P.A., Jordan, D., Aguilar, E., Alonso, R.., and Valentine, R.M., 1980, Engineering geology in Puerto Rico: Annual Meeting, Association of Engineering Geologists Field Trip no. 6, p. 2–7.

Crouse, C.B., and Hengesh, J.V., 2003, Probabilistic seismic hazard analysis for Puerto Rico: Seismological Society of America, 73rd Annual Meeting Program with Abstracts, Seismological Research Letters, v. 74, no. 2, p. 206.

Ellis, S.R., 1976, History of dredging and filling of lagoons in the San Juan area, Puerto Rico: U.S. Geological Survey Water Resources Investigations no. 38-76, 25 p.

Hitchcock, C.S., Lloyd, R.C., and Haydon, W.D., 1999, Mapping liquefaction hazards in Simi Valley, Ventura County, California: Association of Engineering Geologists, Environmental and Engineering Geosciences, v. 5, no. 4, p. 441–458.

Horsfield, W.T., 1975, Quaternary vertical movements in the Greater Antilles: Geological Society of America Bulletin, v. 86, p. 933–938.

Idriss, I.M., and Boulanger, R.W., 2004, Semi-empirical procedures for evaluating liquefaction potential during earthquakes: 11th International Conference on Soil Dynamics and Earthquake Engineering and 3rd International Conference on Earthquake Geotechnical Engineering, January 7–9, 2004: Berkeley, University of California, p. 1–21.

Ishihara, K., and Yoshimine, M., 1992, Evaluation of settlements in sand deposits following liquefaction during earthquakes: Soils and Foundations, v. 32, no. 1, p. 173–188.

Mann, P., Calais, E., Ruegg, J.C., DeMets, C., Jansma, P.E., and Mattioli, G.S., 2002, Oblique collision in the northeastern Caribbean from GPS measurements and geological observations: Tectonics, v. 21, no. 6, p. 1057, doi: 10.1029/2001TC001304.

Masson, D.G., and Scanlon, K.M., 1991, The neotectonic setting of Puerto Rico: Geological Society of America Bulletin, v. 103, p. 144–154, doi: 10.1130/0016-7606(1991)103<0144:TNSOPR>2.3.CO;2.

McCann, W.R., 1985, Earthquake hazards of Puerto Rico and the Virgin Islands: U.S. Geological Survey Open File Report 85-731, p. 53–72.

McCann, W.R., 2002, Microearthquake data elucidate details of the Caribbean subduction zone: Seismological Research Letters, v. 73, no. 1, p. 25–32.

Mitchell, J.K., 1999, Fundamentals of soil behavior: New York, John Wiley & Sons, Inc., 437 p.

Monroe, W.H., 1968, High-level Quaternary beach deposits in northwestern Puerto Rico: U.S. Geological Survey Professional Paper 600-C, p. C140–C143.

Monroe, W.H., 1980, Geology of the middle Tertiary formations of Puerto Rico: USGS Professional Paper 953, 93 p.

Molinelli-Freytes, J.M., 1985, Generalized earthquake induced geologic hazards map for the San Juan Metropolitan area, Commonwealth of Puerto Rico: U.S. Department of Natural Resources, Planning Resources Area, unpublished map, scale 1:40,000.

Mueller, C.S., Frankel, A.D., Petersen, M.D., and Leyendecker, E.V., 2003, Documentation for 2003 USGS seismic hazard maps for Puerto Rico and the U.S. Virgin Islands, http://eqhazmaps.cr.usgs.gov/html/prvi2003.html (accessed February 2005).

Pease, M.H., Jr, and Monroe, W.H., 1977, Geologic map of the San Juan Quadrangle, Puerto Rico: U.S. Geological Survey Miscellaneous Geologic Investigations Map I-1010, scale 1:20,000.

Pyke, R., 2003, Discussion of "Liquefaction Resistance of Soils: Summary Report from the 1996 NCEER and 1998 NCEER/NSF Workshops on Evaluation of Liquefaction Resistance of Soils" by Youd, T.L., Idriss, I.M. Andrus, R.D. Arango, I., Castro, G., Christian, J.T., Dobry, R., Liam Finn, W.D.L., Harder, L.F., Jr., Hynes, M.E., Ishihara, K., Koester, J.P., Liao, S.S.C., Marcuson, W.F., III, Martin, G.R., Mitchell, J.K., Moriwaki, Y., Power, M.S., Robertson, P.K., Seed, R.B., Stokoe, K.H., II": ASCE Journal of Geotechnical and Geoenvironmental Engineering, v. 129, no. 3, p. 283–284, doi: 10.1061/(ASCE)1090-0241(2003)129:3(283).

Reid, H., and Taber, S., 1919, The Porto Rico earthquake of 1918, with descriptions of earlier earthquakes: Washington, D.C., Report of the Earthquake Investigation Commission, House of Representatives document 269, 74 p.

Seed, R.B., and Harder, L.F., 1990. SPT-based analysis of cyclic pore pressure generation and undrained residual strength, *in* Duncan, M.J., ed., H. Bolton Seed Memorial Symposium Proceedings, May 1990: Vancouver, B.C., Canada, BiTech Publishers, p. 351–376.

Seed, H.B., and Idriss, I.M., 1971a, Influence of soil conditions on building damage potential during earthquakes: Journal of the Structural Division, American Society of Civil Engineers, v. 97, no. ST2, Proceedings Paper 7909, p. 639–663.

Seed, H.B., and Idriss, I.M., 1971b, Simplified procedure for evaluating soil liquefaction potential: Proceedings of the American Society of Civil Engineers, Journal of the Soil Mechanics and Foundations Division, v. 93, no. SM9, p. 1249–1273.

Seed, H.B., Idriss, I.M., and Arango, I., 1983, Evaluation of liquefaction potential using field performance data: American Society of Civil Engineers, Journal of Geotechnical Engineering, v. 109, no. 3, p. 458–482.

Seed, R.B., Cetin, K.O., Moss, R.E.S., Kammerer, A.M., Wu., J., Pestana, J.M., and Riemer, M.F., 2002, Recent advances in soil liquefaction engineering and seismic site response evaluation: Rolla, Missouri, University of Missouri, Paper no. SPL-2.

Taggert, B.E., and Joyce, J., 1989, Radiometrically dated marine terraces on northwestern Puerto Rico, *in* Larue, D.K. and Draper, G., eds., Transactions of the 12th Caribbean Conference, St. Croix, U.S. Virgin Islands: Coral Gables, Florida, Miami Geological Society, p. 248–256.

Tinsley, J.C., Youd, T.L., Perkins, D.M., and Chen, A.T.F., 1985. Evaluating liquefaction potential, *in* Ziony, J.I., ed., Evaluating earthquake hazards in the Los Angeles Region—an earth-science perspective: United States Geological Survey Professional Paper 1360, p. 263–316.

U.S. Geological Survey, 1962, February 16, 1962, aerial photographs GS-VAIH XF-6745 5-92 to 5-101; 5-120 to 5-130; 5-80 to 5-89; and 5-6 to 5-9: U.S. Geological Survey.

Youd, T.L., 1991, Mapping of earthquake-induced liquefaction for seismic zonation, *in* Proceedings of the Fourth International Conference on Seismic Zonation, August 1991: Stanford, California, Stanford University, Earthquake Engineering Research Institute, p. 111–138.

Youd, T.L., and Perkins, D.M., 1978, Mapping liquefaction-induced ground failure potential: Journal of Geotechnical Engineering, v. 104, no. 4, p. 433–446.

Youd, T.L., Idriss, I.M., Andrus, R.D., Arango, I., Castro, G., Christian, J.T., Dobry, R., Liam Finn, W.D.L., Harder, L.F., Jr., Hynes, M.E., Ishihara, K., Koester, J.P., Liao, S.S.C., Marcuson, W.F., III, Martin, G.R., Mitchell, J.K., Moriwaki, Y., Power, M.S., Robertson, P.K., Seed, R.B., and Stokoe, K.H., II, 2001, Liquefaction resistance of soils: Summary report from the 1996 NCEER and 1998 NCEER/NSF workshops on evaluation of liquefaction resistance of soils, ASCE: Journal of Geotechnical and Geoenvironmental Engineering, v. 127, no. 10, p. 817–833, doi: 10.1061/(ASCE)1090-0241(2001)127:10(817).

Youd, T.L., Idriss, I.M., Andrus, R.D., Arango, I., Castro, G., Christian, J.T., Dobry, R., Liam Finn, W.D.L., Harder, L.F., Jr., Hynes, M.E., Ishihara, K., Koester, J.P., Liao, S.S.C., Marcuson, W.F., III, Martin, G.R., Mitchell, J.K., Moriwaki, Y., Power, M.S., Robertson, P.K., Seed, R.B., and Stokoe, K.H., II, 2003, Closure to "Liquefaction resistance of soils: summary report from the 1996 NCEER and 1998 NCEER/NSF workshops on evaluation of liquefaction resistance of soils," ASCE: Journal of Geotechnical and Geoenvironmental Engineering, v. 129, no. 3, p. 284–286, doi: 10.1061/(ASCE)1090-0241(2003)129:3(284).

Manuscript Accepted by the Society August 18, 2004

Printed in the USA

Geological Society of America
Special Paper 385
2005

Liquefaction induced by historic and prehistoric earthquakes in western Puerto Rico

Martitia P. Tuttle*
Kathleen Dyer-Williams
M. Tuttle & Associates, 128 Tibbetts Lane, Georgetown, Maine 04548, USA

Eugene S. Schweig
U.S. Geological Survey, 3876 Central Ave., Ste. 2, Memphis, Tennessee 38152-3050, USA

Carol S. Prentice
U.S. Geological Survey, 345 Middlefield Rd., MS 977, Menlo Park, California 94025, USA

Juan Carlos Moya
EcoGeo, LLC, 8503 Brock Cr., Austin, Texas 78745, USA

Kathleen B. Tucker
CERI, University of Memphis, Memphis, Tennessee 38152, USA

ABSTRACT

Dozens of liquefaction features in western Puerto Rico probably formed during at least three large earthquakes since A.D. 1300. Many of the features formed during the 1918 moment magnitude (M) 7.3 event and the 1670 event, which may have been as large as M 7 and centered in the Añasco River Valley. Liquefaction features along Río Culebrinas, and possibly a few along Río Grande de Añasco, appear to have formed ca. A.D. 1300–1508 as the result of a M ≥ 6.5 earthquake. We conducted reconnaissance along Río Culebrinas, Río Grande de Añasco, and Río Guanajibo, where we found and studied numerous liquefaction features, dated organic samples occurring in association with liquefaction features, and performed liquefaction potential analysis with geotechnical data previously collected along the three rivers. Our ongoing study will provide additional information regarding the age and size distribution of liquefaction features along the western, northern, and eastern coasts and will help to improve estimates of the timing, source areas, and magnitudes of earthquakes that struck Puerto Rico during the late Holocene.

Keywords: paleoseismology, paleoliquefaction, Quaternary fault behavior.

*mptuttle@earthlink.net

Tuttle, M.P., Dyer-Williams, K., Schweig, E.S., Prentice, C.S., Moya, J.C., and Tucker, K.B., 2005, Liquefaction induced by historic and prehistoric earthquakes in western Puerto Rico *in* Mann, P., ed., Active tectonics and seismic hazards of Puerto Rico, the Virgin Islands, and offshore areas: Geological Society of America Special Paper 385, p. 263–276. For permission to copy, contact editing@geosociety.org.

INTRODUCTION

Significance

Puerto Rico, located within the tectonically active zone between the North America and Caribbean plates, is recognized as having a significant seismic hazard (Fig. 1; Asencio, 1980; McCann, 1985; Moya and McCann, 1992; Moya, 1999). Several large to very large earthquakes have struck Puerto Rico in the past 400 years. These include a moment magnitude (**M**) 7.5 event in 1943 located northwest of Puerto Rico, a **M** 7.3 event in 1918 centered in the Mona Passage, a **M** ~7.3 event in 1867 in the Anegada Passage, a **M** ~7.3 event in 1787 possibly related to the Puerto Rico Trench, and a **M** ~6 event in 1670 in western Puerto Rico (Fig. 1; McCann, 1985; Panagiotopoulos, 1995; Dolan et al., 1998; McCann et al., this volume; Doser et al., this volume). The 1918 earthquake reportedly induced liquefaction in the Añasco River Valley and generated a tsunami that struck the west coast of Puerto Rico, killing at least 114 persons and causing $4 million in damage (Fig. 2; Reid and Taber, 1919). The 1918 earthquake may also have induced liquefaction near Aguadilla to the north (Moya and McCann, 1991). Because the west coast is much more heavily populated now (486,800 in the Mayagüez-Aquadilla area according to the World-Gazetteer.com) than it was earlier in the century, a repeat of a 1918-type event would likely cause considerably more deaths and damage. Given its current population of almost 4 million with most people living in coastal areas subject to tsunami inundation and liquefaction, Puerto Rico is at significant risk from future large earthquakes.

Previous Related Work

Offshore structures that are potential sources of very large earthquakes include the Puerto Rico Trench, North Puerto Rico Slope fault zone, South Puerto Rico Slope fault zone, Septentrional fault zone, and faults associated with the Mona Passage, Virgin Islands trough, and Muertos trough (Fig. 1; Grindlay et al., 1997; Dolan et al., 1998). Offshore geophysical studies are under way to better define the active traces of these fault zones, particularly those along the western and southern coasts of Puerto Rico (see Grindlay et al., 2000, this volume). Possible onshore sources of damaging earthquakes include the Great Southern Puerto Rico fault zone and Great Northern Puerto Rico fault zone (Figs. 1 and 2). Shallow seismicity occurs below the central portion of the Great Southern Puerto Rico fault zone and in the southwestern corner of the island (Asencio, 1980; McCann, 1985; Joyce et al., 1987), suggesting that structures in these areas may be seismogenic. For much of the island, the youngest exposed rocks are Tertiary in age (Monroe, 1968; Glover, 1971; Seiders et al., 1972), making it difficult to determine the recency of faulting or ground shaking. Consequently, investigations of onshore faults have been limited (McCann, 1985; Moya and McCann, 1991). Field investigations of the Great Southern and Great Northern Puerto Rico fault zones found that they were active during the Tertiary and report predominantly thrust and left-lateral displacement (Glover and Mattson, 1960; Erikson et al., 1991). Other investigations conducted in the 1970s suggested that there might be evidence for late Quaternary deformation in the southern part of the island (Geomatrix Consultants, 1988). More recently, Prentice et al. (2000) identified and trenched the Lajas fault along the southern margin of the Lajas Valley and found evidence for Holocene movement (Fig. 2). With few exceptions, however, there is insufficient information to accurately assess the earthquake potential of onshore or offshore faults.

Paleoseismology, or the study of fault rupture and ground shaking as preserved in the late Quaternary geologic record, offers a time window of thousands of years that is well-suited for studying the behavior of fault systems and estimating rates of earthquake occurrence (Tuttle, 2001). Only a few paleoseismic studies have been conducted in the northern Caribbean but they have been successful at identifying active fault zones and liquefaction features resulting from strong ground shaking. On Hispaniola, paleoseismic studies of the Septentrional fault zone, a major left-lateral strike-slip fault related to the North America–Caribbean plate boundary, have estimated a slip rate of 9 ± 3 mm/yr and found evidence for a large earthquake ca. A.D. 1200 (Fig. 1; Prentice et al., 1993; Mann et al., 1998; Prentice et al., 2003). A study targeting areas where the Septentrional fault zone is buried by Holocene fluvial deposits found numerous historic and prehistoric liquefaction features. This paleoliquefaction study concluded that rupture of the fault ca. A.D. 1200 may have extended farther east than previously thought generating a M ~8 earthquake (Tuttle et al., 2003). These studies contribute to the understanding of earthquake hazards of Hispaniola and demonstrate the potential usefulness of paleoseismology in the tectonically active northeastern Caribbean.

Only one paleoliquefaction study has been previously conducted in Puerto Rico. Moya (1998), working with archaeologists from the Puerto Rico Institute of Culture, found sand dikes and sills intruding a cultural horizon and features at the Barrio Quemados. The archaeological site, thought to have been occupied from A.D. 1200 to 1500, is located a few kilometers east of Mayagüez on the fluvial plain of the Yaquez River (Fig. 2). Moya (1998) interpreted the sand dikes and sills as earthquake-induced liquefaction features that post-date the occupation of the site and that may have formed during the 1670 earthquake. Highly liquefiable sediments are known to occur in the San Juan area (Molinelli, 1985). Coastal plain sediments similar to those in San Juan have been mapped along most of the northern and southern coasts and in river valleys along the western and eastern coasts (Monroe, 1976). Liquefiable sediments have been identified in all the major river valleys of western Puerto Rico from geologic and geotechnical data (Moya and McCann, 1991). Site investigations, including measurement of shear wave velocities, have also found fluvial deposits in western Puerto Rico to be susceptible to liquefaction (Macari, 1994). Therefore, liquefiable sediments are present in river valleys along the coasts of Puerto Rico. If these sediments have been subjected to strong ground shaking, they

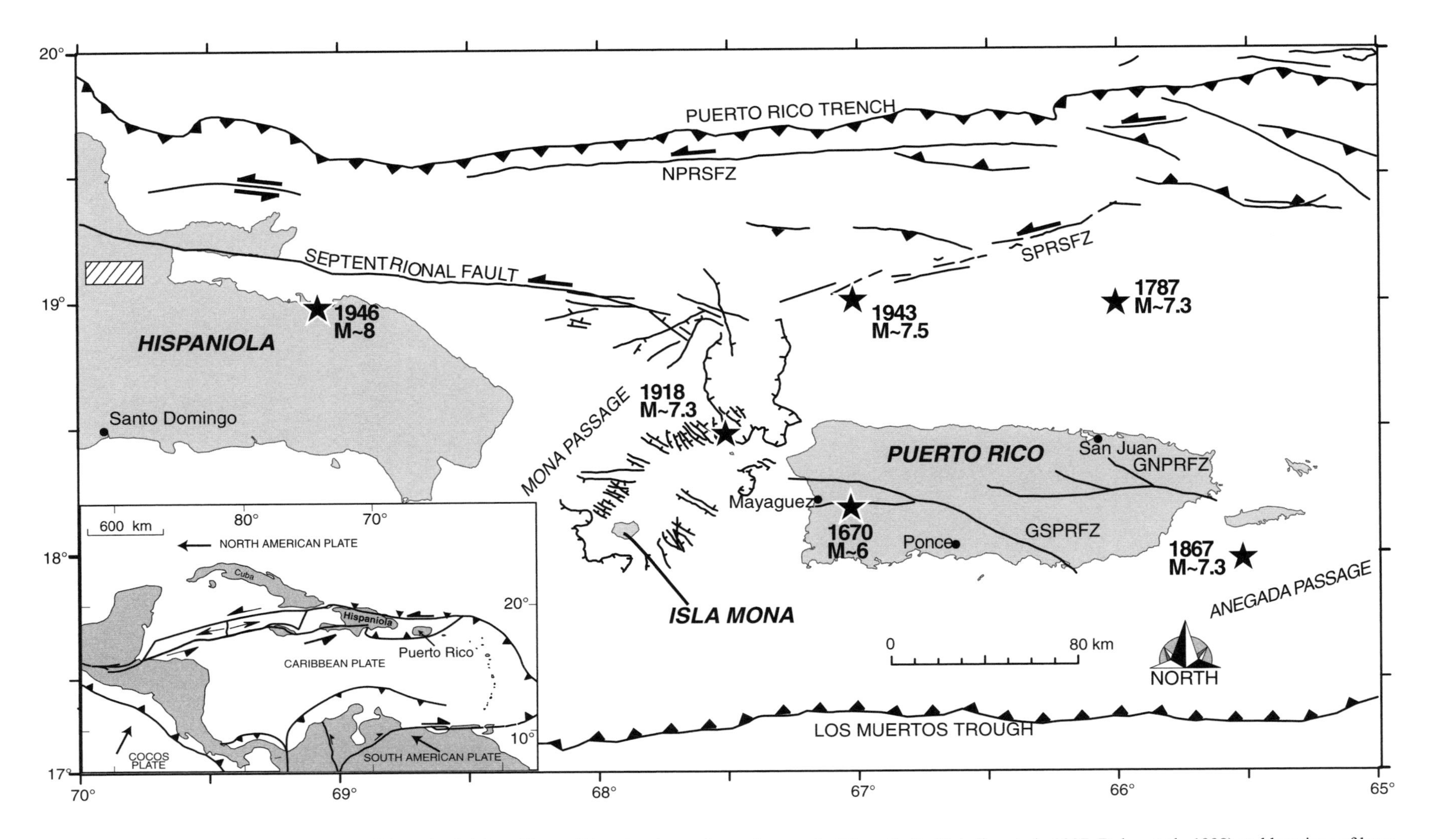

Figure 1. Location map of northeastern Caribbean in vicinity of Puerto Rico showing major onshore and offshore faults (Grindlay et al., 1997; Dolan et al., 1998) and locations of large historical earthquakes (1986 MIDAS catalogue; Dolan et al., 1998). GSPRFZ—Great Southern Puerto Rico fault zone; GNPRFZ—Great Northern Puerto Rico fault zone; NPRSFZ—Northern Puerto Rico Slope fault zone; SPRSFZ—Southern Puerto Rico Slope fault zone. Hispaniola study area, where liquefaction features may be related to 1946 (**M**) ~8 event and to 2–4 closely timed (**M**) 7–8 earthquakes ca. A.D. 1200, indicated by rectangle (Tuttle et al., 2003). Inset map shows plate-tectonic setting of greater Caribbean region.

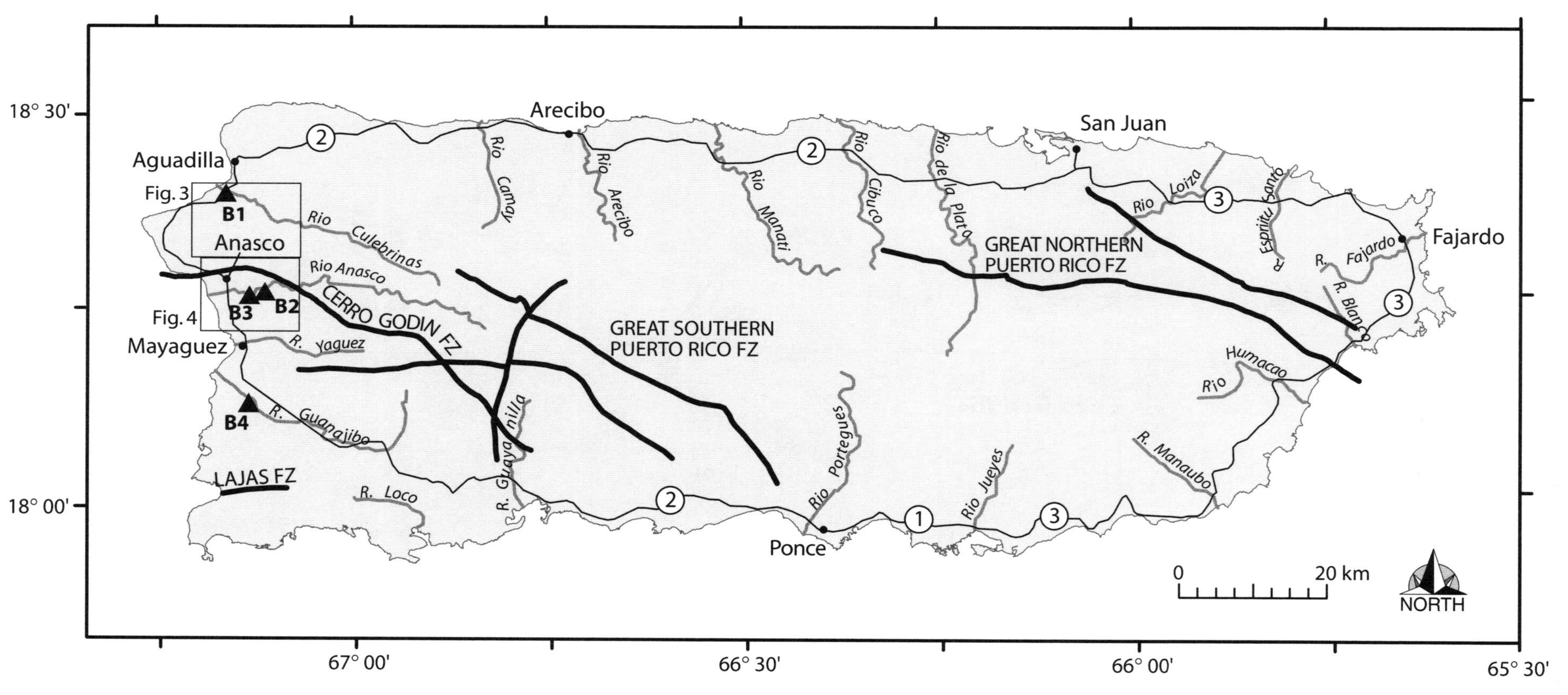

Figure 2. Map of Puerto Rico showing principal faults (Larue and Ryan, 1998; Jolly et al., 1998; Lao-Davila et al., 2000; Meltzer and Almy, 2000; Prentice et al., 2000), major rivers, borehole sites, labeled B1, B2, B3, and B4, and areas of Figures 3 and 4, labeled F3 and F4. FZ—fault zone.

may have liquefied, resulting in liquefaction features preserved in Holocene deposits.

Study Objectives

This paper presents results of an ongoing paleoliquefaction study in coastal areas of Puerto Rico. The purpose of the study is to help constrain the source areas of historic earthquakes and to estimate the timing, source areas, and magnitudes of prehistoric earthquakes. Toward this end, we have conducted (1) reconnaissance for, and documentation of, liquefaction features along three rivers, Río Culebrinas, Río Grande de Añasco, and Río Guanajibo, along the western coast of Puerto Rico, (2) radiocarbon dating of some of the liquefaction features we have found, and (3) liquefaction potential analysis of fluvial deposits.

METHODS OF INVESTIGATION

River Reconnaissance

During reconnaissance conducted in January and February 2000, we examined cutbanks along 14 km of Río Culebrinas, 11 km of Río Grande de Añasco, and 10 km of Río Guanajibo for earthquake-induced liquefaction features and other soft-sediment deformation structures related to ground shaking or faulting (Figs. 2, 3, and 4). Due to higher than normal river levels, cutbank exposure along Río Culebrinas, Río Grande de Añasco, and Río Guanajibo was limited to 4–5 m, 4 m, and 3 m, respectively. We found no evidence of faulting in late Holocene deposits; however, we did find and study many earthquake-induced liquefaction features (Tuttle et al., 2000). In addition, organic material was collected from host sediments for radiocarbon dating purposes. For sand dikes, we tried to find datable material close to their uppermost terminations. For sand blows, we tried to find datable material both above and below the vented deposit in order to establish minimum and maximum age constraints. All samples were later reviewed and the most useful ones for estimating ages of liquefaction features selected for radiocarbon dating by Beta Analytic, Inc.

Liquefaction Potential Analysis

In preparation for liquefaction potential analysis, we compiled borehole data, including blow counts, previously collected by the Puerto Rico Department of Transportation and Public Works at two bridge crossings of Río Grande de Añasco and one crossing each of Río Culebrinas and Río Guanajibo (Figs. 2, 3, and 4). The blow count (N), a measure of soil density, has been empirically related to liquefaction of sediment during actual earthquakes, with liquefaction susceptibility generally increasing with decreasing blow counts. The blow count values, taken from borehole logs, are the number of hammer blows, with a 140-lb hammer dropped a distance of 30 inches, required to advance a 2-inch split-barrel sampler approximately one foot (American Society for Testing and Materials, 1983). Simply comparing blow counts for the three rivers, sediments along the Río Grande de Añasco (N 2–15) appear to be the most susceptible to liquefaction; whereas sediments along the Río Guanajibo (N 17–25) appear to be least susceptible. It is important to take into account these differences in liquefaction susceptibility when interpreting the areal distribution of liquefaction features. We would have preferred to use borehole data collected at selected liquefaction sites, but in situ geotechnical testing was not possible given the scope of this study.

We applied the revised, simplified procedure (Seed and Idriss, 1982; Youd and Idriss, 1997) to evaluate liquefaction potential of fluvial sediments along Río Culebrinas, Río Grande de Añasco, and Río Guanajibo for various scenario earthquakes. The earthquakes we considered include the 1918 earthquake of **M** 7.3 located in the Mona Passage; the 1670 earthquake of **M** 7.0 generated by the Cerro Goden fault zone; the 1670 earthquake of **M** 6.5 or 7.5 generated by the Lajas fault zone; and a prehistoric (ca. A.D. 1300–1508) earthquake produced by an unknown fault in the Culebrinas River Valley, the Cerro Goden fault zone along the northern margin of the Añasco River Valley, or an offshore fault in the vicinity of the 1918 earthquake. For each borehole site, we determined whether representative sandy layers below the water table would be likely to liquefy during the scenario earthquakes. Estimates of peak ground acceleration for the earthquakes are based on ground-motion relations developed for California (Boore et al., 1997). This assessment would benefit from additional analysis using attenuation relations developed specifically for Puerto Rico or the northeastern Caribbean.

RESULTS OF INVESTIGATIONS

River Reconnaissance

Results of the search for and dating of liquefaction features in western Puerto Rico are summarized in Tables 1 and 2 and illustrated on Figures 3 and 4. Along Río Culebrinas, we found twenty-seven liquefaction features at ten sites. Liquefaction features include sand dikes up to 23 cm wide and one, possibly two, sand blow deposits. Along Río Grande de Añasco, we found thirty liquefaction features at eighteen sites, including sand dikes up to 16 cm wide and two, possibly five, sand blows. The five largest dikes occur at sites 2, 10, 11, 17, and 18 located between 3.5 and 4.5 km from the coast. Several of the liquefaction features are discussed below.

Río Culebrinas

At site Río Culebrinas 1 (RC1), several dikes extend into the base of a paleosol characterized by soil structure and organic accumulation (Fig. 5). The largest dike is discontinuous as it crosses the paleosol but appears to widen to form a vent structure within the top of the paleosol and to broaden into the base of an overlying sand layer. The sand layer is of

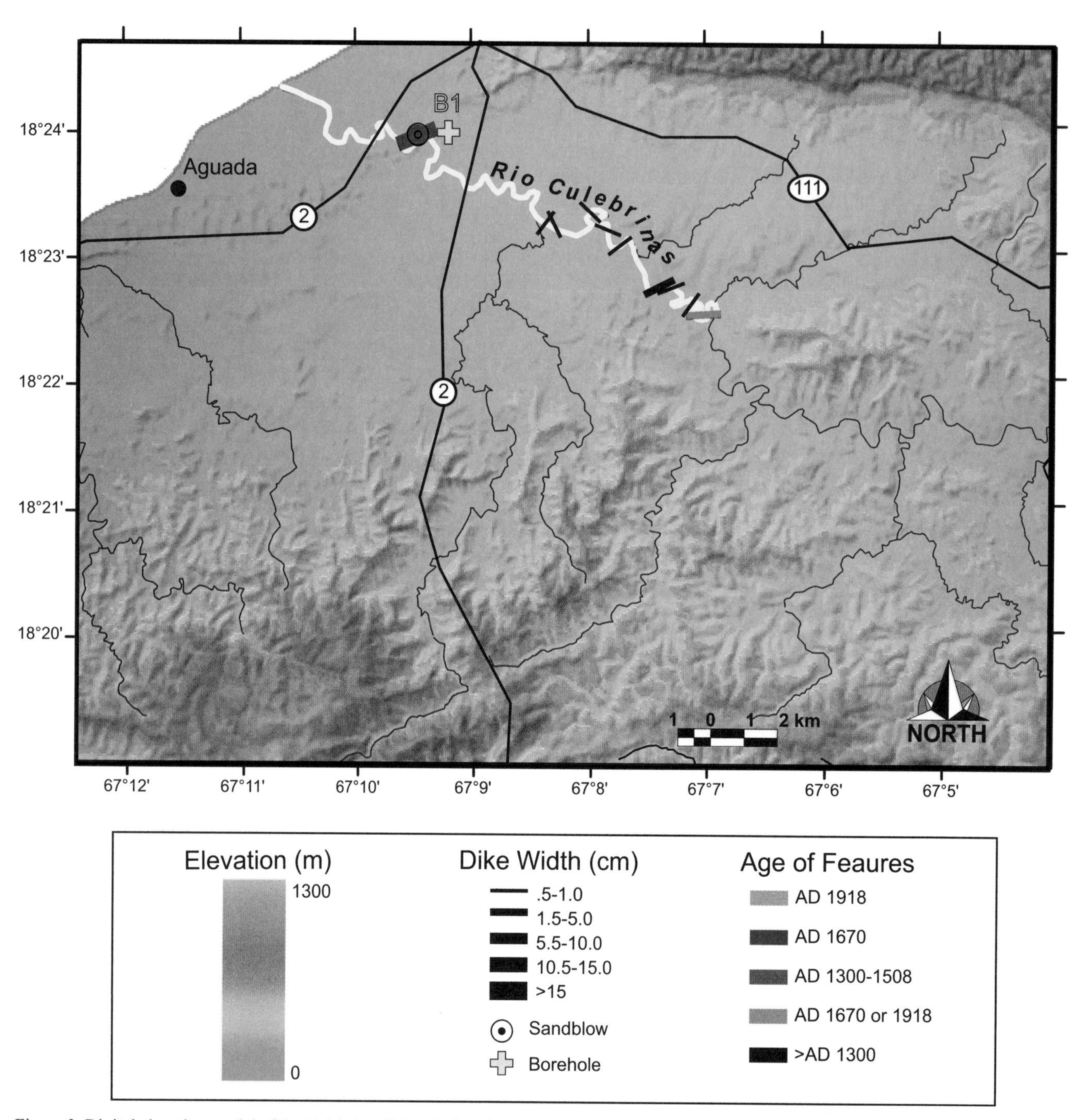

Figure 3. Digital elevation model of the Culebrinas River Valley showing locations, sizes, and estimated ages of earthquake-induced liquefaction features. Yellow line indicates portion of river surveyed. Borehole location also indicated. See Figure 2 for location of area shown.

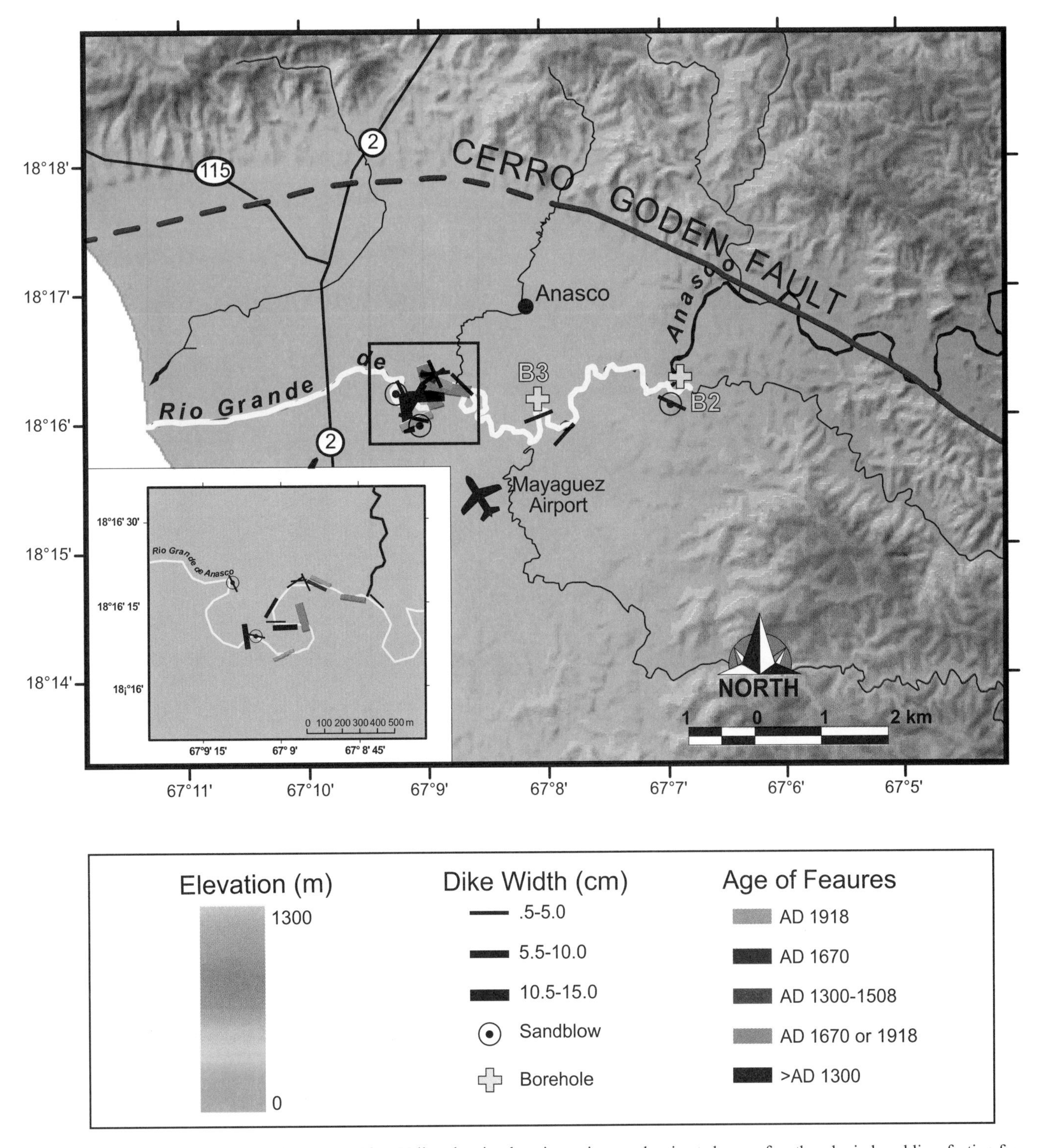

Figure 4. Digital elevation model of the Añasco River Valley showing locations, sizes, and estimated ages of earthquake-induced liquefaction features. Borehole location also indicated. Inset is enlargement of area with numerous liquefaction features. See Figure 2 for location of area shown.

limited lateral extent and pinches out away from the sand dike. The sand layer is interpreted as a sand blow that was deposited on the paleosol that would have been at the surface at the time. Radiocarbon dating of charcoal located 10 cm below the base of the sand blow and within the paleosol yielded a calibrated age of A.D. 1300–1370, 1380–1430 (Table 2). The charcoal sample provides a close maximum age for the liquefaction features and indicates that they formed after A.D. 1300, and possibly after A.D. 1430. Since there is no large earthquake recorded by the Spanish following colonization in A.D. 1508 and prior to the earthquake in A.D. 1670 (McCann et al., this volume), the event that induced liquefaction at this site is estimated to have occurred between A.D. 1300 and A.D. 1508.

At site Río Culebrinas 2 (RC2), several dikes filled with pebbly, coarse sand are exposed in the river cutbank. The largest dike (6 cm wide) extends ~3 m above the river level where it crosscuts a paleosol and broadens into the base of a pebbly sand layer that pinches out away from the sand dike. The sand layer is interpreted as a sand blow. A piece of charcoal was collected from the paleosol 14 cm below the overlying sand blow. Radiocarbon dating of the sample indicates that the sand blow and related sand dike formed after A.D. 1640 (Table 2). Radiocarbon dating at sites RC3 and RC6, where only sand dikes were found, indicates that these liquefaction features formed since A.D. 1300. It appears that at least a 700 yr record of fluvial sedimentation and strong ground shaking is preserved in the upper 4–5 m of sediments along the Río Culebrinas. By inference, we estimate that liquefaction

TABLE 1. RESULTS OF RECONNAISSANCE AND RADIOCARBON DATING OF LIQUEFACTION FEATURES

Site name	Latitude (decimal degrees)	Longitude (decimal degrees)	Thickness of sand blows (cm)	Width of sand dikes (cm)	Strike and dip of largest sand dikes	Preliminary age estimate of features (A.D.)
Río Culebrinas						
1	18.401183	67.158133	20	23, 10, 7, 3, 3, 2	N68°E, 80°NW N38°E, 78°NW N83°E, 75°NW	1300–1508
2	18.377583	67.117267		6, 2	N88°E, vertical	1670
3	18.378450	67.119000		4, 1.5	N38°E, vertical	>1300
4	18.380700	67.121883		4.5	N68°E, 78°NW	>1300
5	18.381117	67.123750		8	N64°E, vertical	>1300
6	18.386333	67.129100		3, 2, 1.5, 1.5, 1.5	N53°E, 81°NW	>1300
7	18.388417	67.130917		3	N68°W, vertical	>1300
8	18.390667	67.133450		5	N47°W, 87°NE	>1300
9	18.389150	67.138500		1.5	N27°W, vertical	>1300
10	18.389333	67.139833		3, 2, 1, 0.5, 0.5	N38°E, vertical	>1300
Río Grande de Añasco						
1	18.271940	67.153040	8	3.5	N27°W, vertical	>1300
2	18.269710	67.152690		13, 1	N7°W, vertical	>1300
3	18.268790	67.151670	0.5	3, 2.5	N77°W, vertical	>1300
4	18.268650	67.151420		8, 4.5	N66°E, 86°SE N23°E, vertical	1918
5	18.271600	67.115200				>1300
6	18.270300	67.115800	3.5	5	N67°W, 80°SW	>1300
7	18.266333	67.130733		3, 1, 1	N43°E, 82°SE	>1300
8	18.268700	67.134333		3	N71°E, 80°NW	>1300
9	18.272983	67.144783		3	N45°W, vertical	>1300
10	18.272950	67.146450		13, 4, 1	N82°W, vertical N27°W, 88°SW	1670 or 1918
11	18.274050	67.148350	12	9, 6	N68°W, 70°NE N57°W, 81°NE	1670 and 1918
12	18.274333	67.148600		4	N27°W, vertical	>1300
13	18.274467	67.148667		2, 1	N83°E, 72°NW	>1300
14	18.274000	67.148983		4	N57°E, vertical	>1300
15	18.271433	67.151567		6	N33°E, vertical	>1300
16	18.271567	67.149800	6	1	N89°E, vertical	>1300
17	18.271783	67.150117		15	N87°W, vertical	>1300
18	18.273150	67.149233		16	N14°W, vertical	1670 or 1918
Río Guanajibo						
1	18.138833	67.141083		3	N48°E, 80°SE	>1300
2	18.156933	67.165833		0.5	N87°W, vertical	>1300

features documented at eight other sites along the river, for which we have no radiocarbon dating data, formed since A.D. 1300.

Río Grande de Añasco

At site Río Grande de Añasco 11 (RGA11), several sand dikes extend 0.63 m above the river level and broaden slightly into the base of a 12-cm-thick sand layer (Fig. 6). The sand layer fines and thins away from the sand dikes and portions of the sand layer appear to have been eroded. These characteristics suggest that the layer is a sand blow deposit. In addition, there is a second generation of sand dikes that crosscuts the sand blow deposit. One of the dikes extends to 1.13 m above the river level (Fig. 7). Charcoal collected 10 cm below the sand blow deposit indicates that the first generation of liquefaction features formed after A.D. 1640; whereas, charcoal collected above the sand blow and 10 cm below the tip of a younger sand dike indicates that the second generation liquefaction features formed after A.D. 1670 (Table 2). Therefore, it appears that both generations of liquefaction features formed in the past 400 yr.

TABLE 2. RESULTS OF RADIOCARBON DATING OF SAMPLES COLLECTED AT LIQUEFACTION SITES ALONG RÍO CULEBRINAS, RÍO GRANDE DE AÑASCO, AND RÍO GUANAJIBO

Site sample	Lab sample	$^{13}C/^{12}C$ ratio	^{14}C age (1σ, yr B.P.)*	Cal yr A.D./B.C. (2σ)†	Probability distribution	Sample description	Earthquake age constraint
(1) RC1-C3	Beta-129692	−21.7	560 ± 40	A.D. 1300–1370 A.D. 1380–1430	0.526 0.474	charcoal 10 cm below sand blow; from paleosol; 1.4 m awl	>A.D. 1300
(2) RC1-W1	Beta-146689	−26.4	160 ± 80	A.D. 1530–1540 A.D. 1640–1960	0.008 0.992	possibly recent deposit; wood ~1 m below top of dike; just awl	
(3) RC2-C1	Beta-146690	−26.5	200 ± 40	A.D. 1640–1700 A.D. 1720–1810 A.D. 1830–1880 A.D. 1920–1950	0.267 0.537 0.049 0.146	charcoal 14 cm below top of dike; 2.86 m awl	>A.D. 1640
(4) RC3-C2	Beta-146691	−16.2	600 ± 40	A.D. 1300–1410	1.000	charcoal adjacent to top of sand dike; 1.39 m awl	>A.D. 1300
(5) RC6-C3	Beta-146692	−18.2	460 ± 40	A.D. 1330–1340 A.D. 1400–1490 A.D. 1500–1510 A.D. 1600–1610	0.001 0.985 0.001 0.013	charcoal 27 cm below top of upper-most sand dike; 2.28 m awl	>A.D. 1300
(6) RGA2-C2	Beta-129693	−27.2	100 ± 40	A.D. 1680–1740 A.D. 1750–1760 A.D. 1800–1940 A.D. 1950–1955	0.290 0.025 0.657 0.028	charcoal few cm above top of dike; 1.91 m awl	
(7) RGA4-C1	Beta-129691	−11.4	110 ± 40	A.D. 1680–1760 A.D. 1770–1780 A.D. 1800–1940 A.D. 1950	0.336 0.005 0.632 0.027	charcoal 6 cm below top of dike; 1.6 m awl	>A.D. 1680
(8) RGA4-C3	Beta-129694	−17.8	420 ± 40	A.D. 1420–1520 A.D. 1570–1630	0.843 0.157	charcoal 23 cm below top of dike; 1.43 m awl	>A.D. 1420
(9) RGA4-O4	Beta-146693	−26.7	90 ± 40	A.D. 1680–1740 A.D. 1750–1760 A.D. 1800–1940 A.D. 1950	0.282 0.009 0.678 0.032	possibly recent deposit; leaves 1.77 m below top of dike; just awl	
(10) RGA10-C1	Beta-146694	−28.5	190 ± 40	A.D. 1640–1700 A.D. 1720–1820 A.D. 1830–1880 A.D. 1910–1950	0.235 0.532 0.074 0.158	charcoal 40 cm below top of dike; 0.4 m awl	>A.D. 1640
(11) RGA11-C3	Beta-146695	−28.4	130 ± 40	A.D. 1670–1780 A.D. 1800–1910 A.D. 1910–1940 A.D. 1950	0.405 0.429 0.147 0.019	charcoal 10 cm below top of younger sand dikes; 1.03 m awl	>A.D. 1670
(12) RGA11-C5	Beta-146696	−28.0	190 ± 40	A.D. 1640–1700 A.D. 1720–1820 A.D. 1830–1880 A.D. 1910–1950	0.235 0.532 0.074 0.158	charcoal 10 cm below sand blow; 0.53 m awl	>A.D. 1640
(13) RGA18-C1	Beta-146697	−31.4	200 ± 50	A.D. 1530–1540 A.D. 1640–1710 A.D. 1720–1890 A.D. 1910–1950	0.004 0.268 0.583 0.145	charcoal 1.1 m below top of sand dike; 0.2 m awl	>A.D. 1640

Note: RC—Río Culebrinas; RGA—Río Grande de Añasco; RG—Río Guanajibo; awl—above water level.
*Conventional radiocarbon ages in years before present (1950) determined by Beta Analytic, Inc.
†Calibrated calendar age ranges determined with CALIB Rev. 4.3, Method B and rounded to the nearest decade (Stuiver et al., 1998).

Figure 5. Photograph of sand dike and related sand blow at site RC1. Charcoal from a paleosol crosscut by sand dike and overlain by sand blow provides close maximum age of A.D. 1300–1430.

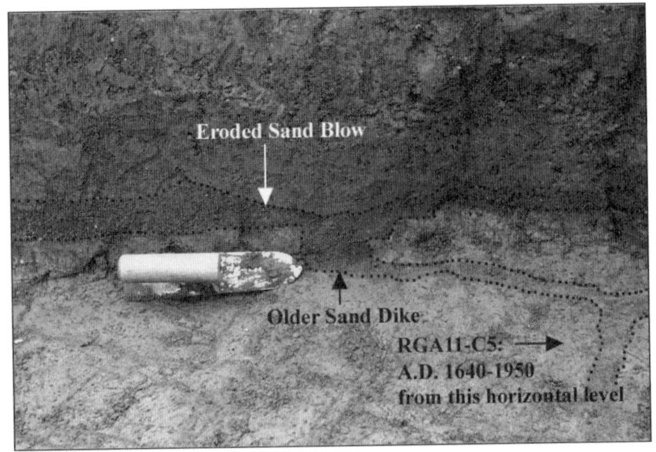

Figure 6. Photograph of sand dike and related sand blow at site RGA11. Upper portion of sand blow appears eroded.

At nearby site RGA10, a 13-cm-wide sand dike extends 0.8 m above the river level and terminates within a paleosol. Charcoal collected 40 cm below the tip of the dike indicates that it formed after A.D. 1640 (Table 2). Similarly, at RGA18, a 16-cm-wide dike extends 1.3 m above the river. Charcoal collected 1.1 m below the tip of the dike indicates that it formed after A.D. 1530, and more likely after A.D. 1640. At RGA4, two sand dikes extend 1.66 m and 1.40 m above the river level. Pieces of charcoal collected 6 cm and 23 cm below the tip of the uppermost sand dike yield stratigraphically consistent results and indicate that the dikes formed after A.D. 1680. It appears that at least a 600–700 yr geological record is preserved in the upper 4 m of cutbanks along the Río Grande de Añasco. We infer that liquefaction features documented at other sites along the river, but for which we have no radiocarbon data, formed since A.D. 1300 (Fig. 8).

Río Guanajibo

Along Río Guanajibo located ~6 km south of Mayagüez, we found only two small dikes at two different sites (Fig. 2; Table 1). The dikes are filled with medium sand and fine sandy silt. No organic material was found at either liquefaction site that could help to constrain the age of the features. Based on dating of materials along Río Grande de Añasco, deposits exposed along the Río Guanajibo are probably less than 700 yr old. Therefore, we infer that the two sand dikes along Río Guanajibo also formed since A.D. 1300. It appears that earthquake-induced liquefaction was less extensive and less severe along the Río Guanajibo than along rivers farther north.

Results of Liquefaction Potential Analysis

The results of liquefaction potential analysis for the scenario earthquakes described above are presented in Tables 3–6. The 1918 **M** 7.3 earthquake, thought to have been centered in the Mona Passage, reportedly induced liquefaction in the Añasco River Valley (Reid and Taber, 1919). The results of our analysis are consistent with this observation (Table 3). However, it seems unlikely that the 1918 earthquake induced liquefaction along Río Culebrinas. This event would have had to be about **M** 7.5 and located closer to shore to do so (Table 6). Therefore, the post–A.D. 1640 liquefaction features on the Culebrinas probably formed during the 1670 earthquake, as did some of the features along the Añasco. The 1670 earthquake is thought to have been centered in southwestern Puerto Rico and to have induced liquefaction in the Yaquez River Valley (Moya, 1998). If the 1670 earthquake were centered in the Lajas Valley, a location of Holocene-age faulting (Prentice et al., 2000), it would have to be of at least **M** 6.5 to induce liquefaction along the Añasco (Table 4). Such an event, however, would not induce liquefaction along the Culebrinas, even if it were of **M** 7.5. If generated by the Cerro Goden fault zone, the 1670 earthquake would have to be of **M** 7 to induce liquefaction along both the Añasco and the Culebrinas (Table 5). Therefore, the 1670 earthquake may have been of **M** ~7 and more likely centered in or near the Añasco River Valley than in the Lajas Valley.

With the available data, we cannot yet resolve the source area of an earthquake that induced liquefaction along Río Culebrinas between A.D. 1300–1508. Our analysis suggests that if the earthquake were generated by a fault in the Culebrinas River

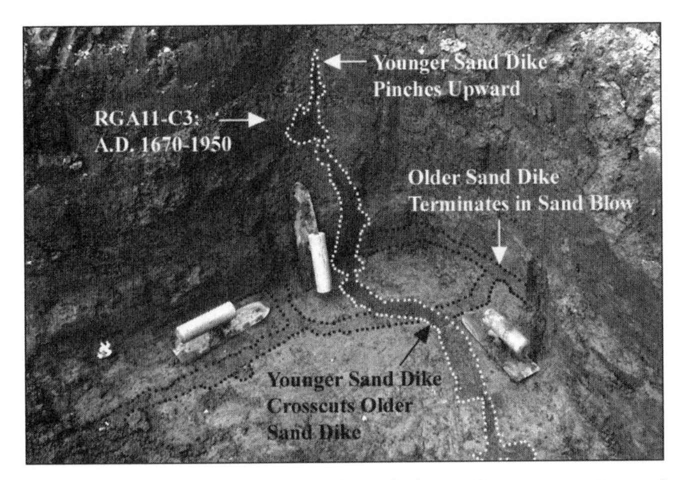

Figure 7. Photograph of crosscutting relations of two generations of sand dikes at site RGA11.

Figure 8. Photograph of small sand dike and related sand blow at site RGA16.

Valley, it could have been as small as **M** 6.5 and still induce liquefaction locally (Table 6). If generated by the Cerro Goden fault zone, the event would have to be of **M** ≥ 7. Alternatively, if produced by an offshore source like that of the 1918 earthquake, the A.D. 1300–1508 event would have to be at least **M** 7.5 and located within 30 km of the shoreline to induce liquefaction along Río Culebrinas (Table 6).

SUMMARY AND CONCLUSIONS

During river reconnaissance in western Puerto Rico, we found and studied twenty-seven liquefaction features along Río Culebrinas, thirty features along Río Grande de Añasco, and two features along Río Guanajibo. The liquefaction features included a few small sand blows and many small to medium-sized sand dikes. Many of the liquefaction features probably formed during the 1918 or 1670 earthquakes (Tables 1 and 2). Those liquefaction features for which we have maximum age constraint of A.D. 1640 could have formed during either of the 1670 or 1918 events (see Table 7 for a summary of our interpretations). Those features with maximum age constraint of 1670 or later probably formed during the 1918 earthquake. At least one liquefaction feature on the Río Culebrinas formed between A.D. 1300 and 1508. Although additional earthquakes cannot be ruled out, other liquefaction features documented along the three rivers probably formed during one of these three earthquakes.

Liquefaction potential analysis of sandy sediments along Río Culebrinas, Río Grande de Añasco, and Río Guanajibo was used to evaluate several scenario earthquakes (Table 7). The results of our analysis are consistent with the observation that the 1918 earthquake induced liquefaction in the Añasco River Valley. Our analysis suggests that the 1670 earthquake, which appears to have produced liquefaction features along both Río Grande de Añasco and Río Culebrinas, may have been of **M** ~7 and located in or near the Añasco River Valley. The source and magnitude of an event ca. A.D. 1300–1508 that induced liquefaction along Río Culebrinas is uncertain, but our analysis indicates that, even if it were a local event, it was probably of **M** ≥ 6.5.

Reconnaissance was conducted when river levels were not especially low. Liquefaction features related to prehistoric earthquakes may occur deeper in the section than was exposed at the time of reconnaissance. Therefore, we think that it would be worthwhile to resurvey at least portions of Río Culebrinas, Río Grande de Añasco, and Río Guanajibo when water levels are lower than they were in January and February 2000. Reconnaissance of river cutbanks and liquefaction potential analysis of fluvial sediments along the northern and eastern coasts of Puerto Rico are currently under way. Additional information regarding the age and size distribution of liquefaction features will help to improve estimates of the timing, source area, and magnitudes of earthquakes that struck Puerto Rico during the late Holocene.

ACKNOWLEDGMENTS

This research is supported by U.S. Geological Survey grant 143400HQGR0004. The views and conclusions present in this document are those of the authors and should not be interpreted as necessarily representing the official policies, either expressed or implied, of the U.S. Government. We are grateful for assistance provided by Pedro Díaz, Maria Irizari, and Matt Larsen at the San Juan Office of the U.S. Geological Survey; Ing. Luis Sandoval and Chief Soils Engineer Edgardo Pagan at the Puerto Rico Highway and Transportation Authority; Raul Cabán and Daniel Cruz Lorenzo, Cuerpo de Vigilantes, Agua-

TABLE 3. THE 1918 EARTHQUAKE OF M 7.3

Site/borehole (map reference)	Magnitude @ distance (km)	amax	Depth to susceptible sediment (ft)	Blow count $N_{1(60)}$	Cyclic stress ratio	Results*
Culebrinas RT-2 (B1)	M7.3@40	0.15	20	16	0.12	N
Añasco RT-406 (B2)	M7.3@50	0.13	35	7	0.09	L
	M7.3@50	0.13	40	6	0.09	L
	M7.3@50	0.13	15	1	0.06	N
	M7.3@50	0.13	15	5	0.06	L
	M7.3@50	0.13	20	8	0.07	N
	M7.3@50	0.13	10	10	0.04	N
	M7.3@50	0.13	15	2	0.06	N
	M7.3@50	0.13	15	13	0.14	N
	M7.3@50	0.13	20	12	0.09	N
	M7.3@50	0.13	15	15	0.15	N
Añasco RT-430 (B3)	M7.3@40	0.15	20	3	0.08	L
	M7.3@40	0.15	25	13	0.11	N
	M7.3@40	0.15	5	13	0.06	N
	M7.3@40	0.15	7	10	0.08	N
	M7.3@40	0.15	10	4	0.09	L
	M7.3@40	0.15	35	4	0.11	L
	M7.3@40	0.15	45	6	0.11	L
	M7.3@40	0.15	15	6	0.11	L
	M7.3@40	0.15	20	12	0.11	N
Guanajibo RT-114 (B4)	M7.3@50	0.13	17	2	0.08	L
	M7.3@50	0.13	25	15	0.09	N
	M7.3@50	0.13	17	8	0.07	N
	M7.3@50	0.13	20	19	0.08	N

*L—liquefaction likely; M—liquefaction marginal; N—liquefaction not likely.

TABLE 4. THE 1670 EARTHQUAKE IF M 6.5 OR M 7.5 AND PRODUCED BY THE LAJAS VALLEY FAULT ZONE

Site/borehole (map reference)	Magnitude @ distance (km)	amax	Depth to susceptible sediment (ft)	Blow count $N_{1(60)}$	Cyclic stress ratio	Results*
Culebrinas RT-2 (B1)	M6.5@40	0.10	20	16	0.08	N
	M7.5@40	0.17	20	16	0.14	N
Añasco RT-406 (B2)	M6.5@30	0.12	35	7	0.09	N
	M6.5@30	0.12	40	6	0.08	N
	M6.5@30	0.12	10	5	0.05	N
	M6.5@30	0.12	15	1	0.06	N
	M6.5@30	0.12	15	5	0.06	N
	M6.5@30	0.12	20	8	0.06	N
	M6.5@30	0.12	10	10	0.04	N
	M6.5@30	0.12	15	2	0.05	N
	M6.5@30	0.12	15	13	0.14	N
	M6.5@30	0.12	20	12	0.09	N
	M6.5@30	0.12	15	15	0.14	N
Añasco RT-430 (B3)	M6.5@30	0.12	20	3	0.07	N
	M6.5@30	0.12	25	13	0.09	L
	M6.5@30	0.12	5	13	0.05	N
	M6.5@30	0.12	7	10	0.06	N
	M6.5@30	0.12	10	4	0.08	L
	M6.5@30	0.12	35	4	0.09	L
	M6.5@30	0.12	45	6	0.09	L
	M6.5@30	0.12	15	6	0.09	L
	M6.5@30	0.12	20	12	0.09	L
Guanajibo RT-114 (B4)	M6.5@15	0.21	17	2	0.13	L
	M6.5@15	0.21	25	15	0.14	N
	M6.5@15	0.21	17	8	0.12	M
	M6.5@15	0.21	20	19	0.12	N

*L—liquefaction likely; M—liquefaction marginal; N—liquefaction not likely.

TABLE 5. THE 1670 EARTHQUAKE IF M 7 AND PRODUCED BY THE CERRO GODEN FAULT ZONE

Site/borehole (map reference)	Magnitude @ distance (km)	amax	Depth to susceptible sediment (ft)	Blow count $N_{1(60)}$	Cyclic stress ratio	Results*
Culebrinas RT-2 (B1)	M7@15	0.27	20	16	0.22	L
Añasco RT-406 (B2)	M7@5	0.49	35	7	0.34	L
	M7@5	0.49	40	6	0.33	L
	M7@5	0.49	10	5	0.18	L
	M7@5	0.49	15	1	0.22	L
	M7@5	0.49	15	5	0.22	L
	M7@5	0.49	20	8	0.25	L
	M7@5	0.49	10	10	0.17	L
	M7@5	0.49	15	2	0.21	L
	M7@5	0.49	15	13	0.54	L
	M7@5	0.49	20	12	0.35	L
	M7@5	0.49	15	15	0.55	L
Añasco RT-430 (B3)	M7@5	0.49	20	3	0.27	L
	M7@5	0.49	25	13	0.36	L
	M7@5	0.49	5	13	0.20	L
	M7@5	0.49	7	10	0.25	L
	M7@5	0.49	10	4	0.29	L
	M7@5	0.49	35	4	0.36	L
	M7@5	0.49	45	6	0.34	L
	M7@5	0.49	15	6	0.33	L
	M7@5	0.49	20	12	0.36	L
Guanajibo RT-114 (B4)	M7@15	0.27	17	2	0.17	L
	M7@15	0.27	25	15	0.19	N
	M7@15	0.27	17	8	0.15	L
	M7@15	0.27	20	19	0.16	N

*L—liquefaction likely; M—liquefaction marginal; N—liquefaction not likely.

TABLE 6. THE PREHISTORIC EARTHQUAKE THAT INDUCED LIQUEFACTION ALONG THE RIO CULEBRINAS

Earthquake source	Magnitude @ distance (km)	amax	Depth to susceptible sediment (ft)	Blow count $N_{1(60)}$	Cyclic stress ratio	Results*
Local source in the Rio Culebrinas River Valley	M6.25@5	0.33	20	16	0.27	N
	M6.5@5	0.37	20	16	0.30	L
Cerro Goden fault zone	M6.75@15	0.24	20	16	0.19	N
	M7@15	0.27	20	16	0.22	L
1918 source area	M7.3@30	0.19	20	16	0.15	N
	M7.3@40	0.15	20	16	0.12	N
	M7.5@30	0.21	20	16	0.17	L
	M7.5@40	0.17	20	16	0.14	N

*L—liquefaction likely; M—liquefaction marginal; N—liquefaction not likely.

TABLE 7. SUMMARY OF INTERPRETATIONS FROM FIELD OBSERVATIONS AND ANALYSIS OF SCENARIO EARTHQUAKES

River names	Mona Passage M ~7.3 1918	Cerro Godin fault zone M ~7.0 1670	Northwestern Puerto Rico M $\geq$ 6.5 1300–1508
Río Culebrinas	No Liquefaction	Liquefaction	Liquefaction
Río Grande de Añasco	Liquefaction	Liquefaction	No Liquefaction
Rio Yaquez	Unknown	Liquefaction	Unknown
Río Guanajibo	Liquefaction for at least one of these events		

dilla; Julio Roman, Mayor; and Martín Conception, Director of Civil Defense, Aguada. Paul Mann, John Sims, and John Tinsley provided thorough and constructive reviews that improved this manuscript.

REFERENCES CITED

American Society for Testing and Materials, 1983, Annual Book of ASTM Standards, Section 4, Construction, Standard method for penetration test and split-barrel sampling of soils, designation D1586, p. 297–299.

Asencio, E., 1980, Western Puerto Rico seismicity: U.S. Geological Survey Open-file Report 80-192, p. 135.

Boore, D.M., Joyner, W.M., and Fumal, T.E., 1997, Equations for estimating horizontal response spectra and peak accelerations from western North American earthquakes: A summary of recent work: Seismological Research Letters, v. 68, no. 1 p. 128–153.

Dolan, J.F., Mullins, H.T., and Wald, D.J., 1998, Active tectonics of the north-central Caribbean: Oblique collision, strain partitioning, and opposing subducted slabs, *in* Dolan, J.F., and Mann, P., eds., Active strike-slip and collisional tectonics in the northern Caribbean plate collisional zone: Geological Society of America Special Paper 326, p. 1–61.

Erikson, J., Pindell, J., and Larue, D.K., 1991, Fault zone deformational constraints on Paleogene tectonic evolution in southern Puerto Rico: Geophysical Research Letters, v. 18, p. 569–572.

Geomatrix Consultants, 1988, Earthquake ground motions for the Portugues Dam, Puerto Rico, Geological-seismological evaluation to assess potential hazards: Department of the Army, Jacksonville District, Corps of Engineers, Jacksonville, Florida.

Glover, L., III, 1971, Geology of the Coamo area, Puerto Rico, and its relation to the volcanic arc-trench association: U.S. Geological Survey Professional Paper 636, 102 p.

Glover, L., III, and Mattson, P.H., 1960, Successive thrust and transcurrent faulting during the early Tertiary in south-central Puerto Rico, *in* Short papers in the geological sciences: U.S. Geological Survey Professional Paper 400-B, p. 363–365.

Grindlay, N.R., Mann, P., and Dolan, J., 1997, Researchers investigate submarine faults north of Puerto Rico: Eos (Transactions, American Geophysical Union), v. 78, p. 404.

Grindlay, N.R., Abrams, L.J., Mann, P., and Del Greco, L., 2000, A high-resolution sidescan and seismic survey reveals evidence of late Holocene fault activity offshore western and southern Puerto Rico: Eos (Transactions, American Geophysical Union), Annual Fall Meeting, p. F1181.

Jolly, W.T., Lidiak, E.G., Schellekens, J.H., and Santos, H., 1998, Volcanism, tectonics, and stratigraphic correlations in Puerto Rico, *in* Lidiak, E.G., and Larue, D.K., eds., Tectonics and Geochemistry of the Northeastern Caribbean: Geological Society of America Special Paper 322, p. 1–34.

Joyce, J., McCann, W.R., and Lithgow, C., 1987, Onland active faulting in the Puerto Rico platelet: Eos (Transactions, American Geophysical Union), v. 68, p. 1483.

Lao-Davila, D.A., Mann, P., Prentice, C.S., and Draper, G., 2000, Late Quaternary activity of the Cerro Goden fault zone, transpressional uplift of the La Cedena Range, and their possible relation to the opening of the Mona Rift, Western Puerto Rico: Eos (Transactions, American Geophysical Union), Annual Fall Meeting, p. F1181.

Larue, D.K., and Ryan, H.F., 1998, Seismic reflection profiles of the Puerto Rico Trench: Shortening between the North American and Caribbean plates, *in* Lidiak, E.G., and Larue, D.K., eds., Tectonics and Geochemistry of the Northeastern Caribbean: Geological Society of America Special Paper 322, p. 193–210.

Macari, J.E., 1994, A field study in support of the assessment for liquefaction and soil amplification in western Puerto Rico: Puerto Rico Earthquake Safety Commission, 35 p.

Mann, P., Prentice, C.S., Burr, G., Peña, L.R., and Taylor, F.W., 1998, Tectonic geomorphology and paleoseismicity of the Septentrional fault system, Dominican Republic, *in* Dolan, J.F., and Mann, P., eds., Active strike-slip and collisional tectonics of the northern Caribbean plate boundary zone: Geological Society of America Special Paper 326, p. 63–123.

McCann, W.R., 1985, On the earthquake hazards of Puerto Rico and the Virgin Islands: Seismological Society of America Bulletin, v. 75, p. 251–262.

Meltzer, A., and Almy, C., 2000, Fault structure and earthquake potential, Lajas Valley, SW Puerto Rico: Eos (Transactions, American Geophysical Union), Annual Fall Meeting, p. F1181.

Monroe, W.H., 1968, The age of the Puerto Rico Trench: Geological Society of America Bulletin, v. 79, p. 153–162.

Monroe, W.H., 1976, The karst landforms of Puerto Rico: U.S. Geological Survey Professional Paper 899, 69 p.

Molinelli, J., 1985, Earthquake vulnerability study for the metropolitan area of San Juan, Puerto Rico: Consultant report to the Department of Natural Resources, San Juan, Puerto Rico, 66 p.

Moya, J.C., 1998, The neotectonics of western Puerto Rico [Ph.D. thesis]: Boulder, Colorado, University of Colorado, 132 p.

Moya, J.C., 1999, Results from neotectonic studies in western Puerto Rico: Geological Society of America Penrose Conference, Subduction to strike-slip transitions along plate boundaries, Dominican Republic, abstracts (http://www.uncwil.edu/people/grindlayn/penrose.html).

Moya, J.C., and McCann, W.R., 1991, Earthquake vulnerability study of Mayaguez, western Puerto Rico: Cooperative Agreement, Earthquake Safety Commission of Puerto Rico—Federal Emergency Management Agency, Internal Report 91-1: FEMAPR-0012, 66 p.

Moya, J.C., and McCann, W.R., 1992, Earthquake vulnerability study of the Mayaguez area, western Puerto Rico: Comision de Seguridad contra Terremotos, Department of Natural Resources, Puerto Rico, 43 p.

Panagiotopoulos, D.G., 1995, Long-term earthquake prediction in Central America and Caribbean Sea based on time and magnitude-predictable model: Bulletin of the Seismological Society of America, v. 85, n. 4, p. 1190–1201.

Prentice, C.S., Mann, P., Taylor, F.W., Burr, G., and Valastro, S., Jr., 1993, Paleoseismicity of the North American–Caribbean plate boundary (Septentrional fault): Dominican Republic: Geology, v. 21, p. 49–52, doi: 10.1130/0091-7613(1993)0212.3.CO;2.

Prentice, C.S., Mann, P., and Burr, G., 2000, Prehistoric earthquakes associated with a Late Quaternary fault in the Lajas Valley, Southwestern Puerto Rico: Eos (Transactions, American Geophysical Union), Annual Fall Meeting, p. F1182.

Prentice, C.S., Mann, P., Pena, L.R., and Burr, G., 2003, Slip rate and earthquake recurrence along the central Septentrional fault, North American–Caribbean plate boundary, Dominican Republic: Journal of Geophysical Research, v. 108, n. B3, p. 2149.

Reid, H., and Taber, S., 1919, The Puerto Rico earthquakes of October–November 1918: Bulletin of the Seismological Society of America, v. 9, p. 95–127.

Seed, H.B., and Idriss, I.M., 1982, Ground motions and soil liquefaction during earthquakes: Berkeley, Earthquake Engineering Research Institute, 134 p.

Seiders, V.M., Briggs, R.P., and Glover, L., III, 1972, Geology of Isla Desecheo, Puerto Rico, with notes on the Great Southern Puerto Rico Fault Zone and Quaternary stillstands of the sea: U.S. Geological Survey Professional Paper 739, 22 p.

Stuiver, M., Reimer, P.J., and Braziunas, T.F., 1998, High-precision radiocarbon age calibration for terrestrial and marine samples: Radiocarbon, v. 40, p. 1127–1151.

Tuttle, M.P., 2001, The use of liquefaction features in paleoseismology: Lessons learned in the New Madrid seismic zone, central United States: Journal of Seismology, v. 5, p. 361–380, doi: 10.1023/A:1011423525258.

Tuttle, M.P., Dyer-Williams, K., Schweig, E.S., Prentice, C.S., and Moya, J.C., 2000, Liquefaction induced by historic and prehistoric earthquakes in western Puerto Rico: Eos (Transactions, American Geophysical Union), Annual Fall Meeting, p. F1183.

Tuttle, M.P., Prentice, C.S., Dyer-Williams, K., Pena, L., and Burr, G., 2003, Late Holocene liquefaction features in the Dominican Republic: A powerful tool for earthquake hazard assessment: Bulletin of the Seismological Society of America, v. 93, no. 1, p. 27–46.

Youd, T.L., and Idriss, I.M., editors, 1997, Evaluation of liquefaction resistance of soils: National Center for Earthquake Engineering and Research, Technical Report, v. NCEER-97-0022, 40 p.

Manuscript Accepted by the Society August 18, 2004

Printed in the USA

Geological Society of America
Special Paper 385
2005

Earthquake-induced liquefaction potential in western Puerto Rico using GIS technology

Emir José Macari*
University of Texas at Brownsville, Brownsville, Texas 78520, USA

Laureano R. Hoyos*
University of Texas at Arlington, Arlington, Texas 76019, USA

ABSTRACT

Seismic hazard and risk analyses play a major role in identifying the potential consequences of an earthquake both in relation to existing facilities as well as in the planning and location of new structures. Such analyses must include consideration of several geological and geotechnical hazards and thus a large number of input factors are required. The resulting databases may be quite large and require an appropriate environment to optimize the risk evaluation procedures. Advances in computer-based Geographic Information Systems (GIS) have provided a technology which is ideally suited to fulfill the needs of earthquake hazard analyses. The overall objective of this study was to integrate a variety of analysis procedures for identifying and mapping geotechnical hazards through the use of GIS technology. This effort complements the comprehensive GIS database developed by the U.S. Geological Survey in Puerto Rico and will benefit emergency planning agencies, private industry, engineers, architects, and insurance agencies, among many others. This study was a pilot program that deals with the development of a GIS database of geotechnical properties for western Puerto Rico and the evaluation of liquefaction potential based on the Liquefaction Potential Index (LPI) method. The results are preliminary in nature and more detailed comprehensive studies are needed to accurately assess the liquefaction potential in western Puerto Rico.

Keywords: liquefaction, seismic hazard analyses, geotechnical hazard analyses, geographic information systems.

INTRODUCTION

The Island of Puerto Rico in the Greater Antilles is an area of significant seismicity and earthquake damage potential. The historical seismic records dating back to the 1500s indicate that several major earthquakes have struck the region. Field investigations have also revealed paleo-liquefaction features, which suggest prehistoric earthquake shaking (McCann, 1993; Tuttle et al., 2000). Existing historical seismic records contain evidence that many large and destructive earthquakes have occurred within 100 km of Puerto Rico (Asencio, 1980). Over the years, the rapid increase in population density and urbanization of the island has

*E-mails: emacari@utb.edu, lhoyos@ce.uta.edu.

Macari, E.J., and Hoyos, L.R., 2005, Earthquake-induced liquefaction potential in western Puerto Rico using GIS technology, *in* Mann, P., ed., Active tectonics and seismic hazards of Puerto Rico, the Virgin Islands, and offshore areas: Geological Society of America Special Paper 385, p. 277–287. For permission to copy, contact editing@geosociety.org.

made it imperative to perform an assessment of the seismic hazard for this region. Such an assessment can be obtained by applying known techniques of seismic hazard analyses to the region based on available geotechnical and geophysical information. Based on historical earthquake records in the region, earthquake events can be simulated and, hence, an assessment of damage potential may be performed.

Recent advances in computer-based Geographic Information Systems (GIS) have provided a technology which is ideally suited to fulfill the needs of earthquake hazard analyses. The overall objective of this ongoing research project is to integrate a variety of analysis procedures for identifying and mapping geotechnical hazards through the use of GIS. The focus of the work is on: (1) the selection and coding of hazard analysis procedures that incorporate the spatial nature of the subsurface data; (2) the parallel usage of these procedures along with a range of spatial analysis systems in a comprehensive framework; and (3) implementation of the framework in a pilot study to assess the potential for a range of earthquake-induced hazards in western Puerto Rico.

GEOLOGIC ENVIRONMENT OF THE MAYAGÜEZ AREA

The Island of Puerto Rico is situated ~220 km southeast of Florida between the Caribbean Sea and the Atlantic Ocean. Roughly rectangular in shape, the island measures ~170 km long (east-west) and 60 km wide (north-south), as shown in Figure 1A. It has a total surface area, including outlying islands, of 8760 km^2, and a population of 3.8 million (U.S. Census Bureau, 2000). The central portion of Puerto Rico is mountainous but the bulk of the island's population lives on the low-lying flat coastal plains. Coastal plain sediments consist primarily of loose, clean, fine sands and silts, but also contain thick deposits of soft clays. These coastal areas, especially in the western portions of the island, are extremely vulnerable to earthquake-induced hazards because of their propensity to liquefaction and seismic wave amplification. The coastal regions were, hence, the focus of this study.

Northwestern Puerto Rico consists of Miocene-age limestone and surface deposits of sandy clays and sands. From ~5 km south of the city of Aguadilla to ~8 km north of the city of Mayagüez, lower Tertiary and Cretaceous formations of limestone, siltstone, and quartz diorite can be found, as illustrated in Figure 1B (Beinroth, 1969). The major western coastal plain of Puerto Rico begins south of these formations. The coastal plain is nearly flat, and the water table consistently occurs within 2 m of the surface. The shore deposits in the Mayagüez area are mainly quartzitic beach sands. These beach deposits interfinger with alluvial deposits near the Rio Grande and Rio Cañas rivers, and continue several miles inland. The city of Mayagüez is founded on loose, alluvial, cohesionless deposits with occasional denser silt deposits from the Alto Rio, Rio Cañas, and Rio Guanajibo rivers. These deposits extend in a southwestern direction from Mayagüez, generally following the Rio Guanajibo and Rio Rosario. The hills south of Rio Guanajibo, toward Cabo Rojo, are older sandy clay deposits to 100 m depths. The southwestern corner of the island consists of loose cohesionless deposits that extend westward along southern Puerto Rico (Beinroth, 1969).

SEISMICITY OF PUERTO RICO

Existing historical records contain evidence that many large and destructive earthquakes have occurred within 100 km of Puerto Rico. Four strong earthquakes have affected the island since the early 1500s (Molinelli, 2000). The most recent of these occurred on 11 October 1918. The epicenter was located northwest of the city of Aguadilla in the Mona Canyon, between the Island of Puerto Rico and the Dominican Republic. This earthquake had an estimated magnitude (M_s) of 7.5 on the Richter scale and was accompanied by a 6 m tsunami. The earthquake killed ~116 people and caused more than 4 million dollars of damage to houses, factories, public buildings, and bridges (Reid and Taber, 1919).

On 18 November 1867, a strong earthquake occurred with an approximate magnitude of 7.5 on the Richter scale. The epicenter was located in the Anegada Passage, between Puerto Rico and St. Croix, Virgin Islands. The strongest earthquake that has affected Puerto Rico since the early 1500s occurred on 2 May 1787. This earthquake may have been as large as magnitude 8 on the Richter scale. Its epicenter was possibly to the north, in the Puerto Rico Trench. The quake was felt very strongly all across the island, demolishing the Arecibo church and El Rosario and La Concepción monasteries, and causing extensive damage to the churches at Bayamón, Toa Baja, and Mayagüez. The other strong earthquake occurred in 1670, severely affecting the area of San German District in western Puerto Rico (Molinelli, 2000).

The seismic activity is concentrated in the following zones around the Island of Puerto Rico (Tuttle et al., this volume):

(1) the Puerto Rico Trench to the north;

(2) the Sombrero Basin to the east northeast;

(3) the Mona Canyon and Mona Passage to the west-northwest;

(4) the Anegada Passage to the east; and

(5) the Muertos trough to the south.

Puerto Rico is situated in the hanging wall of an oblique subduction zone at the boundary between the North America and Caribbean plates. The Puerto Rico Trench, ~50 km north of the island, is believed to be the deepest portion of the Atlantic Ocean. The trench, formed by the subduction of the North America plate under the Caribbean plate and the Puerto Rico platelet, is considered to be the source of the magnitude 8 earthquake of 1787, along with frequent seismic activity of $M_s > 6–7$ in the twentieth century (McCann, 1986). South of Puerto Rico is the Muertos trough, which forms the boundary of the Puerto Rico platelet and the Caribbean plate. To the east of the island lies the Anegada trough. This has been reported to be the source of an $M_s = 7.5$ earthquake in 1867, which affected the eastern part of the island (Asencio, 1980; McCann and Sykes, 1984; McCann, 1986).

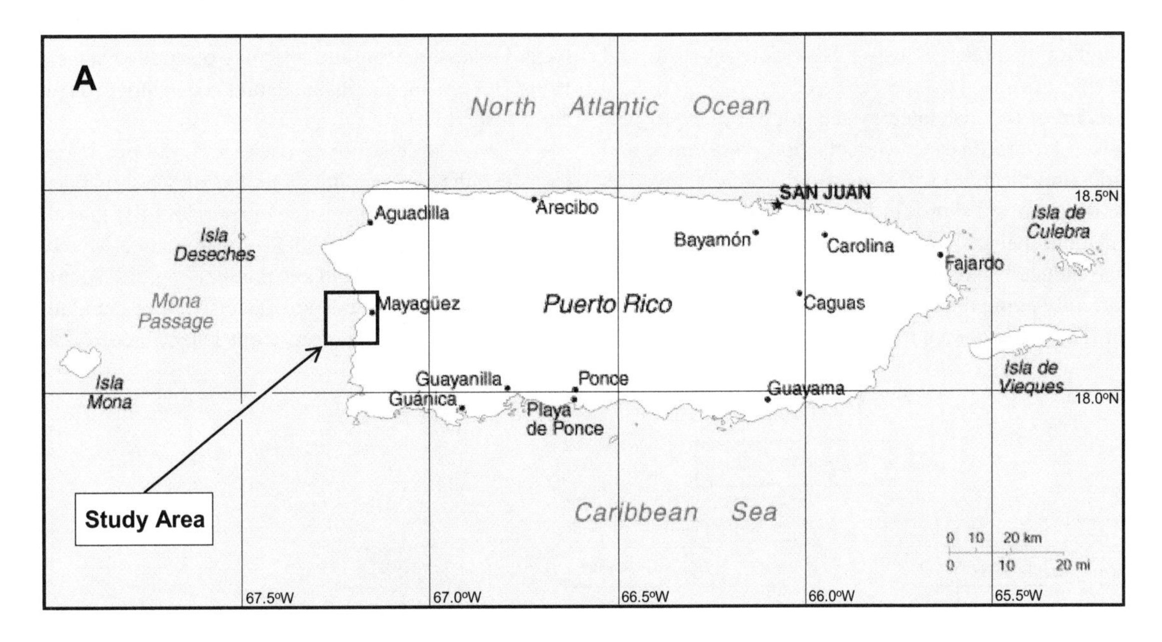

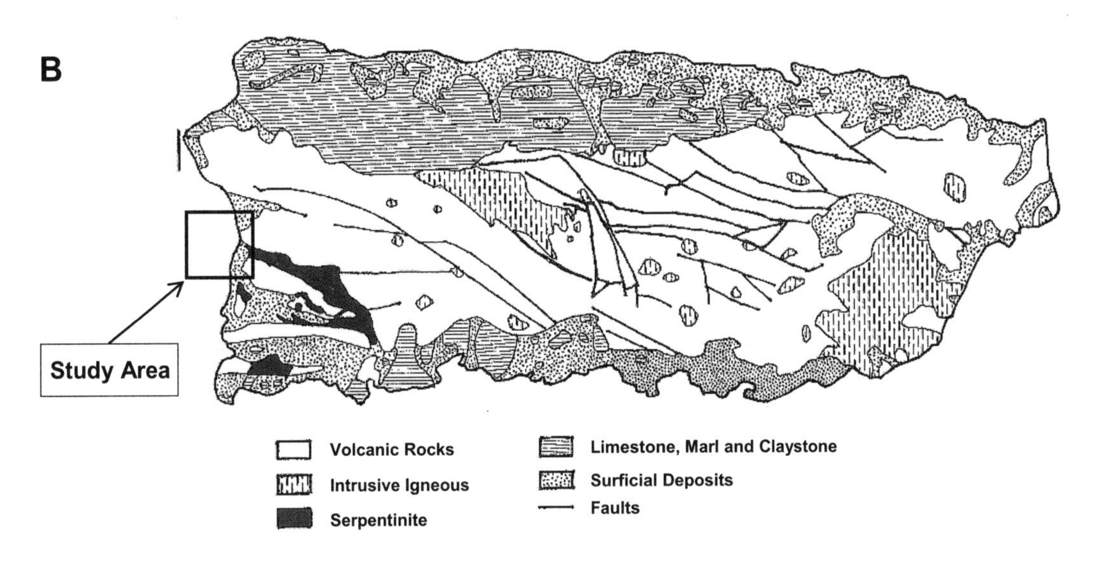

Figure 1. The Island of Puerto Rico. (A) Map and study area. (B) Geologic environment (Beinroth, 1969).

Of special interest to the western portion of Puerto Rico is the Mona Canyon, also known as the Mona Passage. This canyon, representing the separation of Puerto Rico and Hispaniola islands, is associated with the Southern Puerto Rico Fault Zone trending southeast through the Puerto Rico mainland. High seismic activity continues in the Mona Canyon, which is thought to be the source of the 1918 (M_s = 7.5) earthquake. Small earthquakes continue to be recorded in Puerto Rico, consistent with its many sources of seismic shaking. The Seismic Network at the University of Puerto Rico indicates that the number of earthquakes in the region occur at an approximate rate of 300–400 per year (University of Puerto Rico, 2001).

INVESTIGATIONS UNDERTAKEN

A generalized GIS-based framework, that is independent of the geographic location, has been developed to produce earthquake hazards micro-zonations. Hence, the framework may be readily applied at any other location. The portability of this tool is attractive in that it may be updated as analytical procedures are modified, refined, and/or extended, or as additional data concerning site and response characteristics become available. To date, the PC-supported framework integrates GIS software (ArcInfo), visualization software (BOSS GMS), a geostatistical package (GeoEAS), and a series of custom algorithms to compute liquefaction potential using both deterministic and probabilistic routines.

The structure of the integrated system for various geotechnical earthquake hazard analyses (liquefaction, landslides, and ground motion amplification) within a spatial setting, as well as the data-flow diagram, are shown in Figure 2. Programming and integration of additional analytical routines to assess the potential for earthquake-induced landsliding and/or ground motion amplification is currently being undertaken to extend the functionality of the system. The methodology framework was designed to aide hazard assessment studies and may be applied at a regional level using geological, geophysical, and geotechnical databases (Luna and Frost, 1998).

The multiple software package developed herein was used to assess the liquefaction potential of western Puerto Rico by inputting a simulated seismic event of similar magnitude to that of the 1918 western Puerto Rico earthquake, which devastated the city of Mayagüez and surrounding regions. As more detailed analyses are performed, the results of the simulation may be compared to those reported from paleo-liquefaction studies in

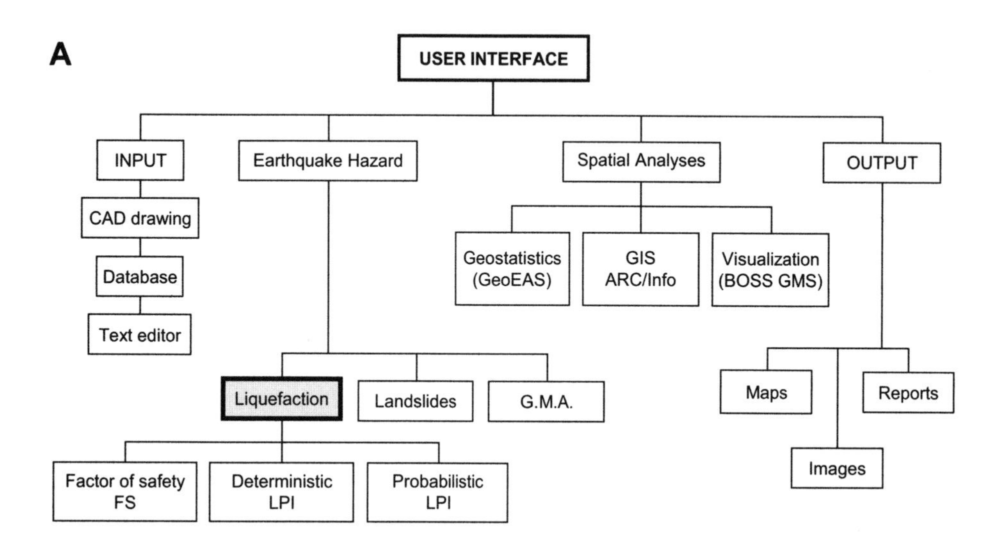

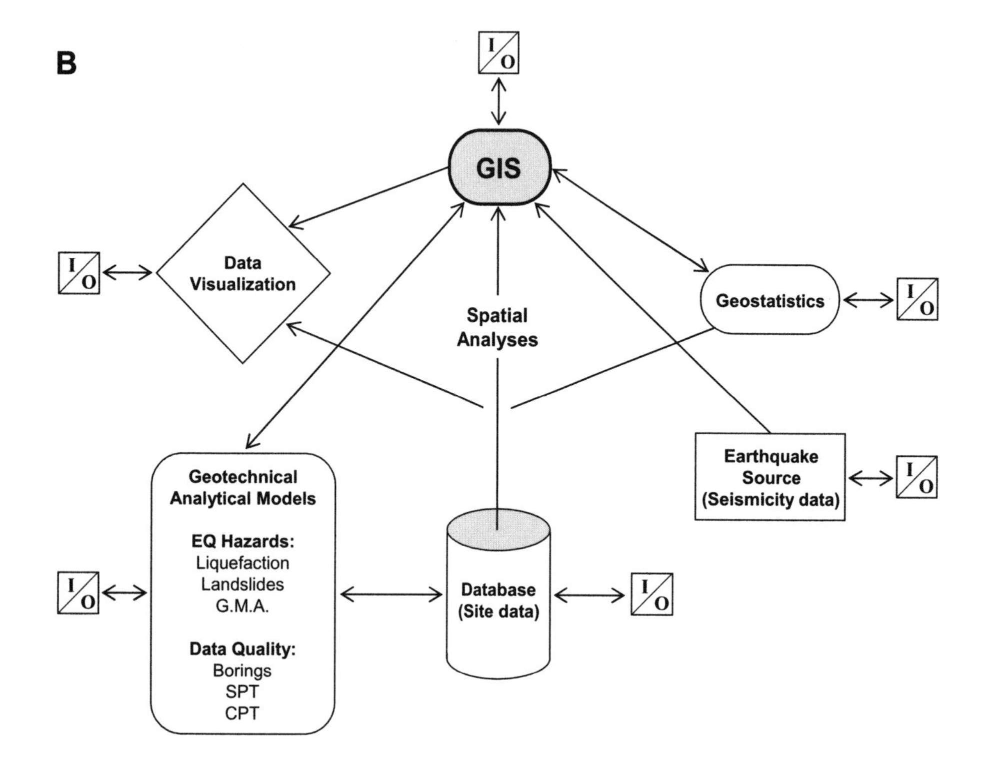

Figure 2. Overview of integrated system. (A) Analysis structure. (B) Data-flow diagram. LPI—Liquefaction Potential Index; G.M.A.—ground motion amplification; SPT—standard penetration test; CPT—cone penetration test.

the region (Tuttle et al., 2000; Doser et al., this volume; Tuttle et al., this volume). Site specific information, either from recent in situ damage assessments or from studies of historical records and paleo-liquefaction studies, may be used to further calibrate the model and help to fine tune similar earthquake hazard assessment studies in the future.

LIQUEFACTION POTENTIAL INDEX METHOD

Liquefaction

Soil liquefaction is considered to be the sudden loss of effective stress (particle-to-particle or inter-granular stresses) in a saturated sandy sediment due to an increase in excess pore water pressure resulting from applied cyclic loads (Seed and Idriss, 1971). Liquefaction may lead to extensive settlement and lateral spreading that often results in serious damage to foundations and buildings as well as other buried infrastructure (Youd et al., 2002). The liquefaction potential of a specific site may be evaluated both qualitatively and quantitatively. Qualitative methods involve mapping techniques aimed at defining seismicity and ground failure susceptibility. Seismic zonation may be performed by overlaying these maps with maps of geologic features (such as faults, topography, sediment layering, etc.), depth to ground water, and infrastructure to obtain a liquefaction hazard map.

Quantitative methods are based on geotechnical models that predict the behavior of soils under an applied loading. These models are calibrated based on parameters obtained from in situ and laboratory tests (Dobry et al., 1981; Stokoe et al., 1988). These and most other methods discard the spatial component of data (x-y coordinates), and simply lump the available data, calculate an average or a low-bound case, and then analyze and/or evaluate as a single case for the entire site. The Liquefaction Potential Index (LPI) method, instead, suggests evaluating liquefaction potential at the location where the data was collected, and interpreting the spatial distribution of these results using the spatial analysis and geostatistical capabilities of a Geographic Information System (GIS).

The LPI method, developed by Iwasaki et al. (1982), has received special attention in recent years. LPI is a weighted integration of the factor of safety against liquefaction (FS_L) within the uppermost 20 m of sediment beneath a site. The method is very well suited for a GIS to evaluate the likelihood that liquefaction may or may not occur in a given region. The method makes an assessment on the susceptibility to liquefaction based on the resistance to penetration measurements (standard penetration test [SPT] or cone penetration test [CPT]) made with depth, and finally calculates the index related to that location on the ground surface. The location may then be spatially referenced by determining the geographic or plan coordinates and, together with all the boring information, the LPI becomes another attribute attached to the spatial element.

The index represents the location at the ground surface because the weighting function used by the LPI has its maximum value at the ground surface, giving more weight to $(1 - FS_L)$ values at shallow depths. If liquefaction is to occur at a greater depth, the effect on ground infrastructure will be less than that if it were to occur at shallower depths (Luna and Frost, 1998).

Similar approaches, based on CPT data, have been reported by others, including Bartlett and Youd (1995) and, more recently, Youd et al. (2002). However, the LPI method, originally based on the work by Seed and Idriss (1971), has been proven to produce reasonable predictions. A recent study by Toprak and Holzer (2003) investigated the correlation of LPI with surface manifestations of liquefaction during the 1989 Loma Prieta earthquake and found them to be in close agreement.

LPI Analytical Approach

In the GIS-based LPI method developed herein, the distribution of the probability of liquefaction at a site is mapped in the form of LPI contours generated by using the results of all SPT or CPT soundings as discrete values. These LPI contours separate areas expected to experience different degrees of liquefaction. Moreover, once the LPI contour map is generated, other useful GIS-based spatial analysis operations may be performed, such as, overlay with other GIS coverages containing the geospatial locations and attribute databases of civil engineering facilities, lifelines, or populations at risk.

Figure 3 illustrates schematically the calculation of the LPI based on in situ SPT or CPT data at one sounding location. The LPI can be then expressed as follows:

$$\mathrm{LPI} = \int_0^{\infty} w(z)\left[1 - FS_L(z)\right]dz \qquad (1)$$

where $w(z)$ = weighting function, which linearly decreases with depth from a value of 10 at the surface to 0 at a depth of 20 m; FS_L = factor of safety against liquefaction, which is defined as the R/L ratio at depth z, where R is the resistance to liquefaction and L is the dynamic applied load; and dz = unit thickness of the sediment layer under consideration, which are, typically, looser granular sediments such as sands or silts.

In Equation (1), the dynamic applied load, L, induced by an earthquake at a depth z can be defined as follows (Seed and Idriss, 1971):

$$L = 0.65 \frac{A_{max}}{g} \frac{\sigma_o}{\sigma'_o} r_d \qquad (2)$$

where A_{max} = peak ground acceleration (or PGA); g = acceleration due to gravity; σ_o = total overburden pressure at depth z; σ'_o = effective overburden pressure at depth z; and r_d = stress reduction factor, which decreases with depth from a value of 1 at the surface to ~0.5 at a depth of 100 m.

The resistance to liquefaction, R, in Equation (1), also known as the cyclic stress ratio (CSR), is a function of the earthquake magnitude, M_s, and the sediment strength, among other factors, and can be obtained by several methods. The method used herein is based on an empirical correlation originally

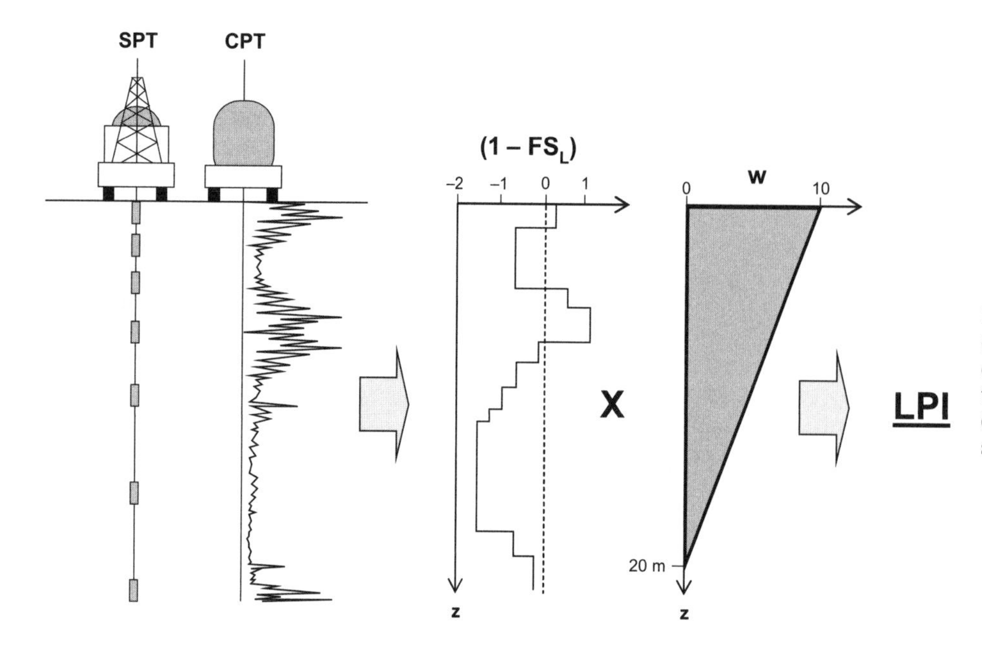

Figure 3. Liquefaction Potential Index (LPI) calculation based on standard penetration test (SPT) or cone penetration test (CPT) data at one sounding location (Luna and Frost, 1998). FS_L—factor of safety against liquefaction.

presented by Seed et al. (1983), and later modified by Shinozuka (1990), that includes the sediment strength as measured by the standard penetration resistance value and the earthquake magnitude, as follows:

$$R = 0.012C_M(N_1)_{60} \longrightarrow \text{for } (N_1)_{60} < 25$$
$$R = \left\{0.0056[(N_1)_{60}]^2 - 0.268(N_1)_{60} + 3.5\right\} \longrightarrow \text{for } (N_1)_{60} > 25 \quad (3)$$

where C_M = coefficient accounting for the number of representative loading cycles, which in turn is a function of M_s; and $(N_1)_{60}$ = standard penetration resistance value (SPT N-value) normalized to an effective overburden pressure of 1 ton per square foot (tsf) and to an effective impact energy delivered to the drill rod equal to 60% of the ideal free-fall energy. Figure 4 illustrates the calculation process for the factor of safety against liquefaction (FS_L) at any depth z, for an $M_s = 7.5$ earthquake with peak ground acceleration (PGA) = 0.10 g's.

The LPI values obtained from the methodology outlined above were divided into four major ranges or levels of expected liquefaction-induced damage, as shown in Table 1.

GIS–GEOTECHNICAL DATABASE FOR WESTERN PUERTO RICO

The first stage of the database development process for the western Puerto Rico region involved primarily extensive collection of geotechnical boring logs from various agencies. Most of the boring logs obtained included standard penetration test results. They were provided by the Puerto Rico Department of Transportation, local engineering consulting firms, and the University of Puerto Rico–Mayagüez campus.

The second stage of the database development process involved an electronic interpretation and digitization of the boring logs into GIS. Details from each boring log were first entered manually into an electronic spreadsheet. In addition to standard penetration test and cone penetration test values with depth, the type of soil, the basic engineering tests performed, and the location of the boring log were also noted. The spreadsheet data was then read by MATLAB and ArcInfo software to generate attribute databases for each boring log recorded. Each boring log was then digitized into ArcInfo using an HP digital plotter and, finally, a GIS coverage was created with all digitized boring logs.

Several other data coverage layers were obtained from the U.S. Geological Survey, as shown in Table 2. These layers included urban zones, population zones, highways, topography, and other lifelines, which were ultimately overlaid with LPI results to generate liquefaction hazard maps for the region.

INTEGRATION OF MATLAB, GIS-ARCINFO, AND LPI METHOD

The generation of LPI hazard maps involves three main steps: (1) generation of boring log database, (2) development of LPI algorithm, and (3) generation of LPI contours using geostatistical tools. These steps are briefly summarized as follows:

Step 1: Generation of boring log database. Boring log databases were generated by converting boring logs information obtained in hard-copy form into electronic spreadsheets. The spreadsheets contained the following information: (1) depth, (2) SPT value, (3) water table information, (4) soil description, and (5) engineering properties.

Step 2: Development of LPI algorithm. Using the electronic information given above, a MATLAB routine was written

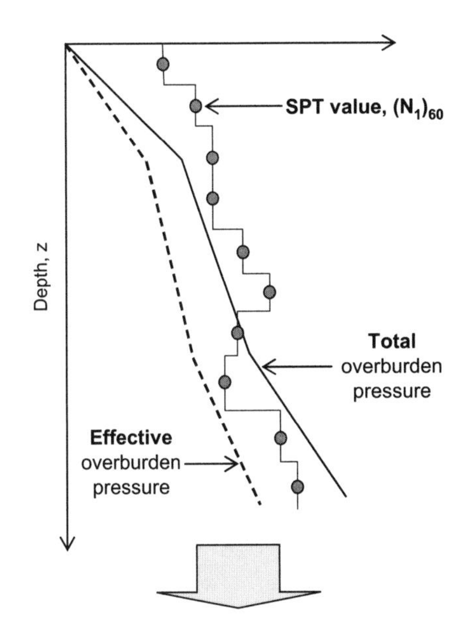

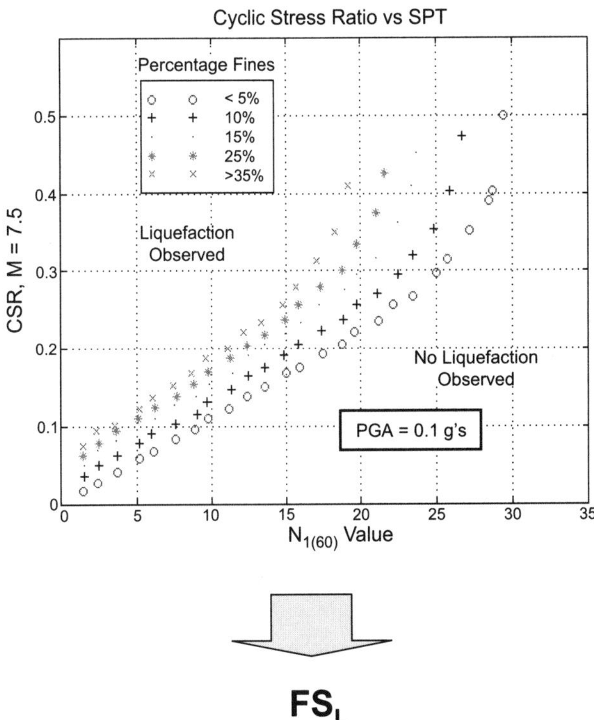

Figure 4. Calculation of factor of safety against liquefaction (Luna and Frost, 1998). SPT—standard penetration test, CSR—cyclic stress ratio, PGA—peak ground acceleration.

in scripting language, essentially by taking the input data from each boring log and computing the LPI for the boring using an LPI algorithm based on Equations (1), (2), and (3) above, and the schematic representations shown in Figures 3 and 4.

Step 3: Generation of LPI contours using geostatistical tools. After obtaining the LPI values for each boring location, the corresponding attribute table containing the coordinated locations of each boring log can be exported from ArcInfo. A new field containing the corresponding LPI value for each boring log was then generated. Using GeoEAS, a U.S. Environmental Protection Agency–developed geostatistical software, it is possible to estimate the LPI values at locations where there is no specific boring log information via geostatistical interpolation techniques.

Using the procedures outlined above, it is possible to obtain a liquefaction hazard map for the western Puerto Rico region. The liquefaction hazard maps that were generated consist of several LPI contour maps showing LPI results over the region for a given earthquake magnitude and varying levels of ground motion acceleration. Using these liquefaction hazard maps, it is possible to identify regions of low, moderate, and high liquefaction potential in the western Puerto Rico region. The same liquefaction hazard maps generation process can be utilized for earthquakes of any given magnitude, M_s, and PGA.

TABLE 1. LIQUEFACTION POTENTIAL INDEX CATEGORIES

S. no.	LPI range	Description
1	0	No damage
2	0 < LPI ≤ 5	Minor Damage
3	5 < LPI ≤ 15	Moderate Damage
4	LPI > 15	Major Damage

TABLE 2. GIS COVERAGES FOR WESTERN PUERTO RICO

S. no.	Coverage name	Coverage description
1	Borings	Boring-log records and location
2	Central_latt	Lattice grid for central la plata quad
3	Mabd01	Mayagüez borders
4	Mabd01_utm	Mayagüez borders in UTM system
5	Magnis_03	Mayagüez geographic name information systems
6	Malu77	Mayagüez lot units
7	Mapl01	Mayagüez pipelines
8	Mard01	Mayagüez roads
9	Masl01	Mayagüez stream lines
10	Mast01	Mayagüez streams
11	Matb_03	Mayagüez tiger blocks
12	Matics	Mayagüez tics
13	Matin	Mayagüez triangular irregular network
14	Matp01	Mayagüez topography
15	Matt_03	Mayagüez tiger tracks
16	Mayagüez_latt	Lattice grid for Mayagüez quad
17	Medium_duty	Medium-duty roads
18	Old_rroad	Old rail roads
19	Pr_2	Heavy-duty roads
20	Rincon_latt	Lattice grid for Rincon quad
21	Rosario_latt	Lattice grid for Rosario quad

RESULTS AND HAZARD MAPS

For the purpose of this study, several scenarios were considered for a given earthquake as shown in Table 3. The maximum credible earthquake for the western coast of Puerto Rico has a magnitude $M_s = 7.5$ and an associated PGA = 0.15 g's (McCann, 1993). The results presented in Table 4 correspond to an earthquake of magnitude $M_s = 7.5$ and PGA varying from 0.05 to 0.15 g's. Once generated, the liquefaction hazard maps may be overlaid over existing digitized coverage data, such as lifelines (e.g., pipelines, highways, bridges, etc.) or important civil infrastructure (e.g., schools, hospitals, police stations, etc.). Such a hazard map may help local and federal officials in emergency planning and response, and may also assist urban development officials involved with city planning and expansion.

Figure 5 shows an elevation map of the western Puerto Rico region outlining the study area. As noted earlier in this document, the bulk of the island's population lives on the low-lying flat coastal plains where sediments consist primarily of loose, clean, fine sands and silty sands, which make these areas extremely

TABLE 3. MATLAB–LIQUEFACTION POTENTIAL INDEX RUN SCENARIOS

Run no.	Magnitude (M_s)	Peak ground acceleration (g's)
1	7.5	0.05
2	7.5	0.10
3	7.5	0.15

TABLE 4. SUMMARY OF VARIOUS MATLAB–LIQUEFACTION POTENTIAL INDEX RUNS

Run no.	PGA (g's)	Maximum LPI
1	0.05	17.6
2	0.10	46.8
3	0.15	62.2

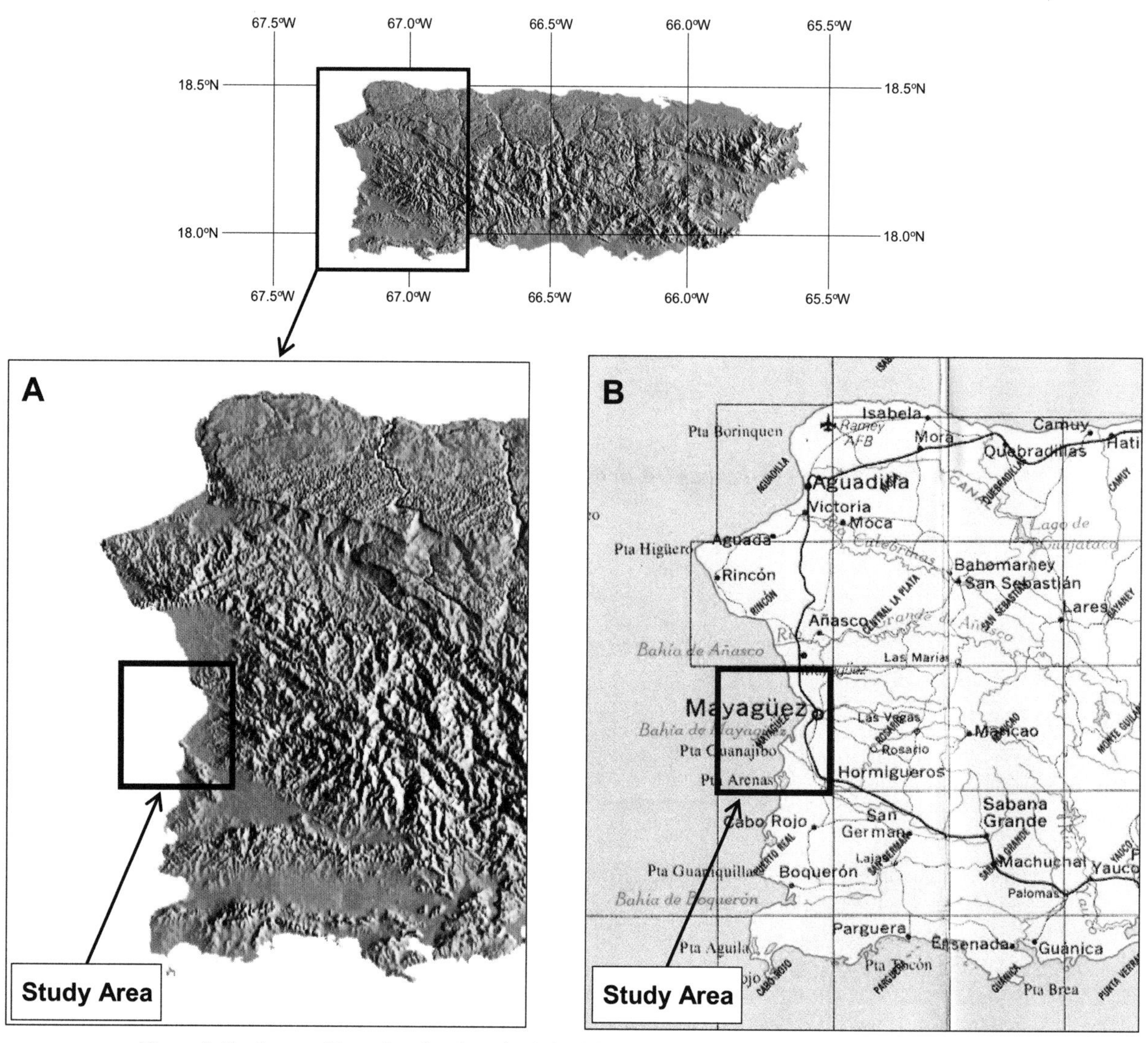

Figure 5. Study area, Mayagüez Quadrangle: (A) USGS elevation map, (B) USGS quadrangles map.

vulnerable to earthquake-induced liquefaction, especially in the western coast of the island (Fig. 6).

Figures 6A, 6B, and 6C show the LPI contour maps for earthquake-induced liquefaction hazard of areas near and within the city of Mayagüez, overlain with PR-2 highway coverage. Figure 6A, for a PGA of 0.05 g's, shows most of the study region with low probability of liquefaction (LPI < 5), some regions with moderate probability of liquefaction (5 < LPI < 15), and a small region with a high likelihood of liquefaction (LPI > 15).

This scenario quickly changes with increasing values of peak ground motion acceleration, PGA. Figure 6C shows the liquefaction hazard map for a PGA of 0.15 g's. As may be seen from this figure, there are more regions with high probability of liquefaction (LPI > 15). However, a much more detailed study is needed in order to properly identify those regions where earthquake-induced liquefaction is to be expected for a seismic event of such magnitude (7.5) and peak ground acceleration (0.15 g's).

CONCLUSIONS AND FUTURE WORK

The LPI method (Iwasaki et al., 1982) was implemented using MATLAB software. This software was integrated with an electronic spreadsheet, a GIS software (ArcInfo), and a geostatistical package (GeoEAS) in order to produce earthquake-induced liquefaction hazard maps. The resulting maps identify areas near and within the city of Mayagüez in the western Puerto Rico region where liquefaction is likely to occur during the maximum credible earthquake for the western coast of Puerto Rico, having a magnitude (M_s) of 7.5 and associated peak ground acceleration of 0.15 g's. These areas, and particularly urban areas, should be investigated more thoroughly to determine the risk to existing civil engineering infrastructure and lifelines.

Using additional GIS coverage layers, it is possible to identify the buildings and critical lifelines located within the trouble spots, which may assist in hazard assessment processes and implementation of suitable mitigation practices. The analysis presented herein can be further extended and refined by using more populated data sets from additional boring logs, which can be obtained from private consulting agencies in the region, along with more complex and sophisticated ground response models. All GIS coverages used in the present work (Table 2), such as roads and pipelines, had a scale of 1:63 K; for better results a more detailed scale (at least 1:20 K) should be used.

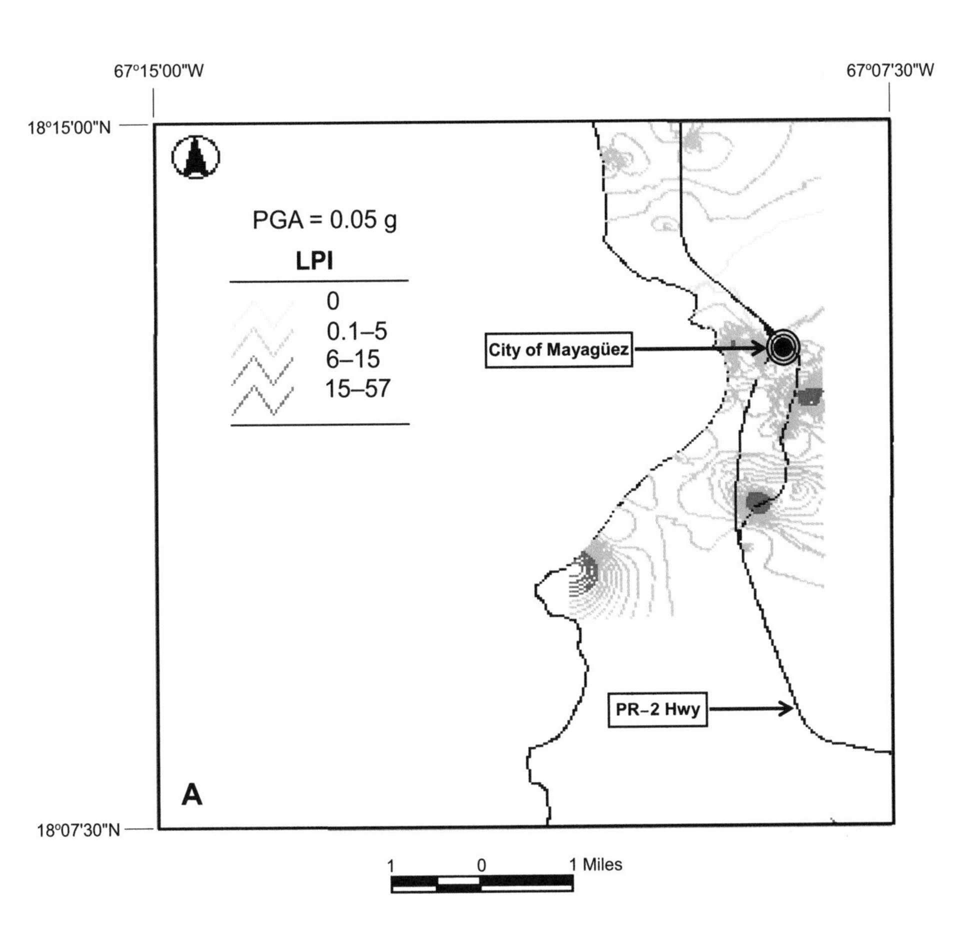

Figure 6 (*on this and following page*). Mayagüez Quadrangle liquefaction hazard maps. PGA—peak ground acceleration. (A) PGA = 0.05 g's, (B) PGA = 0.10 g's, (C) PGA = 0.15 g's.

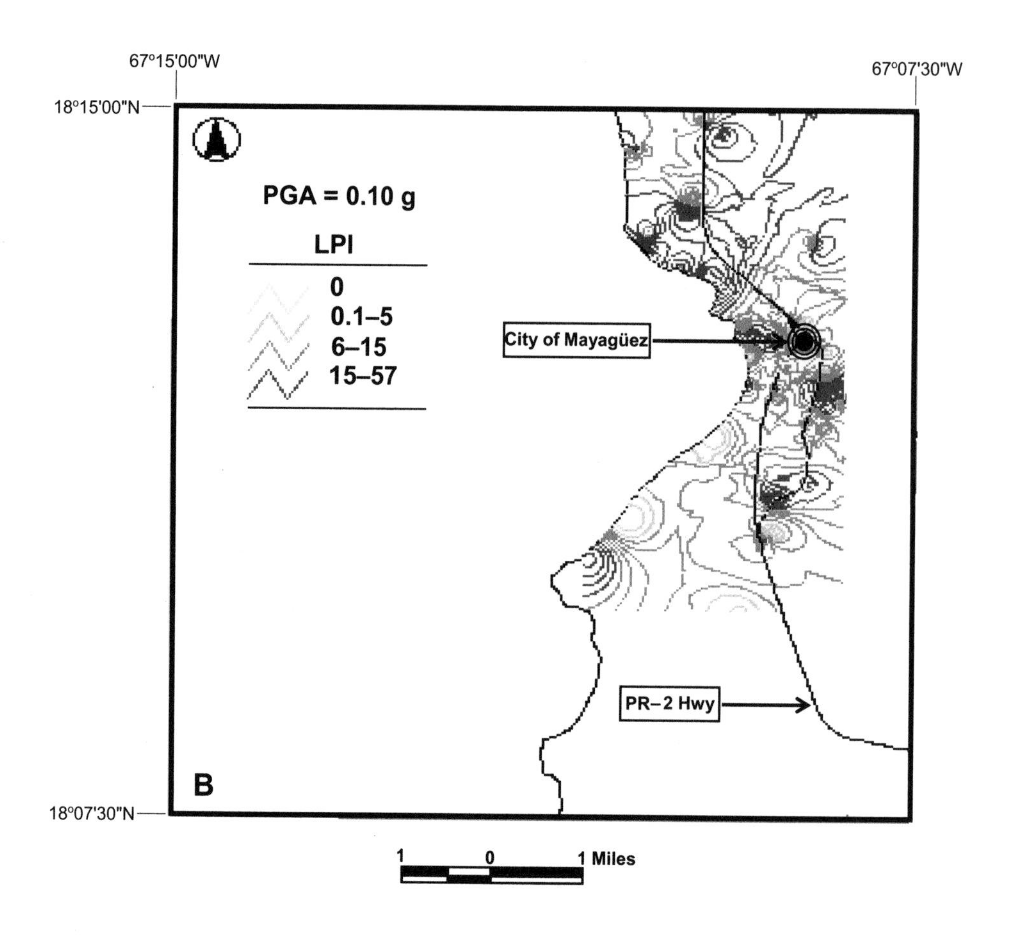

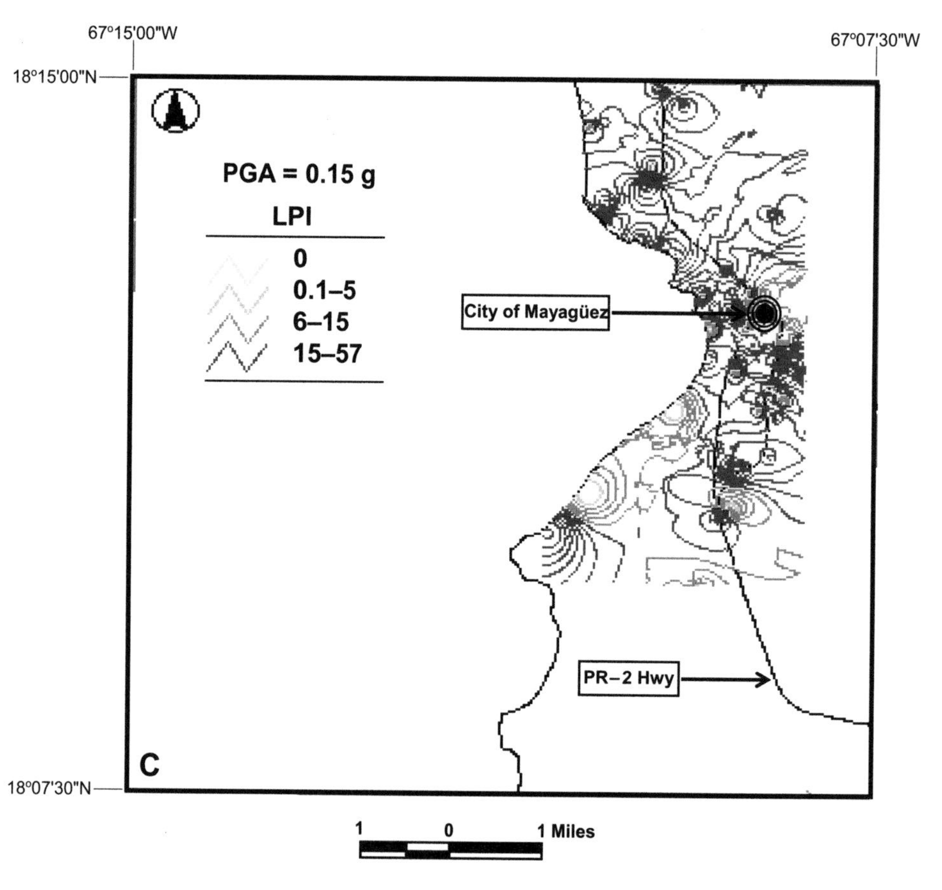

Figure 6 (*continued*).

ACKNOWLEDGMENTS

Funding for this study was provided by the U.S. Geological Survey Grant 1434-HQ-97-GR-03049. The authors greatly acknowledge this support. The authors would also like to express their appreciation to John E. Parks and Matthew C. Larsen of the USGS Puerto Rico office for all their help in providing much of the GIS-ArcInfo coverage data used in this study. The authors are also grateful to the reviewers for their beneficial comments, and especially to Paul Mann of the University of Texas, who worked diligently for over a year to ensure the quality of this publication.

REFERENCES CITED

Asencio, E., 1980, Western Puerto Rico seismicity: U.S. Department of the Interior Geological Survey, Open-File Report 80-192.

Bartlett, S.F., and Youd, T.L., 1995, Empirical prediction of liquefaction-induced lateral spread: ASCE Journal of Geotechnical Engineering, v. 121, no. 4, p. 316–329, doi: 10.1061/(ASCE)0733-9410(1995)121:4(316).

Beinroth, F.H., 1969, An outline of the geology of Puerto Rico, Bulletin # 213, Agricultural Experiment Station, Rio Piedras, Puerto Rico, 00928.

Dobry, R., Stokoe, R.H., Ladd, R.S., and Youd, T.L., 1981, Liquefaction susceptibility from S-wave velocity: Proceedings, session on in situ testing to evaluate liquefaction susceptibility: ASCE National Convention, St. Louis, Missouri.

Iwasaki, T., Tokida, K., Tatsuoka, F., Watanabe, S., Yasuda, S., and Sato, H., 1982, Microzonation for Soil Liquefaction Potential using the Simplified Model: Proceedings of the 3rd International Earthquake Microzonation Conference, Seattle, p. 1319–1330.

Luna, R., and Frost, J.D., 1998, Spatial liquefaction analysis system: ASCE Journal of Computing in Civil Engineering, v. 12, no. 1, p. 48–56, doi: 10.1061/(ASCE)0887-3801(1998)12:1(48).

McCann, W.R., 1993, Seismic hazard map for Puerto Rico: Report prepared for the Seismic Safety Commission of Puerto Rico.

McCann, W.R., 1986, Historic earthquakes and the earthquake hazard of Puerto Rico, Proceedings of the Workshop on Assessment of Geologic Hazard and Risk in Puerto Rico, May 14–16, 1986: U.S. Department of the Interior Geological Survey, Open-File Report 87-008.

McCann, W.R., and Sykes, L.R., 1984, Subduction of aseismic ridges beneath the Caribbean Plate: Implications for the tectonics and seismic potential of the northeastern Caribbean: Journal of Geophysical Research, v. 89, p. 4493–4519.

Molinelli, J., 2000, Terremotos: Puerto Rico Civil Defense and Federal Emergency Management Agency (FEMA) Report.

Reid, H.R., and Taber, S., 1919, Puerto Rico earthquake of 1918, with descriptions of earlier earthquakes: Report of the Earthquake Investigation Commission, House of Representatives, Document 269, Washington, D.C., 74 p.

Seed, H.B. and Idriss, I.M., 1971, Simplified procedure for evaluating liquefaction potential: ASCE Journal of Soils and Foundations Division, 97(SM9), p. 1249–1272.

Seed, H.B., Idriss, I.M., and Arango, I.A., 1983, Evaluating liquefaction potential using field performance data: ASCE Journal of Geotechnical Engineering Division, v. 109, no. 3, p. 458–482.

Shinozuka, M., 1990, Crude oil transmission study—Probabilistic models for liquefaction: National Center for Earthquake Engineering Research Bulletin, p. 3–8.

Stokoe, K.H., Roesset, J.M., Bierschawale, J.G., and Aouad, M., 1988, Liquefaction potential of sands from shear wave velocity: Proceedings of the Ninth World Conference on Earthquake Engineering, Tokyo-Kyoto, Japan.

Toprak, S., and Holzer, T.L., 2003, Liquefaction Potential Index: A field assessment: ASCE Journal of Geotechnical and Geoenvironmental Engineering, v. 129, no. 4, p. 315–322.

Tuttle, M., Dyer-Williams, K., Schweig, E., Prentice, C., and Moya, J., 2000, Liquefaction induced by historic and prehistoric earthquakes in western Puerto Rico: Eos (Transactions, American Geophysical Union), Annual Fall Meeting, v. 81, no. 48, p. 1183.

U.S. Census Bureau, 2000, Census 2000 Redistricting Data (Public Law 94–171): U.S. Department of Interior Publication.

University of Puerto Rico, 2001, Preliminary locations of earthquakes recorded near Puerto Rico: Seismic Bulletin, Department of Geology, University of Puerto Rico Seismic Network, Mayagüez, Puerto Rico.

Youd, T.L., Hansen, C.M., and Bartlett, S.F., 2002, Revised MLR equations for prediction of lateral spread displacement: ASCE Journal of Geotechnical and Geoenvironmental Engineering, v. 128, p. 1007–1017.

Manuscript Accepted by the Society August 18, 2004

Index

A

B

C

D

N

O

P

Q

R